PRESS INFORMATION

NASA SPACE SHUTTLE
TRANSPORTATION SYSTEM
MANUAL January 1984

NASA Space Transportation Systems
Operations Office, MO, NASA,
Washington, D.C., 20546

 Rockwell International

Integration and Satellite Systems Division
12214 Lakewood Boulevard
Downey, CA, 90241

©2011 Periscope Film LLC All Rights Reserved
ISBN #978-1-935700-84-5

Contents

	Page
Space Transportation System	1
Space Shuttle Program	3
Shuttle Requirements	3
Background and Status	6
Launch Sites	7
Mission Profile	9
Aborts	13
Orbiter Ground Turnaround	15
Payloads	17
Space Shuttle Flights	20
Solid-Rocket Boosters	21
External Tank	46
Main Propulsion System	51
External Tank	54
Orbiter Propellant Management Subsystem	59
Propellant Management Gaseous Propellant Collection and Supply Network	63
Propellant Management Valves	63
Orbiter Helium Subsystem	64
Space Shuttle Main Engines	70
POGO Suppression System	74
SSME Controller	75
Malfunction Detection	81
Orbiter Hydraulic Systems	83
Thrust Vector Control	84
Helium, Oxidizer and Fuel Flow Sequence	88
Orbiter-External Tank Separation System	102
Space Shuttle Coordinate System	107
Space Shuttle Spacecraft Structures	109
Orbiter Structure	109
Forward Fuselage	111
Crew Compartment	114
Forward Fuselage and Crew Compartment Windows	129
Wing	131
Mid Fuselage	135
Payload Bay Doors	137

	Page
Aft Fuselage	148
Vertical Tail	154
Cargo Bay Stowage Assembly	157
Passive Thermal Control System	157
Purge, Vent, and Drain System	159
Space Shuttle Spacecraft Systems	163
Thermal Protection System	163
Orbital Maneuvering System	178
Helium Pressurization	180
Propellant Storage and Distribution	183
Engine Bipropellant Valve Assembly	184
Engine	188
OMS Thrusting Sequence	188
Thermal Control	188
Engine Gimbal and Control Assembly	189
OMS Payload Bay Kit	190
OMS to RCS Interconnect	190
OMS to RCS Gauging Sequence	191
Abort Control Sequences	191
OMS Engine Fault Detection and Identification	193
OMS Actuator FDI	194
Reaction Control System	195
Pressurization System	198
Propellant System	200
RCS Quantity Monitor	201
Engine Propellant Feed	202
RCS Engines	205
Heaters	206
RCS Jet Selection	207
Electrical Power System	214
Power Reactant Storage and Distribution System	215
Fuel Cell Powerplants	223
Electrical Power Distribution and Control	230
Environmental Control and Life Support System	241
Atmospheric Revitalization and Control Subsystem	242
Cabin Air	251

Contents

	Page
Water Coolant Loop Subsystem	254
Active Thermal Control Subsystem	261
Food, Water and Waste Management	271
Airlock Support Subsystem	283
Personal Egress Air Pack	302
Antigravity Suit	303
Auxiliary Power Unit	306
Water Spray Boiler	318
Hydraulic System	322
Landing Gear System	333
Main Landing Gear Brakes	339
Nose Wheel Steering	342
Avionics Systems	345
Data Processing System	346
Communications	357
S-Band System	363
Ku-Band System	370
Audio System	370
UHF System	378
Instrumentation	380
Master Timing Unit	383
Operational Recorders	384
Payload Recorder	385
Modular Auxiliary Data System	387
Ku-Band Rendezvous Radar	392
Text and Graphics	392
Equipment Location	392
Television	392
EMU-TV	397

	Page
Payload Deployment and Retrieval System	399
Payload Retention Mechanisms	411
Navigation Aids	415
Guidance, Navigation and Control	438
Caution/Warning System	501
Smoke Detection and Fire Suppression	504
Head Up Display	505
Displays and Controls	508
Space Transportation System Payloads	511
Payload Assist Module	511
Inertial Upper Stage	516
Spacelab	535
Space Telescope	546
Getaway Special	548
Space Transportation System Background Information	551
Orbiter Approach and Landing Test Program	551
Space Shuttle Flight Test Program	552
Shuttle Carrier Aircraft	553
Space Shuttle Chronology	554
Rockwell Space Shuttle Management	569
Space Transportation System Facilities	574
Space Transportation System Glossary	588
Space Transportation System Contractors	592

 # Space Transportation System

The preamble for this STS Press Information book is relatively simple in context:

> The economic and social benefits accruing from development of the spatial environment...abundant power and thermal control in a near absolute weightless and vacuum condition...can only be constrained by lack of imagination or not understanding that the principle ingrediant of progress is growth.

———————————————

The Shuttle spacecraft is a "magnificent flying machine." These words were first voiced by the two astronaut crews who flew the *Enterprise* on the STS program's Approach and Landing Test (ALT) series of flights. Several years later the same words were echoed by each crew flying the *Columbia* in its first orbital missions, and then re-echoed by each crew of *Challenger*....as will they for *Discovery* and *Atlantis*.

The Shuttle spacecraft was designed by Rockwell International for the National Aeronautics and Space Administration (NASA) as a flexible "cargo carrier" for travel into near-Earth orbit...with a return to landing as a reusable system capable of repeatable flights without requiring major overhaul.

The first half-dozen flights of *Columbia* and *Challenger* proved a capability even beyond that design, providing the potential for a foundation leading into the first of several decades of the 21st century.

In its first flights, the Shuttle spacecraft proved to be an extremely stable orbiting platform, not only providing an excellent launch site for satellites destined for boosting to higher orbits and interplanetary missions, but it proved to be an outstanding housing for the most precise of scientific instrumentation.

With the Remote Manipulator System and its 50 foot-long arm—containing shoulder, elbow, wrist-type action—STS mission specialists efficiently and expertly could extract cargo from the spacecraft's cargo bay which is as long as a six story building is high. And, it was not only possible to deploy cargo in this fashion, but to release it and later retrieve it from orbital journey.

The potentially expansive "surface" of what we may accomplish in space, is just being scratched. We know that our Earth-bound environment has a "gravity" constraint on the production and processsing of materials. Materials are layered, they cannot mix, there is friction and motion in our ground-based processing environment resulting in contamination in glass, separation in chemicals and defects in electronic materials.

Space Transportation System

The experiments and demonstrations conducted in the Apollo, Skylab and Apollo/Soyuz Test Project flights explored the benefits of processing in zero-gravity. Materials processing offers a great potential from our space environment.

In the STS flights, a Monodisperse Latex Reactor (MLR) experiment designed to study feasibility of producing latex spheres larger than two or three microns in size...the largest produced in the one-G atmosphere. A micron is one-millionth of a meter or about 25,000 microns are needed to stretch out an inch. In space, the result of several flights, latex spheres up to 20 microns in diameter have been grown. These latex microspheres have major medical and industrial research applications... measuring pores in the wall of human intestines for cancer research, or the eyes for glacoma investigations; as a carrier of drugs and radioactive isotops for treatment of tumors or measuring blood flow in the body for heart research.

A Continuous Flow Electrophoresis System experiment (CFES) flown in several STS missions has successfully demonstrated that biological materials can be separated in quantities 700 times greater in space than on Earth....and achieve four times greater purification. Scientists are continuing to discover that certain treatment, and even cures, are possible by using specific cells, hormones or proteins. Unfortunately, those materials are produced only in extremely small quantities on Earth. But, in space..........

The fullest possible development in space of materials—physically, thermally, chemically, optically and electrically—could afford major breakthroughs in medicine, electronics, energy and instrumentation.

The potential of space is unlimited....for science.

Immediately, is the realistic knowledge and use of remote sensing devices from space. From the stable platform some 150 miles above Earth, the world's resources may be monitored. And with the Shuttle spacecraft's huge cargo capacity and mild launch environment, a variety of satellites may be placed in orbit including some which could not be launched previously because of size, shape, weight or sensitivity to launch forces.

Malfunctioning satellites can be repaired in space by Shuttle crewmen or the satellite can be returned in the cargo bay for repair on Earth. The Shuttle spacecraft also serves as a launch platform for higher-orbit satellites or for interplanetary craft.

Eventually, the spacecraft can transport elements of large structures to orbit and support the construction crews in space. Such large structures could include satellite power systems, permanent space stations, multi-satellite service platforms, and orbiting space power stations.

A special NASA study team recently reviewed the U.S. space program and identified specific contributions to the welfare of mankind it could make in the next 25 years. These contributions—compiled in the team's report, *Outlook for Space*—range from weather and communications through manufacturing-processing to earth-oriented activities. In the last category, a grouping of contributions aimed at responding to "basic human needs," the use of satellites and space laboratories was seen helping meet the following needs:

- **Food production, forestry management**—Forecasting of global crop production and water availability, land use assessment, timber inventory, assessment of marine resources and rangeland.

- **Environmental protection**—Large-scale weather forecasting, climate prediction, weather modification experiments, stratospheric changes and effects, water quality monitoring, global marine weather forecasting.

Space Transportation System

- **Protection of life and property**—Storm tracking, tropospheric pollutant monitoring, hazard forecasting, communication-navigation, earthquake prediction, control of harmful insects.

- **Energy and mineral exploration**—Solar power relay to earth from satellite power stations, disposal of hazardous waste in space, world geologic atlas.

- **Information transfer**—Domestic, intercontinental, and personal communications.

- **Use of space environment**—Basic physics and chemistry, material science, commercial inorganic processing, biological material research, effects of gravity on terrestrial life, physiology and disease processes.

- **Earth science**—Earth's magnetic field, crustal dynamics, ocean interior and dynamics, dynamics and energetics of lower atmosphere, ionosphere-magnetosphere coupling, and structure, chemistry, and dynamics of stratosphere/mesosphere.

SPACE SHUTTLE PROGRAM

The Space Shuttle is developed by the National Aeronautics and Space Administration. NASA coordinates and manages the Space Transportation System (NASA's name for the overall Shuttle program) including inter-governmental agency requirements and international and joint projects. NASA also oversees the launch and space flight requirements for civilian and commercial use.

The Space Shuttle system consists of four primary elements: an orbiter spacecraft, two solid-rocket boosters, an external tank to house fuel and oxidizer, and the three main engines.

The orbiter is built by Rockwell International's Space Transportation and Systems Group, Downey, CA, facility and also has contractual responsibility for integration of the overall Space Transporation System. Both orbiter and integration contracts are under the direction of NASA's Johnson Space Center (JSC) in Houston, TX.

The solid-rocket booster motors are built by the Wasatch Division of Thiokol Corp., the external tank by Martin Marietta Corp., and the Shuttle main engines by Rockwell's Rocketdyne Division. These contracts are under the direction of NASA's George C. Marshall Space Flight Center (MSFC) in Huntsville, AL.

SHUTTLE REQUIREMENTS

The Shuttle will transport into near earth orbit 100 to 600 nautical miles (115 to 690 statute miles) up to 29,484 kilograms (65,000 pounds) of cargo. This cargo (called payload) is carried in a bay 4.57 meters (15 feet) in diameter and 18 meters (60 feet) long. It can bring back from space cargo weighing a total of 14,515 kilograms (32,000 pounds).

Major system requirements are that the orbiter and the two solid-rocket boosters be reusable, and that the orbiter have a 160-hour turnaround time (1972 design concept requirement); that is, be ready to return to space 160 hours (two weeks, based on 80-hour workweeks) after landing from the previous mission.

Other features of the Shuttle:

- The orbiter normally will carry a flight crew of four, plus three additional passengers. A total of 10 persons could be carried under emergency conditions.

- The basic mission is seven days in space; with additional supplies, a 30-day mission is possible.

Space Transportation System

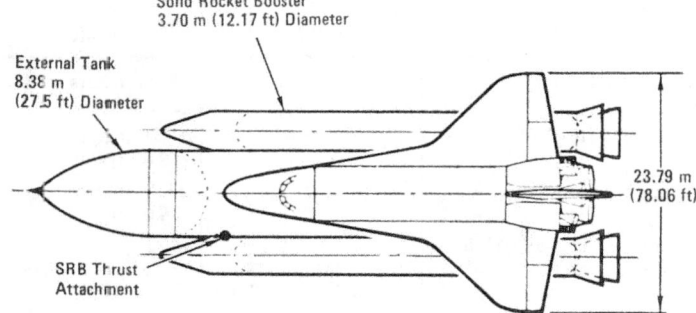

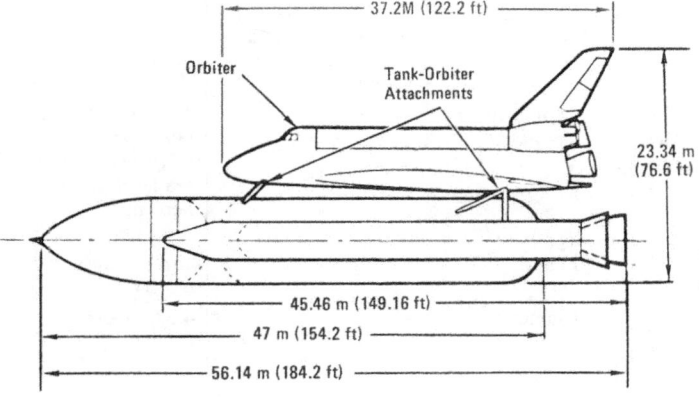

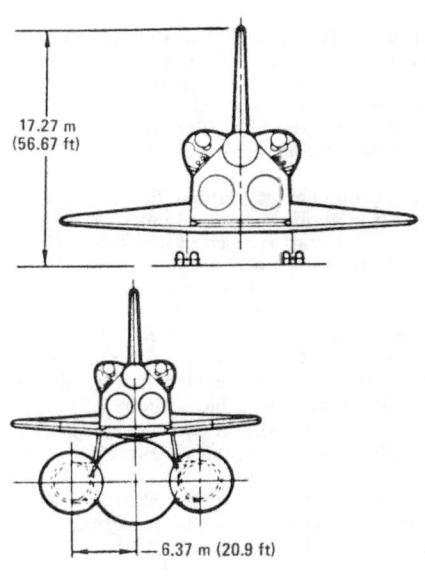

Orbiter	. . .	68,040 kg (150,000 lb) Dry*
SRB (2)	. . .	584,692 kg (1,289,004 lb) Ea.
ET	. . .	738,717 kg (1,628,565 lb)
SSME	. . .	1,751,622 Newtons (393,800 lb)
		104 percent thrust at sea level

*Plus Payload and Consumables
Weights Approximate

Space Shuttle Statistics

 **Space Transportation System**

- The crew compartment has a shirtsleeve environment, and the acceleration load is never greater than 3 g's.

- The Shuttle can be on launch pad standby for up to 24 hours, and can be launched from standby within 2 hours.

- In its return to earth, the orbiter has a cross-range maneuvering capability of 1,100 nautical miles (1,265 statute miles).

The Shuttle is launched in an upright position, with thrust provided by the three main engines and the two solid-rocket boosters. After about two minutes, at an altitude of about 24 nautical miles (28 miles), the two boosters are spent and are separated from the orbiter. They fall into the ocean at predetermined points and are recovered for reuse.

The main engines continue firing for about eight minutes, cutting off at about 59 nautical miles (68 statute miles) altitude just before the craft is inserted into orbit. The external tank is separated. It follows a ballistic trajectory back into a remote area of the ocean but is not recovered.

Two smaller liquid rocket engines (orbital maneuvering system) are used to put the orbiter into orbit, for maneuvers while in orbit, and for slowing the vehicle for reentry. The spacecraft's velocity in orbit is about 7,743 meters per second (25,405 feet per second); the deorbit velocity decrease is approximately 91 meters per second (300 feet per second).

After reentry, the unpowered orbiter glides to earth and lands on a runway like an airplane. Normal touchdown speed is 184 to 196 knots (213 to 226 mph).

There are two launch sites for the Shuttle: Kennedy Space Center (KSC), FL, and Vandenberg Air Force Base, CA. The orbiter normally will land at the site from which it was launched.

Basic Shuttle characteristics:

	Overall Shuttle	Orbiter
Length	56.14 meters (184.2 ft)	37.24 meters (122.2 ft)
Height	23.34 meters (76.6 ft)	17.27 meters (56.67 ft)
Wingspan	—	23.79 meters (78.06 ft)
Approx weight Gross liftoff	2,041,200 kilograms (4.5 million pounds)	—
Landing with payload	—	96,163 kilograms (212,000 pounds)
Thrust (sea level) Solid-rocket boosters	14,409,740 Newtons (3,239,600 pounds) of thrust each in vacuum	—
Orbiter main engines	—	1,751,622 Newtons (393,800 pounds) of thrust each at sea level at 104 percent
Cargo bay: Length	—	18.28 meters (60 ft)
Diameter	—	4.57 meters (15 ft)

Space Transportation System

BACKGROUND AND STATUS

On July 26, 1972, NASA selected Rockwell's Space Transportation and Systems Group in Downey, CA, as the industrial contractor for design, development, test, and evaluation (DDT&E) of the orbiter. The contract called for fabrication and testing of two orbiter spacecraft, a full-scale structural test article, and a main propulsion test article. The award followed years of NASA and Air Force studies on definition and feasibility of a reusable space transportation system.

NASA previously (March 31, 1972) had selected Rockwell's Rocketdyne Division to design and develop the Space Shuttle main engines. Contracts followed to Martin Marietta for the external tank (August 16, 1973) and Thiokol's Wasatch Division for the solid-rocket boosters (June 27, 1974).

In addition to the orbiter DDT&E contract, Rockwell's Space Transportation and Systems Group was given contractual responsibility as industrial integrator for the overall Shuttle system.

Rockwell International's Launch Operations, part of the Space Transportation and Systems Group Shuttle Integration and Satellite Systems Division, Seal Beach, CA, was under contract to NASA's Kennedy Space Center, FL, for turnaround, processing, pre-launch testing, launch and recovery operations from STS-1 through the STS-11 missions.

The first orbiter spacecraft, *Enterprise* (OV—orbiter vehicle—101), was rolled out on Sept. 17, 1976. On Jan. 31, 1977, it was transported overland from Rockwell's assembly facility at Palmdale, CA, to the Dryden Flight Research Facility at Edwards Air Force Base (36 miles) for the approach and landing test (ALT) program.

The ALT program was conducted from February through November, 1977, and demonstrated that the orbiter could fly in the atmosphere and land as an airplane. The program consisted of:

- Five "captive" flights in which the orbiter, unmanned, was mounted atop a specifically modified 747 Shuttle carrier aircraft (SCA).

- Three manned captive flights in which two-man astronaut crews operated the orbiter control systems.

- Five "free" flights, in which the orbiter was released from the SCA and maneuvered to a landing at Edwards. In the first four such flights the landing was on a dry lake bed; in the fifth, the landing was on Edwards' main concrete runway under conditions simulating a return from space. The last two free flights were made without the tail cone, which is the spacecraft's configuration during an actual landing from earth orbit.

On March 13, 1978, the *Enterprise* was ferried atop the SCA to NASA's Marshall Space Flight Center in Huntsville, Ala., to undergo a series of mated vertical ground vibration tests. These were completed in March, 1979. On April 10, 1979, the *Enterprise* was ferried to the Kennedy Space Center, mated with the external tank and solid rocket boosters, and transported via the mobile launch platform to launch complex 39A. At launch complex 39A, the *Enterprise* served as a practice and launch complex fit check verification tool representing the flight vehicles. It was ferried back to NASA's Dryden Flight Research Facility at Edwards AFB, CA, on August 16, 1979, and then returned overland to Rockwell's Palmdale final assembly facility on October 30, 1979. Certain components were refurbished for use on flight vehicles being assembled at Palmdale. The *Enterprise* was then returned overland to NASA's Dryden Flight Research Facility on September 6, 1981. In the May-June 1983 time period, *Enterprise* was ferried to the Paris, France air show, as well as to Germany, Italy, England, and Canada and returned to the Dryden Flight Research Center.

 Space Transportation System

It will eventually be used as a practice and fit check verification tool at Vandenberg AFB. The *Enterprise* was built as a test vehicle and is not equipped for space flight.

The second orbiter (OV-102), *Columbia*, was the first to fly into space. It was transported overland on March 8, 1979, from Palmdale to Edwards for mating atop the SCA and ferrying to Kennedy Space Center, FL. It arrived at KSC on March 25, 1979 to begin preparations for the first flight into space.

The structural test article, after 11 months extensive testing was modified at Rockwell's final assembly facility at Palmdale to become the second orbiter available for operational missions. It was redesignated OV-099, the *Challenger*.

The main propulsion test article (MPTA-098) consisted of an orbiter aft fuselage, a truss arrangement which simulated the orbiter mid fuselage, and the Shuttle main propulsion system (three main engines and the external tank). This test structure is at the National Space Technology Laboratory in Mississippi. A series of static firings were conducted in 1978 through 1981 in support of the first flight into space.

On Jan. 29, 1979, NASA contracted with Rockwell for the manufacture of two additional orbiters, OV-103 and OV-104 (*Discovery* and *Atlantis*), conversion of the STA to space flight configuration (*Challenger*), and modification of *Columbia* from its developmental configuration to that required for operational flights.

LAUNCH SITES

During normal operations, Shuttle launchings will be from both KSC and the Western Test Range (Vandenberg AFB). Shuttle missions destined for equatorial orbital trajectories will be launched from KSC and those requiring polar orbital planes will be launched from WTR.

NASA named the first four orbiter spacecraft after famous sailing ships. In the order they will become operational, they are:

- *Columbia* **(OV-102)**, after a sailing frigate launched in 1836, one of the first Navy ships to circumnavigate the globe. Columbia also was the name of the Apollo 11 command module which carried Neil Armstrong, Michael Collins, and Edward (Buzz) Aldrin on the first lunar landing mission, July 20, 1969.

- *Challenger* **(OV-099)**, also a Navy ship, which from 1872 to 1876 made a prolonged exploration of the Atlantic and Pacific Oceans. It also was used in the Apollo program, for the Apollo 17 lunar module.

- *Discovery* **(OV-103)**, after two ships, Henry Hudson's which in 1610-11 attempted to search for a northwest passage between Atlantic and Pacific oceans and instead discovered Hudson Bay, and Captain Cook's which discovered the Hawaiian Islands and explored southern Alaska and western Canada.

- *Atlantis* **(OV-104)**, after a two-masted ketch operated for the Woods Hole Oceanographic Institute from 1930 to 1966, which traveled more than half a million miles in ocean research.

Space Transportation System

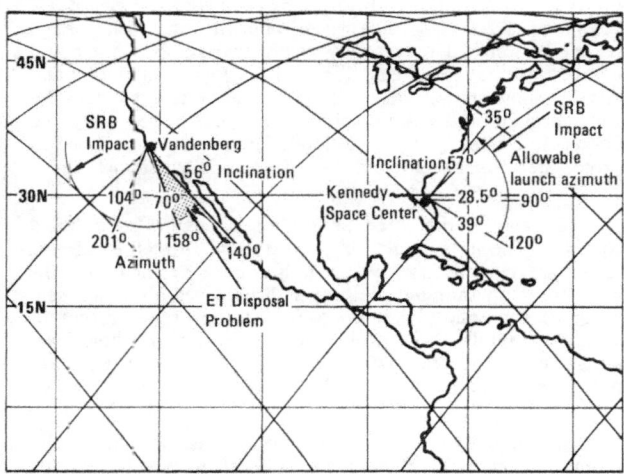

Space Shuttle Launch Sites

Orbital mechanics and the complexities of mission requirements, plus safety and the possibility of infringement on foreign air and land space, prohibit polar orbit launches from KSC.

KSC launches have an allowable path no less than 35 degrees northeast and no greater than 120 degrees southeast. These are azimuth degree readings based on due east from KSC as 90 degrees.

A 35-degree azimuth launch places the spacecraft in an orbital inclination of 57 degrees. This means the spacecraft in its orbital trajectories around the earth will never exceed an earth latitude higher or lower than 57 degrees north or south of the equator.

A launch path from KSC at an azimuth of 120 degrees will place spacecraft in an orbital inclination of 39 degrees (it will be above or below 39 degrees north or south of the equator).

These two azimuths 35 and 120 degrees—represent the launch limits from KSC. Any azimuth angles further north or south would launch a spacecraft over a habitable land mass, adversely affect safety provisions for abort or vehicle separation conditions, or would be undesirable because of the possibility that the solid rocket boosters or external tank could land on foreign land or sea space.

Launches from the Western Test Range have an allowable launch path suitable for polar insertions south, southwest, and southeast.

The launch limits at WTR are 201 and 158 degrees. At a 201-degree launch azimuth, the spacecrft would be orbiting at a 104-degree inclination. Zero degree would be due north of the launch site and the orbital trajectory would be within 14 degrees east or west of the north-south pole meridian. At a launch azimuth of 158 degrees, the spacecraft would be orbiting at a 70-degree inclination, the trajectory would be within 20 degrees east or west of the polar meridian. Similar to KSC, the WTR has allowable launch azimuths which do not pass over habitable areas or affect safety, abort, separation, and political considerations.

Mission requirements and payload weight penalties also are major factors in selecting two launch sites.

The earth rotates from west to east at a speed of approximately 900 nautical miles per hour (1,035 miles per hour). A launch to the east uses the earth's rotation somewhat as a springboard. This means, for example, that the Shuttle can carry a 29,484-kilogram (65,000-pound) payload from a KSC launch, but only 18,144 kilograms (40,000 pounds) with a

Space Transportation System

launch inclination of 90 degrees from WTR. Incidentally, the earth's rotational rate is also the reason the orbiter has a cross-range capability of 1,100 nautical miles (1,265 statute miles) to provide the abort once around (AOA) capability in polar orbit launches.

Attempting to launch and place a spacecraft in polar orbit from KSC to avoid habitable land mass would be uneconomical because the Shuttle's payload would be reduced severely—down to 7,711 kilograms (17,000 pounds). A northerly launch into polar orbit 8 to 20 degrees azimuth—would necessitate a path over a land mass, and most safety, abort, and political constraints would have to be waived. This prohibits polar orbit launches from KSC.

The following orbital insertion inclinations and payload weights exemplify the Shuttle's capabilities:

1. Equatorial orbit from KSC (for low earth orbit, geosynchronous orbit, or interplanetary escape)—at an orbital inclination of 28.5 degrees from KSC, maximum payload weight is 29,484 kilograms (65,000 pounds); at an inclination of 57 degrees, maximum payload weight is 25,855 kilograms (57,000 pounds).

2. Polar orbit from WTR—at an orbital inclination of 90 degrees, maximum payload weight is 18,144 kilograms (40,000 pounds) or 14,515 kilograms (32,000 pounds) at an inclination of 104 degrees.

MISSION PROFILE

In the launch configuration, the orbiter and two solid-rocket boosters are attached to the external tank and all are in a vertical position (nose up) on the launch pad. Each solid-rocket booster is attached at its aft skirt to the mobile launch platform by four bolts.

Emergency exit for the flight crew while on the launch pad up to 30 seconds prior to liftoff is by slide wire. There are five 365-meter (1,200-foot) slide wires, each with one basket. Each basket is designed to carry normally two persons but could handle three. The baskets, 1.5 meters (5 feet) in diameter and 1,066 millimeters (42 inches) deep, are suspended beneath the slide mechanism by four cables. The slide wires carry the baskets to a bunker at ground level. The bunker is designed to protect personnel even from an explosion on the launch pad.

At launch, the three Shuttle main engines—fed liquid hydrogen fuel and liquid oxygen oxidizer from the external tank—are ignited first. When it has been verified that the engines are operating at the proper thrust level, a signal is sent to ignite the solid-rocket boosters. At the proper thrust-to-weight ratio, initiators (small explosives) at the eight holddown bolts are fired to release the Shuttle for liftoff. All this takes only a few seconds.

Maximum dynamic pressure (max q) is reached early in the ascent, nominally at 10,241 meters (33,600 feet) at 60 seconds after liftoff.

Approximately a minute later (two minutes into the ascent phase), the two solid-rocket boosters have consumed their propellant and are jettisoned from the external tank. This is triggered by a separation signal from the orbiter.

The boosters briefly continue to ascend, while small motors fire to carry them away from the Shuttle. The boosters then turn and descend, and at a predetermined altitude, parachutes are deployed to decelerate them for a safe splashdown in the ocean. Splashdown will occur approximately 141 nautical miles (162 statute miles) from the launch site. The boosters are recovered and reused.

Space Transportation System

Shuttle Mission Profile

MAIN ENGINE CUTOFF, EXTERNAL TANK SEPARATION

Altitude: 59 nmi (68 miles); velocity: 7,796 m/s (25,581 f/s, 17,440 m/h) about 8 minutes after launch (just before orbit insertion)

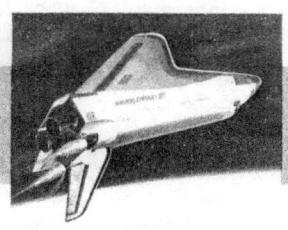

ORBIT INSERTION AND CIRCULARIZATION

Altitude varies according to mission

ORBITAL OPERATIONS

Mission from 7 to 30 days; 100 to 600 nmi (115 to 690 miles) orbits; 7,743 m/s (25,405 f/s, 17,321 m/h)

SRB SEPARATION

Altitude: 24 nmi (28 miles); velocity: 1,383 m/s (4,538 f/s, 3,094 m/h) 2 minutes after launch

DEORBIT

Velocity decreased nominal 91 m/s (300 f/s, 204 m/h) from earth orbit operations

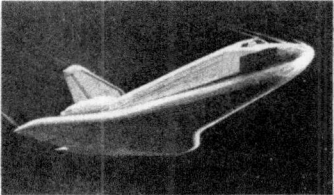

LAUNCH

Maximum dynamic pressure at 10,241 meters (33,600 ft); about 60 seconds after launch

MAINTENANCE Two-week turnaround (14 days — 160 hours)

LANDING Touchdown speed 184 to 196 knots (213 to 226 m/h)

f/s = feet per second   m/s = meters per second   m/h = miles per hour

Space Transportation System

Meanwhile, the orbiter and external tank continue to ascend, using the thrust of the three main engines. Approximately eight minutes after launch and just short of orbital velocity, the three engines are shut down (main engine cutoff—MECO) and the external tank is jettisoned on command from the orbiter.

The external tank continues on a ballistic trajectory and enters the atmosphere, where it disintegrates. Its projected impact is in the Indian Ocean (except for 57 degree inclinations), in the case of equatorial orbits (KSC launch), and in the extreme southern Pacific Ocean, in the case of WTR launch.

Two orbital maneuvering systems (OMS) at the aft end of the orbiter are used in a two-step firing, to complete insertion into earth orbit, and to circularize the spacecraft's orbit. Forward and aft reaction control system (RCS) thrusters provide attitude control (pitch, yaw, and roll) of the orbiter, as well as any minor translation maneuvers along a given axis. The orbiter is designed to operate in earth orbit between 100 to 600 nautical miles (115 to 690 statute miles).

At completion of the orbital operations (from 1 to 30 days), the orbiter is oriented to a tail-first attitude. The two OMS engines are then used to slow the vehicle for deorbit.

The RCS thrusters then turn the orbiter nose forward for entry. These thrusters continue to control the orbiter until atmospheric density is sufficient for the pitch and roll aerodynamic control surfaces to become effective.

Entry is considered to occur at 121,920 meters (400,000 feet) altitude approximately 4,400 nautical miles (5,063 statute miles) from the landing site and at approximately 7,620 meters per second (25,000 feet per second) velocity.

At 121,920 meters the spacecraft is maneuvered to zero degrees roll and yaw (wings level) and a predetermined angle of attack for entry. In the KSC flights, the angle of attack is 40 degrees; in WTR flights, it will be between 28 degrees and 38 degrees. The flight control system issues the commands to roll, pitch, and yaw RCS jets for rate damping.

The forward RCS jets are inhibited at 121,920 meters, and the aft RCS jets maneuver the spacecraft until a dynamic pressure of 0.929 meters squared (10 pounds per square foot) is sensed, which is when the orbiter's ailerons become effective. The aft RCS roll jets are then deactivated. At a dynamic pressure of 1.8 meters squared (20 pounds per square foot), the orbiter's elevators become active and the aft RCS pitch jets are deactiviated. The orbiter's speed brake is used below Mach 10 to induce a more positive downward elevator trim deflection. At Mach 3.5, the rudder becomes activated and the aft RCS yaw jets are deactivated at 13,716 meters (45,000 feet).

Entry guidance must dissipate the tremendous amount of energy the orbiter posseses when it enters the earth's atmosphere to assure that the orbiter does not either burn up (entry angle too steep) or skip out of the atmosphere (entry angle too shallow) and that the orbiter is properly positioned to reach the desired touchdown point.

During entry, energy is dissipated by the atmospheric drag on the orbiter's surface. Higher atmospheric drag levels enable faster energy dissipation with a steeper trajectory. Normally, the angle of attack and roll angle enable the atmospheric drag of any flight vehicle to be controlled. However, for the orbiter, angle of attack was rejected because it creates surface temperatures above the design specification. The angle of attack scheduled during entry is loaded into the orbiter computer as a function of relative velocity, leaving roll angle for energy control. Increasing the roll angle decreases the vertical component of lift, causing a higher sink rate and energy dissipation rate. Increasing the roll rate does raise the surface temperature of the orbiter, but not nearly as drastically as an equal angle of attack command.

Space Transportation System

If the orbiter is low on energy (current range-to-go much greater than nominal at current velocity) entry guidance will command lower than nominal drag levels. If the orbiter has too much energy (current range-to-go much less than nominal at the current velocity), entry guidance will command higher than nominal drag levels to dissipate the extra energy.

Roll angle is used to control cross-range. Azimuth error is the angle between the plane containing the orbiter's position vector and the heading alignment cylinder tangency point and the plane containing the orbiter's position vector and velocity vector. When the azimuth error exceeds a computer-loaded number, the orbiter's roll angle is reversed.

Thus, descent rate and down ranging are controlled by bank angle. The steeper the bank angle, the greater the descent rate and the greater the drag; conversely, the minimum drag attitude is wings level. Cross-range is controlled by bank reversals.

The entry thermal control phase is designed to keep the backface temperatures within the design limits. A constant heating rate is established until below 5,791 meters per second (19,000 feet per second).

The equilibrium glide phase shifts the orbiter from the rapidly increasing drag levels of the temperature control phase to the constant drag level of the constant drag phase. The equilibrium glide flight is defined as flight in which the flight path angle, the angle between the local horizontal and the local velocity vector, remains constant. Equilibrium glide flight provides the maximum downrange capability. It lasts until the drag acceleration reaches 33 feet per second squared.

The constant drag phase begins at that point. In the KSC flights the angle of attack is initially 40 degrees but it begins to ramp down in this phase to approximately 36 degrees by the end of this phase. In WTR flights, it will be between 28 degrees and 38 degrees.

The transition phase is where the angle of attack continues to ramp down, reaching the approximately 14-degree angle of attack at entry terminal area energy management (TAEM) interface, approximately 25,298 meters (83,000 feet) altitude, 762 meters per second (2,500 feet per second), Mach 2.5 and 52 nautical miles (59 statute miles) from the landing runway. Control is then transferred to TAEM guidance.

During the entry phases described, the orbiter's roll commands keep the orbiter on the drag profile and control crossrange.

TAEM guidance steers the orbiter to the nearest of two heading alignment cylinders (HAC's) whose radii are 5,480 meters (18,000 feet), which are located tangent to and on either side of the runway centerline on the runway centerline on the approach end. In TAEM guidance, excess energy is dissipated with an S turn and the speed brake can be utilized to modify drag, L/D (lift/drag) ratio, and flight path angle in high energy conditions. This increases the ground track range as the orbiter turns away from the nearest HAC until sufficient energy is dissipated to allow a normal approach and landing guidance phase capture, which begins at 3,048 meters (10,000 feet) altitude. The orbiter also can be flown near the velocity for maximum lift over drag or wings level for the range stretch case. The spacecraft slows to subsonic velocity at approximately 14,935 meters (49,000 feet) altitude, about 22 nautical miles (25.3 statute miles) from the landing site.

At TAEM acquisition, the orbiter is turned until it is aimed at a point tangent to the nearest HAC and continues until it reaches the point, WP-1 (way point one). At WP-1, the TAEM heading alignment phase begins. The HAC is followed until landing runway alignment plus or minus 20 degrees has been achieved. In the TAEM prefinal phase, the orbiter leaves the HAC, pitches down to acquire the steep glide slope, increases airspeed, and banks to acquire the runway centerline and continues until on the runway centerline, on the outer glide slope,

 **Space Transportation System**

and on airspeed. The approach and landing guidance phase begins with the completion of the TAEM prefinal phase and ends when the spacecraft comes to a complete stop on the runway.

The approach and landing trajectory capture phase begins at the TAEM interface and continues to guidance lock on to the steep outer glide slope. The approach and landing phase begins at about 3,048 meters (10,000 feet) altitude at an equivalent air speed (EAS) of 290 plus or minus 12 knots, 6.9 nautical miles (7.9 statute miles) from touchdown. Autoland guidance is initiated at this point to guide the orbiter to the minus 20° glide slope (which is over seven times that of a commercial airliner's approach) aimed at a target 0.86 nautical mile (one statute mile) in front of the runway. The spacecraft speed brake is positioned to hold the proper velocity. The descent rate in the later portion of TAEM and approach and landing is greater than 3,048 meters (10,000 feet) per minute (approximately 20 times higher rate of descent than a commercial airliner's standard three-degree instrument approach angle).

At 533 meters (1,750 feet) above ground level, a pre-flare maneuver is started to position the spacecraft for a 1.5-degree glideslope in preparation for landing with the speed brake positioned as required. The flight crew deploys the landing gear at this point.

The final phase reduces the sink rate of the spacecraft to less than 2.7 meters per second (9 feet per second). Touchdown occurs approximately 762 meters (2,500 feet) past the runway threshold at a speed of 184 to 196 knots (213 to 226 mph).

ABORTS

The Shuttle has three abort alternatives, depending on when it becomes necessary. These are to return to launch site (RTLS), abort once around (AOA), and to abort to orbit (ATO).

RETURN TO LAUNCH SITE. This mode will be used in the event of a main engine failure between liftoff and the point at which the next abort mode (AOA) is available. RTLS will not begin until the solid-rocket boosters complete their normal thrusting period and are jettisoned, as in a normal ascent.

The Space Shuttle (orbiter and external tank) continues to thrust downrange, with the two remaining main engines, the two OMS, and the four aft +X RCS thrusters firing, until the remaining propellant for the main engines equals the amount required to reverse the direction of flight.

A pitch-around (plus pitch) maneuver is then performed at approximately 5 degrees per second; this places the orbiter and external tank in a "heads-up" attitude, pointing back toward the launch site. Main engine cutoff is commanded when altitude, attitude, flight path angle, heading, weight, and velocity/range conditions combine for acceptable orbiter-external tank separation [tank impact no closer than 24 nautical miles (28 statute miles) from the U.S. coast] and orbiter glides to the launch site runway.

ABORT ONCE AROUND. This mode will be used from approximately two minutes after normal solid-rocket booster separation to the point at which the abort-to-orbit mode becomes available. Again, this abort would occur in the event of a main engine failure.

The Space Shuttle vehicle continues to thrust with the remaining main engines and the OMS and aft RCS +X thrusters. The OMS and RCS thrusting periods terminate when the amount of propellant remaining in these two systems will support OMS thrusting periods after MECO.

Main engine cutoff is followed by jettisoning of the external tank. The OMS thrusters are fired after jettisoning the external tank to obtain an apogee of an intermediate orbit. The second firing of the OMS places the spacecraft into a suborbital

Space Transportation System

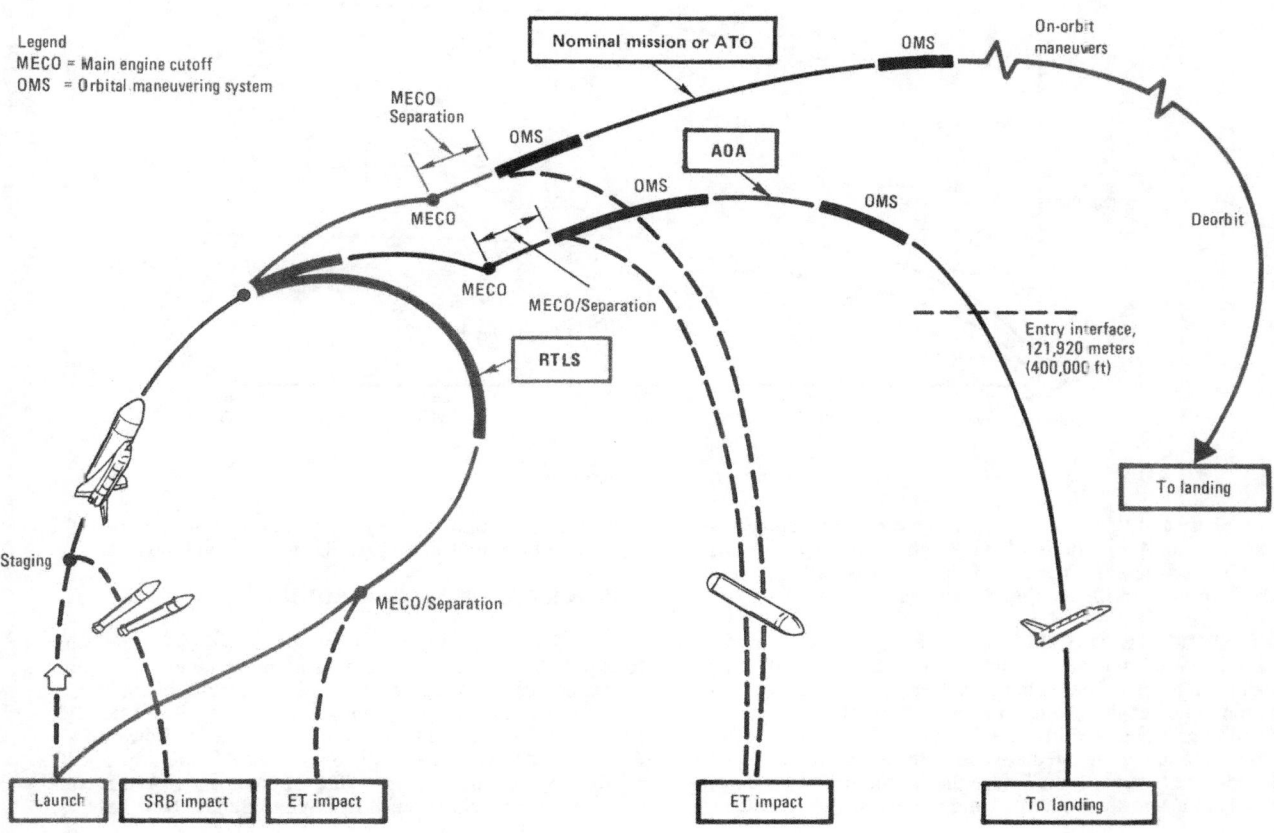

Abort and Normal Mission Profile

Space Transportation System

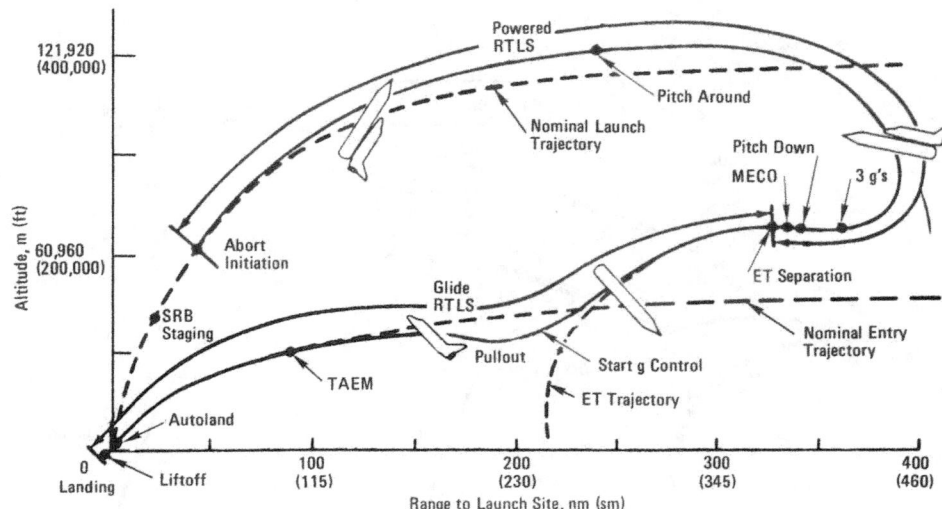

Typical RTLS Abort Profile

coast phase and "free return" orbit for the desired entry interface. The flight conditions—range, flight path angle, headings, and velocity—at entry resulting from this orbit will enable the orbiter to glide to the landing site runway.

ABORT TO ORBIT. This mode begins after the AOA point is passed and also would occur in the event of a main engine failure. The Space Shuttle continues to thrust with the remaining main engines to main engine cutoff and external tank jettisoning. The OMS thrusters fire twice, to insert the orbiter into orbit and then to circularize the orbit. The orbit coast time altitude and the coast time before the deorbit maneuver depend on when the abort was initiated and the mission. Alternate missions may be planned in case of an ATO orbit. The deorbit, entry, and landing would be similar to a normal mission.

ORBITER GROUND TURNAROUND

NASA's Kennedy Space Center, FL, had contracted with Rockwell International for the turnaround maintenance and processing as well as spacecraft recovery ground operations at all primary and contingency landing sites for the STS-1 through STS-11 missions. In addition to the prime sites at Edwards Air Force Base, CA, and the Kennedy Space Center, FL, KSC recovery operations are responsible for landing activities at contingency sites at Northrup Strip, White Sands, New Mexico;

Space Transportation System

ROTA Naval Air Station, Spain; Dakar Senegal; Kadena Air Force Base, Okinawa, and Hickman Air Force Base, HI.

Columbia landed on the dry lakebed runway No. 23 at Edwards AFB, CA (Dryden Flight Research Facility) on the first two Earth orbital flights (STS-1 and -2) and the alternate landing strip at the Army's White Sands Missile Range, NM (Northrup strip No. 17) was used for STS-3. The fourth and fifth flights landed on Edwards AFB concrete runway 22. The first flight of *Challenger* landed on concrete runway 22 at Edwards AFB (STS-6), the second flight of *Challenger* landed on runway 15 at Edwards AFB a the third flight of *Challenger* landed on concrete runway 22 at Edwards AFB. The sixth flight of *Columbia* landed on runway 17 at Edwards AFB.

The spacecraft recovery operations at Edwards Air Force Base or eventually at the Kennedy Space Center is supported by approximately 160 Launch Operations team members. Ground team members wearing self contained atmospheric protective ensemble (SCAPE) suits that protect them from toxic chemicals will approach the spacecraft as soon as it stops rolling. The ground team members will take sensor measurements to insure the atmosphere in the vicinity of the spacecraft is not explosive. In the event of propellant leaks, a wind machine truck carrying a large fan will be moved into the area to create a turbulent air flow that will break up gas concentrations and reduce the potential for an explosion.

An air conditioning purge unit is attached to the orbiter so cool air can be directed through the orbiter's aft fuselage, payload bay, forward fuselage, wings, vertical stabilizer, and orbital maneuvering system/reaction control system pods to dissipate the heat of entry. This heat, if not dissipated, will "soak" to the orbiter systems within 15 minutes of landing.

A second ground cooling unit is connected to the spacecraft Freon coolant loops to provide cooling for the flight crew and avionics during post landing and system checks. The spacecraft fuel cells remain powered up at this time. The flight crew will then exit the spacecraft and a ground crew will power down the spacecraft.

Within one to two hours the spacecraft and ground support equipment convoy will be ready to move the spacecraft to the service area at NASA's Dryden Flight Research Center at Edwards. After detailed inspection and preparations at DFRC, the spacecraft is ferried atop the Shuttle Carrier Aircraft to the Kennedy Space Center.

When the spacecraft lands and completes its runout at the Kennedy Space Center, the same procedures as at Edwards Air Force Base are accomplished, with the exception being that it is expected that only one hour will be required before the spacecraft and convoy is ready to move to the Orbiter Processing Facility (OPF).

The orbiter should be refurbished and readied for another launch in the shortest possible time. Short turnaround decreases the maintenance cost (part of the cost per flight), decreases the number of orbiters and support elements needed, and increases the utilization rate of each orbiter.

The spacecraft is towed to the Orbiter Processing Facility where it is safed (fuel and oxidizer systems drained, tanks purged, and ordnance removed). The OMS and RCS pods are removed and reinstalled, if required, and other vehicle maintenance performed. The payload may be installed in the Orbiter Processing Facility and spacecraft functioning verified.

The spacecraft is then towed to the Vehicle Assembly Building (VAB), and mated to the external tank. These elements were stacked and mated on the mobile launch platform while the orbiter was refurbished. Shuttle connections and the integrated vehicle are checked and ordnance is installed.

Space Transportation System

The mobile launch platform moves the entire Space Shuttle system on four crawlers to the launch pad, where connections are made and servicing, checkout, activities begin. If the payload was not installed in the Orbiter Processing Facility, it will be installed at the launch pad followed by pre-launch activities.

In the event of a landing at an alternate site, a crew of about eight team members will move to the landing site to assist the astronaut crew in preparing the orbiter for loading aboard the Shuttle Carrier Aircraft for transport back to the Kennedy Space Center. If the landing is outside the U.S., personnel at the contingency landing sites will be provided minimum training on safe handling of the orbiter with emphasis on crash rescue training, how to tow the orbiter to a safe area, and prevention of propellant configuration.

Space Shuttle flights from the Western Test Range, CA, will utilize the Vandenberg Launch Facility (SL6) which was built for but never used for the manned orbital laboratory program. This facility will be modified for Space Transportation System use.

The runway at Vandenberg will be strengthened and lengthened from 2,438 meters (8,000 feet) to 3,657 meters (12,000 feet) to accommodate the orbiter returning from space.

Shuttle buildup at Vandenberg will differ from the NASA Kennedy Space Center plan in that the integration-on-pad technique will be employed. Solid rocket boosters will start on-the-pad buildup followed by the external tank. The orbiter will then be mated to the external tank.

The orbiter maintenance and checkout facility at Vandenberg will be used for orbiter processing. It will also provide an area for processing security classified payloads. SL6 includes the launch mount, access tower, mobile service tower, launch control tower, payload preparation room, payload changeout room, solid rocket booster refurbishment facility, solid rocket booster disassembly facility and liquid hydrogen and liquid oxygen storage tanks.

The launch processing system will be similar to the one at the Kennedy Space Center.

PAYLOADS

The Space Shuttle is a transportation system. What it carries to earth orbit and back is its reason for existence. The Shuttle orbiter's 18.28-meter-long (60-foot) cargo bay will take hundreds of payloads into space.

All of NASA's centers are vitally concerned with the Shuttle's payload capabilities. In the first 11 years of operation through 1991, more than 400 Shuttle missions will be flown, each carrying one or more payloads. Estimates are that nearly every federal agency—as well as universities, the scientific community, and private industry—will have Shuttle payloads during the rest of this century.

Payloads already carried aboard the orbiter include the Tracking and Data Relay Satellites, communications satellites, Spacelab, and Department of Defense non-weapon military payloads. In addition to the payloads already carried into space as well as additional payloads of similar types in the future flights, additional payloads include the Space Telescope, multimission modular spacecraft and Long Duration Exposure Facility. The Department of Defense will total about 25 percent of the Shuttle missions.

Satellites that require high orbital altitudes will be carried aboard the orbiter with a payload assist module (PAM) or inertial upper stage (IUS) or other type of booster stage. The satellite and upper stage will be deployed together from the

 **Space Transportation System**

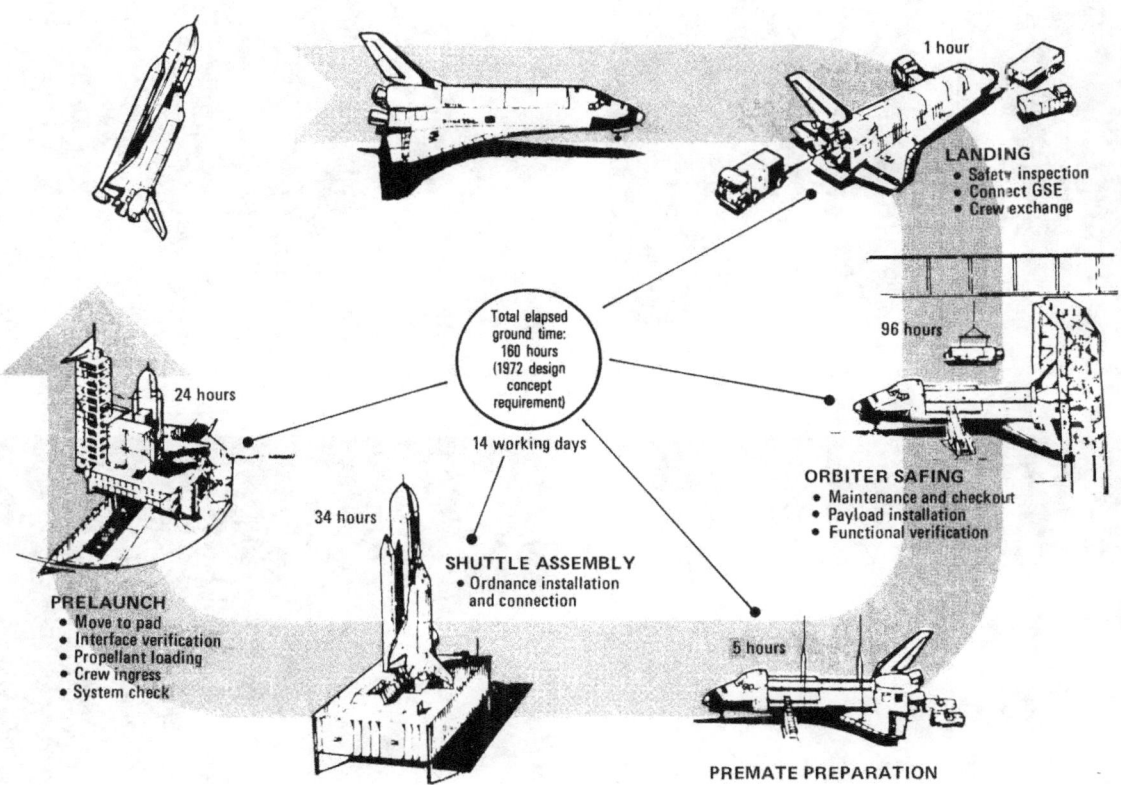

Ground Turnaround Sequence

 **Space Transportation System**

Satellite Business Systems Deployment with Payload Assist Module
STS-5

Tracking Data Relay Satellite Deployment with Inertial Upper Stage
STS-6

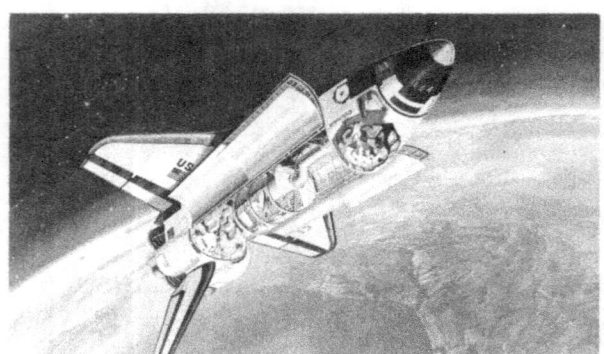
Spacelab-1

Space Telescope

Space Transportation System

payload bay, and the upper stage is used to boost the satellite to the required orbit. Weather, communication, and navigation satellites, as well as deep space probes, would require upper stages.

SPACE SHUTTLE FLIGHTS

In preparation for the first flight into space of Orbiter 102 (*Columbia*), tens of thousands of hours of tests and simulations were expended at government and contractor facilities throughout the nation to qualify all the structures, flight equipment, and computer programs software.

Major test programs included the 13 flights of the Approach and Landing Test (ALT) program at NASA's Dryden Flight Research Center, Edwards Air Force Base, CA. The main propulsion test article—consisting of the orbiter aft fuselage, three Space Shuttle main engines, external tank, and truss arrangement to simulate the mid fuselage—qualified the main propulsion system at the National Space Technology Laboratory in Mississippi. The solid rocket booster's solid-propellant rocket motor was qualified at Thiokol Chemical Corp., near Brigham City, Utah. The orbiter reaction control and orbital maneuvering systems were qualified at NASA's White Sands Test Facility near Las Cruces, NM Flight simulation and avionics testing qualified the Space Shuttle avionics at JSC and at Rockwell's Space Transportation and Systems Group in Downey, CA. The orbiter full-scale structure was tested at Lockheed's facility in Palmdale, CA. The orbiter vehicle (OV-101), external tank and solid rocket boosters completed a fit check and checkout with the mobile launch platform and launch complex 39A at KSC.

The initial four flights of *Columbia* were launched from KSC and verified design and operational capability of the Shuttle and all of the ground-based monitoring, communications, and support systems. The first flight was structured to minimize risks and complexity and was two days in duration. Because of a shutdown fuel cell, the second flight was also two days in duration and contained the OSTA-1 (Office of Space Terrestrial and Applications) pallet and the RMS (remote manipulator system). The third flight contained the OSS-1 (Office of Space Science) pallet and the RMS and was scheduled for seven days, however due to weather conditions at the White Sands Missile Range, NM, lakebed landing strip, the mission was extended one day with the eventual landing at the White Sands Missile Range, NM lakebed runway. The fourth flight was a seven day mission and developed and demonstrated mission and payload capabilities with a Department of Defense (DOD) payload and was the first landing on a concrete runway, No. 22 at Edwards Air Force Base. The fifth flight of *Columbia* got down to the "real business" and purpose of the Space Transportation System, hauling cargo into earth orbit and successfully deploying two commercial communication satellites, Satellite Business System (SBS)-C and a Canadian communication satellite ANIK-C and after five days landed on concrete runway 22 at Edwards Air Force Base. *Columbia*'s next flight, carried the European Space Agency (ESA) Spacelab-1 into earth orbit with a flight crew of six. In 1984 the *Columbia* will be returned to Rockwell's Palmdale, CA, plant for major modification which includes installation of "operational systems."

Challenger was the first "operational Shuttle spacecraft." All of its on-board systems were qualified to operate for a minimum of 100 missions without major overhaul. Although *Columbia*, on its STS-5 flight, was "billed" as the first operational mission, the onboard spacecraft systems did not have "operational certification." Structural fabrication of *Challenger* at Rockwell International's California plant actually got underway in 1975, about a year earlier than the origin of *Columbia*. In transitional periods which took *Challenger* from structural assembly to becoming the Structural Test Article (STA-099), then back into a manufacturing modification period

Space Transportation System

for uprating to "flight worthy" status (OV-099), to final assembly and checkout, rollout, ferry flight to Kennedy Space Center, prelaunch testing and its first flight into earth orbit.

Discovery was delivered to the Kennedy Space Center in November, 1983. At Rockwell's Palmdale facility, *Atlantis* is in final assembly and checkout and is scheduled for delivery late 1984.

KSC launch operations has responsibility for all mating, prelaunch testing, and launch control ground activities until the Space Shuttle vehicle has cleared the umbilical tower; then responsibility is turned over to JSC's Mission Control Center in Houston. This responsibility includes earth entry and approach and landing until runout completion, at which time the orbiter is handed over to the post-landing operations at the landing site for turnaround and relaunch. KSC also will process the solid rocket boosters and external tank for launch, as well as for recycling the solid rocket boosters for reuse.

SOLID-ROCKET BOOSTERS

The two solid-rocket boosters (SRB's) provide the main thrust to lift the Space Shuttle off the pad and up to an altitude of about 45,720 meters (150,000 feet), 24 nautical miles (28 statute miles). In addition, the two SRB's carry the entire weight of the external tank and orbiter and transmit the weight load through their structure to the mobile launch platform. Each booster has a thrust (sea level) of 14,409,740 Newtons (3,239,600 pounds) at launch. They are ignited after the three Shuttle main engine thrust level is verified. The two SRB's provide 71.4 percent of the thrust at lift-off and during first stage ascent. After SRB separation, 75 seconds later, SRB apogee occurs at an altitude of 67,056 meters (220,000 feet), 35 nautical miles (41 statute miles). SRB impact occurs in the ocean around 122 nautical miles (141 statute miles) downrange.

The SRB's are the largest solid-propellant motors ever flown and the first designed for reuse. Each is 45.46 meters (149.16 feet) long and 3.70 meters (12.17 feet) in diameter.

Prior to the STS-6 mission, each solid rocket booster weighed approximately 586,504 kilograms (1,293,000 pounds) at launch. The propellant weight for each solid rocket motor was approximately 502,588 kilograms (1,108,000 pounds). The inert weight of each solid rocket booster was approximately 83,916 kilograms (185,000 pounds). The initial thrust of each solid rocket booster was approximately 13,811,040 Newtons (3,150,000 pounds).

Beginning with the STS-6 mission, each solid rocket booster weighed approximately 586,051 kilograms (1,292,000 pounds) at launch. The propellant weight for each solid rocket motor is approximately 503,949 kilograms (1,111,000 pounds). The inert weight of each solid rocket booster is approximately 82,101 kilograms (181,000 pounds). This 1,814 kilogram (4,000 pound) weight reduction for each solid rocket motor was achieved by reducing the thickness of each solid rocket motor steel casing approximately two to four hundreths of an inch. Areas that are reduced were the cylindrical, attach, and stiffener segments. The initial thrust for each solid rocket booster remains the same as prior to STS-6.

Beginning with the STS-8 mission, each solid rocket

Space Transportation System

booster utilizes new high performance solid rocket motors with the lightweight shaved casings used beginning with STS-6. The propellant and inert weight of each solid rocket booster remains the same as those beginning with the STS-6 mission. The high performance solid rocket motors provide an initial thrust increase to approximately 14,678,400 Newtons (3,300,000 pounds). The increase in thrust was achieved by lengthening the exit cone of the solid rocket motor nozzles by 254 millimeters (10 inches) and decreasing the solid rocket motor nozzles throat diameter by 101 millimeters (4 inches) which increases the velocity of the solid rocket motor gases as they exit through the nozzle. Also, some of the solid rocket motor propellant inhibitor used in the four segments in each solid rocket motor is omitted, thus causing the propellant to burn faster.

In 1985, the solid rocket booster steel shaved skin casings will be replaced with filament wound casings using the high performance motors. The filament wound casings result in a decrease of the inert weight to approximately 67,132 kilograms (148,000 pounds) for each solid rocket booster which results in a weight reduction of approximately 146,784 Newtons (33,000 pounds) for each solid rocket booster to that of the steel shaved skin casings. The propellant weight for each solid rocket booster remains the same as those beginning with STS-6, the thrust remains the same as those, beginning with STS-8, and the nozzle and propellant inhibitor is the same as those, beginning with STS-8. The total eight with the filament wound casings for each solid rocket booster will then be approximately 571,082 kilograms (1,259,000 pounds) at launch. The solid rocket boosters are designed for 20 reuses.

Primary elements of each booster are the motor (including case, propellant, igniter, and nozzle), structure, separation systems, operational flight instrumentation, recovery avionics, pyrotechnics, deceleration system, thrust vector control system and range safety destruct system.

Each booster is attached to the external tank at the SRB's aft frame by two lateral sway braces and a diagonal attachment. The forward end of each SRB is attached to the external tank at the forward end of the SRB forward skirt. On the launch pad, each booster also is attached to the mobile launch platform at the aft skirt by four bolts which are severed by small explosives at liftoff.

The propellant mixture in each SRB motor consists of an ammonium perchlorate (oxidizer, 69.6 percent by weight), aluminum (fuel, 16 percent), iron oxide (a catalyst, 0.4 percent), a polymer (a binder that holds the mixture together, 12.04 percent), and an epoxy curing agent (1.96 percent). The propellant is an 11-point star-shaped perforation in the forward motor segment and a double-truncated-cone perforation in each of the aft segments and aft closure. This configuration provides high thrust at ignition, then reduces the thrust by approximately a third 50 seconds after liftoff to prevent overstressing of the vehicle during maximum dynamic pressure (max q).

The SRB's are interchangeable. They are used as matched pairs and each is made up of four solid rocket motor segments. The pairs are matched by loading each of the four motor segments in pairs from the same batches of propellant ingredients to minimize any thrust imbalance. The segmented casing design give maximum flexibility in fabrication and ease of transportation and handling. Each segment is shipped to the launch site on a heavy duty rail car with a specially built cover.

The nozzle expansion ratio of each booster beginning with the STS-8 mission is 7:79. The nozzle is gimbaled for thrust vector (direction) control. Each SRB has its own redundant auxiliary power units and hydraulic pumps. The all-axis gimbaling capability is 8 degrees. Each nozzle has a carbon cloth liner which erodes and chars during firing. The nozzle is a convergent-divergent, movable design in which an aft pivot-point flexible bearing is the gimbal mechanism.

 **Space Transportation System**

The cone-shaped aft skirt reacts the aft loads between the SRB and the mobile launch platform. The four aft separation motors are mounted on the skirt. The aft section contains avionics, thrust vector control system which consists of two auxiliary power units and hydraulic pumps, hydraulic systems, and nozzle extension jettison system.

The forward section of each booster contains avionics, sequencer, forward separation motors, nose cone separation system, drogue and main parachutes, recovery beacon, recovery light, and a range safety system.

Each SRB has two integrated electronic assemblies, one forward and one aft. The forward assembly jettisons the nozzle after burnout, initiates release of the nose cap and frustum, detaches the parachutes, and turns on the recovery aids. The aft assembly, mounted in the external tank/SRB attach ring, connects with the forward assembly and the orbiter avionics systems for SRB ignition commands and nozzle thrust vector control. Each integrated electronic assembly has a multiplexer/demultiplexer which sends or receives more than one message, signal, or unit of information on a single communication channel.

Eight booster separation motors (four in the nose frustum and four in the aft skirt) of each SRB thrust for 1.02 second at SRB separation from the external tank. Each solid-rocket separation motor is 789 millimeters (31.1 inches) long and 325 millimeters (12.8 inches) in diameter.

Location aids are provided for each SRB, frustum/drogue chute, and main parachutes. These include a transmitter, antenna, strobe/converter, battery, and salt water switch electronics. The location aids are designed for a minimum operating life of 72 hours and are considered usable up to 20 times by refurbishment. The flashing light is an exception. It has an operating life of 280 hours. The battery is used only once.

The SRB nozzle extensions are not recovered.

The recovery crew retrieves the SRB's, frustum/drogue chutes, and main parachutes. The nozzles are plugged, dewatered, and towed back to the launch site. Each booster is removed from the water and components disassembled and washed with fresh and deionized water to limit salt water corrosion. The motor segments, igniter, and nozzle are shipped back to Thiokol for refurbishment.

Each SRB incorporates range safety system which includes a battery power source, receiver/decoder, antennas, and ordnance.

HOLD DOWN POSTS. Each solid rocket booster (SRB) has four hold-down posts that fit into corresponding support posts on the mobile launch platform. Hold-down bolts hold the SRB and launch platform posts together. Each bolt has a nut at each end, but only the top nut is frangible. It contains two NASA standard initiators (NSI's), which are ignited at solid rocket motor ignition commands.

When the two NSI's are ignited at each hold down, the hold-down bolt travels downward because of the release of tension in the bolt (pre-tensioned prior to launch), NSI gas pressure, and gravity. The bolt is stopped by the stud deceleration stand, which contains sand. The SRB bolt is 711 millimeters (28 inches) long and is 88 millimeters (3.5 inches) in diameter.

The solid rocket motor ignition commands are issued by the orbiter's computers through the master event controllers (MEC's) to the hold-down pyrotechnic initiator controllers (PIC's) on the mobile launch platform. They provide the ignition to the hold-down NSI's. The launch processing system (LPS) monitors the SRB hold-down PIC's for low voltage during the last 16 seconds before launch. PIC low voltage will initiate a launch hold.

 **Space Transportation System**

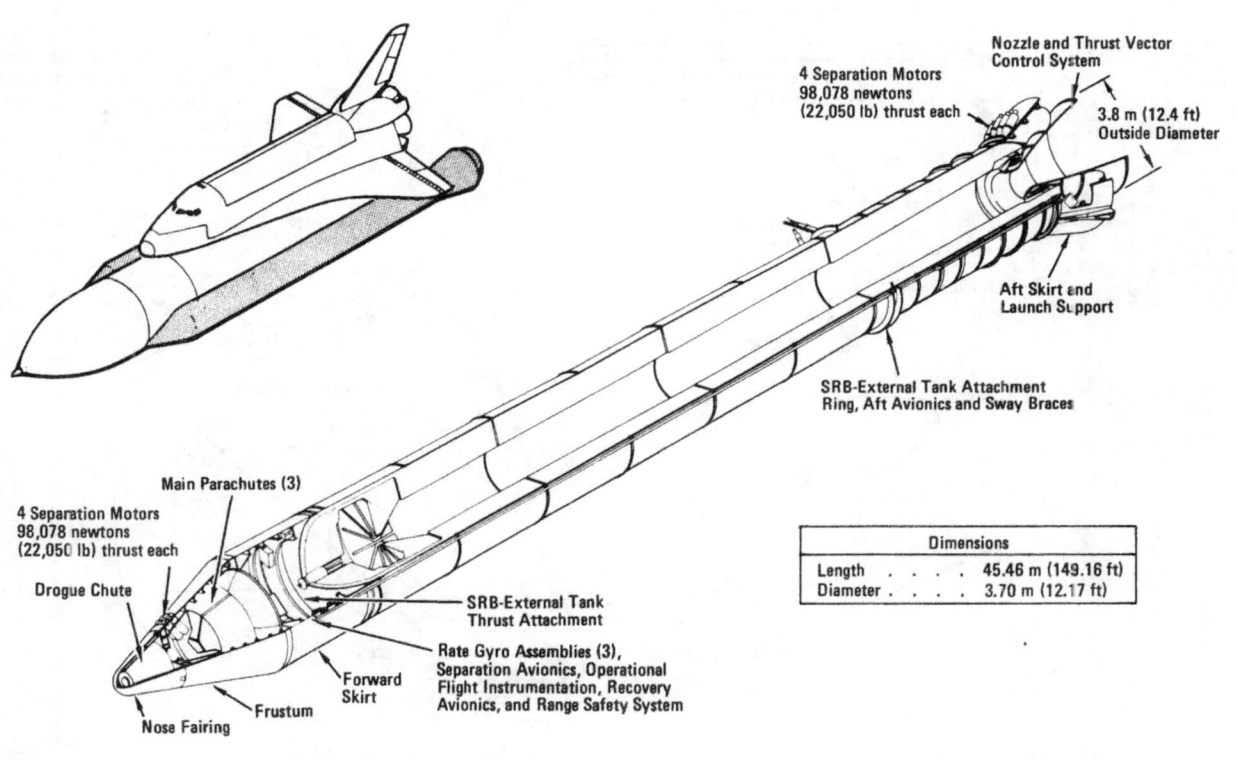

Solid Rocket Booster

Space Transportation System

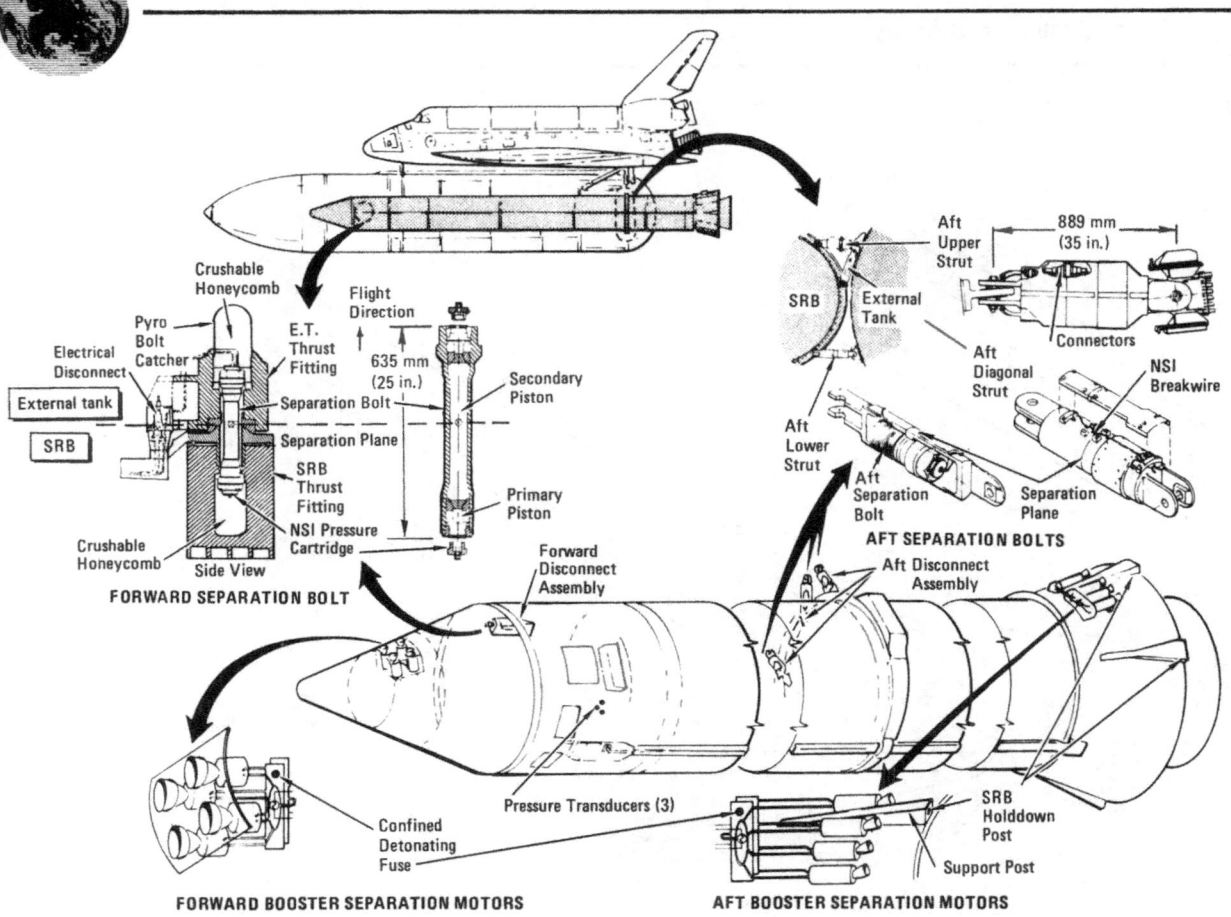

Separation System Elements

Space Transportation System

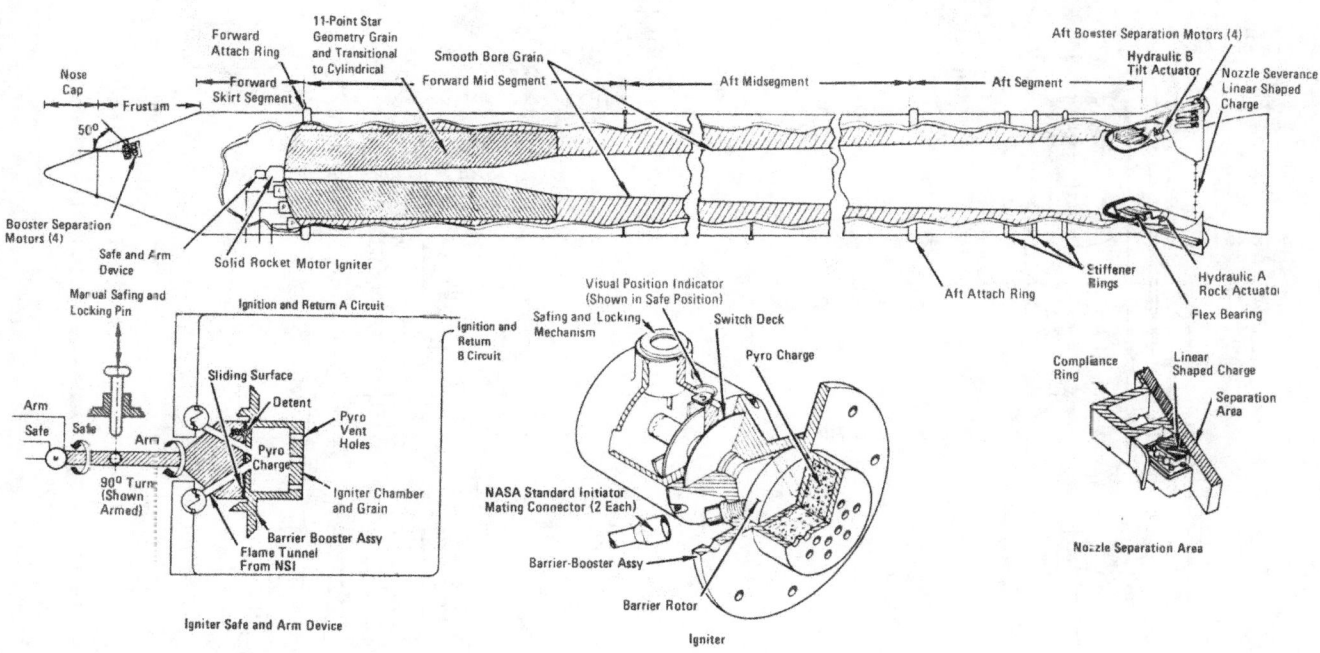

Motor Segments, Safe and Arm Device, Igniter, and Nozzle Severance

Space Transportation System

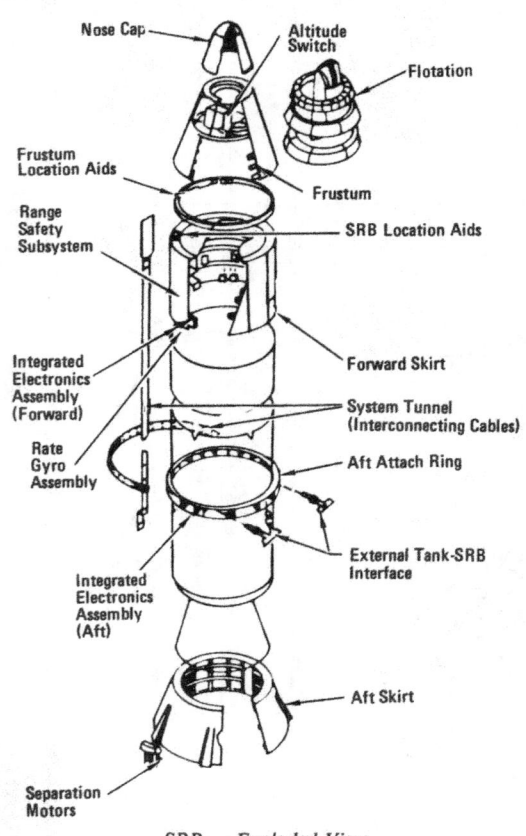

SRB — Exploded View

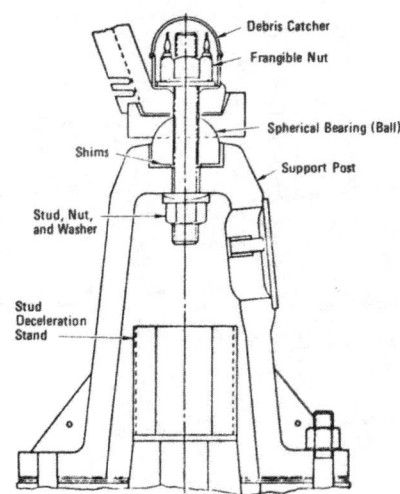

SRB Holddown Configuration

SRB Support/Hold-Down Post

 **Space Transportation System**

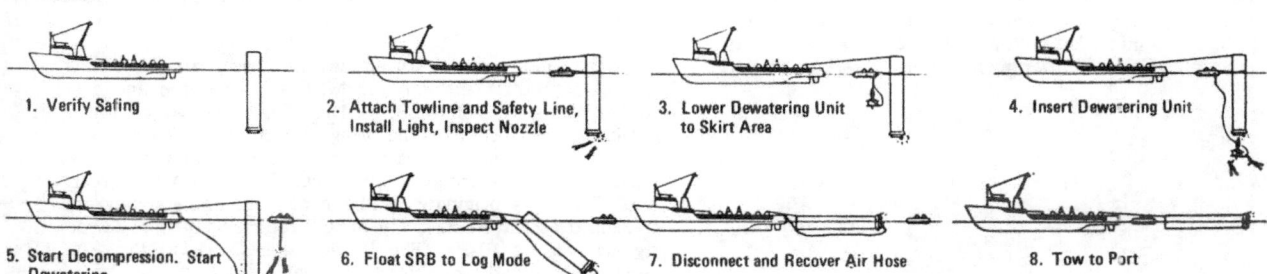

SRB Recovery Procedure

SRB Dewatering Units

UTC Freedom

Space Transportation System

SRB IGNITION. SRB ignition can occur only when a manual lock pin from each SRB safe and arm device has been removed. The ground crew removes the pin during pre-launch activities. The solid rocket motor ignition commands are issued when the three Space Shuttle main engines (SSME's) are at or above 90 percent rated thrust, no SSME fail and/or SRB ignition PIC low voltage is indicated, and there are no holds from the LPS.

The solid rocket motor ignition commands are sent by the orbiter computers through the MEC's to the safe and arm device NSI's in each SRB. A PIC single-channel capacitor discharge device controls the firing of each pyrotechnic device. Three signals must be present simultaneously for the PIC to generate the pyro firing output. These signals — Arm, Fire 1, and Fire 2 — originate in the orbiter computers and are transmitted to the MEC's. The MEC's reformats them to 28 vdc signals for the PIC's. The "arm" signal charges the PIC capacitor to 40 vdc (minimum of 20 vdc).

The "arm" signal causes a barrier rotor to move into a position from which redundant NSI's fire through a thin barrier seal down a flame tunnel. This ignites a pyro booster charge,

UTC Liberty

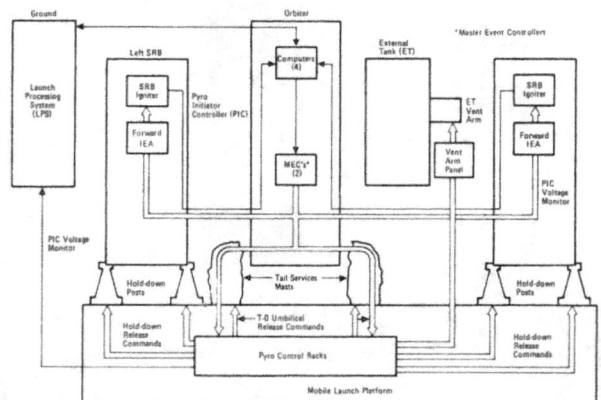

Solid Rocket Ignition, Hold-Down Release, and Umbilical Retract Commands

Space Transportation System

which is retained in the safe arm device behind a perforated plate. The booster charge ignites the propellant in the igniter initiator, and combustion products of this propellant ignite the solid rocket motor initiator, which fires down the length of the solid rocket motor igniting the solid rocket motor propellant.

The computer launch sequence also controls certain critical main propulsion system valves and monitors the engine-ready indications from the main engines. The MPS start commands are issued by the onboard computers at T minus 6.6 seconds (staggered start — engine three, engine two, engine one — all approximately within one-fourth of a second) and the sequence monitors the thrust buildup of each engine. All three engines must reach the required 90 percent thrust within 3 seconds or an orderly shutdown is commanded up to 4.6 seconds and safing functions are initiated.

Normal thrust build-up to the required 90-percent thrust level will result in the engines begin commanded to the liftoff position at T minus 3 seconds as well as the Fire 1 command being issued to arm the SRB's. At T minus 3 seconds, the vehicle base bending load modes are allowed to initialize (movement of approximately 650 millimeters [25-1/2 inches] measured at the tip of the external tank - with movement towards the external tank).

At T-0, the two SRB's are ignited, under command of the four onboard computers; separation of the four explosive bolts on each SRB is initiated (each bolt is 711 millimeters - 28 inches-long and 88 millimeters - 3.5 inches - in diameter); the two T-0 umbilicals (one on each side of the spacecraft) are retracted; the onboard master timing unit, event timer, and mission event timers are started; the three SSME's are at 104 percent and the ground launch sequence is terminated.

The solid rocket motor thrust profile is tailored to reduce thrust during the maximum dynamic pressure (max q) region.

ELECTRICAL POWER DISTRIBUTION. Electrical power distribution in each SRB consists of orbiter-supplied main dc bus power to each SRB via SRB buses A, B, and C. The orbiter main dc buses A, B, and C supply main dc bus power to the respective SRB buses A, B, and C. In addition, orbiter main dc bus C supplies backup power to SRB buses A and B, and orbiter bus B supplies backup power to SRB bus C. This electrical power distribution arrangement allows all SRB buses to remain powered in the event one orbiter main bus fails.

The nominal dc voltage is 28 vdc with an upper limit of 32 vdc and lower limit of 24 vdc.

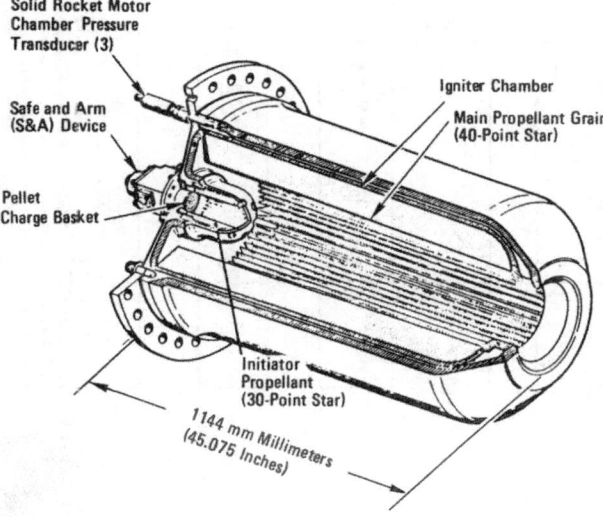

Solid Rocket Motor Igniter

 **Space Transportation System**

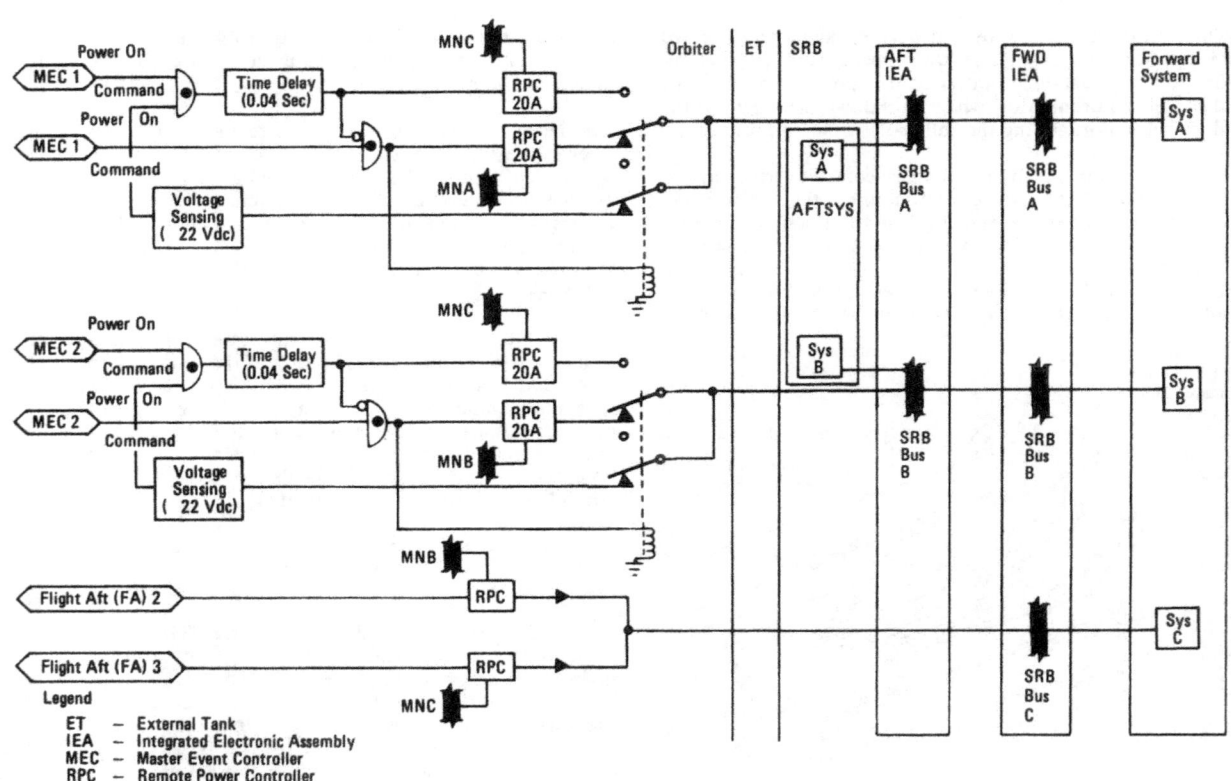

SRB Electrical Power Distribution

Space Transportation System

HYDRAULIC POWER UNITS. There are two self-contained, independent hydraulic power units (HPU) on each SRB. Each HPU consists of an auxiliary unit (APU), fuel supply module (FSM), hydraulic pump, hydraulic reservoir, and hydraulic fluid manifold assembly. The APU's are hydrazine-fueled and generate mechanical shaft power to a hydraulic pump that produces hydraulic pressure for the SRB hydraulic system. The two separate HPU's and two hydraulic systems are located on the aft end of each SRB, between the SRB nozzle and aft skirt. The HPU components are mounted on the aft skirt between the rock and tilt actuators. The two systems operate from T minus 28 seconds until SRB jettison. The two independent hydraulic systems are connected to the rock and tilt servoactuators.

The APU controller electronics are located in the SRB aft integrated electronic assemblies (IEA's) on the aft external tank attach rings.

The APU's and their fuel systems are isolated from each other. Each fuel supply module (tank) contains 9.9 kilograms (22 pounds) of hydrazine. The fuel tank is pressurized with gaseous nitrogen at 20,700 millimeters of mercury (mmHg) (400 psi) which provides the force to expel (positive expulsion) the fuel from the tank to the fuel distribution line, maintaining a positive fuel supply to the APU throughout its operation.

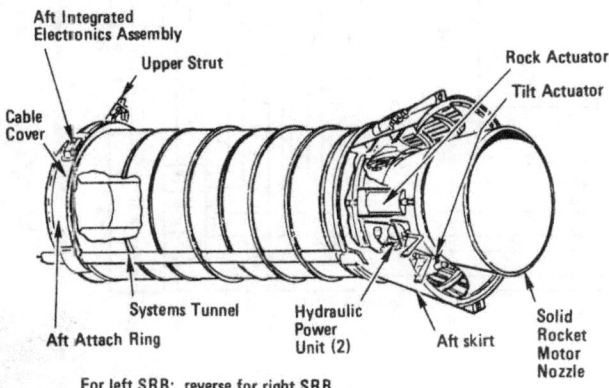

SRB Thrust Vector Control Component Location

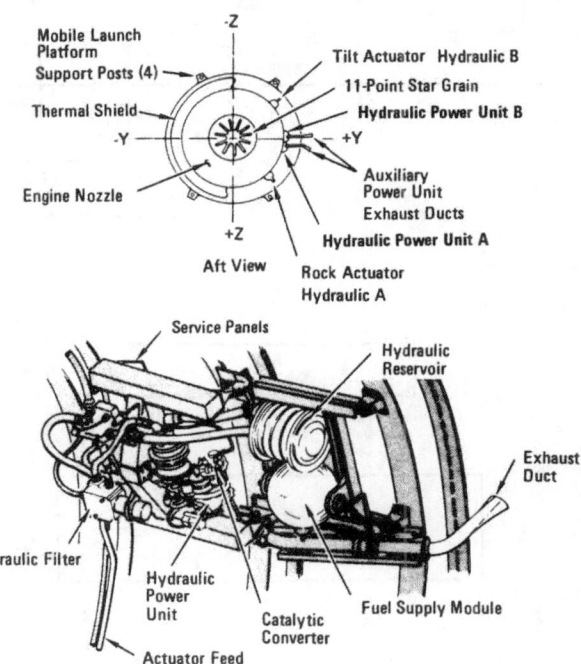

SRB Thrust Vector Control Component Location

 **Space Transportation System**

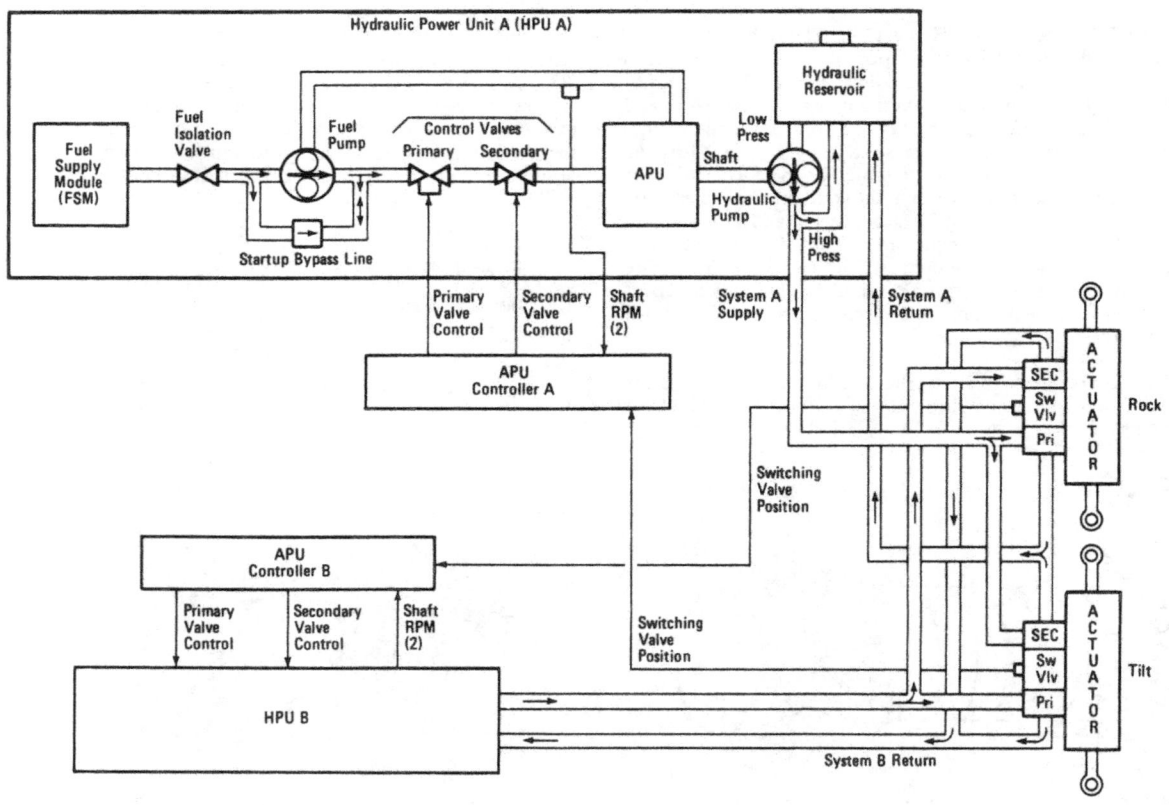

Hydraulic Power Unit System

Space Transportation System

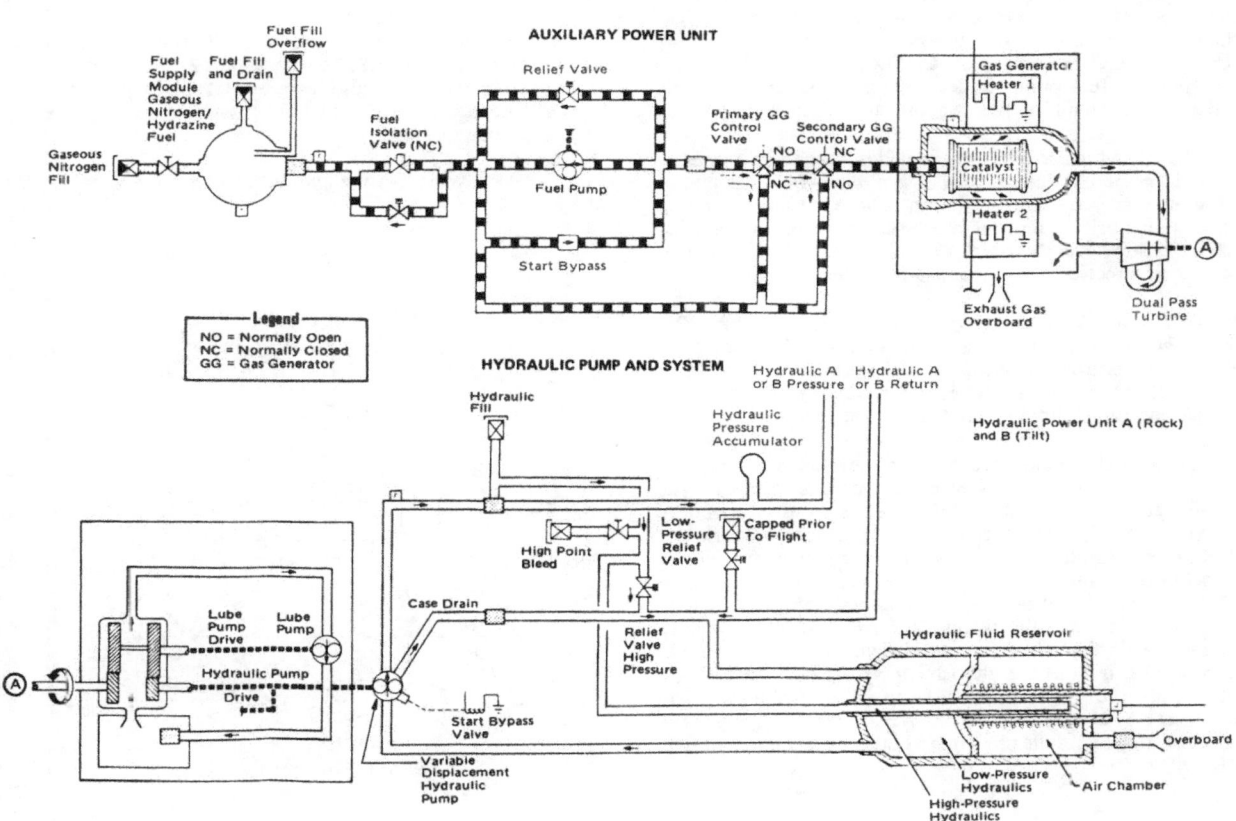

Auxiliary Power Unit and Hydraulic Pump System

 **Space Transportation System**

The fuel isolation valve is opened at APU startup to allow fuel to flow to the APU fuel pump and control valves, then to the gas generator. The gas generator catalytic action decomposes the fuel and creates a hot gas and feeds the hot gas exhaust product to the APU two-stage gas turbine. Fuel flows primarily through the startup bypass line until the APU speed is such that the fuel pump outlet pressure is greater than the bypass line. Then all the fuel is supplied to the fuel pump.

The APU turbine assembly provides mechanical power to the APU gearbox. The gearbox drives the APU fuel pump, hydraulic pump, and lube oil pump. The APU lube oil pump provides oil lubrication of the gearbox. The turbine exhaust of each APU flows over the exterior of the gas generator, cooling it and is then directed overboard through an exhaust duct.

When the APU speed reaches 100-percent, the APU primary control valve closes, and the APU speed is controlled by the APU controller electronics. If the primary control valve logic fails to the "open" state, the secondary control valve assumes control of the APU at 112-percent speed.

Each HPU is connected to both servoactuators on that SRB. One HPU serves as the primary hydraulic source for the servoactuator and the other HPU serves as the secondary hydraulics for the servoactuator. Each servoactuator has a switching valve that allows the secondary hydraulics to power the actuator if the primary hydraulic pressure drops below 106,087 mmHg (2,050 psi). A switch contact on the switching valve will close when the switching valve is in the secondary position. When the valve is closed, a signal is sent to the APU controller that inhibits the 100-percent APU speed control logic and enables the 112-percent APU speed control logic. The 100-percent APU speed enables one APU/HPU to supply sufficient operating hydraulic pressure to both servoactuators of that SRB.

The APU 100-percent speed corresponds to 72,000 rpm of the APU, 110-percent to 79,200 rpm, and 112-percent to 80,640 rpm.

The hydraulic pump speed is 3,600 rpm and supplies hydraulic pressure of 157,837 plus or minus 2,587 mmHg (3,050 plus or minus 50 psi). A high-pressure relief valve provides over-pressure protection to the hydraulic system and relieves at 194,062 mmHg (3,750 psi).

The APU's/HPU's and hydraulic systems are reusable for 20 missions.

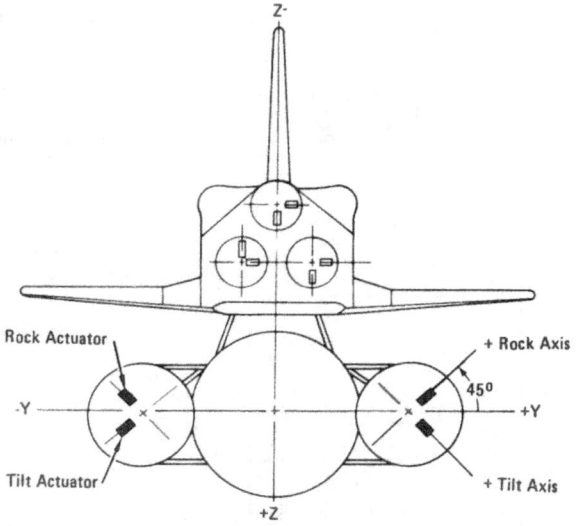

SRB Actuator Orientation

Space Transportation System

THRUST VECTOR CONTROL. Each SRB has two hydraulic gimbal servoactuators: one for rock and one for tilt. The servoactuators provide the force and control to gimbal the nozzle for thrust vector control.

The Space Shuttle ascent thrust vector control (ATVC) portion of the flight control system directs the thrust of the three Shuttle main engines and the two SRB nozzles to control Shuttle attitude and trajectory during liftoff and ascent. Commands from the guidance system are transmitted to the ATVC drivers, which transmit signals proportional to the commands to each servoactuator of the main engines and SRB's. Four independent flight control system channels and four ATVC channels control six main engine and four SRB ATVC drivers, with each driver controlling one hydraulic port on each main and SRB servoactuator.

Each SRB servoactuator consists of four independent, two-stage servovalves that receive signals from the drivers. Each servovalve controls one power spool in each actuator, which positions an actuator ram and the nozzle to control the direction of thrust.

The four servovalves in each actuator provide a force-summed majority voting arrangement to position the power spool. With four identical commands to the four servovalves, the actuator force sum action prevents a single erroneous command from affecting power ram motion. If the erroneous command persists for more than a predetermined time, differential pressure sensing activates a selector valve to isolate and remove the defective servovalve hydraulic pressure, permitting the remaining channels and servovalves to control the actuator ram spool.

Failure monitors are provided for each channel to indicate which channel has been bypassed. An isolation valve on each channel provides the capability of resetting a failed or bypassed channel.

Each actuator ram is equipped with transducers for position feedback to the thrust vector control system. Within each servoactuator ram is a splashdown load relief assembly to cushion the nozzle at water splashdown and prevent damage to the nozzle flexible bearing.

SRB RATE GYRO ASSEMBLIES. Each SRB contains three rate gyro assemblies (RGA's), with each RGA containing one pitch and one yaw gyro. These provide an output proportional to angular rates about the pitch and yaw axis to the orbiter computers and guidance, navigation, and control system during first-stage ascent flight in conjunction with the orbiter roll rate gyros until SRB separation. At SRB separation, a switchover is made from the SRB RGA's to the orbiter RGA's.

The SRB RGA rates pass through the flight aft multiplexer/demultiplexers (MDM's) to the orbiter computers. The RGA rates are then mid-value selected in redundancy management (RM) to provide one rate from each SRB pitch and yaw rate gyro (left and right SRB) to the user software. The RGA's are designed for 20 missions.

SRB SEPARATION. SRB separation is initiated when the three solid rocket motor chamber pressure transducers are processed in the redundancy management middle value select and the head-end chamber pressure of both SRB's is less than or equal to 2,587 mmHg (50 psi). A backup cue is the time-elapsed from booster ignition.

The separation sequence is initiated, commanding the thrust vector control actuators to the null position and putting the main propulsion system into a second-stage configuration (at 0.8 second from sequence intialization) that ensures the thrust of each SRB is less than 444,800 newtons (100,000 pounds). Orbiter yaw attitude is held for four seconds, while SRB thrust drops to less than 266,880 newtons (60,000 pounds).

 **Space Transportation System**

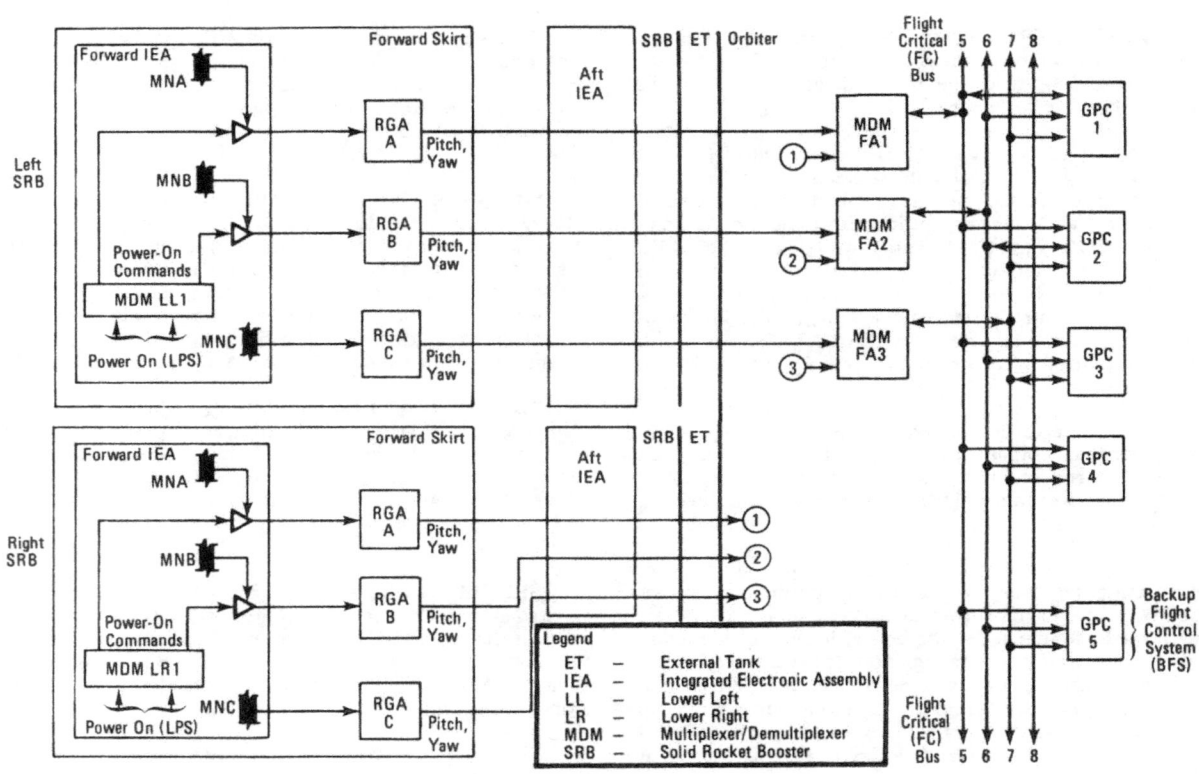

SRB Rate Gyro Assembly (RGA)

Space Transportation System

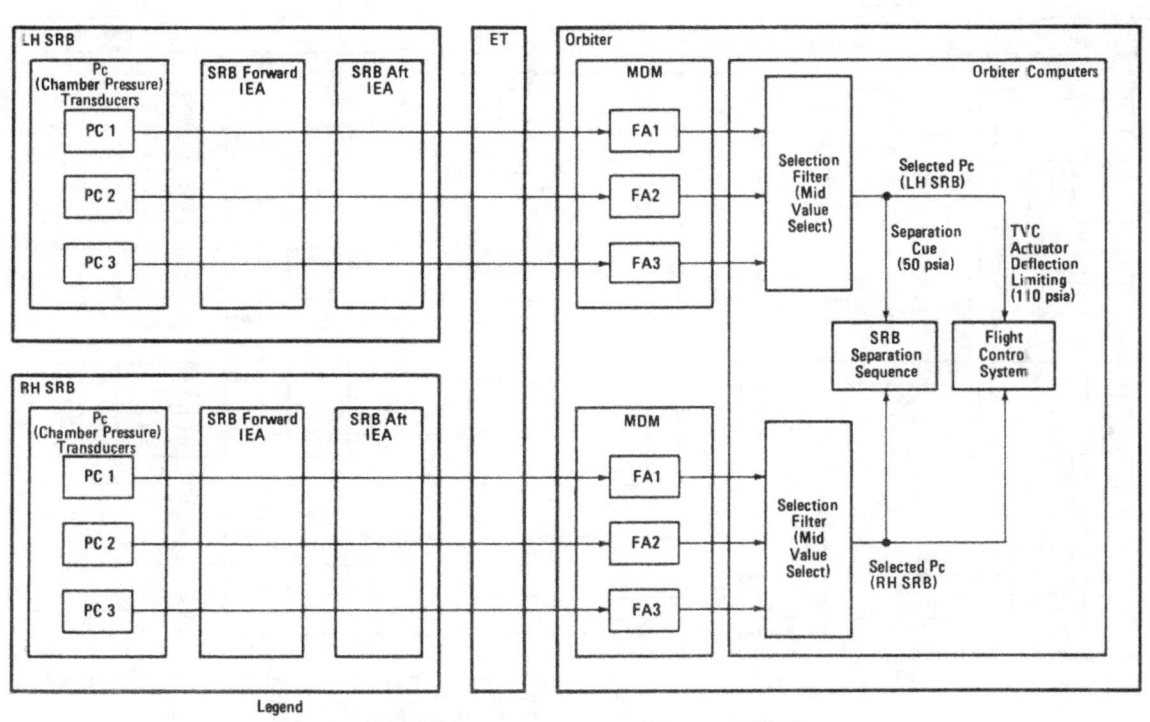

Solid Rocket Motor Chamber Pressure Data Flow

Space Transportation System

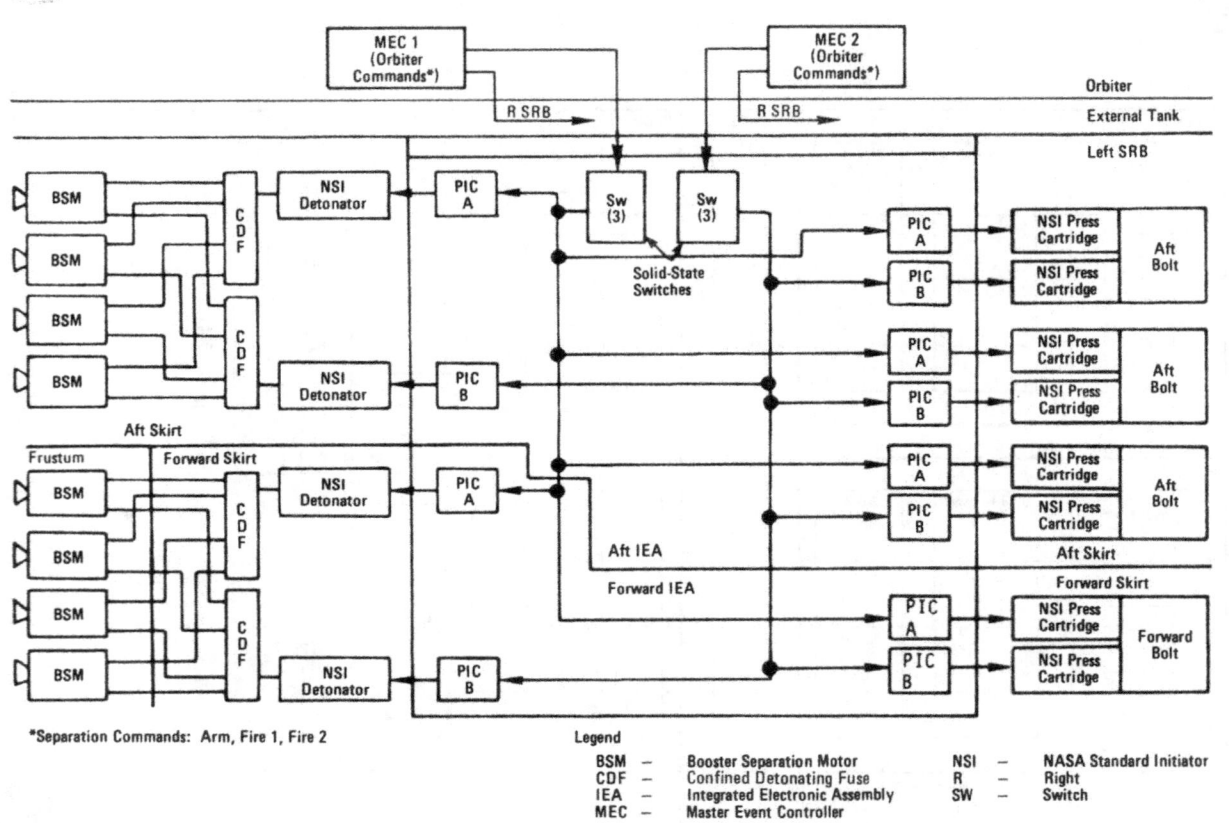

SRB Separation System

Space Transportation System

The SRB's separate from the external tank within 30 milliseconds of the ordnance firing command.

The forward attachment point consists of a ball (SRB) and socket (ET) held together by one bolt. The bolt contains one NSI at each end. It is noted that the forward attachment point also carries the range safety system (RSS) cross-strap wiring connecting each SRB RSS and the ET RSS with each other.

The aft attachment points consist of three separate struts: upper, diagonal, and lower. Each strut contains one bolt with an NSI at each end. The upper strut also carries the umbilical interface between its SRB and the external tank and on to the orbiter.

There are four booster separation motors (BSM's) on each end of each SRB. The BSM's separate the SRB's from the external tank. The solid rocket motors in each cluster of four are ignited by firing redundant NSI's into redundant confined detonating fuse manifolds.

The separation commands issued from the orbiter by the SRB separation sequence initiate the redundant NSI's in each bolt and ignite the BSM's to effect a clean separation.

RANGE SAFETY SYSTEM. The Shuttle vehicle has three range safety systems (RSS's). One is located in each SRB and one in the external tank. Any one, or all three, is capable of receiving two command messages (arm and fire) transmitted from the ground station. The RSS is used only when the Shuttle vehicle violates a launch trajectory red line.

An RSS consists of two antenna couplers, command receivers and command decoders, a dual distributor, a safe and arm device with two NSI's, two confined detonating fuse (CDF) manifolds, and two linear shaped charges (LSC's).

The antenna couplers provide the proper impedance for radio frequency (RF) and ground support equipment (GSE) commands. The command receivers are tuned to RSS command frequencies and provide the input signal to the distributors when an RSS command is sent. The command decoders use a coded plug to prevent any RF signal other than the proper RF signal from getting into the distributors. The distributors contain the logic to supply valid destruct commands to the RSS pyrotechnics.

The NSI's provide the spark to ignite the CDF, which, in turn ignites the LSC for Shuttle vehicle destruction. The safe and arm device provides mechanical isolation between the NSI's and the CDF during prelaunch and during the SRB separation sequence.

The first message called arm allows the onboard logic to enable a destruct and illuminates a light on the flight deck display and control panel at the CDR and PILOT station. The second message transmitted is the fire command.

The SRB distributors in the SRB's and the ET are cross-strapped together. Thus, one arm and/or destruct signal received by one SRB will provide the arm and/or destruct signals to the other SRB and the external tank.

Electrical power from the RSS battery in each SRB is routed to RSS system A. The recovery battery in each SRB is used to power RSS system B, as well as the recovery system in that SRB. The SRB RSS is powered down during the separation sequence, and the recovery system for that SRB is powered up.

SRB DESCENT AND RECOVERY. During the descent portion of the trajectory, the SRB's achieve a nose-up trim condition.

 **Space Transportation System**

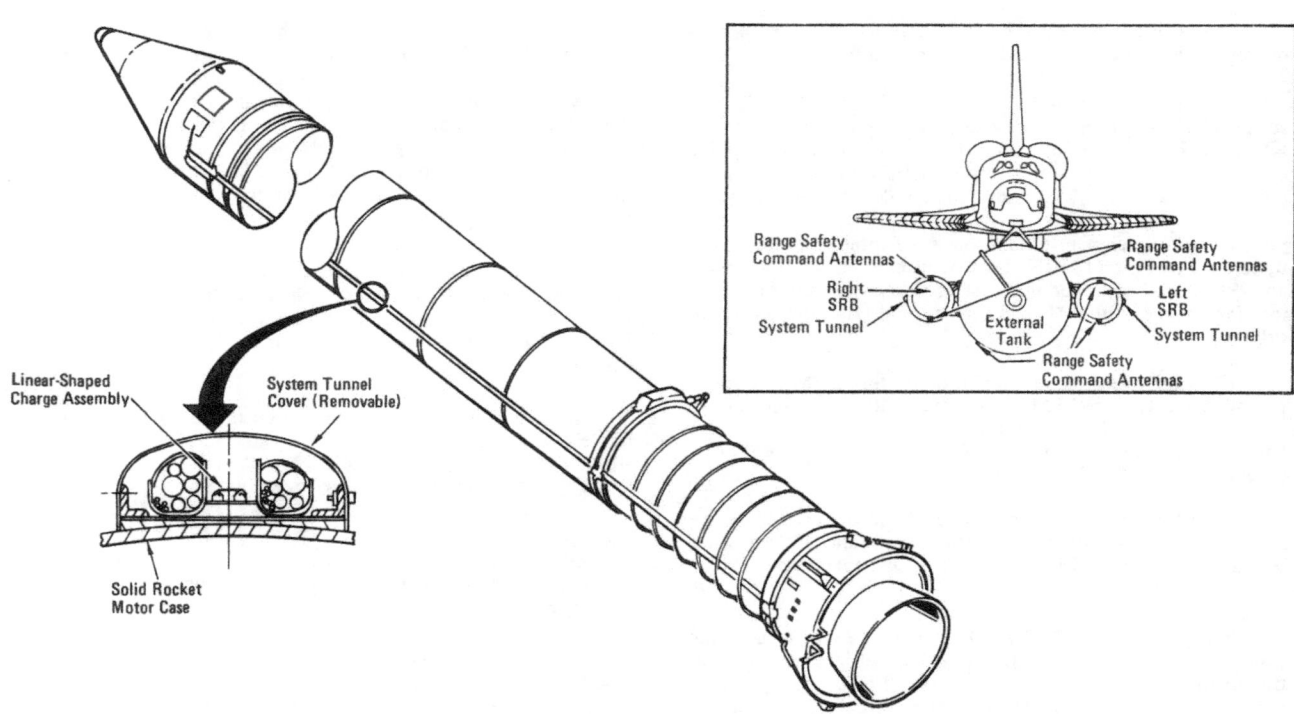

SRB Range Safety System

Space Transportation System

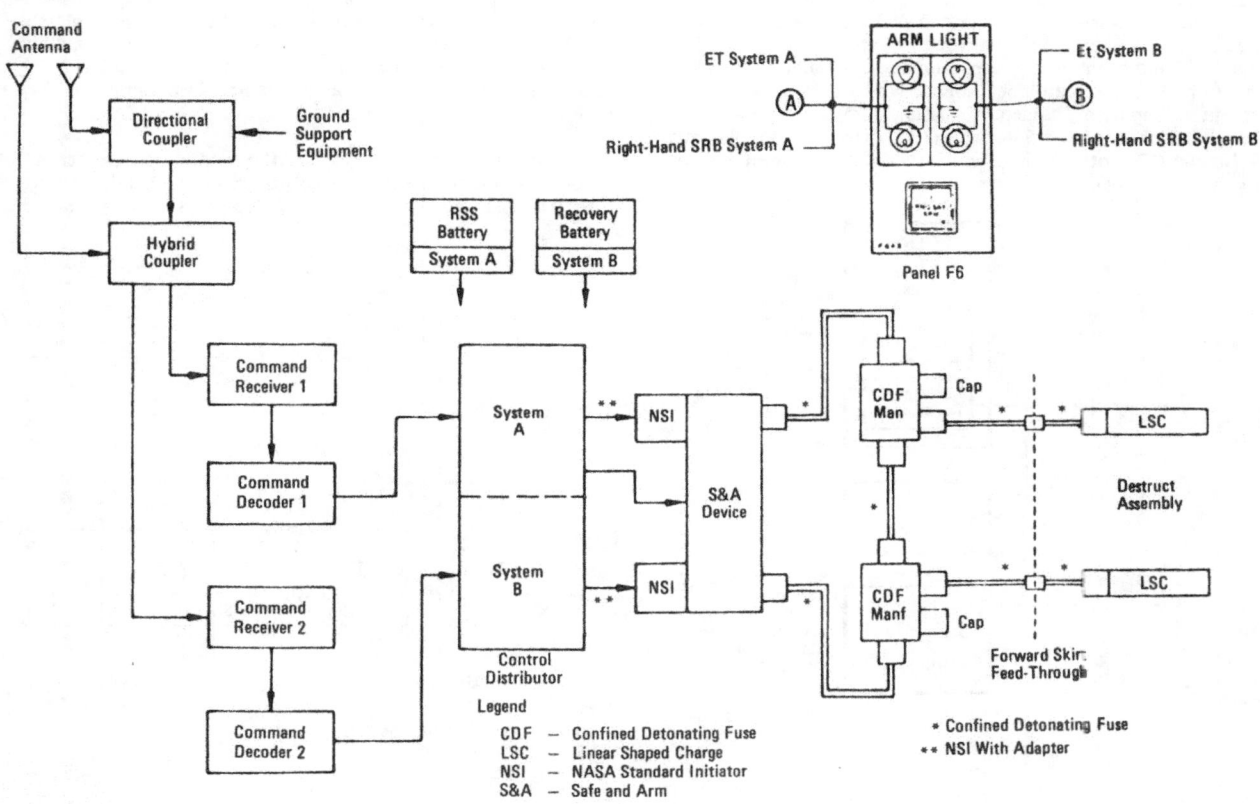

SRB Range Safety System (RSS)

Space Transportation System

Approximately 220 seconds after separation and when the altitude switch senses an atmospheric pressure corresponding to approximately 4,693 meters (15,400 feet), a nose cap initiator fires, and the energy is sent through confined detonating fuse manifold and assemblies to three thrusters on the top ring of the frustum. The nose cap is jettisoned, which fires a pyrotechnic charge to deploy a 3.5-meter-diameter (11.5-foot) conical, ribbon, pilot chute with a 5.4-meter (18-foot) suspension line and 9.7-meter (32-foot) riser line, which then strips the pilot bag and extracts the drogue parachute.

The drogue chute, 16 meters (54 feet) in diameter with 12 30-meter (100-foot) suspension lines, slows the booster's descent. The drogue disreefs to full inflation approximately 234.3 seconds after separation at an altitude of approximately 2,834 meters (9,300 feet). It withstands a load of 122,472 kilograms (270,000 pounds) and weights 544 kilograms (1,200 pounds).

Approximately 241.4 seconds after separation, another initiator fires, and its energy is sent through a confined detonating fuse to a linear-shaped charge, which separates the frustum and

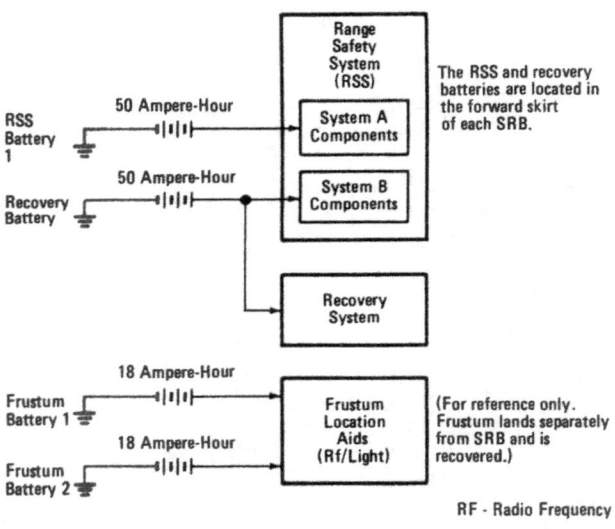

SRB Battery Power

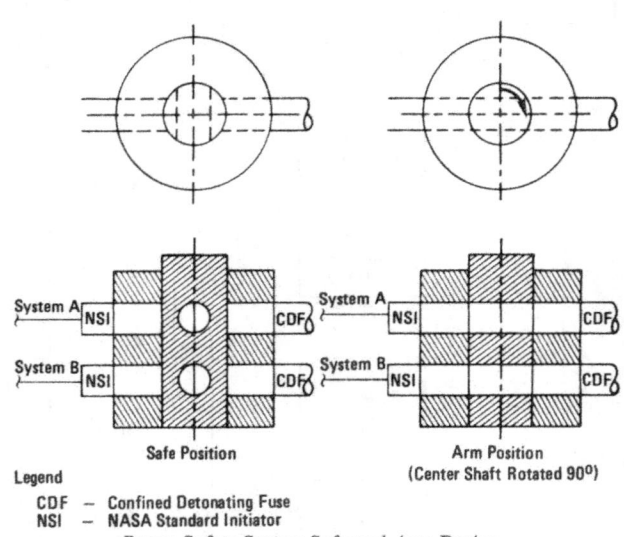

Legend
CDF — Confined Detonating Fuse
NSI — NASA Standard Initiator

Range Safety System Safe and Arm Device (Simplified)

Space Transportation System

Separation	T = 0 SEC
Apogee	T = 70 SEC
Nose Cap Jettison	T = 220 SEC
Nose Cap and Pilot Parachute Bag	
Pilot Parachute Deploys Drogue Parachute	
Pilot/Drogue Deployment	
Drogue Parachute Inflates to 1st Reefed Condition	T = 223.3 SEC
Drogue Disreefs to Full Inflation	T = 234.3 SEC
Drogue and Frustum Deploy With Main Packs	T = 241.4 SEC
Main Disreef to 1st Reefed Condition	T = 244.5 SEC
2nd Reefed Condition	T = 253.7 SEC
Full Inflation	T = 261.9 SEC
Nozzle Severance	
Frustum and Drogue Impact at 18 mps (60 fps)	
Water Impact Chutes Detach at Impact Deploying Tow Pendant	T = 281 SEC
Parachute Flotation	
Tow Pendant	
Sonar Beacons	
"Break" Lanyard	
Sonar Beacons	
Tow Pendant Floats on Surface	

SRB Descent and Splash Down

 **Space Transportation System**

drogue chute and strips the main parachute bags. This occurs at about 2,011 meters (6,600 feet).

The three main parachutes also are conical ribbon types. They are 35 meters (115 feet) in diameter with 48 40-meter (132-foot) suspension lines and 8 12-meter (40-foot) riser lines. At approximately 261.9 seconds and an altitude of 6,707 meters (2,200 feet), the three main parachutes reach full open. Each withstands a load of 81,648 kilograms (180,000 pounds) and weighs 680 kilograms (1,500 pounds). The parachutes are made of nylon ribbons about 50 millimeters (2 inches) wide. The horizontal, vertical, and radial ribbons are about as thick as canvas but structural members are about as thick as military webb belts.

At approximately 20 seconds after deployment of main chutes, a linear-shaped charge severs the nozzle extension.

The velocity of each SRB at water impact, about 281 seconds after separation, is approximately 26 meters (87 feet) per second. With the SRB descending aft end first, air in the now empty (burned out) motor casing is trapped and compressed, causing the booster to float with the forward end out of the water after splashdown.

At impact, the main parachutes are released, and a towing pendant is deployed. The pendant may also be deployed manually by the SRB recovery crew. The parachutes are designed for reuse on 10 missions.

The radio transponder in each SRB has a range of 8.9 nautical miles (10.35 statute miles), and the flashing light has a night-time range of 4.9 nautical miles (5.75 statute miles), with 10 power optical aids with wave heights of 2.9 meters (9.5 feet). The flashing light is activated upon water impact.

Each frustum/drogue chute has an RF transmitter with a range of 8.9 nautical miles (10.35 statute miles) and a flashing light with a night-time range of 4.9 nautical miles (5.75 statute miles) with wave heights of 2.9 meters (9.5 feet). The flashing light is activated upon water impact.

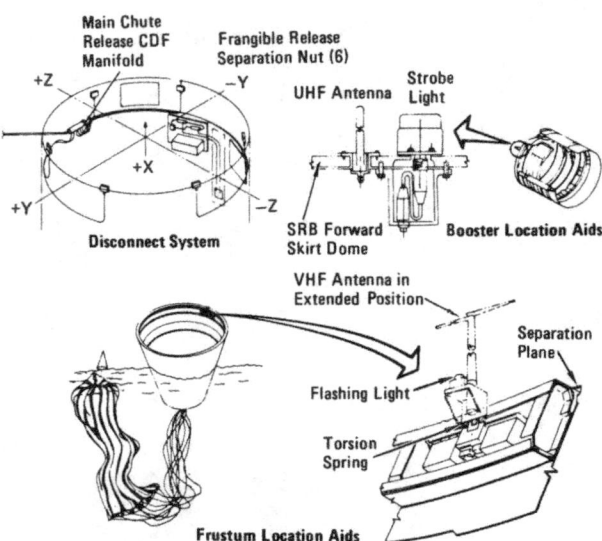

SRB Parachute Disconnect and Recovery Aids

The main parachutes have an RF transmitter with a range of 8.9 nautical miles (10.35 statute miles) and a flashing light with a night-time range of 4.9 nautical miles (5.75 statute miles) with wave heights of 2.9 meters (9.5 feet). The flashing light is activated upon water impact. In addition, sonar beacons are used with a range of 2.9 nautical miles (3.45 statute miles).

Space Transportation System

Various parameters of SRB operation are monitored and displayed on the orbiter flight deck control and display panel and are transmitted to ground telemetry.

CONTRACTORS. Thiokol Chemical Corp., Wasatch Division, Brigham City, Utah, is prime contractor for the solid-rocket booster motors. Other contractors include McDonnell Douglas Astronautics Co., Huntington Beach, CA (SRB structures); United Space Boosters Inc., Sunnyvale, CA (SRB checkout, assembly, launch, and refurbishment, except for motors); Pioneer Parachute Co., Manchester, CT (parachutes); Abex Corp., Oxnard, CA (hydraulic pumps); Arde Inc., Mahwah, NJ (hydrazine fuel modules); Arkwin Industries Inc., Westbury, NY (hydraulic reservoirs); Aydin Vector Division, Newtown, PA (integrated electronic assemblies); Bendix Corp., Teterboro, NJ (integrated electronic assemblies); Consolidated Controls Corp., El Segundo, CA (fuel isolation valves, hydrazine); Eldec Corp., Lynnwood, WA (integrated electronic assemblies); Explosive Technology, Fairfield, CA (CDF manifolds); Martin Marietta, Denver, CO (pyro initiator controllers); Moog Inc., East Aurora, NY (servoactuators); Sperry Rand Flight Systems, Phoenix, AZ (multiplexers/demultiplexers); Teledyne, Lewisburg, TN (location aid transmitters); United Technology Corp., Sunnyvale, CA (separation motors); Sundstrand, Rockford, IL (auxiliary power units); Motorola Inc., Scottsdale, AZ (range safety receivers).

EXTERNAL TANK

The external tank contains the liquid hydrogen fuel and liquid oxygen oxidizer and supplies them under pressure to the three main engines in the orbiter during liftoff and ascent. When the main engines are shut down the external tank is jettisoned, enters the earth's atmosphere, breaks up, and impacts in a remote ocean area. It is not recovered.

The largest and heaviest (when loaded) element of the Space Shuttle, the external tank, has three major components: the forward liquid oxygen tank, an unpressurized intertank which contains most of the electrical components, and the aft liquid hydrogen tank. The external tank is 47 meters (154.2 feet) long and has a diameter of 8.38 meters (27.5 feet).

Beginning with the STS-6 mission, a lightweight external tank is utilized. Although each future tank may vary slightly, each will weigh approximately 30,096 kilograms (66,000 pounds) inert. The final heavy weight tank flown on STS-7 weighed approximately 34,927 kilograms (77,000 pounds) inert. For each pound of weight reduced from the ET, the cargo carrying capability of the Space Shuttle spacecraft increases almost 0.4 kilograms (one pound). The weight reduction was accomplished by eliminating portions of stringers (structural stiffeners running the length of the hydrogen tank), using fewer stiffener rings and by modifying major frames in the hydrogen tank. Also, significant portions of the tank are milled differently to reduce thickness, and the weight of the ET's aft solid rocket booster attachments were reduced by using a stronger, yet lighter and less expensive titanium alloy. Earlier several hundred kilograms (pounds) were eliminated by deleting the antigeyser line. The line paralleled the oxygen feedline and provided a circulation path for liquid oxygen to reduce accumulation of gaseous oxygen in the feedline while the oxygen tank was being filled prior to launch. After assessing propellant loading data from ground tests and the first few Space Shuttle missions, the antigeyser line was removed from the STS-5 and subsequent missions. The total length and diameter of the ET remains unchanged due to the weight reduction.

 **Space Transportation System**

The external tank is attached to the orbiter at one forward attachment point and two aft points. In the aft attachment area, there are also umbilicals which carry fluids, gases, electrical signals, and electrical power between the tank and the orbiter. Electrical signals and controls between the orbiter and the two solid-rocket boosters also are routed through those umbilicals.

The LO_2 tank is an aluminum monocoque structure composed of a fusion-welded assembly of preformed, chemmilled gores, panels, machined fittings, and ring chords. It operates in a pressure range of 1,035 to 1,138 mmHg (20 to 22 psi). The tank contains antislosh and antivortex provisions to minimize liquid residuals and for damping fluid motion. The tank feeds into a 431 millimeter (17 inch) diameter feedline which conveys

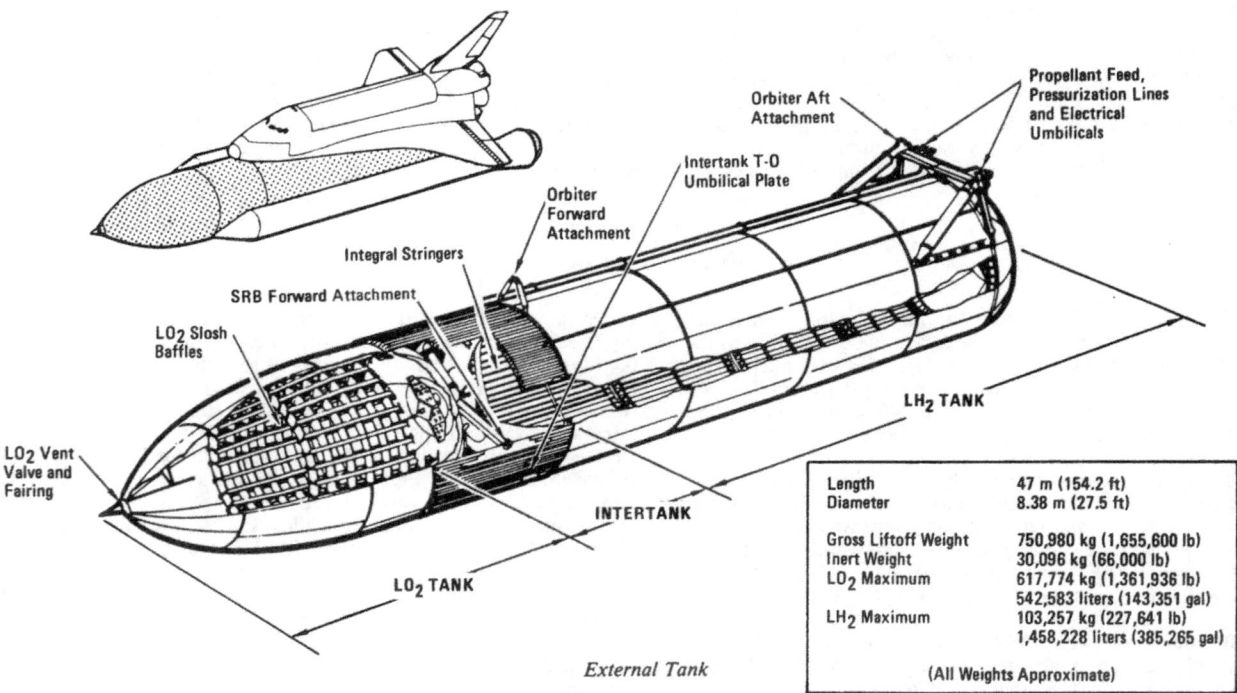

External Tank

Length	47 m (154.2 ft)
Diameter	8.38 m (27.5 ft)
Gross Liftoff Weight	750,980 kg (1,655,600 lb)
Inert Weight	30,096 kg (66,000 lb)
LO_2 Maximum	617,774 kg (1,361,936 lb)
	542,583 liters (143,351 gal)
LH_2 Maximum	103,257 kg (227,641 lb)
	1,458,228 liters (385,265 gal)
(All Weights Approximate)	

Space Transportation System

the LO_2 through the intertank, then external to the ET to the aft right hand ET/orbiter disconnect umbilical. The 431 millimeter (17 inch) diameter feedline permits LO_2 flow of approximately 1,264 kilograms (2,787 pounds) per second with the Space Shuttle Main Engines (SSME's) operating at 104 present or permits a maximum flow of 71,979 liters per minute (19,017 gallons per minute). The LO_2 tank's double wedge nose cone reduces drag, heating, contains the vehicle's ascent air data system, and serves as a lightning rod. The LO_2 tank volume is 559 cubic meters (19,786 cubic feet). Its diameter is 8,407 millimeters (331 inches), its length is 15,036 millimeters (592 inches) and the empty weight of 5,443 kilograms (12,000 pounds).

The intertank is a steel/aluminum semi-monocoque cylindrical structural with flanges on each end for joining the LO_2 and LH_2 tanks. The intertank houses the ET instrumentation components and provides an umbilical plate that interfaces with the ground facility arm for purge gas supply, hazardous gas detection, and hydrogen gas boiloff during ground operations. It consists of mechanically joined skin, stringers, and machined panels of aluminum alloy. The intertank is vented during flight. The intertank contains the forward SRB-external tank attach thrust beam and fittings which distribute the SRB loads to the LO_2 and LH_2 tanks. The intertank is 6,858 millimeters (270 inches) long 8,407 millimeters (331 inches) in diameter and weighs 5,488 kilograms (12,100 pounds).

The LH_2 tank is an aluminum semi-monocoque structure of fusion-welded barrel sections, five ring frames, and forward and aft ellipsoidal domes. Its operating pressure range is 1,656 to 1,759 mmHg (32 to 34 psi). The tank contains an antivortex baffle and siphon outlet to transmit the LH_2 from the tank through a 431-millimeter-diameter (17 inch) line to the left aft umbilical. The LH_2 feedline flow rate is 210 kilgrams (465 pounds) per second with the SSME's at 104 percent or permits a maximum flow of 184,420 liters per minute (48,724 gallons per minute). The LH_2 tank provides at its forward end the ET/orbiter forward attach pod strut and at its aft end the two ET/

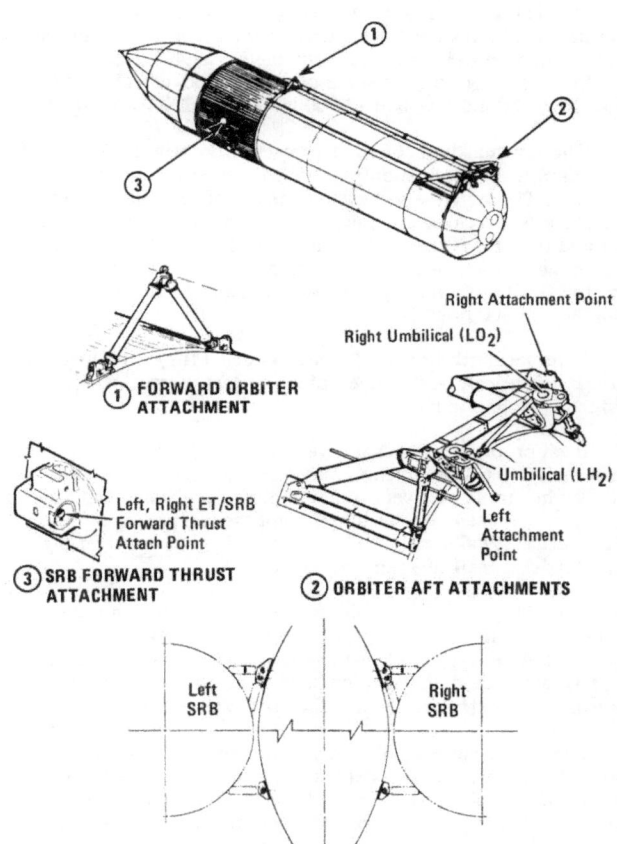

Orbiter — External Tank Attachment Points

 **Space Transportation System**

orbiter aft attach ball fittings, as well as the aft SRB-external tank stabilizing strut attach ball fittings, as well as the aft SRB-external tank stabilizing strut attachments. The LH_2 tank is 8,407 millimeters (331 inches) in diameter, 29,464 millimeters (1160 inches) long, has a volume of 1,514 cubic meters (53,518 cubic feet) and a dry weight 12,700 kilograms (28,000 pounds).

The external tank thermal protection system consists of foam applied or sprayed-on premolded insulation and ablator materials. The system also includes the use of phenolic thermal insulators to preclude air liquification. Thermal isolators are required for LH_2 tank attachments to preclude the liquification of air exposed metallic attachments and to reduce heat flow into the LH_2. The thermal protection system weight is 2,177 kilograms (4,800 pounds).

External hardware, ET/orbiter attach fittings, umbilical fittings, electrical, and range safety system weight is 4,127 kilograms (9,100 pounds).

Each propellant tank has a vent and relief valve at its forward end. This dual-function valve can be opened by GSE-supplied helium for the vent function during prelaunch and can open during flight when the ullage (empty space) pressure of the LH_2 reaches 1,966 mmHg (38 psi) or the ullage pressure of the LO_2 tank reaches 1,293 mmHg (25 psi).

The LO_2 tank contains a separate pyrotechnically-operated propulsive, tumble vent valve at its forward end. At separation, the LO_2 tumble vent valve is opened, providing impulse to assist in the separation maneuver and more positive control of the entry aerodynamics of the external tank.

There are eight propellant depletion sensors, four each for fuel and oxidizer. The fuel depletion sensors are located in the bottom of the fuel tank. The oxidizer sensors are mounted in the orbiter LO_2 feedline manifold downstream of the feedline disconnect. During main engine thrusting, the orbiter computers constantly compute the instantaneous mass of the vehicle due to the usage of the propellants. Normally, main engine cutoff (MECO) is based on a predetermined velocity; however, if any two of the fuel or oxidizer sensors sense a dry condition, the engines will be shut down.

Location of the LO_2 sensors allow the maximum amount of oxidizer to be consumed in the engines, while allowing sufficient time to shut down the engines before the oxidizer pumps cavitate (run dry). In addition, 498 kilograms (1,100 pounds) of LH_2 are loaded over and above that required by 6:1 oxidizer/fuel engine mixture ratio. This assures that main engine cutoff from the depletion sensors is fuel-rich; oxidizer-rich engine shutdowns can cause burning and severe erosion of engine components.

Four pressure transducers located at the top end of the LO_2 and LH_2 tanks monitor the ullage pressures.

Each of the two aft external tank umbilical plate mates with a corresponding plate on the orbiter. The plates help maintain alignment among the umbilicals. Physical strength at the umbilical plates is provided by bolting corresponding umbilical plates together. When the orbiter computers command external tank separation, the bolts are severed by pyrotechnic devices.

The external tank has five propellant umbilicals that interface with orbiter umbilicals: two for the LO_2 tank and three for the LH_2 tank. One of the LO_2 tank umbilicals is for liquid oxygen, the other for gaseous oxygen. The LH_2 tank umbilical has two for liquid and one for gas. The smaller-diameter LH_2 umbilical is a recirculation umbilical used only during LH_2 chilldown sequence in prelaunch.

The external tank also has two electrical umbilicals, which carry electrical power from the orbiter to the tank and the two solid-rocket boosters and provide information from the SRB's and external tank to the orbiter.

Space Transportation System

A swing-arm-mounted cap to the fixed service structure covers the oxygen tank vent on top of the external tank during the countdown and is retracted about two minutes prior to liftoff. The cap will siphon off oxygen vapor that threatened to form large ice on the external tank, thus protecting the orbiter's thermal protection system during launch.

A range safety system provides a means for destructively dispersing the propellants. It includes a battery power source, a receiver/decoder, antennas, and ordnance.

Various parameters are monitored and displayed on the flight deck display and control panel and are transmitted to the ground.

The contractor for the external tank is Martin Marietta Aerospace, Denver, CO, the tank is manufactured at Michoud, LA, Motorola Inc., Scottsdale, AZ, is the contractor for range safety receivers.

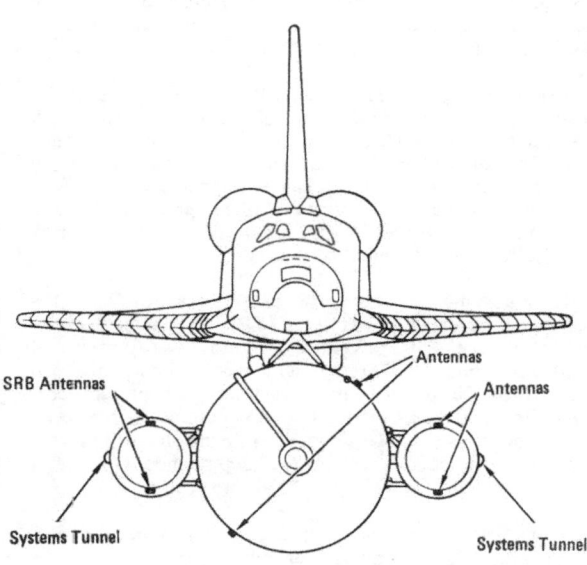

External Tank Range Safety System Antennas

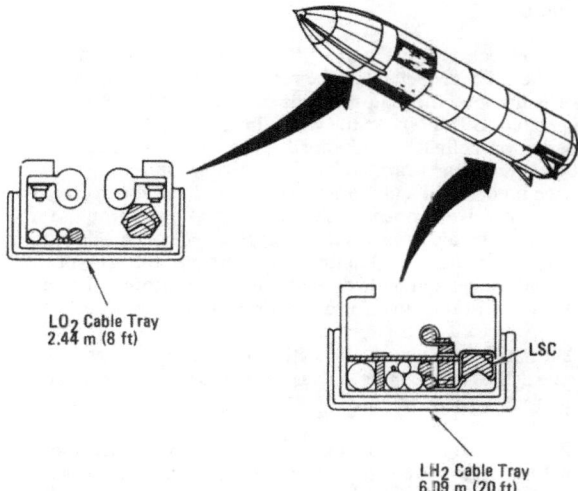

External Tank Range Safety System Linear Shaped Changes

 Space Transportation System

MAIN PROPULSION SYSTEM

The main propulsion system (MPS), assisted by the two solid rocket boosters (SRB's) during the initial phases of the ascent trajectory, provides the velocity increment (ΔV) from liftoff to a predetermined velocity increment (ΔV) prior to orbit insertion. After the two SRB's are expended and jettisoned, the MPS continues to thrust until the predetermined velocity is achieved. At that time main engine cutoff (MECO) is initiated. The external tank (ET) is jettisoned, and the orbital maneuvering system (OMS) is ignited to provide the final ΔV for orbital insertion. The magnitude of the ΔV supplied by the OMS depends on payload weight, mission trajectory, and system limitations.

Coincident with the start of the OMS thrusting maneuver (settles the MPS propellants), the remaining liquid oxygen propellant is dumped through the bells of the three Space Shuttle main engines (SSME's). At the same time, the remaining liquid hydrogen propellant is dumped overboard through the hydrogen fill and drain valves for six seconds, then the hydrogen inboard fill and drain valve is closed and the hydrogen recirculation valve is opened. The hydrogen flows through the hydrogen engine bleed valves to the orbiter hydrogen MPS line between the inboard and outboard hydrogen fill and drain valves, and the remaining hydrogen is dumped through the outboard fill and drain valve for approximately 100 seconds.

During on-orbit operations, the flight crew will vacuuminert the MPS by opening valves that allow remaining propellants to be vented to space.

Prior to entry, the flight crew repressurizes the MPS propellant lines with helium to prevent contaminants from being drawn into the lines during entry. It is also used to purge the spacecraft aft fuselage. The last activity of the MPS occurs at the end of landing rollout. At that time the helium remaining in onboard helium storage tanks is released into the MPS to provide an inert atmosphere for safety.

The MPS consists of the following major subsystems: three SSME's, three SSME controllers, external tank, orbiter MPS propellant management subsystems and helium susbsystem, four ascent thrust vector control (ATVC) units, and six SSME hydraulic servoactuators.

The main engines are reusable, high-performance liquid-propellant rocket engines with variable thrust. The propellant fuel is liquid hydrogen (LH_2) and the oxidizer is liquid oxygen (LO_2). The propellant is carried in separate tanks in the external tank and supplied to the main engines under pressure. Each engine can be gimbaled plus or minus 10.5 degrees in pitch and plus or minus 8.5 degrees in yaw for thrust vector control by hydraulically powered gimbal actuators.

The main engines can be throttled over a range of 65 to 109 percent of their rated power level (RPL) in one-percent increments. A value of 100 percent corresponds to a thrust level of 1,668,000 newtons (375,000 pounds) at sea level and 2,090,560 newtons (470,000 pounds) in a vacuum. 104 percent corresponds to 1,751,622 newtons (393,800 pounds, at sea level, 2,174,182 newtons (488,800 pounds) in a vacuum. 109 percent corresponds to 1,856,150 newtons (417,300 pounds) at sea level 12,282,936 newtons (513,250 pounds) in a vacuum.

At sea level, the engine throttling range is reduced due to flow separation in the nozzle, prohibiting operation of the engine at its 65-percent throttle setting, referred to as minimum power level (MPL). All three main engines receive the same throttle command at the same time. Normally these come automatically from the general-purpose computers through the engine controllers. During certain contingency situations,

Space Transportation System

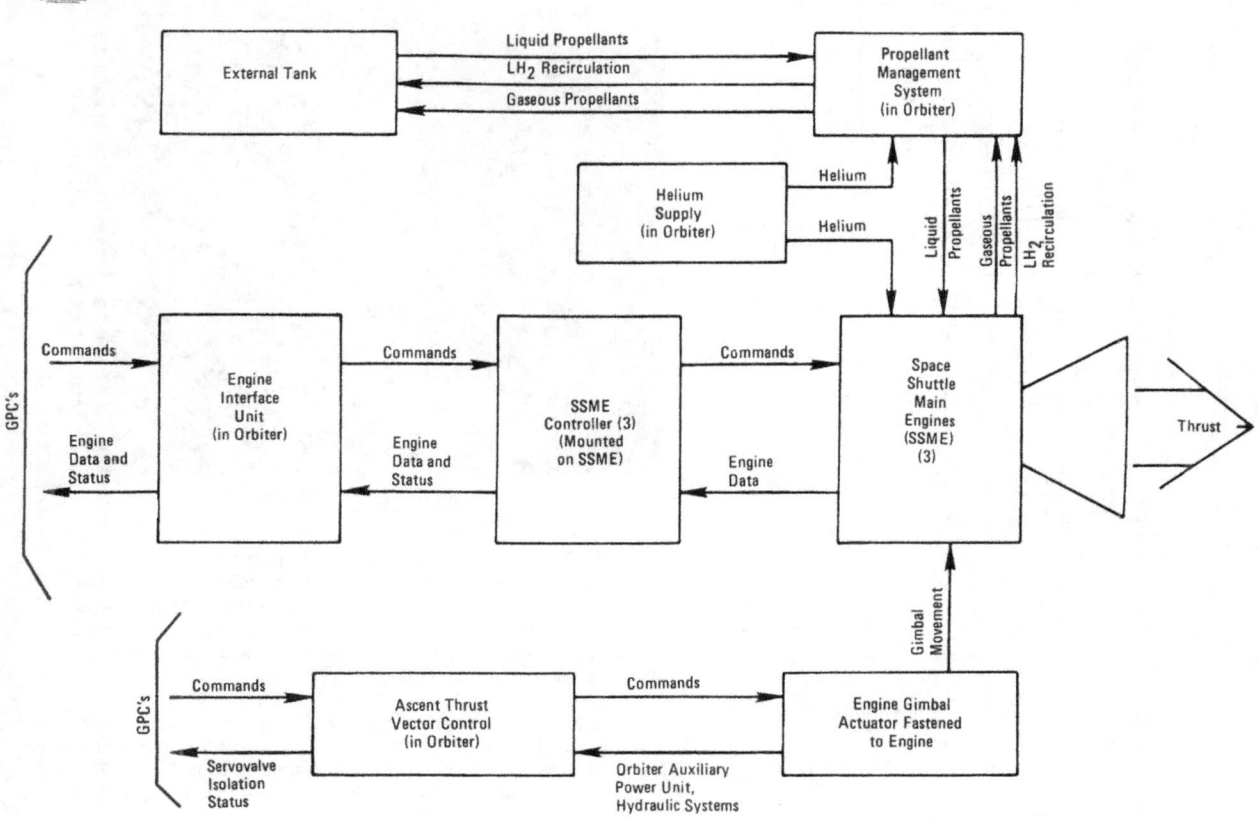

MPS Subsystem

Space Transportation System

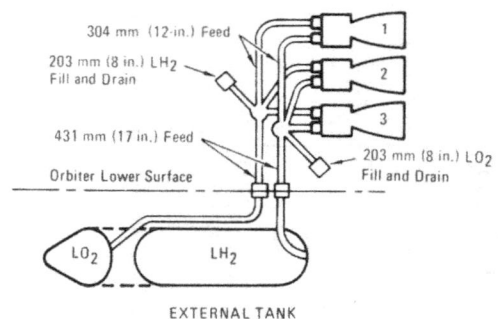

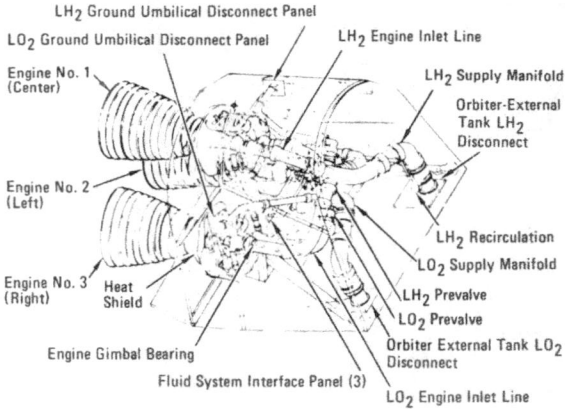

Main Propulsion System

Space Shuttle Main Engine

manual control of engine throttling is possible by use of the speed brake/engine throttle controller handle. The throttling ability reduces vehicle loads during maximum aerodynamic pressure, limits vehicle acceleration to 3 g's maximum during boost, and makes it possible to abort with all main engines thrusting or one engine out.

Each engine is designed for 7-1/2 hours of operation over a life span of 55 starts. Throughout the throttling range, the LO_2-LH_2 mixture is 6:1. Each nozzle area ratio is 77:5:1. The engines are 4.2 meters (14 feet) long and 2.4 meters (8 feet) in diameter at the nozzle exit.

Space Transportation System

The SSME controllers are a digital computer system electronic packages mounted on the SSME's. They operate in conjunction with engine sensors, valve actuators, and spark igniters to provide a self-contained system for monitoring engine control, checkout, and status. Each controller is attached to the forward end of the SSME. Engine data and status collected by each controller from the SSME are transmitted to the engine interface unit (EIU), which is mounted in the orbiter. There is one EIU for each SSME.

The EIU transmitts commands from the orbiter computers to the main engine controller. When engine data and status are received by the EIU, the data are held in a buffer until an orbiter computer request for data is received by the EIU.

It is noted, that each of the three orbiter auxiliary power units (APU's), provide mechanical shaft power through a gear train to drive a hydraulic pump. Each of the three hydraulic pumps, provide hydraulic pressure to its respective hydraulic system. The three orbiter hydraulic systems provide hydraulic pressure to position the SSME servoactuators for thrust vector control during the ascent phase of the mission in addition to other functions in the main propulsion system.

The ascent thrust vector control (ATVC) receives commands from the orbiter computers and sends commands to the engine gimbal actuators. The ATVC's are an electronics package (four in all) mounted in the orbiter's aft fuselage avionics bays. Hydraulic isolation commands are directed to the engine gimbal actuators that have faulty servovalves as a result of excessive pressure. In conjunction with this, a servovalve isolation signal is transmitted to the computers.

The SSME hydraulic servoactuators are used to gimbal the main engine. There are two actuators per SSME, one for pitch motion and one for yaw motion. They convert electrical commands received from the orbiter computers and position servovalves which direct hydraulic pressure to a piston that converts the pressure into a mechanical force that is used to gimbal the SSME's. Hydraulic presure status for each servovalve is transmitted to the ATVC.

The orbiter MPS propellant management subsystem consists of manifolds, distribution lines, and valves by which the liquid propellants pass from the ET to the SSME's and the gaseous propellants pass from the SSME's to the ET. The SSME's gaseous propellants provide ET pressurization. All the valves in the propellant managment subsystems are under direct control of the orbiter computers and are either electrically or pneumatically actuated.

The orbiter MPS helium subsystem consists of a series of helium supply tanks and regulators, check valves, distribution lines, and control valves. The subsystem supplies helium used within the engine for purging the high pressure oxidizer turbopump intermediate seal, preburner oxidizer domes and actuating valves during emergency pneumatic shutdown. The balance of the helium is used to actuate all the pneumatically operated valves within the propellant management subsystem and to pressurize the propellant lines prior to reentry.

EXTERNAL TANK

The ET aft attachment struts connect the orbiter through a number of umbilicals that carry fluids, gases, signals, and electrical power between the tank and the orbiter. These umbilicals are collected into two bundles. Each bundle is permanently connected to the ET at one end physically supported by one of the rear attachment struts and terminated in an umbilical plate. Each ET umbilical plate mates with a corresponding orbiter umbilical plate. The plates help to maintain alignment among the various connecting components. Physical strength at the joint is provided by bolting corresponding umbilcal plates together. When the onboard computers command ET separation, the bolts are severed by pyrotechnic devices.

54

 **Space Transportation System**

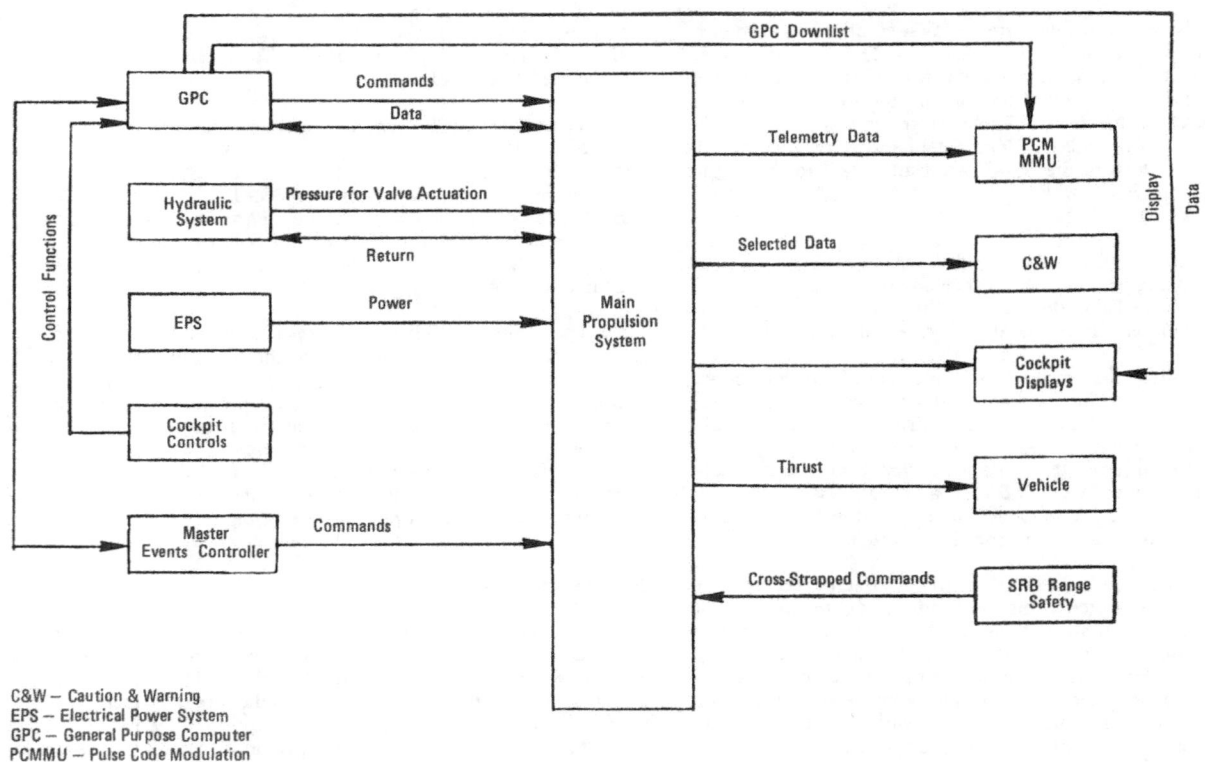

C&W – Caution & Warning
EPS – Electrical Power System
GPC – General Purpose Computer
PCMMU – Pulse Code Modulation
         Master Unit

MPS Interfaces With Other Vehicle Systems

Space Transportation System

At the forward end of each ET propellant tank is a vent and relief valve. This valve is a dual function valve: it can be opened by GSE-supplied helium (vent) or excessive tank pressure (relief). The vent is available only in prelaunch; once lift-off has occurred, only the relief function is operable. The LO_2 (liquid oxygen) tank will relieve at an ullage pressure of 1,293 millimeters of mercury (mmHg) (25 psi). The LH_2 (liquid hydrogen) tanks will relieve at an ullage pressure of 1,966 mmHg (38 psi). The flight crew has no control over the position of the vent and relief valves in either prelaunch or ascent. Normal tank ullage pressure range during ascent for the LH_2 tank is 1,656 to 1,759 mmHg (32 to 34 psi) and for the LO_2 tank is 1,035 to 1,138 mmHg (20 to 22 psi). During prelaunch, the LH_2 tank is pressurized to 2,282 mmHg (44.1 psi) to meet the start requirement of the SSME low-pressure fuel turbopump (LPFT).

In addition to the vent and relief valve, the LO_2 tank has a tumble vent valve that is opened at the ET separation sequence. The thrust force provided by opening the valve imparts an angular velocity to the ET to assist in the separtion maneuver and provide more positive control of the ET reentry aerodynamics.

There are eight propellant depletion sensors. Four of them sense fuel depletion, and four sense oxidizer depletion. The fuel depletions sensors are mounted in the ET LO_2 feedline manifold downstream of the tank. During prelaunch, the launch processing system (LPS) tests each propellant depletion sensor. If any are found to be in a failed condition, the LPS sets a flag in the computer SSME OPS (operational sequence) sequence logic which will instruct the computer to ignore the output of that particular sensor. During SSME engine thrusting, the computer constantly computes the instantaneous mass of the vehicle due to propellant usage from the ET. Vehicle mass constantly decreases. When the computed vehicle mass matches a predetermined initialized loaded value, the computer arms the propellant depletion sensors. After this time, if any two of the

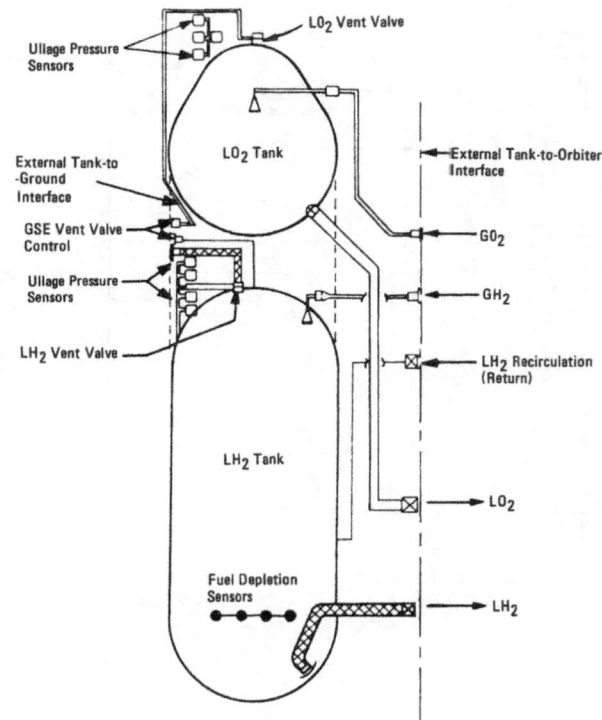

External Tank Schematic

"good" fuel depletion sensors (those not flagged during prelaunch) or any two of the "good" oxidizer depletion sensors indicate "dry" condition, the computers issue a MECO command. This type of MECO is the backup to the nominal

 **Space Transportation System**

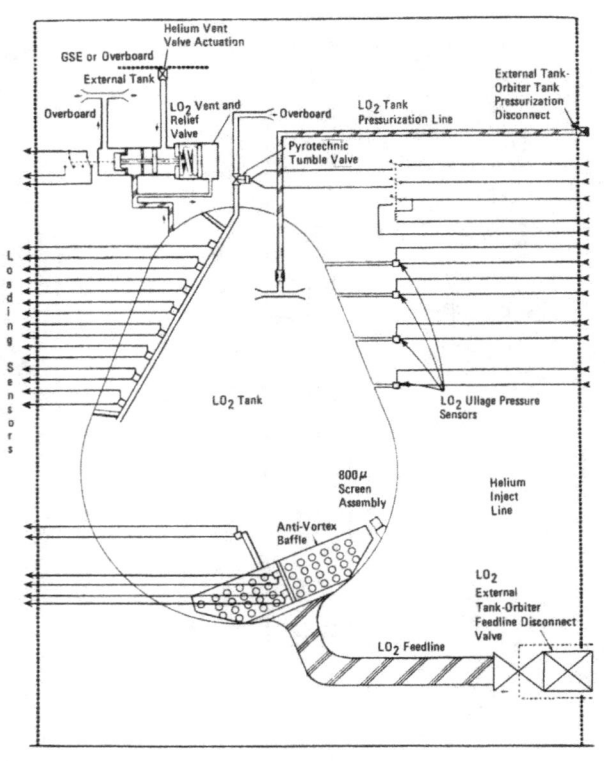

External Tank Schematic, LO$_2$ Tank Detail

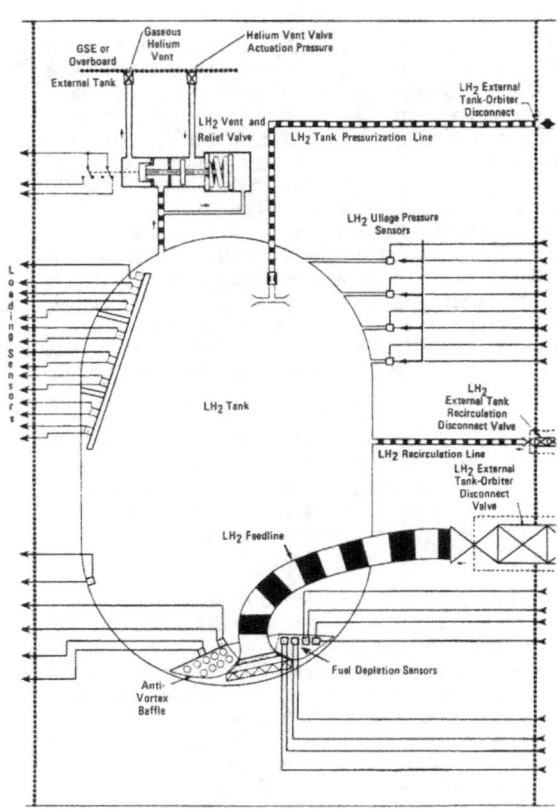

External Tank Schematic, LH$_2$ Tank Detail

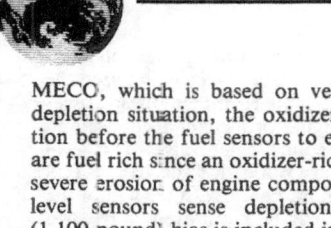

Space Transportation System

MECO, which is based on vehicle velocity. In a propellant depletion situation, the oxidizer-level sensors sense the condition before the fuel sensors to ensure that all depletion cutoffs are fuel rich since an oxidizer-rich cutoff can cause burning and severe erosion of engine components. To ensure that oxidizer-level sensors sense depletion first, a plus 498-kilogram (1,100-pound) bias is included in the amount of LH_2 loaded in the ET. This amount is in excess of that dictated by the 6:1 (oxidizer/fuel) mixture ratio. The position of the oxidizer propellant depletion sensors allows the maximum amount of oxidizer to be consumed in the engines and allows sufficient time to cut off the engines before the oxidizer turbopumps cavitate (run dry).

Four ullage pressure transducers are located at the top end of each propellant tank (LO_2 and LH_2). One of the four is considered a spare and is normally off line. During prelaunch, the GSE will normally check out the four transducers and if one of the three active transducers is determined to be "bad," that transducer can be taken off line and the output of the spare transducer selected. The flight crew can also perform this operation after lift-off, via the computer keyboard; however, because of the short time involved from lift-off to MECO, this would probably be impractical. The three active ullage pressure sensors provide outputs for cathode-ray-tube (CRT) display and control of ullage pressure within their particular propellant tanks. For CRT display, computer processing selects the middle valve output of the three transducers and displays this single value. For ullage pressure control, all three outputs are used.

The ET has five propellant disconnects: two for the LO_2 tank and three for the LH_2 tank. One of the LO_2 tank propellant umbilicals carries LO_2, and the other carries gaseous oxygen (GO_2). The LH_2 tank has two disconnects that carry LH_2 and one that carries gaseous hydrogen (GH_2). The ET LH_2 recirculation disconnect is the smaller of the two disconnects that carry LH_2 and is used only during the LH_2 chilldown sequence in prelaunch.

The ET has two electrical umbilicals, each made of many smaller electrical cables. These cables carry electrical power from the orbiter to the ET and the two SRB's and bring back to the orbiter telemetry from the SRB and ET. The operational instrumentation (OI) telemetry that comes back from the SRB's is conditioned, digitized, and multiplexed in the SRB's themselves. The ET OI measurements that return to the orbiter are the "raw" transducer outputs and must be processed within the orbiter telemetry system.

Oxidizer flow during launch is GSE-supplied LO_2 and is loaded through the orbiter propellant management subsystem. It exits the orbiter at the LO_2 feedline disconnect valve and flows through the LO_2 umbilical into the LO_2 tank. Once the loading is completed, the LO_2 vent and relief valve in the ET is closed and the LO_2 tank is pressurized to 1,086 mmHg (21 psi) by GSE-supplied helium injected into the GO_2 pressurization line. During SSME thrusting, the LO_2 flows out of the ET, through the ET LO_2 umbilical, and enters the orbiter at the LO_2 feedline disconnect valve. Pressurization in the tank is maintained by GO_2 tapped from the three SSME's and supplied to the LO_2 tank through the ET GO_2 umbilical. The interface between the ET GO_2 umbilical and the orbiter is a self-sealing quick disconnect. The ET LO_2 umbilical, the ET GO_2 umbilical, and one ET electrical umbilical are grouped together in a single bundle and connect to the orbiter at the right-hand orbiter umbilical plate.

Fuel flow during prelaunch is GSE-supplied LH_2 and is loaded through the orbiter propellant management subsystem. It exits the orbiter at the LH_2 feedline disconnect valve and flows through the ET LH_2 umbilical in the LH_2 tank. Once loading is complete, the LH_2 vent and relief valve in the ET is closed, and the LH_2 tank is pressurized to 2,199 mmHg (42.5 psi) by GSE-supplied helium injected into the GH_2 pressurization line. Approximately 45 minutes before SSME start, three electrically powered LH_2 pumps in the orbiter begin to circulate the LH_2 out of the ET, through the three SSME's, back to the

Space Transportation System

ET via a special ET LH$_2$ recirculation umbilical. This recirculation chills down all the LH$_2$ lines between the ET and the high-pressure fuel turbopump (HPFT) in the SSME so that the path is free of any GH$_2$ bubbles and is at the proper temperature for engine start. LH$_2$ recirculation ends approximately six seconds before engine start. During engine thrusting, LH$_2$ flows out of the ET and through the ET LH$_2$ umbilical and enters the orbiter at the LH$_2$ feedline disconnect valve. Tank pressurization is maintained by GH$_2$ tapped from the three SSME's and supplied to the LH$_2$ tank through the ET GH$_2$ umbilical. The interface between the ET GH$_2$ umbilical, and the orbiter is a self-sealing quick disconnect. The ET LH$_2$ umbilical, the ET LH$_2$ recirculation umbilical, the ET GH$_2$ umbilical, and one electrical umbilical are grouped together in a single bundle and connected to the orbiter at the left-hand orbiter umbilical plate.

ORBITER PROPELLANT MANAGEMENT SUBSYSTEM

During engine thrusting, propellants under ET tank pressure flow from the ET to the orbiter through two umbilicals: one for LH$_2$ and the other for LO$_2$. Within the orbiter, the propellants pass through a system of manifolds, distribution lines, and valves to the main engines. The valves within the orbiter are under direct control of the orbiter computers and are electrically or pneumatically actuated. This system also provides a path that allows gases tapped from the three SSME's to flow back to the ET through two gas umbilicals to maintain pressures in the fuel and oxidizer tanks. During prelaunch this system is used to control the loading of propellants in the ET. During orbit it controls propellant dump, vacuum inerting, and, prior to entry, system repressurization.

The two 431-millimeter-diameter (17-in.) propellant feedline manifolds in the orbiter, one for LO$_2$ and one for LH$_2$, have a feedline disconnect valve at one end and two fill and drain valves, one inboard, and one outboard, connected in series at the other end. The feedline manifolds connect to the ET liquid propellant umbilicals at the feedline disconnect valve and to either GSE liquid propellant umbilicals (prelaunch only) or overboard at the outboard fill and drain valves. Between the feedline disconnect valves and the inboard fill and drain valves are three outlets for the three engine propellant feedlines and one outlet for the propellant feedline relief valve. The LH$_2$ feedline manifold contains an extra outlet for the LH$_2$ RTLS (return to launch site) feedline dump line. Pressures within the LO$_2$ and LH$_2$ feedline manifolds can be monitored on the two engine manifold (ENG MANF) meters on panel F7 or on the CRT display.

The six 304-millimeter-diameter (12-in.) feedlines in the orbiter, three for LO$_2$ and three for LH$_2$, connect the LO$_2$ feedline manifold to each of the three engine low-pressure oxidizer turbopump (LPOT) inlets and connect the LH$_2$ feedline manifold to each of the three engine low-pressure fuel turbopump (LPFT) inlets. There is one prevalve in each of the six engine propellant feedlines. The prevalves are designated as LEFT, CENTER, and RIGHT engine LO$_2$ prevalve and LEFT, CENTER, and RIGHT engine LH$_2$ prevalve. Each prevalve has its own switch on panel R4. Each switch has a GPC, OPEN, and CLOSE position. The GPC position allows automatic computer control of each prevalve, and the OPEN and CLOSE positions give the flight crew direct manual control of each valve.

There are two 25.4-millimeter-diameter (1-in.) relief valves in the orbiter propellant management subsystem: one for LO$_2$ and one for LH$_2$. Each line connects to one of the propellant feedline manifolds at one end and to an overload port at the other end. Each line contains a relief valve and an electrically actuated relief isolation valve. The isolation valve is mounted in series with, and upstream of, the relief valve. Flow through the relief line and relief valve is enabled by deenergizing the normally open isolation valve, allowing it to open. The position of the two relief isolation valves is controlled by its individual FEEDLINE RLF (relief) ISOL (isolation) switches on panel R4. Normally these switches are left in the GPC position. With the

Space Transportation System

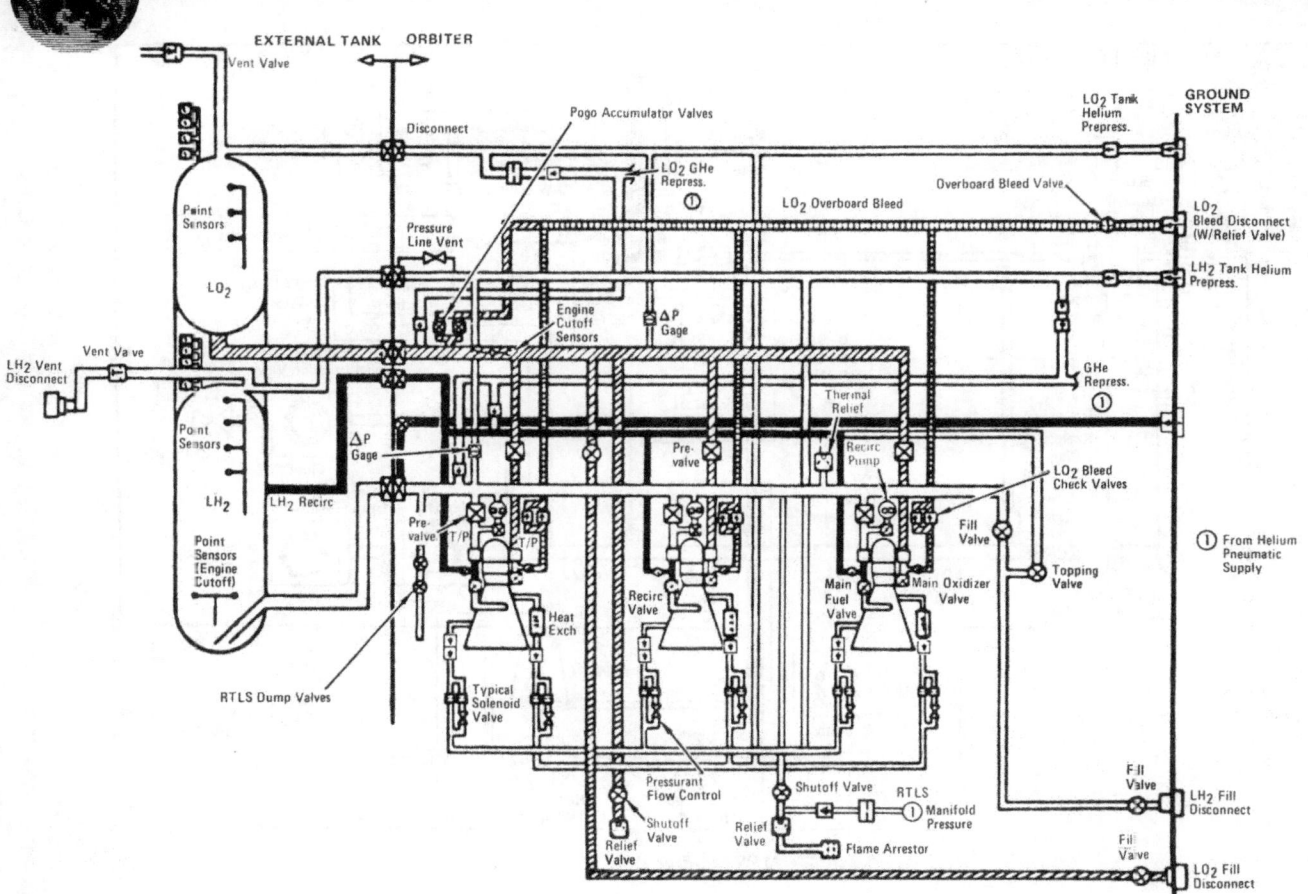

MPS Fluid System

Space Transportation System

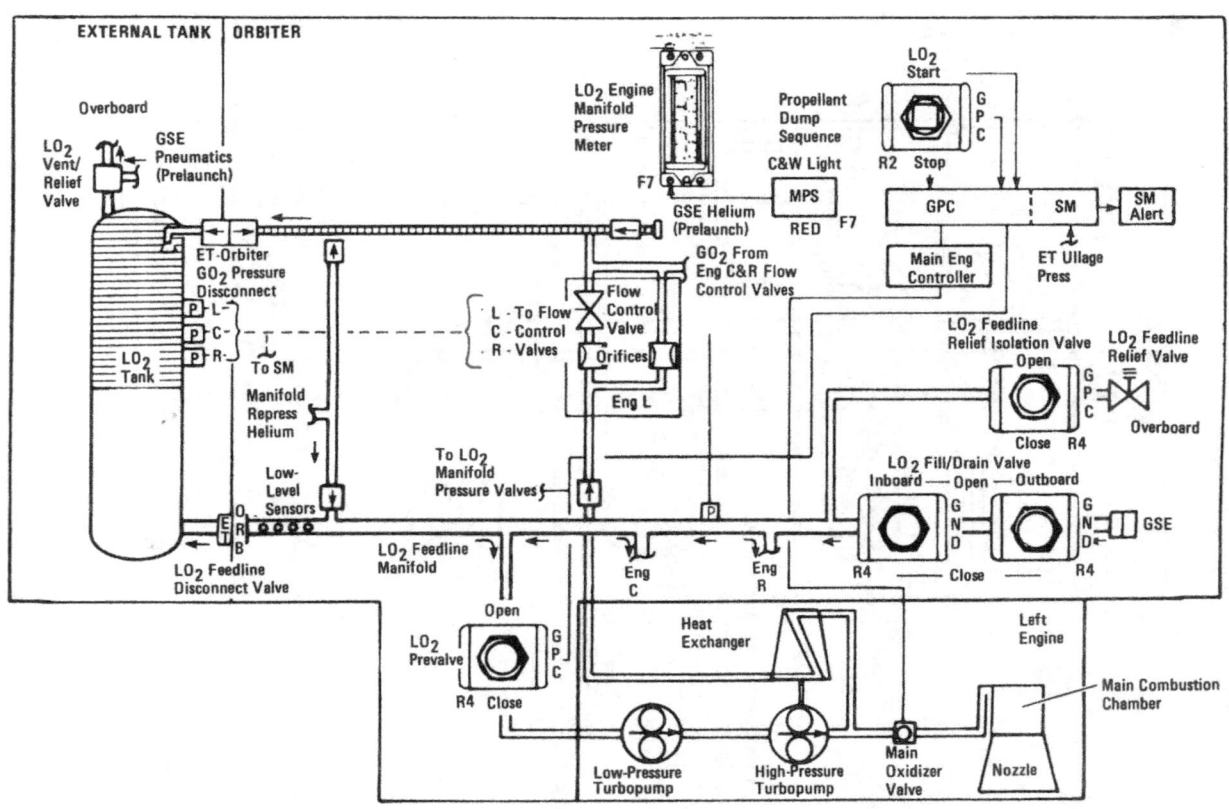

MPS Oxidizer Flow

Space Transportation System

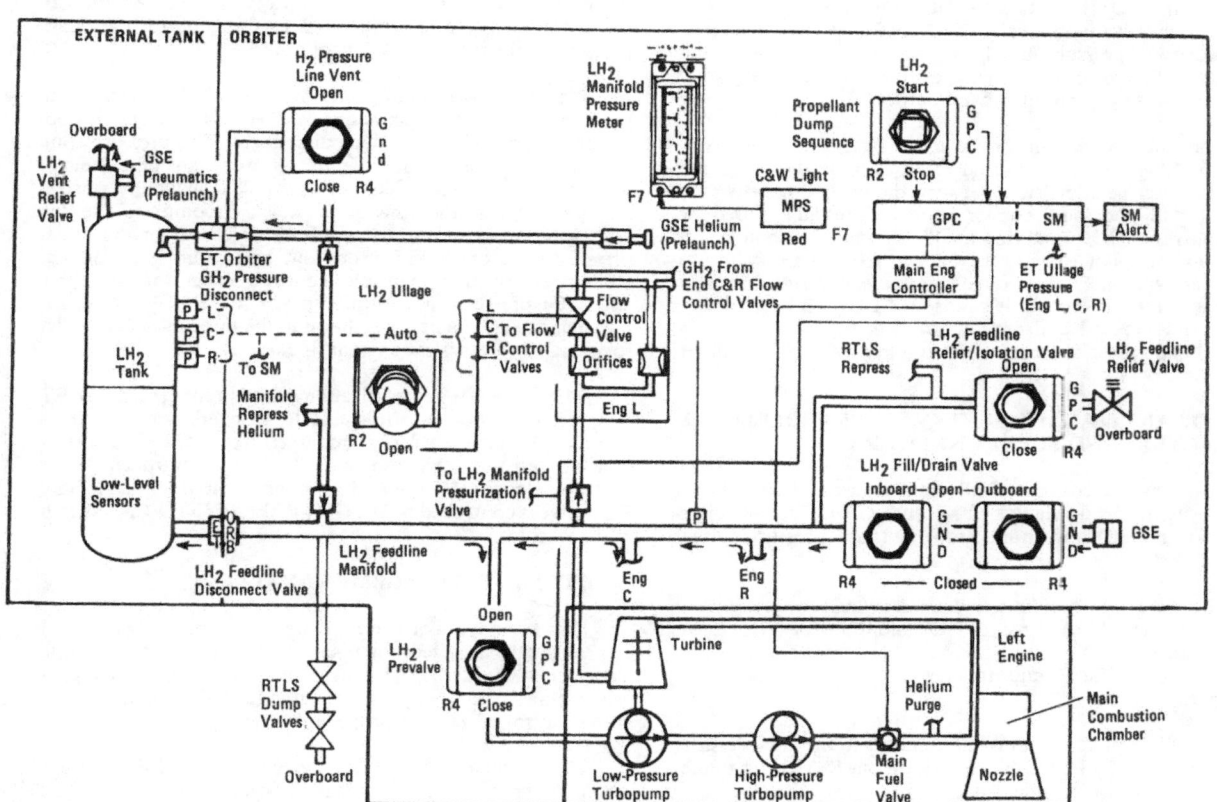

MPS Fuel Flow

Space Transportation System

two switches in the GPC position, both relief isolation valves are opened automatically after MECO. Each switch has an OPEN and CLOSE position, which allows the flight crew to control each valve individually. The relief lines and valves prevent excessive pressure buildups, generated by heatup and expansion of the propellants in the feedline manifolds, by allowing the pressure to be vented overboard.

The single 50-millimeter-diameter (2-in.) line which connects the LH_2 feedline manifold to an overboard port on the outer aft fuselage's left side, between the orbital maneuvering system (OMS) pod and the upper surface of the wing, is used for dumping residual LH_2 during an RTLS abort. In non-RTLS situations, the pilot can use the backup LH_2 dump switch to open these valves. Flow through the line is controlled by two series-connected, normally closed, LH_2 RTLS dump valves—one inboard and one outboard in the line. The LH_2 RTLS dump valves are controlled automatically by computer commands.

PROPELLANT MANAGEMENT GASEOUS PROPELLANT COLLECTION AND SUPPLY NETWORK

This network consists of all the lines used to collect and supply gaseous propellants, GO_2 and GH_2, for all three SSME's to the ET to maintain propellant tank pressure during engine thrusting.

There are six 16-millimeter-diameter (0.63 in.) pressurization lines in the orbiter: three for GO_2 and three for GH_2. Each GO_2 pressurization line connects to the oxidizer heat exchanger outlet on one SSME and to the GO_2 ET pressurization manifold.

Each GH_2 pressurization line connects to the low-pressure fuel turbopump (LPFT) turbine outlet on one SSME and to the GH_2 ET pressurization manifold.

Each of the six pressurization output lines contains two orifices connected in parallel and one flow control valve connected in series with one of the orifices. The combination of orifices and flow control valves is used to control the ullage pressure in the two ET propellant tanks.

There are two 50-millimeter-diameter (2 in.) manifolds in the orbiter: one for GO_2 and one for GH_2. At each end of both manifolds are self-sealing quick disconnects. The pressurization manifolds connect to the ET gaseous propellant umbilicals at one set of quick disconnects and to the GSE helium pressurization umbilicals at the other set of quick disconnects. The two GSE helium pressurization umbilicals are used for the initial pressurization of the ET propellant tanks during prelaunch. Each pressurization manifold contains inlets for the three engine ET pressurization output lines. The ET GH_2 pressurization manifold contains, in addition to the three inlets, an outlet for the GH_2 pressurization vent line.

The single GH_2 pressurization vent line connects to the ET GH_2 pressurization manifold and to an overboard port. This line is used exclusively for vacuum inerting the GH_2 pressurization lines on orbit. Flow through the line is controlled by the normally closed GH_2 pressurization line vent valve in the line. This valve is controlled by the GH_2 PRESS LINE VENT switch on panel R4.

PROPELLANT MANAGEMENT VALVES

Two basic types of valve are used: pneumatically actuated and electrically actuated. Pneumatic valves are used where large loads are encountered, such as in the control of liquid propellant flows. Electrical valves are used for lighter loads such as in the control of gaseous propellant flows.

The pneumatically actuated valves are divided into two subtypes: those that require pneumatic pressure to open and

Space Transportation System

close the valve (type one) and those that are spring loaded to one position and require pneumatic pressure to move to the other position (type two).

Each type one valve actuator is equipped with two electrically actuated solenoid valves. Each of the two solenoid valves controls helium pressure to either an "open" port or a "close" port on the actuator. Energizing the solenoid valve on the "open" port allows helium pressure to open the pneumatic valve. Energizing the solenoid on the "close" port allows helium pressure to close the pneumatic valve. Removing power from a solenoid valve removes helium pressure from the corresponding port of the pneumatic actuator and allows the helium pressure trapped in that side of the actuator to vent overboard. Removing power from both solenoids allows the pneumatic valve to remain in the last commanded position. This type of valve is used for the LO_2 and LH_2 feedline disconnect valve, the three LH_2 and LO_2 prevalves, and the LH_2 and LO_2 inboard and outboard fill/drain valves.

Each type two valve is a single electrically actuated solenoid valve that controls helium pressure to either an "open" or a "close" port on the actuator. Removing power from the solenoid valve removes helium pressure from the corresponding port of the pneumatic actuator and allows helium pressure trapped in that side of the actuator to vent overboard. Spring force takes over and drives the valve to the opposite position. If the spring force drives the valve to the open position, the valve is referred to as a normally open (NO) valve. If the spring force drives the valve to a closed position, the valve is referred to as a normaly closed (NC) valve. This type of valve is used for the LH_2 RTLS inboard dump valves (NC), LH_2 RTLS outboard dump valves (NC), LH_2 feedline relief shutoff valve (NO), and LO_2 feedline relief shutoff valve (NO).

The electrically actuated solenoid valves are spring-loaded to one position and move to the other position when electrical power is applied. These valves are referred to as either normally open (NO) or normally closed (NC) based on their position in the deenergized state. The electrically actuated solenoid valves are the GH_2 pressurization line vent valve (NC), the three GH_2 pressurization flow control valves (NO), and the three GO_2 pressurization flow control valves (NO).

ORBITER HELIUM SUBSYSTEM

The orbiter helium system consists of seven 0.13-cubic-meter (4.7-cubic-foot) helium supply tanks, three 0.48-cubic-meter (17.3 cubic-foot) helium supply tanks, and associated regulators, check valves, distribution lines, and control valves. Four of the 0.13-cubic-meter (4.7-cubic-foot) helium supply tanks are located in the orbiter aft fuselage, and the other three are located below the payload bay liner in the mid fuselage in the area originally reserved for the cryogenic storage tanks of the power reactant and storage tanks for the power reactant and storage distribution system. The three 0.48-cubic-meter (17.3-cubic-foot) helium supply tanks are also located below the payload bay liner in the mid fuselage.

The tanks are of a composite construction consisting of a titanium liner with a fiber glass structural overwrap. The large tanks are 1,023 millimeters (40.3 inches) in diameter and have a dry weight of 123 kilograms (272 pounds). The smaller tanks are 660 millimeters (26 inches) in diameter and have a dry weight of 33 kilograms (73 pounds). The tanks are serviced before liftoff to 217,350 mmHg (4,200 psi).

Each of the larger tanks is plumbed to two of the smaller ones, forming three sets of tanks for the engine helium pneumatic supply system. Each set—one in the mid fuselage, the other in the aft fuselage—supplies helium to one engine and is referred to as left, center, or right engine helium. Each set normally provides helium to only that engine for in-flight purges and provides pressure to that engine for emergency pneumatic shutdown of that engine.

 **Space Transportation System**

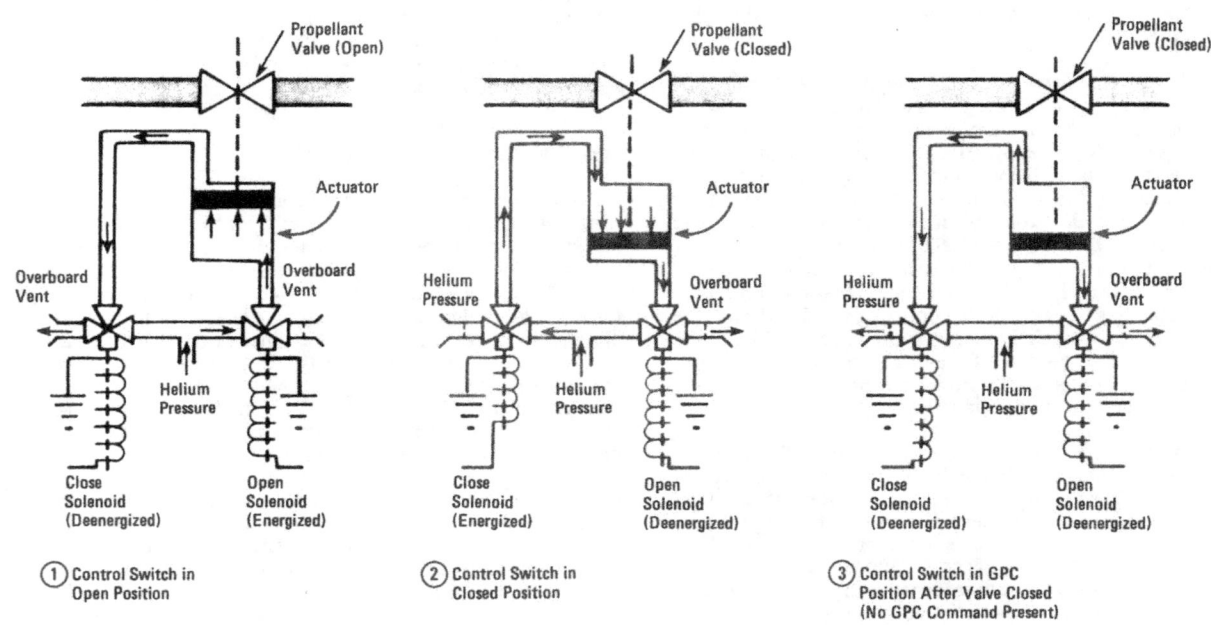

Propellant Management System Typical Pneumatically Actuated Propellant Valve, Type 1

 **Space Transportation System**

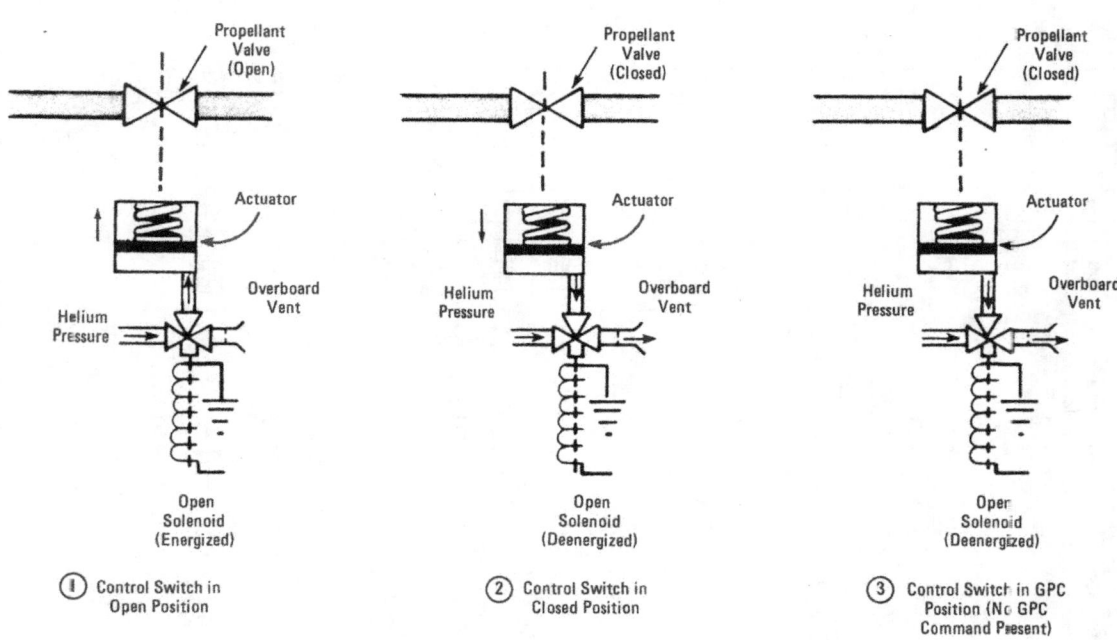

Propellant Management System Typical Pneumatically Actuated Propellant Valve, Type 2

Space Transportation System

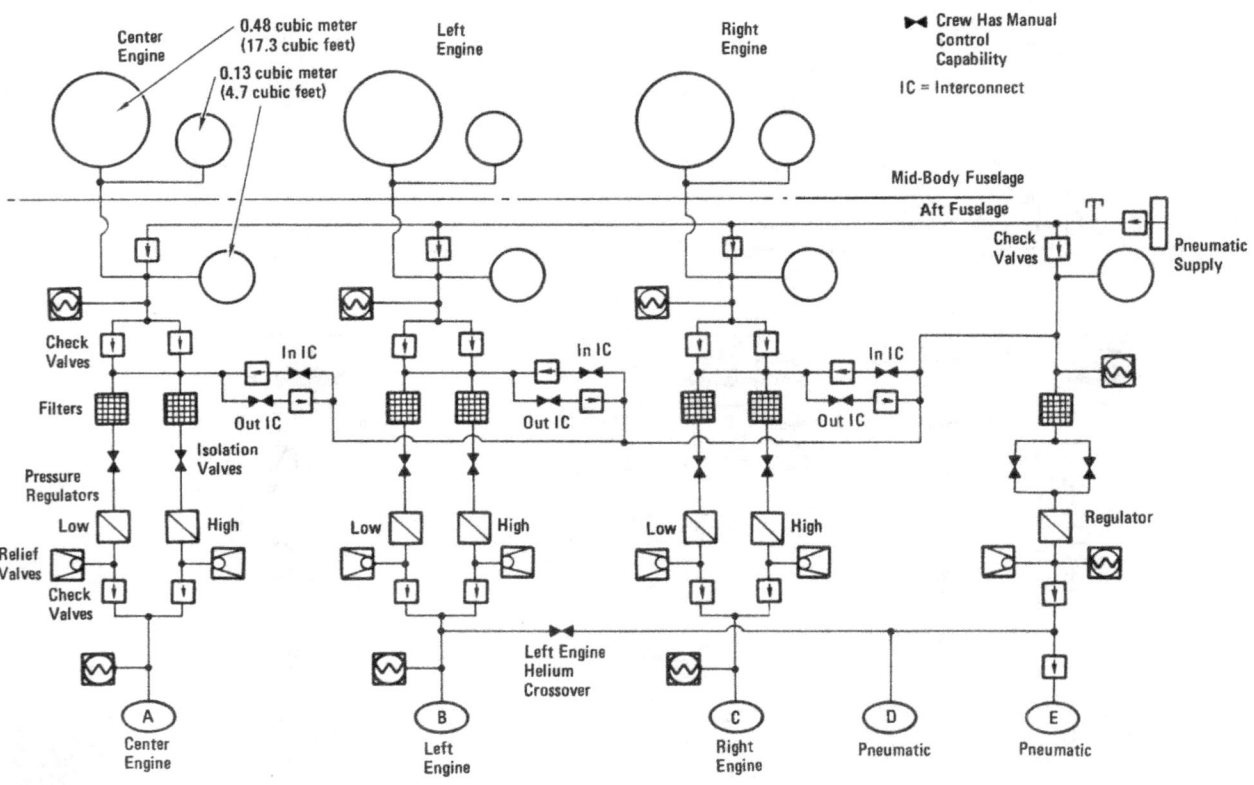

Helium System Storage and Supply

Space Transportation System

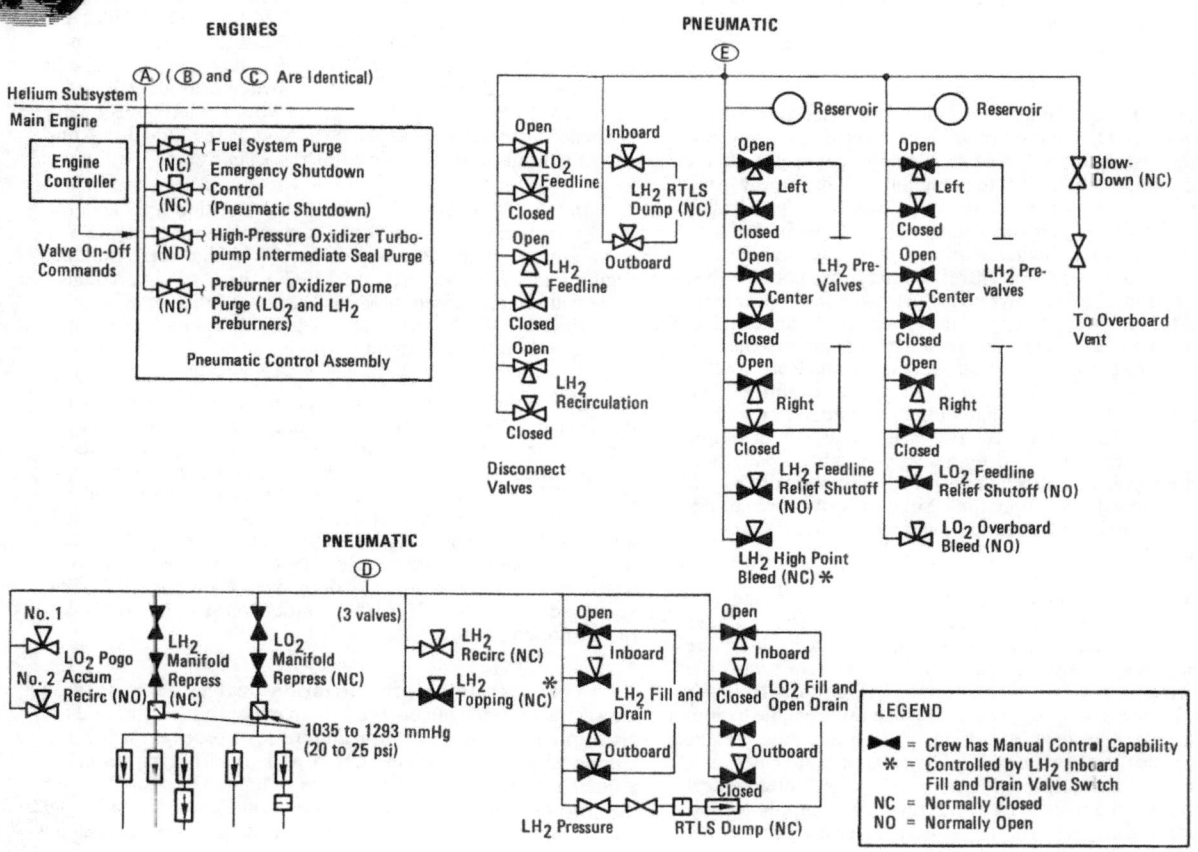

Helium System Distribution

Space Transportation System

The remaining small tank is referred to as pneumatic helium supply. It supplies helium pressure to actuate all pneumatically operated valves in the propellant management subsystem.

There are eight helium supply tank isolation valves grouped in pairs. One pair of valves is connected to each engine helium supply tank cluster and one pair is connected to the pneumatic supply tank. In the engine helium supply tank system, each pair of isolation valves is connected in parallel, with each valve in the pair controlling helium flow through one leg of a dual redundant helium supply circuit. Each helium supply circuit contains two check valves, a filter, an isolation valve, a regulator, and a relief valve. The two isolation valves connected to the pneumatic supply tanks are also connected in parallel; however the rest of the pneumatic supply system consists of a filter, the two isolation valves, a regulator, a relief valve, and a single check valve. Each of the engine helium supply isolation valves can be individually controlled by its own switch on panel R2, whereas the two pneumatic helium supply isolation valves are controlled by a single switch on panel R2. All of these valves can also be controlled automatically by the computers. The helium supply tank isolation valves are spring-loaded to the closed position and pneumatically actuated to the open position.

Each engine helium supply system operates independently until after MECO. The three engine helium supply systems have a helium "out" interconnect valve which connects it to the pneumatic supply system. A check valve is in series with the "out" interconnect valve, which permits helium flow from the engine to the pneumatic supply system. The three engine helium supply systems also have a helium "in" interconnect valve which connects the pneumatic supply system to that engine supply system. The check valve, in series with the "in" interconnect valve, permits helium flow from the pneumatic supply to that engine system. Each pair of interconnect valves is controlled by a single switch on panel R2, which has three positions.

The three switch positions, on panel R2, are IN OPEN/OUT CLOSE, GPC, and IN CLOSE/OUT OPEN. With the switch in the IN OPEN/OUT CLOSE position, the "in" interconnect valve is open and the "out" interconnect valve is closed. The IN CLOSE/OUT OPEN switch position will do the reverse. With the switch in GPC, the "out" interconnect valve is opened automatically at the beginning of the LO_2 dump and closed automatically at the end of the LH_2 dump.

In the RTLS abort, the GPC switch position will cause the "in" interconnect valve to open automatically at MECO and close automatically 20 seconds later. If the corresponding engine was shutdown prior to MECO, its "in" interconnect valve will remain closed at MECO. At any other time, placing the switch in GPC position results in both interconnect valves closing and remaining closed.

The additional interconnect crossover valve between the left engine helium supply and the pneumatic helium supply is used in the event of the pneumatic helium supply regulator failure due to only one regulator in the pneumatic helium supply system. The crossover helium valve would be opened and the pneumatic helium supply tank isolation valves would be closed, allowing the left engine helium supply system to supply helium to the pneumatic helium supply. The crossover helium valve is controlled by its own three-position switch on panel R2. The switch positions are OPEN, GPC, and CLOSE. The GPC position is presently undefined.

The manifold pressurization valves are located downstream of the pneumatic helium pressure regulator and are used to control the flow of helium to the propellant manifolds during nominal propellant dump and manifold repressurization. There are four of these valves grouped in pairs. One pair controls helium pressure to the LO_2 propellant manifolds, and the other pair controls helium pressure to the LH_2 propellant manifold.

Space Transportation System

The LH$_2$ RTLS dump pressurization valves are located downstream of the pneumatic helium pressure regulator and are used to control the pressurization of the LH$_2$ propellant manifolds during an RTLS LH$_2$ dump. There are two of these valves connected in series. Unlike the LH$_2$ manifold pressurization valves, the LH$_2$ RTLS dump pressurization valves cannot be controlled from the flight deck. During an RTLS abort, these valves are opened and closed automatically by GPC commands. An additional difference between the nominal and the LH$_2$ dump is in the routing of the helium and the location at which it enters the LH$_2$ feedline manifold. For the nominal LH$_2$ dump, helium passes through the LH$_2$ manifold pressurization valves and enters the feedline manifold in the vicinity of the LH$_2$ feedline disconnect valve. For the LH$_2$ RTLS dump, helium passes through the RTLS/LH$_2$ dump pressurization valves and enters the feedline manifold in the vicinity of the LH$_2$ inboard fill/drain valve on the inboard side. There is no RTLS LO$_2$ dump pressurization valve since the LO$_2$ manifold is not pressurized during the RTLS LO$_2$ dump.

Each engine helium supply tank cluster has two pressure regulators operating in parallel. Each regulator controls pressure in one leg of a dual-redundant helium supply circuit and is designated either a high regulator or low regulator. The high regulator is set to provide a nominal outlet pressure of 41,400 mmHg (800 psi); the low regulator is set to provide a nominal outlet pressure of 37,001 mmHg (715 psi). When the system demand is low, only the high regulator will operate. When demand increases, the low regulator will begin to operate, doubling the flow capacity of the system.

The pressure regulator for the pneumatic helium supply system is not redundant and is set to provide outlet pressure between 37,001 to 34,847 mmHg (715 to 770 psi). Downstream of the regulator are two more regulators: the LH$_2$ manifold pressure regulator and the LO$_2$ manifold pressure regulator. These regulators are used only during MPS propellant dump and manifold repressurization. Both regulators are set to provide outlet pressures between 1,035 to 1,293 mmHg (20 to 25 psi). Flow through the regulators is controlled by the appropriate set of two normally closed manifold pressurization valves.

Downstream of each pressure regulator, with the exception of the two manifold repressurization regulators, is a relief valve. The valve protects the downstream helium distribution lines from overpressurization in the event the associated regulator fails fully open. The two relief valves in each engine helium supply are set at different values, reflecting the differences in the settings of the corresponding low and high regulators. The relief valve in the pneumatic helium supply circuit relieves at 41,400 to 43,987 mmHg (800 to 850 psi) and reseats at 40,623 mmHg (785 psi).

There is one pneumatic control assembly (PCA) manifold on each of the three SSME's. The PCA is essentially a manifold pressurized by of the engine helium supply systems and contains solenoid valves to control and direct pressure to perform various essential functions. The valves are energized by discrete on/off commands from the output electronics of the associated SSME controller. Functions controlled by the PCA include in-flight purges of the fuel system high-pressure oxidizer turbo-pump (HPOT) intermediate seal cavity and preburner oxidizer domes, and pneumatic shutdown.

SPACE SHUTTLE MAIN ENGINES

Oxidizer from the external tank enters the orbiter at the LO$_2$ feedline disconnect valve, then into the orbiter LO$_2$ feedline manifold. There it branches out into three parallel paths, one to each engine. In each branch an LO$_2$ prevalve must be opened to permit flow to the low-pressure oxidizer turbopump (LPOT).

The LPOT is an axial flow pump driven by a six-stage turbine powered by LO$_2$. It boosts the LO$_2$ pressure from 5.175

Space Transportation System

mmHg (100 psia) to 21,838 mmHg (422 psia). The flow from the LPOT is supplied to the high-pressure oxidizer turbopump (HPOT). During engine operation, the pressure boost permits the HPOT to operate at high speeds without cavitation. The LPOT operates at approximatley 5,150 rpm. The LPOT is approximately 457 by 457 millimeters (18 by 18 inches) and is connected to the vehicle propellant ducting and supported in a fixed position by the orbiter structure.

The HPOT consists of two single-stage centrifugal pumps (main pump and preburner pump) mounted on a common shaft and driven by a two-stage hot gas turbine. The main pump boosts the LO_2 pressure from 21,838 mmHg (422 psia) to 222,525 mmHg (4,300 psia) while operating at approximately 27,940 rpm. The HPOT discharge flow splits into several paths, one of which is routed to drive the LPOT turbine. Another path is routed to and through the main oxidizer valve and enters into

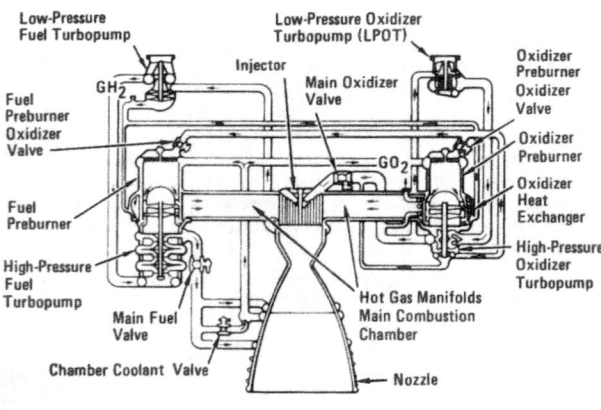

Main Engine Flow

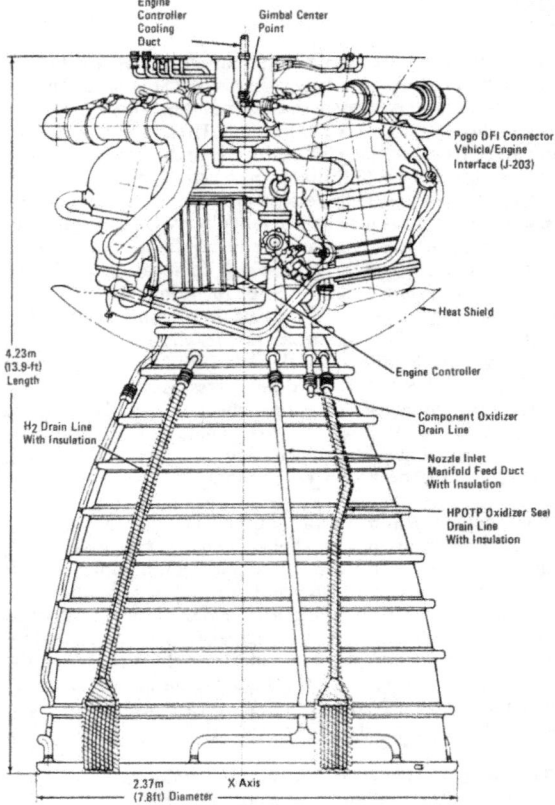

SSME Controller Side

71

the engine main combustion chamber. Another small flow path is tapped off and sent to the oxidizer heat exchanger. Before the LO_2 enters the oxidizer heat exchanger, it flows through an anti-flood valve that prevents LO_2 from entering the heat exchanger until sufficient heat is present to convert the LO_2 to gas. The heat exchanger utilizes the heat contained in the discharge gases from the HPOT turbine to convert the LO_2 to gas. The gas is then sent to a manifold, where it joins gas from the other engines and is sent to the external tank to pressurize the LO_2 tank. Another path enters the HPOT second-stage (preburner pump) to boost the LO_2 pressure from 222,525 mmHg (4,300 psia) to 383,985 mmHg (7,420 psia) and passes through the oxidizer preburner oxidizer valve (OPOV) into the oxidizer preburner (OPB) and through the fuel preburner oxidizer valve (FPOV) and into the fuel preburner (FPB). The HPOT is approximately 609 by 914 millimeters (24 by 36 inches). The HPOT is flange attached to the hot gas manifold.

Fuel enters the orbiter at the LH_2 feedline disconnect valve, then into the orbiter GH_2 feedline manifold, and branches out into three parallel paths to each engine. In each LH_2 branch is an LH_2 prevalve, which, when open, permits LH_2 flow to the low-pressure fuel turbopump (LPFT).

The LPFT is an axial-flow pump driven by a two-stage turbine powered by GH_2. It boosts LH_2 pressure from 1,552 mmHg (30 psia) to 14,283 mmHg (276 psia) and supplies it to the high-pressure fuel turbopump (HPFT). During engine operation, the pressure boost provided by the LPFT permits the HPFT to operate at high speeds without cavitation. The LPFT operates at approximately 15,290 rpm. The LPFT is approximately 457 by 609 millimeters (18 by 24 inches). The LPFT is connected to the vehicle propellant ducting and is supported in a fixed position by the orbiter structure 180 degrees from the LPOT.

The HPFT is a three-stage centrifugal pump driven by a two-stage hot-gas turbine. It boosts LH_2 pressure from 14,283 mmHg (276 psia) to 337,151 mmHg (6,515 psia). The HPFT operates at approximately 30,015 rpm. The discharge flow from the turbopump is routed to and through the main valve and then splits into three flow paths. One path is through the jacket of the main combustion chamber, where the hydrogen is used for cooling the chamber walls. It is then routed from the main combustion chamber to the LPFT, where it is used to drive the LPFT turbine. A small portion of the flow from the LPFT is then directed to a common manifold from all three engines to form a single path to the external tank to maintain LH_2 tank pressurization. The remaining hydrogen passes between the inner and outer walls to cool the hot-gas manifold and is discharged into the main combustion chamber. The second hydrogen flow path from the main fuel valve is through the engine nozzle (for cooling the nozzle). It then joins the third flow path from the chamber coolant valve. The combined flow is then directed to the fuel and oxidizer preburner. The HPFT is approximately 558 by 1,117 millimeters (22 by 44 inches). The HPFT is flange-attached to the hot-gas manifold.

The oxidizer (OPB) and fuel (FPB) preburners are welded to the hot-gas manifold. The fuel and oxidizer enter each preburner, where they are mixed so that efficient combustion can occur. The augmented spark igniter is a small combustion chamber located in the center of the injector of each preburner. The two dual spark igniters, which are activated by the controller, are used during the engine-start sequence to initiate combustion in each preburner. They are turned off after approximately three seconds as the combustion process is self-sustaining. The preburners produce the fuel-rich hot gas which passes through the turbines to generate the power to operate the high-pressure turbopumps. The oxidizer preburner outflow drives the turbine which is connected to the HPOT and oxidizer preburner pump. The fuel preburner outflow drives the turbine which is connected to the HPFT.

 **Space Transportation System**

The HPOT turbine and HPOT pumps are mounted on a common shaft. Mixing of the fuel-rich hot gas in the turbine section and the LO_2 in the main pump could create a hazard. To prevent this, the two sections are separated by a cavity that is continuously purged with the MPS engine helium supply during engine operation. Two seals minimize leakage into the cavity. One is between the turbine section and the cavity and other is between the pump section and cavity. Loss of helium pressure to this cavity results in automatic engine shutdown.

Speed of the HPOT and HPFT turbines depends on the position of the coresonding oxidizer and fuel preburner oxidizer valves. These valves are positioned by the engine controller, which uses them to throttle the flow of LO_2 to the preburners and thus control engine thrust. The oxidizer and fuel preburner valves increase or decrease the LO_2 and LH_2 flow, thus increasing or decreasing preburner chamber pressure, HPOT and HPFT turbine speed, and LO_2 and GH_2 flow into the main combustion chamber, increasing or decreasing the engine thrust, thus throttling the engine. The oxidizer and fuel preburner valves operate together to throttle the engine and maintain the engine mixture ratio at constant 6:1.

The main oxidizer valve (MOV) and main fuel valve (MFV) control LO_2 and LH_2 flow into the engine and are controlled by each engine controller. When an engine is operating, the main valves are fully open.

A coolant control valve (CCV) is mounted on the combustion chamber coolant bypass duct of each engine. The engine controller regulates the amount of GH_2 allowed to bypass the nozzle coolant loop, thus controlling its temperature. The chamber coolant valve is 100-percent open before engine start and after shutdown. During engine operation, it will be 100-percent open for throttle settings of 100 to 109 percent for minimum cooling. For throttle settings between 65 to 100 percent, its position will range from 66.4- to 100-percent open for maximum cooling.

Each engine main combustion chamber (MCC) receives fuel-rich hot gas from a hot-gas manifold cooling circuit. The GH_2 and LO_2 enter the chamber at the injector, which mixes the propellants. A small augmented spark igniter (ASI) chamber is located in the center of the injector. The dual-redundant igniter is used during the engine start sequence to initiate combustion. After approximately three seconds, the igniters are turned off, and the combustion process is self-sustaining. The main injector and dome assembly is welded to the hot-gas manifold. The main combustion chamber is bolted to the hot-gas manifold.

The inner surface of each combustion chamber, as well as the inner surface of each nozzle, is cooled by GH_2 flowing through coolant passages. The nozzle assembly is a bell-shaped extension bolted to the main combustion chamber. The nozzle is 2,870 millimeters (113 inches) long and has an exit outside diameter of 2,387 millimeters (94 inches). An engine-to-orbiter-heat-shield support ring is welded to the forward end of the nozzle and is the engine attach point for the orbiter-supplied heat shield. Thermal protection for the nozzles is necessary because of the exposure portions of the nozzles experience during the launch, ascent, on-orbit, and entry phases of a mission. The insulation consists of four layers of metallic batting covered with a metallic foil and screening.

The five propellant valves on each engine (oxidizer preburner oxidizer, fuel preburner oxidizer, main oxidizer, main fuel, and chamber coolant) are hydraulically actuated and controlled by electrical signals from the controller. They can be fully closed by using the MPS engine helium supply system as a backup actuation system.

The low-pressure oxygen (LPOT) and fuel (LPFT) turbopump are mounted 180 degrees apart on the orbiter aft fuselage thrust structure. The lines from the low-pressure turbopumps to the high-pressure turbopumps contain flexible bellows that enable the low-pressure turbopumps to remain stationary while

Space Transportation System

the rest of the engine is gimbaled for thrust vector control. The LH$_2$ line from the LPFT to the HPFT is insulated to prevent the formation of liquid air.

The main oxidizer and fuel valves also are used after shutdown. During a propellant dump, the main oxidizer valve will be opened to allow residual LO$_2$ to be dumped overboard through the engine and the LH$_2$ dumped through the LH$_2$ fill and drain valves overboard. After the dump is completed, the valves close and remain closed for the remainder of the mission.

The gimbal bearing is bolted to the main injector and dome assembly and is the thrust interface between the engine and orbiter. The bearing assembly is approximately 287 by 355 millimeters (11.3 by 14 inches).

Overall an SSME weighs approximately 3,129 kilograms (6,900 pounds).

POGO SUPPRESSION SYSTEM

A pogo suppression system prevens the transmission of low-frequency flow oscillations into the high-pressure turbopump and ultimately prevents main combustion chamber pressure (engine thrust) oscillation. Flow oscillations transmitted from the Space Shuttle vehicle are suppressed by a partially filled gas accumulator, which is flange-attached to the high-pressure oxidizer turbopump inlet duct.

The system consists of a 0.01-cubic-meter (0.6-cubic-foot) accumulator with an internal standpipe, helium precharge valve package, gaseous oxygen supply valve package, and two recirculation isolation valves (one being located on the orbiter).

During engine start, the accumulator is charged with helium 1.5 seconds after the start command to provide pogo protection until the engine heat exchanger is operational and gaseous oxygen is available.

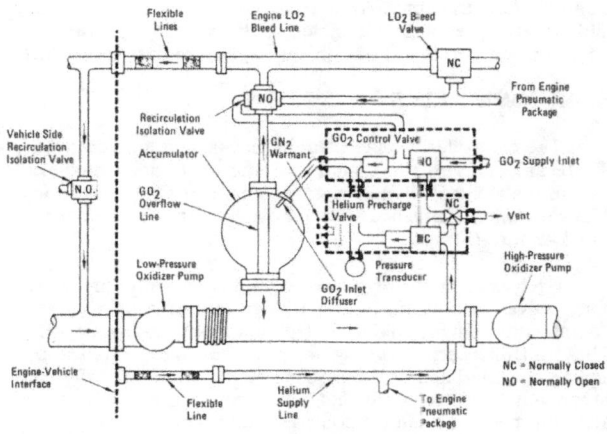

Pogo Suppression System

The accumulator is partially chilled by LO$_2$ during the engine chilldown operation. The accumulator fills to the overflow standpipe line inlet level, which is sufficient to preclude gas ingestion at start.

During engine operation, the accumulator is charged with a continuous gaseous oxygen flow maintained at a rate governed by the engine operation point.

The liquid level in the accumulator is controlled by the overflow standpipe line in the accumulator, which is orificed to regulate the gaseous oxygen overflow over the engine operating power level. The system is sized to provide sufficient gaseous oxygen replenish at the minimum gaseous oxygen flow rate and to permit sufficient gaseous oxygen overflow at the maximum

Space Transportation System

decreasing pressure transient in the low-pressure oxidizer turbopump discharge duct. At all other conditions, excess gaseous and liquid oxygen are recirculated back to the low-pressure oxidizer turbopump inlet through the engine oxidizer bleed duct.

SSME CONTROLLER

The controller is an electronics package mounted on each SSME and contains two digital computers with associated electronics that control all main engine components and operations. The controller is attached to the main combustion chamber by shock-mount fittings.

Each controller operates in conjunction with engine sensors, valves, actuators, and spark ignitors to provide a self-contained system for engine control, checkout, and monitoring. The controller provides engine flight readiness verification, engine start and shutdown sequencing, closed-loop thrust and propellant mixture ratio control, sensor excitation, valve actuator and spark ignitor control signals, engine performance limit monitoring, onboard engine checkout and response to vehicle commands and transmission of engine status, and performance and maintenance data.

Each engine controller receives engine commands transmitted by the orbiter GPC's through an engine interface unit (EIU) dedicated to that engine controller. The engine controller provides its own commands to the main engine components. Engine data are sent to the engine controller, where the data are stored in a vehicle data table (VDT) in the controller's computer memory. Controller status compiled by the engine controller's computer are also added to the vehicle data table. The vehicle data table is periodically output by the controller to the EIU for transmission to the orbiter GPC's.

The EIU is a specialized multiplexer/demultiplexer (MDM) that interfaces with the orbiter GPC's and with the engine controller. When engine commands are received by the

- Computer Control of Engine
- Input Power
  115-volt, 400-Hz, 3-phase ac
  28-volt dc
  490 watts (Standby)
  600 watts (Mainstage)
- Size: 596 x 368 x 431 millimeters
  (23.5 x 14.5 x 17 inches)
- Heat Transfer
  Forced Air-Cooling (Ground Checkout)
  Convection Cooling (Engine Operation)

SSME Controller

EIU, the data are held in a buffer until the EIU receives an orbiter GPC request for data. The EIU then sends data to each orbiter GPC. Each EIU is dedicated to one SSME and communicates only with the engine controller that controls the SSME. The EIU's have no interface with each other.

 **Space Transportation System**

The controller provides responsive control of engine thrust and mixture ratio throughout the digital computer in the controller, updating the instructions to the engine control elements 50 times per second (every 20 milliseconds). Engine reliability is enhanced by a dual redundant system that allows normal operation after the first failure and a fail-safe shutdown after a second failure. High-reliability electronic parts are used throughout the controller.

The digital computer is programmable, allowing modification of engine control equations and constants by change to the stored program (software). The controller is packaged in a sealed, pressurized chassis and is cooled by convection heat transfer through pin fins as part of the main chassis. The electronics are distributed on functional modules with special thermal and vibration protection.

The controller is divided into five subsystems: input electronics, output electronics, computer interface electronics, digital computer, and power supply electronics. Each subsystems is duplicated to provide dual-redundant capability.

The input electronics receive data from all engine sensors, condition the signals, and convert to digital values for processing by the digital computer. Engine control sensors are dual-redundant, and maintenance data sensors are non-redundant.

The output electronics convert the computer digital control commands into voltages suitable for powering the engine spark igniters, the off/on valves, and the engine propellant valve actuators.

The computer interface electronics control the flow of data within the controller, input data to the computer, and computer output commands to the output electronics. They also provide the controller interface with the vehicle engine electronics interface unit for receiving engine commands which are triple-redundant channels from the vehicle and transmission of engine status and data through dual-redundant channels to the vehicle. The computer interface electronics includes the watchdog timers that determine which channel of the dual redundant mechanization is in control.

The digital computer is an internally stored, general-purpose, digital computer that provides the computational capability necessary for all engine control functions. The memory has a program storage capacity of 16,384 data and instruction words (17-bit words; 16 bits for program use, one bit for parity).

The power supply electronics convert the 115-V, 3-phase, 400-Hz vehicle power to the individual power supply voltage levels required by the engine control system and monitor the level of power supply channel operation within satisfactory limits.

Each orbiter GPC, operating in a redundant set, will issue engine commands to the engine interface units for transmission to its corresponding engine controllers. Each orbiter GPC has an SSME system operating program (SOP) application software residing in it. The engine commands are output over its assigned flight control (FC) data bus (a total of four GPC's outputting over four FC data buses). Therefore, each EIU will receive four commands. The nominal ascent configuration has orbiter GPC's 1, 2, 3, and 4 outputting on the FC data buses 5, 6, 7, and 8 respectively. Each FC data bus is connected to one multiplexer interface adapter (MIA) in each EIU.

The EIU checks the received engine commands for transmission errors. If there are none, the EIU passes the validated engine commands on to the controller interface assemblies, which output the validated engine commands to the engine controller. An engine command that does not pass validation is not sent to the controller interface unit (CIA). Instead it is dead-ended in the engine interface unit's multiplexer interface unit (MIU). Commands that come through MIA's

Space Transportation System

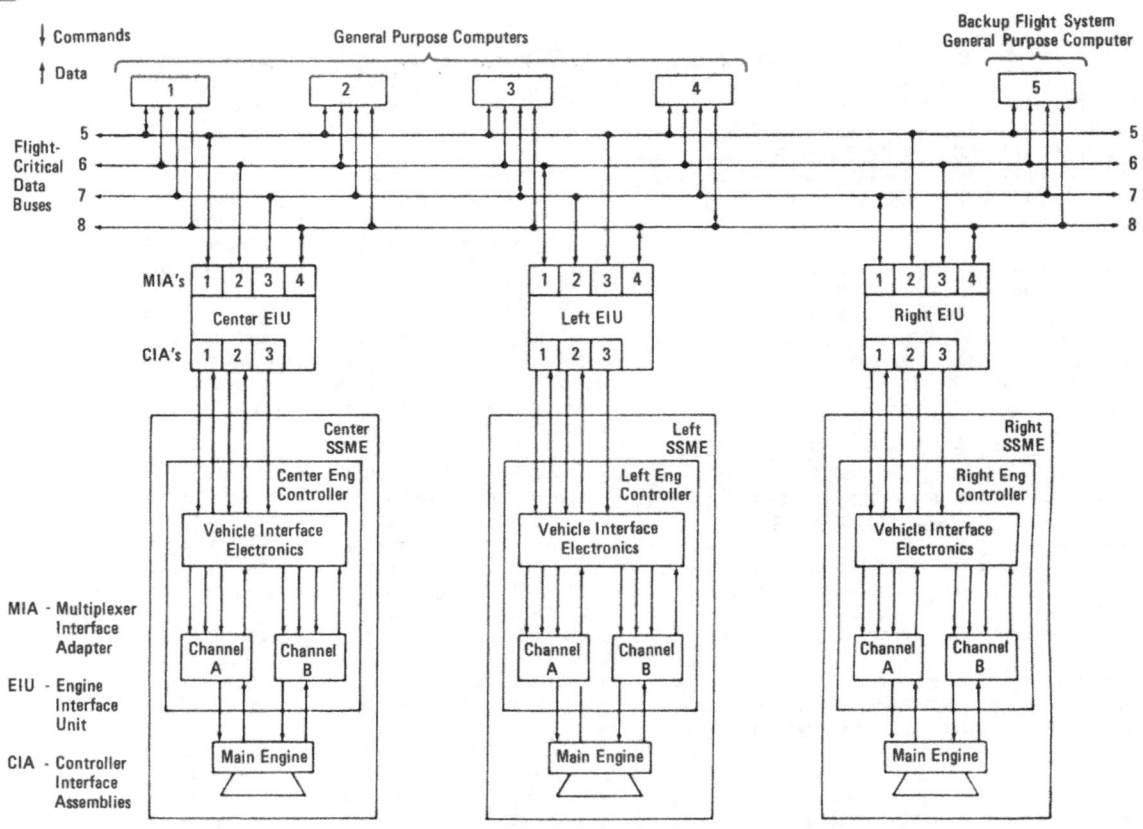

General Purpose Computer

Space Transportation System

one and two are sent to CIA's one and two, respectively. Commands that come to MIA's three and four pass through a CIA three data select logic. This logic outputs the command that arrives at the logic first, from either MIA three or four. The other command is dead-ended in CIA three select logic. The selected command is output through CIA three. In this manner, the EIU reduces the four commands sent to the EIU to three commands output by the EIU.

The engine controller vehicle interface electronics (VIE) receive the three engine commands output by its EIU, check for transmission errors (hardware validation), and send controller hardware-validated engine commands to the Controller A and B electronics. Normally, Channel A electronics is in control, with Channel B electronics active, but not in control. If Channel A fails, Channel B will assume control. If Channel B subsequently fails, the engine controller will shut down the engine pneumatically. If two or three commands pass voting, the engine controller will issue its own commands to accomplish the function commanded by the orbiter GPC's. If command voting fails, two or all three commands fail, the engine controller will maintain the last command that passed voting.

The backup flight system (BFS) orbiter GPC 5 contains an SSME hardware applications software interface program (HIP). When the four orbiter GPC's (1, 2, 3, and 4) are in control, the BFS GPC 5 does no commanding. When the BFS GPC 5 is in control, the BFS sends commands to, and requests data from, the EIU's, and in this configuration, the four orbiter GPC's (1, 2, 3, and 4) neither command nor listen. The BFS, when engaged, allows GPC 5 to command FC buses 5, 6, 7, and 8 for main engine control through the SSME HIP. The SSME HIP performs the same main engine command functions as the SSME SOP. The command flow through the EIU's and engine controllers remains the same for BFS engaged as for the four-GPC redundant set.

The engine controller provides all the SSME data to the orbiter GPC's. Sensors in the engine supply pressures, temperatures, flow rates, turbopump speeds, valve position, and engine servovalve actuator positions to the engine controller. The engine controller assembles these data into a vehicle data table and adds status data of its own to the vehicle data table. The vehicle data tables output Channels A and B go to the vehicle interface electronics for transmission to the EIU's. The vehicle interface electroncis output over both data paths. The data paths are called primary and secondary. Channel A vehicle data table is normally sent over both primary and secondary control (Channel A has failed), then the vehicle electronics interface outputs Channel B vehicle data table over both primary and secondary data paths.

The vehicle data table is sent by the controller to the EIU. There are only two data paths versus three command paths between the engine controller and the EIU. The data path that interfaces with controller interface adapter (CIA) one is called primary data. The path that interfaces with CIA two is called secondary data. Primary and secondary data are held in buffers until the orbiter GPC's send a data request command to the EIU's. The GPC's request both primary and secondary data. Primary data will be output only through multiplexer interface adapter one on each EIU. Secondary data will be output only through multiplexer interface adapter four on each EIU.

During prelaunch, the orbiter GPC's will look at both primary and secondary data. Loss of either primary or secondary data will result in data path failure and either engine ignition inhibit or launch pad shutdown of all three SSME's.

At T-O, the orbiter GPC's will request both primary and secondary data from each EIU. For no failures, only primary data are looked at. If there is a loss of primary data (which can occur between the engine controller Channel A electronics and the SSME SOP), then the secondary data are looked at.

 **Space Transportation System**

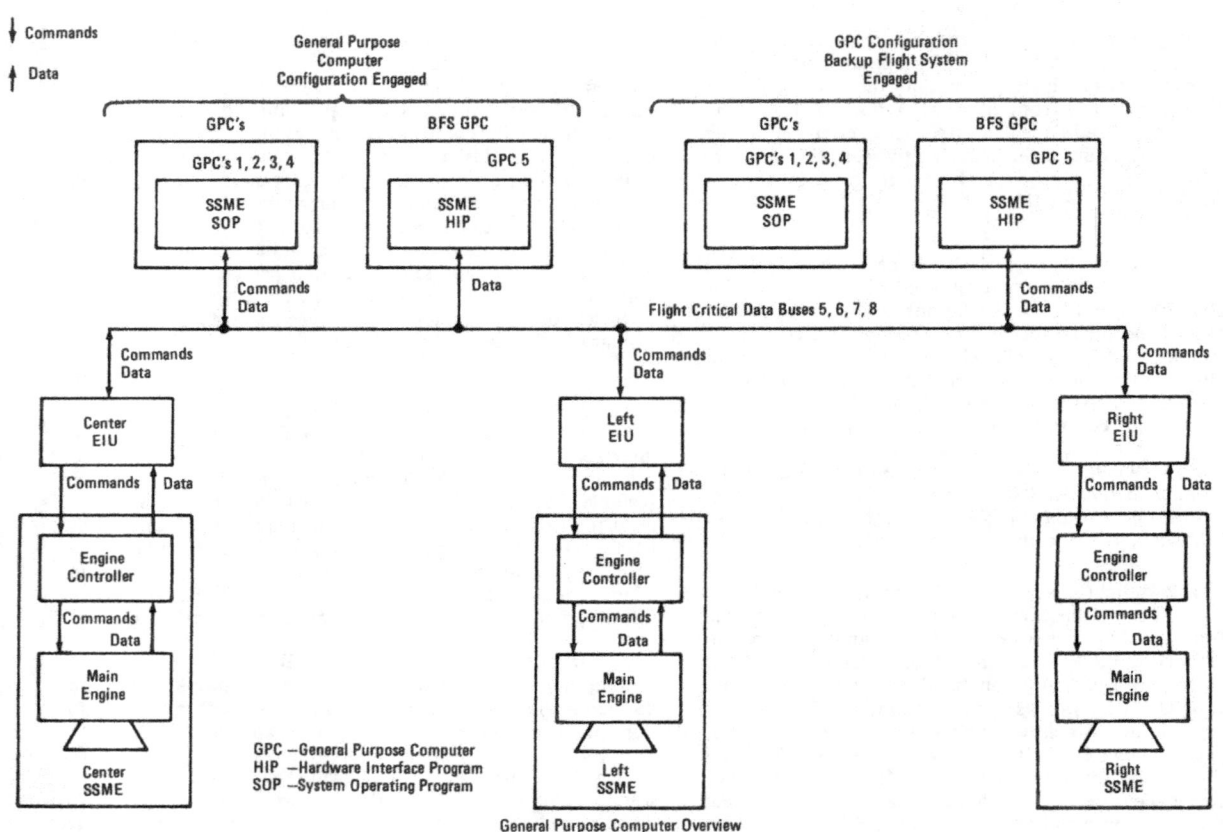

General Purpose Computer Overview

Space Transportation System

There are two primary written engine controller computer software programs: the flight operational program and the test operational program. The flight operational program is an on-line, real-time, process-control program that processes inputs from engine sensors; controls the operation of the engine servovalves, actuators, solenoids, and spark igniters; accepts and processes vehicle commands; provides and transmits data to the vehicle; and provides checkout and monitoring capabilites. The test operational program supports engine testing and, functionally, is similar to the flight operational program, but differs with respect to implementation. The computer software programs are modular and are defined as computer program components, which consist of a data base organized into tables and 15 computer program components. During application of the computer program components, the program performs data processing functions for failure detection and status to the vehicle. As system operation progresses through an operating phase, different combinations of control functions are operative at different times. These combinations within a phase are defined as operating modes.

The checkout phase, operational programs, is initiated to begin active control monitoring or checkout. The standby mode in this phase is a waiting mode of controller operation during which active control sequence operations are in process. Monitoring functions that do not affect engine hardware status are continually active during the mode. Such functions include processing of vehicle commands, status update, and controller self-test. During checkout, data and instructions can be loaded into the engine controller computer memory. This permits updating of software program and data as necessary to proceed with engine firing operations or checkout operations. Also in this phase, a component checkout mode, consisting of checkout or engine leak test, is performed on an individual engine system component.

The start preparation phase consists of system purges and propellant conditioning, which are performed in preparation for engine start. Purge sequence number one mode is the first purge sequence, including oxidizer system and intermediate seal purge operation. Purge sequence number two mode is the second purge sequence, including fuel system purge operation, and continuation of purges initiated in purge sequence number one. Purge sequence number three mode includes propellant recirculation (bleed valve operation). The fourth purge sequence mode includes fuel system purge after propellant drop. The engine ready mode is when proper engine thermal conditions for start have been attained and other criteria for start have been satisfied, including a continuation of purge sequence mode four.

The start phase covers operations for starting or engine firing, beginning with scheduled open-loop operation of propellant valves. The start initiation mode includes all functions prior to ignition confirmed and closing of the thrust control loop. The thrust buildup mode detects ignition by monitoring main combustion chamber pressure and that closed-loop thrust buildup sequencing is in progress.

The mainstage phase is automatically entered upon successful completion of the start phase. The normal control mode has initiated mixture ratio control and thrust control is operating normally. In case of a malfunction, the electrical lock mode will be activated. In that mode engine propellant valves are electrically held in a fixed configuration, and all control loop communications are suspended. There is also the hydraulic lockup mode, in which all fail-safe valves are deactivated to hydraulically hold the propellant valves in a fixed configuration, and all control loop functions are suspended.

The shutdown phase covers operations to reduce main combustion chamber pressure and drive all valves closed to effect full engine shutdown. Throttling to minimum power level (MPL) mode is that portion of the shutdown in progress at a programmed shutdown thrust reference level above MPL. Valve schedule throttling mode is that stage in the shutdown sequence

Space Transportation System

where the programmed thrust reference has decreased below MPL. Propellant valves closed, is that stage in the shutdown sequence following closure of all liquid propellant valves, and the shutdown purge is activated and verification sequences are in progress. The fail-safe pneumatic mode is when the fail-safe pneumatic shutdown is used.

The post-shutdown phase represents the state of the SSME and engine controller at completion of engine firing. The standby mode is a waiting mode of controller operations, with functions identical to those of standby during checkout and is the normal mode of post-shutdown entered after completion of the shutdown phase. The terminate sequence mode terminates a purge sequence by a command from the vehicle. All propellant valves are closed, and all solenoid and torque motor valves are deenergized.

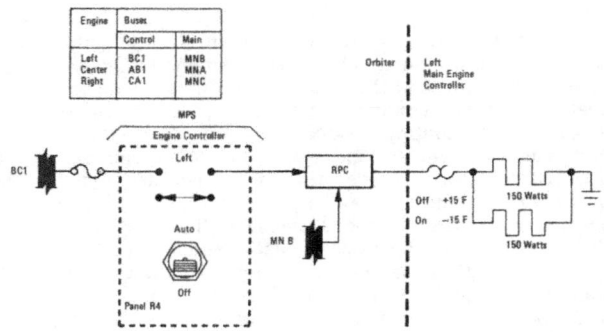

Main Engine Controller Heater Power

Each controller utilizes AC power provided by the MPS ENGINE POWER LEFT, CTR, RIGHT switches on panel R2.

Each controller has electrical heaters that are powered by main bus power through a remote power controller (RPC). The RPC is controlled by the MAIN PROPULSION SYSTEM ENGINE CNTRL HTR LEFT, CTR, RIGHT switches on panel R4. The heaters are not normally used until after MECO and provide survival heat for the remainder of the mission.

Each controller is 596 by 368 by 431 millimeters (23.5 by 14.5 by 17 inches) and weighs 95 kilograms (211 pounds).

MALFUNCTION DETECTION

There are three separate means of detecting malfunctions within the main propulsion system: engine controller, caution and warning system, and GPC.

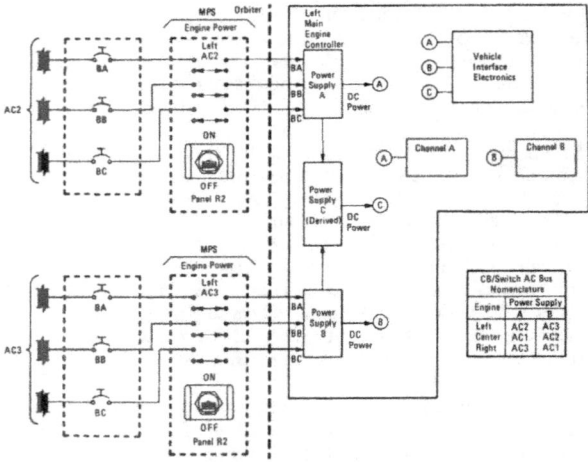

Main Engine Controller Power

Space Transportation System

The engine controller, through its network of sensors, has access to numerous engine operating parameter. A group of these parameters has been designated as critical operating parameters, and, for these parameters, special limits have been defined and are hard-wired and are limit-sensed within the caution and warning system. If a violation of any limit is detected, the caution and warning system will illuminate the red MPS (main propulsion system) caution and warning on panel F7. The MPS red caution and warning light will illuminate due to an MPS engine LO_2 manifold pressure above 12,885 mmHg (249 psia), MPS engine LH_2 manifold pressure below 1,449 or above 3,105 mmHg (below 28 or above 60 psia), MPS helium pressure center, left, or right below 5,951 mmHg (1,150 psia), MPS helium regulated pressure center, left, or right above 42,435 mmHg (820 psia), MPS helium dp/dt, left, center, or right above 1,500 mmHg (29 psia). It is noted that the flight crew can monitor on panel F7 the MPS PRESS HELIUM PNEU, L, C, R TANK or REG in conjunction with the TANK or REG position of the switch on panel F7. In addition, the MPS PRESS ENG MANF LO_2, LH_2 can also be monitored on panel F7. A number of the conditions will require crew action: the MPS engine LH_2 manifold pressure below the minimum setting will require the flight crew to pressurize the external LH_2 tanks using LH_2 ULLAGE PRESS switch on panel R2 to OPEN, and a low helium pressure may require the flight crew to interconnect the pneumatics helium tank and the engine helium tanks using the MPS He INTERCONNECT valve switches on panel R2 for which engine helium system is affected.

The engine controller also has a self-test feature allowing it to detect certain malfunctions involving its own sensors and control devices. For each of the three engines, there is an amber (lower half) MAIN ENGINE STATUS LEFT, CRT, RIGHT light on panel F7. A light will illuminate when the corresponding engine helium pressure is below 59,512 mmHg (1,150 psia) or regulated helium pressure is above 42,435 mmHg (820 psia).

The amber (lower half) of the MAIN ENGINE STATUS LEFT, CTR, RIGHT light on panel F7 may also be illuminated by the SSME SOP (GPC detected malfunctions). The amber light may be illuminated due to an electronic hold, hydraulic lockup, loss of two or more command channels or command reject between the GPC to the SSME controller, or loss of both data channels between the SSME controller to the GPC of the corresponding engine. In an electronic hold for the affected SSME, the fuel flow rate and Pc (chamber pressure) each have four sensors grouped in two pairs, and the loss of data from both pairs of its four sensors (to fail both sensors, it is only necessary to fail one sensor in each pair) will result in the propellant valve actuators being maintained electronically in the positions existing at the time of the second sensor failure. In the case of either the hydraulic lockup or an electronic hold, all engine throttling capability for the affected engine is lost, thus subsequent throttling commands to that engine will not change the thrust level.

The red (upper half) of the MAIN ENGINE STATUS LEFT, CTR, RIGHT light on panel F7 will illuminate if the corresponding engine high pressure oxidizer turbine discharge temperature is above 1,900°R, main combustion chamber pressure is below 51,750 mmHg (1,000 psia), high-pressure oxidizer turbopump intermediate seal purge pressure is below 8,797 or above 33,637 mmHg (below 170 or above 650 psia), high-pressure oxidizer turbopump secondary seal purge pressure is below 258 or above 4,398 mmHg (below 5 or above 85 psia), or flight acceleration safety cutoff system is at 10g. Because of the rapidity with which it is possible to exceed these limits, the engine controller has been programmed to sense the limits and automaticlly cut off the engine if the limits are exceeded. Although a shutdown as a result of the operating limit violation is normally automatic, the flight crew can, if necessary, inhibit an automatic shutdown through the use of the MAIN ENGINE LIMIT SHUT DN switch on panel C3. The LIMIT SHUT DN

 Space Transportation System

switch has three positions: ENABLE, AUTO, and INHIBIT. The ENABLE position allows only the first engine that violates operating limits to be shut down automatically, and if either of the two remaining engines subsequently violates operating limits, it will be inhibited from automatic shutdown. The INHIBIT position inhibits all automatic shutdowns. The MAIN ENGINE SHUTDOWN LEFT, CTR, RIGHT pushbuttons on panel C3 have spring loaded covers (guards). When the guard is raised and the pushbutton is depressed, the corresponding engine shuts down immediately.

The backup caution and warning processing of the orbiter GPC's can detect certain specified out-of-limit or fault conditions of the MPS. The BACKUP C/W ALARM on panel F7 illuminates, a fault message appears on all CRT displays, and an audio alarm sounds if the MPS engine LO_2 manifold pressure is at 0 mmHg or above 1,500 mmHg (0 or above 29 psia), MPS engine LH_2 manifold pressure is below 1,552 or above 2,380 mmHg (below 30 or above 46 psia), MPS helium pressure L, C, or R are below 5,951 mmHg (1,150 psia), or MPS regulated helium pressure L, C, or R are above 42,435 mmHg (820 psia). This is identical to the parameter limit sensed by the caution and warning system; thus the MPS red light on panel F7 will also be illuminated.

The SM ALERT indicator on panel F7 will illuminate, a fault message will appear on all CRT displays, and an audio alarm will be sounded when MPS malfunctions/conditions are detected by the orbiter GPC's SSME SOP or special systems monitoring processing. The first four conditions are detected by the SSME SOP and are identical to those which illuminated the amber (lower half) light of the respective MAIN ENGINE STATUS light on panel F7 due to electronic hold, hydraulic lockup, loss of two or more command channels or command reject between the GPC to the SSME controller, or loss of both data channels between the SSME controller to the orbiter GPC. The last four conditions are special systems monitoring processing and illuminate the SM ALERT light on panel F7, sound an audio alarm, and provide a fault message on all CRT's due to ET LH_2 ullage pressure below 1,552 or above 2,380 mmHg (below 30 or above 46 psia) or ET LO_2 ullage pressure at 0 mmHg or above 1,500 mmHg (0 or above 29 psia). (It is noted that the MAIN ENGINE STATUS LIGHTS on panel F7 will not illuminate.)

ORBITER HYDRAULIC SYSTEMS

The three orbiter hydraulic systems provide hydraulic pressure to the main propulsion system for thrust vector control (TVC) and to actuate engine valves on each SSME.

The three hydraulic supply systems are distributed to the main propulsion system thrust vector control (MPS/TVC) valves. These valves are controlled by HYDRAULICS MPS/TVC 1, 2, 3 switches on panel R4. A valve is opened by its respective switch being positioned to OPEN on panel R4. The talkback indicator located above each switch indicates OP or CL for open and close.

When the three MPS/TVC hydraulic isolation valves are opened, hydraulic pressure actuates the engine main fuel valve, main oxidizer valve, fuel preburner oxidizer valve, oxidizer preburner oxidizer valve, and chamber coolant valve. All hydraulic actuated engine valves on an engine receive hydraulic pressure from the same hydraulic system. The left engine valves are actuated from hydraulic system 2, the center engine valves are actuated from hydraulic system 1, and the right engine valves are actuated from hydraulic system 3. Each engine valve actuator is controlled by dual redundant signals; Channel A/engine servo valve 1 and Channel B/engine servo valve 2 from that engine controller electronics. As a backup, all hydraulic actuated engine valves on an engine are supplied with helium pressure from the helium subsystem engine helium tank supply system, left, center, and right. In the event of hydraulic lockup, helium pressure is used to actuate that engine's propellant valves to their fully closed position upon engine shutdown.

Space Transportation System

Hydraulic lockup is a condition in which all of the propellant valves on an engine are hydraulically locked in a fixed position. This is a build-in protective response of the MPS propellant valve actuator/control circuit. It goes into effect any time that low hydraulic pressure or loss of control of one or more propellant valve actuators renders closed-loop control of engine thrust or mixture ratio impossible. Hydraulic lockup allows an engine to continue to thrust in a safe manner under conditions that normally would require that the engine be shutdown; however, the affected engine will continue to operate at the throttle level in effect at the time hydraulic lockup occurred. Once an engine is in a hydraulic lockup conditions, any subsequent shutoff commands, whether nominal or premature, will cause a pneumatic helium shutdown. Hydraulic lockup does not affect the capability of the engine controller to monitor critical operating parameters or issue an automatic shutdown if an operating limit is out of tolerance; however, the engine shutdown would be accomplished pneumatically.

The three MPS/TVC valves must also be opened to supply hydraulic pressure to the six SSME TVC actuators. There are two servoactuators per SSME: one for yaw and one for pitch. Each actuator is fastened to the orbiter thrust structure and to the powerhead of one of the three SSME's. The two actuators per each SSME provides attitude control and trajectory shaping by gimbaling of the SSME's in conjunction with the SRB's in first-stage ascent and without the SRB's in second-stage ascent. Each SSME servoactuator receives hydraulic pressure from two of the three orbiter hydraulic systems; one system as a primary system and the other as a standby system. Each servoactuator has its own hydraulic switching valve. The switching valve receives hydraulic pressure from two of the three orbiter hydraulic systems and provides a single source pressure to the actuator. Normally, the primary hydraulic supply is directed to the actuator; however, if the primary system were to fail and lose hydraulic pressure, the switching valve would automatically switch over to the standby system, and the actuator would continue to function on the standby system. The left engine pitch actuator utilizes hydraulic system 2 as the primary and hydraulic system 1 as the standby. The yaw actuator utilizes hydraulic system 1 as the primary and hydraulic system 2 as the standby. The center engine pitch actuator utilizes hydraulic system 1 as the primary and hydraulic system 3. And the yaw actuator utilizes hydraulic system 3 as the primary and hydraulic system 1 as the standby. The right engine pitch actuator utilizes hydraulic system 3 as the primary and hydraulic system 2 as the standby, and the yaw actuator utilizes hydraulic system 2 as the primary and hydraulic system 3 as the standby.

The hydraulic systems are distributed among the actuators and engine valves to equalize the hydraulic work load among the three systems.

During on-orbit operations, the HYDRAULIC MPS/TVC ISOL VLV SYS1, SYS2, SYS3 switches on panel R4 are positioned to CLOSE to protect against hydraulic leaks downstream of these valves. In addition, there is no requirement to gimbal the SSME's from the stow position. During on-orbit operations when the MPS/TVC valves are closed, the hydraulic pressure and return lines within each MPS/TVC are interconnected to enable hydraulic fluid circulation for thermal conditioning.

THRUST VECTOR CONTROL

The Space Shuttle ascent thrust vector control (ATVC) portion of the flight control system directs the thrust of the three main engines and two SRB nozzles to control attitude and trajectory during liftoff and first stage ascent and without the SRB's in second stage ascent.

The ATVC is an avionics hardware package that provides gimbal commands and fault detection for each hydraulic gimbal actuator. The MPS ATVC's are located in the three aft avionics bays in the orbiter aft fuselage and are cooled by coldplates and the Freon-21 system. The associated flight aft (FA) multiplexer/demultiplexers (MDM's) are also located in the aft avionics bays.

Space Transportation System

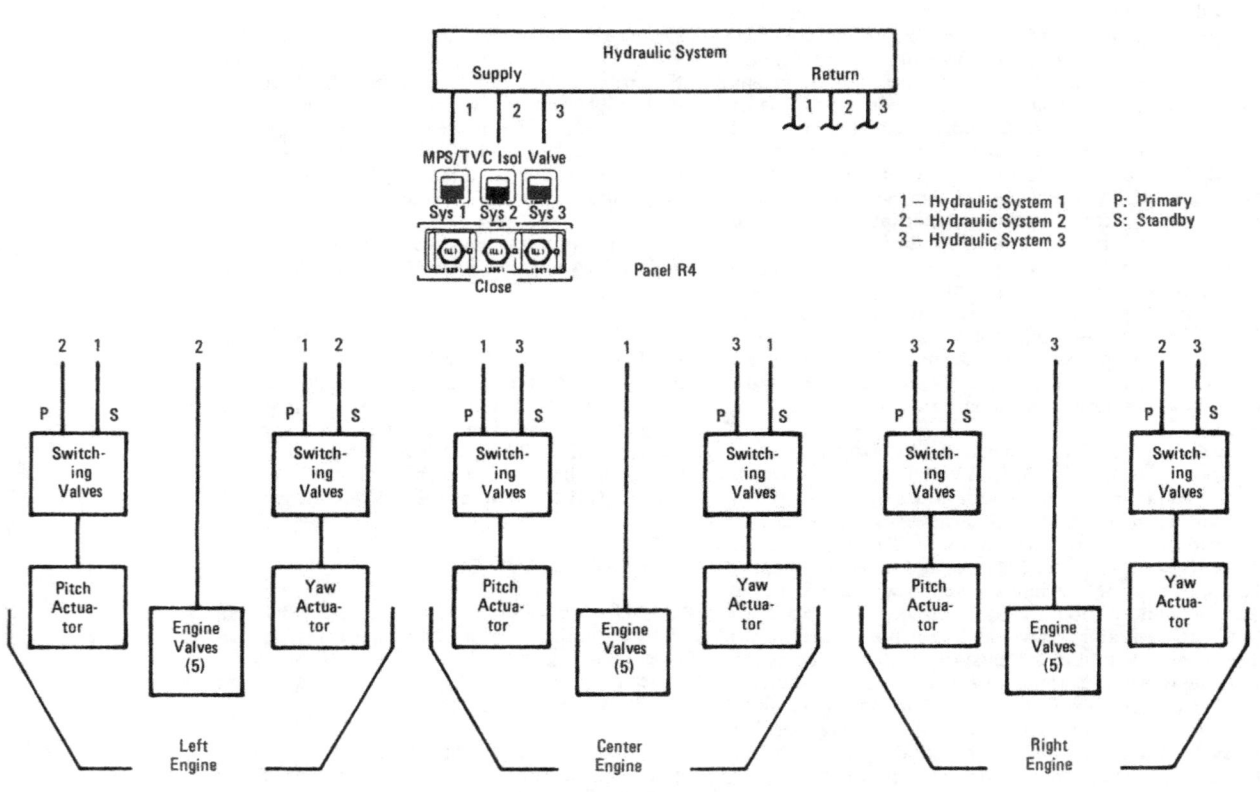

MPS Hydraulic System Distribution

Space Transportation System

The MPS TVC command flow starts in the GPC's, where the flight control system generates the TVC position commands, and is terminated at the SSME servoactuators, where the actuators gimbal the SSME's in response to the commands. All the MPS TVC position commands generated by the flight control system are issued to MPS TVC CMD (command) SOP (system operating command), where they are processed and disbursed to their corresponding flight aft MDM's. The flight aft MDM's separate these linear discrete commands and disburse them to the ATVC channel, which generates equivalent command analog voltages for each command issued. These voltages are, in turn, sent to the servoactuators commanding the SSME hydraulic actuator to extend or retract, thus gimbaling the main engine to which it is fastened.

There are six MPS TVC actuators that respond to command voltages issued from four ATVC channels. Each ATVC channel has six MPS drivers and four SRB drivers. Each actuator receives four identical command voltages from four different MPS drivers, each located in different ATVC channels.

Each main engine servoactuator consists of four independent, two-stage servovalves, which receive signals from the drivers. Each servovalve controls one power spool in each actuator, which positions an actuator ram and the engine to control thrust direction.

The four servovalves in each actuator provide a force-summed majority voting arrangement to position the power spool. With four identical commands to the four servovalves, the actuator force sum action prevents a single erroneous command from affecting power ram motion. If the erroneous command persists for more than a predetermined time, differential pressure-sensing activates an isolation driver, which energizes an isolation valve, which isolates and removes the defective servovalve hydraulic pressure, permitting the remaining channels and servovalves to control the actuator ram spool providing that the FCS CHANNEL 1, 2, 3, 4 switches on panel C3 are in the AUTO position. A second failure would isolate and remove the defective servovalve hydraulic pressure in the same manner as the first failure, leaving only two remaining channels.

Failure monitors are provided for each channel on the CRT and backup caution and warning light to indicate which channel has been bypassed for the MPS and/or SRB. If the FCS CHANNEL 1, 2, 3, or 4 switch on panel C3 is positioned to the OFF position, that ATVC channel is isolated from that servovalve on all MPS and SRB actuators. The OVERRIDE position of FCS CHANNEL 1, 2, 3, 4 switch on panel C3 inhibits the isolation valve driver from energizing the isolation valve for its respective channel and provides the capability of resetting a failed or bypassed channel.

The ATVC 1, 2, 3, 4 power switches are located on panel 017. The ON position enables the respective ATVC channel, and the OFF position disables that respective channel.

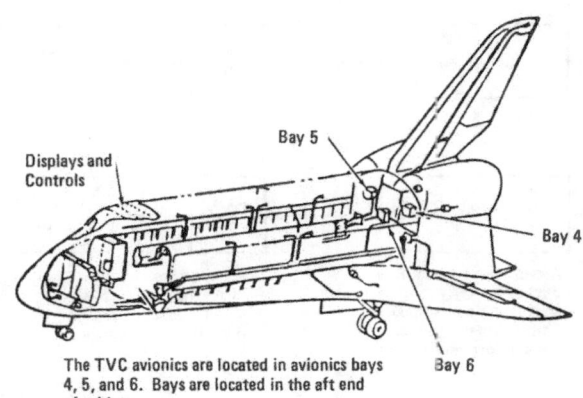

The TVC avionics are located in avionics bays 4, 5, and 6. Bays are located in the aft end of orbiter.

Location of Orbiter Avionics Bays

Space Transportation System

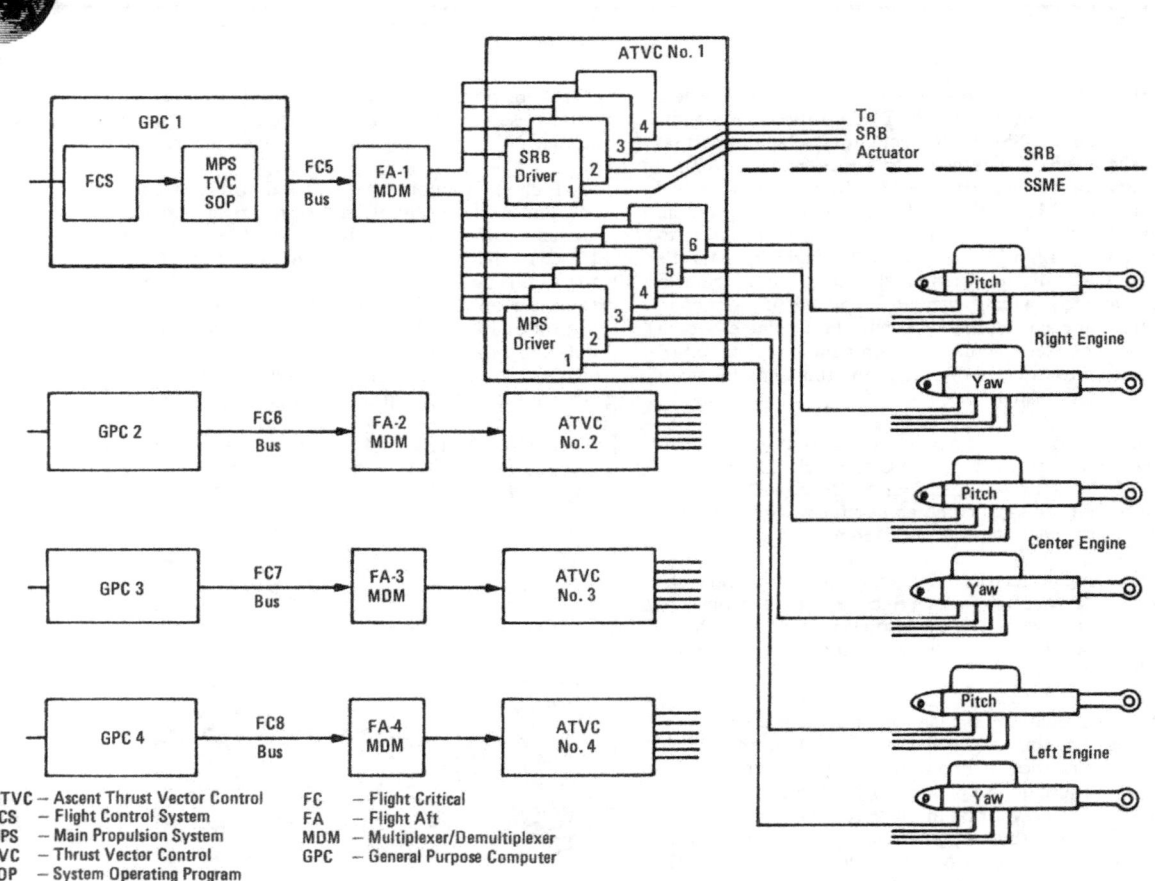

Command Flow Overview

Space Transportation System

Each actuator ram is equipped with transducers for position feedback to the TVC system.

The SSME servoactuators change each SSME's thrust vector direction as needed during the flight sequence. The three pitch actuators gimbal the engine up or down a maximum of 10 degrees 30 minutes from the installed null position. The three yaw actuators gimbal the engine left or right a maximum of 8 degrees 30 minutes from the installed position. The installed null position for the left and right main engines is 10 degrees up from the X-axis in a negative Z direction and 3 degrees 30 minutes outboard from an engine center line parallel to the X-axis. The center engine's installed null position is 16 degrees above the X-axis for pitch and on the X-axis for yaw. When any engine is installed in the null position, the other engines cannot have an engine-to-engine collision with that engine.

The gimbal rate is 10 degrees per second minimum, 20 degrees per second maximum.

There are three actuator sizes for the main engines. For the one upper pitch actuator, piston area is 160 centimeters squared (24.8 square inches), stroke is 274 millimeters (10.8 inches), peak flow is 189 liters per minute (50 gallons per minute), and weight is 120 kilograms (265 pounds). Piston area of the two lower pitch actuators is 129 centimeters squared (20 square inches), stroke is 274 millimeters (10.8 inches), peak flow is 170 liters per minute (45 gallons per minute), and weight is 111 kilograms (245 pounds). For all three yaw actuators, piston area is 129 centimeters squared (20 square inches), stroke is 223 millimeters (8.8 inches), peak flow is 170 liters per minute (45 gallons per minute), and weight is 108 kilograms (240 pounds).

HELIUM, OXIDIZER, AND FUEL FLOW SEQUENCE

At T minus 05:15:00, the fast-full portion of the LO_2 and LH_2 loading sequence begins under control of the launch processing system (LPS).

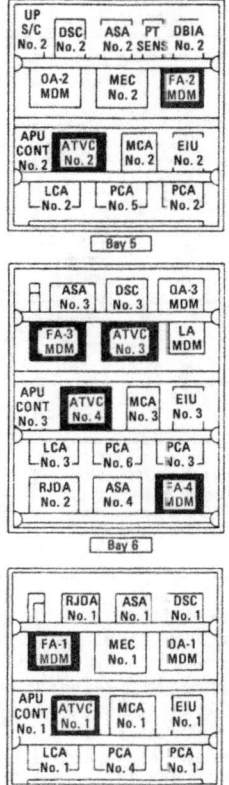

Avionics Bays 4, 5, and 6

Space Transportation System

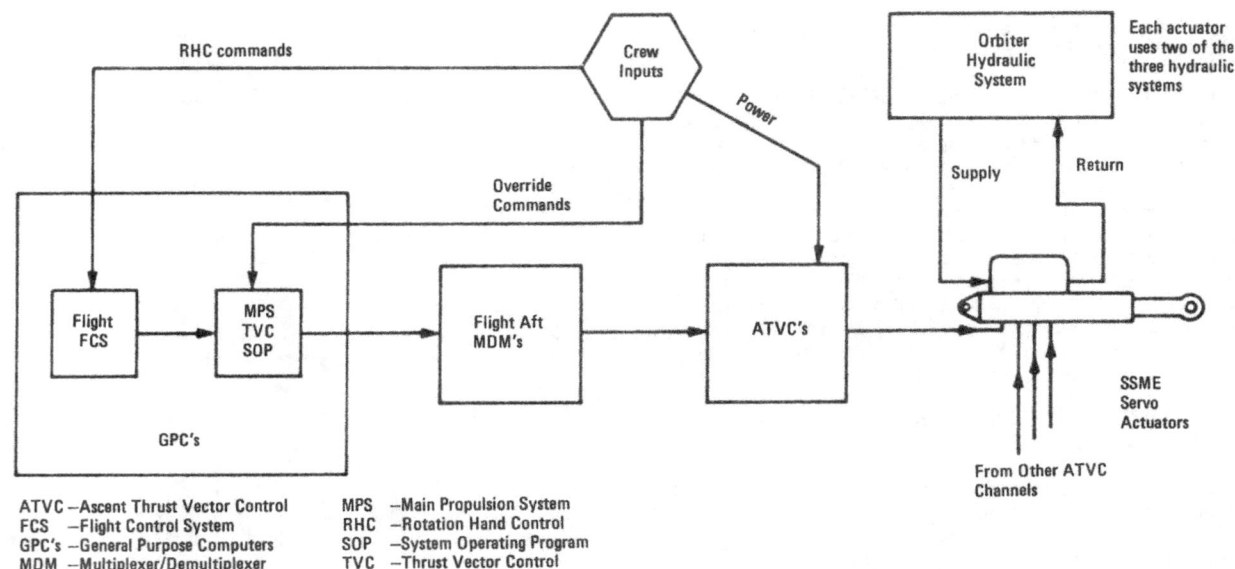

SSME Thrust Vector Control Interface Flow

At T minus 03:45:00, the LH$_2$ tank fast fill is completed to 98 percent and a slow topping off process is begun and stabilizes to 100 percent. At T minus 03:30:00, the LO$_2$ fast fill is complete. At T minus 03:15:00 LH$_2$ replenishment begins and at T minus 03:10:00 LO$_2$ replenishment begins and are completed at T minus 02:30:00. At T minus 02:05:00, the LH$_2$ chilldown sequence is initiated by the LPS. It opens the LH$_2$ recirculation valves and starts the LH$_2$ recirculation pumps. At T minus 02:03:54, as part of the LH$_2$ chilldown sequence, the LH$_2$ prevalves are closed and remain closed until T minus 00:00:09.3.

During prelaunch, the pneumatic helium supply provides pressure to operate the LO$_2$ and LH$_2$ prevalves, the LO$_2$ and LH$_2$ inboard fill/drain valves, and the LO$_2$ and LH$_2$ outboard fill/drain valves. The three engine helium supply systems are not used at this time.

When the flight crew enters the orbiter, all ten helium supply tanks are partially pressurized to approximately 103,500 mmHg (2,000 psi). The filling of the helium tanks to their full pressure begins at T minus 02:10:00. Regulated helium pressure

Space Transportation System

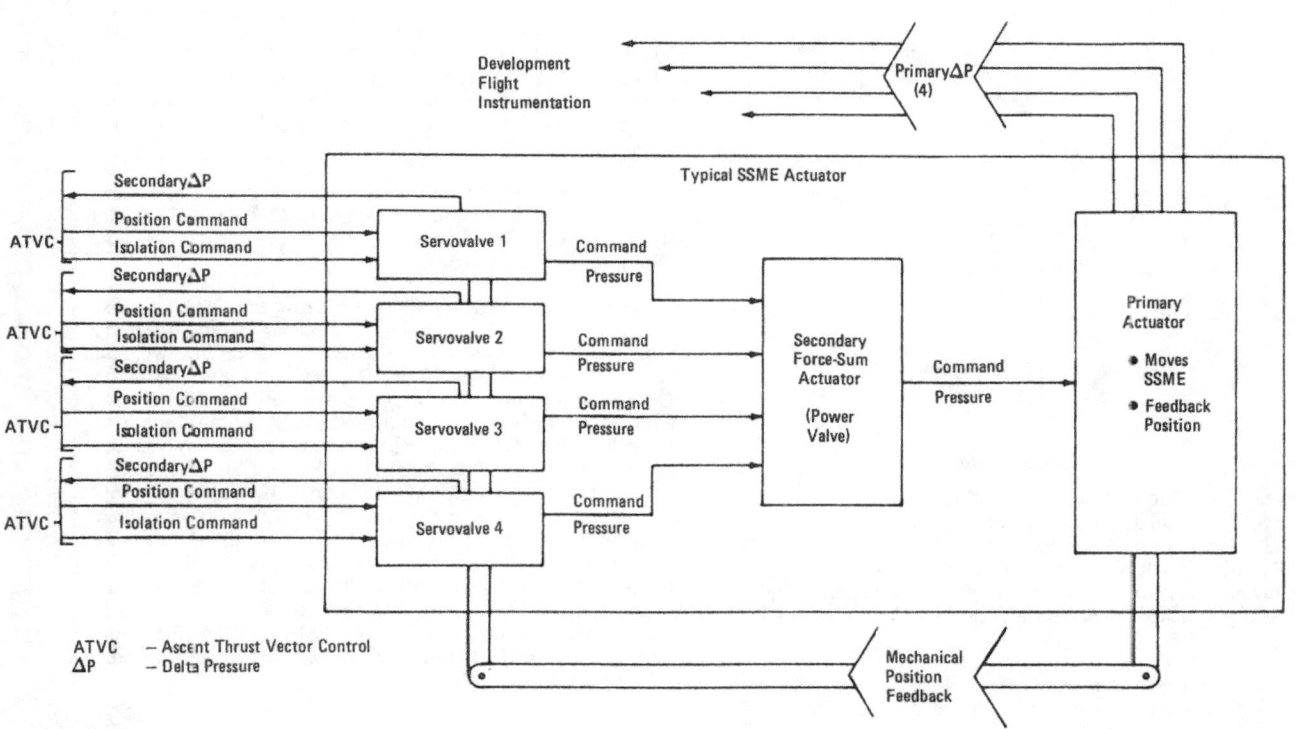

Typical SSME Actuator

Space Transportation System

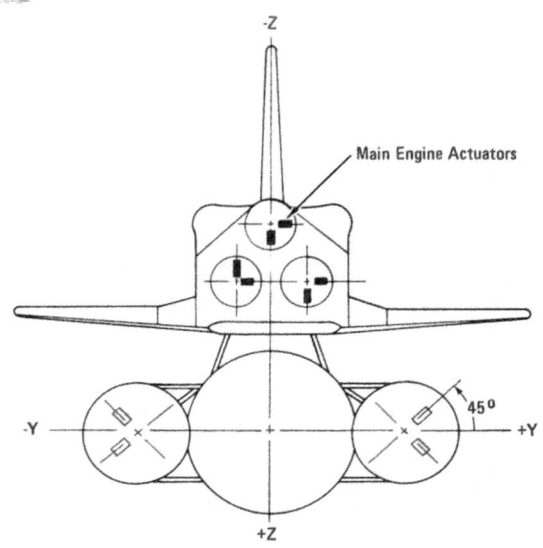

Actuator Orientation — Main Engines

is between 37,000 to 40,603 mmHg (175 to 785 psi). Helium supply tank and regulated pressures are monitored on the MPS PRESS, PNEU, L, C, R meters on panel F7. The MPS PRESS TANK, REG switch position on panel F7 selects either supply or regulated pressures to be displayed on the meters. Engine helium and regulated pressures are also available on the CRT display.

When the flight crew enters the orbiter, the eight MPS He ISOLATION A and B switches on panel R2 are in the GPC position, the MPS PNEUMATICS L ENG TO XOVR and He ISOL switches on panel R2 are in the GPC position, and the MPS He INTERCONNECT LEFT, CTR, RIGHT switches on panel R2 are in the GPC position. With the switches in these positions, the eight helium isolation valves are in the open position, and the left engine crossover and the six helium interconnect valves are in the closed position.

At T minus 00:16:00, one of the first actions by the flight crew is to place the six MPS He ISOLATION A and B switches and MPS PNEUMATICS He ISOL switch on panel R2 to the OPEN position. This will not change the position of the helium isolation valves; however, it inhibits the LPS control of valve position.

At T minus 01:00:00, the helium supply tanks are pressurized to their final liftoff value of 232,875 mmHg (4,500 psi). This process is a gradual one to prevent excessive heat buildup in the supply tanks.

During prelaunch, liquid oxygen from the GSE is loaded through the GSE/LO_2 umbilical and passes through the LO_2 outboard fill/drain valve, the LO_2 inboard fill/drain valve, and the LO_2 feedline manifold. The LO_2 then exits the orbiter at the LO_2 feedline disconnect valve and passes through the ET LO_2 umbilical into the LO_2 tank. During loading, the LO_2 tank vent and relief valves are open to prevent pressure buildup in the tank due to LO_2 loading, the MAIN PROPULSION SYSTEM PROPELLANT FILL/DRAIN LO_2 OUTBD and INBD switches on panel R4 are in the GND (ground) position, which allows the LPS to control the positions of these valves as required. When LO_2 loading is completed, the LPS will first command the LO_2 inboard fill/drain valve closed. The LO_2 in the line between the inboard and outboard fill/drain valves is then allowed to drain back into the GSE, and the LPS commands the outboard fill/drain valve closed. Maximum fill time is 75 minutes for each fluid.

Also during prelaunch, liquid hydrogen supplied from the GSE LH_2 umbilical passes through the LH_2 outboard fill/drain valve, the LH_2 inboard fill/drain valve, and the LH_2 feedline

Space Transportation System

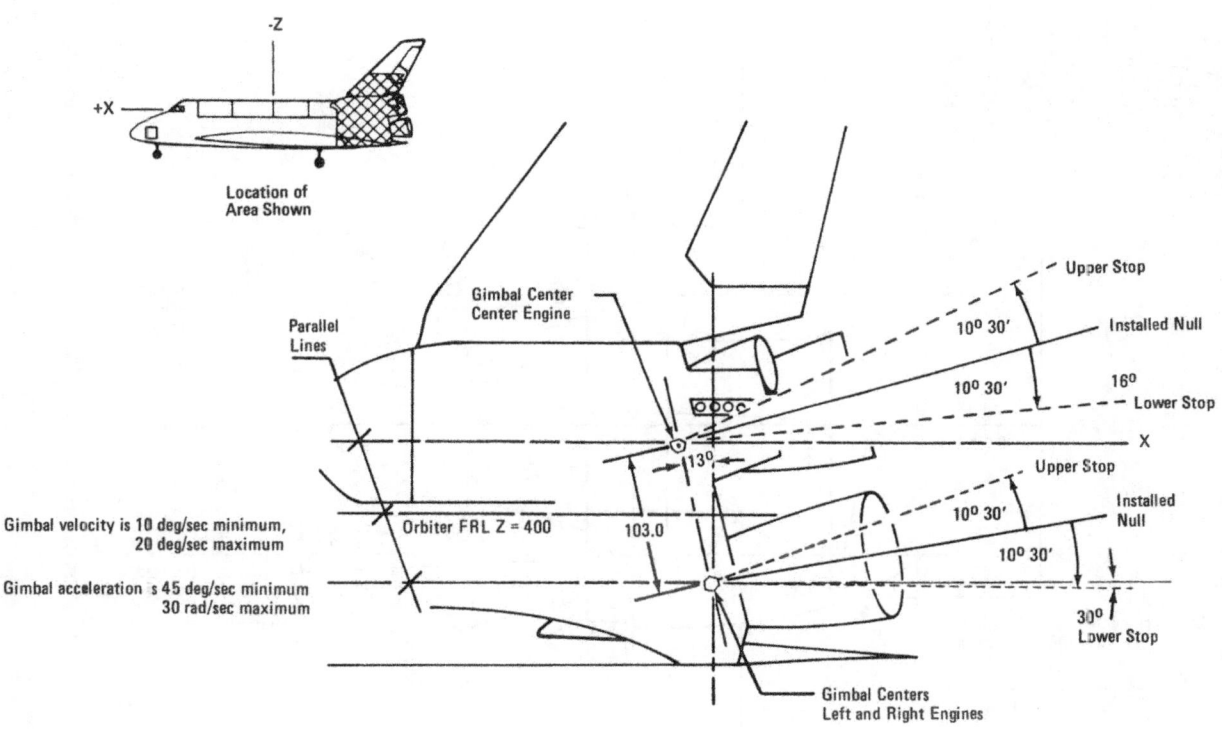

SSME Gimbal Pitch Axis

Space Transportation System

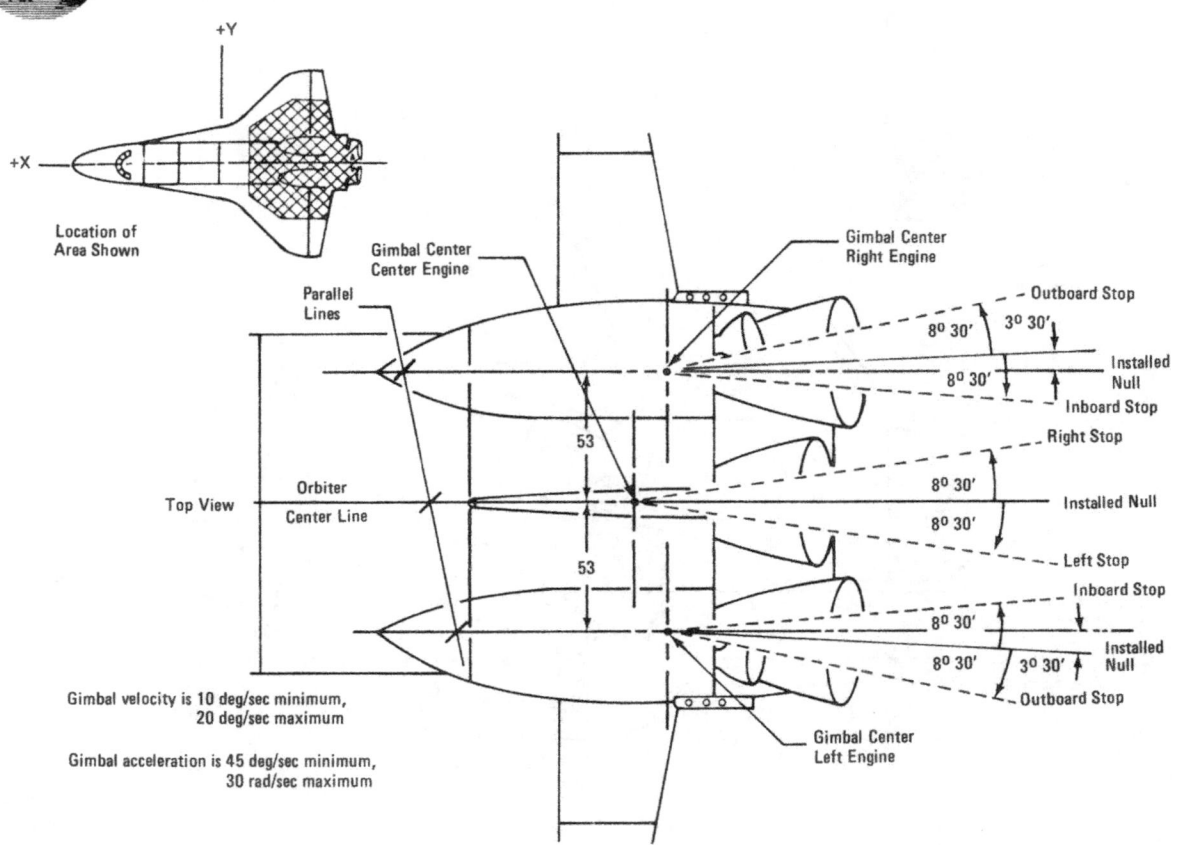

SSME Gimbal Yaw Axis

Space Transportation System

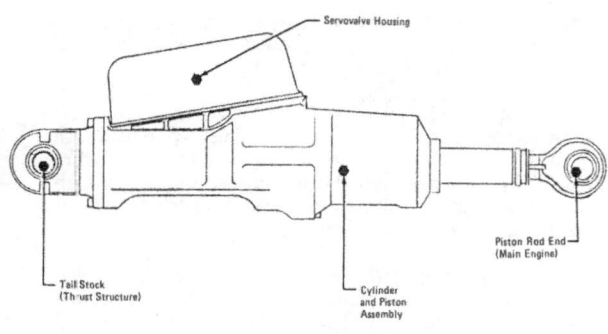

Main Engine Electrohydraulic Servoactuator

manifold. The LH2 then exits the orbiter at the LH2 feedline disconnect valve and passes throught the ET LH2 umbilical into the LH2 tanks. During loading, the LH2 tank vent valve is left open to prevent pressure buildup in tank due to LH2 boiloff. During LH2 loading, the MAIN PROPULSION SYSTEM PROPELLANT FILL/DRAIN LH2 INBD and OUTBD switches on panel R4 are in the GND position, which allows the LPS to control the position of these valves as required.

At T minus 00:04:00 minutes seconds, the fuel system purge begins, followed at T minus 00:03:25 with the beginning of the engine gimbal tests. During the MPS engine gimbal test, each gimbal actuator is operated through a canned profile of extensions and retractions. If all actuators function satisfactorily, the engines are gimbaled to predefined positions at T minus 00:02:15. The engines will remain in these positions until engine ignition. In the predefined start positions, the engines are gimbaled in an outward direction (away from one another) so that the engine start transient will not cause the engine bells to contact one another during the start sequence.

At T minus 00:02:55 minutes-seconds, the LPS closes the LO2 tank vent valve and the tank is pressurized to 1,086 mmHg (21 psi) with GSE-supplied helium. The LO2 tank pressure can be monitored on the MPS PRESS ENG MANF LO2 meter on panel F7 as well as on the CRT. An LO2 tank pressure of 1,086 mmHg (21 psi) corresonds to an LO2 engine manifold pressure of 5,433 mmHg (105 psia).

At T minus 00:01:57, the LPS closes the LH2 tank vent valve, and the tank is pressurized to 2,277 mmHg (44 pisa) with GSE-supplied helium. LH2 tank pressure is monitored on the MPS PRESS ENG MANF LH2 meter on panel F7 as well as on the CRT display. An LH2 tank pressure of 2,277 mmHg (44 psia) corresponds to an LH2 engine manifold pressure of 2,326 mmHg (44.96 psia).

At T minus 00:00:31, the onboard redundant set launch sequence is enabled by the LPS. From this point on, all sequencing is performed by the orbiter GPC's in the redundant set, based on the onboard clock time. The GPC's however, still respond to hold, resume count, and recycle commands from the LPS.

At T minus 00:00:16, the GPC's will begin to issue arming commands for the SRB ignition pyro initiator controller's (PIC's), the hold-down release PIC's, and T-0 umbilical release PIC's.

At T minus 00:00:09.3, the engine chilldown sequence is completed and the GPC's command the LH2 prevalves open—the LO2 prevalves were open during loading to permit engine chilldown. The MAIN PROPULSION SYSTEM LO2 and LH2 PREVALVE LEFT, CTR, RIGHT switches on panel R4 are in the GPC position.

At T minus 00:00:16, helium flows out of the nine helium supply tanks through the helium isolation valves, the regulators, and check valves and enters the engine at the inlet of the

Space Transportation System

pneumatic control assembly (PCA). The PCA is a manifold containing solenoid valves to control and direct helium pressure under the control of the engine controller to perform various essential functions. The valves are energized by discrete on/off commands from the output electronics of the engine controller. One essential function from T minus 00:00:04 to MECO plus six seconds is the purging of the high-pressure oxidizer turbopump (HPOT) intermediate seal cavity. This cavity is between two seals, one of which contains the hot fuel-rich gas in the oxidizer turbine and the other the liquid oxygen in the oxidizer turbopump. Leakage through one of both of the seals and mixing of the propellants could result in a catastrophic explosion. Continuous overload purge of this area prevents the propellants from mixing as they are dumped overboard through drain lines. The PCA also functions as an emergency backup for closing the engine propellant valves with helium pressure. In a normal engine shutdown, the engine propellant valves are hydraulically actuated.

At T minus 00:00:06.6, the GPC's issue the engine start command, opening the main oxidizer valve in each engine. Between main engine oxidizer valve opening and MECO, LO_2 flows out of the ET, through the ET LO_2 umbilical and the LO_2 feedline disconnect valves, and into the LO_2 umbilical and the LO_2 feedline disconnect valves and into the LO_2 feedline manifold. From this manifold, LO_2 is distributed to the engines through the three engine LO_2 feedlines. In each line, LO_2 passes through the prevalve and enters the main engine at the inlet to the low-pressure oxidizer turbopump (LPOT). In the engine, a small portion of the LO_2 is diverted into the oxidizer heat exchanger. In the heat exchanger, heat generated by the high-pressure oxidizer turbopump (HPOT) is used to convert liquid oxygen into gaseous oxygen (GO_2). This GO_2 is then directed back to the ET to maintain oxidizer tank pressure. The GO_2 flowback to the ET begins at the outlet of the heat exchanger. From this point, GO_2 passes through a check valve, then splits into two paths, each containing a flow control orifice. One of these paths also contains a valve which is normally controlled by one of three pressure transducers located in the LO_2 tank. Downstream of the two flow control orifices and the pressure control valves, the GO_2 lines empty into the ET GO_2 pressurization manifold. This single line exits the orbiter at the GO_2 pressurization disconnect and passes through the ET GO_2 umbilical into the top of the LO_2 tank.

At T minus 00:00:06.6, if the PIC voltages are within limits and all three engine controllers are indicating engine ready, the GPC's will issue the engine start commands to the three SSME's. If the PIC conditions are not met in four seconds, the engine start commands will not be issued, and the GPC's will proceed a countdown hold condition.

At T minus 00:00:06.6, the GPC's issue the engine start command, opening the main fuel valve in the engine. Between main fuel valve opening and MECO, LH_2 flows out of the ET through the ET LH_2 umbilical and the LH_2 flows out of the ET through the ET LH_2 umbilical and the LH_2 feedline disconnect valve into the LH_2 feedline manifold. From this manifold, LH_2 is distributed to the engines through the three engine LH_2 feedlines. In each line, LH_2 passes through the prevalve and enters the main engine at the inlet to the low-pressure fuel turbopump (LPFT). In the engine, the LH_2 is used to cool various engine components and in the process is converted to gaseous hydrogen (GH_2). The majority of this GH_2 is burned in the engine; a smaller portion is directed back to the ET to maintain LH_2 tank pressure. GH_2 flowback to the ET begins at the turbine outlet of the LPFT. GH_2 tapped from this line first passes through a check valve and then splits into two paths, each containing a flow control orifice. One of these paths also contains a valve normally controlled by one of three pressure transducers located in the LH_2 tank.

If all three SSME's reach 90 percent of their rated thrust at T minus 00:00:03 seconds after the ignition command was issued, and approximately 3 seconds have elapsed from the time when all engines are at 90 percent to permit vehicle base bending

Space Transportation System

load models to return to a minimum, the GPC's will issue the commands to fire the SRB ignition PIC's, the hold-down release PIC's, and the T-0 umbilical release PIC's. At T-0 liftoff occurs almost immediately because of the extreme rapid thrust buildup by the SRB's.

If one or more of the three SSME's do not reach 90 percent of their rated thrust in less than 4.6 seconds from the SSME ignition command, all SSME's will be shutdown, the SRB's will not be ignited, and a countdown hold condition will exist. Between liftoff and MECO, as long as the SSME's perform nominally, all MPS sequencing and control functions will be executed automatically by the GPC's. During this time period, the flight crew will monitor MPS performance, back up automatic functions, if required, and provide manual inputs in the event of MPS malfunctions.

Beginning at T-0, the SSME gimbal actuators, which were locked in their special pre-ignition positions, are first commanded to their null positions for SRB start and then allowed to operate as needed for thrust vector control.

During ascent, LH_2 tanks pressure is maintained between 1,707 and 1,811 mmHg (33 and 35 psia) by the orifices in the two lines and the action of the flow control valve. When pressure in the LH_2 tank reaches 1,811 mmHg (35 psia), the valve closes, and when pressure drops below 1,707 mmHg (33 psia), the valve opens. Tank pressures greater than 1,966 mmHg (38 psia) will cause the tank to relieve through the tank vent valve. If tank pressure falls below a 1,707 mmHg (33 psia) (LH_2 engine manifold pressure less than 1,759 mmHg (34 psia) or rises above 1,811 mmHg (35 psia), the flight crew will position the MPS LH_2 ULLAGE PRESS switch on panel R2 to the OPEN position. This allows the three flow control valves to go to the full-open position. Normally the MPS LH_2 ULLAGE PRESS switch is in the AUTO position. Downstream of the two flow control orifices and the flow control valves, the GH_2 lines empty into the ET GH_2 pressurization manifold. This single line then exits the orbiter at the GH_2 pressurization disconnect and passes through the ET GH_2 umbilical into the top of the LH_2 tank. During ascent LO_2 tank pressure is maintained between 1,035 and 1,138 mmHg (20 and 22 psia) by the orifices in the two lines and the action of the flow control valve. When pressure in the tanks reaches 1,138 mmHg (22 psia), the valves closes. When pressure drops below 1,035 mmHg (20 psia), the valve opens. LO_2 tank pressure greater than 1,293 mmHg (25 psia) will cause the tank to relieve through the tank vent and relief valves.

The SSME thrust level setting is dependent upon the flight, may be 100 percent or 104 percent or some high-orbit missions and missions involving heavy payloads may require the maximum thrust setting of 109 percent. The initial thrust level is normally maintained until approximately 31 seconds into the mission, when the GPC's throttle the engines to a lower thrust percent setting which is dependent upon that flight percent. The purpose of the lower throttle setting is to minimize structrual loading when the orbiter is passing through the region of maximum aerodynamic pressure (max Q). This normally occurs around 63 seconds mission elapsed time. At approximately 65 seconds, the engines are once again throttled to the applicable higher percent and remain at that setting.

The SRB's burn out at approximately 00:02 minutes mission elapsed time and are separated from the orbiter by GPC command, via the mission events controller and the SRB separation PIC's. The flight crew can initiate SRB separation manually in the event of an automatic sequence failure.

Beginning at approximately 00:07:40 mission elapsed time, the engines are throttled back to maintain vehicle acceleration at, or less than, three gs. The 3-g limit is an operational limit devised to prevent physical stresses on the flight crew. Approximately eight seconds before MECO, the engines are throttled back to 65 percent.

Space Transportation System

Although MECO is based on the attainment of a specified velocity, the engine can also be shutdown due to LO_2 depletion before the specified velocity of MECO is reached. LO_2 depletion is sensed by four oxidizer depletion sensors located in the LO_2 feedline manifold. Any two of the four sensors indicating a dry condition will cause the GPC's to issue a MECO command to the engine controller.

The engine can also be shutdown by LH_2 depletion should it occur before the specified velocity is reached. LH_2 depletion is sensed by four fuel depletion sensors located in the bottom of the LH_2 tank. Any two of the four sensors indicating a dry condition will cause the GPC's to issue a MECO command to the engine controller.

Once MECO has been confirmed, the GPC's will execute the ET separation sequence. The ET separation sequence requires approximately 11 seconds and includes arming of the ET separation PIC's, closure of the LO_2 and LH_2 prevalves, firing of the ET tumble system pyrotechnic, LH_2 and LO_2 feedline disconnect valve closure, gimbaling of the SSME's to the MPS propellant dump position (full down), ET signal conditioners power off (deadfacing), firing of umbilical unlatch pyrotechnics, and retraction of the umbilical plate hydraulically.

At this point, the GPC's check for ET separation inhibits. If the vehicle's pitch, roll, and yaw rates are not less than 0.2 degrees/second, automatic ET separation is inhibited. If these conditions are met, the GPC's will issue the commands to the ET separation pyrotechnics. In manual ET separation or return to launch site (RTLS) aborts, the inhibits are overriden.

At separation, the orbiter begins a reaction control system minus-Z translation separation maneuver to move it away from the ET. This maneuver takes approximately 13 seconds and results in a negative-Z delta component of approximately three meters per second (11 feet per second).

Once MECO has occurred due to the specified velocity attainment or due to LO_2 or LH_2 sensor depletion and before ET separation, the GPC's will isolate the orbiter LH_2 feedline from the ET by closing the LH_2 feedline disconnect valve (actually two valves, one on each side of the separation interface) and the LO_2 feedline disconnect valve (also two valves, one on each side of the separation interface). At orbiter/ET separation, the GO_2 feedline is sealed at the separation interface by action of the self-sealing quick-disconnect as is the GH_2 feedline.

The MPS pneumatic control assembly on each SSME provides an emergency backup mode of closing the engine propellant valves pneumatically by helium pressure. The normal engine shutdown of the engine propellant valves is by hydraulic actuation.

At MECO the GPC's will open the LO_2 feedline relief isolation valve, allowing any pressure buildup generated by oxidizer trapped in the orbiter LO_2 feedline manifold to be vented overboard through the relief valve, providing that the MAIN PROPULSION SYSTEM FEEDLINE RLF ISOL LO_2 switch on panel R4 is in the GPC position. The GPC's will also open the LH_2 feedline relief isolation valve, allowing any pressure buildup of fuel trapped in the orbiter LH_2 feedline manifold to be vented overboard through the relief valve, providing that the MAIN PROPULSION SYSTEM FEEDLINE RLF ISOL LH_2 switch on panel R4 is in the GPC position.

At MECO, the pneumatic control assembly for each engine performs a 6-second purge of the engine preburner oxidizer domes and a 2-second post-charge of the pogo accumulator. This purge ensures that no residual propellant remains in these areas to cause an unsafe condition and to prevent a water hammer effect in the LO_2 manifolds of the SSME. This helium usage and the high-pressure oxidizer turbopump (HPOT) intermediate seal cavity purge can be observed on the MPS HELIUM L, C, R meters on panel F7 and are also available on the CRT.

Space Transportation System

At MECO, the RTLS LH_2 dump valves will be opened for 30 seconds to ensure that the LH_2 manifold pressure does not result in operation of the LH_2 feedline relief valve.

After the 6-second purge is completed, the GPC's interconnect the pneumatic helium and engine helium supply system by opening the three "out" interconnect valves providing that the MPS He INTERCONNECT LEFT, CENTER, RIGHT switches on panel R2 are in the GPC position. This connects all ten helium supply tanks to the common manifold and ensures there is sufficient helium available to perform the LO_2 and LH_2 propellant dumps, which is required after ET separation.

After ET separation, there are still approximately 771 kilograms (1,700 pounds) of propellants trapped in the SSME's and an additional 1,678 kilograms (3,700 pounds) of propellants trapped in the orbiter propellant management system. This 2,449 kilograms (5,400 pounds) of propellant represents an overall center of gravity (c.g.) shift for the orbiter of approximately 17 centimeters (7 inches). Non-nominal center-of-gravity locations can create major guidance problems during reentry. The residual LO_2, by far the heavier of the two propellants, poses the greatest impact on c.g. travel. The greatest hazard from the trapped LH_2 occurs during reentry when any LH_2/GH_2 remaining in the propellant lines may combine with atmospheric oxygen to form a potentially explosive mixture. In addition, unless dumped overboard, the trapped propellants will sporadically outgas through the orbiter LO_2 and LH_2 feedline relief valves, causing vehicle accelerations which are of such a low-level that they cannot be sensed by onboard guidance, yet represent a significant source of navigation error when applied over an entire mission. Outgassing propellants are also a potential source of contamination for scientific experiments contained in the payload bay.

At approximately 11 seconds after MECO, the ET separates from the orbiter. Approximately 109 seconds later, the orbital maneuvering (OMS) 1 thrusting begins. Coincident with the start of the OMS-1 thrusting, the GPC's automatically initiate the LO_2 dump, providing that the MPS PRPLT DUMP SEQUENCE LO_2 switch on panel R2 is in the GPC position. The GPC commands the two LO_2 manifold repressurization valves open, providing that the MAIN PROPULSION SYSTEM MANF PRESS LO_2 switch on panel R4 is in the GPC position and command each engine controller to open that SSME main oxidizer valve. And the GPC commands the three LO_2 prevalves open, providing that the MAIN PROPULSION SYSTEM LO_2 PREVALVES LEFT, CTR, RIGHT are in the GPC position. The LO_2 trapped in the orbiter feedline manifold is expelled under pressure from the helium subsystem through the engine bell of each SSME. If the MAIN PROPULSION, SYSTEM MANF PRESS LO_2 switch on Panel R4 is left in the GPC position, the LO_2 dump continues for 105 seconds. At the end of this period, the GPC's automatically terminate the dump by closing the two LO_2 manifold repressurization valves, wait ten seconds and then close the three LO_2 prevalves and command that engine controller to close that engine main oxidizer valve.

If necessary, the crew can perform the LO_2 dump manually by utilizing the START and STOP positions of the MPS PRPLT DUMP SEQUENCE LO_2 switch on panel R2. When the LO_2 dump is performed manually, all valve opening and closing sequences are still automatic. Positioning the MPS PRPLT DUMP SEQUENCE LO_2 switch to the START position causes the GPC's to immediately begin commanding all of the required valves to open automatically and in the proper sequence. The LO_2 dump would continue as long as the switch is in the START position. Placing the switch to the STOP position would cause the GPC's to begin commanding all of the required valves closed automatically and in the proper sequence. The earliest time a manual LO_2 dump can be performed is MECO plus ten seconds since the SSME's require a cooldown of at least ten seconds after MECO.

The GPC's software, MPS dump sequence will automatically initiate the LO_2 dump at one time only—at initiation of the OMS-1 thrusting period. If the MPS PRPLT DUMP

Space Transportation System

SEQUENCE LO$_2$ switch on panel R2 is not in the GPC position at that time, the LO$_2$ dump must be initiated manually. In addition, once the LO$_2$ dump has been initiated and the MPS PRPLT DUMP SEQUENCE LO$_2$ switch is placed in the STOP position, the GPC's will no longer monitor any of the positions of this switch. For this reason, the LO$_2$ dump cannot be reinitiated, manually or automatically.

Simultaneously with LO$_2$ dump, the GPC's automatically initiate the MPS LH$_2$ dump, providing that the MPS PRPLT DUMP SEQUENCE LH$_2$ switch on panel R2 is in the GPC position. The GPC's command each engine controller to command a ten-second helium purge of that engine fuel lines downstream of the main engine fuel valves, command the LH$_2$ manifold repressurization valve open, providing that the MAIN PROPULSION MANF PRESS LH$_2$ switch on panel R4 is in the GPC position, and command the two LH$_2$ fill/drain valve (inboard-outboard) open.

The LH$_2$ trapped in the orbiter feedline manifold is expelled under pressure from the helium subsystem through the LH$_2$ fill and drain valves overboard for six seconds. The inboard fill and drain valve is then closed, the recirculation valves are opened, and flow is through the engine bleed valves into the orbiter MPS between the inboard and outboard fill and drain valve and overboard through the outboard fill and drain valve for approximately 180 seconds.

If necessary, the flight crew can perform the LH$_2$ dump manually by utilizing the START and STOP positions of the MPS PRPLT DUMP SEQUENCE LH$_2$ switch on panel R2. When the LH$_2$ dump is performed manually, all valve opening and closing sequences are still automatic. Placing the MPS PRPLT DUMP SEQUENCE switch in the START position will cause the GPC's immediately to begin commanding all the required valves open automatically and in the proper sequence. The LH$_2$ dump will continue for as long as the switch is in the START position. Placing the switch in the STOP position causes the GPC's to begin commanding all of the required valves closed automatically and in the proper sequence.

At the end of the LO$_2$ and LH$_2$ dumps, the GPC's close the helium-out interconnect valves and all of the supply tank isolation valves, providing that the MPS He ISOLATION LEFT CTR, RIGHT A and B, PNEUMATIC He ISOL, and He INTERCONNECT LEFT, CTR, RIGHT switches on panel R2 are in the GPC position.

After the LO$_2$ and LH$_2$ dumps are completed, the SSME's are gimbaled to their entry positions. In the entry positions, the engine bells are moved inward (toward one another) to reduce aerodynamic heating during entry.

At approximately 00:19:00 mission elapsed time and after the MPS LO$_2$ and LH$_2$ dumps, the flight crew will initiate the procedure for vacuum inerting the orbiter LO$_2$ and LH$_2$ lines. Vacuum inerting allows any traces of LO$_2$ or LH$_2$ remaining after the propelllant dumps to be vented into space.

The LO$_2$ vacuum inerting is accomplished by opening the three LO$_2$ prevalves, the LO$_2$ inboard fill/drain valve, and the LO$_2$ outboard fill/drain valve. These valves are opened by placing the MAIN PROPULSION SYSTEM LO$_2$ PREVALVES LEFT, CTR, RIGHT and the PROPELLANT FILL/DRAIN LO$_2$ OUTBD, INBD switch on panel R4 to the OPEN position.

Helium is provided for valve actuation through the two pneumatic helium isolation valves, and these valves are closed by the GPC's at the end of the MPS LH$_2$ dump. The MPS PNEUMATIC He ISOL switch on panel R2 is positioned to OPEN. If additional helium is required to open and/or close the fill/drain valves, it can be obtained by opening the engine helium supply isolation valves and the out helium interconnect valves. These are controlled by placing the MPS He ISOLATION LEFT CTR, RIGHT A and B on panel R2 to OPEN and placing MPS He INTERCONNECT LEFT, CTR,

Space Transportation System

RIGHT switches on panel R2 to INCLOSE/OUT position. These valves were also closed by the GPC at the end of the MPS LH2 dump.

For LH2 vacuum inerting, the LH2 inboard and outboard fill/drain valves are opened by placing the MAIN PROPULSION SYSTEM PROPELLANT FILL/DRAIN LH2 OUTBD, INBD switches on panel R4 to OPEN. The ET GH2 pressurization manifold will also be vacuum-inerted by opening the H2 pressurization line vent valve, by placing the MAIN PROPULSION SYSTEM H2 LINE VENT switch on panel R4 to OPEN.

As a result, both LO2 lines and LH2 lines will be inerted simultaneously. Approximately 30 minutes is allowed for vacuum inerting. At the end of the 30 minutes, the flight crew will close the LO2 outboard fill/drain valve by placing the MAIN PROPULSION SYSTEM PROPELLANT FILL/DRAIN LO2 switch on panel R4 to CLOSE. The three LO2 prevalves and LO2 inboard fill/drain valve will be left open. To conserve electrical power after completion of the LO2 vacuum inerting sequence, the MAIN PROPULSION SYSTEM PREVALVES LO2 LEFT, CTR, RIGHT switches on panel R4 are positioned to GPC, and the MAIN PROPULSION SYSTEM PROPELLANT FILL/DRAIN LO2 OUTBD, INBD switches on panel R4 are placed to the GRD position. These switch positions remove power from the opening and closing solenoids of the corresponding valves and, because the valves are pneumatically actuated, they remain in their last commanded position. At the end of the same 30-minute period, the LH2 outboard fill/drain valve and the H2 pressurization line vent valve are closed by positioning the MAIN PROPULSION SYSTEM PROPELLANT FILL/DRAIN LH2 OUTBD switch on panel R4 to CLOSE and the MAIN PROPULSION SYSTEM H2 PRESS LINE VENT switch on panel R4 to CLOSE. The three LH2 prevalves and the LH2 inboard fill/drain valve are left open. The MAIN PROPULSION SYSTEM PREVALVES LEFT, CTR, RIGHT switches on panel R4 will be positioned to GPC, also for power conservation, and the MAIN PROPULSION SYSTEM PROPELLANT FILL/DRAIN LH2 INBD, OUTBD and H2 PRESS LINE VENT switches on panel R4 are positioned to the GRD position to conserve power. The H2 pressurization line vent valve is electrically activated; however, it is normally closed (spring-loaded to the closed position), and removing power from the valve solenoid leaves the valve closed.

After vacuum inerting, the helium isolation valves and interconnect valves are closed by placing the MPS He ISOLATION LEFT, CTR, RIGHT, PNEUMATICS He ISOL on panel R2 to the CLOSE position and the He INTERCONNECT LEFT, CTR, RIGHT switches on panel R2 to GPC. This ensures that the helium supply tanks are isolated from any leakage in the downstream lines during orbital operations.

The electrical power is also turned off to each engine controller and engine interface unit, and the engine controller heaters are placed in the AUTO mode.

During the early portion of the entry timeline, the propellant feedline manifolds and the ET pressurization lines are repressurized with helium from the helium subsystem. This prevents atmospheric contamination from being drawn into the manifolds and feedlines during entry. Removing contamination from the manifolds or feedlines can be a long and costly process since it involves disassembly of the affected part. Manifold repressurization is a total manual sequence performed by the flight crew.

Manifold repressurization is enabled by opening the LO2 and LH2 manifold pressurization valves. These valves are controlled by the MAIN PROPULSION SYSTEM PRESS LO2 and LH2 switches on panel R4. The pilot, strapped into his seat, may not be able to reach these switches, so before he is strapped in, the LO2 and LH2 MANF PRESS switches are positioned to OPEN.

Space Transportation System

Later in the entry timeline, the pneumatic helium isolation valves will be opened to allow helium flow through the manifold pressurization valves and into the feedline manifolds and pressurization lines by placing the MPS PNEUMATICS He ISOL switch on panel R2 to OPEN. In addition, the engine helium isolation valves and the out-helium interconnect valves are also opened to ensure there is sufficient helium to perform the pressurization by placing the MPS He ISOLATION LEFT, CTR, RIGHT A and B switches on panel R2 to OPEN and placing the MPS He INTERCONNECT LEFT, CTR, RIGHT to INCLOSE/OPEN OUT. Because of overboard leakage through the SSME's all of the helium subsystems valves opened during the manifold repressurization sequence will be left open until vehicle touchdown and rollout; thus any helium loss caused by overboard leakage will be immediately replenished from the helium supply tanks.

The aft compartment is purged by helium from the MPS system. The engine valve actuation pneumatic plumbing also has two orficed valves located in the aft compartment. These valves are controlled by the GPC's. They are opened by the GPC's at the same time that the repressurization valves are actuated.

If MECO is preceded by a return to launch site (RTLS) abort, the subsequent MPS LO_2 dump will begin ten seconds after the ET separation command is issued, and, simultaneously, the LH_2 dump will also begin. The LO_2 and LH_2 dump is initiated and terminated automatically by the GPC's regardless of the MPS PRPLT DUMP SEQUENCE LO_2 and LH_2 switch positions on panel R2.

Because of the geometry of the LO_2 feedline manifold and its orientation to the orbiter's acceleration vector during RTLS abort, little LO_2 is actually lost overboard during the LO_2 dump. For this reason, the RTLS LO_2 dump is considered more of a venting process than a dump. In the nominal LO_2 dump, the GPC's terminate the dump 180 seconds or at 30,480 meters (100,000 feet), whichever occurs first, after initiation. The LO_2 is dumped through the bell of each SSME; however, each engine is gimbaled to the entry stow position rather than the normal dump position. The LO_2 feedline manifold is not pressurized in this mode and the two LO_2 manifold repressurization valves remain closed throughout the entire dump. No vacuum inerting or system repressurization is performed in an RTLS abort. The MAIN PROPULSION SYSTEM PREVALVES LO_2, LEFT, CTR RIGHT switches panel R4 are in the GPC position, and the GPC's command that engine controller to open each engine main oxidizer valve for the dump.

In the RTLS mode, the LH_2 dump is initiated and terminated automatically by the GPC's and simultaneously with the LO_2 dump regardless of the position of the MPS PRPLT DUMP SEQUENCE LH_2 switch on panel R2. The LH_2 dump opens the two RTLS dump valves and the two RTLS dump valves and the two RTLS manifold repressurization valves. The LH_2 trapped in the feedline manifold is expelled, under pressure from the helium subsystem, through a special opening on the port side of the orbiter between the wing and the orbital maneuvering system (OMS) pod. The GPC's will terminate the LH_2 dump automatically when the orbiter reaches 30,480 meters (100,000 feet). At that time, the inboard and outboard RTLS dump valves are closed, and the two RTLS manifold repressurization valves are closed. No vacuum inerting or system repressurization is performed following an RTLS LH_2 dump.

CONTRACTORS. The Rocketdyne Division of Rockwell International, Canoga Park, CA, is prime contractor for the Space Shuttle main engines. Other contractors include Aeroflex Laboratories, Plainview, NY (MPS vibration mounts); Airite Division, Sargent Industries, El Segundo, CA (MPS surge pressure receiver); Ametek Calmec, Pico Rivera, CA (1-1/2 inch and 2-inch LO_2 and LH_2 shutoff valve, 4-inch LH_2 disconnect, 2-inch GH_2/GO_2 disconnect); Ametek Straza, El Cajon, CA (8-inch LH_2/LO_2 fill/drain, 2-inch and 4-inch El Cajon, CA LH_2 recirculation lines, high point bleed line manifold, gimbal joint); Arrowhead Products, Division of Federal Mogul, Los

Space Transportation System

Alamitos, CA (12 to 17-inch diameter LO_2 and LH_2 feedlines, flexible purge gas connector); Astech, Santa Ana, CA (MPS heat shield); Brunswick, Lincoln, NE (17.3-cubic-foot and 4.7 cubic foot capacity helium tanks); Brunswick-Circle Seal, Anaheim, CA (helium check valves, GO_2 and GH_2 1-inch helium pressurization line, 3/8-inch LH_2 relief valve, engine isolation check valves); Brunswick-Wintec, El Segundo, CA (helium filter); Coast Metal Craft, Compton, CA (metal flex hose); Conrac Corp., West Caldwell, NJ (engine interface unit); Consolidated Controls, El Segundo, CA (oxygen primary flow control hydraulic valve, hydrogen/oxygen pressurant flow control valves, 20-psi helium regulator, 850-psi helium relief valve, 750-psi helium regulator); Fairchild Stratos, Manhattan Beach, CA (12-inch prevalves, 1-1/2 inch LO_2 disconnect, 8-inch LO_2 and LH_2 fill and drain valves and GN_2 and GH_2 disconnects); Gulton Industries, Costa Mesa, CA (pogo pressure transducer); K-West, Westminister, CA (LO_2 and LH_2 external tank ullage pressure signal conditioner, MPS differential pressure transducer and electronics propellant head pressure); Megatek, Van Nuys, CA (MPS line flange cryo seals); Moog, Inc., East Aurora, NY (main engine gimbal actuators); Parker Hannifan Corp., Irvine, CA (1-inch relief isolation valves, pogo check valves, 17-inch LH_2 and LO_2 disconnects, 8-inch LO_2 and LH_2 disconnects, LO_2 and LH_2 relief valves); Simmonds Precision Instruments, Vergennes, VT (LO_2 and LH_2 point sensors and electronics); Sterer Engineering, Los Angeles, CA (main engine hydraulic solenoid shutoff valve); Whittaker Corp., North Hollywood, CA (750/20-psi helium regulator); Wright Components, Inc. Clifton Springs, NJ (two-way pneumatic solenoid valve, three-way helium solenoid valve, hydraulic latching solenoid valve).

ORBITER-EXTERNAL TANK SEPARATION SYSTEM

This system separates the orbiter from the external tank at three structural points and two umbilicals. Separation of the orbiter and external tank occur just before insertion, and is triggered automatically by the orbiter computers. Orbiter crewmen also can initiate separation.

The forward structural attachment consists of a shear bolt unit mounted in a spherical bearing. The bolt separates at a specified break area when two pressure cartridges are initiated. The pressure from one or both cartridges drives a tandem piston to shear the bolt with the second piston acting as a hole plugger to fill the cavity left by the sheared bolt. A centering mechanism rotates the unit from the displacement position to a centered position aligning the bearing flush with the adjacent thermal protection system moldline.

The aft structural attachment consists of two special bolts attaching the external tank strut hemisphere to the orbiter left and right-side cavities by pyro-actuated frangible nuts. At separation the frangible nuts are split by a booster cartridge initiated by a detonator cartridge. The attach bolts are driven by the separation forces and a spring into a cavity in the tank strut. The frangible nut, cartridge fragments and hot gases are contained within a cover assembly and a hole plugger isolates the fragments within the container.

The orbiter-external tank umbilical plate separation consists of right and left assemblies. Each assembly contains three dual-detonator frangible nut/bolt combinations that hold the orbiter and external tank umbilical plates together during mated flight. Each bolt has a retraction spring which, after release of the nut, retracts the bolt to the external tank side of the interface. On the orbiter side, each frangible nut with its detonators is enclosed in debris container that contains nut fragments and hot gases generated by operation of the detonators, either of which will fracture the nut. Each orbiter umbilical plate has three hydraulic retractors which, after release of the three frangible nut/bolt combinations retract the plate approximately

Space Transportation System

66 millimeters (2-1/2 inches). The retraction disconnects the orbiter-external tank electrical umbilical in the first 12.7 millimeters (0.5 inch) of travel, releases the fluids between the disconnected LO_2 and LH_2 shutoff valves, and (as a secondary function of the retract motion through linkage) closes the LO_2 and LH_2 main feedline disconnect valves. A dedicated MPS pneumatic helium supply is the primary MPS valve closing system. Each orbiter umbilical plate has three bungees that hold the plate in the lateral position after separation from the external tank umbilical plate.

An electromechanical actuation system operates two separate doors to cover the left and right umbilical cavities after the external tank is jettisoned and umbilical plates are retracted. Normal operation of the doors by the flight crew is manual. Each door is approximately 1,270 millimeters (50 inches) square.

The doors are held fully open during liftoff and ascent by two centerline latches, one forward and one aft, each latch engaging both doors. After tank separation two redundant ac

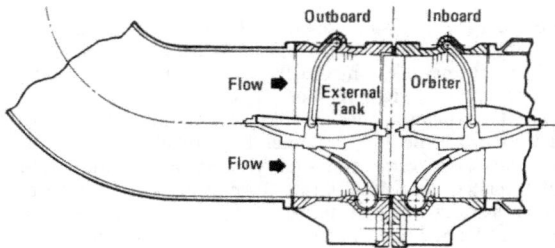

External Tank-Orbiter Disconnect Valves

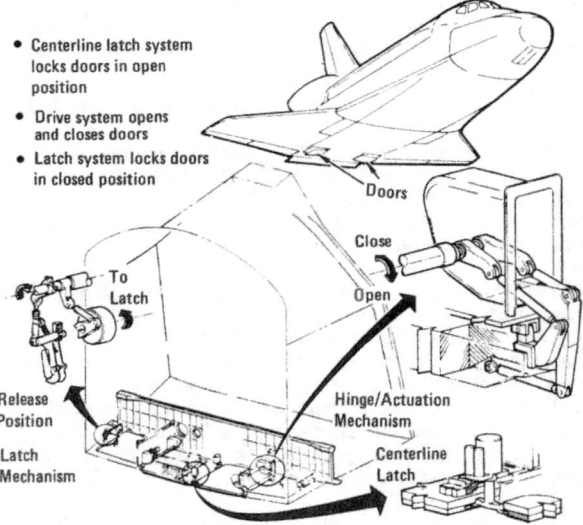

Orbiter/External Tank Umbilical Door Operational Sequence

Orbiter Umbilical Doors

Space Transportation System

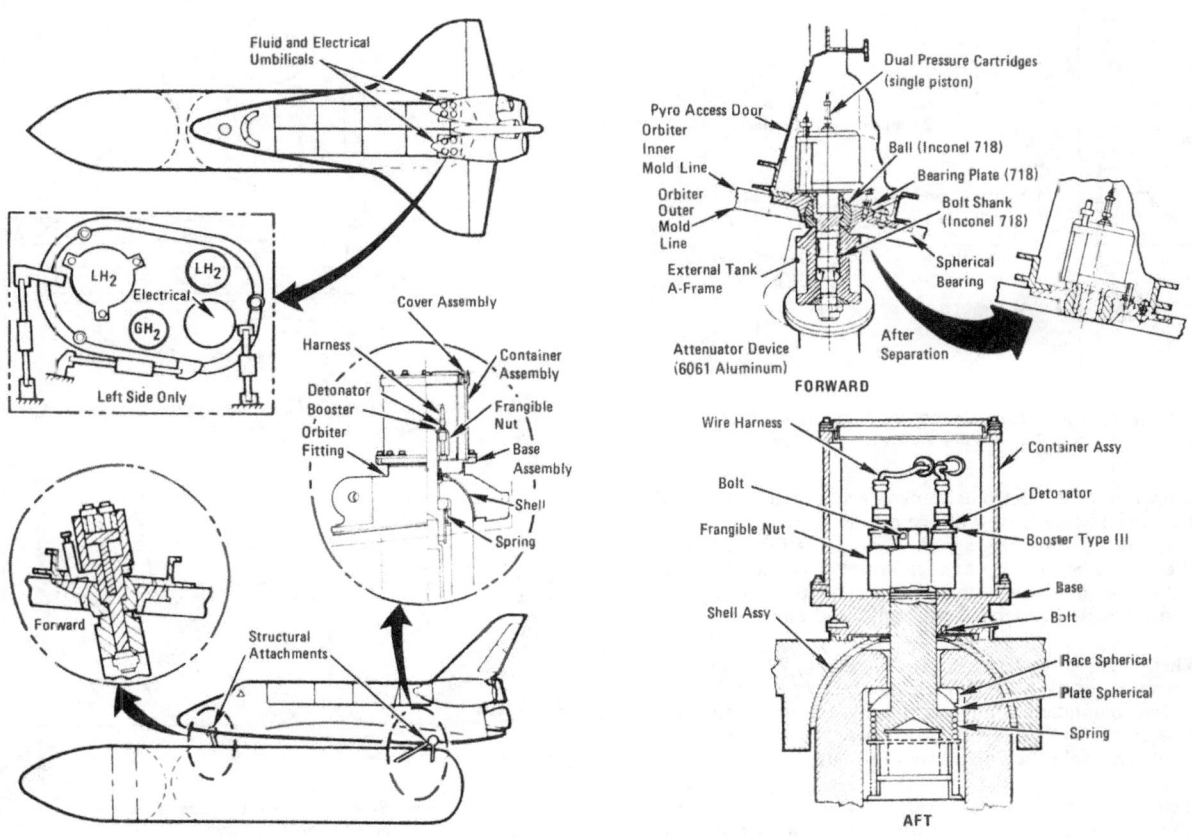

Orbiter — External Tank Separation System

Space Transportation System

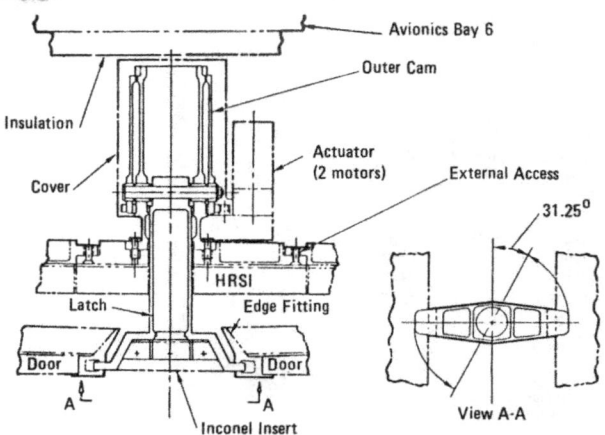

*Orbiter/External Tank Umbilical Closeout Doors —
Centerline Latch*

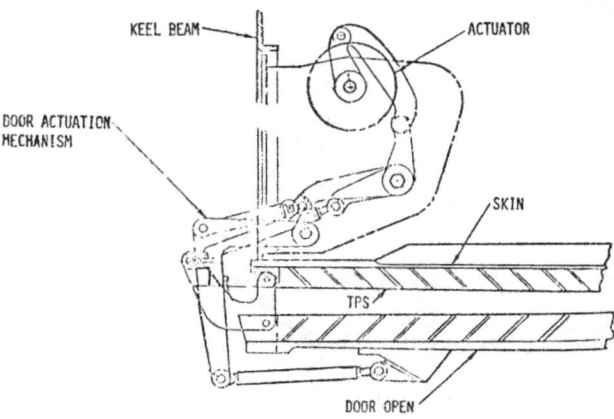

*Orbiter/External Tank Umbilical Closeout Doors —
Hinge/Actuation System*

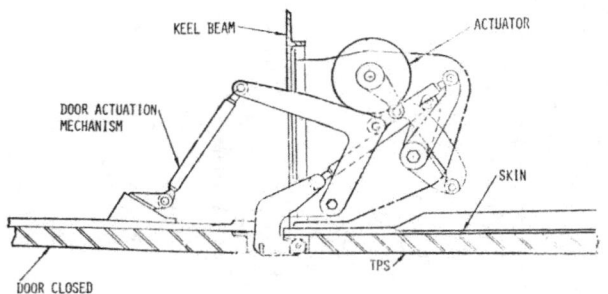

*Orbiter/External Tank Umbilical Closeout Doors —
Hinge/Actuation System*

motors operate each electromechanical actuator to rotate and retract the latch blade flush with the orbiter thermal protection system moldline. The hinges and door actuation system is located on the inboard edge of each door. Two redundant ac motors operate an electromechanical actuator to open or close each door through a system of bellcranks and pushrods.

When the door is within 50 millimeters (2 inches) of the closed position ready-to-latch indicators activate the uplatch system. Two redundant ac motors provide the power to drive three uplatch hooks that engage rollers near the outboard edge of the door and hold the door closed for entry.

Each umbilical door is covered with reusable surface insulation. An aerothermal barrier which requires 310 mmHg (6

Space Transportation System

pounds per square inch) to compress is incorporated on each door to seal the door tiles with adjacent tiles.

The right and left umbilical doors are controlled by the ET UMBILICAL DOOR switches located on the flight deck display and control panel R2. The right and left umbilical doors are controlled automatically in RTLS aborts and are controlled manually by the flight crew upon completion of the orbital maneuvering system (OMS) thrusting period in all other sequences. It is noted that the automatic mode can be used to provide a backup to the manual sequence.

The manual sequence is enabled by positioning the ET UMBILICAL DOOR MODE switch on panel R2 to MAN (manual). The STOW position of the ET UMBILICAL DOOR CENTERLINE LATCH switch on panel R2 releases the two umbilical door centerline latches and stows both centerline latches. The ET UMBILICAL DOOR CENTERLINE LATCH talkback indicator indicates STO (stow) when the latches complete their motion in six seconds. The GRD (ground) position of the ET UMBILICAL DOOR CENTERLINE switch is used to isolate STOW command stimuli from motor control assembly relays. Prior to release of the centerline latches the ET UMBILICAL DOOR CENTERLINE LATCH talkback indicates BARBERPOLE.

The manual control of opening or closing the right and left umbilical door is controlled by the respective LEFT and RIGHT ET UMBILICAL DOOR OPEN, CLOSE, OFF switch on panel R2. The CLOSE position commands the respective door closed in approximately 24 seconds and when the door reaches its limit of travel, the electrical motors are automatically turned off. The LEFT and RIGHT ET UMBILICAL DOOR talkback indicator above the respective switch on panel R2 indicates OP (open) prior to door closure, BARBERPOLE when the door is in transit, and CL (close) when two of the three ready-to-latch switches for that door has sensed door closure. The OFF position of the respective switch, removes electrical power from the manual door closure sequence. The OPEN position of the respective switch is used for ground operations.

The manual control of latching or releasing the right and left umbilical door is controlled by the respective LEFT and RIGHT ET UMBILICAL DOOR LATCH, RELEASE, OFF, LATCH switch on panel R2. The LATCH position commands the respective door latches to their latched positions in approximately six seconds, providing two of the three ready-to-latch switches have sensed door position and the latches hold the door closed. The LEFT and RIGHT ET UMBILICAL DOOR, talkback indicator above the respective switch on panel R2, indicates REL (release) during ascent, BARBERPOLE when that door is in transit, and LAT (latch) when that door is in its latched position. The OFF position of the respective switch, removes electrical power from the door latch closure sequence. The REL (release) position of the respective switch is used for ground operations.

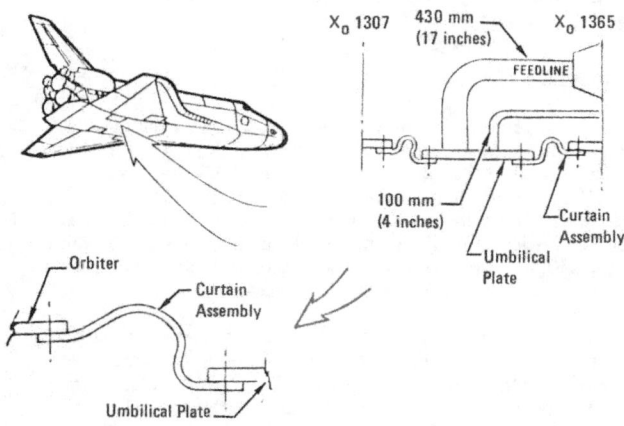

Orbiter/External Tank Umbilical Closeout Curtain

Space Transportation System

The automatic sequence of the left and right umbilical doors occurs during the return-to-launch site (RTLS) abort. The automatic sequence can be performed with the ET UMB DOOR MODE switch on panel R2 in the manual or GPC position. Two seconds after external tank (ET) separation the centerline latches release the doors and the latches are stowed. The ET UMBILICAL DOOR CENTERLINE LATCH talkback indicator will indicate STO when the centerline latches complete their motion at ET separation plus eight seconds. The left and right umbilical doors are closed and the ET UMBILICAL DOOR LEFT and RIGHT DOOR talkback indicates CL at ET separation plus 32 seconds. The left and right umbilical door latches, latch the doors closed and the ET UMBILICAL DOOR LEFT and RIGHT talkback indicates LAT at ET separation plus 38 seconds.

A curtain is installed at each of the orbiter/external tank umbilicals. After ET separation, the residual LO_2 is dumped through the three Space Shuttle Main Engines and the residual LH_2 is dumped overboard. The umbilical curtain prevents hazardous gases (GO_2, GH_2) from entering into the orbiter aft fuselage through the orbiter umbilical openings prior to umbilical door closure. The curtain also acts as a seal during the ascent phase of the mission to permit the aft fuselage to vent through the orbiter purge and vent system, thereby providing protection to the orbiter aft bulkhead at station X_O 1307.

Various parameters are monitored and displayed on the flight deck control panel and CRT (cathode ray tube) and transmitted by telemetry.

Contractors for the separation system include Hoover Electric, Los Angeles (external tank umbilical centerline latch and actuator, umbilical door actuator, and umbilical door latch actuator), U.S. Bearing, Chatsworth, CA (external tank-orbiter spherical bearing); Bertea Corp., Irvine, CA (umbilical retractor actuator); Space Ordnance System Division, Trans Technology Corp., Saugus, CA (orbiter-external tank separation bolt/cartridge detonator assembly, 3/4-inch frangible nut orbiter-external tank umbilical separation and 2-1/2 inch frangible nut/pyro components in orbiter/external tank aft attach separation system).

SPACE SHUTTLE COORDINATE SYSTEM

The Space Shuttle coordinate reference system is a means of locating specific points on the Shuttle. The system is measured in inches and decimal places: X_O designates the longitudinal (forward and aft) axis, Y_O the lateral (inboard and outboard) axis, and Z_O the vertical (up and down) axis. The subscript O indicates orbiter; similar reference systems are used for the external tank (T), solid rocket booster (B), and overall Space Shuttle System (S).

In each coordinate system, the X-axis zero point is located forward of the nose tip; that is, the orbiter nose tip location is 236 inches aft of the zero point (at X_O 236); the external tank nose cap tip location is at X_T 322.5, and the solid rocket booster (SRB) nose tip location is at X_B 200. In the orbiter, the horizontal X_O, Y_O reference plane is located at Z_O 400, which is 336.5 inches above the external tank horizontal X_T, Y_T reference plane located at Z_T 400. The solid rocket booster horizontal X_B, Y_B reference plane is located at Z_B 0 and coincident with the external tank horizontal plane at Z_T 400. The solid rocket booster vertical X_B, Z_B planes are located at $+Y_S$ 250.5 and $-Y_S$ 250.5. Also, note that the orbiter, external tank, and Shuttle system center X, Z planes coincide.

From the X = 0 point, aft is positive, and forward is negative for all coordinate systems. Looking forward, each Shuttle element Y-axis point right of the center plane (starboard) is positive and left of center (port) is negative. The Z axis of each point within all elements is positive with Z = 0 located

Space Transportation System

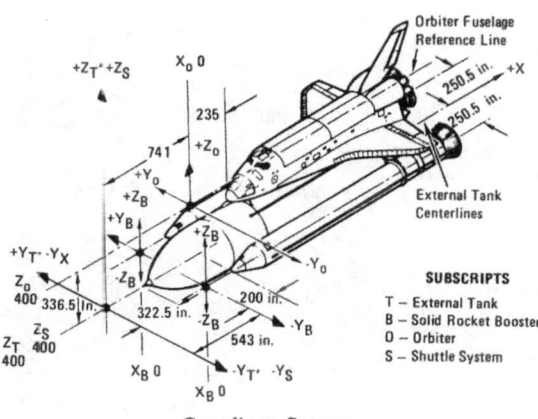

Coordinate Systems

below the element, except for the SRB's, in which each Z-coordinate point below the SRB X_B, Y_B reference plane is negative and each point above that plane is positive.

The Shuttle system and Shuttle element coordinate systems are related as follows: The external tank X_T 0 point coincides with X_S 0, the SRB X_B 0 point is located 543 inches aft, and the orbiter Y_O, Z_O reference plane is 741 inches aft of X_S 0.

Space Shuttle Spacecraft Structures

ORBITER STRUCTURE

The orbiter structure is divided into major sections: the forward fuselage, which consists of upper and lower sections that fit clam-like around a pressurized crew compartment; wings; mid fuselage; payload bay doors; aft fuselage; and the vertical tail. The majority of the structures are of conventional aluminum construction protected by reusable surface insulation.

The forward fuselage structure is composed of 2024 aluminum alloy skin-stringer panels, frames, and bulkheads.

The crew compartment, which is supported within the forward fuselage at four attachment points, is welded to create a pressure-tight vessel. The three-level compartment has a side hatch for normal passage and a hatch from the airlock into the payload bay for extra-vehicular activity (EVA) and intravehicular activity (IVA).

The mid fuselage is an 18.28-meter (60-foot) section of primary load-carrying structure between the forward and aft fuselages. It includes the wing carry-through structure and the payload bay doors. The skins are integral-machined aluminum panels and aluminum honeycomb sandwich panels. The frames are constructed as a combination of aluminum panels with riveted or machined integral stiffeners and a truss structure center section. The upper half of the mid fuselage consists of structural payload bay doors hinged along the side and split at the top centerline. The doors are graphite epoxy frames and honeycomb panel construction.

The aft fuselage includes a truss-type internal structure of diffusion-bonded elements that transfers the main engine thrust loads to the mid fuselage and external tank. The aft fuselage external surface is of standard construction except for the removable orbital maneuvering system/reaction control system

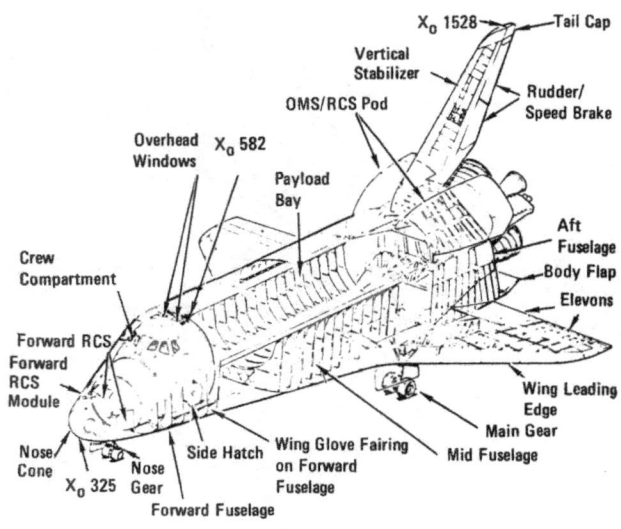

Orbiter Structure

(OMS/RCS) pods. They are constructed of graphite epoxy skins and frames. An aluminum bulkhead heat shield with reusable insulation at the rear of the orbiter protects the main engines.

The wing is constructed of conventional aluminum alloy. Corrugated spar web, truss-type ribs and riveted skin-stringer and honeycomb covers are used. The elevons are of aluminum honeycomb and are split into two segments to minimize hinge binding and interaction with the wing.

Space Shuttle Spacecraft Structures

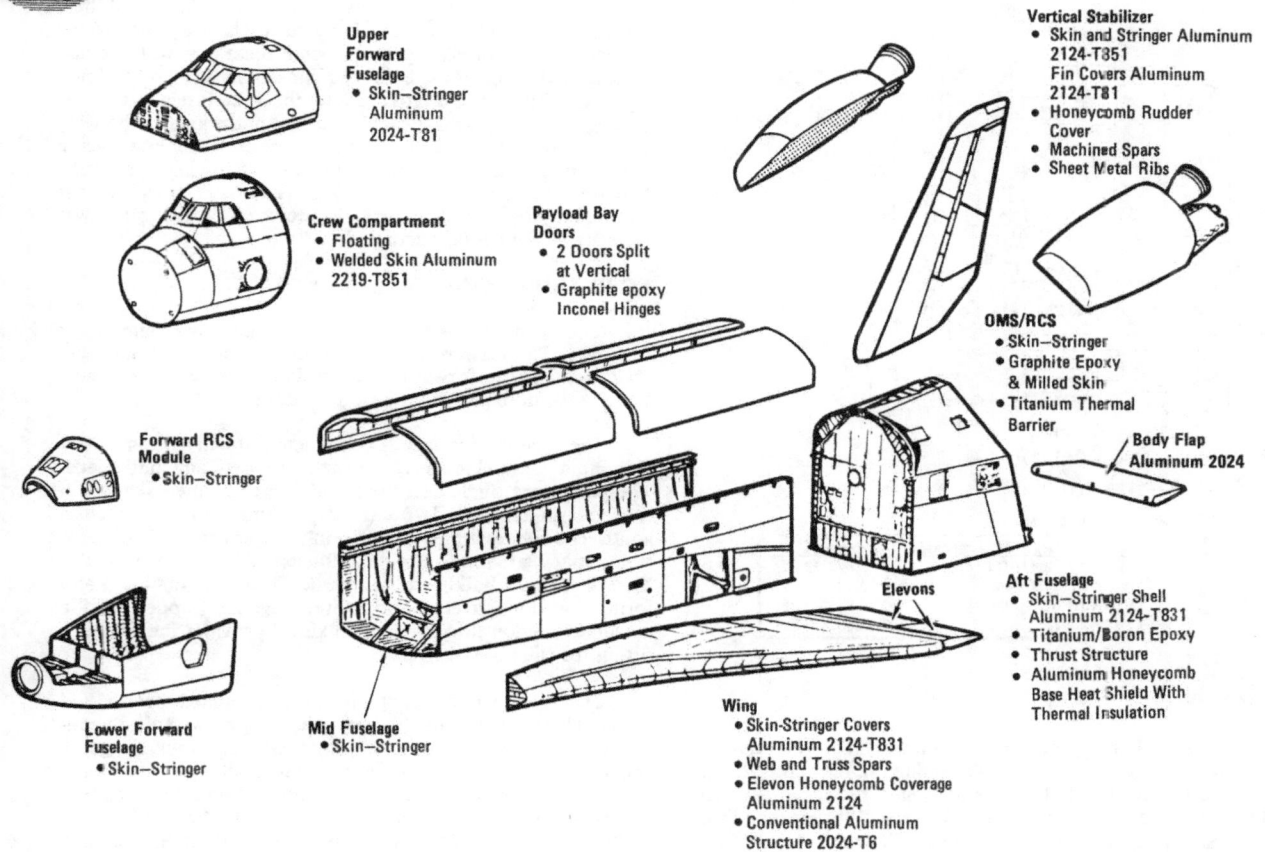

Orbiter Structural Elements

Space Shuttle Spacecraft Structures

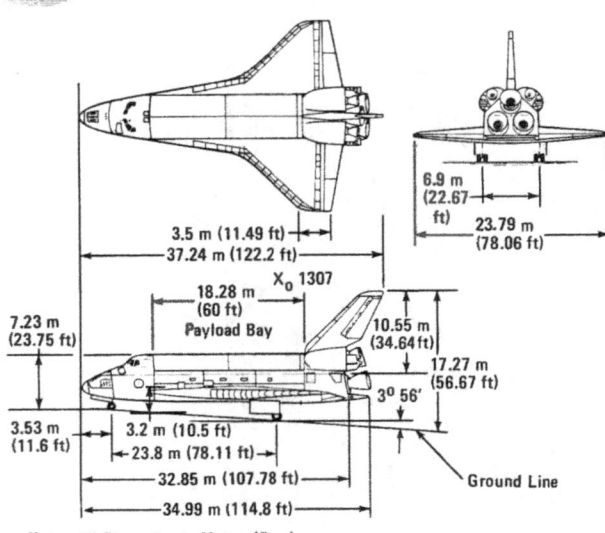

Orbiter Dimensions

The vertical tail, a conventional aluminum alloy structure, is a two-spar, multi-rib, integrally machined skin assembly. The tail is attached to the aft fuselage by bolted fittings at the two main spars. The rudder/speed brake assembly is divided into upper and lower sections. Each is also split longitudinally and individually actuated to serve as both rudder and speed brake.

These major structural assemblies are mated and held together with rivets and bolts. The mid fuselage is joined to the forward and aft fuselage primarily with shear ties, with the mid fuselage overlapping the bulkhead caps at stations X_O 582 and X_O 1307. The wing is attached to the mid fuselage and aft fuselage primarily with shear ties, except in the area of the wing carry-through. There the upper panels are attached with tension bolts. The vertical tail is attached to the aft fuselage with bolts that work in both shear and tension. The body flap, which has aluminum honeycomb covers, is attached to the aft lower fuselage by four rotary actuators.

FORWARD FUSELAGE

The forward fuselage consists of the upper and lower fuselages. The forward fuselage houses the crew compartment and supports the forward RCS module, nose cap, nose gear wheel well, nose gear, and nose gear doors.

The forward fuselage is constructed of conventional 2024 aluminum alloy skin-stringer panels, frames, and bulkheads. The panels are single curvature and stretch-formed skins with riveted stringers spaced 76.2 to 127 millimeters (3 to 5 inches) apart. The frames are riveted to skin-stringer panels. The major frame spacing is 762 to 914.4 millimeters (30 to 36 inches). The Y_O 378 forward bulkhead is constructed of flat aluminum and formed sections riveted and bolted together (upper) and a machined section (lower). The bulkhead provides the interface fitting for the nose section.

The nose section contains large machined beams and struts. The structure for the nose landing gear wheel well consists of two support beams, two upper closeout webs, drag link support struts, nose landing gear strut and actuator attachment fittings, and the nose landing gear door fittings. The two (left and right) landing gear doors are attached by hinge fittings in the nose section. The doors are constructed of aluminum alloy honeycomb. The left door is wider than the right, although both

Space Shuttle Spacecraft Structures

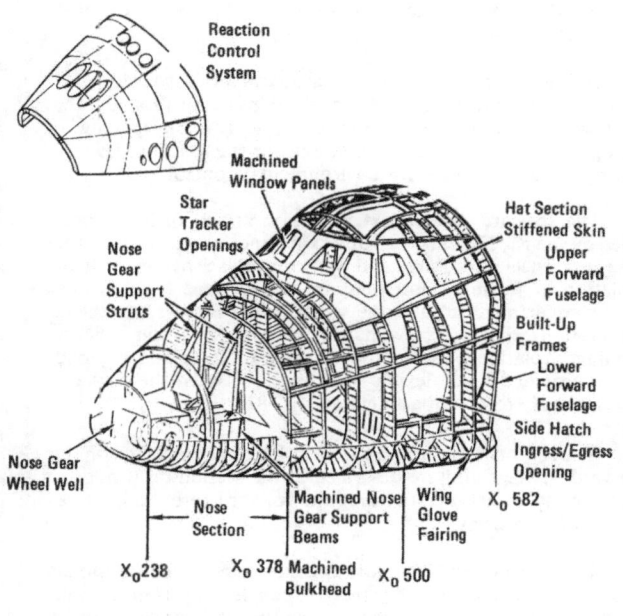

- Aluminum Construction
- Riveted Skin—Stringer/Frame Structure

Forward Fuselage Structure

Upper Forward Fuselage

Lower Forward Fuselage

are the same length. Each door has an up-latch fitting at the forward and aft ends to lock the doors closed when the gear is retracted. Each door has a pressure seal in addition to a thermal barrier. Ballast provisions are provided in the nose wheel well and on the X_O 378 bulkhead for weight and center of gravity

Space Shuttle Spacecraft Structures

Upper Forward Fuselage Being Mated to Lower Forward Fuselage

control. The ballast is lead and the provisions in the nose wheel well will accommodate 612.36 kilograms (1,350 pounds) and the X_O 378 bulkhead will accommodate a maximum of 1,206.57 kilograms (2,660 pounds).

The forward fuselage carries the basic body bending loads (a tendency to change the radius of a curvature of the body) and reacts nose landing gear loads.

The forward fuselage will be covered with the reusable thermal protection system, except for the six windshields, two overhead windows, side hatch window, and some of the area around the forward RCS engines. The nose cap is also a reusable thermal protection system. It is constructed of reinforced carbon-carbon and will have thermal barriers at the nose cap-structure interface.

In the forward fuselage skin are structural provisions for installation of antennas, the deployable air data probes, and the door eyelet openings for the two star trackers. An opening in the upper forward fuselage is required for star tracker viewing. The opening has a door for environmental control.

The forward orbiter/external tank attach fitting is provided by the X_O 378 bulkhead and skin panel structure aft of the nose gear wheel well. Purge and vent control is provided with installation of flexible boots between the forward fuselage and crew compartment around the windshield windows, overhead observation window, crew hatch window, and star tracker opening. Isolation between the forward fuselage and payload bay is provided by a flexible membrane between the forward fuselage and crew compartment at X_O 582.

The six forward outer-pane windshields are installed on the forward fuselage and are described in the section on windows. The window structural frames in the forward fuselage are five-axis machined parts.

The forward RCS module is constructed of conventional 2024 aluminum alloy skin-stringer panels and frames. The panels are composed of single-curvature and stretch-formed skins with riveted stringers. The frames are riveted to the skin-stringer panels. The forward reaction control system module is secured to the forward fuselage nose section and forward bulkhead of the forward fuselage with 16 fasteners, which permit installation and removal of the forward RCS module. The components of the forward RCS are mounted and attached to the forward RCS module. The forward RCS module will be covered with reusable thermal protection in addition to thermal

Space Shuttle Spacecraft Structures

Forward RCS Module

barriers installed around the forward RCS module and engine interface and forward RCS module interface-attachment area to the forward fuselage.

The forward fuselage and forward RCS module are built by Rockwell's Space Transportation and Systems Group, Shuttle Orbiter Division, Downey, CA.

CREW COMPARTMENT

The three-level crew compartment is constructed of 2219 aluminum alloy plate with integral stiffening stringers and internal framing welded together to create a pressure-tight vessel. The compartment has a side hatch for normal ingress and egress, a hatch into the airlock from the mid deck, and a hatch through the aft bulkhead into the payload bay for extravehicular activity, and payload bay access.

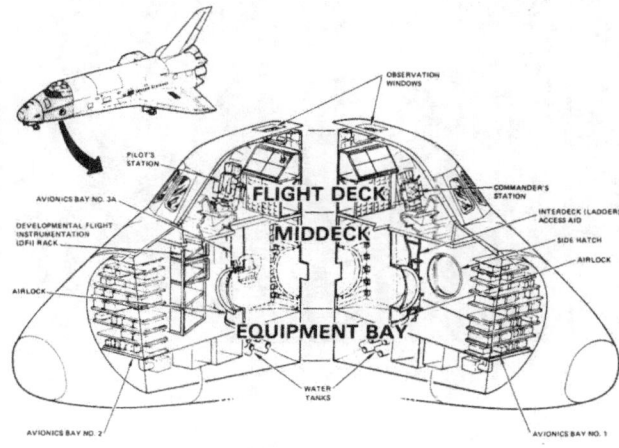

Crew Compartment and Arrangement

Redundant pressure window panes are provided in the windshield and in the overhead, aft viewing, and side hatch windows; they are described in the window section. The approximately 300 penetrations in the pressure shell are sealed with plates and fittings. A large removable panel in the aft bulkhead provides access to the crew compartment interior during initial fabrication and assembly and provides for airlock installation and removal. Equipment supported in the compartment includes the environmental control life and support system (ECLSS), avionics, guidance and navigation (G&N), displays and controls (D&C), navigation star tracker base, and crew accommodations for sleeping, waste management, seats, and galley.

The crew compartment is supported within the forward fuselage at only four attach points to minimize the thermal conductivity between them. The two major attach points are at the

Space Shuttle Spacecraft Structures

Crew Compartment

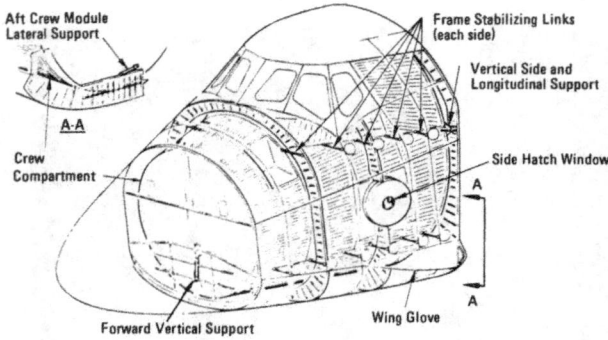

Crew Compartment-Forward Fuselage Interface

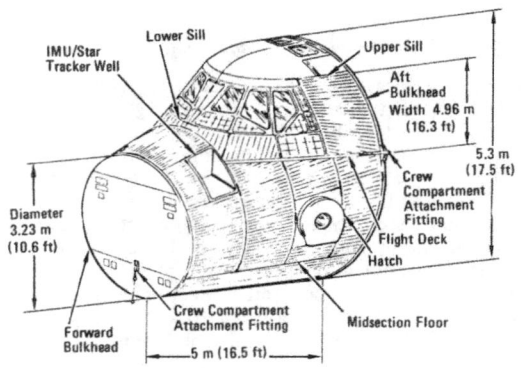

Crew Compartment

aft end of the crew compartment at the flight deck section floor level. The vertical load reaction link is on the centerline of the forward bulkhead. The lateral load reaction links are on the lower segment of the aft bulkhead.

The compartment is configured to accommodate a crew of four on the flight deck and three in the mid deck. The crew cabin arrangement consists of a flight deck, mid deck, and lower level equipment bay.

The crew compartment is pressurized to 760 plus or minus 10 millimeters of mercury (mmHg) (14.7 plus or minus 0.2 psia) and is maintained at an 80 percent nitrogen and 20 percent oxygen composition by the ECLSS, which provides a shirtsleeve environment for the flight crew. The crew compartment is designed for 828 mmHg (16 psia).

The crew compartment volume with the airlock in the mid deck is 66 cubic meters (2,325 cubic feet). If the airlock is in the

Space Shuttle Spacecraft Structures

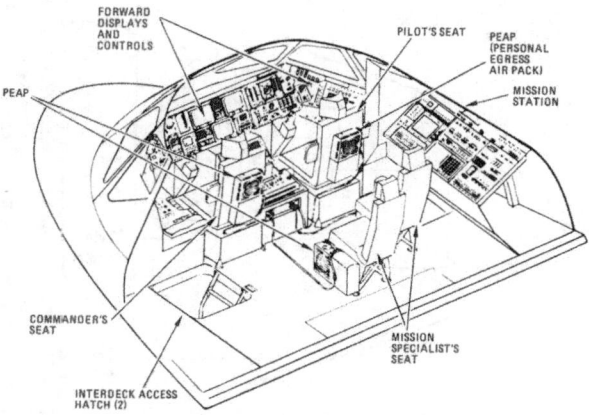

Crew Compartment Flight Deck

payload bay, the crew compartment cabin volume is 74 cubic meters (2,625 cubic feet).

The flight deck is the uppermost compartment of the cabin. The commander and pilot work stations are positioned side by side in the forward portion of the flight deck. These stations have controls and displays for maintaining autonomous control of the vehicle throughout all mission phases. Directly behind and to the sides of the commander and pilot centerline are the mission specialist seats. The mission specialist station is located on the right side of the orbiter and has controls and displays for monitoring systems, communication management, payload operation management, and payload/orbiter interface operations. The payload specialist station is located on the left side of the orbiter and contains controls and displays.

The forward flight deck including the center console and seats is approximately 2.2 meters squared (24 square feet) and with the side console control and displays another approximate 0.3 meters squared (3.5 square feet) is added. If the center console is subtracted from the 2.2 meters squared (24 square feet) this would amount to approximately 0.4 meters squared (5.2 square feet).

The aft flight deck is approximately 3.7 meters squared (40 square feet).

Between these two aft stations are the on-orbit pilot and payload handling stations. These rearward facing stations are unoccupied during launch. Orbital operation visibility is provided by overhead and aft viewing windows. The orbital stations contain displays and controls for executing attitude or translational maneuvers for terminal-phase rendezvous, stationkeeping and docking, and payload deployment and retrieval. The mission specialists are designated the payload manipulator. The commander and/or pilot would be the rendezvous and docking operator.

The commanders and pilots seat system will have controls for electrically adjusting the seat fore, aft, up, and down and for support during vertical launch and horizontal flight and tilt for back angle positioning. It also has an inertia reel, which allows mobility for performing tasks, and capability for locking the seat. The seat will accommodate mounting of a rotational controller and stowage of inflight and emergency equipment.

The seats have removable cushions and pads. Specialist seats are not adjustable, but have mounting provisions for emergency equipment, communication, and biomedical monitoring and controls and mechanization to release the seat from the flight and/or mid-deck for stowage during zero-g orbital flight. The specialists' seats will have restraint devices and controls to lock and unlock the seat back for tilt change and removable cushions and pads. The seat is 647 millimeters (25.5 inches) in length and 393 millimeters (15.5 inches) wide and when folded for stowage is 279 millimeters (11 inches) in height.

Space Shuttle Spacecraft Structures

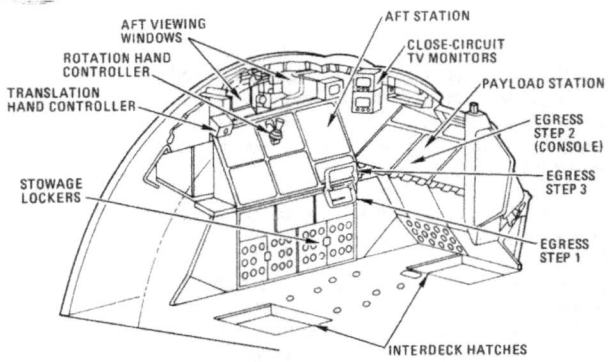

Aft Flight Station

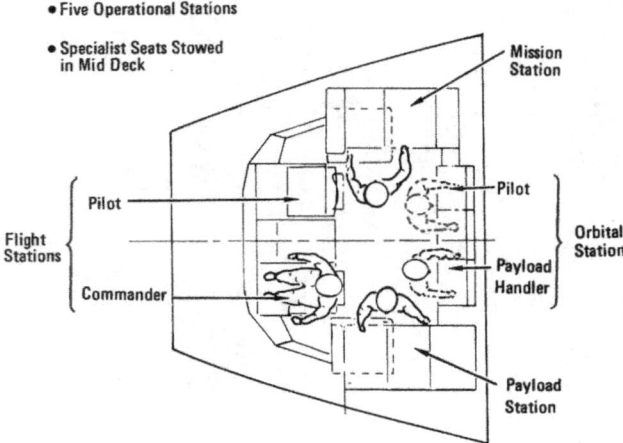

Flight Deck Station — Orbital Operations

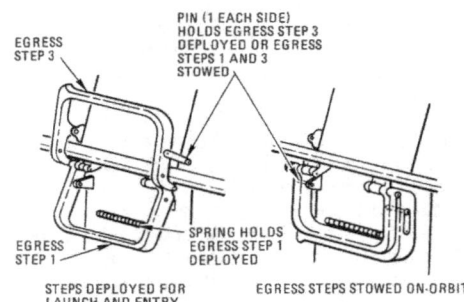

Egress Steps to Left Hand (port) Overhead Window

The left-hand overhead window will provide an emergency exit route.

Directly beneath the flight deck is the mid deck. Access to the mid deck-flight deck is by two interdeck openings 660 by 711 millimeters (26 by 28 inches). A ladder attached to the left interdeck access allows easy passage in one-g conditions. The mid deck provides crew accommodations. Three avionics equipment bays are also located here. The two forward avionics bays utilize the complete width of the cabin and extend into the mid deck 990 millimeters (39 inches) from the forward bulkhead. The aft bay extends into the mid deck 990 millimeters (39 inches) from the aft bulkhead on the right side of the airlock. Just forward of the waste management system is the side hatch. The mid deck completely stripped is approximately 14.5 meters squared (160 square feet). The gross mobility area is approximately 9.29 meters squared (100 square feet).

The side hatch in the mid deck is used for normal crew entrance/exit and may be operated from within the crew cabin

Space Shuttle Spacecraft Structures

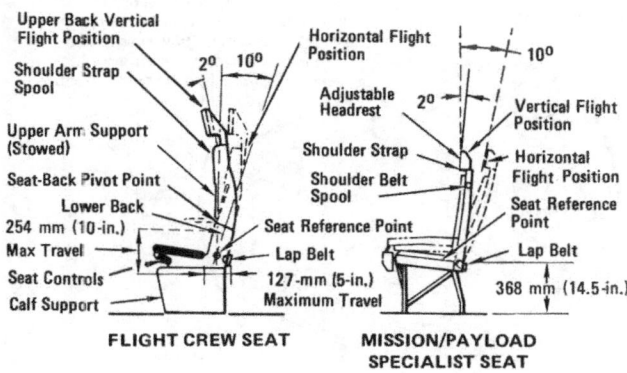

Flight Seats

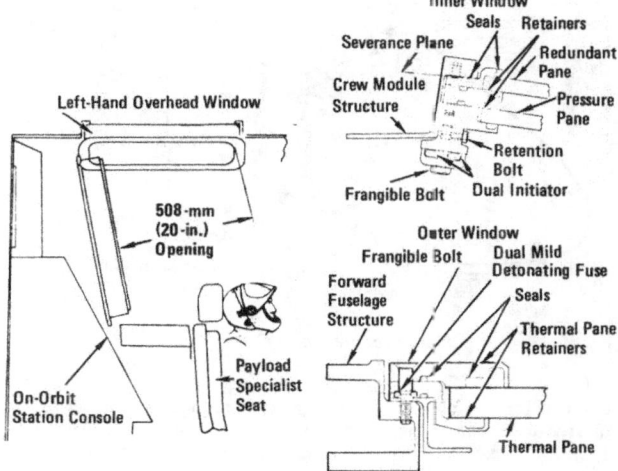

Flight Deck Emergency Egress Window Exit

mid deck or externally. The pressure side hatch is assembled to the crew cabin tunnel through hinges, torque tube, and support fittings. The side hatch opens outwardly 90 degrees down with the orbiter horizontal or 90 degrees sideways with the orbiter vertical. The side hatch is 1,016 millimeters (40 inches) in diameter. It has a 254 millimeter-diameter (10-inch) clear-view window in the center of the hatch. The window consists of three panes of glass. The side hatch seal provides a pressure seal that is compressed by the side hatch latch mechanisms when the hatch is locked closed. A thermal barrier of Inconel wire mesh spring with a ceramic fiber braided sleeve is installed between the reusable surface insulation tiles on the forward fuselage and the side hatch. The total weight of the side hatch is 129 kilograms (294 pounds).

Sleep stations and a galley or a payload in lieu of galley can be installed in the mid deck; in addition, three seats can be installed on the mid deck floor of the same type as the specialist seats on the flight deck. Three additional seats can be installed on the mid deck for rescue missions if the sleep stations are removed.

The mid deck also provides a stowage volume of 3.96 cubic meters (140 cubic feet). Accommodations are included for dining, sleeping, maintenance, exercising, and data management. The floor contains removable panels that provide access to the ECLSS equipment and stowage beneath the floor in the equipment bay. On the orbiter centerline, just aft of the forward avionics equipment bay, an opening in the ceiling provides access to the inertial measurement units.

Stowage modular lockers are used to store the flight crew's personal gear, mission-necessary equipment, and experiments

Space Shuttle Spacecraft Structures

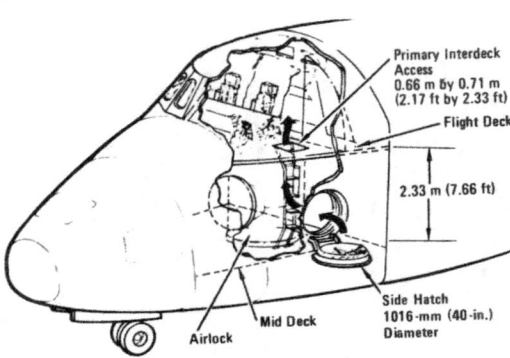

Normal Ingress and Egress

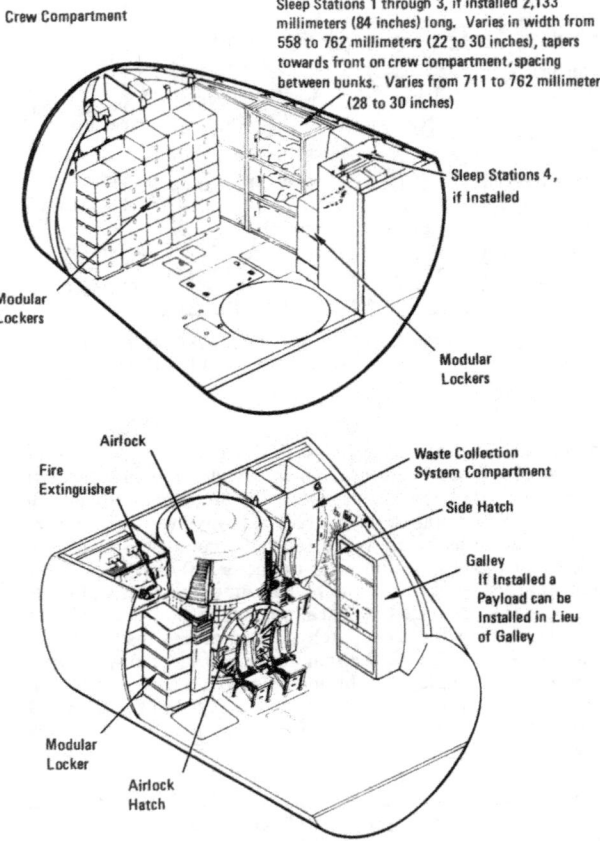

Crew Compartment Mid Deck

use sandwich panels of Kevlar/epoxy and nonmetallic core. This reduced the weight by 83 percent compared to all aluminum lockers. This is a reduction of approximately 68 kilograms (150 pounds). There are 42 identical boxes 279 by 457 by 533 millimeters (11 by 18 by 21 inches) which can be used for stowage.

An airlock is located in the mid deck. The airlock and airlock hatches permit extra vehicular activities (EVA) flight crew members to transfer from the mid deck crew compartment into the payload bay without depressurizing the orbiter crew cabin.

Normally, two extravehicular mobility units (EMU's) are stowed in the mid-deck. The EMU's are an integrated space suit assembly and life support system which provides the capability for the flight crew members to leave the pressurized orbiter crew cabin and work outside the cabin in space.

Space Shuttle Spacecraft Structures

Avionics Bays 1 and 2

Airlock Internal View

Airlock View From Mid Deck

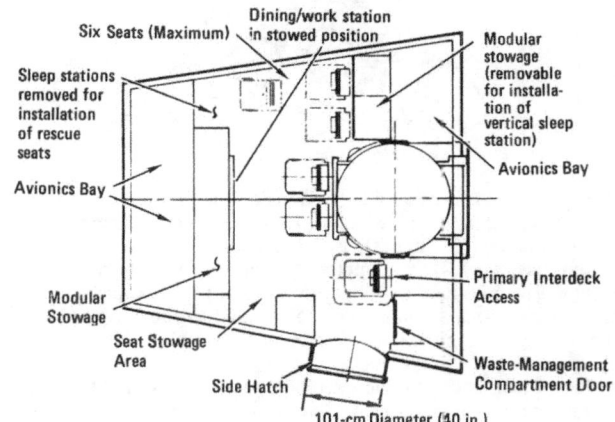
Mid-Deck-Rescue Configuration

 # Space Shuttle Spacecraft Structures

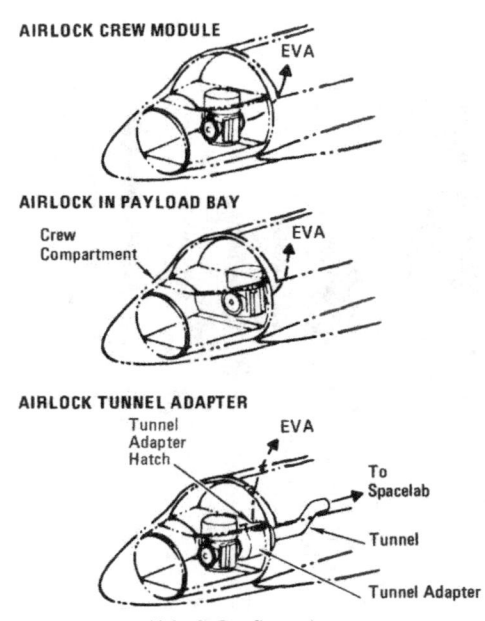

Airlock Configuration

The airlock is normally located inside the mid-deck of the spacecraft's pressurized crew cabin. It has an inside diameter of 1,600 millimeters (63 inches), is 2,108 millimeters (83 inches) long, and has two 1,016 millimeter (40 inch) diameter D-shaped openings, 914 millimeters (36 inches) across, plus two pressure sealing hatches and a complement of airlock support systems. The airlock volume is 4.24 cubic meters (150 cubic feet).

The airlock is sized to accommodate two fully suited flight crew members simultaneously. The airlock support provides airlock depressurization and repressurization, EVA equipment recharge, liquid cooled garment water cooling, EVA equipment checkout, donning and communications. All EVA gear, checkout panel, and recharge stations are located against the internal walls of the airlock.

The airlock hatches are mounted on the airlock. The inner hatch is mounted on the exterior of the airlock (orbiter crew cabin mid-deck side) and opens in the mid-deck. The inner hatch isolates the airlock from the orbiter crew cabin. The outer hatch is mounted in the interior of the airlock and opens in the airlock. The outer hatch isolates the airlock from the unpressurized payload bay when closed and permits the EVA crew members to exit from the airlock to the payload bay when open.

Airlock repressurization is controllable from inside the orbiter crew cabin mid-deck and from inside the airlock. It is performed by equalizing the airlock and cabin pressure with airlock hatch-mounted equalization valves mounted on the inner hatch. Depressurization of the airlock is controlled from inside the airlock. The airlock is depressurized by venting the airlock pressure overboard. The two D-shaped airlock hatches are installed to open toward the primary pressure source, the orbiter crew cabin, to achieve pressure assist sealing when closed.

Each hatch has six interconnected latches with a gearbox/actuator, a window, a hinge mechanism and hold-open device, a differential pressure gage on each side, and two equalization valves.

The window in each airlock hatch is 101 millimeters (4 inches) in diameter. The window is used for crew observation from the cabin/airlock and the airlock/payload bay. The dual window panes are made of polycarbonate plastic and mounted directly to the hatch using bolts fastened through the panes. Each hatch window has dual pressure seals with seal grooves located in the hatch.

Space Shuttle Spacecraft Structures

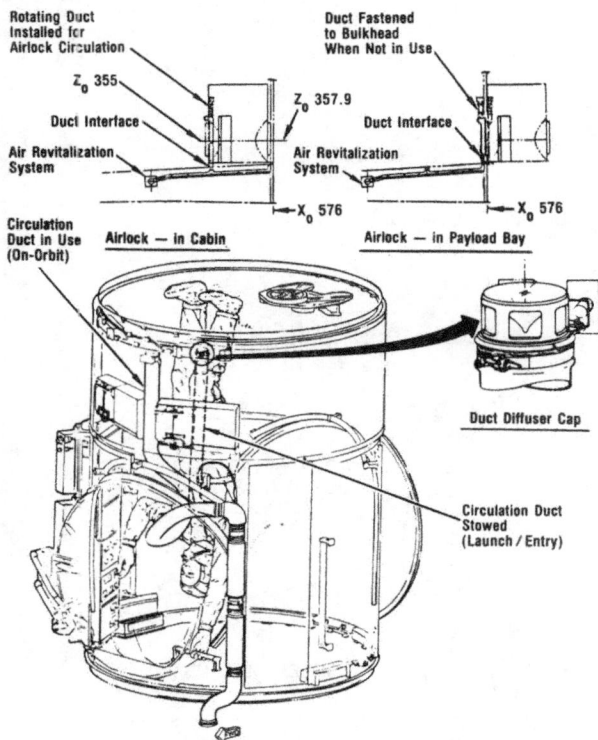

Airlock in Mid Deck Without Tunnel Adapter

Each airlock hatch has dual pressure seals to maintain pressure integrity for the airlock. One seal is mounted on the airlock hatch and the other on the airlock structure. A leak check quick disconnect is installed between the hatch and the airlock pressure seals to verify hatch pressure integrity prior to the flight.

The gearbox with latch mechanism on each hatch allows the flight crew to open and/or close the hatch during transfers and EVA operation. The gearbox and the latches are mounted on the low pressure side of each hatch, with a gearbox handle installed on both sides to permit operation from either side of the hatch.

Three of the six latches on each hatch are double acting. They have cam surfaces which force the sealing surfaces apart when the latches are opened, thereby acting as crew assist devices. The latches are interconnected with "push-pull" rods and an idler bellcrank installed between the rods for pivoting the rods. Self-aligning dual rotating bearings are used on the rods for attachment to the bellcranks and the latches. The gearbox and hatch open support struts are also connected to the latching system, using the same rod/bellcrank and bearing system. To latch or unlatch the hatch, a rotation of 440 degrees on the gearbox handle is required.

The hatch actuator/gearbox is used to provide the mechanical advantage to open/close the latches. The hatch actuator lock lever requires a force of 35 to 44 Newtons (8 to 10 pounds) through an angle of 180 degrees to unlatch the actuator. A rotation of 440 degrees minimum with a force of 133 Newtons (30 pounds) maximum applied to the actuator handle is required to operate the latches to their fully unlatched positions.

The hinge mechanism for each hatch permits a minimum opening sweep into the airlock or the crew cabin mid-deck. The inner hatch (airlock to crew cabin) is pulled/pushed forward to the crew cabin approximately 152 millimeters (6 inches). The hatch pivots up and to the starboard (right) side. Positive locks are provided to hold the hatch in both an intermediate and a full

Space Shuttle Spacecraft Structures

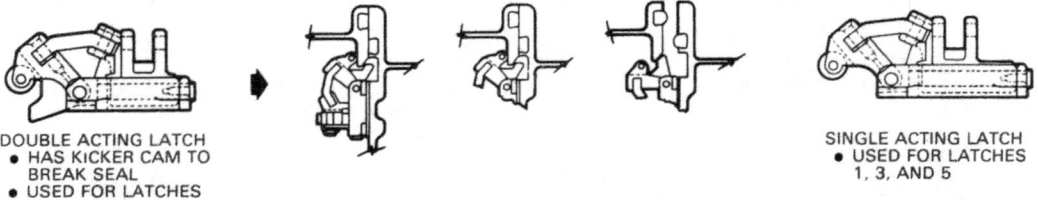

DOUBLE ACTING LATCH
- HAS KICKER CAM TO BREAK SEAL
- USED FOR LATCHES 2, 4, AND 6

SINGLE ACTING LATCH
- USED FOR LATCHES 1, 3, AND 5

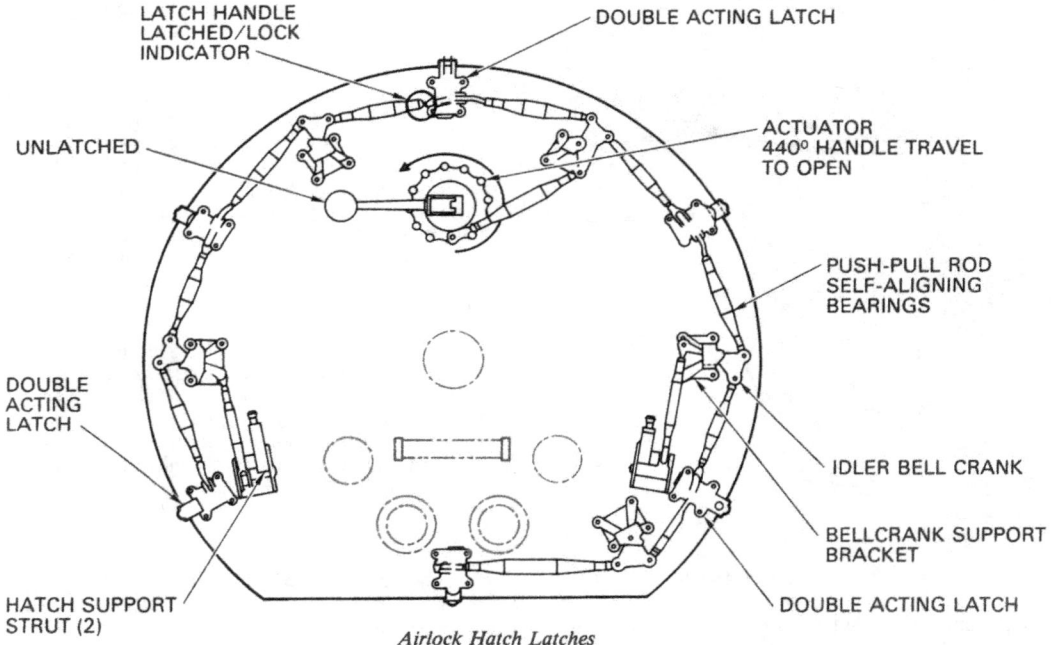

Airlock Hatch Latches

Space Shuttle Spacecraft Structures

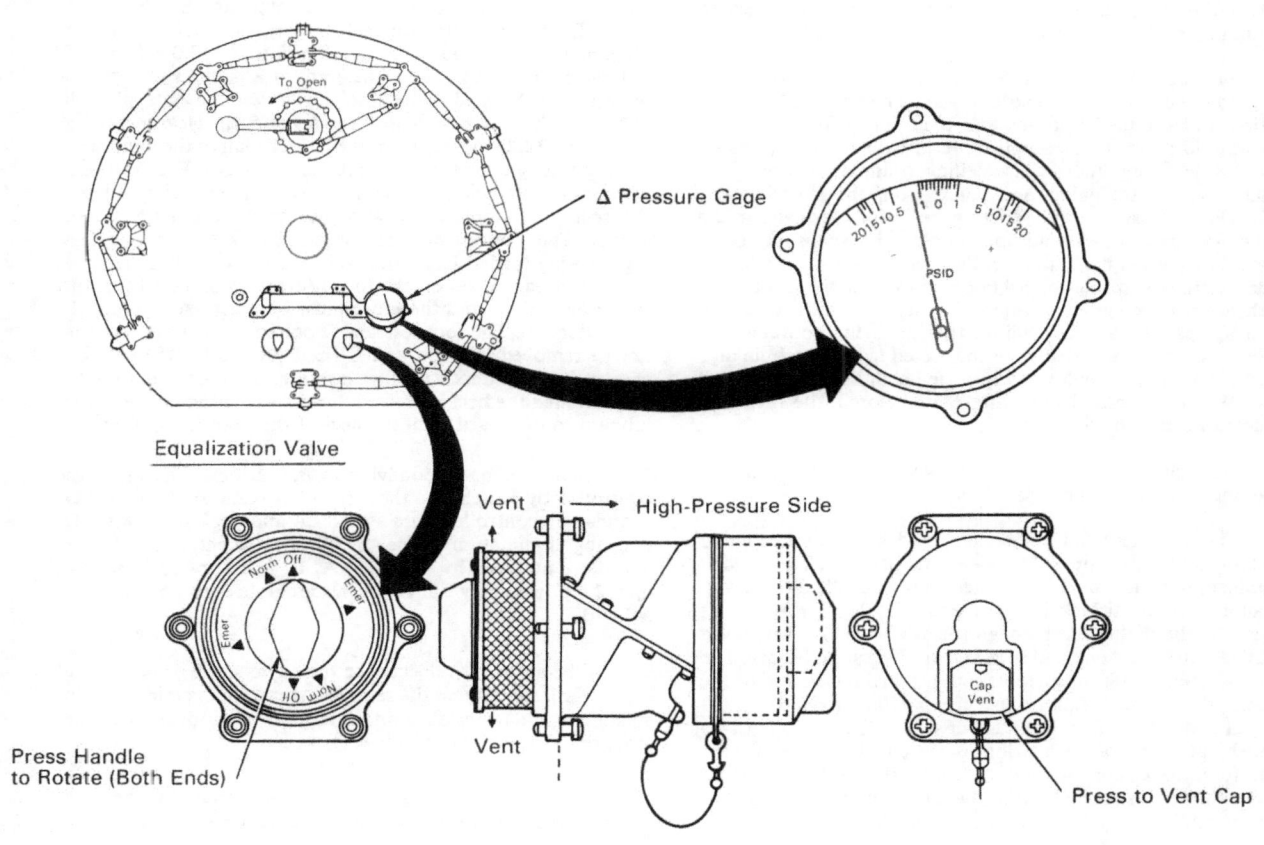

Airlock Repressurization

Space Shuttle Spacecraft Structures

open position. To release the lock, a spring-loaded handle is provided on the latch hold-open bracket. Friction is also provided in the linkage to prevent the hatch from moving if released during any part of the swing.

The outer hatch (in airlock to payload bay) opens and closes to the contour of the airlock wall. The hatch is hinged to be first pulled into the airlock and then pulled forward at the bottom and rotated down until it rests with the low pressure (outer) side facing the airlock ceiling (mid-deck floor). The linkage mechanism guides the hatch from the closed/open, open/closed position with friction restraint throughout the stroke. The hatch has a hold-open hook which snaps into place over a flange when the hatch is fully open. The hook is released by depressing the spring-loaded hook handle and by pushing the hatch toward the closed position. To support and protect the hatch against the airlock ceiling, the hatch incorporates two deployable struts. The struts are connected to the hatch linkage mechanism and are deployed when the hatch linkage is rotated open. When the hatch latches are rotated closed, the struts are retracted against the hatch.

The airlock hatches can be removed in-flight from the hinge mechanism via pip pins, if required.

Airlock air circulation system provides conditioned air to the airlock during non-EVA operation periods. The airlock revitalization system duct is attached to the outside airlock wall at launch. Upon airlock hatch opening in-flight, the duct is rotated by the flight crew through the cabin/airlock hatch and installed into the airlock and held in place by a strap holder. The duct has a removable air diffuser cap installed on the end of the flexible duct which can adjust the airflow from 0 to 97 kilograms per hour (216 pounds per hour). The duct must be rotated out of the airlock prior to closing the cabin/airlock hatch for airlock depressurization. During the EVA preparation period, the duct is rotated out of the airlock and can be used as supplemental air circulation in the mid-deck.

To assist the crew member in pre- and post-EVA operations, the airlock incorporates handrails and foot restraints. Handrails are located alongside the avionics and ECLSS panels. A handhold is mounted on each side of the hatches. They are aluminum alloy and oval configurations 19.05 by 33.52 millimeters (0.75 by 1.32 inches) and are painted yellow. The handrails are bonded to the airlock walls with an epoxyphenolic adhesive. Each handrail provides a handgrip clearance of 57 millimeters (2.25 inches) from the airlock wall to the handrail to allow gripping operations in a pressurized glove. Foot restraints are installed on the airlock floor nearer the payload bay side and the ceiling handhold is installed nearer the cabin side of the airlock. The foot restraints can be rotated 360 degrees by releasing a spring-loaded latch and will lock in every 90 degrees. A rotation release knob on the foot restraint is designed for shirt sleeve operation, and therefore must be positioned before the suit is donned. The foot restraint is bolted to the floor and cannot be removed in flight and is sized for the EMU boot. The crew member ingresses by first inserting the foot under the toe bar and then the heel is pressed down by rotating the heel from inboard to outboard until the heel of the boot is captured.

There are four floodlights in the airlock. The lights are controlled by switches in the airlock on panel AW18A; light 2 can also be controlled by a switch on mid-deck panel M013Q, allowing illumination of the airlock prior to entry. Lights 1, 3, and 4 are powered by buses MNA, B, and C respectively and light 2 is powered by ESS1BC. The circuit breakers are on panel ML86B.

If the airlock is relocated to the payload bay from the mid deck, it will function in the same manner as in the mid deck. Insulation would be installed on the airlock exterior for protection from the extreme temperatures of space.

For Spacelab missions, the airlock remains in the crew compartment mid deck and a tunnel adapter is installed in the

Space Shuttle Spacecraft Structures

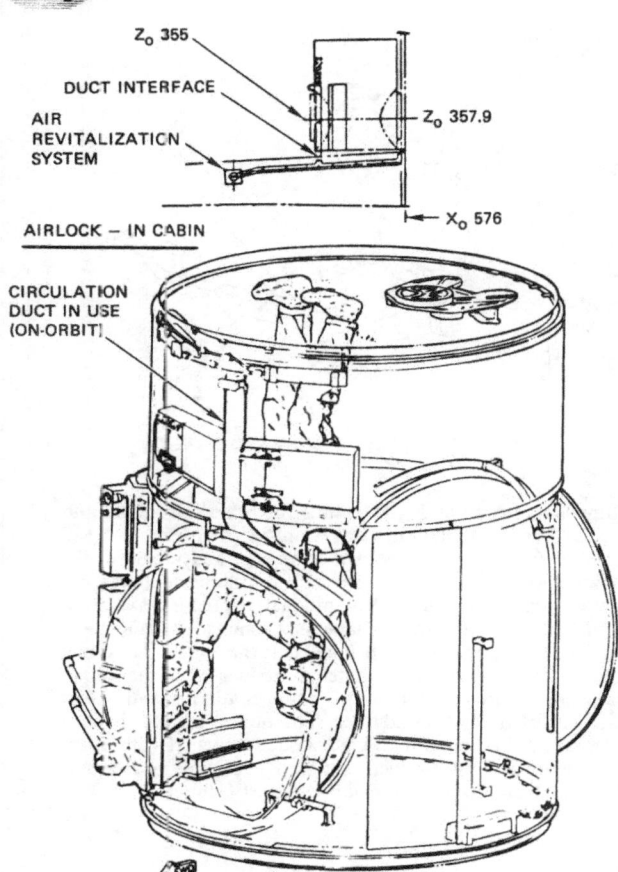

Airlock With Tunnel Adapter for Spacelab

Tunnel Adapter

payload bay which mates with the airlock and the Spacelab tunnel.

The airlock, tunnel adapter, hatches, tunnel extension and tunnel permits the flight crew members to transfer from the spacecraft pressurized mid deck crew compartment into Spacelab in a pressurized shirt sleeve environment.

In addition, the airlock, tunnel adapter and hatches permit the EVA flight crew members to transfer from the airlock/tunnel adapter in the space suit assembly into the payload bay without depressurizing the spacecraft crew cabin and Spacelab.

The tunnel adapter is located in the payload bay and is attached to the airlock at orbiter station X_o 576 and attached to

Space Shuttle Spacecraft Structures

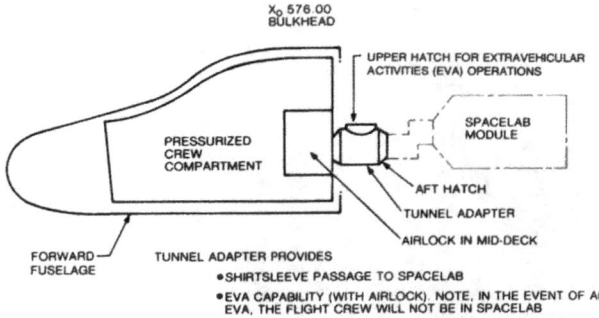

Airlock/Tunnel Adapter

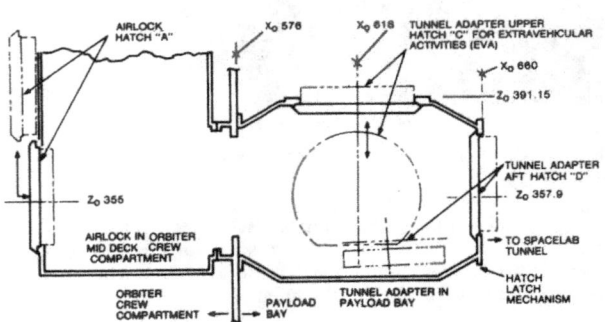

Airlock and Tunnel Adapter Hatch Mechanical Systems

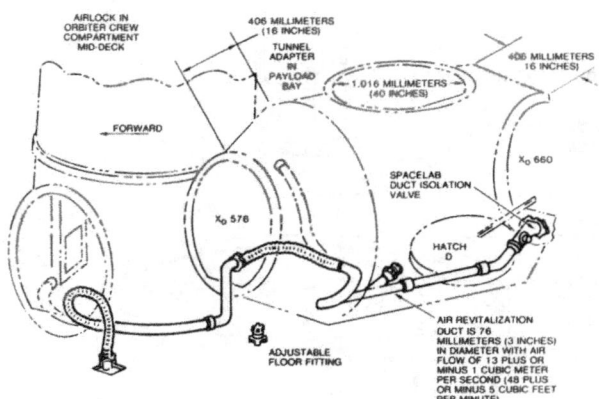

Environmental Control Life Support System (ECLSS) Air Circulation Duct Routing

the tunnel extension at X_O 660, thus the Spacelab tunnel and Spacelab. The tunnel adapter has an inside diameter of 1,600 millimeters (63 inches) at the widest section and tapers in the cone area at each end, to two 1,016 millimeter (40 inch) diameter D-shaped openings, 914 millimeters (36 inches) across. A 1,016 millimeter (40 inch) diameter D-shaped opening, 914 millimeters (36 inches) across is located at the top of the tunnel adapter. Two pressure sealing hatches are located in the tunnel adapter, one at the upper area of the tunnel adapter and one at the aft end of the tunnel adapter. The tunnel adapter is constructed of 2219 aluminum and is a welded structure with 60 by 60 millimeter (2.4 by 2.4 inch) exposed structural ribs on exterior surface and an external waffle skin stiffening.

Space Shuttle Spacecraft Structures

The hatch located on the mid-deck side of the airlock is mounted on the exterior of the airlock and opens into the mid-deck. This hatch isolates the airlock from the spacecraft crew cabin. The hatch located in the tunnel adapter aft end isolates the tunnel adapter/airlock from the tunnel extension, tunnel and Spacelab. This hatch opens into the tunnel adapter. The hatch located in the tunnel adapter at the upper D-shaped opening isolates the airlock/tunnel adapter from the unpressurized payload bay when closed and permits the EVA crew members to exit from the airlock/tunnel adapter to the payload bay when open. This hatch opens into the tunnel adapter.

Airlock repressurization is controllable from inside the orbiter crew cabin mid-deck and from inside the airlock. It is performed by equalizing the airlock and cabin pressure with airlock hatch-mounted equalization valves mounted on the inner hatch. Depressurization of the airlock is controlled from inside the airlock. The airlock is depressurized by venting the airlock pressure overboard. The airlock hatch is installed to open toward the primary pressure source, the orbiter crew cabin, to achieve pressure assist sealing when closed. The two hatches in the tunnel adapter are also installed to open toward the primary pressure source, the orbiter crew cabin, to achieve pressure assist sealing when closed.

Each hatch has six interconnected latches (with the exception of the aft hatch which has 17) with a gearbox/actuator, a window, a hinge mechanism and hold-open device, a differential pressure gage on each side, and two equalization valves.

The window in each hatch is 101 millimeters (4 inches) in diameter. The window is used for crew observation from the cabin/airlock, tunnel adapter to tunnel, and tunnel adapter to payload bays. The dual window panes are made of polycarbonate plastic and mounted directly to the hatch using bolts fastened through the panes. Each hatch window has dual pressure seals with seal grooves located in the hatch.

Each hatch has dual pressure seals to maintain pressure integrity. One seal is mounted on the hatch and the other on the structure. A leak check quick disconnect is installed between the hatch and the pressure seals to verify hatch pressure integrity prior to flight.

The gearbox with latch mechanism on each hatch allows the flight crew to open and/or close the hatch during transfers and EVA operations. The gearbox and the latches are mounted on the low pressure side of each hatch, with a gearbox handle installed on both sides to permit operation from either side of the hatch.

Three of the six latches on each hatch are double acting (with the exception of the aft hatch which has two). They have cam surfaces which force the sealing surfaces apart when the latches are opened, thereby acting as crew assist devices. The latches are interconnected with "push-pull" rods and an idler bellcrank installed between the rods for pivoting the rods. Self-aligning dual rotating bearings are used on the rods for attachment to the bellcranks and the latches. The gearbox and hatch open support struts are also connected to the latching system, using the same rod/bellcrank and bearing system. To latch or unlatch the hatch, a rotation of 440 degrees on the gearbox handle is required.

The hatch actuator/gearbox is used to provide the mechanical advantage to open/close the latches. The hatch actuator lock lever requires a force of 35 to 44 Newtons (8 to 10 pounds) through an angle of 180 degrees to unlatch the actuator. A rotation of 440 degrees minimum with a force of 133 Newtons (30 pounds) maximum applied to the actuator handle is required to operate the latches to their fully unlatched positions.

The hinge mechanism for each hatch permits a minimum opening sweep into the tunnel adapter or the spacecraft crew

Space Shuttle Spacecraft Structures

cabin mid-deck. The airlock crew cabin hatch in the mid deck is pulled/pushed forward to the mid deck approximately 152 millimeters (6 inches). The hatch pivots up and to the starboard (right) side. Positive locks are provided to hold the latch in both an intermediate and a full open position. To release the lock, a spring-loaded handle is provided on the latch hold-open bracket. Friction is also provided in the linkage to prevent the hatch from moving if released during any part of the swing.

The aft hatch is hinged to be first pulled into the tunnel adapter and then pulled forward at the bottom. The top of the hatch is rotated towards the tunnel and downward until the hatch rests with the Spacelab side facing the tunnel adapter floor. The linkage mechanism guides the hatch from the closed/open, open/closed position with friction restraint throughout the stroke. The hatch is held in the open position by straps and velcro.

The upper (EVA) hatch in the tunnel adapter opens and closes to the port (left) wall of the tunnel adapter. The hatch is hinged to be first pulled into the tunnel adapter and then pulled forward at the hinge area and rotated down until it rests against the port wall of the tunnel adapter. The linkage mechanism guides the hatch from the closed/open, open/closed position with friction restraint throughout the stroke. The hatch is held in the open position by straps and velcro.

The hatches can be removed in flight from the hinge mechanisms via pip pins, if required.

The equipment bay houses the major components of the waste management and atmospheric revitalization systems such as pumps, fans, lithium hydroxide (LiOH) absorbers, heat exchangers, and ducting. This compartment provides stowage space for LiOH canisters and five separate spaces for crew equipment stowage with a volume of 0.8 cubic meters (29.92 cubic feet).

The crew compartment and stowage modular lockers are built by Rockwell's Space Transportation and Systems Group, Shuttle Orbiter Division, Downey, CA. The crew seat contractor is AMI, Colorado Springs, CO.

FORWARD FUSELAGE AND CREW COMPARTMENT WINDOWS

The orbiter windows provide visibility for entry, landing, and on-orbit operations. For atmospheric flight, the flight crew needs forward, left, and right viewing areas. On-orbit mission phases require visibility for rendezvous, docking, and payload handling.

The six windows located at the forward flight deck commander and pilot stations provide the forward, left, and right viewing. The two overhead windows and two payload viewing windows at the aft station location on the flight deck provide rendezvous, docking, and payload viewing. A window in the mid deck side hatch is for viewing by a crew person from that position.

The six planform-shaped forward windows are the largest pieces of glass ever produced in the optical quality for see-through viewing. Each consists of three individual panes. The innermost pane is constructed of tempered aluminosilicate glass to withstand the crew compartment pressure. It is 15.7 millimeters (5/8 of an inch) thick. Aluminosilicate glass is a low expansion glass that can be tempered to provide maximum mechanical strength. The exterior of this pane, called a pressure pane, is coated with a red reflector coating to reflect the infrared (heat portion) rays while transmitting the visible spectrum.

The center pane is constructed of low expansion fused silica glass due to its high optical quality and excellent thermal shock resistance. This pane is 33 millimeters (1.3 inches) thick.

Space Shuttle Spacecraft Structures

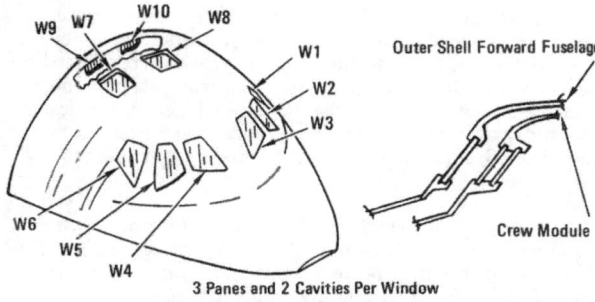

Flight Deck Windows

The exterior and interior are coated with a high efficiency, anti-reflection coating to improve visible light transmission. This pane is redundant to the pressure pane. These windows withstand a proof pressure of 445,050 mmHg (8,600 psi) at 115 °C (240 °F) and 0.017 relative humidity. This pane is also a redundant pane to the outer thermal pane.

The outer pane's exterior is of the same material as the center pane and is 15.7 millimeters (5/8 of an inch) thick. The exterior is uncoated, but the interior is coated with the high efficiency anti-reflective coating. The outer surface withstands approximately 482 °C (900 °F) and the interior surface withstands approximately 426 °C (800 °F).

Each of the forward six windows' outer panes measures 1,066.8 millimeters (42 inches) diagonally, and the center and inner panes each measure 889 millimeters (35 inches) diagonally. The outer panes of the forward six windows are mounted and attached to the forward fuselage. The center and inner panes are mounted and attached to the crew compartment. Redundant seals are employed for each window. No sealing/bonding compounds are used.

The two overhead windows at the flight deck aft station are identical in construction to the six forward windows except for thickness. The inner and center panes are 11.4 millimeters (0.45 of an inch) thick, and the outer pane is 17.2 millimeters (0.68 of an inch) thick. The outer pane is attached and mounted to the forward fuselage, and the center and inner panes are mounted and attached to the crew compartment. The two overhead window's clear view area is 508 by 508 millimeters (20 by 20 inches). The left-hand overhead window will provide the crew members with an emergency egress. Three egress steps are provided at the aft flight station for access to the overhead window. The inner and center panes will open into the crew cabin, and the outer pane will open up and over the top of the orbiter. This provides an emergency exit area of 508 by 508 millimeters (20 by 20 inches).

On the aft flight deck, each of the two windows for viewing the payload bay consists of only two panes of glass identical to the six forward windows' inner and center pane. The outer thermal pane is not installed. Each pane is 7.6 millimeters (0.30 of an inch) thick. The windows are 370 by 279 millimeters (14-1/2 by 11 inches). Both panes are attached and mounted to the crew compartment.

The side hatch viewing window consists of three panes of glass identical to the six forward windows. The inner pane is 289.5 millimeters (11.4 inches) in diameter and 6.35 millimeters (1/4 of an inch) thick. The center pane is 289.5 millimeters (11.4 inches) in diameter and 12.7 millimeters (1/2 inch) thick. The outer pane is 381 millimeters (15 inches) in diameter and 7.62 millimeters (3/10 of an inch) thick.

Each window has shade/filter covers to reduce the light entering the cabin. The shade/filters are stowed until required. Attachment mechanisms and devices are provided for the installation at each window on the flight deck and the mid deck side hatch window.

Space Shuttle Spacecraft Structures

Contractor for the windows is Corning Glass Co., Corning, NY.

WING

The wing is an aerodynamic lifting surface that provides conventional lift and control for the orbiter. The left and right wings consist of the wing glove, the intermediate section which includes the main landing gear well, the torque box, the forward spar for the mounting of the reusable reinforced carbon-carbon leading edge structure thermal protection system, the wing/elevon interface, the elevon seal panels, and the elevons.

The wing is constructed of conventional aluminum alloy with a multi-rib and spar arrangement with skin-stringer-stiffened covers or honeycomb skin covers. Each wing is approximately 18.28 meters (60 feet) long at the fuselage intersection, with a maximum thickness of 1.52 meters (5 feet).

The forward wing box is an extension of the basic wing that aerodynamically blends the wing leading edge into the mid fuselage wing glove. The forward wing box is of a conventional design of aluminum multi-ribs, aluminum tubes, and tubular struts. The upper and lower wing skin panels are stiffened aluminum. The leading edge spar is of corrugated aluminum construction.

The intermediate wing section consists of the conventional aluminum multi-ribs and aluminum tubes. The upper and lower skin covers are of aluminum honeycomb. A portion of the lower wing surface skin panel is made up of the main landing gear door. The intermediate section houses the main landing gear compartment and reacts a portion of the main landing gear loads. A structural rib supports the outboard main landing gear door hinges and the main landing gear trunnion and drag link. The support for the inboard main landing gear trunnion and drag link attachment is provided by the mid fuselage. The main landing gear door is a conventional aluminum honeycomb configuration.

The four major spars are of corrugated aluminum to minimize thermal loads. The forward spar provides the attachment of the thermal protection system reusable reinforced carbon-carbon leading edge structure. The rear spar provides the attachment interfaces for the elevons, hinged upper seal panels, and associated hydraulic and electrical system components. The upper and lower wing skin panels are of aluminum stiffened skins.

The elevons provide orbiter flight control during atmospheric flight. The two-piece elevons are of conventional aluminum multi-rib and beam construction with aluminum honeycomb skins for compatibility with the acoustic environment and thermal interaction. The elevons are divided into two segments for each wing and each segment is supported by three hinges. Attachment to the flight control system hydraulic actuators is along the forward extremity of each elevon, and all hinge moments are reacted at these points. Each elevon travels 40 degrees up and 25 degrees down.

The transition area on the upper surface between the torque box and the movable elevon consists of a series of hinged panels which provide a closeout of the wing-to-elevon cavity. These panels are of Inconel honeycomb sandwich construction outboard of wing station Y_w 312.5 and of titanium honeycomb sandwich construction inboard of wing station Y_w 312.5. The upper leading edge of each elevon incorporates titanium rub strips. The rub strips are of titanium honeycomb construction and are not covered with the thermal protection system reusable surface insulation. The rub strips provide the sealing surface area for the elevon seal panels.

The exposed areas of the wings, main landing gear doors, and elevons are covered with the thermal protection system

Space Shuttle Spacecraft Structures

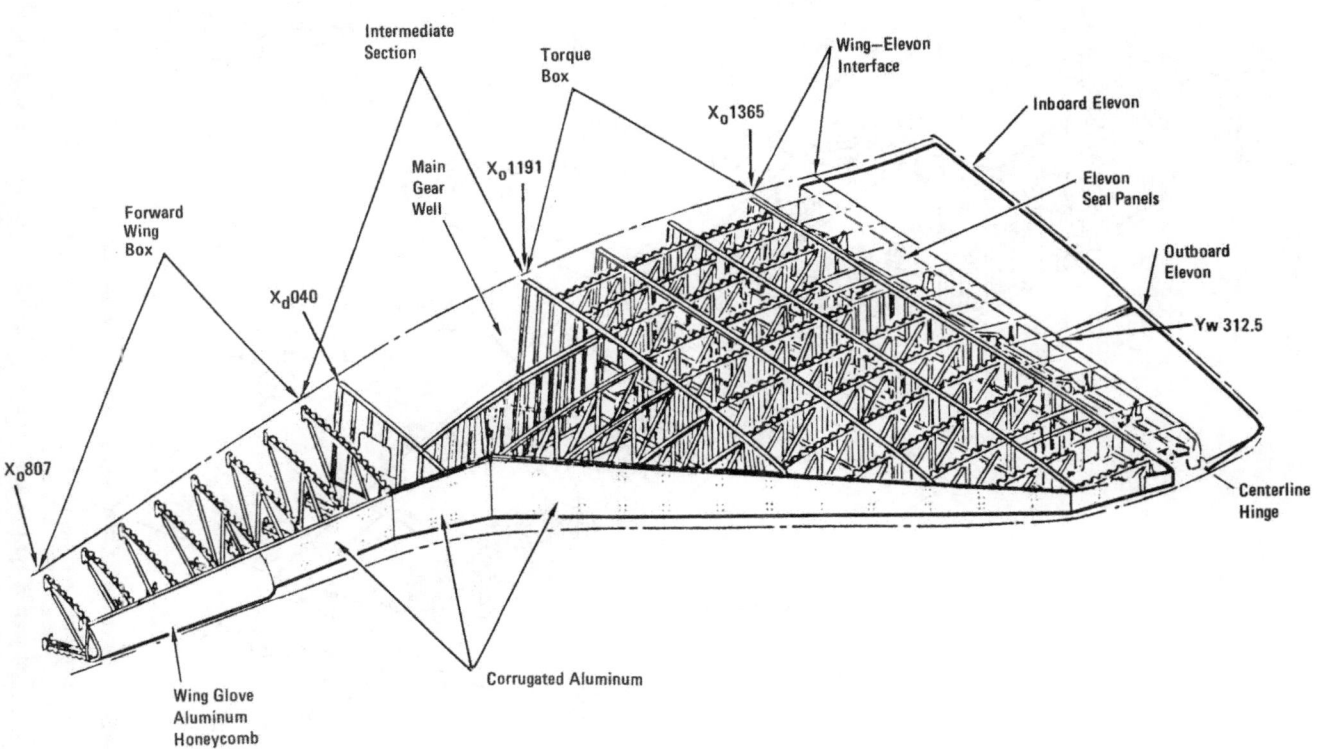

Wing Configuration

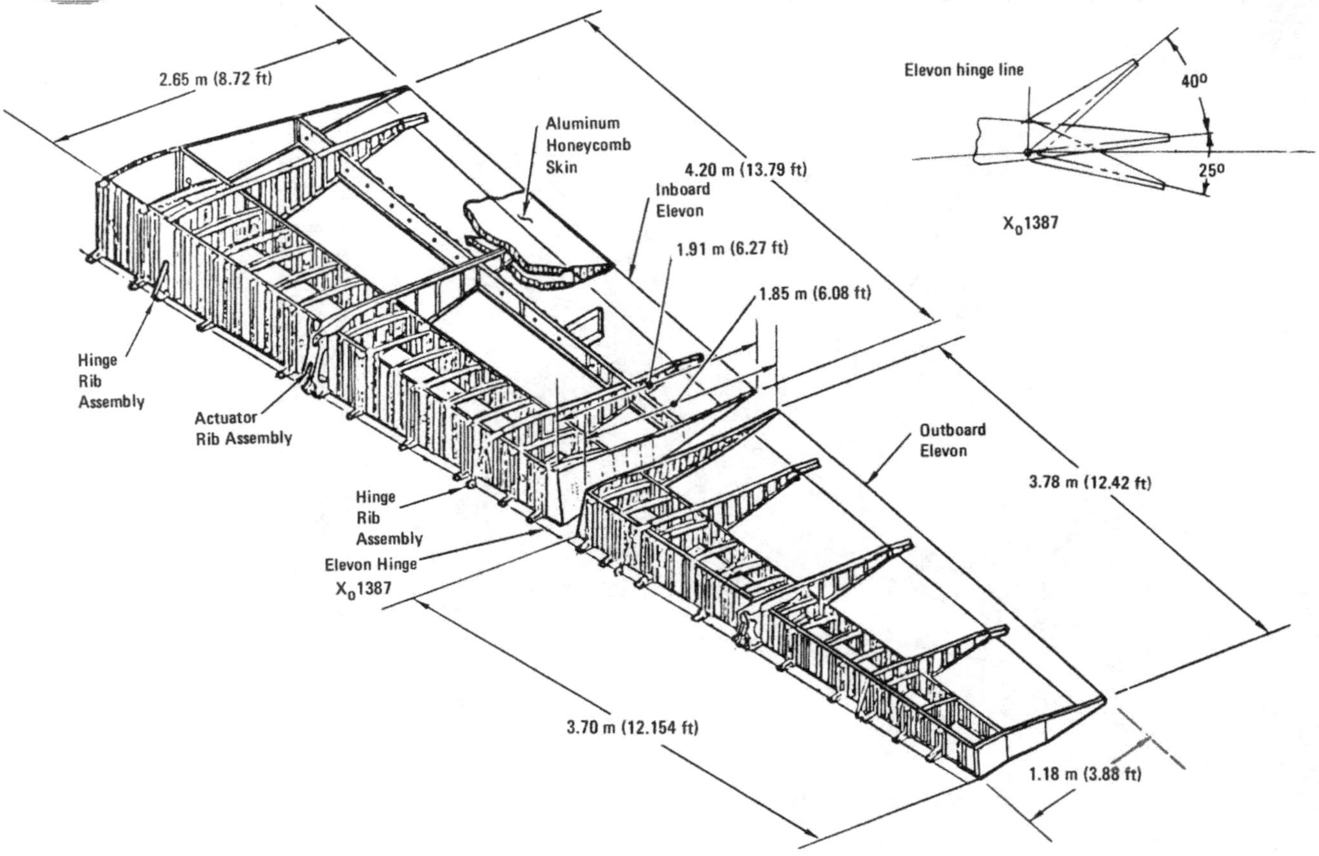

Space Shuttle Spacecraft Structures

Elevon Construction

Space Shuttle Spacecraft Structures

Left Wing

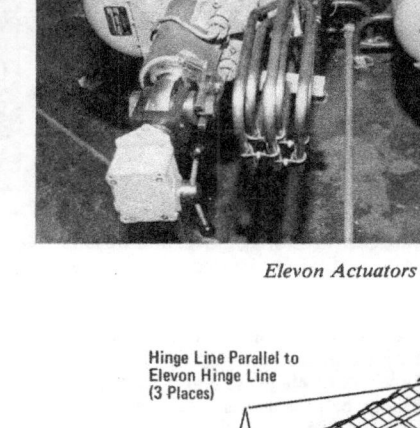

Elevon Actuators

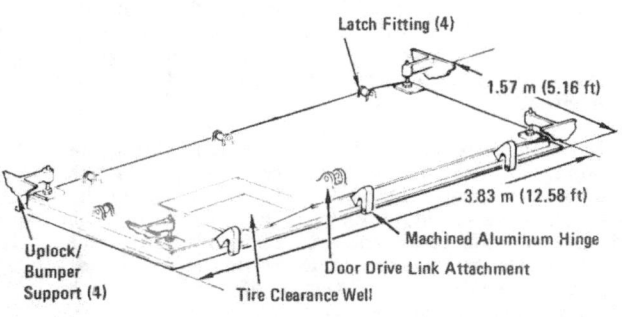

Main Landing Gear Door Construction

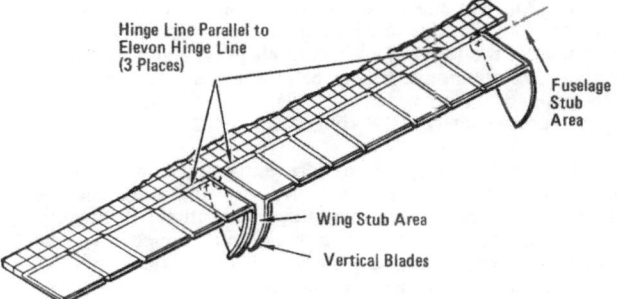

Wing Seal Panel Arrangement

Space Shuttle Spacecraft Structures

reusable surface insulation materials except for the elevon seal panels.

Thermal seals are provided on the elevon lower cove area along with thermal spring seals on the upper rub panels. Pressure seals and thermal barriers are provided on the main landing gear doors.

The wing is attached to the fuselage with a tension bolt splice along the upper surface. A shear splice along the lower surface in the area of the fuselage carry-through completes attachment interface.

The wing, elevons, and main landing gear door contractor is Grumman Corp., Bethpage, NY.

MID FUSELAGE

The mid fuselage structure interfaces with the forward fuselage, aft fuselage, and wings. It supports the payload bay doors, hinges, tiedown fittings, forward wing glove, and various orbiter system components, in addition to forming the payload bay area.

The forward and aft ends of the mid fuselage are open, with reinforced skin and longerons interfacing with the bulkheads of the forward and aft fuselage. The mid fuselage is primarily an aluminum structure 18.28 meters (60 feet) long, 5.18 meters (17 feet) wide, and 3.96 meters (13 feet) high. It weighs approximately 6,124 kilograms (13,502 pounds).

The mid fuselage skins are integrally machined by numerical control. The panels above the wing glove and the wings for the forward eight bays have longitudinal T stringers. The aft five bays have aluminum honeycomb panels. The side skins in the shadow of the wing are also numerically control machined but have vertical stiffeners.

Mid Fuselage

There are 12 main frame assemblies that stabilize the mid fuselage structure. The assemblies consist of vertical side elements and horizontal elements. The side elements are machined, whereas the horizontal elements are boron/aluminum tubes with bonded titanium end fittings, which substantially reduce weight. This reduced the weight by 49 percent. This is a reduction of approximately 138 kilograms (305 pounds).

In the upper portion of the mid fuselage are the sill and door longerons. The machined sill longerons not only are the primary body bending elements, but also take the longitudinal loads from payloads in the payload bay. Attached to the payload bay door longerons and associated structure are the 13 payload bay door hinges. These hinges provide the vertical reaction from the payload bay doors and five of the hinges react the payload bay door shears. The sill longeron also provides in the

# Space Shuttle Spacecraft Structures

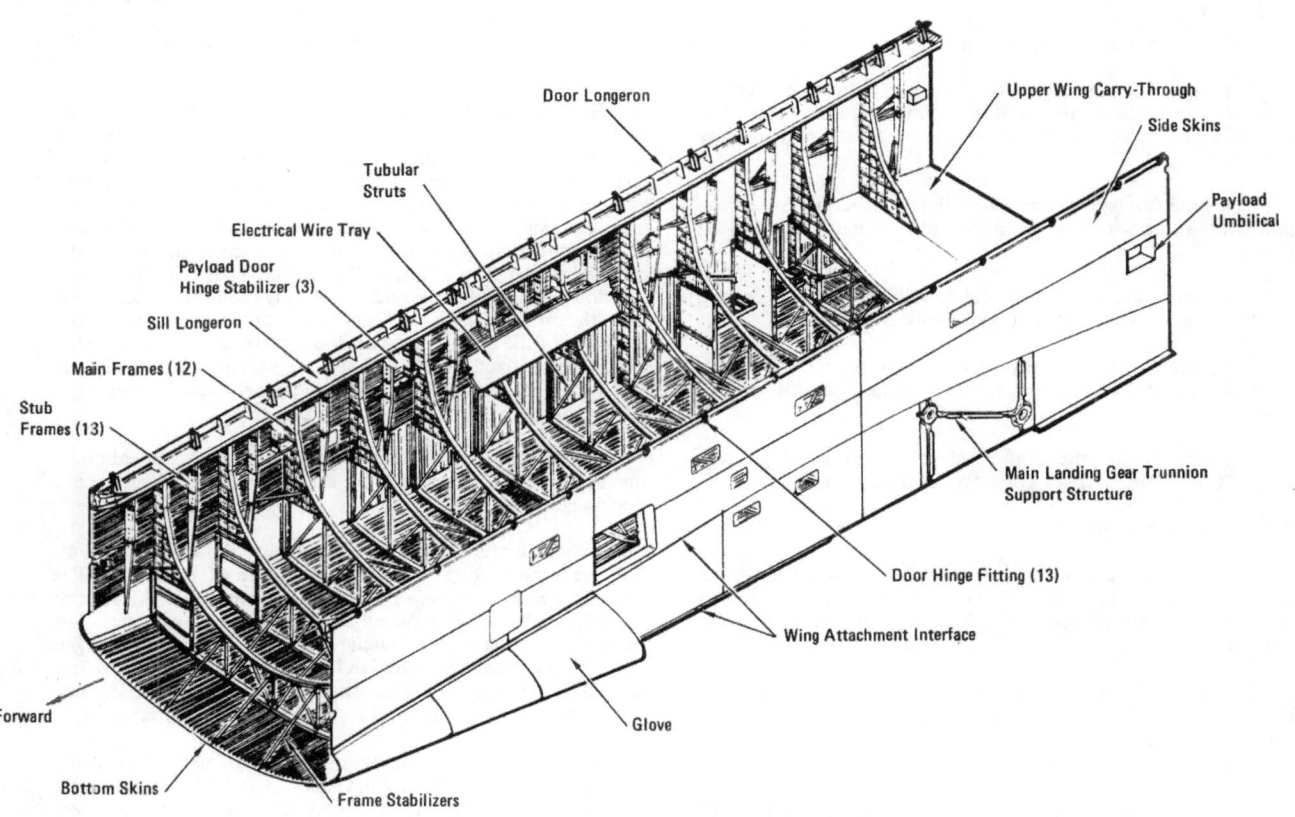

Mid Fuselage Structure

Space Shuttle Spacecraft Structures

base support for the payload bay manipulator arm or arms and its stowage provisions, the Ku-band rendezvous antenna and antenna base support and its stowage provisions, and the payload bay door actuation system.

The side wall forward of the wing carry-through structure provides the inboard support for the main landing gear. The total lateral landing gear loads are reacted by the mid fuselage structure.

The mid fuselage also supports the two electrical wire trays that contain the wiring between the crew compartment and aft fuselage.

Plumbing and wiring in the lower portion of the mid fuselage are supported by fiberglass milk stools.

The remainder of the exposed areas of the mid-fuselage is covered with the thermal protection system reusable surface insulation.

Contractor for the mid fuselage is General Dynamics Corp., Convair Aerospace Division, San Diego, CA.

PAYLOAD BAY DOORS

The payload bay doors consist of left- and right-hand doors hinged at each side of the mid fuselage and latched mechanically at the forward and aft fuselage and at the split top centerline.

Each door hinges on 13 Inconel 718 external hinges (five shear and eight idlers). The lower half of each hinge attaches to the mid fuselage sill longeron. The hinges rotate on bearings with dual rotational surfaces.

Each door actuation system provides the mechanism to drive each door side to the open or closed position. Each mechanism consists of an electromechanical power drive unit and six rotary gear actuators. The actuators are connected by torque tubes to each other, and to the power drive unit. Linkages transmit torque from the rotary actuators to the doors.

The forward 9.14-meter (30-foot) sections of the left- and right-hand doors incorporate radiators that can be deployed and are hinged and latched to the door inner surface which rejects the excess heat of the Freon-21 coolant loops from both sides of the radiator panels when the doors are open. An electromechanical actuation system on the door unlatches and deploys the radiators when open and latches and stows the radiators when closed. The radiators may be left in the stowed position for a given flight and would only radiate the excess heat from the one side. Fixed radiator panels are installed on the forward end of the aft payload bay doors and radiate from one side only. Kitted, fixed radiator panels may be installed on the aft end of the aft payload bay doors when required by a specific mission and also radiate from only one side.

During payload bay door closure, the crew optical alignment sight (COAS) is used at the aft flight deck station to check door alignment.

When closed, the doors are latched to the forward and aft bulkheads and along the upper centerline of the doors. The latching system consists of 16 bulkhead latches (eight aft and eight forward) and 16 payload bay door centerline latches. The forward and aft bulkhead latches are in groups of four ganged latch hooks. The centerline latches are also in groups of four ganged latches. Each gang incorporates four latches, bellcranks, pushrods, levers, rollers, and an electromechanical actuator.

The actuators, actuator output arm and active latch mechanisms are mounted on the forward and aft doors. Passive latch rollers, one for each payload bay door latch hook, are mounted on the forward and aft bulkheads. The four hooks in

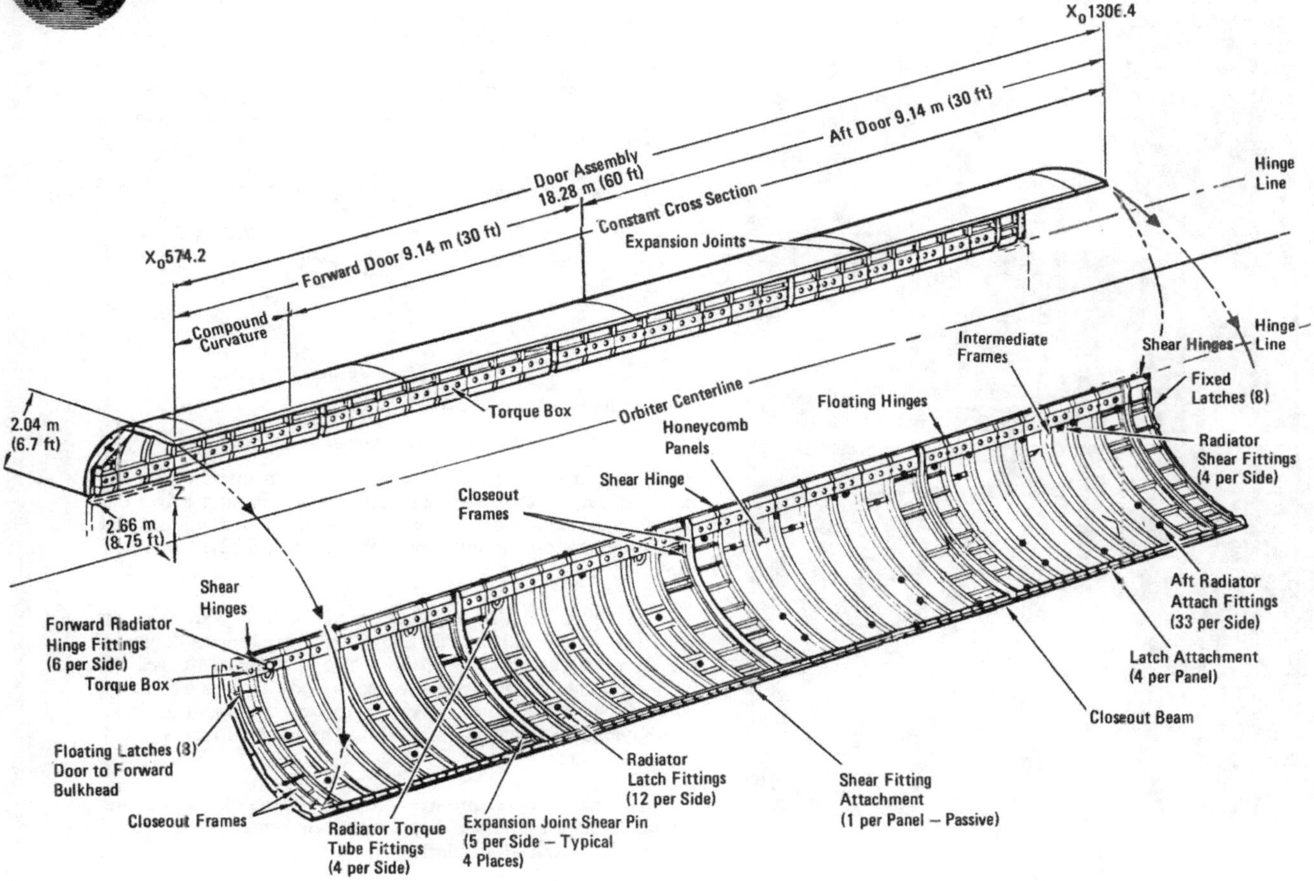

Space Shuttle Spacecraft Structures

Payload Bay Doors

Space Shuttle Spacecraft Structures

Payload Bay Door

- Payload Bay Door Drive System
  Opens/Closes Payload Bay Doors
- Payload Bay Door Latch System
  Locks Payload Bay Doors
- Radiator Deployment System
  Deploys/Retracts Radiators
- Radiator Latch System
  Locks/Supports Radiators

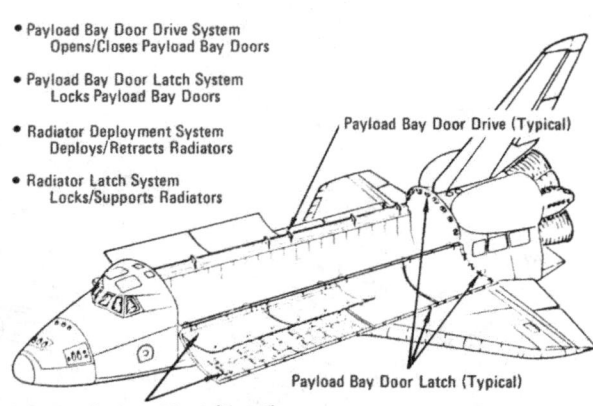

Payload Bay Doors and Radiators Mechanisms

each latch group are linked by a mechanism operated by the actuator output arm. During unlatching the actuator drives the output arm, which disengages all four latch hooks from the passive rollers on the bulkhead. The process is reversed when the doors close.

When the payload bay doors are closed, they are fixed at the aft fuselage bulkhead and allowed to move longitudinally at the forward fuselage. The doors also accommodate vehicle torsional loads (a force which causes a body such as a shaft to twist about its longitudinal axis), aerodynamic pressure loads, and payload bay vent lag pressures. The payload bay is not a pressurized area.

Thermal and pressure seals are used to close the gaps at the forward and aft fuselage interface, door centerline, and circumferential expansion joints.

Space Shuttle Spacecraft Structures

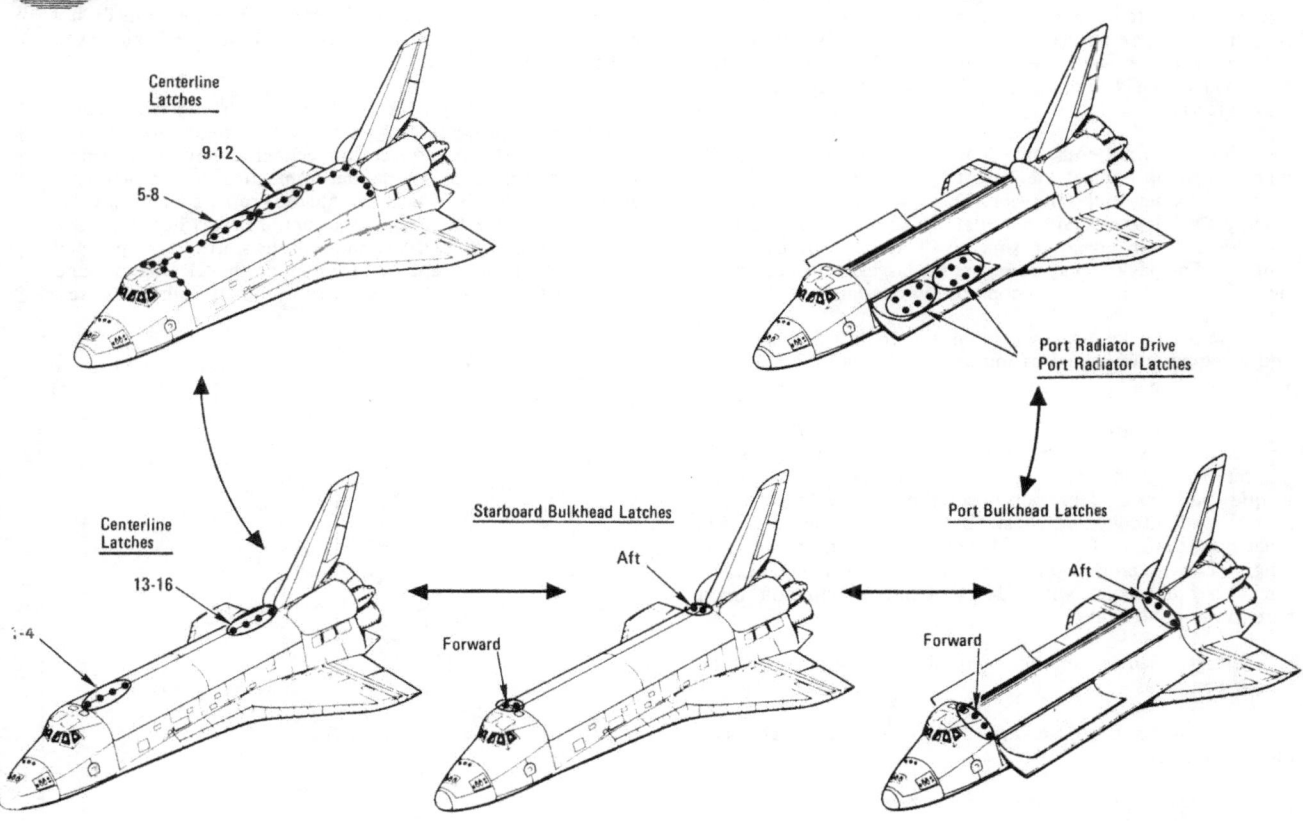

Payload Bay Doors and Radiator Latches

Space Shuttle Spacecraft Structures

The doors are 18.28 meters (60 feet) long. Each consists of five segments interconnected by expansion joints. The chord of each half of these curved doors is approximately 3.04 meters (10 feet) and the doors are 4.57 meters (15 feet) in diameter. The surface area is approximately 148.64 meters squared (1,600 square feet).

The doors are constructed of graphite/epoxy composite material, which reduces the weight by 23 percent over that of aluminum honeycomb sandwich. This is a reduction of approximately 408 kilograms (900 pounds), which brings the weight of the doors down to approximately 1,480 kilograms (3,264 pounds). The payload bay doors are the largest aerospace structure to be constructed from composite material.

The composite doors will withstand 163 dB acoustic noise and a temperature range of minus 112° to plus 57°C (minus 170° to plus 135°F).

The doors are made up of subassemblies consisting of graphite/epoxy honeycomb sandwich panel, solid graphite/epoxy laminate frames, expansion joint frames, torque box, seal depressor, centerline beam intercostals, gussets, end fittings, and clips. There are also aluminum 2024 shear pins, titanium fittings and Inconel 718 floating and shear hinges. The assembly is joined by mechanical fasteners. Lightning strike protection is provided by bonding aluminum mesh wire to the outer skin.

Extra-vehicular activity handholds are attached in the torque box areas.

The payload bay doors are covered with the reusable surface insulation.

The port (left-hand) door with attached systems weighs approximately 1,077 kilograms (2,375 pounds) and the starboard weighs about 1,149 kilograms (2,535 pounds). The starboard door contains the centerline latch active mechanisms, which accounts for the weight difference. These weights do not include the radiator panel system, which adds 377 kilograms (833 pounds) per door.

The payload bay door power and control system consists of the data processing system (DPS), midmotor control assemblies (MMCA) and electro-mechanical actuator assembly electrical motors. Auto or manual commands from the DPS are sent to the MMCA's, which activate the proper actuator motors to open or close latches and doors. Microswitch feedback signals are sent to the DPS to indicate the status of the payload bay door (PBD) latch and drive system on the CRT display and to operate the PBD status indicator on the aft flight deck crew

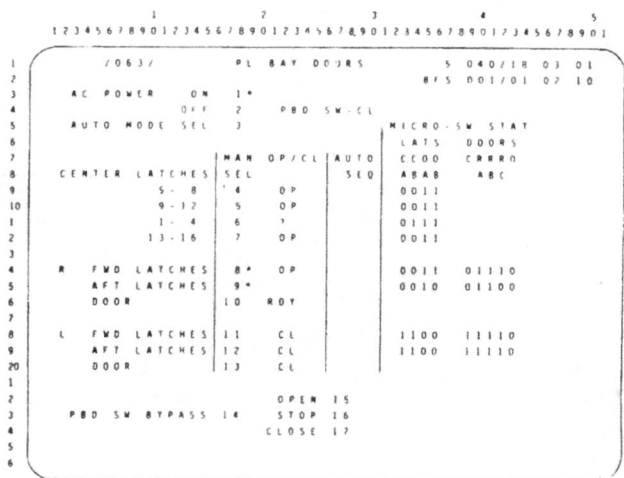

Payload Bay Doors Deployment and Closure System

Space Shuttle Spacecraft Structures

display and control panel. The status of the payload bay door limit switches is also displayed on the CRT.

The forward and aft bulkhead latches are in groups of four ganged latch hooks. Each group of hooks is opened or closed by an electro-mechanical actuator consisting of two redundant three phase ac reversible electric motors operated by the OPEN/CLOSE/STOP switch on Panel R13.

During latching, bulkhead switch module striker arms come into contact with the doors when they are nearly closed.

The ready-to-latch switches activate bulkhead latch electrical motors. When the payload bay doors are fully closed, switches on the forward and aft bulkheads turn off the payload bay door electrical drive motors.

The payload bay door bulkhead latch groups can be opened and closed automatically in a predetermined sequence or manually by individual latch groups through the computer keyboard. In the automatic mode, the forward and aft bulkhead latches operate simultaneously.

A PL (payload) BAY DOORS indicator on Panel R13 will show whether the payload bay doors are closed or open only when they are operated automatically. The indicator will remain in its original state when the manual mode is used.

Payload Bay Doors Ready to Latch Switch Module

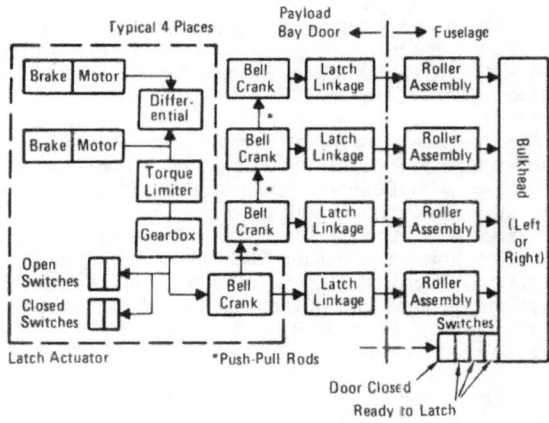

Payload Bay Doors Bulkhead Latches Mechanical Block Diagram

Space Shuttle Spacecraft Structures

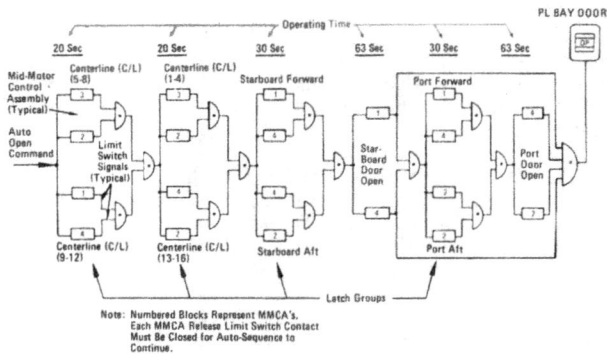

Payload Bay Doors Auto-Open Signal Flow Simplified

The payload bay door drive motors are automatically turned off if closing takes more than two times the normal 63 seconds. This prevents damage to the payload bay door drive system if a switch fails. A two-out-of-three voting logic of the ready-to-latch switches precludes premature start signals to the bulkhead latch drive motors.

The bulkhead latch drive motors are turned off automatically only if the operating time is more than twice the normal 30 seconds. If only one bulkhead electrical drive motor operates, 60 seconds are required to open or close the bulkhead latches. Each MMCA receives commands from the DPS and is turned on by limit switch closure. Each has its own timer set to twice the normal operating time. This allows enough time for single motor operation of a bulkhead latch group without causing a sequence fail signal PLB (payload bay) DOORS CRT message and SM ALERT.

All limit switch contact closures are sent to the DPS and are shown on the PBD (payload bay door) CRT display under MICRO-SW (switch) STAT (status). The flight crew can observe the change in the status of these switches. Microswitch status is also transmitted to telemetry.

Torque limiters in the bulkhead latch groups allow slippage if a limit switch fails to turn off an electrical drive motor or a mechanism jams during latch operations, thus preventing damage to the motors or mechanisms.

Extra vehicular activity (EVA) disconnect points in the bulkhead latch group mechanisms are provided in case the mechanism jams when the doors close. This permits crew members to close the doors manually from inside the payload bay.

The starboard payload bay doors must be opened first and closed last because of the arrangement of the centerline latch mechanism and the structural and seal overlap.

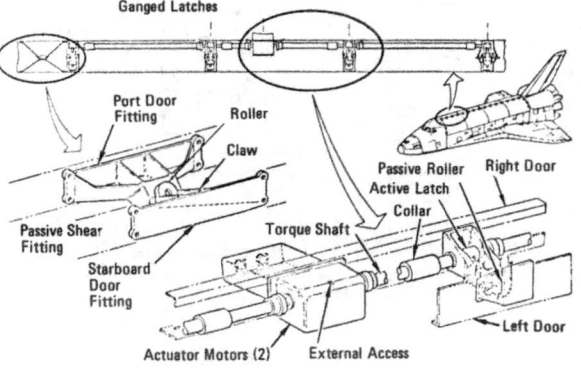

Payload Bay Door Upper Centerline Latch System

143

Space Shuttle Spacecraft Structures

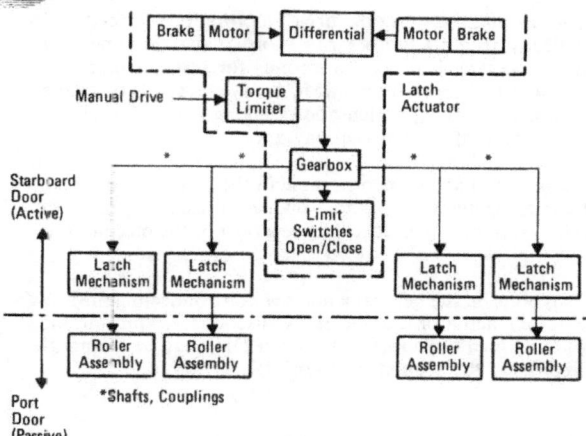

Payload Bay Doors Centerline Latches Mechanical Block Diagram

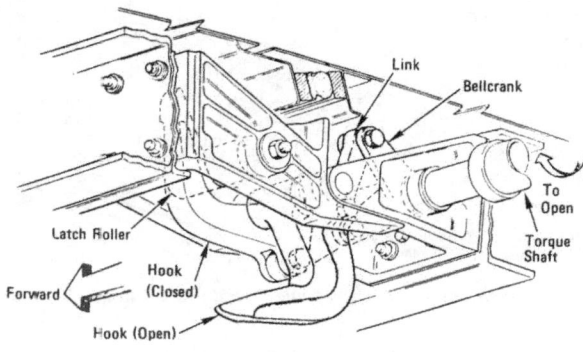

Centerline Door Latch Hook (typical) open Position

During centerline latching, the electrical drive motors turn the rotary shaft, bellcrank, and link, causing the hook to engage the passive roller. Alignment rollers on the starboard doors eliminate overlapping of the closed doors caused by thermal distortion. All 16 centerline hook assemblies contain the alignment rollers. The passive shear fittings in each centerline latch group align the closing doors and cause the fore and aft shear loads to react once the doors are closed.

The centerline latch groups can be opened and closed automatically in a predetermined sequence manually or by single individual latch groups through the computer keyboard.

The centerline latch drive motors are turned off automatically if they operate more than twice the normal 20 seconds. If only one motor operates, it takes 40 seconds to open or close the centerline latches. Each MMCA receives commands from the DPS and is turned on by limit switch closure, each has its own timer set to twice the normal operating time. This allows enough time for single-motor operation of a centerline latch group without causing a sequence fail signal and computer alert.

Torque limiters in the centerline latch groups allow slippage if limit switches fail to turn off an electrical drive motor or the mechanisms jam.

EVA disconnects in a centerline latch group can be used to isolate a jammed latch from the group.

The payload bay doors are driven by a rotary actuator consisting of two electrical three-phase reversible ac motors per power drive unit (PDU). There is one PDU for the starboard doors and one for the port doors.

The PDU drives a 16-meter (55-foot) long torque shaft. The shaft turns the rotary actuators, which causes the push rod, bellcrank, and link to push the doors open. The same arrangement pulls the doors closed. Limit switches on each drive system turn off drive motors when the doors are open.

Space Shuttle Spacecraft Structures

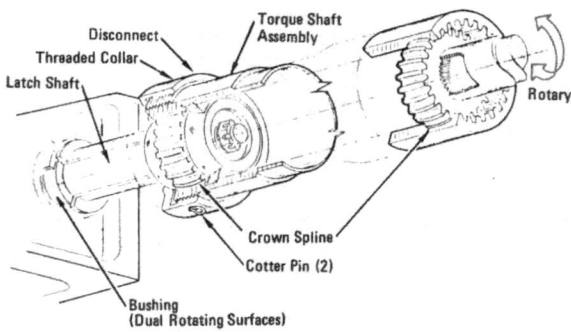

Extra-Vehicular Activity (EVA) Disconnect of Torque Shaft From Actuator and Latch Assembly

The payload bay door drive motors are turned off automatically if both motors run more than two times the normal 63 seconds. It takes 126 seconds for just one motor to open or close the doors. Each MMCA times is set to twice the normal operating time, which allows enough time for single-motor operation of the payload bay door.

Torque limiters are incorporated in the rotary actuators to avoid damaging the drive motors or mechanisms if limit switches fail to turn off an electrical drive motor or the mechanisms jam.

Two bolts on the bellcrank and the bolt connecting the link to the rotary actuator can be EVA disconnect points if the linkage fails when the doors close. The PDU's can be disengaged manually on the ground or on orbit.

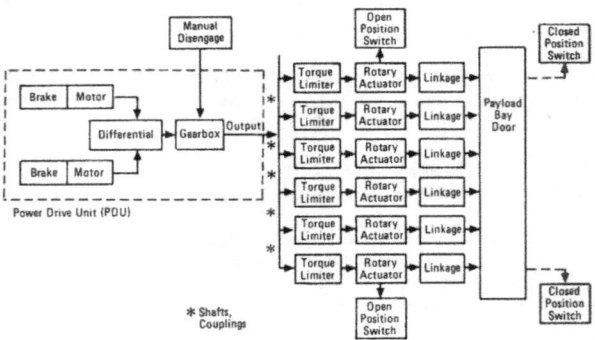

Payload Bay Door Drive System Control Mechanical Block Diagram

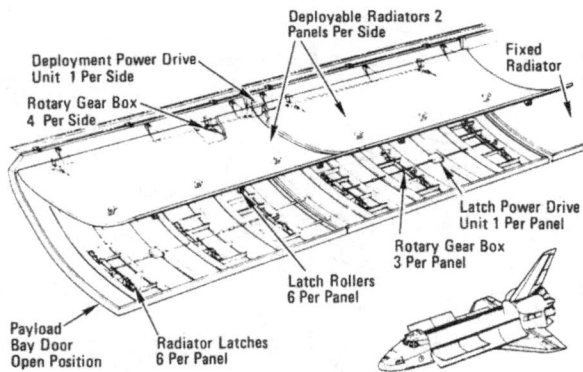

Payload Bay Door, Deployable Radiator Mechanism

Space Shuttle Spacecraft Structures

The payload bay doors open through an angle of 175.5 degrees.

Two radiator panels on each forward payload bay door can be deployed when the doors are opened on orbit and stowed when the doors are closed before entry or left in the stowed position for a given flight. Freon-21 Coolant Loop No. 1 flows through the left-hand radiator panels and the No. 2 loop flows through the right-hand panels. On orbit the panels radiate excess heat collected by the Freon-21 coolant loops from heat exchangers and cold plates throughout the orbiter. Coolant flows through the radiators from aft to forward. The radiator panels mounted on the forward end of the aft payload bay doors are fixed to the bay doors.

The radiator deploy and stow operation is controlled manually from the aft flight deck crew display and control panel. The PL BAY MECH (payload bay mechanisms) PWR switches and the RADIATOR LATCH and RADIATOR CONTROL SYS switches control the panels. Four indicators show the radiator latch and deploy status.

When the payload bay doors are fully open, the SYS1 and SYS2 PL BAY MECH PWR switches are turned to ON and the RADIATOR LATCH CONTROL switches to RELEASE. In approximately 26 seconds, the status indicators will show REL (release). The RADIATOR CONTROL SYS A and SYS B switches are turned to DEPLOY, and in approximately 43 seconds, the indicators will show DEP (deploy). The stow sequence is reversed. The electrical power and control system for the deployable radiators is designed to permit the loss of one latch or radiator control switch or one PL BAY MECH PWR SYS1 or SYS2 switch and not affect the operation of the radiators.

Each deployable radiator panel is secured to the payload bay door in the stowed position by six ganged latches. Two electrical drive motors latch or unlatch the six latches in each panel simultaneously. The motors can be reversed by the LATCH and UNLATCH switches on Panel R13. Limit switches turn off the motors when the latches are opened or closed. Each electrical drive motor is controlled by a separate set of switches. A differential within the PDU allows dual- or single-motor operation. Latching or unlatching the radiators takes approximately 26 seconds with both motors operating or 52 seconds with one motor.

The electric drive motors rotate torque shafts, and the shafts turn the rotary actuators, which operate push rods that latch or unlatch the hooks. The linkages and latches are attached to the payload bay doors and passive rollers are attached to the radiator panels. Torque limiters in the PDU prevent damage to the system in the event of jamming or binding during system operation.

Deploy systems on the port and starboard sides drive the radiators away from the payload bar doors and retract them.

Each deploy system uses two reversible electric motors to operate the power drive unit (PDU). The motors are not turned on until the MMCA's have received two signals from the radiator panel latch drives. This prevents inadvertent deployment of the radiators while still latched. As the rotary actuator shaft turns, the deployment crank and link straighten out and push the radiator panels away from the payload bay door to the deployed position. Stowing the radiators is the reverse operation. Limit switches turn off the electric drive motors when the radiators are deployed or stowed. Both port and starboard radiators are deployed or stowed simultaneously. The radiators deploy 35.5 degrees from the payload bay doors in 43 seconds with both electrical motors operating or in 86 seconds with one electrical motor operating. The DEPLOY/STOW switches on Panel R13 allow single- or dual-motor operation.

Torque limiters in the PDU prevent damage in the event of jamming or binding during operation.

Space Shuttle Spacecraft Structures

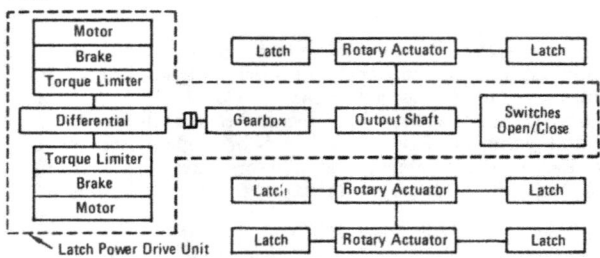

Deployable Radiator Latch System Block Diagram

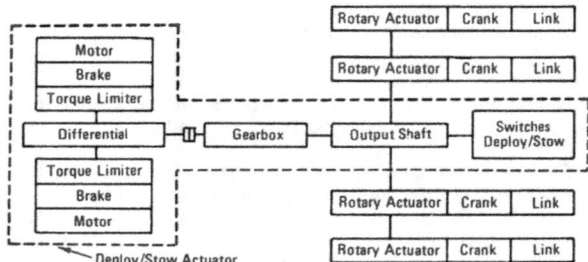

Deployable Radiator Deploy/Stow Control

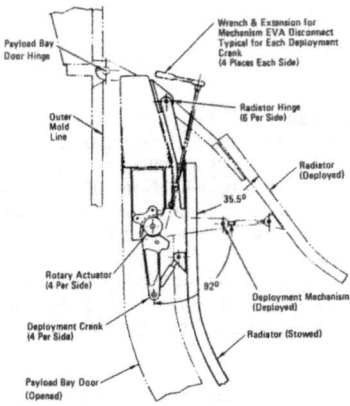

Payload Bay Door, Deployable Radiator Deployed

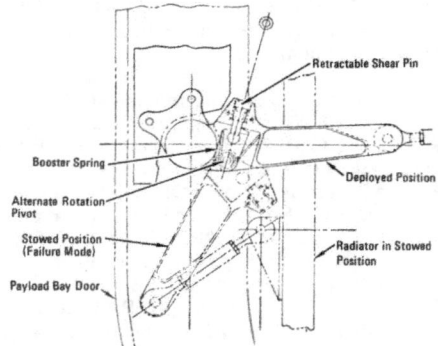

Payload Bay Door, Deployable Radiator Mechanism Disconnect Features

Each rotary crank can be disengaged from the rotary actuator (via EVA operations) by retracting a shear pin. Retraction allows the crank to rotate about an alternate pivot and permits the crew to stow the panels if the system fails. If the PDU fails, all four shear pins must be removed to allow manual stowing of the radiators. The pins are accessible when the radiators are fully deployed. No disengagement is planned if the radiators fail to deploy.

Space Shuttle Spacecraft Structures

Contractors are Rockwell's Tulsa Division, Tulsa, OK (payload bay doors); Curtiss Wright, Caldwell, NJ (payload bay door power drive unit, rotary actuators, drive shafts, torque tubes and couplings, radiator deploy/latch actuator and latch mechanism); Hoover Electric, Los Angeles, CA (payload bay door electromechanical rotary actuators); Vought Corp., Dallas, TX (radiators); Rockwell's Space Transportation and Systems Group, Shuttle Orbiter Division, Downey, CA (latches and linkages).

AFT FUSELAGE

The aft fuselage consists of an outer shell, thrust structure, and internal secondary structure. It is approximately 5.48 meters (18 feet) long, 6.70 meters (22 feet) wide, and 6.09 meters (20 feet) high.

The aft fuselage supports and interfaces with the left-hand and right-hand aft OMS/RCS pods, the wing aft spar, mid fuselage, orbiter/external tank rear attachments, Space Shuttle main engines, heat shield, body flap, vertical tail, and two T-0 launch umbilical panels.

The aft fuselage provides the load path to the mid fuselage main longerons, main wing spar continuity across the forward bulkhead of the aft fuselage, structural support for the body flap, structural housing around all internal systems for protection from operational environments (pressure, thermal, and acoustic), and for controlled internal pressures during flight.

The forward bulkhead closes off the aft fuselage from the mid fuselage and is composed of machined and beaded sheet metal aluminum segments. The upper portion of the bulkhead attaches to the front spar of the vertical tail.

The internal thrust structure supports the three main engines. The upper section of the thrust structure supports the

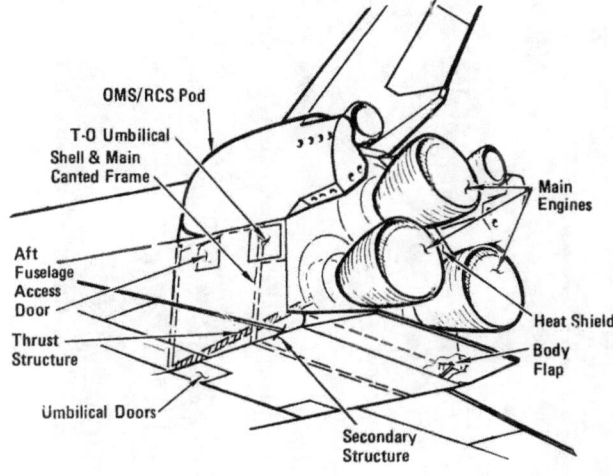

Aft Fuselage Structure

upper main engine, and the lower section of the thrust structure supports the two lower main engines. The internal thrust structure includes the main engines, load reaction truss structures, engine interface fittings, and the actuator support structure and supports the main engines, low pressure turbopumps, and propellant lines. The orbiter/external tank two aft attach points interface at the longeron fittings.

The internal thrust structure is composed mainly of 28 machined, diffusion-bonded truss members. In diffusion bonding, titanium strips are bonded together under heat, pressure, and time. This fuses the titanium strips into a single, hollow homogenous mass that is lighter and stronger than a forged

 ## Space Shuttle Spacecraft Structures

Aft and Mid Fuselage

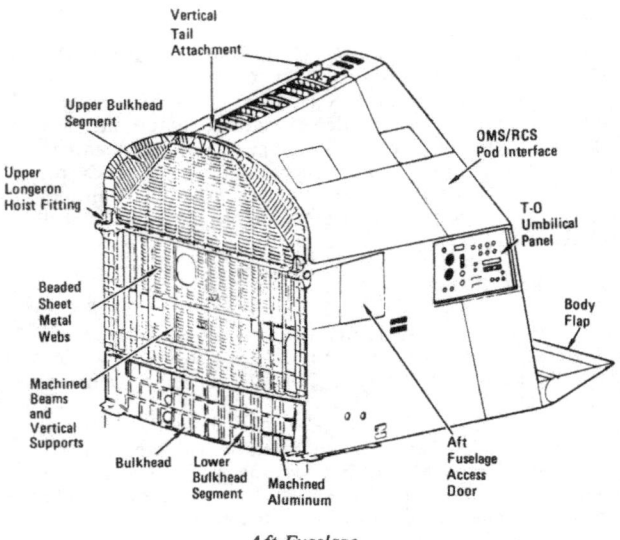

Aft Fuselage

part. In looking at the cross section of a diffusion bond, one sees no weld line. It is an homogenous parent metal, yet composed of pieces joined by diffusion bonding. In selected areas, the titanium construction is reinforced with boron/epoxy tubular struts to minimize weight and add stiffness. This reduced the weight by 21 percent. This is a reduction of approximately 408 kilograms (900 pounds).

The upper thrust structure of the aft fuselage is of integral machined aluminum construction with aluminum frames, except for the vertical fin support frame, which is titanium. The skin panels are integrally machined aluminum and attach to each side of the vertical fin to react drag and torsion loading.

The outer shell of the aft fuselage is constructed of integral machined aluminum. Various penetrations are provided in the shell for access to installed systems.

Space Shuttle Spacecraft Structures

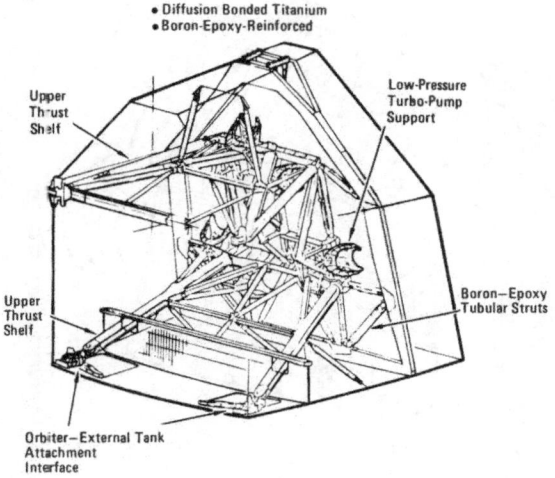

Aft Fuselage Internal Structure

The secondary structure of the aft fuselage is of conventional aluminum construction, except that some titanium and fiberglass is used for thermal isolation of equipment. The aft fuselage secondary structures consists of brackets, build-up webs, truss members, and machined fittings as required by system loading and support constraints. Certain system components, such as the avionics shelves, are shock-mounted to the secondary structure. The secondary structure includes support provisions for the auxiliary power units, hydraulics, ammonia boiler, flash evaporator, and electrical wire runs.

The two external tank umbilical areas interface with the external tank two aft attach points of the orbiter and the external tank liquid oxygen, liquid hydrogen feed lines, as well as electrical wire runs. The umbilicals are retracted, and the umbilical areas are closed off after external tank separation by an electromechanically operated beryllium door at each umbilical. Thermal barriers are employed at each umbilical door. The exposed area of each door when closed is covered with reusable surface insulation.

The aft fuselage heat shield and seal provides a closeout of the orbiter aft base area. The aft heat shield consists of a base heat shield of machined aluminum. Attached to the base heat shield are domes of honeycomb construction which support flexible and sliding seal assemblies. The engine-mounted heat shield is of Inconel honeycomb construction and is removable for access to the main engine power heads. The heat shield is covered with reusable surface insulation except for the Inconel segments. The exposed areas of the aft fuselage also are covered with reusable surface insulation.

The OMS/RCS left- and right-hand pods are attached to the upper aft fuselage left and right sides. Each pod is fabricated primarily of graphite/epoxy composite and aluminum. Each pod is 6.45 meters (21.8 feet) long and 3.46 meters (11.37 feet) wide at its aft end and 2.56 meters (8.41 feet) wide at its forward end, with a surface area of approximately 40.41 meters squared (435 square feet). Each pod is divided into two compartments: the OMS and the RCS housing. Each pod houses all the OMS propulsion components and RCS propulsion components. Each pod is attached to the aft fuselage with 11 bolts. The pod skin panels are graphite/epoxy honeycomb sandwich. The forward and aft bulkhead, aft tank support bulkhead, and floor truss beam are machined aluminum 2124. The centerline beam is 2024 aluminum sheet with titanium stiffness and graphite/epoxy frames. The frames are graphite/epoxy. The OMS thrust structure is conventional 2124 aluminum construction. The cross braces are aluminum tubing, and the attach fittings at the forward and aft fittings are 2124 aluminim. The intermediate fittings are corrosion-resistant steel. The RCS housing, which attaches to the OMS pod structure, contains the RCS thrusters

Space Shuttle Spacecraft Structures

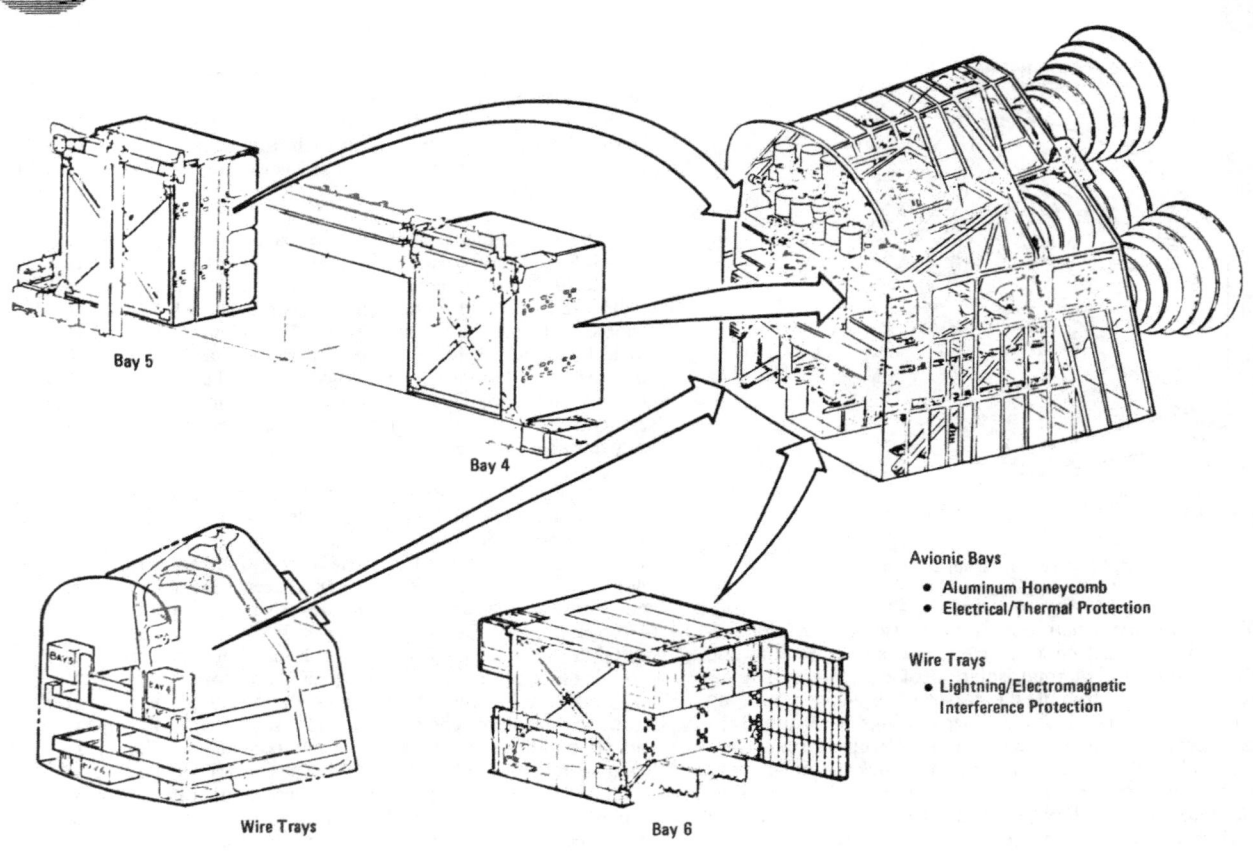

Aft Fuselage Avionics Provisions

Space Shuttle Spacecraft Structures

Aft Avionics Bays

and associated propellant feed lines. The RCS housing is of aluminum sheet metal construction, including flat outer skins, and the curved outer skin panels are graphite/epoxy honeycomb sandwich. Access to the OMS and RCS and attach points are provided by 24 doors in the skins.

The two graphite/epoxy pods per spacecraft reduces the weight by 10 percent. This is a reduction of approximately 204 kilograms (450 pounds). The pods will withstand 162 dB acoustic noise and a temperature range from minus 112°C to plus 57°C (minus 170° to plus 135°F).

The exposed areas of the OMS/RCS pods are covered with reusable surface insulation and a pressure and thermal seal

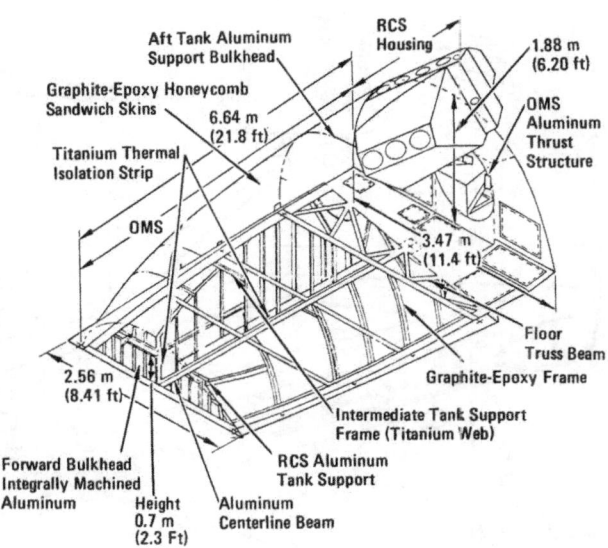

Aft OMS/RCS Pod

is installed at the OMS/RCS pod aft fuselage interface. Thermal barriers also are installed and they interface with the RCS thrusters and reusable surface insulation.

The body flap thermally shields the three main engines during entry and provides the orbiter with pitch control trim during its atmospheric flight after entry.

The body flap is an aluminum structure consisting of ribs, spars, skin panels, and a trailing edge assembly. The main upper

Space Shuttle Spacecraft Structures

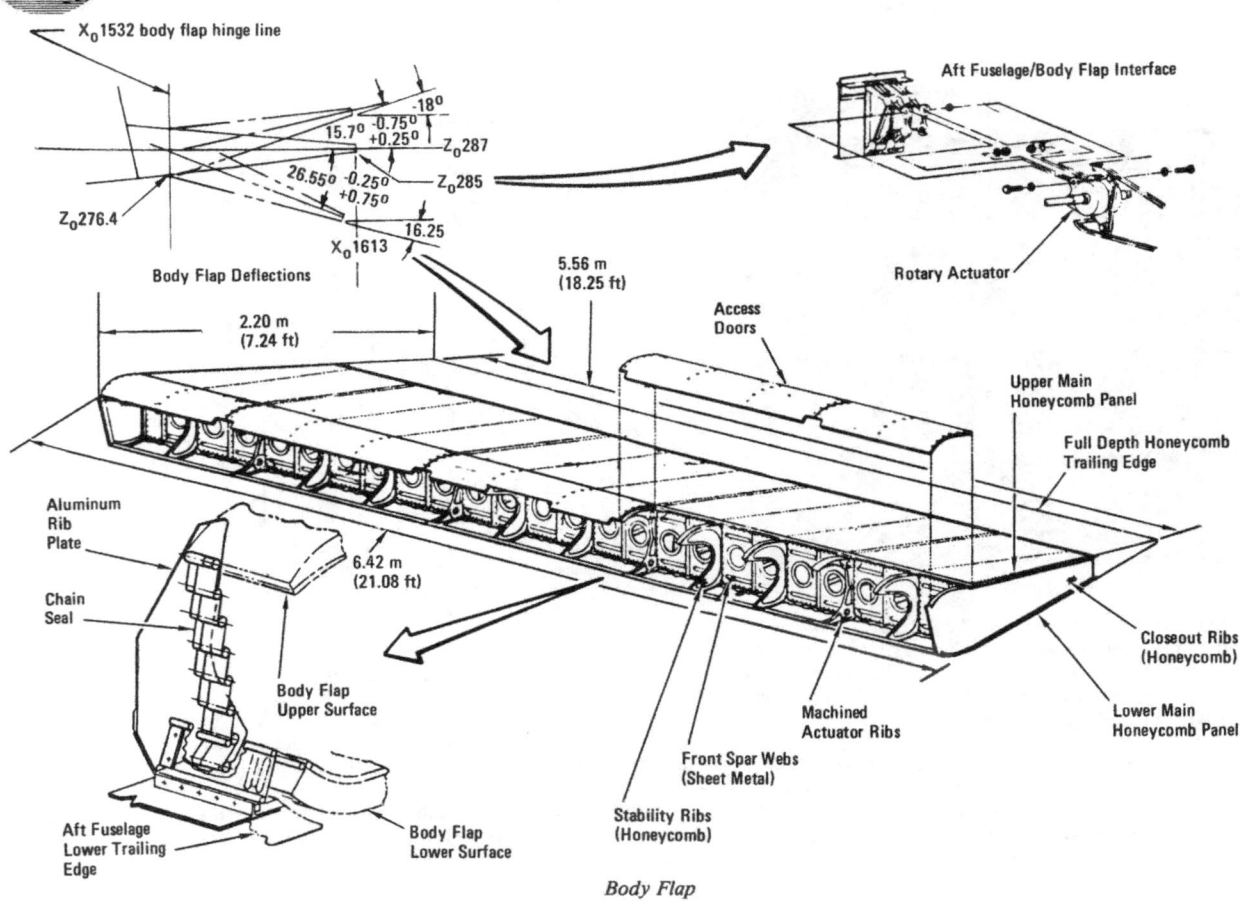

Body Flap

Space Shuttle Spacecraft Structures

and lower forward honeycomb skin panels are joined to the ribs, spars, and honeycomb trailing edge with structural fasteners. The removable upper forward honeycomb skin panels complete the body flap structure.

The upper skin panels aft of the forward spar and the entire lower skin panels are mechanically attached to the ribs. The forward upper skin consists of five removable access panels attached to the ribs with quick-release fasteners. The four integral machined aluminum actuator ribs provide the aft fuselage interface through self-aligning bearings. Two bearings are located in each rib for attachment to the four rotary actuators located in the aft fuselage, which are controlled by the flight control system and the hydraulically actuated rotary actuators. The remaining ribs consist of eight stability ribs and two closeout ribs constructed of chemically milled aluminum webs bonded to aluminum honeycomb core. The forward spar web is of chemically milled sheets with flanged holes and stiffened beads. The spar web is riveted to the ribs. The trailing edge includes the rear spar, which is composed of piano-hinge half-cap angles, chemically milled skins, honeycomb aluminum core, closeouts, and plates. The trailing edge attaches to the upper and lower forward panels by the piano-hinge halves and hinge pins. Two moisture drain lines and one hydraulic fluid drain line penetrate the trailing edge honeycomb core for horizontal and vertical drainage.

The body flap is covered with reusable surface insulation and an articulating pressure and thermal seal to its forward cover area on the lower surface of the body flap to block heat and air flow from the structures.

The aft fuselage also is built by Rockwell's Space Transportation and Systems Group, Shuttle Orbiter Division, Downey, CA. The OMS/RCS pods are built by McDonnell Douglas, St. Louis, MO. The body flap is built by Rockwell's Columbus, OH division.

VERTICAL TAIL

The vertical tail consists of a structural fin surface., the rudder/speed brake surface, a tip, and a lower trailing edge. The rudder splits into two halves to serve as a speed brake.

The vertical tail structure fin is made of aluminum. Integral machined skins and strings, ribs, and two machined spars make up the main torque box. The fin is attached by two tension tie bolts at the root of the front spar of the vertical tail to the forward bulkhead of the aft fuselage and by eight shear

Vertical Tail

154

Space Shuttle Spacecraft Structures

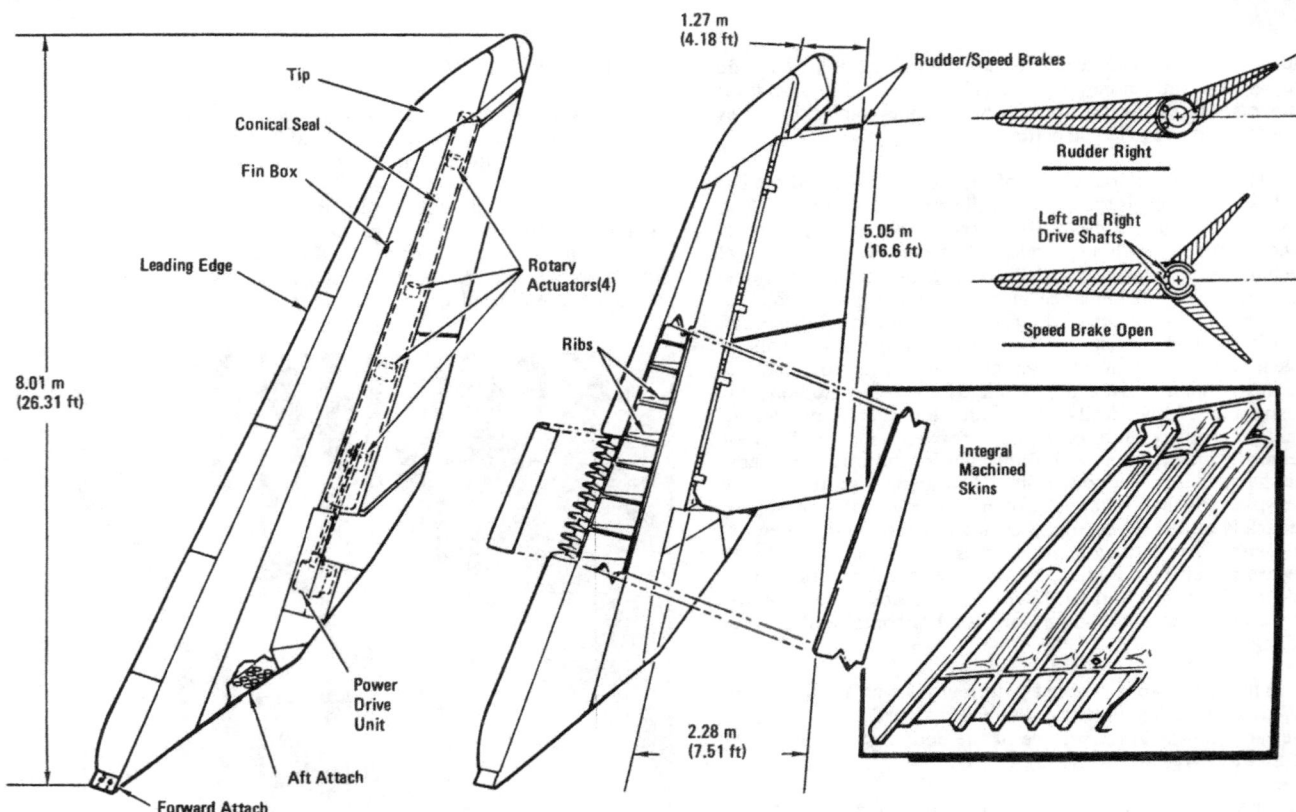

Vertical Tail

Space Shuttle Spacecraft Structures

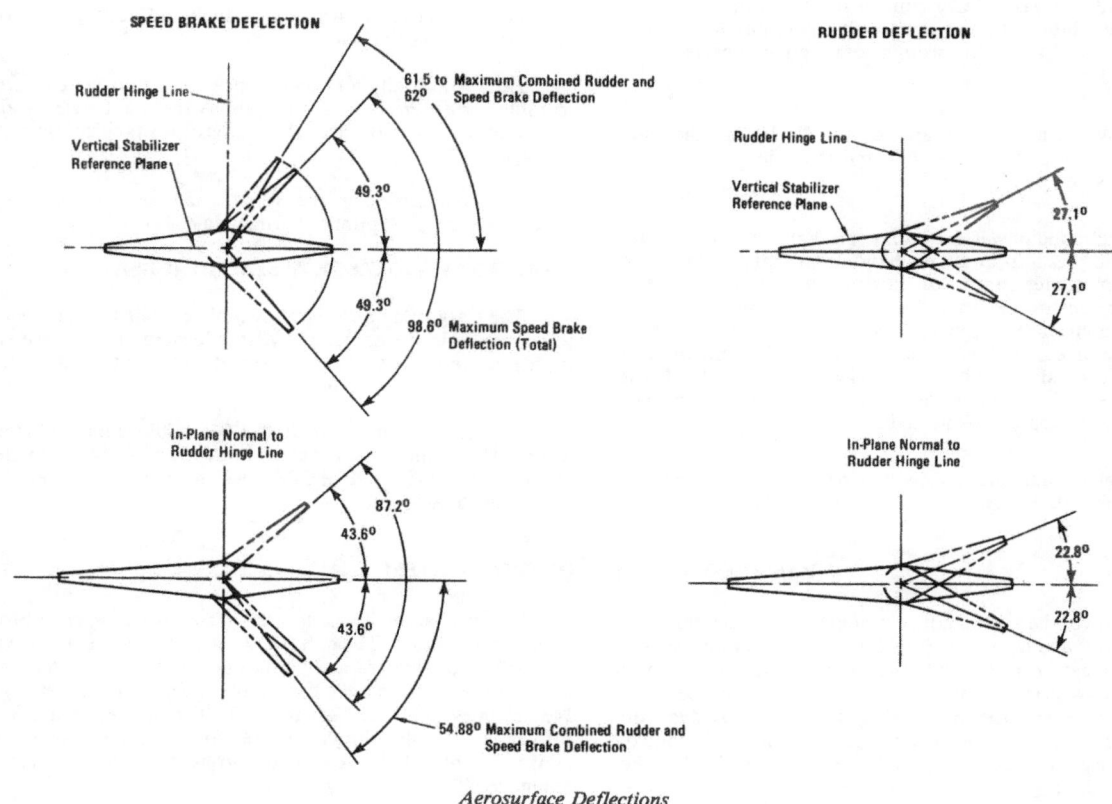

Aerosurface Deflections

Space Shuttle Spacecraft Structures

bolts at the root of the vertical tail rear spar to the upper structural surface of the aft fuselage.

The rudder/speed brake control surface is made of conventional aluminum ribs and spars with aluminum honeycomb skin panels and is attached through rotating hinge parts to the vertical tail fin.

The lower trailing edge area of the fin houses the rudder/speed brake power drive unit and is made of aluminum honeycomb skin.

The hydraulic power drive unit/mechanical rotary actuation system drives left- and right-hand drive shafts in the same direction for rudder control of plus or minus 27 degrees. For speed brake control, the drive shafts turn in opposite directions for a maximum of 49.3 degrees each. The rotary drive actions are also combined for joint rudder/speed brake control. The hydraulic power drive unit is controlled by the orbiter flight control system. A maximum deflection rate of 10 degrees per second control capability is available.

The vertical tail structure is designed for 163 dB acoustic environment with a maximum temperature of 176 °C (350 °F).

An Inconel honeycomb conical seal houses the rotary actuators and provides a pressure and thermal seal which withstands a maximum of 648 °C (1,200 °F).

The split halves of the rudder panels and trailing edge contain a thermal barrier seal.

The vertical tail and rudder/speed brake are covered with reusable surface insulation materials. A thermal barrier is also employed at the interface of the vertical stabilizer and aft fuselage.

The contractor for the vertical tail and rudder/speed brake is Fairchild Republic, Farmingdale, NY.

CARGO BAY STOWAGE ASSEMBLY (CBSA)

The Cargo Bay Stowage Assembly contains miscellaneous tools for use in the payload bay. It is located on the starboard (right) side of the payload bay forward, between Orbiter Station $X_O = 589$ and $X_O = 636$.

The CBSA is approximately 1,066 millimeters (42 inches) wide, 609 millimeters (24 inches) in depth and 914 millimeters (36 inches) in height. The CBSA weight is approximately 259 kilograms (573 pounds).

PASSIVE THERMAL CONTROL SYSTEM

A passive thermal control system helps maintain the orbiter systems and components within their temperature limits. This system uses orbiter heat sources and heat sinks and is supplemented by insulation blankets, thermal coatings, and thermal isolation methods. Heaters are provided on components and systems where passive thermal control techniques are not adequate. (The heaters are described in the various systems.)

The insulation blankets are of two basic types: fibrous bulk and multilayer. The bulk blankets are fibrous material with a density of 0.9 kilogram (2 pounds per cubic foot) with a sewn cover of reinforced acrylic film Kapton. The cover material has 145,317 holes per meters squared (13,500 holes per square foot) for venting. Acrylic film tape is used for cutouts, patching, and reinforcements. Tufts are used throughout the blankets to minimize billowing during venting.

Space Shuttle Spacecraft Structures

The muiltilayer blankets are constructed of alternate layers of perforated acrylic film Kapton reflectors and Dacron net separators for a total of 16 reflector layers, with the two cover halves counting as two layers. Covers, tufting, and acrylic film tape are similar to the bulk blankets.

The contractors are Hi-Temp Insulation, Inc., Camarillo, CA (fibrous insulation); Scheldahl, Northfield, MN (cover materials and inner layers); Apex Mills, Los Angeles, CA (separators).

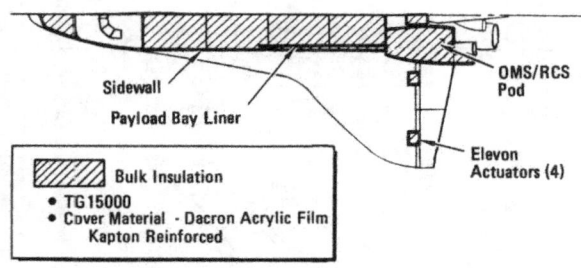

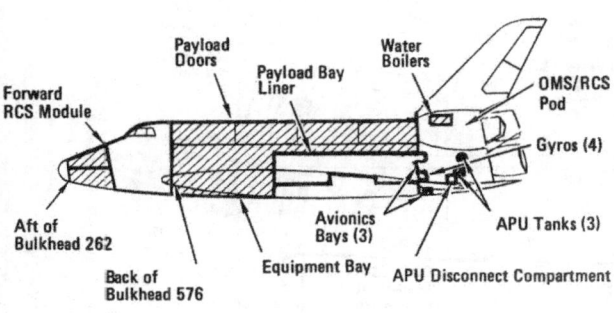

Bulk Insulation

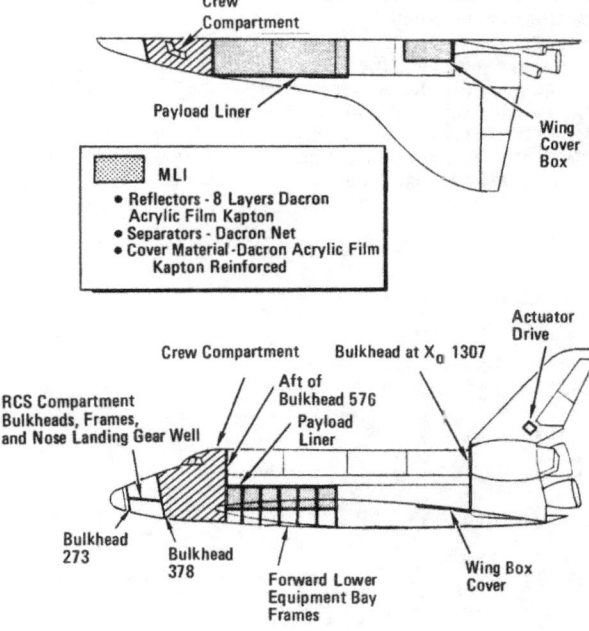

Multilayer Insulation

Space Shuttle Spacecraft Structures

PURGE, VENT, AND DRAIN SYSTEM

The purge, vent, and drain systems accomplish the following: provide the unpressurized compartments of the orbiter with an air purge that thermally conditions system components and prevents hazardous gas accumulation, vent compartments during ascent and take in air during descent to minimize differential pressures, and drain trapped fluids and condition the window cavities to maintain visibility.

The purge system carries conditioned gas from ground support equipment to the orbiter cavities via the T-O umbilical disconnect during preflight and postflight phases. Purge gas is provided to three separate sets of distribution plumbing: (1) the forward fuselage, OMS/RCS pods, wings, and vertical stabilizer, (2) mid fuselage, and (3) the aft fuselage.

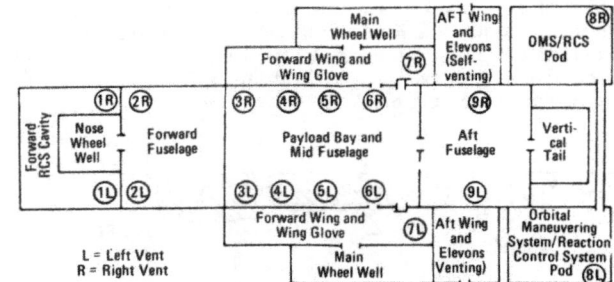

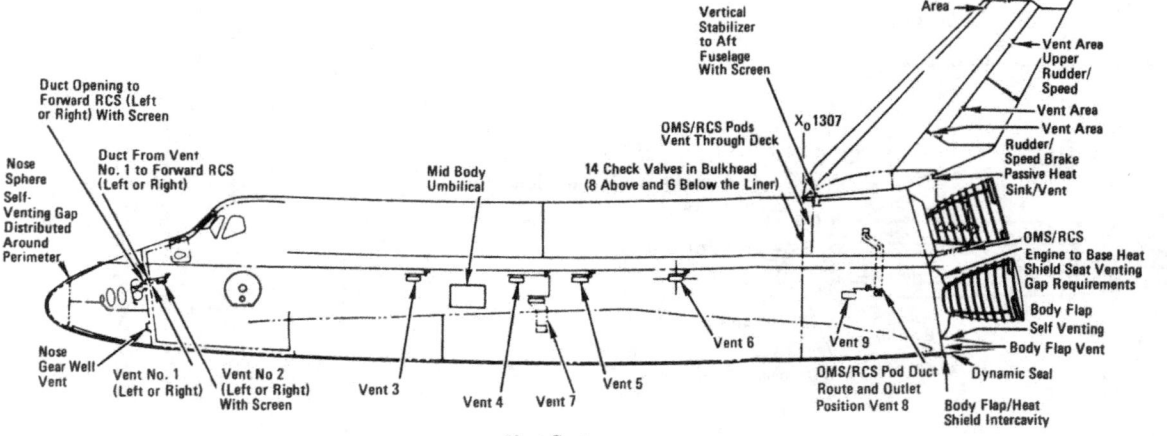

Vent System

Space Shuttle Spacecraft Systems

The vent system provides the flow area for control of pressure during purge, depressurization during ascent, molecular venting in orbit, and repressurization during descent. There are 18 vent ports in the fuselage skin that serve specific orbiter cavities. The purge and vent outlets are located and sized to vent the cavities within structural, hazard, and purge limitations. Vent doors are operated by electromechanical actuators and are sequenced during the mission to protect against gas ingestion, high acoustic levels, and entry heating.

The drain system provides flow paths and systems to drain or remove accumulated water. The paths are provided through a series of limber holes that allow drainage to the lowest point for removal. Locations that cannot be served by limber holes are evacuated through a series of tubes and disconnects by ground support equipment.

The purge and vent ducting is Kevlar/epoxy (115 pieces up to 279.4 millimeters [11 inches] in diameter) which replaced fiberglass or aluminum ducts. This reduced the weight by 33 percent. This is a reduction of approximately 90 kilograms (200 pounds).

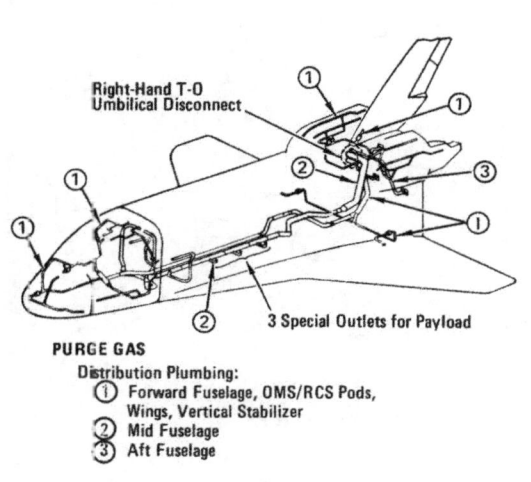

Purge Gas System

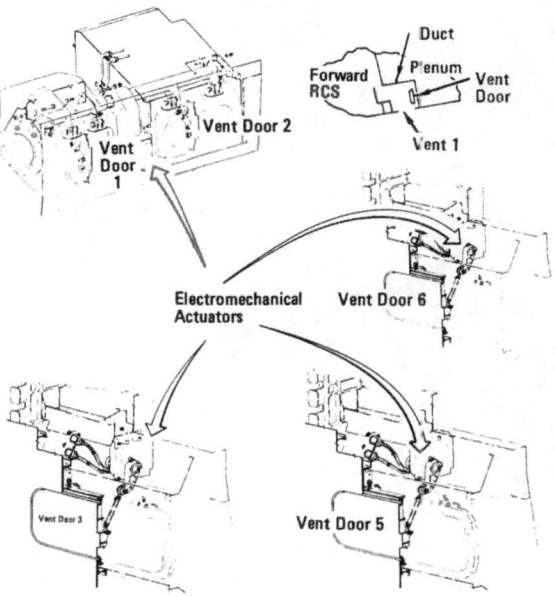

Vent Doors (1, 2, 3, 5 & 6)

# Space Shuttle Spacecraft Structures

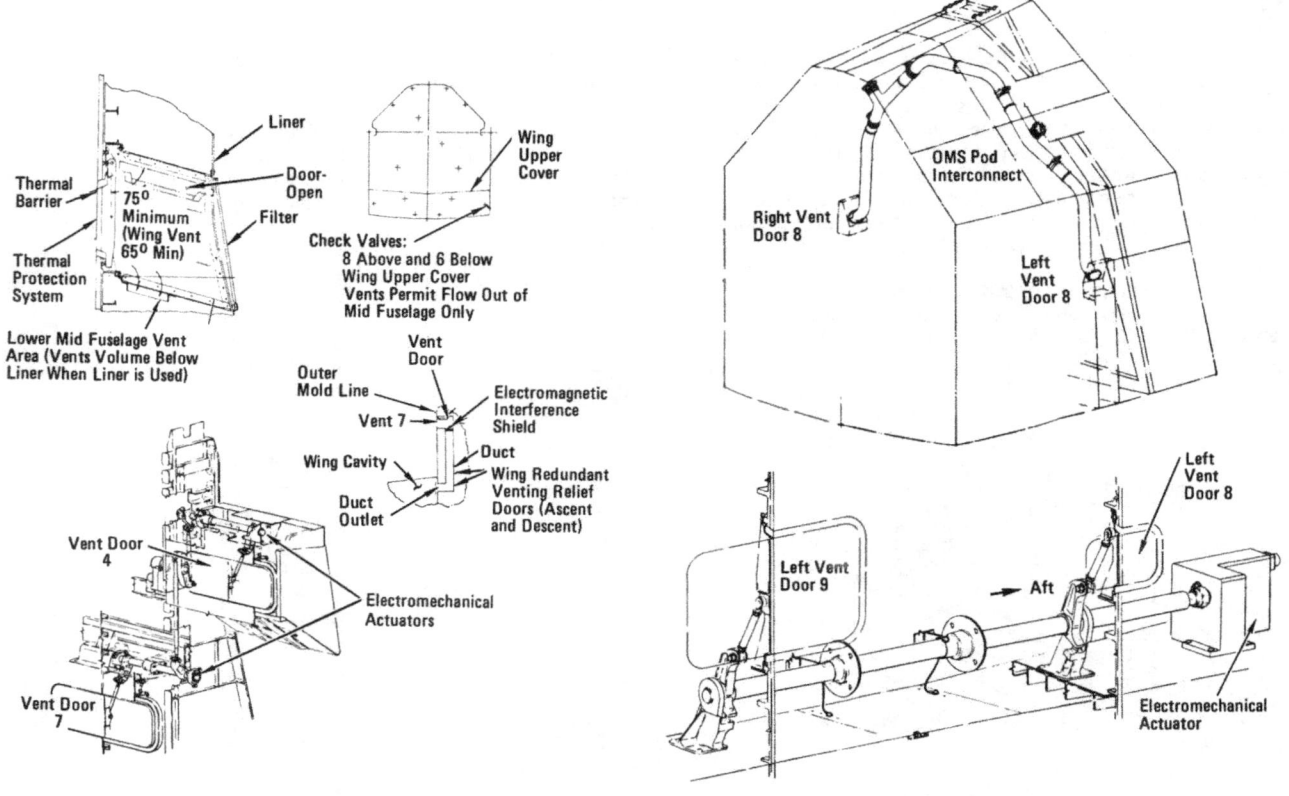

Vent Doors (4 & 7)

Vent Doors (8 & 9)

Space Shuttle Spacecraft Structures

The window cavity conditioning system prevents moisture ingress into the windshields and overhead window and payload viewing window cavities and provides for depressurization and repressurization of these cavities during flight. This system also provides the purge conditioning (drying) of these areas during ground operations. The side hatch window is self-contained.

The hazardous gas detection system detects hazardous levels of explosive or toxic gases. The onboard orbiter sample lines duct the compartment gases to the ground support equipment at the T-O right-hand umbilical panel and to the ground-based mass spectrometer for analysis at the launch pad.

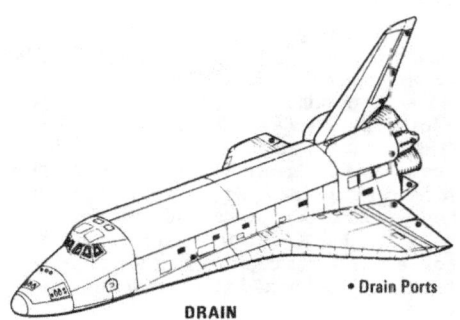

Drain System

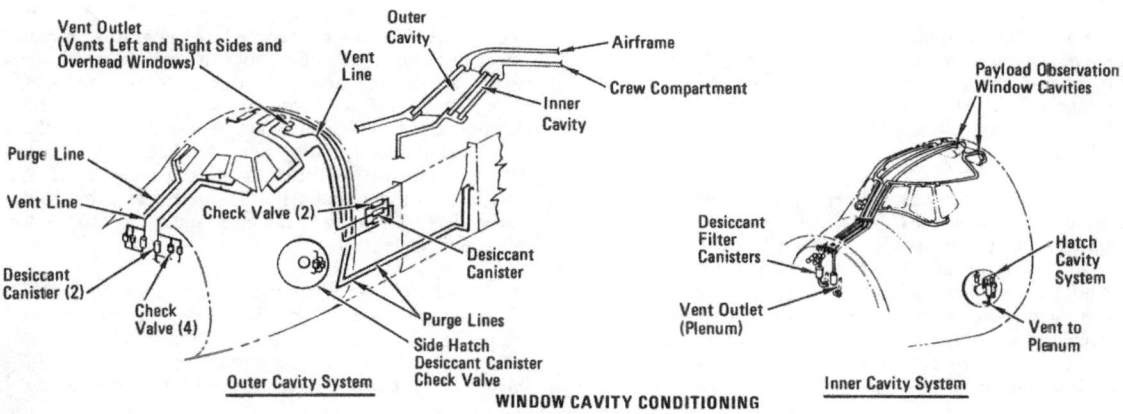

Window Cavity Conditioning System

Space Shuttle Spacecraft Systems

THERMAL PROTECTION SYSTEM

The thermal protection system (TPS) consists of materials applied externally to the primary structural shell of the orbiter to maintain the orbiter airframe outer skin within acceptable temperature limits. The skins are constructed primarily of aluminum and/or graphite epoxy. During entry they must be protected from temperatures above 176 °C (350 °F). The TPS materials must be capable of performing a minimum of 100 missions in which temperatures will range from a minus 156 °C (minus 250 °F) in the cold soak of space to reentry temperatures that will reach nearly 1,648 °C (3,000 °F) on the wing leading edge and the nose cap. Interior compartment temperatures are controlled by internal insulation and heaters and through purging techniques.

The TPS is a passive system selected for stability at high temperatures and weight efficiency. Its applications are as follows:

1. Coated Nomex felt reusable surface insulation (FRSI) is used where temperatures are less than 371 °C (700 °F). FRSI is used on the upper payload bay doors, mid and aft fuselage sides, upper wing, and a portion of the orbital maneuvering system/reaction control system (OMS/RCS) pods.

2. On Orbiter 102, the *Columbia,* low-temperature reusable surface insulation (LRSI) tiles are used where temperatures go below 648 °C (1,200 °F) and above 371 °C (700 °F) nominal. These areas are the lower portion of payload bay doors; forward, mid, and aft fuselage; upper wing; vertical tail and a portion of the OMS/RCS pods. These tiles have a white surface coating in color which provides better thermal characteristics on orbit.

On Orbiter 099, the *Challenger,* some of the LRSI tiles on the orbital maneuvering system/reaction control system (OMS/RCS) pods are replaced with a sewn composite blanket, advanced flexible reusable surface insulation (AFRSI), a quilted fabric blanket that improves producibility and durability, reduces fabrication and installation cost, reduces installation schedule time, and results in a weight reduction. Subsequent orbiters would also use AFRSI to replace the majority of the LRSI tiles.

3. On Orbiter 102, the *Columbia,* high-temperature reusable surface insulation (HRSI) tiles are used where temperatures are below 1,260 °C (2,300 °F) and above 648 °C (1,200 °F). The areas are the forward fuselage, lower mid fuselage, lower wing, selected areas of the vertical tail, a portion of the OMS/RCS pods, and around the forward fuselage windows. The HRSI has two different densities: one weighs 4 kilograms per cubic meter (9 pounds per cubic foot) and is used in all areas except around the nose and main landing gear doors, nose cap interface, wing leading edge, reinforced carbon-carbon/HRSI interface, external tank umbilical doors, vent doors, and vertical stabilizer leading edge. Those areas use HRSI tiles with a density of 9.9 kilograms per cubic meter (22 pounds per cubic foot). These tiles have a black surface coating necessary for entry emittance.

On Orbiter 103 and subsequent orbiters, some of the HRSI 22-pounds-per-cubic-foot tiles may be replaced with fibrous refractory composite insulation (FRCI) HRSI tiles. When all 22 pounds-per-cubic-foot tiles have been used, FRCI-12 tiles will be used, and on

Space Shuttle Spacecraft Systems

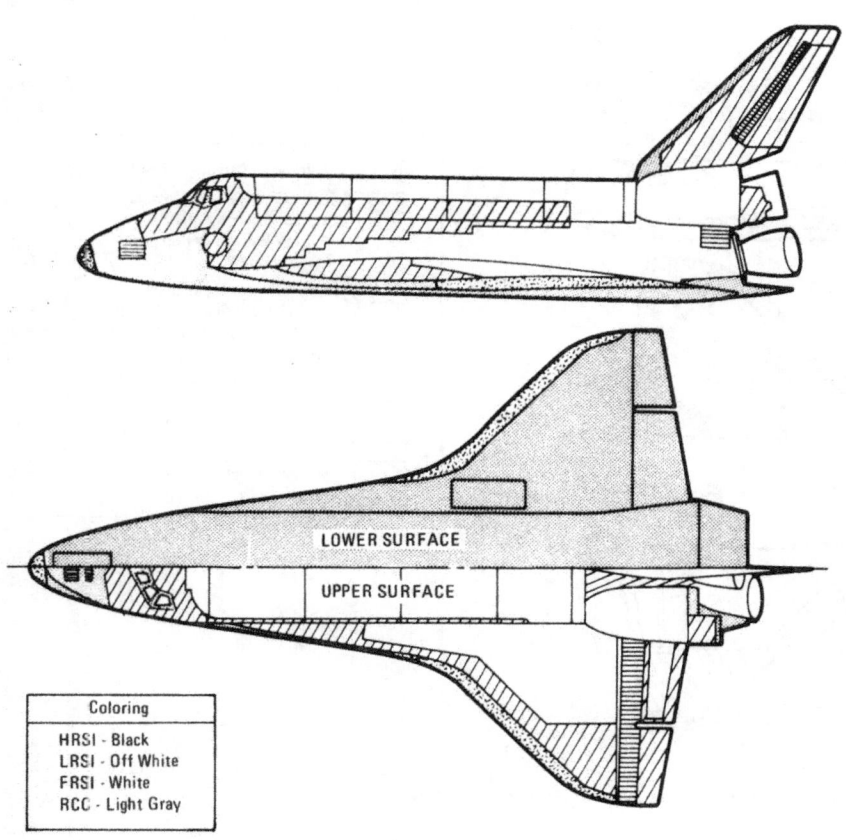

Thermal Protection System, Orbiter 102

 **Space Shuttle Spacecraft Systems**

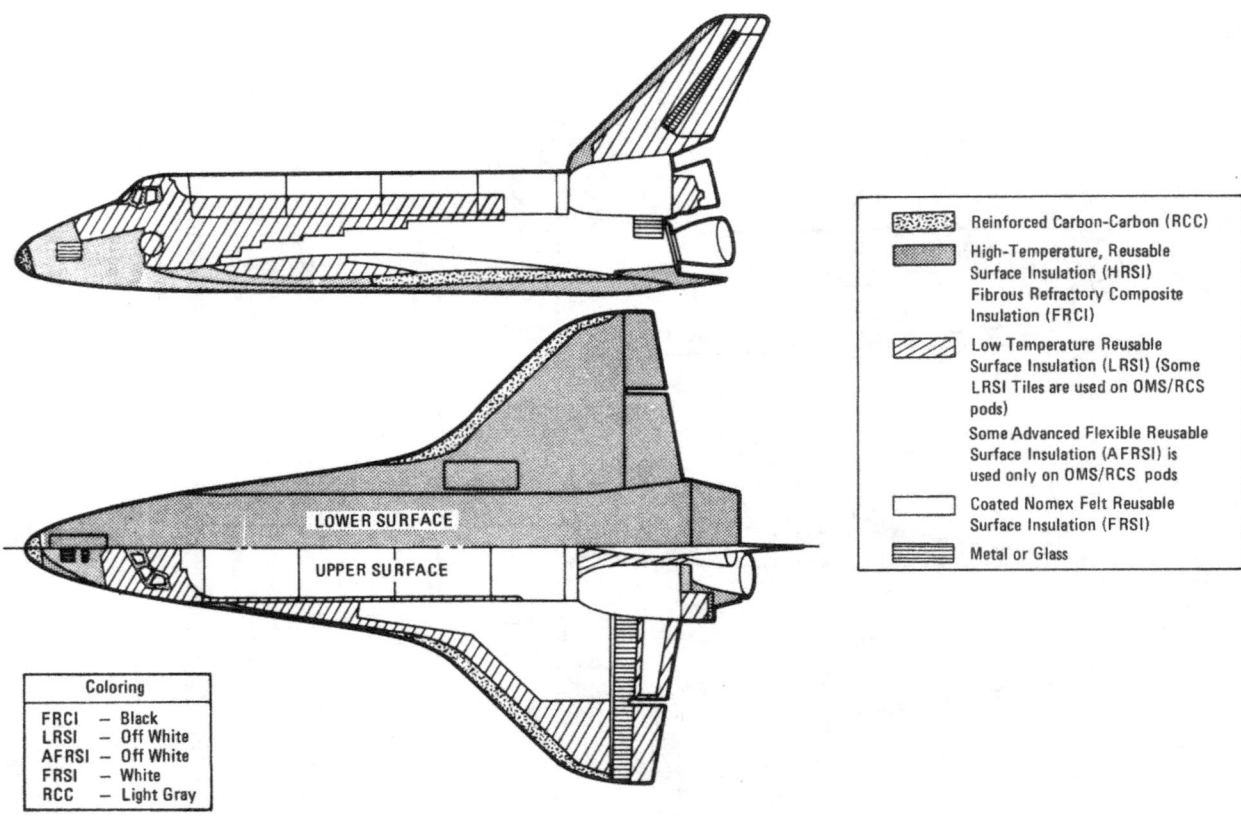

Thermal Protection System, Orbiter 099 and Subsequent Orbiters

Space Shuttle Spacecraft Systems

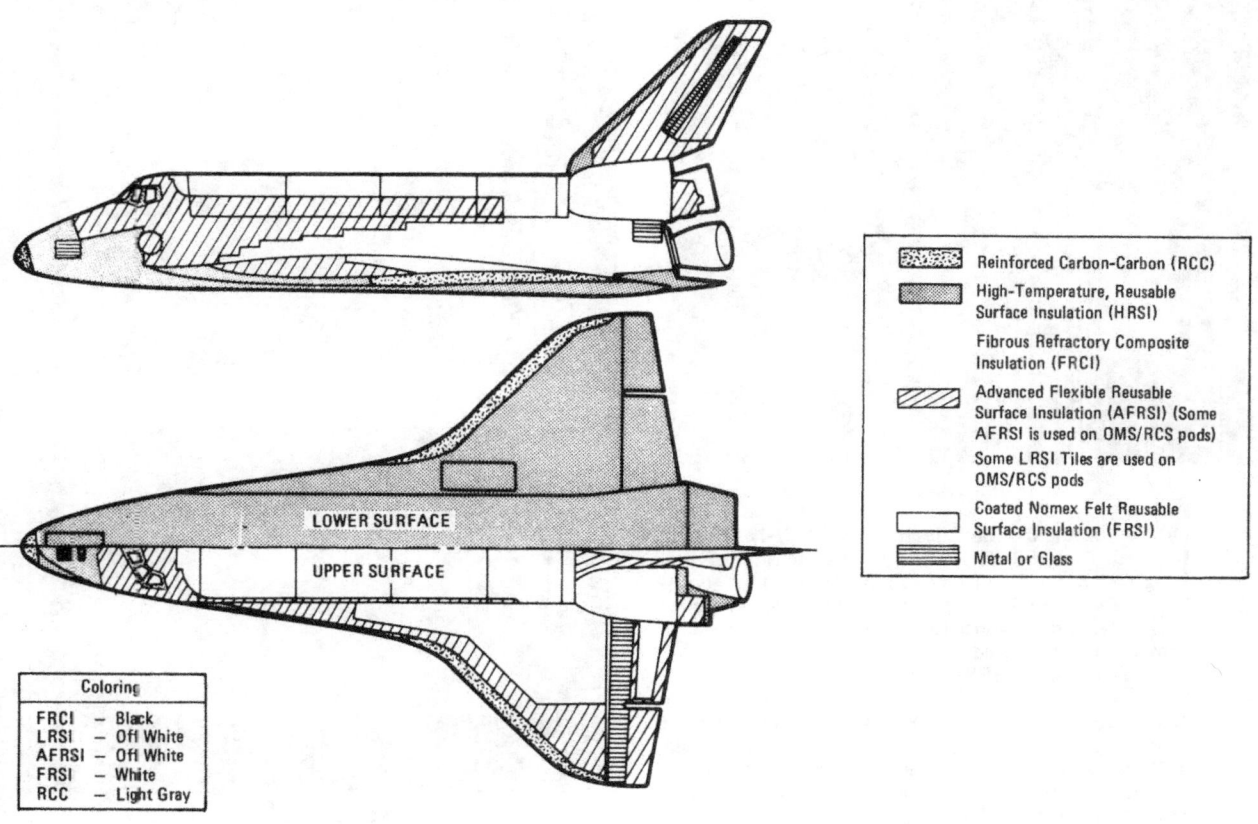

Thermal Protection System, Orbiter 103 and Subsequent Orbiters

Space Shuttle Spacecraft Systems

HRSI Tiles

RCC Wing

RCC Nose Cap

Orbiter 102 and 099, any tiles replaced FRCI-12 tiles will be used. The FRCI-12 tiles have a density of 5.4 kilograms per cubic meter (12 pounds per cubic foot). The FRCI HRSI tiles have improved strength, durability, and resistance to coating cracking. They also provide a weight reduction from that of the 99.8 percent-pure silica HRSI tiles.

4. Reinforced carbon-carbon (RCC) is used on the wing leading edge and nose cap, where temperatures exceed 1,260 °C (2,300 °F). RCC is also used in the immediate area around the forward orbiter-external tank structural attachment on the orbiter.

Space Shuttle Spacecraft Systems

Additional materials are used in special areas. These are:

1. Thermal panes are used for the windows.

2. Metal is used for the forward reaction control system fairings and elevon seal panels on the upper wing elevon interface.

3. A combination of white and black-pigmented silica cloth for thermal barriers and gap fillers is installed around operable penetrations such as main and nose landing gear doors egress/ingress side hatch, umbilical doors, elevons, forward RCS module and RCS thrusters, vent doors, payload bay doors, rudder/speed brake, and OMS/RCS pod and RCS thrusters, and gaps between tiles in high differential pressure areas.

The TPS has been designed for ease of maintenance and for flexibility of ground and flight operations while satisfying its primary function of maintaining acceptable airframe outer skin temperatures.

The FRSI Nomex felt varies in thickness from 4.06 millimeters (0.160 inch) to 10.1 millimeters (0.40 inch) thick and consists of sheets 0.9 by 1.2 meters (3 to 4 feet) (except for closeout areas) bonded directly to the orbiter exterior. The felt is coated with a white pigmented silicone elastomer to waterproof the felt and to provide the required thermal and optical properties. The FRSI provides an emittance of 0.8 and solar absorptance of 0.32. FRSI covers nearly 50 percent of the orbiter's upper surface.

The felt is a basic Nomex (aramid) fiber. The fibers are two denier 76.2 millimeters (3 inches) long and crimped. The fibers are loaded into a carding machine, which untangles the clumps of fibers and "combs" them to make a tenuous mass of lengthwise-oriented, relatively parallel fibers called a web. The cross-lapped web is fed into a loom, where it is lightly needled into a batt. Generally, two such batts are placed face to face, where they are needled together to form felt. The felt is then subjected to a multineedle-pass process until the desired strength is reached. The needled felt is then calendered to stabilize thickness 4.06 millimeters (0.16 inch) to 10.1 millimeters (0.40 inch) by passing it through heated rollers at selected pressures. The calendered material is then heat-set at approximately 260 °C (500 °F) to thermally stabilize the felt.

The FRSI is bonded to the orbiter surface by a room temperature vulcanizing silicon adhesive. The silicon adhesive glue is applied at 0.20 millimeter (0.008 inch) thick. The very thin glue line reduces weight and minimizes the thermal expansion during temperatures of 112 °C (500 °F) at the glue line (entry) and temperatures below minus 223 °C (minus 170 °F) on orbit. The orbiter structure could be as low as minus 121 °C (minus 250 °F). The FRSI bond is cured at room temperature, with vacuum bags used to apply pressure.

On Orbiter 102, the HRSI tiles are nominally 152.4 by 152.4 millimeters (6 by 6 inches). The tiles are made of a low-density (lightweight), high-purity silica (glass) 99.8 percent amorphous fiber (fibers derived from common sand, one to two mils thick) insulation that is made rigid by ceramic (clay) bonding. Ninety percent of the tile is void, and ten percent is material that results in the tiles weighing 4 kilograms per cubic meter (9 pounds per cubic foot). However, the areas around the nose and main landing gear doors, external tank umbilical doors, vent doors, and vertical stabilizer leading edge utilize tiles that have a density of 9.9 kilograms per cubic meter (22 pounds per cubic foot). A slurry containing fibers mixed with water is frame-cast to form soft, porous blocks to which collodial silica binder solution is added. When sintered, a rigid block is produced, which is then cut into quarters and then machined to the precise dimensions required for individual tiles.

The HRSI tiles will vary in thickness: 25.4 millimeters (1 inch) to 127 millimeters (5 inches) to minimize the orbiter weight

Space Shuttle Spacecraft Systems

and not permit the orbiter structure to see more than 176 °C (350 °F). The tiles will vary slightly in sizes and shapes at the closeout areas. The tile thickness also provides adequate on-orbit space cold soak protection, and the tiles must withstand repeated heating and cooling, plus extreme acoustic environments (165 decibels at launch) in some local areas. Resistance to thermal shock is very good. The material can be taken from a 1,260 °C (2,300 °F) oven and immersed in cold water without damage. Surface heat dissipates so quickly that an uncoated tile can be held by its edges with an ungloved hand seconds after removal from the oven and while the tile interior still glows red hot. The HRSI tiles are coated on the top and sides with a glass mixture formed by mixing tetra-silicide with boro-silicate glass in a powder with a liquid carrier and sprayed on the tile to a coating thickness of 16 to 18 mils. The coated HRSI tiles are then placed in an oven and heated to a temperature of 1,260 °C (2,300 °F). This results in a black waterproof glossy coating covering the tile which has a surface emittance of 0.85 and a solar absorptance of about 0.85. In addition, the silica fibers are treated with a silicone resin after the ceramic coating heating process to provide bulk waterproofing.

Over 23,000 HRSI tiles are used on the bottom portion of the orbiter and nose areas of the orbiter in addition to other selected areas of the orbiter. The HRSI tiles cannot withstand airframe load deformation; therefore, stress isolation is necessary between the tiles and the orbiter structure. This isolation is provided by strain isolation pads (SIP's). The SIP's isolate the HRSI tiles from the orbiter's structural deflections, expansions, and acoustic excitation. They thereby prevent stress failure in the tiles. The SIP is a thermal isolator. The SIP's are made of the same Nomex felt material as the FRSI and are either 2.28 millimeters (0.09 inch) or 4.06 millimeters (0.16 inch) thick. The SIP's are bonded to the tiles, and the SIP/tile assemblies are bonded to the orbiter structure by the same room temperature vulcanizing process as in the FRSI.

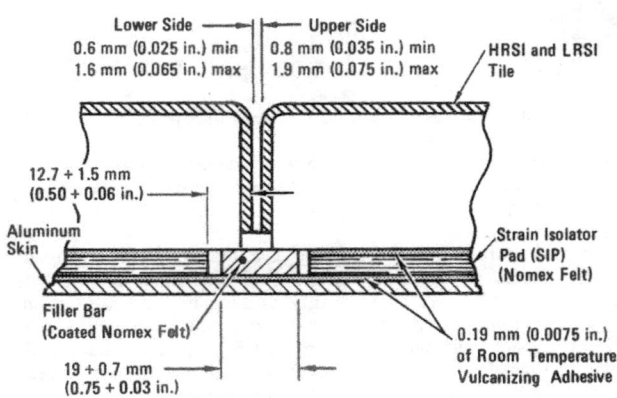

HRSI and LRSI Tile Interface, Orbiter 102 and 099

Since the HRSI tiles thermally expand or contract very little compared to the orbiter structure, it is necessary to leave gaps of 25 to 65 mils between the tiles to prevent tile-to-tile contact. Insulation is required in the bottom of the gap between tiles and is of the same Nomex felt material as FRSI. It is referred to as a filler bar. The material is 2.28 millimeters (0.09 inch) or 4.0 millimeters (0.16 inch) thick material cut into strips 19.0 millimeters (0.75 inch) wide and bonded to the structure at the same time as the HRSI SIP pads. The filler bar is waterproof and temperature resistant up to approximately 426 °C (800 °F), top side exposure.

The LRSI and HRSI tiles that were removed from Orbiter 102 at the Kennedy Space Center were replaced with tiles that are densified. The densification process was required due to fiber polarization in the SIP manufacture, which resulted in

Space Shuttle Spacecraft Systems

stress concentrations on the SIP/tile bond interface. The densification process utilizes a Ludox AS, which is an ammonia-stabilized binder. When mixed with silica slip particles, it becomes a cement. When mixed with water, it dries to a finished hard surface. A silica-tetraboride coloring agent is mixed with the compound for penetration identification. The pigmented Ludox slip slurry is brush-painted with several coats on the tile and allowed to air-dry for 24 hours. A heat treatment and other processing follow prior to installation. The densification coating only penetrates the tiles about 2.79 millimeters (0.11 inch) to 1.77 millimeters (0.07 inch), yet the strength and stiffness of the tile/SIP system are increased by a factor of two. Densified tiles are used on Orbiter 099 and subsequent orbiters.

The FRCI-12 HRSI tiles were developed by NASA's Ames Research Center, Mountain View, CA, and are manufactured by Lockheed Missiles and Space Division, Sunnyvale, CA, the same manufacturer of the original 99.8 percent-pure silica HRSI tiles.

The FRCI-12 HRSI tiles are a higher strength tile derived by adding AB312 (alumino boro silicate fiber) called Nextel developed by the 3M Company in St. Paul, MN to the 80 percent pure silica tile slurry. The Nextel activates boron fusion and figuratively welds the micron-size fibers of pure silica into a rigid structure during sintering in a high-temperature furnace. The resulting composite fiber refractory material composed of 20 percent Nextel and 80 percent silica fiber has entirely different physical properties than the original HRSI 99.8-percent pure silica tiles. The Nextel, with an expansion coefficient 10 times that of the 99.8-percent pure silica, acts like a preshrunk concrete reinforcing bar in the fiber matrix.

One characteristic of the FRCI-12 HRSI tiles places the reaction cured glass (black) coating into a compression as it is cured. This sharply reduces the sensitivity to cracking of coating during handling and operations. In addition to the improved coating compatibility, the FRCI-12 tiles are about 10 percent lighter than the HRSI 99.8-percent pure silica tiles. This reduces weight about 10 percent. The FRCI-12 HRSI tiles have also demonstrated a tensile (pull) strength at least three times greater than that of the HRSI 99.8-percent pure silica tiles and a use temperature approximately 37 °C (100 °F) higher.

The FRCI-12 HRSI tile manufacturing process is essentially the same as the 99.8-percent pure silica HRSI tiles, the only change being in the "wet end" prebinding of the slurry before it is cast. It also requires a higher sintering temperature. When dried, a rigid block is produced as in the 99.8-percent pure silica HRSI tile. These blocks will be cut into quarters and then machined to the precise dimensions required for each tile. The FRCI-12 tiles are the same 152.4- by 152.4-millimeters (6 by 6 inch) size tiles and vary in thickness from 25.4 millimeters (1 inch) to 127 millimeters (5 inches). They will also vary in size and shape at the closeout areas as in the 99.8-percent pure silica HRSI tiles. The FRCI-12 HRSI tiles are bonded to the orbiter essentially the same as the 99.8-percent silica HRSI tiles.

On Orbiter 102, the *Columbia*, the 99.8-percent pure silica LRSI tiles are the same construction and have the same basic functions as the 99.8-percent pure silica HRSI tiles. They are thinner [5.0 to 35 millimeters (0.2 to 1.4 inches)] than the HRSI tiles to minimize orbiter weight. The 99.8-percent pure silica LRSI tiles are manufactured in the same manner as the 99.8-percent pure silica HRSI tiles, except that the tiles are 203 by 203 millimeters (8 by 8 inches) and a white optical and moisture-resistant coating is applied ten mils thick to the top and side of the LRSI tiles. The white coating provides additional on-orbit thermal control for the orbiter. The coating comprises primarily silica compounds with shiny aluminum oxide to obtain optical properties. The coated 99.8-percent pure silica LRSI tiles are treated with bulk waterproofing similar to the 99.8-percent pure silica HRSI tiles. The 99.8-percent pure silica LRSI tiles are installed on the orbiter in the same manner as the

Space Shuttle Spacecraft Systems

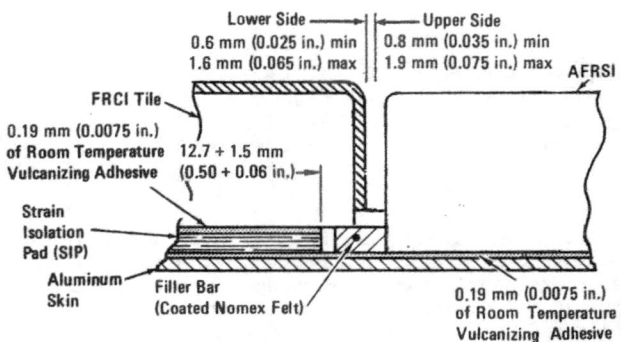

FRCI Tile and AFRSI Interface, Orbiter 103 and Subsequent

99.8-percent pure silica HRSI tiles (room temperature vulcanizing, SIP, filler bar, and vacuum bond-cure). The 99.8-percent LRSI has a surface emittance of 0.8 and a solar absorptance of 0.32. More than 6,000 99.8-percent pure silica LRSI tiles are used.

On Orbiter 099, the *Challenger,* some of the OMS/RCS pods 99.8-percent pure silica LRSI tiles are replaced with an advanced flexible reusable surface insulation (AFRSI) quilted fabric blanket; on Orbiter 103 and subsequent orbiters all the LRSI tiles except for some areas on the OMS/RCS pods are replaced with AFRSI. The AFRSI consists of a low density fibrous, silica batting which is made up of high-purity silica (glass), 99.8-percent amorphous silica fibers (fibers derived from common sand and 1 to 2 mils thick). This batting is sandwiched between an outer woven silica high temperature fabric and an inner lower temperature woven glass fabric. The composite is sewn with silica thread, which provides a quilt like look. The AFRSI blankets are treated with a material to provide water-repellency. The AFRSI composite density is approximately 4.9 kilograms per cubic meter (11 pounds per cubic foot) and varies in thickness from 11 to 24 millimeters (0.45 to 0.95 inch). The blankets are bonded directly to the spacecraft by a room temperature vulcanizing (RTV) silicon adhesive. The silicon adhesive is applied at a thickness of 5.08 millimeters (0.20 inch). The very thin glue line reduces weight and minimizes the thermal expansion during temperature changes. The sewn quilted fabric blanket is obtained in 0.9 by 0.9 meter (3 by 3 feet) squares at the proper thickness by Johns Manville. Rockwell then cuts the blanket, if required, to fit and bonds the blanket to the spacecraft. Rockwell manufactures and sews the closeout blankets.

The nose cap and the leading edges of the orbiter wing utilize reinforced carbon-carbon (RCC) panels to maintain airfoil shape above 1,260°C (2,300°F). The wing leading edge is made up of 44 RCC panels (22 each wing), whereas the nose cap is one piece. RCC fabrication begins with a nylon cloth graphitized and impregnated with a phenolic resin. This impregnated cloth is layed up as a laminate and cured in an autoclave. After cure, the laminate is pyrolized (taking the resin out) at high temperature to convert the resin to carbon. The part is then impregnated with furfural alcohol in a vacuum chamber, then cured and pyrolized again to convert furfural alcohol to carbon. This process is repeated three times until the required reinforced carbon-carbon is achieved.

For it to become an oxidation-resistant coating, the material is packed in a retort with a dry pack material made up of a mixture of alumina, silicon, and silicon carbide. The retort is placed in a furnace, and the coating process takes place in argon with a stepped-time-temperature cycle up to 1,760°C (3,200°F). A diffusion reaction occurs between the dry pack and carbon-carbon in which the outer layers of the carbon-carbon are converted to silicon carbides (whiteish-grey color) with no thickness increase. It is this silicon-carbide coating which protects the carbon-carbon from oxidation. Further oxidation resistance is provided by impregnation of a coated RCC

Space Shuttle Spacecraft Systems

part with tetraethyl-ortho silicate (TEOS). The RCC laminate is superior to a sandwich design because it is light in weight and rugged and it promotes internal cross radiation from the hot stagnation region to cooler areas, thus reducing stagnation temperatures and thermal gradients around the leading edge. The operating range of RCC is from minus 121 °C (minus 250 °F) to about 1,648 °C (3,000 °F). The RCC is highly resistant to fatigue loading, which will be experienced during ascent and entry.

The RCC panels are mechanically attached to the wing with a series of floating joints to reduce loading on the panels due to wing deflections. The seal between each wing leading edge panel is referred to as a "tee" seal. The "tee" seals allow lateral motion and for thermal expansion differences between the RCC and the orbiter wing cooler structure, behind the leading edge. In addition, it prevents the direct flow of hot boundary layer gases into the wing leading edge cavity during entry. The "tee" seals are constructed of RCC.

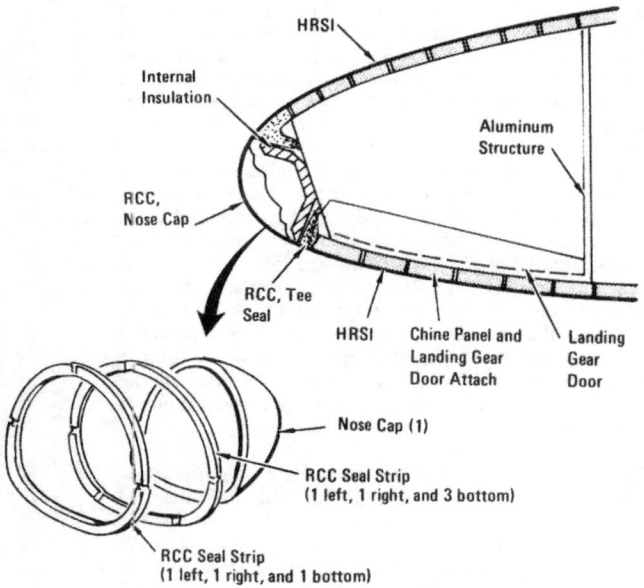

RCC Nose Cap, Orbiter 102 and 099

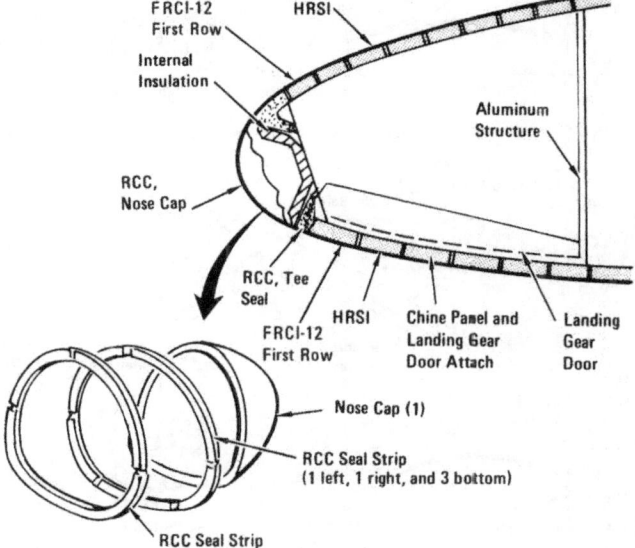

RCC Nose Cap, Orbiter 103 and Subsequent

172

Space Shuttle Spacecraft Systems

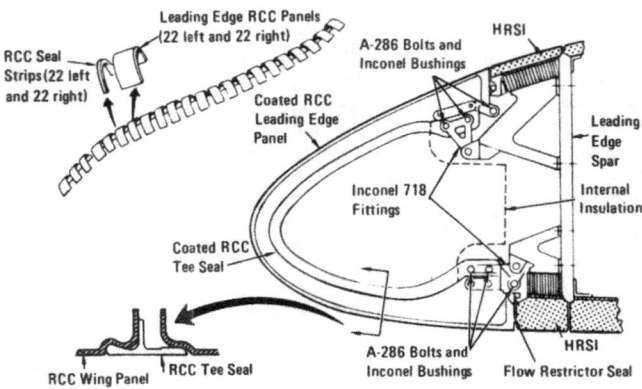

RCC Wing Leading Edge, Orbiters 102 and 099

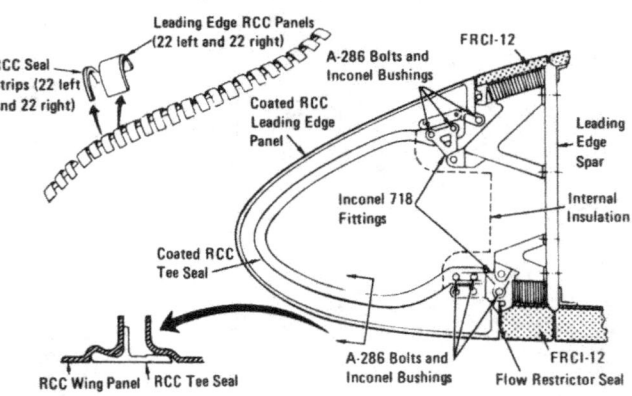

RCC Wing Leading Edge, Orbiter 103 and Subsequent

Since carbon is not a good insulator, the adjacent aluminum and the metallic attachments are protected from exceeding temperature limits by internal insulation. Inconel 718 and A-286 fittings are bolted to flanges on the RCC components and attached to the aluminum wing spars and nose bulkhead. Inconel-covered dynaflex insulation protects the metallic attach fittings and spar from the heat emitted from the inside surface of the RCC wing panels. The nose cap thermal insulation utilizes a blanket made from ceramic fibers and filled with silica fibers and some 99.8-percent pure silica HRSI tiles or FRCI-12 tiles on Orbiter 103 and subsequent orbiters to protect the forward fuselage from the heat emitted from the hot inside surface of the RCC. Approximately six HRSI tiles were removed from the surrounding area of the forward orbiter/ET attach point and an AB312 ceramic cloth blanket is placed on the forward fuselage in this area and RCC is placed over the blanket and attached by metal standoffs.

Thermal barriers are utilized in the closeout between various components of the orbiter and the TPS such as the forward RCS and aft RCS, rudder/speed brake, nose and main landing gear doors, crew hatch, vent doors, external tank umbilical doors, vertical stabilizer/aft fuselage interface, payload bay doors, wing leading edge RCC/HRSI interface, and nose cap/HRSI interface. The various materials utilized are white AB312 ceramic alumina boro silica fibers or black pigmented AB312 ceramic fiber cloth braided around an inner tubular spring made from Inconel X750 wire with silica fibers within the tube, alumina mat, quartz thread, and Macor-machinable ceramic.

Where surface pressure gradients would cause cross-flow of boundary layer air within the intertile gaps, tile gap fillers are provided to minimize increased heating within the gaps. The tile gap filler materials consist of white AB312 ceramic alumina boro silica fibers or black pigmented AB312 ceramic fiber cloth cover with alumina fibers within the cover and are used around

Space Shuttle Spacecraft Systems

the leading edge of the forward fuselage nose cap, windshields and side hatch, wing, vertical stabilizer and trailing edge of elevons, vertical stabilizer, rudder/speed brake, body flap, and heat shield of the Shuttle's main engines.

Fused silica threaded inserts and plugs are used in the tiles to provide access through the tiles to remove door or access panel attachments.

The contractors are Vought Corporation, Dallas, TX (RCC); Lockheed Missiles and Space Co., Inc., Sunnyvale, CA (HRSI and LRSI tiles and HRSI-FRCI-12 tiles); Albany International Research Co., Dedham, MA (Nomex felt); General Electric, Waterford, NY (room temperature vulcanizing adhesive); 3M Company, St. Paul, MN (AB312 fibers); Santa Fe Textiles, Santa Ana, CA (Inconel 750 wire spring and fabric

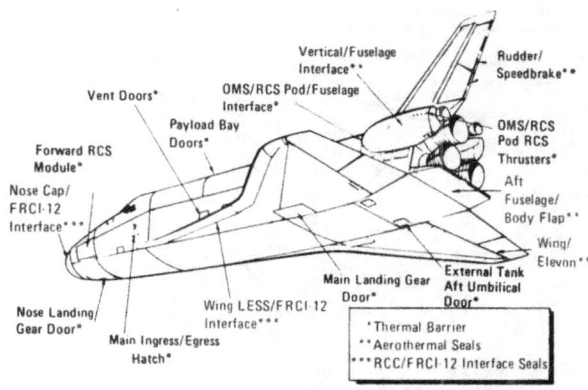

Thermal Barriers, Orbiter 103 and Subsequent

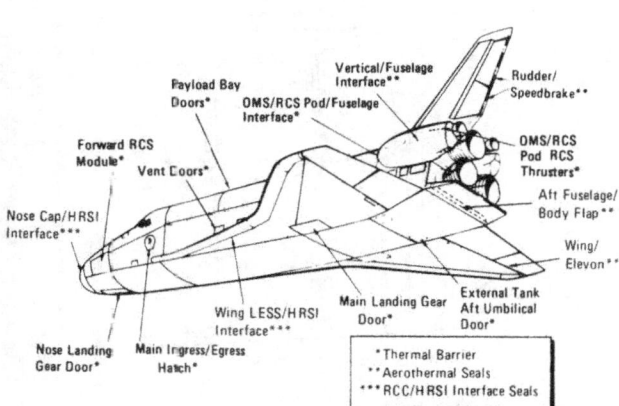

Thermal Barriers, Orbiter 102 and 099

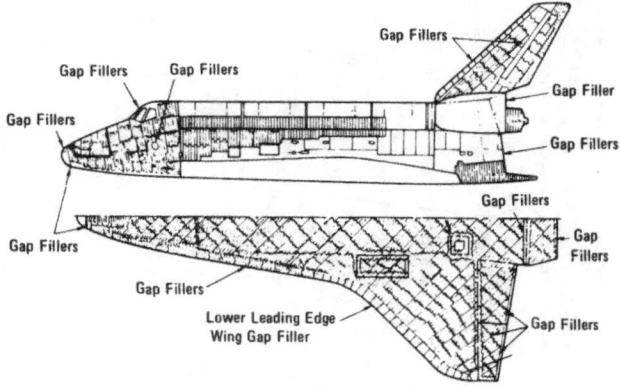

TPS Tile Gap Filler Locations

Space Shuttle Spacecraft Systems

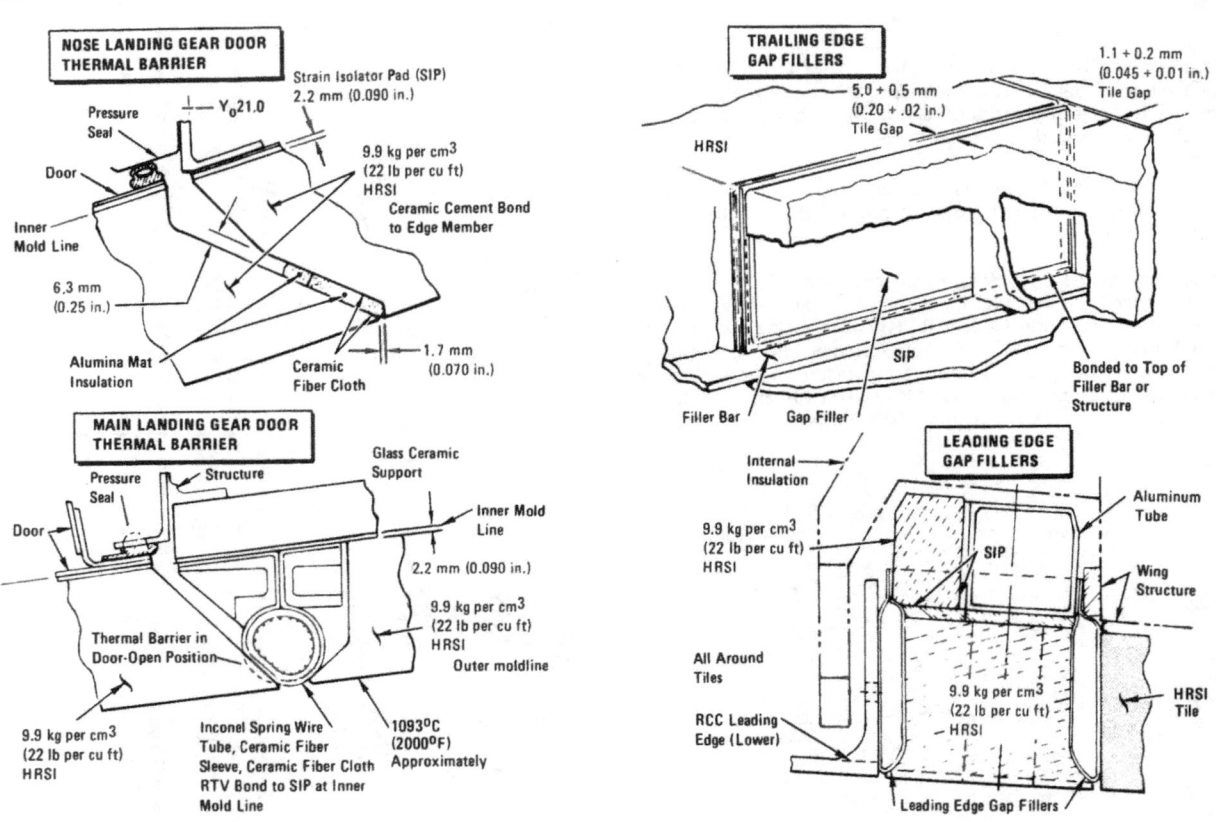

Typical TPS Gap Fillers and Thermal Barriers, Orbiter 102 and 099

Space Shuttle Spacecraft Systems

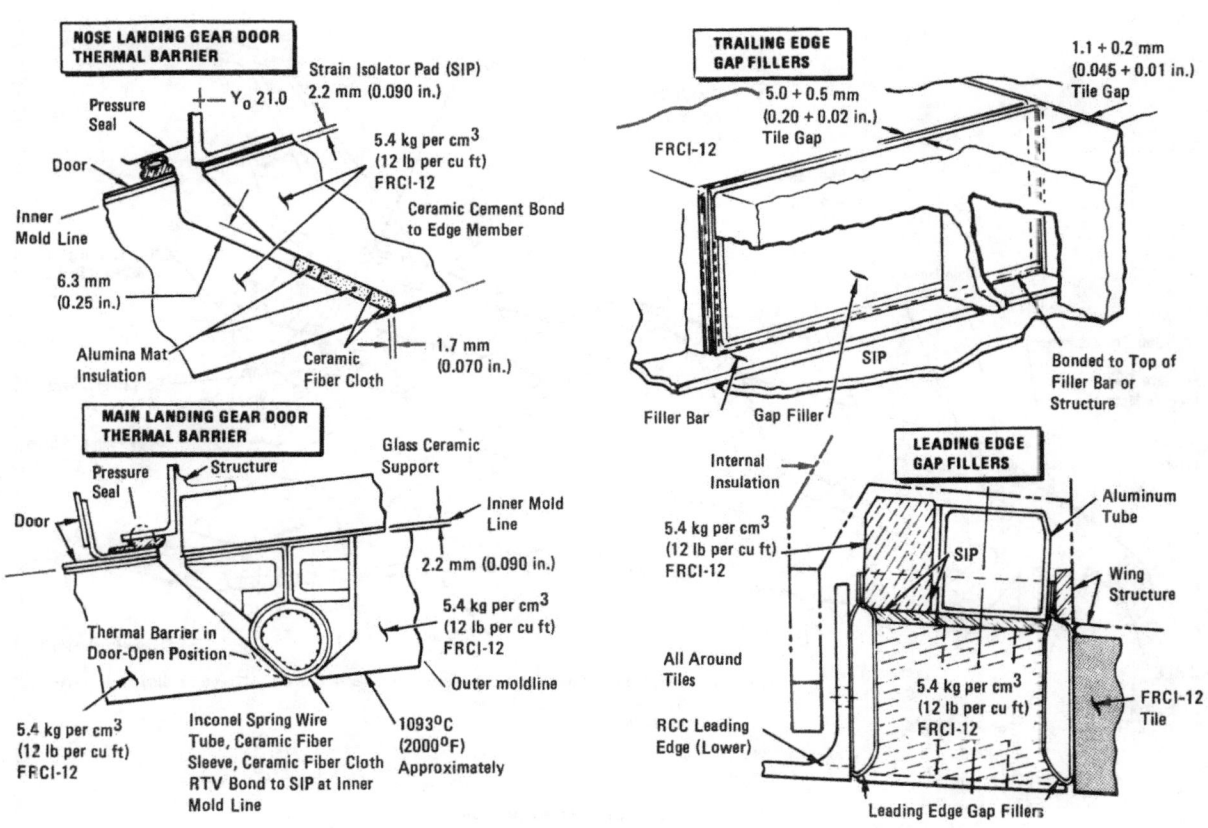

Typical TPS Gap Fillers and Thermal Barriers, Orbiter 103 and Subsequent

 **Space Shuttle Spacecraft Systems**

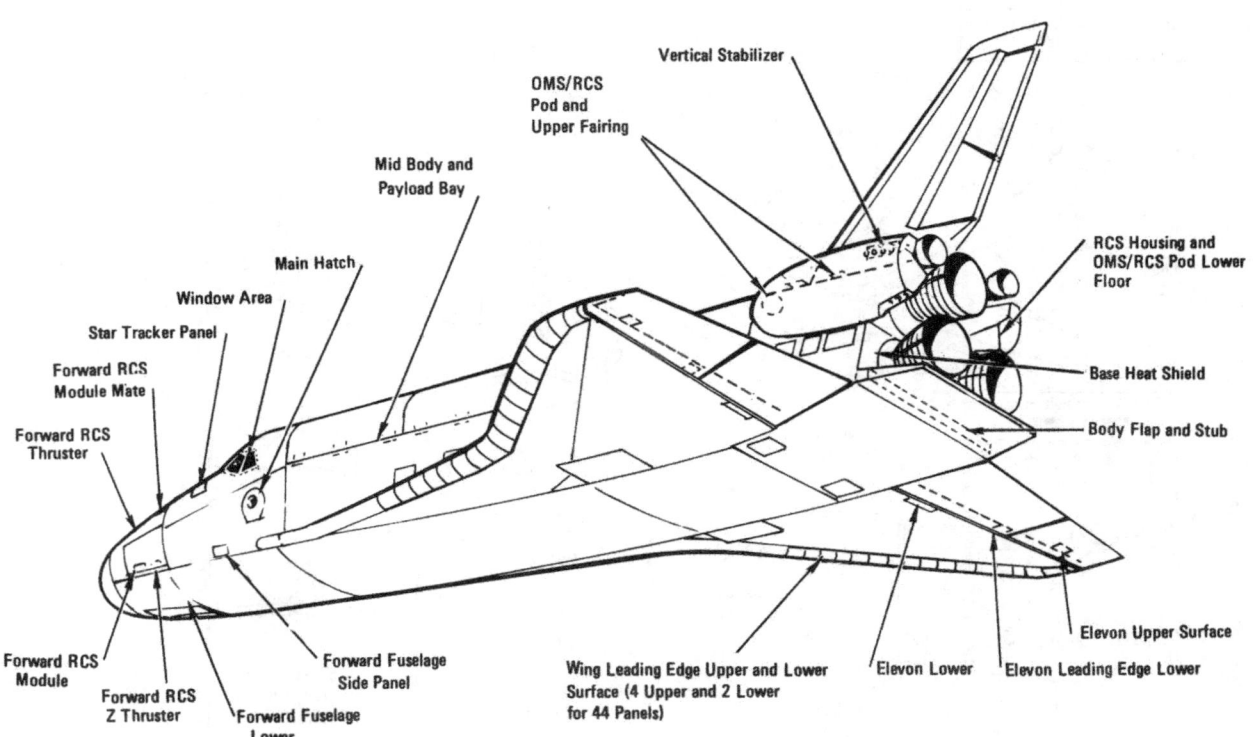

Fused Silica Insert and Plug Locations

Space Shuttle Spacecraft Systems

sleeving); ICI United States, Inc., Wilmington, DE (alumina mat); J.P. Stevens Co., Log Angeles (quartz thread); Corning Glass Works, Corning, NY (Macor-machinable glass ceramic); Velcro Corp., NY, NY (Velcro hooks and loops); Prodesco, Perkasie, PA (fibrous pile-S glass); Johns Manville, Waterville, OH (high purity silica glass); and Johns Manville, Manville, NJ (advanced flexible reusable surface insulation quilted fabric blanket).

ORBITAL MANEUVERING SYSTEM

The orbital maneuvering system (OMS) provides the thrust for orbit insertion, orbit circularization, orbit transfer, rendezvous and deorbit. The OMS is housed in two independent pods located on each side of the orbiter's aft fuselage. The pods also house the aft reaction control system and are referred to as the orbital maneuvering system/reaction control system (OMS/RCS) pods. Each pod contains one OMS engine and the hardware needed to pressurize, store, and distribute the propellants to perform the velocity maneuvers. The two pods provide the redundancy for the OMS.

The first OMS thrusting (OMS-1 insertion) is used to raise the orbiter's orbit to a predetermined elliptical orbit after main engine cut-off (MECO). The crew has the ability to modify the targeting data in the computers by inserting new values via the CRT keyboard unit.

The OMS-2 thrusting period occurs near the apogee of the orbit established by OMS-1 thrusting, and is used to circularize the orbit. The targeting data for this thrusting period also is selected before the mission and also can be modified by the flight crew if necessary.

Additional OMS thrusting periods are performed according to specific mission requirements: to modify the orbit to perform a rendezvous to deploy a payload, or to transfer to another orbit.

Deorbit thrusting normally is performed with the two OMS engines. Target data for the deorbit maneuver is computed by the ground and loaded in the GPC's via uplink. This data also is voiced to the flight crew for verification of loaded values. After verifying the correct deorbit data, the flight crew initiates the OMS gimbal test via item entry in the CRT/keyboard unit. Once the test is complete, the flight crew maneuvers the spacecraft to the desired deorbit thrusting attitude, using the rotational hand controller and RCS thrusters. Upon the completion of the thrusting period the OMS engines are disarmed and any residual velocity is nulled by the use of the translation hand controller (THC) and RCS thrusters.

The system in each pod consists of a high pressure gaseous helium storage tank, pressure regulation system, pressure relief valves, fuel tank, oxidizer tank, propellant distribution system, OMS engine, plus 12 RCS primary and two RCS vernier engines. (The RCS is described in the RCS section.)

In each of the two OMS pods, gaseous helium pressure is supplied to the oxidizer and fuel tanks, which, in turn, supply the oxidizer and fuel under pressure to that OMS engine. The propellants are nitrogen tetroxide as the oxidizer and monomethylhydrazine as the fuel. The propellants are earth-storable and hypergolic (ignite upon contact with each other). The propellants are directed to the OMS engine, where they ignite (hypergolic), producing a hot gas, thus thrust.

 **Space Shuttle Spacecraft Systems**

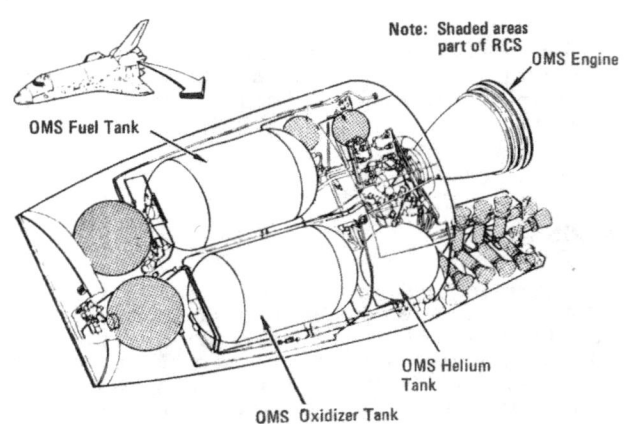

Orbital Maneuvering System

When the fuel reaches the engine, it is directed first through the combustion chamber walls (this regeneratively cools the chamber walls) and then into the engine injector. The oxidizer goes directly into engine injector. The nozzle extension is radiant-cooled and constructed of aluminum alloy.

Each OMS engine provides 26,688 Newtons (6,000 pounds) of thrust. The oxidizer-to-fuel mixture ratio is 1:65, the expansion ratio of the nozzle exit to the throat is 55:1, the dry weight of the engine is 117 kilograms (260 pounds), and the chamber pressure of the engine is 6,468 millimeters of mercury (mmHg) (125 psia).

Each OMS engine is reusable for 100 missions and capable of 1,000 starts and 15 hours of cumulative firing. The minimum firing duration of an OMS engine is two seconds. The OMS may be utilized to provide thrust above 21,336 meters (70,000 feet). For velocity changes of between 0.9 and 1.8 meters per second (3 and 6 feet per second), the normal operating mode is one OMS engine.

Each engine has two electromechanical gimbal actuators, which control the OMS engine thrust direction in pitch and yaw (thrust vector control). The OMS engines can be used singularly by directing the thrust vector through the orbiter center of

Space Shuttle Spacecraft Systems

gravity or together by directing the thrust vector of each engine parallel to the other. It is noted that during a two-OMS-engine thrusting period, the RCS will come into operation only in the event the OMS gimbal exceeds gimbal rate or gimbal limits and should not normally come into operation during the OMS thrust period. However, during a one-OMS engine thrusting period, roll RCS is required. The pitch and yaw actuators are identical except for the stroke length and contain redundant channels (active and standby), which couple to a common drive assembly.

The OMS/RCS pods are designed to be reused for up to 100 missions with only minor repair, refurbishment, and maintenance. The pods are removable to facilitate orbiter turnaround from a mission, if required.

HELIUM PRESSURIZATION

The pressurization system consists of helium tanks, helium pressurization valves, two dual-pressure regulator assemblies, parallel vapor isolation valves, dual-series/parallel check valve assemblies, and pressure relief valves.

The helium storage tank in each pod consists of a titanium liner with a fiberglass structural overwrap. This increases safety and decreases the weight of the tank 32 percent over that of conventional tanks. The helium tank is 1,021 millimeters (40.2 inches) in diameter with a volume of 0.48 cubic meter (17.03 cubic feet) minimum. Its dry weight is 123 kg (272 pounds). The helium tank operating pressure range is 248,400 to 23,805 millimeters of mercury (mmHg) (4,800 to 460 psia) with a maximum operating limit of 252,281 mmHg (4,875 psia) at 37 °C (100 °F).

A pressure sensor downstream of each helium tank in each pod monitors the helium source pressure and transmits it to the "N_2, He, Kit He" switch on Panel F7. In the He position the left and right OMS helium pressure is displayed on the "OMS Press Left, Right" meters. This pressure also is transmitted for display on the CRT.

The two helium pressurization valves in each pod permit helium source pressure to the propellant tanks or isolate the helium from the propellant tanks. The parallel paths in each pod assure helium flow to the propellant tanks of that pod. The helium valves are continuous-duty and solenoid-operated. The valves are energized open and spring-loaded closed. The "OMS He Press/Vapor Isol" switches on Panel 08 permit automatic or manual control of the valves. With the switches in the GPC position, the valves are automatically controlled by the GPC during an engine thrusting sequence. The valves are controlled manually by placing the switches to the OPEN or CLOSE positions.

The pressure regulators reduce the helium source pressure to the desired working pressure. Pressure regulation is accomplished by two pressure-regulating assemblies, one downstream of each helium pressurization valve. Each assembly contains primary and secondary regulators in series and a flow limiter. The primary regulator is normally the controlling regulator. The secondary regulator is normally open during a dynamic flow condition. The secondary regulator will not become the controlling regulator until the primary regulator allows a higher pressure than normal, thus allowing the secondary regulator to assume control. All regulator assemblies are in reference to a bellows assembly that is vented to ambient. The primary regulator outlet pressure at normal flow is 13,041 to 13,558 mmHg (252 to 262 psig) and at high flow is 12,782 mmHg (247 psig) minimum, with lockup at 13,765 mmHg (266 psig). The secondary regulator outlet pressure at normal flow is 13,403 to 13,920 mmHg (259 to 269 psig) and at high flow is 13,144 mmHg (254 psig) minimum, with lockup at 14,127 mmHg (273 psig). The flow limiter restricts the flow to a maximum of 29 cubic meters (1,040 cubic feet) per minute and to a minimum of 8 cubic meters (304 cubic feet) per minute.

Space Shuttle Spacecraft Systems

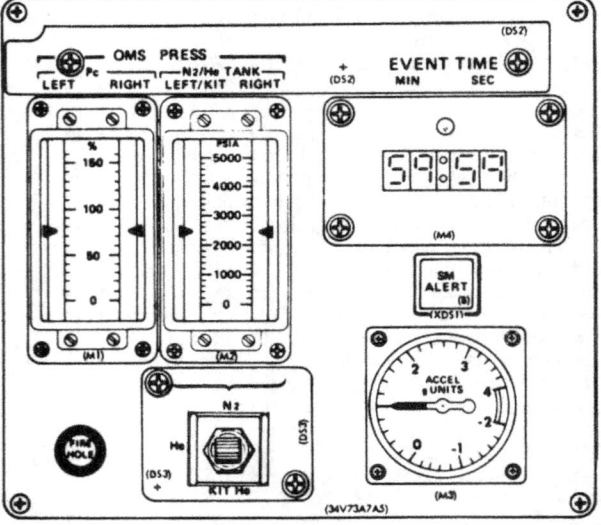

Panel F7

The vapor isolation valves located in the oxidizer pressurization line to the oxidizer tank prevent oxidizer vapor from migrating upstream to the pressure regulators, which could degrade the pressure regulators. These are low pressure, two-position, two-way, solenoid-operated valves. The valves are energized open and spring-loaded closed. They can be commanded manually or automatically by the positioning of the "He Press/Vapor Isol" switches on Panel 08. When either of the two switches A or B are in the OPEN position, both vapor isolation valves are energized open and when both switches A or B are in the CLOSE position, both vapor isolation valves, are closed. When the A and B switches are in the GPC positions, the GPC opens or closes the valves automatically.

The check valve assembly in each parallel path contains four independent check valves connected in a series-parallel configuration provide a positive-checking action against a reverse flow of propellant liquid or vapor and the parallel path permits redundant paths of helium to be directed to the propellant tanks. Filters are incorporated in the inlet of each check valve assembly.

Two-pressure sensors in the helium pressurization line upstream of the fuel and oxidizer tanks monitor the regulated tank pressure and transmit it to the RCS/OMS PRESS rotary switch on Panel 03. When in the OMS PRPLNT position, the left and right fuel and oxidizer pressure are displayed. If the tank pressure is lower than 12,109 mmHg (234 psia) or above 14,697 mmHg (284 psia), the applicable left or right OMS red caution/warning light on Panel F7 will illuminate. These pressures also are transmitted for display on the CRT.

The relief valves in each pressurization path limit excessive pressure in the propellant tanks. Each pressure relief valve consists of a relief valve, a burst diaphragm, and a filter. In the event of excessive pressure caused by helium or propellant vapor, the diaphragm ruptures and the relief valve opens and

Space Shuttle Spacecraft Systems

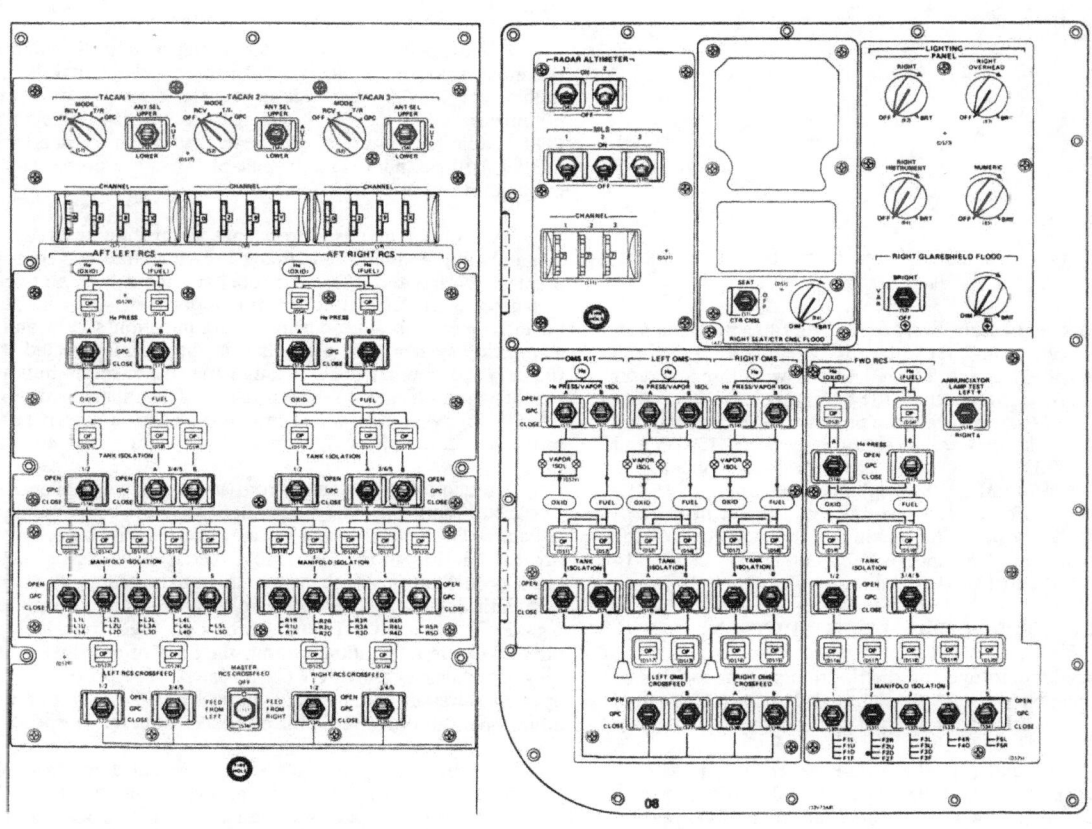

Panel 07 *Panel 08*

Space Shuttle Spacecraft Systems

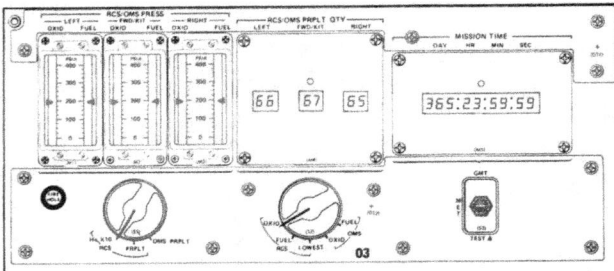

Panel 03

vents the excessive pressure overboard. The filter prevents particulates from the non-fragmentation type diaphragm from entering the relief valve seat. The relief valve will close and reset after the pressure has returned to the operating level. The burst diaphragm is used to provide a more positive seal of helium than the relief valve. The diaphragm ruptures between 15,680 and 16,197 mmHg (303 to 313 psig). The relief valve opens at a minimum of 14,697 mmHg (284 psig) and a maximum of 16,197 mmHg (313 psig). The relief valve minimum reseat pressure is 14,490 mmHg (280 psig). The maximum flow capacity of the relief valve at 15°C (60°F) and 16,197 mmHg (313 psig) is 14 cubic meters (520 cubic feet) per minute.

PROPELLANT STORAGE AND DISTRIBUTION

The propellant storage and distribution system consists of one fuel and one oxidizer tank in each pod. It also contains propellant feed, interconnect lines, and isolation valves.

The OMS integral propellant tankage of both pods provides the capability of a 304 meter-per-second (1,000-foot-per-second) velocity change with a 29,484-kilogram (65,000-pound) payload in the orbiter's payload bay. An OMS pod crossfeed line allows use of the propellants in both pods for operation of either OMS engine.

The propellant supply is contained in domed cylindrical titanium tanks within each pod. Each propellant tank is 2,448 millimeters (96.38 inches) long with a diameter of 1,247.1 millimeters (49.1 inches) and a volume of 0.54 cubic meters (89.89 cubic feet) unpressurized. The dry weight of each tank is 113 kg (250 pounds). The propellant tanks are pressurized by the helium system.

Each tank contains a propellant acquisition and retention assembly in the aft end. Each tank is divided into two compartments: forward and aft. The propellant acquisition and retention assembly is located in the aft compartment and consists of an intermediate bulkhead with communication screen and an acquisition system. The propellant in the tank is directed from the forward compartment through the intermediate bulkhead via the communication screen into the aft compartment during OMS velocity maneuvers. The communication screen retains propellant in the aft compartment during zero-g conditions.

The acquisition assembly consists of four stub galleries and a collector manifold. The stub galleries acquire wall bound propellant at OMS starting and during RCS velocity maneuvers to prevent gas ingestion. The stub galleries have screens which allow propellant flow while preventing gas ingestion. The collector manifold is connected to the stub galleries and also contains a gas arrestor screen to further prevent gas ingestion which permits OMS engine ignition without the need of a propellant settling maneuver employing RCS thrusters. The propellant tank operating pressure limit is 12,937 mmHg (250 psi), with a maximum operating pressure limit of 16,197 mmHg (313 psia).

A capacitance gauging system is installed in each OMS propellant tank to measure the propellant in that tank. The system consists of capacitance gauging with a forward and aft probe, and a totalizer. The forward and aft fuel probes use fuel

Space Shuttle Spacecraft Systems

(which is a conductor) as one plate of the capacitor and a glass tube which is metallized on the inside as the other. The forward and aft oxidizer probes use two concentric nickel tubes as the capacitor plates, and oxidizer as the dielectric. (Helium is also a dielectric, but has a different dielectric constant from the oxidizer.) The aft probes in each tank contain a resistive temperature-sensing element to provide correction to variations in fluid density. The bottom of the tanks and the area of the communication screens cannot be measured.

The totalizer receives OMS valve operation information and inputs from the forward and aft probes from each tank, and outputs total and aft quantities, and a low-level quantity output. The inputs from the OMS valves allow control logic in the totalizer to determine when an OMS engine is thrusting and which tanks are being used. The totalizer begins an engine flow rate/time integration process at the start of the OMS thrusting period, which reduces the indicated amount of propellants by a preset estimated rate for the first 14.8 seconds. After 14.8 seconds of OMS thrusting, the probe capacitance gauging system outputs are enabled, which permits the display of actual propellant quantity remaining. The totalizer and capacitance gauging is displayed on the OMS/RCS PRPLNT QTY meters on Panel 03 when the rotary switch is positioned to OMS FUEL or OXID positions.

When the wet or dry analog comparator indicates the forward probe is dry, the ungaugable propellant in the region of the intermediate bulkhead is added to the aft probe output quantity, decreasing the total quantity at a preset rate for 98.15 seconds, and inhibits updates from the probes. After 98.15 seconds of thrusting, the aft probe output inhibit is removed and the aft probe updates the total quantity. When the quantity is less than five percent, the low-level signal is output.

The propellant from each tank in a pod flows through parallel ac motor-driven isolation valves that can be actuated manually by the OPEN or CLOSE position or automatically by the GPC position of the tank isolation A or B switch on Panel 08. The left and right OMS tank isolation A or B valves permit propellant flow to the OMS engine of that pod or isolate propellant flow to the OMS engine. An indicator above each switch on Panel 08 indicates the status of that set of valves (one fuel and one oxidizer valve for set A and set B). OP (open) when both valves are open, barber pole when the valves are in transit or one valve is open and one closed, and CL (close) when both valves are closed.

ENGINE BIPROPELLANT VALVE ASSEMBLY

Each OMS engine receives the pressure-fed propellants at its bipropellant valve assembly. The bipropellant valve assembly consists of a gaseous nitrogen (GN_2) pressure vessel, an engine pressurization valve, a GN_2 regulator, a GN_2 relief valve, a check valve, a line reservoir, engine purge valves with a check valve, solenoid control valves, actuators, and bipropellant ball valves.

One GN_2 spherical tank is mounted on each bipropellant valve assembly to supply pressure to that engine pressurization valve. The tank contains enough GN_2 to operate the ball valves and purge the engine 10 times. The nominal GN_2 tank capacity is 983 cubic centimeters (60 cubic inches). The maximum tank operating pressure is 155,250 mmHg (3,000 psi), with a proof pressure of 310,500 mmHg (6,000 psig).

Two pressure sensors downstream of the GN_2 tank monitor the pressure; the N_2, He, Kit He switch on Panel F7, when positioned to N_2, displays tank pressure on the "OMS Press N_2 Tank Left, Right" meters.

The GN_2 engine pressurization valve, one for each engine, are dual-coil solenoid-operated valves. Each valve is energized open and spring-loaded closed. The engine pressurization valve permits GN_2 flow to the control ball valves and purge valves 1 and 2 or blocks the flow. The GN_2 engine pressurization valves

Space Shuttle Spacecraft Systems

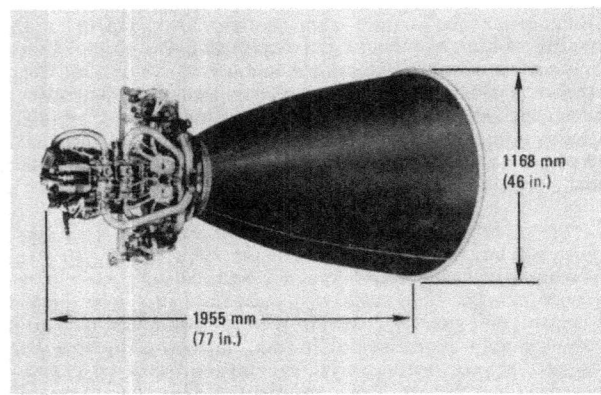

OMS Engine

OMS Engine (Minus Nozzle)

are controlled by the OMS ENG-LEFT RIGHT switches on Panel C3. When the OMS ENG LEFT switch is placed in the ARM PRESS position, the left OMS engine pod's pressurization valve is energized open. When the OMS ENG RIGHT switch is placed in the ARM PRESS position, the right OMS engine pod's pressurization valve is energized open. The GN_2 engine pressurization valve, when energized open, allows GN_2 supply pressure to be directed into a regulator, through a check valve, an in-line reservoir, and to a pair of engine control valves. The engine control valves are controlled by the OMS thrust ON/OFF commands from the GPC's.

A single-stage regulator is installed in each GN_2 pneumatic control system between the GN_2 engine pressurization valve and the engine bipropellant valves. The regulator reduces the GN_2 service pressure to a desired working pressure between 16,301 and 17,336 mmHg (315 and 335 psig).

Each pneumatic control system has a pressure relief valve downstream of the GN_2 regulator. This relief valve limits the pressure applied to the engine control valve in the event of a GN_2 pressure regulator malfunction preventing damage to the engine control valves and bipropellant ball valve actuators. The valve relieves at 23,287 mmHg (450 psig) and resets at 20,700 mmHg (400 psig).

A pressure sensor downstream of the GN_2 regulator transmits the regulated GN_2 pressure to the CRT display.

A check valve downstream of the GN_2 regulator closes to isolate the GN_2 source pressure side of the valve in the event of loss of GN_2 upstream of the check valve.

A 311-cubic centimeter (19-cubic-inch) GN_2 spherical reservoir is installed downstream of the check valve and upstream from the engine control valves. The reservoir accumulates enough GN_2 to provide the pressure necessary to operate the engine bipropellant ball valves one time with the

Space Shuttle Spacecraft Systems

GN$_2$ pressurization valve closed or in the event of loss of pressure in the line upstream of the check valve.

In the GN$_2$ engine pressurization system there are two solenoid-operated, three-way, two-position control valves for actuator control and positioning of the ball valves. Control valve No. 1 controls No. 1 actuator and two ball valves. Control valve No. 2 controls the No. 2 actuator and two ball valves. Each control valve has two solenoid coils; energizing either coil will open the control valve.

The right OMS engine GN$_2$ solenoid control valves 1 and 2 are energized open by computer commands provided that the RIGHT OMS ENG switch on Panel C3 in the ARM or ARM/PRESS position and the RIGHT OMS ENG VLV switch on Panel 016 is on; the valves are de-energized normally when thrust off is commanded or if the RIGHT OMS ENG switch is positioned to OFF. The left OMS engine GN$_2$ solenoid control valves 1 and 2 are controlled in the same manner except through the LEFT OMS ENG switch on panel C3 and the left OMS ENG VLV switch on panel 014.

When the GN$_2$ solenoid control valves are energized open, the regulated GN$_2$ pressure is directed into the two GN$_2$ actuators of that engine. The GN$_2$ acts against the piston in each actuator, overcoming the spring force on the opposite side of each actuator. Each actuator has a rack and pinion gear and the linear motion of the actuator connecting arm is converted into rotary motion which drives two ball valves, one fuel and one oxidizer, to the open position. Each pair of ball valves opens simultaneously. Fuel and oxidizer is then directed to the injector of the engine, where the hypergolic propellants ignite, producing thrust.

Each OMS engine has a sensor that monitors its chamber pressure and transmits it to the OMS PRESS LEFT and RIGHT PC (chamber pressure) meter on Panel F7.

When the computer commands are removed from the solenoid control valves or an engine's OMS ENG switch on Panel or ENG VLV switch on Panel 014/016 is positioned OFF, the solenoid control valves are de-energized removing the regulated GN$_2$ pressure from the actuators and venting the gas in the actuators. The actuator spring forces the actuator piston to move in the opposite direction and the GN$_2$ goes back through the solenoid control valve and is vented overboard. The actuators drive the fuel and oxidizer ball valves closed simultaneously. The series arrangement of ball valves ensure engine thrust termination.

Each actuator incorporates a linear position transducer which supplies information on ball valve positioning to a CRT and also to the OMS gauging system, where gauging processes are initiated upon detecting both valves open or terminated when both valves are closed.

Check valves are installed in the vent port outlet of each GN$_2$ solenoid control valve on the spring pressure side of each actuator to protect the seal of these components from the hard vacuum of space.

Each engine has two engine purge valves in series. These valves, solenoid-operated open and spring-loaded closed are normally energized open after each thrusting period by the GPC's unless inhibited by a crew entry on the MANEUVER CRT display. When the two purge valves of an engine are opened 0.36 second after OMS engine control valve command terminates, GN$_2$ is permitted to flow through the valves and a check valve in the fuel line downstream of the ball valves and out through the engine injector and combustion chamber to space for two seconds. This purges the residual fuel from the combustion chamber and injector of the engine, permitting a safe engine restart. The purge valves are then de-energized and closed. When the purge is completed, the GN$_2$ tank pressurization supply valve is closed, placing the respective OMS ENG

 **Space Shuttle Spacecraft Systems**

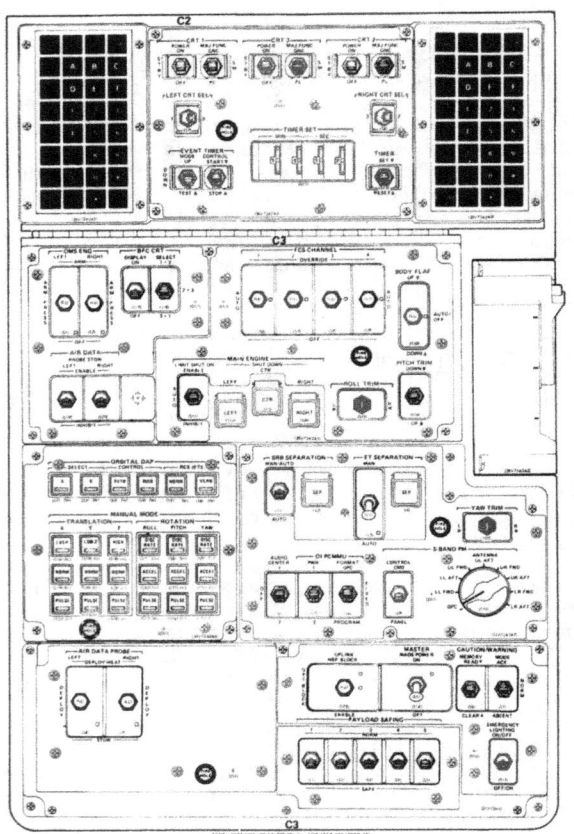
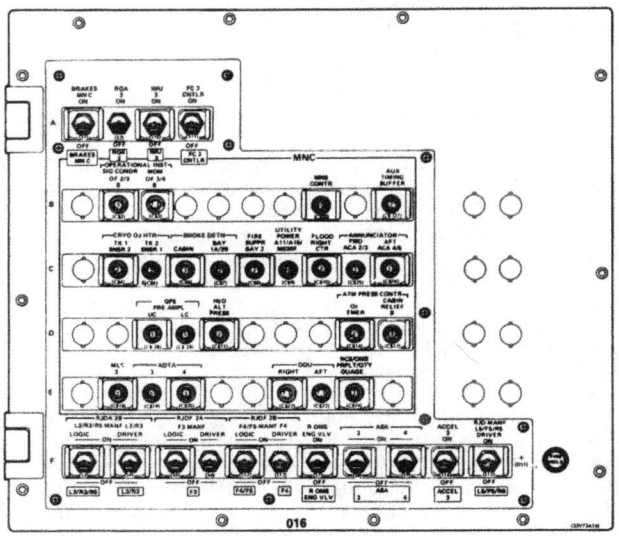

Panel 016

Space Shuttle Spacecraft Systems

switch (Panel C3) to OFF. The check valve downstream of the purge valves prevents the fuel from flowing to the engine purge valves during engine thrusting.

ENGINE

When the fuel reaches the thrust chamber, it is directed first through 102 coolant channels in the combustion chamber wall, providing regenerative cooling to the combustion chamber walls, then to the injector of the engine. The oxidizer is routed directly to the injector. The platelet injector assembly consists of a stack of plates, each with an etched pattern that provides for proper distribution and propellant injection velocity vector. The stack is diffusion-bonded and welded to the back side of the injector. The fuel and oxidizer orifices are positioned so that they will impinge and atomize, causing fuel and oxidizer ignition because of hypergolic reaction.

The contoured nozzle extension is bolted to the aft flange of the combustion chamber. The nozzle extension is made of a columbium alloy and is radiantly cooled.

The nominal flow rate of oxidizer and fuel to each engine is 5.4 kilograms (11.93 pounds) per second and 3.2 kilograms (7.23 pounds) per second, respectively. The vacuum specific impulse of each engine is 313 seconds.

OMS THRUSTING SEQUENCE

The OMS thrusting sequence commands the OMS engines on or off and commands the engine purge function. The flight crew can select, via item entry on the maneuver display, a one- or two-engine thrusting period, and can inhibit the OMS engine purge.

The sequence determines which engines are selected and then provides the necessary computer commands to open the appropriate helium vapor isolation and engine GN_2 solenoid control valves 1 and 2 and sets an engine-on indicator. The sequence will monitor the OMS engine fail flags and, if one or both engines have failed, issue the appropriate OMS cutoff command as soon as the crew has confirmed the failure by placing the "OMS Eng" switch in the Off position. This will then result in the appropriate engine(s) control valve commands being terminated.

In a normal OMS thrusting period, when the OMS cutoff flag is true, the sequence terminates commands to the helium pressurization, helium vapor, and GN_2 engine control valves 1 and 2. If the engine purge sequence is not inhibited, the sequence will check for the left and right engine ARM PRESS signals and, after 0.36 second open the engine GN_2 purge valves for two seconds for the engines which have the ARM PRESS signals present.

THERMAL CONTROL

OMS thermal control is achieved by insulation on the interior surface of the pods that enclose the OMS hardware components and the use of strip heaters. Line-wraparound heaters and insulation control the crossfeed lines. The heaters prevent propellant from freezing in the tanks and lines. The heater system is divided into two areas: the OMS/RCS pods and the aft fuselage crossfeed and bleed lines. Each heater system has two redundant heater systems, A and B, and are controlled by the RCS/OMS HEATERS switches on Panel A14.

Each OMS/RCS pod is divided into nine heater areas. Each of the heater patches in the pods contains an A and B element, and each element has a thermostat which controls the temperature from 12° to 23°C (55° to 75°F). These heater elements are controlled by the LEFT and RIGHT POD switches on Panel A14. Temperature sensors located throughout the pods supply temperature information to the propellant thermal CRT display and telemetry.

Space Shuttle Spacecraft Systems

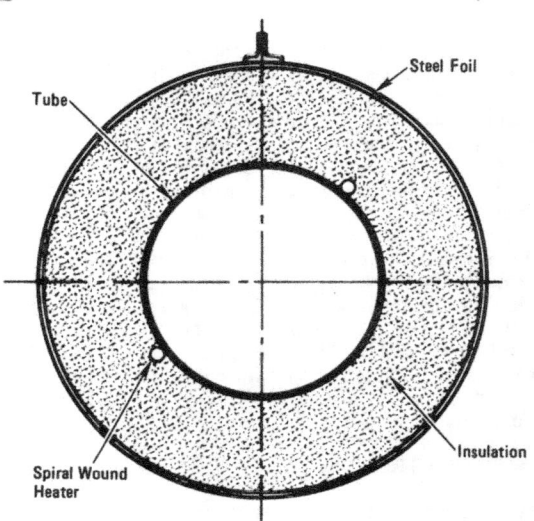

OMS Crossfeed System Tubing Insulation Installation

The crossfeed line thermal control in the aft fuselage is divided into 11 heater areas. Each area is heated in parallel by heater systems A and B, and each area has a control thermostat to maintain temperature at 12 °C minimum (55 °F) to 23 °C (75 °F) maximum. These heater elements are controlled by the respective CRSFD LINES switch on Panel A14. Temperature sensors are located on the control thermostats and on the crossfeed and bleed lines. The temperature sensors supply temperature information on the propellant thermal CRT display and telemetry.

ENGINE GIMBAL AND CONTROL ASSEMBLY

The engine gimbal and control assembly consists of a gimbal ring assembly, a gimbal actuator assembly, and a gimbal actuator controller. The engine gimbal ring assembly and the gimbal actuator assembly provide the OMS thrust vector control by gimbaling the engines in pitch and yaw. Each engine has a pitch actuator and a yaw actuator. Each actuator is extended or retracted by a pair of dual-redundant electric motors and are actuated by GPC control signals.

The gimbal ring assembly contains the two mounting pads to attach the engine to the gimbal ring and the two pads to attach the engine gimbal ring to the orbiter, which transmits the engine thrust through the mounting pads.

The pitch and yaw gimbal actuator assembly for each OMS engine provides the force to gimbal the engines. Each actuator has a primary and secondary servo control unit housed in the gimbal actuator controller. The primary and secondary channels are isolated and are not operated concurrently. Each actuator consists of redundant brushless dc motors, gear train, jack screw and nut-tube assembly, and linear position feedback. A GPC position command signal from the primary servo control unit energizes the primary dc motor, which is coupled with a reduction gear and a no-back device. The output from the primary power train drives the jack screw of the drive assembly, causing the nut-tube to translate (with the secondary power train at idle) and results in angular engine movement, providing linear travel. If the primary power train is inoperative, a GPC position command from the secondary servo control unit energizes the secondary dc motor, and results in angular engine movement, providing linear travel by applying torque to the nut-tube through the spline that extends along the nut-tube for the stroke length of the unit. Rotation of the nut-tube about the stationary jack screw causes the nut-tube to move along the

Space Shuttle Spacecraft Systems

screw. The no-back device prevents backdriving of the standby system.

The electrical interface, power, and electronic control elements for active and standby control channels are assembled in separate enclosures designated as active actuator controller and standby actuator controller. These are mounted on the OMS/RCS pod structure. The active and standby actuator controllers are electrically and mechanically interchangeable.

The gimbal assembly provides control angles of plus or minus six degrees in pitch and plus or minus seven degrees in yaw with clearance provided for an additional one degree for snubbing and tolerances. The engine null position is with the engine nozzles up 15° 49 seconds (as projected in the orbiter XZ plane) and outboard 6° 30 seconds (measured in the 15° 49 minutes plane).

The thrust vector control (TVC) command SOP (software operating procedure) processes and outputs pitch and yaw OMS engine actuator commands, and the actuator power selection discretes. The OMS TVC command SOP is active during OPS (operational sequences), orbit insertion (OMS-1 and OMS-2), orbit coast, deorbit, deorbit coast, and RTLS (return to launch site) abort.

The flight crew can select either the primary or the secondary motors of the pitch and yaw actuators by item entry on the maneuver display, or can select actuators off. The actuator command outputs are selected by the TVC command SOP depending upon the MM (major modes), deorbit maneuver coast, and RTLS abort, CG (center of gravity) TRIM and GIMBAL CHECK flags. The deorbit maneuver coast flag causes the TVC command SOP to output I-loaded valves to command the engines to the entry stowed position. The presence of the RTLS abort and CG trim flags causes the engines to be commanded to a predefined position with the thrust vector through the CG. The MM RTLS flag by itself will cause the engines to be commanded to a stowed position for RTLS (return to launch site) entry. The gimbal check flag causes the engines to be commanded to plus 7° yaw and 6° pitch, then to minus 7° yaw and 6° pitch, then back to 0° yaw and pitch. In the absence of these flags, the TVC command SOP will output the digital autopilot (DAP) gimbal actuator commands to the engine actuators. The backup flight control system (BFS) allows only manual TVC during a thrusting period, but it is otherwise similar.

The OMS TVC feedback SOP monitors the primary and secondary actuator selection discretes from the maneuver display and performs compensation on the selected pitch and yaw actuator feedback data. This data is output to the OMS actuator fault detection and identification (FDI) and to the maneuver display. The OMS TVC feedback SOP is active in OPS orbit insertion OMS-1, OMS-2, orbit coast, deorbit maneuver, and deorbit maneuver coast. The present OMS gimbal positions can be monitored on the maneuver CRT display when this SOP is active and the primary or secondary actuator motors are selected.

OMS PAYLOAD BAY KIT described in previous press reference has not received authorization to build.

OMS-RCS INTERCONNECT

An interconnect between the OMS crossfeed line and the aft RCS manifolds provides the capability for operating the aft RCS using 453 kilograms (1,000 pounds) per pod of OMS propellant for orbital maneuvers. The aft RCS may use OMS propellant from either one or both OMS pods in orbit.

The orbital interconnect sequence is available during orbit operations and orbit checkout.

The flight crew must first configure the following switches (using a feed from the left OMS as an example): position the Aft LEFT RCS TANK ISOLATION 1/2, 3/4/5A, 3/4/5B and

Space Shuttle Spacecraft Systems

AFT RIGHT RCS TANK ISOLATION 1/2, 3/4/5B switches on Panel 07 to CLOSE position; check that talkback indicator above these switches indicates CL, position Aft LEFT RCS CROSSFEED 1/2, 3/4/5 and Aft RIGHT RCS CROSSFEED 1/2, 3/4/5 switches to OPEN, check that indicators show OP, open Left OMS TANK ISOLATION A and B valves (Panel 08) and verify talkback indicators show OP; open LEFT OMS XFD A and B valves and verify indicators show OP; close RIGHT OMS XFD A and B valves and verify indicators show CL, and position Left OMS HE PRESS/VAPOR ISOL VALVE A switch in GPC position. The left OMS to aft RCS interconnect sequence can then be initiated by the item entry on the RCS SPEC display.

The left OMS helium pressure vapor isolation valve A will be commanded open when the left OMS tank (ullage) pressure decays to 12,213 mmHg (236 psig) and the open commands will be terminated 30 seconds later. If the left OMS tank (ullage) pressure remains below 12,213 mmHg (236 psia), the sequence will set an OMS/RCS valve "mis-compare" flag and will set a Class 3 alarm and a CRT fault message. The sequence also will enable the OMS to RCS gauging sequence at the same time.

The flight crew can terminate the sequence and inhibit the OMS to RCS gauging sequence by use of the OMS PRESS ENA-OFF item entry on the RCS SPEC display. The valves can then be reconfigured to their normal position on Panels 07 and 08. The OMS to aft RCS interconnect sequence is not available in the backup flight system (BFS).

OMS TO RCS GAUGING SEQUENCE

The OMS to aft RCS propellant quantities are calculated by burn time integration. Once each cycle, the accumulated aft RCS thruster cycles are used to compute the OMS propellant used since initiation of the gauging. The number of RCS thruster cycles is provided by the RCS command SOP to account for minimum-inpulse firing of the RCS thrusters. The gauging sequence is initiated by item entry of the OMS right or OMS left interconnect on the RCS SPEC CRT display and is terminated by the return to normal item entry.

The gauging sequence maintains a cumulative total of left and right OMS propellant used during OMS to aft RCS interconnects, and displays the cumulative totals as percent of left and right OMS propellant on the RCS SPEC display. The flight crew will be alerted by a Class 3 alarm and a fault message when the total quantity used from either OMS pod exceeds 354 kg (1,000 pounds) or 8.37 percent.

ABORT CONTROL SEQUENCES

The abort control sequence is the software that manages, among other items, the OMS and aft RCS configuration and thrusting periods during ascent aborts to improve performance or to consume OMS and aft RCS propellants for center-of-gravity control.

Pre-mission determined parameters are provided for the OMS and aft RCS thrusting periods during aborts, since the propellant loading and orbiter center of gravity vary with the mission.

The premission parameters for the abort to orbit (ATO) thrusting period are modified during flight based on the vehicle velocity at abort initiation.

The abort once around (AOA) premission parameters are grouped with different values for early or late AOA.

The return to launch site (RTLS) parameters are contained in a single table.

Space Shuttle Spacecraft Systems

The abort control sequence is available in OPS (operational sequences) 1 and 6, and is initiated at SRB separation if selected before then, or at time of selection if after SRB separation.

ATO-AOA ABORTS. The OMS and aft RCS begin thrusting as soon as an ATO or AOA is initiated with one main engine out.

In some cases of an abort, an OMS to aft RCS interconnect is not desired. A parallel aft RCS plus X thrusting period using aft RCS propellant and the four aft RCS plus X thrusters will be performed with the OMS-1 thrusting period to achieve the desired orbit. If a plus X aft RCS thrusting period is required before main engine cutoff (MECO), the abort control sequence will command the four aft plus X RCS jets on if vehicle acceleration is greater than 0.8 g and will monitor the RCS cutoff time to terminate the thrusting period. If an RCS propellant dump (burn) is required before MECO and vehicle acceleration is greater than 1.8 g, the abort control sequence will command an eight aft RCS null thrust and monitor the RCS cutoff time to terminate the thrusting period.

In other abort cases, an OMS to aft RCS interconnect is desired. This thrusting is performed with the OMS and four aft RCS plus X thrusters to consume OMS propellant for orbiter center-of-gravity control and the four aft RCS plus X thrusters are used to consume OMS propellant. More aft RCS jets can be commanded if needed to increase OMS propellant usage, such as OMS propellant dump (burn) where 14 aft RCS null jets can be commanded to thrust to improve orbiter center-of-gravity location.

If the amount of OMS propellant used before MECO results in less than 28 percent of OMS propellants remaining, a 15-second aft RCS ullage thrusting period is made after MECO to provide a positive OMS propellant feed to start the OMS-1 thrusting period.

The OMS to aft RCS interconnect sequence provides for an automatic interconnect of the OMS propellant to the aft RCS when required and reconfigures the propellant feed from the OMS and aft RCS tanks to their normal state after the thrusting periods are completed. The interconnect sequence is initiated by the abort control sequence.

In order to establish a known configuration of the valves, the interconnect sequence terminates the GPC commands to the following valves if they have not been terminated before honoring a request from the abort control sequence: left and right OMS crossfeed A and B valves, aft RCS crossfeed valves, and aft RCS tank isolation valves.

A request from the abort control sequence for an OMS to aft RCS interconnect will sequentially configure the OMS/RCS valves as follows: close the left and right aft RCS propellant tank isolation valves, open the left and right OMS crossfeed A and B valves, and open the left and right aft RCS crossfeed valves. The OMS to aft RCS interconnect complete flag is then set to true.

When the abort control sequence requests a return to normal configuration, all affected OMS/RCS propellant valve commands are removed to establish a known condition and the interconnect sequence will then sequentially configure the valves as follows: close aft RCS crossfeed valves, close left and right OMS crossfeed valves, open aft RCS propellant tank isolation valves. The OMS to aft RCS reconfiguration complete flag is then set to false, and the sequence is terminated.

RETURN TO LAUNCH SITE ABORT. An RTLS abort requires OMS propellant dumping by burning the OMS propellant through both OMS engines and through the 24 aft RCS thrusters to improve abort performance and to achieve an acceptable entry orbiter vehicle weight and center-of-gravity location. The thrusting period is premission determined and depends on the OMS propellant load.

Space Shuttle Spacecraft Systems

The OMS engines first start the thrusting sequence and after the OMS to aft RCS interconnect is complete, the aft RCS thrusters are commanded on. The OMS engines and RCS thrusters then continue their burn for a predetermined period. The interconnect sequence is the same as mentioned in the ATO-AOA aborts. The OMS and aft RCS will begin thrusting at SRB staging if the abort is initiated during the first stage of flight, or immediately upon abort initiation if during second stage.

CONTINGENCY ABORT. A contingency abort is selected automatically at the loss of a second main engine or manually by the flight crew using an item entry on the RTLS TRAJ (trajectory 2) or RTLS TRANS CRT displays. For the contingency aborts the OMS to aft RCS interconnect is performed in a modified manner to allow continuous flow of propellants to the aft RCS jets for vehicle control and to allow contingency rapid dump (burn) of OMS aft RCS propellants. The abort control sequence tracks the total OMS and aft RCS on-time to determine the amount of propellants used.

The request for an interconnect will cause the interconnect sequence to configure the valves sequentially as follows: open the aft RCS crossfeed valves, open the left OMS crossfeed valves A, open the right OMS crossfeed valves B, close the left and right aft RCS tank isolation valves, open the left OMS crossfeed valves B, and open the right OMS crossfeed valves A. The OMS to aft RCS interconnect complete flag will then be set to true.

If the rapid dump (burn) is selected before MECO, the OMS to aft RCS interconnect occurs and both OMS engines and 24 aft RCS jets are commanded to thrust until the desired amount of propellant is consumed. The rapid dump (burn) will be interrupted during external tank separation if the thrusting period is not completed before MECO; otherwise the thrusting period terminates at thrusting time = 0, or if the normal acceleration exceeds a threshold value.

Upon completion of the thrusting period, the OMS to aft RCS configuration flag will be set to false and the sequence will be terminated. A return to normal configuration request by the abort control sequence will cause the interconnect sequence to configure the valves sequentially as follows: open aft RCS propellant tank isolation valves, close the aft RCS crossfeed valves, close the left and right OMS crossfeed A and B valves. The OMS to aft RCS interconnect complete flag will be set to false, and the sequence will be terminated.

OMS ENGINE FAULT DETECTION AND IDENTIFICATION (FDI)

The OMS engine FDI is to detect and identify off-nominal performance of the OMS engine such as off-failures during OMS thrusting periods, on-failures after or before a thrusting period, and high or low engine chamber pressures.

The OMS engine FDI is performed by the redundancy management software, and assumes that the flight crew arms only the OMS engines to be used, and that OMS engines which are not armed cannot be used for thrusting. The OMS engine FDI will be initialized at SRB ignition and terminate after the OMS-1 thrusting period, or in the case of an RTLS abort, will terminate at the transition from RTLS entry to the RTLS landing sequence program. The FDI also will be initiated before each OMS burn and will be terminated after the OMS thrusting period is complete.

The OMS engine FDI uses both a velocity comparison and a chamber pressure (Pc) comparison method to determine a failed on or off engine. The velocity comparison is used only after MECO, since the OMS thrust is small in comparison to MPS thrust before MECO.

The measured velocity increment is compared to a predetermined one-engine and two-engine acceleration threshold value by the redundancy management software to

Space Shuttle Spacecraft Systems

determine the number of engines actually firing. This information along with the assumption that an armed engine is to be used, allows the software to determine if the engine has low thrust or has shut down prematurely.

The Pc comparison test compares a predetermined threshold chamber pressure level to the measured Pc to determine a failed engine (on, off, or low thrust).

The engine-on command and the Pc are used before MECO to determine a failed engine. The velocity indication and the Pc indication are used after MECO to determine a failed engine. If, after MECO, the engine fails the chamber pressure test but passes the velocity test, the engine will be considered as failed. Such a failure would result in illumination of the red RIGHT (or LEFT) OMS caution/warning light on Panel F7, the master alarm, and a fault message. In addition, if an engine fails the Pc and velocity tests, a down arrow is displayed on the maneuver CRT next to the failed engine.

When the flight crew disarms a failed engine by turning the ARM/PRESS switch on Panel C3 to off, a signal is sent to the OMS thrusting sequence to shut down the engine and to signal guidance to reconfigure. Guidance reconfigures and downmodes from two OMS engines to one OMS engine, to four plus X RCS jets.

OMS ACTUATOR FDI

The OMS actuator FDI detects and identifies off-nominal performance of the pitch and yaw gimbal actuators of the OMS engines.

The OMS actuator FDI is divided into two parts. The first determines if the actuator should move from its present position. If so, the second part determines the amount to be moved and whether the desired movement has occurred.

The first part checks the gimbal deflection error (which is the difference between the commanded new position and the last known position) of each actuator and determines whether the actuator should extend or retract, or if it is being driven against a stop. If the actuator is in the desired position, or being driven against a stop, the first part of the process will be repeated. Should the first part determine that the actuator should move, the second part of the actuator FDI process will be performed.

The second part of the actuator FDI process checks the present position of each actuator against its last known position to determine whether the actuator has moved more than a threshold amount. If the actuator has not moved more than this amount, an actuator failure is incremented by one. Each time the actuator subsequently fails this test, the failure is again incremented by one. When the actuator failure counter reaches an I-loaded value of four, the actuator is declared failed and a fault message output. The actuator failure counter is reset to zero any time the actuator passes the threshold test.

The first and second parts of the actuator FDI continue to perform in this manner. The actuator FDI process can detect full-off gimbal failures, and full-on failures indirectly. The full-on failure is determined when the gimbal actuator servo loop determines that the gimbal has extended or retracted too far and commands reverse motion. If no motion occurs, the actuator will be declared failed. The flight crew response to a failed actuator is to select the secondary actuator electronics by item entry on the maneuver CRT display.

The contractors are McDonnell Douglas Astronautics Co., St. Louis, MO (OMS pod assembly and integration); Aerojet Liquid Rocket Co., Sacramento, CA (OMS engine); Aerojet Manufacturing Co., Fullerton, CA (OMS propellant tanks); Aircraft Contours, Los Angeles, CA (OMS pod edge member); Brunswick-Wintec, El Segundo, CA (OMS propellant tank acquisition screen assembly); Consolidated Controls, El Segundo,

Space Shuttle Spacecraft Systems

CA (high- and low-pressure solenoid valves and OMS regulators); Fairchild Stratos, Manhattan Beach, CA (hypergolic servicing couplings); Metal Bellows Co., Chatsworth, CA (alignment bellows); Simmonds Precision Instruments, Vergennes, VT (OMS propellant gauging system); SSP Products, Burbank, CA (gimbal bellows assembly); Tayco Engineering, Long Beach, CA (electrical heaters); AiResearch Manufacturing Co., Torrance, CA (gimbal actuators); Futurecraft Corp., City of Industry, CA (OMS engine valve components); L.A. Gauge, Sun Valley, CA (ball valves); PSM Division of Fansteel, Los Angeles, CA (OMS nozzle extension fabrication); Rexnord Inc., Downers Grove, IL (OMS engine bearings); Sterer Engineering and Manufacturing, Pasadena, CA (OMS engine pressure regulators); Parker-Hannifin, Irvine, CA (OMS propellant tank isolation valves, relief valves, manifold interconnect valves; Rockwell International, Rocketdyne Division, Canoga Park, CA (OMS check valves); Brunswick, Lincoln, NE (OMS helium tanks).

REACTION CONTROL SYSTEM

The forward and aft reaction control system (RCS) on the orbiter provides the thrust for attitude (rotational) maneuvers (pitch, yaw, roll) as well as the thrust for small velocity changes along the orbiter axis (translation maneuvers) when the orbiter is above 21,336 meters (70,000 feet).

The reaction control system (RCS) engines are first used in the mission after main engine cutoff (MECO) to maintain the vehicle in attitude until external tank (ET) separation and to provide at external tank separation a minus-Z translational maneuver of minus 1.2 meters (4 feet) per second with use of the aft RCS pods and forward RCS module. During OMS burns, vehicle attitude is maintained by gimbaling the OMS engines. The RCS engines will come into operation during a two-OMS-engine thrusting period only if the OMS gimbal exceeds gimbal rate or gimbal limits. They should not normally come into operation during the OMS thrust period. However, during a one-OMS-engine thrusting period, roll RCS is required.

After the OMS-1 thrusting period, the orbiter is in attitude hold. The flight crew uses the translation hand control to command translational maneuvers to the RCS and to null the residual velocity. Attitude hold is maintained until time to maneuver to the OMS-2 burn attitude, which is performed manually by the flight crew with the rotational hand controller. The RCS plus X jets can be used to complete either the OMS-1 or OMS-2 thrusting periods or to perform the OMS-2 thrusting period in event of OMS engine failures. In this case, the OMS-to-aft-RCS interconnect capability will be used to feed OMS propellant to the RCS thrusters. In orbit, after the OMS-2 thrusting period is complete, RCS maneuvers in attitude or translation are performed according to the flight plan. For attitude control on orbit, the flight crew can select either primary or vernier jets.

During deorbit, the RCS will nominally be used only to maneuver to the OMS thrusting attitude and to null the remaining delta velocity. If both OMS engines malfunction, the RCS can be used to perform or complete the deorbit maneuver; again in this case the OMS-to-aft-RCS interconnect capability will be selected to feed OMS propellant to the RCS thrusters.

Once the deorbit thrusting period is complete, the orbiter is maneuvered to the entry interface (EI) attitude. The forward RCS module is disabled at EI.

Space Shuttle Spacecraft Systems

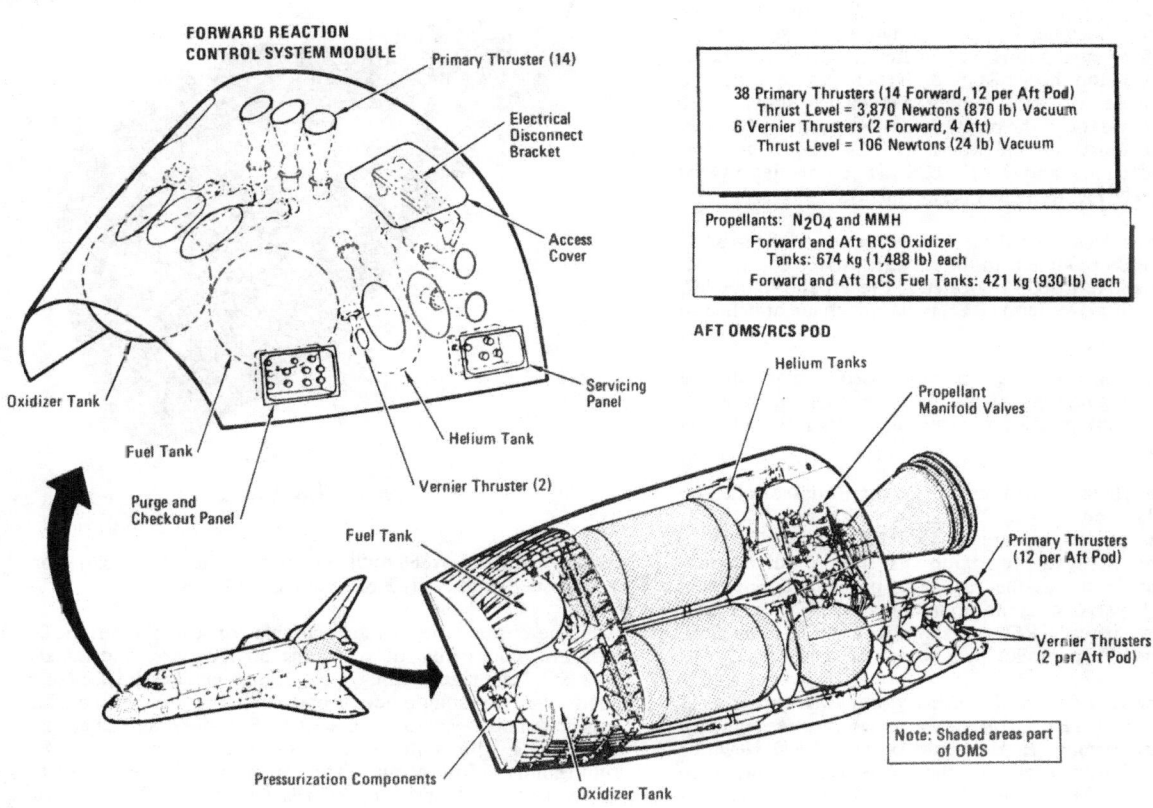

Reaction Control System

Space Shuttle Spacecraft Systems

From EI 121,920 meters (400,000 feet) the orbiter is controlled in roll, pitch, and yaw with the aft RCS thrusters. The orbiter's ailerons become effective at a dynamic pressure of 517 millimeters of mercury (mmHg) per meters squared (10 pounds per square foot), and the aft RCS roll jets are deactivated. At a dynamic pressure of 1.035 mmHg per meters squared (20 pounds per square foot), the orbiter's elevons become effective, and the aft RCS pitch jets are deactivated. The rudder becomes activated at Mach 3.5, and the aft RCS yaw jets are deactivated at approximately 13,716 meters (45,000 feet).

The RCS is located in three different areas of the orbiter. The forward RCS module is in the forward fuselage nose area. The aft (right and left) RCS is located in the left and right OMS (orbital maneuvering system)/RCS pods, which are attached to the aft fuselage.

Each RCS consists of high-pressure gaseous helium storage tanks, pressure regulation systems, pressure relief valves, a fuel and an oxidizer tank, and a propellant distribution system to its RCS engines.

The two helium tanks of each RCS supply gaseous helium pressure to the oxidizer and fuel tanks, which supply the oxidizer and fuel under pressure to the RCS engines. The propellants are nitrogen tetroxide as the oxidizer and monomethylhydrazine as the fuel. The propellants are earth storable and hypergolic (ignite upon contact with each other). The propellants of each RCS are supplied to its RCS engines, where they ignite (hypergolic), producing a hot gas, thus thrust.

The forward RCS has 14 primary and two vernier RCS engines. The aft RCS has 12 primary and two vernier engines in each pod. The primary RCS engines provide 3,870 Newtons (870 pounds) of thrust each, and the vernier RCS engines provide 106 Newtons (24 pounds) of thrust each. The oxidizer-to-fuel ratio for each engine is 1:6. The chamber pressure of the

Primary Engine

primary engines is 7,866 millimeters of mercury (152 psia). For each vernier engine it is 5,692 mmHg (110 psia).

The primary engines are reusable for a minimum of 100 missions and capable of sustaining 50,000 starts and 20,000 seconds of cumulative firing. The primary engines are used in a steady-state thrusting mode of one to 150 seconds, with a contingency of 500 seconds for the aft RCS engines, as well as in a pulse mode with a minimum impulse thrusting time of 80 milliseconds (0.080 second). The expansion ratio (exit area to throat area) of the primary engines ranges from 22:1 to 30:1. The primary engines provide the normal control of the orbiter.

Space Shuttle Spacecraft Systems

Vernier Engine

The vernier engines are also reusable for a minimum of 37 missions and capable of sustaining 500,000 starts and 125,000 seconds of cumulative firings. The vernier engines are used in a steady-state thrusting mode of one to 125 seconds as well as in a pulse mode with a minimum impulse time of 80 milliseconds. The vernier engines are used for finite maneuvers and stationkeeping (or long-time attitude hold). The expansion ratio of the vernier engine ranges from 20:1 to 50:1.

PRESSURIZATION SYSTEM

Each RCS has two helium tanks, isolation valves, pressure regulators, relief valves, and lines for draining and filling.

The helium storage tanks are spherical and of a composite construction consisting of a titanium liner with a Kevlar structural overwrap, which increases safety and decreases the tank weight over conventional titanium tanks. Each helium tank is 475 millimeters (18.71 inches) in diameter with a volume of 49,866 cubic centimeters (3,043 cubic inches). Its dry weight is 10 kilograms (24 pounds). Each helium tank is serviced to 186,000 millimeters of mercury (mmHg) (3,600 psig).

The two helium tanks in each RCS supply gaseous helium individually, one to the fuel tank and one to the oxidizer tank.

There are two helium isolation valves in parallel between the helium tanks and the pressure regulators in each RCS. The helium isolation valves permit the helium source pressure to flow to the propellant tank in the open position and isolate the helium from the propellant tank in the closed position. The helium isolation valves are opened by a solenoid which is momentarily energized until the valve is magnetically latched open. To close the valve, an electric coil surrounding the magnetic latch is energized, allowing spring pressure and helium pressure to force the valve closed.

The helium isolation valves are controlled by the FWD RCS He PRESS A/B switches on Panel 08 and the AFT LEFT RCS He PRESS A/B and AFT RIGHT RCS He PRESS A/B switches on Panel 07. Each switch controls two helium isolation valves, one in the helium oxidizer line and one in the helium fuel line. The switch positions are OPEN, GPC, and CLOSE. Each position is a permanent position, but applies only momentary

Space Shuttle Spacecraft Systems

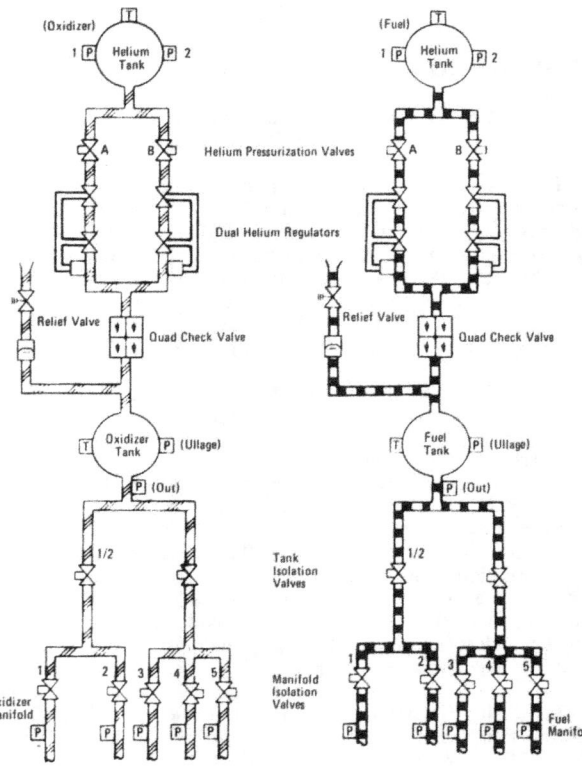

Reaction Control System Pressurization and Propellant Feed System

power to the solenoid due to the logic in the electrical load controller assembly. The GPC position allows the orbiter computer to open or close the valve, and the OPEN and CLOSE positions open and close the valve respectively. A position microswitch in each valve indicates valve position to the controller assembly and controls a position indicator (talkback) above each switch on Panel 07 and 08. When both valves (helium fuel and helium oxidizer) are open, the talkback indicates OP (open), and when both valves are closed the talkback indicates CL (closed). If one valve is open and the other is closed, the talkback indicates barberpole.

The RCS helium supply pressure can be monitored on Panel 03. The rotary switch on Panel 03 positioned to RCS He X10 allows the RCS LEFT, RIGHT, and FWD helium pressures to be displayed on the RCS/OMS PRESS FUEL AND OXID meters on Panel 03.

Helium pressure is regulated by two assemblies, connected in parallel, downstream of each helium isolation valve. The pressure regulators reduce the helium source pressure to the desired working pressure. Each assembly contains two regulators, a primary and a secondary, connected in series. If the primary regulator fails open, the secondary regulator will regulate the pressure. The primary regulates the pressure at 12,523 to 12,834 mmHg (242 to 248 psig). The secondary regulates the pressure at 13,092 to 13,403 mmHg (253 to 259 psig).

Two check valve assemblies in a series-parallel arrangement are located between the pressure regulator assemblies and relief valves. The series arrangement limits the backflow of that propellant vapor and maintains propellant tank pressure integrity in event of an upstream helium leak. The parallel arrangement ensures the flow of helium pressure to the propellant tank if a series check valve fails in the closed position.

Space Shuttle Spacecraft Systems

The helium pressure relief valve assemblies are located between the check valve assemblies and the propellant tanks and vent excessive pressure over-board before it reaches the propellant tanks. Each valve consists of a burst diaphragm, filter, and relief valve. The diaphragm provides a positive seal against helium leakage and will rupture between 16,767 and 17,595 mmHg (324 to 340 psig). The diaphragm is a non-fragmentation type; however, the filter prevents any particles of the burst diaphragm from reaching the relief valve seat. The relief valve relieves at 16,301 mmHg (315 psig) minimum.

PROPELLANT SYSTEM

The propellant system distributes the propellants to the RCS thrusters. Each RCS propellant system consists of the fuel and oxidizer tanks, fuel and oxidizer tank isolation valves, manifold isolation valves, crossfeed valves, and filling, draining, and distribution lines.

Each RCS contains two spherical propellant tanks, one for fuel and one for oxidizer. Each propellant tank in the forward and aft RCS is constructed of titanium and is 9.9 millimeters (0.39 inches) in diameter.

The nominal full load of the forward RCS tanks is 670 kilograms (1,477 pounds) in the oxidizer tank and 421 kilograms (928 pounds) in the fuel tank. The dry weight of the forward tanks is 32 kilograms (70.4 pounds). The aft RCS tanks are over filled and contain a propellant load of 738 kilograms (1,627 pounds) in the oxidizer tank and 448 kilograms (987 pounds) in the fuel tank of each RCS/OMS pod. The dry weight of the aft tanks is 35 kilograms (77 pounds).

Each tank is pressurized with helium, which expels the propellant into an internally mounted surface-tension propellant acquisition screen. It acquires and delivers the propellant to the RCS thrusters on demand. The propellant acquisition device in each RCS propellant tank is required because of the orbiter's orientation during boost, on orbit, and entry and because of the omni-directional acceleration spectrum, which ranges from very high during boost, entry, or abort to very low during orbital operation. The forward RCS propellant tanks have propellant acquisition devices designed to operate primarily in a low-g environment, whereas the aft RCS propellant tanks are designed to operate in both the high and low-g environment. The propellant acquisition devices ensure propellant and pressurant separation during tank operation.

A compartmental tank with individual screen devices in both the upper and lower compartments supplies propellant independant of tank load or orientation. The devices are constructed of stainless steel and mounted in the titanium tank shells. A titanium barrier separates the upper and lower compartments in each tank, and fine stainless steel screens are used in propellant acquisition devices.

At orbiter-ET separation and for orbital operations, propellant flows from the upper compartment bulk region into the channel network to the upper compartment transfer tube and into the lower compartment bulk region. This continues into the upper compartment until gas is ingested into the upper compartment device and transferred to the lower compartment.

The lower compartment of the forward RCS propellant tanks would expel propellant to depletion, as in the case of the upper compartment; however, orbital operations would be terminated with the forward RCS with sufficient propellant reserve to preclude gas ingestion to the forward RCS engines and achieve a 86-percent expulsion efficiency.

During launch, the launch acceleration positions the propellant over the propellant tank outlet. If the only flow path to the lower compartment were through the upper compartment transfer tube, pressure losses due to propellant flow and high-g

Space Shuttle Spacecraft Systems

boost acceleration during the return to launch site (RTLS) abort condition, which requires the aft RCS, would result in helium gas passing directly into the lower compartment early in the expulsion of propellant. This would result in excessive consumption of lower compartment propellant, lower compartment screen breakdown, and helium gas ingestion directly to the RCS engines. To prevent this, a separate subsurface liquid flow path, the abort duct, is provided. It permits propellant flow from the upper to lower compartment, while hydrostatic pressure is utilized to prevent ullage (empty propellant area of tank) gas from passing through the transfer tube into the lower compartment.

The aft RCS propellant tank's lower compartment is not used on orbit, but is required for entry. For entry, the aft RCS tank propellants are positioned 100 degrees away from the tank outlet due to the influence of up to 2.5-g acceleration. As the acceleration builds up, the channel screen in the ullage area of both the upper and lower compartment devices breaks down and ingests gas. During expulsion, the upper compartment propellant may flow through both the transfer tube and the abort duct. When the bulk propellant uncovers the entrance to the abort duct, flow is exclusively through the entry collector to the transfer tube. The abort duct screen remains wet during the balance of the upper compartment expulsion. As entry expulsion continues, propellant is withdrawn from the lower compartment until a 90-percent expulsion efficiency is achieved under this condition.

The aft RCS propellant tanks incorporate the abort duct and entry collector, sumps, and gas traps to ensure proper operation during abort and entry mission phases. Because of these components, the aft RCS propellant tanks are approximately 3.1 kilograms (seven pounds) heavier than the forward RCS propellant tanks. The forward RCS is not required during the abort and entry phases of the mission and does not incorporate these devices.

The left, forward, and right RCS fuel and oxidizer tank ullage pressures can be monitored on Panel 03. When the rotary switch on Panel 03 is positioned to RCS PRPLNT, the left, forward, and right RCS propellant tank pressures are displayed on the RCS/OMS PRESS FUEL, OXID meters on Panel 03.

The left, forward, and right RCS fuel and oxidizer tank ullage pressures will also illuminate the LEFT RCS, FWD RCS, RIGHT RCS red caution/warning respectively if that module ullage tank pressure is below 10,350 mmHg (200 psig) or above 16,146 mmHg (312 psia).

RCS QUANTITY MONITOR

The RCS quantity monitor sequence uses the GPC to calculate the usable percent of fuel and oxidizer in each RCS module (forward, aft left, and aft right). The RCS quantities are computed based upon the pressure, volume, and temperature method, which requires that pressure and temperature measurements be combined with a unique set of constants to calculate the percent remaining in each of the six propellant tanks. Correction factors are included for residual tank propellant. Correction factors are included for residual tank propellant at depletion, gauging inaccuracy, and trapped line propellant. Thus the computed quantity should represent the usable (rather than total) quantity for each module and make it possible to determine if the difference between each pair of tanks in a module exceeds a preset tolerance (leakage detection).

The calculations include effects of helium gas compressibility, helium pressure vessel expansion at high pressure, oxidizer vapor pressure as a function of temperature, and oxidizer and fuel density as a function of temperature and pressure. The sequence assumes that helium that leaves the helium pressure vessel flows to the propellant tanks to replace propellant leaving, and, as a result, the computed quantity remaining in a propellant tank will be decreased by normal usage, propellant leaks, or helium leaks.

Space Shuttle Spacecraft Systems

The left and right aft RCS and forward RCS module quantities are displayed to the flight crew on Panel 03. When the rotary switch on Panel 03 is positioned to RCS fuel or OXID position, the RCS/OMS QTY meters on Panel 03 will indicate in percent the amount of FUEL or OX for the LEFT, FWD, and RIGHT RCS respectively. If the switch is positioned to RCS LOWEST, the gauging system selects whichever is lower (fuel or oxidizer) for display on the RCS/OMS QTY, LEFT, FWD, and RIGHT RCS respectively.

The left, right, and forward RCS quantities are also sent to the cathode-ray tube (CRT), and, in the event of failures, substitution of alternate measurements and the corresponding quantity will be displayed on the CRT. If no substitute is available, the quantity calculation for that tank is suspended with a fault message.

The sequence also provides automatic closure of the high-pressure helium isolation valves on orbit when the propellant tank ullage pressure is above 16,146 mmHg (312 psia) for the left, right, or forward RCS, and the caution/warning red light on Panel F7 is illuminated for the respective FWD RCS, LEFT RCS, or RIGHT RCS, and a fault message is sent to the CRT. When the tank ullage pressure returns below this limit, the close command is removed.

Exceeding a preset absolute difference of 12.6 percent between fuel and oxidizer in a module will illuminate the respective LEFT RCS, RIGHT RCS, or FWD RCS red caution/warning light on Panel F7, activate the backup caution/warning, and cause a fault message to be sent to the CRT. A bias of 12.6 percent is added when a leak is detected so that subsequent leaks in that same module may be detected.

ENGINE PROPELLANT FEED

The propellant tank isolation valves are located between the propellant tanks and the manifold isolation valves and are

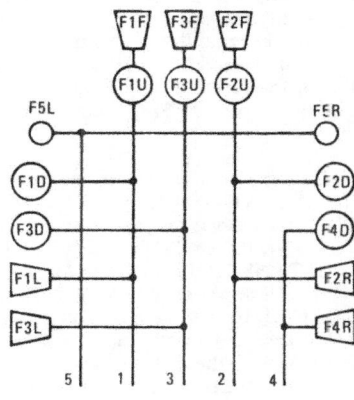

Reaction Control System Propellant Manifold Feed System to Engines

Space Shuttle Spacecraft Systems

used to isolate the propellant tanks from the remainder of the propellant distribution system. The isolation valves are ac-motor-operated and contain a liftoff ball flow control device. One pair of valves (one fuel and one oxidizer valve) isolates the propellant tanks from the 1/2 manifold. Two pairs of valves in parallel, identified as A, B, isolate the propellant tanks from the 3/4/5 manifold line in the forward RCS. One pair of valves isolates the propellant tanks from the 3/4/5 manifold line in the forward RCS.

The forward RCS tank isolation valves are controlled by the FWD RCS TANK ISOLATION 1/2 and 3/4/5 switches on Panel 08. The aft RCS tank isolation valves are controlled by the AFT LEFT RCS TANK ISOLATION 1/2 and 3/4/5 A, B and AFT RIGHT RCS TANK ISOLATION 1/2 and 3/4/5 A, B switches on Panel 07. These are permanent position switches OPEN, GPC, and CLOSE, but only apply signals to the electrical motor controller assembly, which applies power to the valve motor drives. The logic in the motor controller assembly is such that once the valve is in the commanded position, power to the motor is removed. The OPEN/CLOSE switch positions isolate the valves from the computers, and the GPC position permits the orbiter computer to open or close the valves. Each valve contains a position microswitch that indicates valve position to the motor controller assembly and controls a talkback above each switch on Panel 07 and 08. When both valves, fuel and oxidizer, are open, the talkback indicates OP; when closed it indicates CL. The talkback indicates barberpole when there is a mismatch in the oxidizer and fuel valve position or when the valves are in transit.

The manifold isolation valves are between the tank isolation valves and the RCS engines. The manifold isolation valves isolate the propellants from the RCS engines. Manifolds 1, 2, 3, and 4 supply the propellants to the primary RCS thrusters. Manifold 5 supplies the propellants to the RCS vernier thrusters.

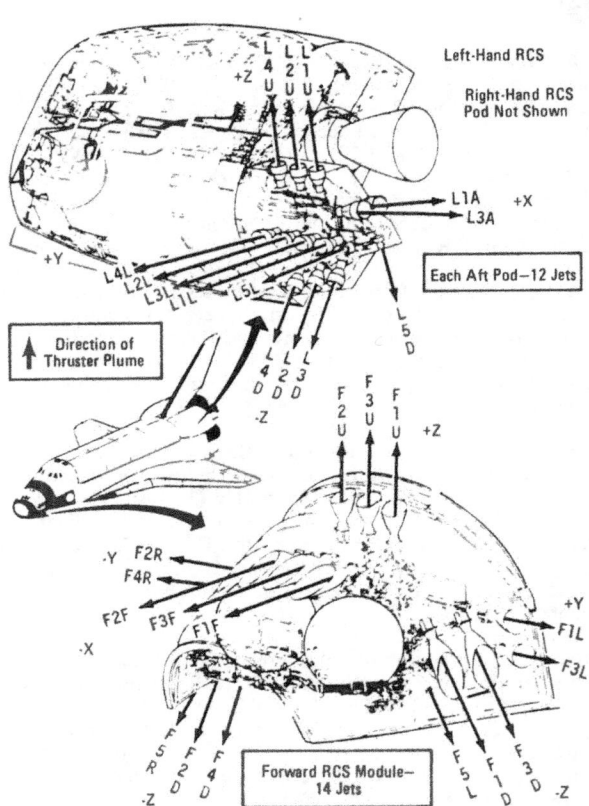

Reaction Control System Engine Location and Identification

203

Space Shuttle Spacecraft Systems

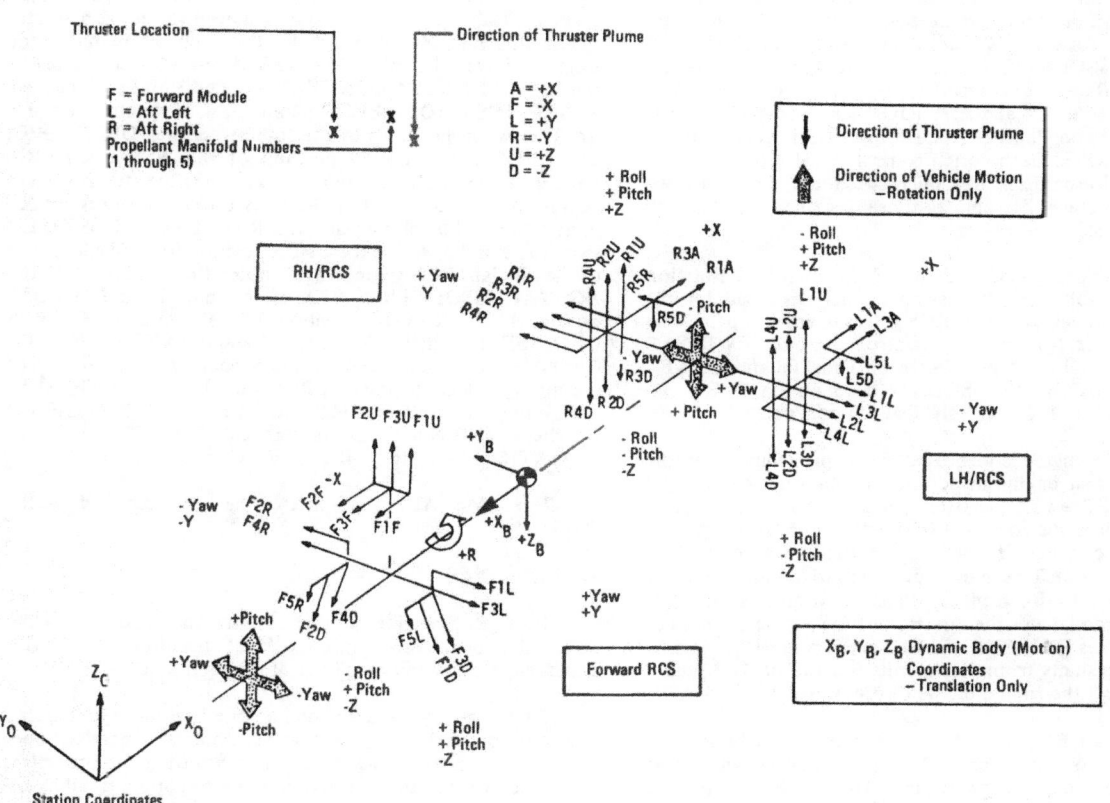

Reaction Control System Engine Identification and Direction of Vehicle Motion

Space Shuttle Spacecraft Systems

The manifold isolation valves in manifolds 1, 2, 3, and 4 are ac-motor-operated valves and contain the same motor assembly switching logic as the tank isolation valves. These valves for the forward RCS are controlled by the FWD RCS MANIFOLD ISOLATION 1, 2, 3, 4 switches on Panel 08 and for the aft RCS are controlled by the AFT LEFT RCS MANIFOLD 1, 2, 3, 4 and AFT RIGHT RCS MANIFOLD 1, 2, 3, 4 switches on Panel 07. The switch positions are OPEN, GPC, and CLOSE. Each switch controls a valve pair, one for fuel and one for oxidizer. The microswitch talkback functions for these valves are the same as those described for the tank isolation valves.

The functioning and control of the manifold isolation valves in manifold 5 are the same as that described for the helium isolation valves. These RCS valves operate in pairs, one oxidizer and one fuel, and are controlled by the FWD RCS MANIFOLD ISOLATION 5 switch on Panel 08 and the AFT LEFT RCS MANIFOLD ISOLATION 5 and AFT RIGHT RCS MANIFOLD ISOLATION 5 switch on Panel 07.

Each RCS engine is identified by the propellant manifold that supplies that engine in addition to the direction of that engine plume. The first identification is a letter, F, L, or R. It designates if it is the forward RCS, left aft RCS, or right aft RCS. The second identification is a number, 1 through 5. It designates the propellant manifold. The third identification is a letter—A (aft), F (forward), L (left), R (right), U (up), D (down)—that designates the direction of the engine plume. For example, engines F2U, F3U, and F1U are forward RCS engines receiving propellants from forward RCS manifold 2, 3, and 1 respectively and the engine plume direction is up.

If either aft RCS pod propellant system must be isolated from its RCS jets, the other aft RCS propellant system can be configured to crossfeed propellant. The aft RCS crossfeed valves which tie the crossfeed manifold into the propellant distribution lines below the tank isolation valves can be configured so that one aft RCS propellant system can feed both left and right RCS pods. The aft RCS crossfeed valves are ac-motor operated, similar in design and identical in operation to the tank isolation valves. The aft RCS crossfeed valves are controlled by the AFT LEFT RCS CROSSFEED 1/2 and 3, 4, 5 and the AFT RIGHT RCS CROSSFEED 1/2 and 3, 4, 5 switches on Panel 07. The switch positions of the four switches is OPEN, GPC, and CLOSE. The OPEN position of the AFT LEFT RCS CROSSFEED switch 1/2 and 3, 4, 5 permits the aft left RCS to supply propellants to aft right RCS crossfeed valves, which must be opened by placing the AFT RIGHT RCS CROSSFEED switch 1/2 and 3, 4, 5 to the OPEN position for propellant flow to the aft right RCS engines. (It is noted that the AFT RIGHT RCS TANK ISOLATION A, B switches must be in the CLOSE position.) The CLOSE position of the AFT LEFT RCS CROSSFEED switch 1/2 and 3, 4, 5 and the AFT RIGHT RCS CROSSFEED isolates the crossfeed capability. The crossfeed of the aft right RCS to the left RCS would be accomplished by positioning the AFT RIGHT and LEFT RCS CROSSFEED switches to OPEN and positioning the AFT LEFT RCS TANK ISOLATION 1/2 and 3, 4, 5 A, B switches to CLOSE.

The OMS/AFT RCS interconnect is described in the OMS.

RCS ENGINES

Each RCS engine contains a fuel and oxidizer injector solenoid control valve, injector head assembly, combustion chamber, nozzle, and an electrical junction box.

Each primary RCS engine has one fuel and one oxidizer solenoid-operated pilot poppet valve, which are energized open by an electrical thrusting command permitting the propellant hydraulic pressure to open the main valve poppet to allow the propellants to flow through the injector into the combustion

Space Shuttle Spacecraft Systems

chamber. There they atomize and ignite (hypergolic) and produce a hot gas, thus thrust. When the thrusting command is terminated, the valves are deenergized and spring-loaded closed.

Each vernier RCS engine has one fuel and one oxidizer solenoid-operated poppet valve, which are energized open by an electrical command. When the on command is terminated, the valves are deenergized and spring-loaded closed.

The primary RCS engine injector head assembly has injector holes arranged in two concentric rings. The outer ring is fuel and the inner ring oxidizer. They are canted toward each other to cause impingement of the hypergolic propellant streams within the combustion chamber. Separate outer fuel injector holes provide cooling of the combustion chamber walls.

The vernier RCS engine injector head assembly has a single pair of fuel and oxidizer injector holes canted to cause impingement of the fuel and oxidizer streams for combustion. The fuel stream is more divergent than the oxidizer stream, which provides cooling of the combustion chamber walls.

The combustion chamber of each RCS engine is constructed of columbium. The nozzle of each RCS engine is tailored to that engine for its structural mated configuration with the forward RCS module or the left and right aft RCS pod. The nozzle is radiation-cooled, and insulation around the combustion chamber and nozzle prevents the excessive heat of 1,093 to 1,315 °C (2,000 to 2,400 °F) from radiating into the orbiter structure.

The electrical junction box for each RCS engine contains an electrical heater, a Pc (chamber pressure) transducer, a leak detection device, and the electrical connections to the propellant valves.

HEATERS

The RCS electrical heaters prevent the injectors and propellants from becoming too cold. Each primary RCS engine contains a 20-watt heater, except the four aft RCS protruding engines, which have 30-watt heaters. Each vernier RCS engine has a 10-watt heater. The forward RCS module contains six heaters mounted on radiation panels in six locations in the module. Each aft OMS/RCS pod is also divided into nine heater zones. Each zone is heated in parallel by an A and B heater system, and the aft RCS thrusting housing contains heaters in the yaw, pitch up, pitch down, vernier, and drain purge panels. The OMS/RCS HEATERS switches are located on Panel A14.

The forward RCS panel heaters are controlled by the FWD RCS AUTO A, B, OFF switch on Panel A14. When the FWD RCS switch is positioned to AUTO A or B position, a thermostat will on either the three port side panels or the three starboard panels. The thermostats are located on helium lines on either side of the forward module. When the temperature of the line reaches a minimum of 13 °C (55 °F), the three respective panels are turned on. When it reaches a maximum of 24 °C (75 °F), they are turned off. There is a minimum 3 °C (6 °F) deadband between the maximum and minimum temperatures. The OFF position removes all electrical power from the heaters.

The aft RCS heaters are controlled by the LEFT POD AUTO A and AUTO B and RIGHT POD AUTO A and AUTO B switches on Panel A14. When the LEFT POD AUTO A or AUTO B switches are positioned to either AUTO A or AUTO B, thermostats will control automatically the nine individual heater zones in each pod. Each heater zone is different, but generally the thermostats control the temperature between 13 °C (55 °F) minimum to 24 °C (75 °F) maximum. The OFF position of the respective switch removes all electrical power from that pod heater system.

Space Shuttle Spacecraft Systems

The forward and aft RCS primary and vernier thruster heaters are controlled by the FWD RCS JET 1, 2, 3, 4, and 5 AUTO or OFF and AFT RCS JET 1, 2, 3, 4, and 5 AUTO or OFF switches on Panel A14. When the FWD or AFT RCS JET 1, 2, 3, 4, and 5 switches are positioned to AUTO, individual thermostats on each thruster will control automatically the individual heaters on each thruster. The primary RCS thruster heaters will turn on between 19° and 24°C (66° to 76°F) and go off between 34° and 43°C (94° to 109°F). The vernier RCS thruster heaters will turn on between 60° and 66°C (140° to 150°F) and go off between 84° and 90°C (184° to 194°F). The OFF position of the respective switch removes all electrical power from that set of thruster heaters. The 1, 2, 3, 4, and 5 designation refers to propellant manifolds with two to four thrusters per manifold.

RCS JET SELECTION

The RCS sends pressure, temperature, and valve position data to the data processing system (DPS) through the flight critical multiplexer/demultiplexers (MDM's) for processing by the orbiter computers. The computers use the data to monitor and display the configuration and status of the RCS. The DPS provides valve configuration commands to the RCS and RCS jet on/off commands to the RCS via the reaction jet drivers (RJD's), aft and forward (RJDA and RJDF) respectively. Data from the RCS through the MDM's are also sent to the pulse code modulation master unit (PCMMU) for incorporation in the downlink to ground telemetry and to the orbiter onboard recorders.

The RJD's AND's the fire commands A and B for an RCS jet. If both are true, they send a voltage to open the RCS jet fuel and oxidizer solenoid valves. This voltage is used to generate the RJD discrete. The fire command A is also sent and used by the RCS redundancy management (RM). The RJD driver and logic power for the aft and forward RJD's are controlled by the RJDA-1A L2/R2 MANF LOGIC and DRIVER, RJDA-2A L4/R4 MANF LOGIC and DRIVER, and RJD-1B F1 MANF LOGIC and DRIVER ON and OFF switches on Panel 014, RJDA-1B L1/L5/R1 MANF LOGIC and DRIVER and RJDA-1A F2 MANF LOGIC and DRIVER ON and OFF switches on Panel 015, and RJDA—2B L3/R3/R5 MANF LOGIC and POWER, RJDA-2A-F3 MANF LOGIC and DRIVER and RJDF-2A F4/F5 MANF LOGIC and DRIVER ON and OFF switches on Panel 016.

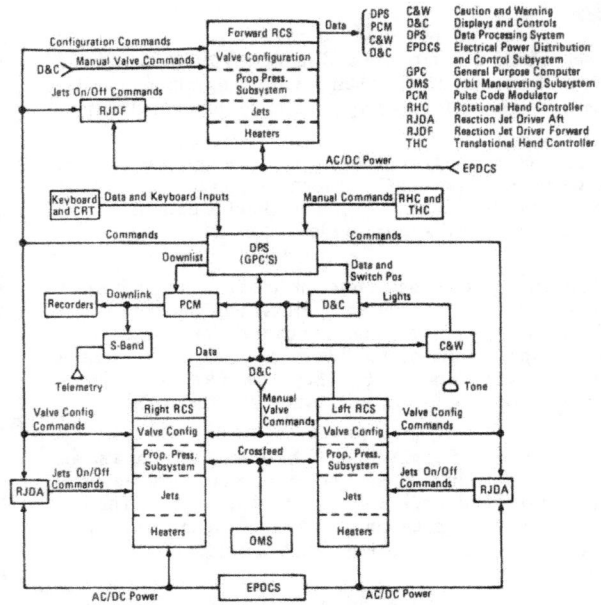

Reaction Control System Overview

Space Shuttle Spacecraft Systems

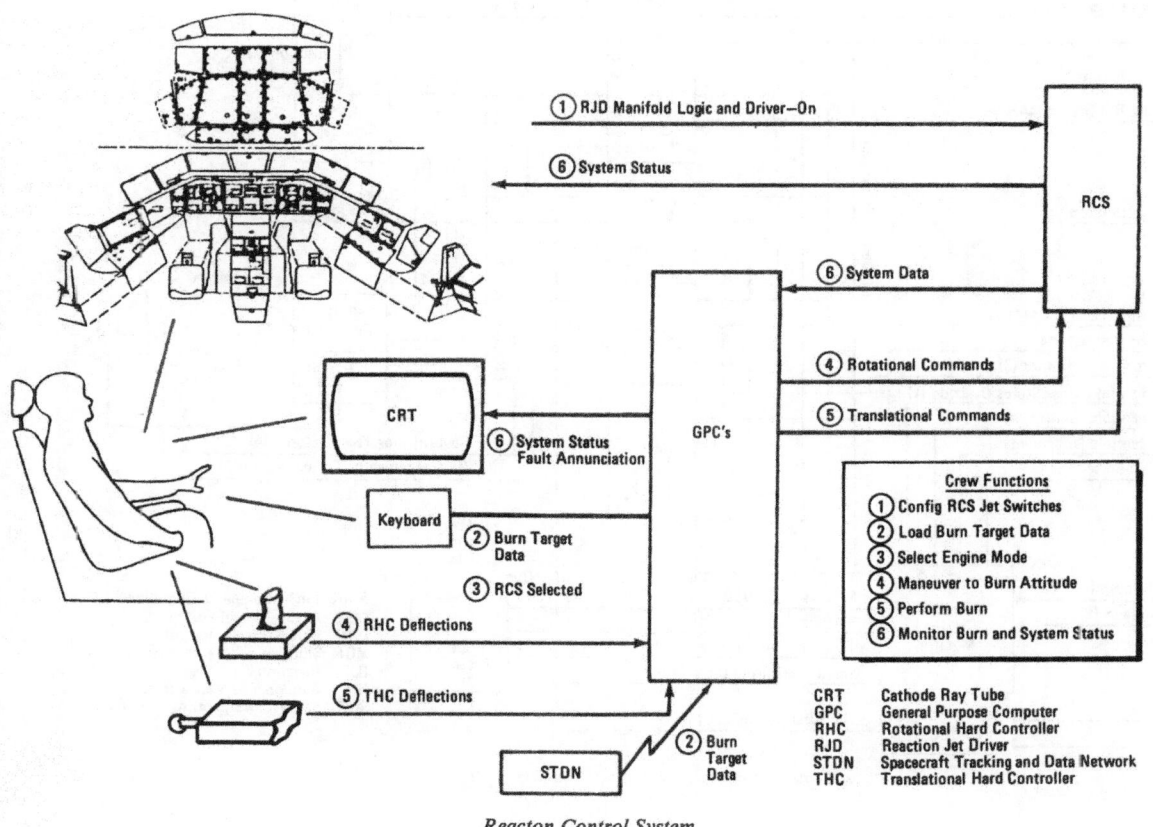

Reacton Control System

Space Shuttle Spacecraft Systems

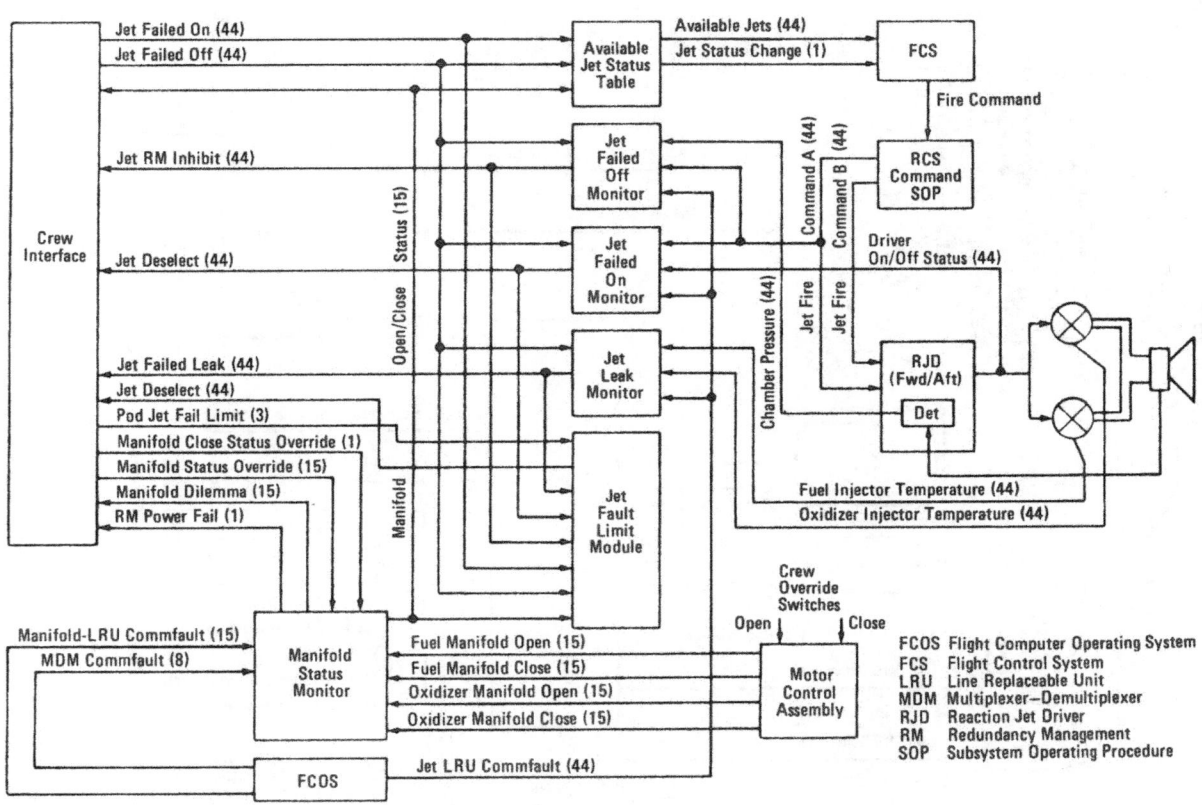

Reaction Control System Redundancy Management

Space Shuttle Spacecraft Systems

The RCS RM monitors the RCS jet's chamber pressure, fuel and oxidizer injector temperatures, RJD on/off output discretes, jet fire commands, and manifold valve status.

The DPS software provides status information on any RCS errors to the RCS RM software. The errors are referred to as communication faults (commfaults). When an RCS error is detected by any orbiter computer, the data on the entire chain are flagged as invalid (commfaulted) for the applications software when present for two consecutive cycles. Commfaults in the RCS RM help prevent the redundant orbiter computers from moding to dissimilar software, to optimize the number of RCS jets available for use, and to prevent the RCS RM from generating additional alerts to the flight control operational software (FCOS). The RCS RM will reconfigure for commfaults regardless of whether the commfault is permanent, transient, or subsequently removed. On subsequent transactions, if the problem is isolated, only the faulty element is flagged as invalid.

The RCS jet-failed-on monitor uses the jet fire command A discretes, the RJD on/off discretes, the jet RM inhibit discretes, and the jet commfault discretes as inputs from each of the 44 jets. The RCS jet-failed-on logic checks for the presence of a RJD-on discrete when no jet fire command A exists. It outputs that the RCS jet has failed on if this calculation is true for three consecutive cycles during any flight phase. It is noted that the consecutive cycles are not affected by commfaults or by cycles in which there are fire commands for the affected RCS jet. However, the three-consecutive-cycle logic will be reset if the non-commanded jet has its RJD output discrete reset to indicate the jet is not firing. A jet-failed-on determination sets the jet-failed-on discrete (even for a minimum jet fire command pulse of 80 milliseconds on and off) and outputs the jet-failed-on to the backup C&W light, the yellow RCS JET light on Panel F7 C&W, a fault message on the CRT, and an audible alarm. These discretes will be reset when the associated RCS jet RM inhibit discrete is reset by the flight crew. A jet-failed-on will not be automatically deselected by the RCS RM, nor will the orbiter digital autopilot (DAP) reconfigure the jet selections.

The RCS jet-failed-on monitor uses the RCS jet fire command A discretes, the jet chamber pressure discretes, the RCS jet RM inhibit discretes, and the jet commfault discretes as inputs from each of the 44 jets. The RCS jet-failed-off logic checks for the absence of the jet chamber pressure discretes when a jet fire command A discrete exists. It outputs that the RCS jet has failed off if true for three consecutive cycles during any mission phase after MECO (main engine cutoff). It is noted that the consecutive cycles are not affected by the commfaults or by cycles in which there are no fire commands for the affected RCS jet. However, the three-consecutive-cycle-logic leading to a failed-off indication must begin anew, if, before the third consecutive cycle is reached, the fire command, and its associated chamber pressure, indicates that the RCS jet has fired. A jet-failed-off determination sets the jet-failed-off discrete (even for a minimum jet fire command pulse of 80 milliseconds on and off) and outputs the jet-failed-off indication to the backup C&W light, the yellow RCS JET light on Panel F7 C&W, a fault message on the CRT, and an audible alarm. The RCS jet-failed-off monitor will be inhibited for the jet that has failed off until the flight crew resets the RM inhibit discrete. The RCS RM will automatically deselect a jet that has failed off, and the DAP will reconfigure the jet selection accordingly. The RCS RM will annunciate a failed-off jet, but not deselect the jet if the jets RM inhibit discrete has been set in advance.

The RCS jet-failed leak monitor uses the RCS jet fuel and oxidizer injector temperatures for each of the 44 jets with the specified temperature of minus 1.1°C (30°F) for the primary and 54°C (130°F) for the vernier jets and declares the RCS jet failed leak if either of the temperatures is less than the specified limit for three consecutive cycles. The RCS jet leak monitor logic checks if either the fuel or oxidizer injector temperature is less than the specified limit. It outputs the RCS jet leak monitor

Space Shuttle Spacecraft Systems

if true for three consecutive cycles leading to the jet failed leak detection will begin anew if the fuel and oxidizer temperatures are both greater than the specified limits before the jet failed leak counter reaches three consecutive cycles. An RCS jet failed leak monitor outputs the RCS jet failed leak to the backup C&W light, the yellow RCS JET light on Panel F7 C&W, a fault message on the CRT, and an audible alarm. The RCS jet failed leak monitor will be inhibited for the jet that has leak detection until the flight crew resets the RCS RM inhibit discrete. The RCS RM will automatically deselect a jet declared leaking, and the DAP will reconfigure the jet selection accordingly. The RCS RM will annunciate a failed leak jet, but not deselect the jet if the jet's RM inhibit discrete has been set in advance.

The RCS fault jet limit module limits the number of jets that can be automatically deselected in response to failures detected by RCS RM. The limits are modifiable by the flight crew input on the RCS SPEC display (RCS F, L, R jet fail limit). This module also reconfigures a jet's availability status. Automatic deselection of a jet occurs if all the following are satisfied: jet detected failed off or leak (jet-on failures do not result in automatic deselection), jet select/deselect status in SELECT, jet's manifold status is OPEN, RM is not inhibited for this jet, jet failure has not been overridden, and the number of automatic deselections of primary jets on that aft RCS pod is less than the associated jet fail limit (no limit on vernier jets), which inhibits the RCS RM for that pod. A jet's status can be changed from deselect to select only by item entry on the RCS SPEC page. Automatic deselection of a jet can be prevented by use of the inhibit item entries on the RCS SPEC page.

The manifold status monitor uses the open and close discretes of the oxidizer and fuel manifold isolation valves (provided by the motor control assemblies) to determine the open/close status, independent of status changes made by the flight crew. The flight crew can override the status of all manifolds on an individual basis by item entries on the RCS SPEC via the manifold status override discretes. The use of the manifold status override feature will not inhibit or modify any of the other functions of the manifold status monitor.

The available jet status table module provides to the flight control systems a list of jets available for use. The available jet status table uses the manifold open/close discretes from the manifold status monitor, and the jet deselect output discretes from the jet fault limit module as inputs and outputs the jet available discretes and the jet status discrete. The available jet status module statuses a jet as available to the flight control system if the jet deselect output discretes and the manifold open/close discretes indicate select and open respectively. The available jet status table will be computed each time the jet status change discrete is true.

The DAP (digital autopilot) jet select module contains default logic in certain instances. When the orbiter is mated to the ET, roll rate default logic inhibits, roll rotation, and yaw commands are normally in the direction of favorable yaw/roll coupling. During insertion with open RCS crossfeed, a seven-aft-RCS-jet limit applies. If negative Z and plus X translation commands are commanded simultaneously, both will be degraded. A limit of five aft RCS jets applies on one side during insertion with RCS crossfeed closed, and plus X is degraded when simultaneous negative Z and plus X and Y translation and yaw rotation commands exceed the demand of five aft RCS jets. During deorbit, only five aft RCS jets can fire simultaneously when the RCS crossfeed is open. If plus X and negative Z translation are commanded simultaneously, plus X translation is given priority.

The DAP jet select module determines which aft (right, left, or both) RCS jets must be turned on in response to the pitch, roll, and yaw jet commands from the entry flight control system. The forward RCS jets are not used during entry. After entry interface, only the four Y-axis and six Z-axis RCS jets on each aft RCS pod are utilized. No X-axis or vernier jets are utilized. The DAP sends the discretes which designate which aft

Space Shuttle Spacecraft Systems

RCS jets are available for firing (a maximum of four RCS jets per pod may be fired) and, during reconfiguration or when the RCS crossfeed valves are open, the maximum combined total number of yaw jets available during certain pitch and roll maneuvers.

During ascent or entry, the DAP jet select logic module, which is located in the flight control system (FCS), receives both RCS rotation and translation commands. By using a table lookup technique, the module outputs 38 jet on/off commands to the RCS command (CMD) system operating program (SOP), which then generates dual fire commands A and B to the individual RCS reaction jet drivers (RJD's) to turn each of the 38 primary RCS jets on or off. The fire commands A and B for each of the 38 primary RCS jets are set equal to the digital autopilot (DAP) RCS commands. Commands are issued to the six RCS vernier similarly during on orbit.

The transition digital autopilot (TRANS DAP) becomes active immediately after MECO (main engine cutoff) and maintains attitude hold in preparation for ET separation. The TRANS DAP controls the spacecraft in response to manual (CSS-control stick steering) or automatic commands during orbit insertion OMS-1 and OMS-2 thrusting periods, orbit coast, orbit checkout, deorbit maneuver, and deorbit maneuver coast. These commands are converted to OMS engine deflections (thrust vector control) during the two OMS insertion thrusting periods and RCS jet firings during the entire insertion phase. RCS commands are issued to support OMS rotations (roll control) when only one OMS engine is utilized, or for rotation, attitude hold, or translation when the OMS engines are not thrusting. The TRANS DAP utilizes attitude feedback and velocity increments from the inertial measurement units (IMU's) via the attitude processor. This feedback information allows the TRANS DAP to operate as a closed loop system for pointing and rotation but not for translation.

The on-orbit DAP and RCS command orbit SOP generates the dual fire commands to the individual RCS jets in response to commands from the flight control system during orbit operations and orbit checkout and set the fire A and fire B commands for each jet equal to the on-orbit DAP RCS commands. The fire A commands are also sent to redundancy management. There are automatic or manual (CSS) rotation modes, manual translation, and primary or vernier (rotation only) RCS capabilities on orbit.

The automatic or guided rotation commands are supplied by the universal pointing processor, and manual (CSS) rotation or translation commands are supplied by the RHC's (rotation hand control) or THC's (translation hand control) respectively. Crew commands from the flight deck forward or aft station are accepted. Three selectable manual (CSS) rotation modes and two selectable translation modes (for X, Y, and Z translations) are provided. The capability to select nose (forward RCS) or tail (aft RCS) only for pitch and/or yaw control is provided by the primary jets. Primary jet roll control is provided only by the aft RCS jets.

The vernier jets are used for tight attitude deadbands and fuel conservation. The loss of one vernier jet results in the loss of the entire vernier mode.

The on-orbit DAP has two sets of I-loaded (initialized) deadbands, DAP A and DAP B. DAP A has a wide deadband and is used for maneuvers that do not require accurate pointing. DAP B has a narrow deadband and is used for maneuvers that require accurate pointing such as IMU alignment.

The entry/landing RCS command SOP generates the dual fire commands to the individual RCS thrusters in response to commands from the flight control system during entry guidance and terminal area energy management and approach/landing

Space Shuttle Spacecraft Systems

and sets the fire A and fire B commands equal to the aerojet DAP commands or the return to launch site abort DAP commands, depending upon which is selected by the flight control system. These commands are sent to the 20 aft RCS Y and Z jets. The fire A commands are also sent to redundancy management.

The aerojet DAP is a set of general equations used to develop effector commands that will control and stabilize the orbiter during its descent to landing from orbital attitude. The aerojet DAP resides in the entry OPS, but is used only during entry and terminal area energy management and approach and landing.

This is accomplished by using either manual (CSS) commands or automatic commands as inputs to the equations. The solution of these equations results in fire commands to the available RCS jets and/or appropriate orbiter aerosurfaces.

The on-orbit and TRANS DAP are also rate command control systems. Sensed body rate feedback is employed for stability augmentation in all three axes. This basic rate system is retained in a complex network of equations whose principal terms are constantly changing to provide the necessary vehicle stability while ensuring that sufficient maneuver capability exists to follow the planned trajectory.

For exoatmosphere flight or flight during the trajectory where certain control surface are rendered ineffective by adverse aerodynamics, a combination of aft RCS jet commands and aerosurface commands is issued. For conventional vehicle flight in the atmosphere, the solution of equations results in deflection commands to the elevons (elevator/aileron), rudder, speedbrake, and body flap. Inputs can consist of automatic commands from entry guidance in the form of attitude, angle of attack, surface position, acceleration commands and manual (CSS) roll, pitch, and yaw rate commands from the flight-crew-operated controllers or a combination of the two since the software channels may be moded independently.

The RCS activity light processing sequence performs the processing necessary to drive the roll, pitch, and yaw indicators lights on Panels F6 and A8 to indicate the presence of an RCS command during entry and terminal area energy management and approach/landing. The indicators are L, R, for roll and yaw left or right, and U, D, for pitch up and down. The RCS commands are stretched by this sequence to provide increased visibility of the indicator lights. Their primary function is to indicate when more than two yaw jets are commanded and when the elevon drive rate is saturated.

From entry interface until the dynamic pressure is greater than 517 millimeters of mercury (mmHg) per meters squared (10 pounds per square foot), the ROLL L (left) and R (roll) lights indicate that left or right roll commands have been issued by the DAP. The minimum light-on duration is extended to allow the light to be seen even for a minimum impulse firings. When dynamic pressure of 517 mmHg per meters squared (10 pounds per square foot) has been sensed, neither ROLL light will illuminate until 2,587 mmHg per meters squared (50 pounds per square foot) has been sensed and two RCS yaw jets are commanded on.

The pitch (U,D) lights indicate up and down pitch jet commands until a dynamic pressure of 1,035 mmHg per meters squared (20 pounds per square foot) is sensed, after which the pitch jets are no longer used. When 2,587 mmHg per meters squared (50 pounds per square foot) is sensed, the PITCH lights assume a new function like the ROLL lights. Both PITCH lights will illuminate whenever the elevon surface drive rate exceeds 20 degrees per second (10 degrees per second if only one hydraulics system is remaining).

Space Shuttle Spacecraft Systems

The yaw (L and R) lights will function as yaw jet command indicators throughout entry until the yaw jets are disabled at approximately 13,716 meters (45,000 feet). The yaw lights have no other function.

The forward RCS and aft (left and right) OMS/RCS pods are removable units to facilitate orbiter turnaround and reusable for a minimum of 100 missions, if required.

The contractors are McDonnell Douglas Astronautics Co., St. Louis, MO (OMS/RCS pod assembly and integration); CCI Corp., Marquardt Co. Division, Van Nuys, CA (primary and vernier thrusters); Brunswick, Lincoln, NE (RCS helium tanks); Consolidated Controls, El Segundo, CA (RCS high-pressure helium valves and low-pressure vernier engine manifold valves, dc); Cox and Co., New York, NY (RCS electrical heaters); Fairchild Stratos, Manhattan Beach, CA (forward and aft RCS helium pressure regulators, couplings nitrogen tetroxide/monomethyl hydrazine, helium fill disconnects); Honeywell, Inc., St. Petersburg, FL (forward and aft RCS reaction jet drivers); Martin Marietta, Denver, CO (forward and aft RCS propellant tanks); Metal Bellows Co., Chatsworth, CA (RCS flexible line assembly); Parker-Hannifin, Irvine, CA (ac motor-operated manifold valves, RCS crossfeed valves, interconnect valves payload, hypergolic couplings, and manually operated OMS/RCS valves); Rockwell International, Rocketdyne Division, Canoga Park, CA (RCS check valves); Brunswick-Wintec, Los Angeles, CA (filters).

ELECTRICAL POWER SYSTEM

The electrical power system (EPS) consists of three subsystems: power reactant storage and distribution (PRSD), fuel cell powerplants (FCP) (the electrical power generation), and electrical power distribution and control (EPDC).

The PRSD subsystem stores and supplies the reactants (hydrogen and oxygen) to the three fuel cell powerplants, which generate the electrical power during all mission phases; in addition, oxygen is supplied to the environmental control and life support system (ECLSS) for crew cabin pressurization. The hydrogen and oxygen are stored in their respective storage tanks at cryogenic temperatures and supercritical pressures. The EPDC subsystem distributes the fuel cell powerplant direct current and converts direct current to alternating current and distributes alternating current to all Space Shuttle electrical equipment throughout all mission phases.

The three fuel cell powerplants, through a chemical reaction, generate all of the 28-volt dc electrical power for the Space Shuttle after launch. Electrical power before launch is supplied in conjunction with ground support equipment at the launch pad. Each of the three fuel cell powerplants consists of a power section, where the chemical reaction occurs, and a compact accessory section attached to the power section, which controls and monitors the power section performance. The three fuel cell powerplants are individually coupled to the reactant (hydrogen and oxygen) distribution subsystem, the heat rejection subsystem the potable water storage subsystem, and the EPDC subsystem. The fuel cell powerplants generate heat and water as a byproduct. The heat is directed to fuel cell heat exchangers, where excess heat is rejected to the Freon coolant loops. The water is directed to the potable water storage subsystem.

The dc power generated by the fuel cells is routed to a three-bus system that distributes dc power to the forward, mid, and aft orbiter sections for equipment in these areas. The three main dc buses—MN (main) A, MNB, and MNC—are the prime source of power for the orbiter's dc loads. Each of the three dc

Space Shuttle Spacecraft Systems

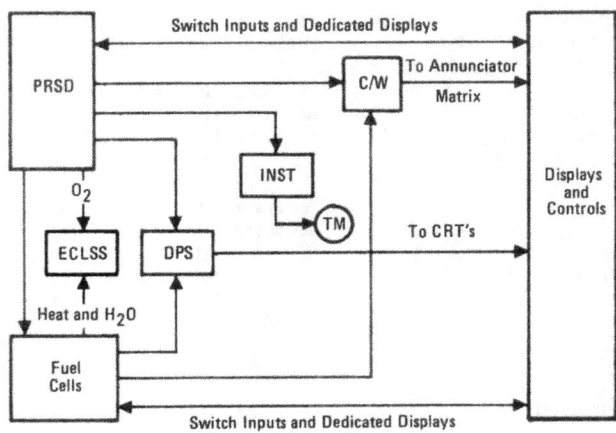

PRSD Interfaces

Fuel Cell Powerplant

main buses supplies power to three solid (static), single-phase inverters, which constitute one inverter system; thus, the nine inverters convert dc power to ac power for distribution to three ac buses, AC1, AC2, and AC3. They are the source power for the orbiter's ac loads.

The EPDC subsystem controls and distributes electrical power (ac and dc) to the orbiter subsystems, the solid-rocket boosters, the external tank, and payloads. Power distribution is controlled and distributed by assemblies. Each assembly is a storage area or box for electrical components such as remote switching devices, buses, resistors, diodes, and fuses. Each assembly usually contains a power bus or buses and remote switching devices for distributing bus power to subsystems located in its area.

POWER REACTANT STORAGE AND DISTRIBUTION (PRSDS)

Hydrogen and oxygen are stored in a supercritical condition in double-walled, thermally insulated, spherical tanks with a vacuum annulus between the inner pressure vessel and outer shell of the tank. Each tank has multi-layer thermal insulation and heaters to add energy to the reactants during depletion for pressure control. Each tank has capacitance quality sensing capability for measuring quality remaining.

The hydrogen and oxygen tanks are grouped in sets consisting of one hydrogen and one oxygen tank. The amount of tank sets installed is dependent upon that specific mission requirement. The baseline tank sets are two sets of tanks. Up to

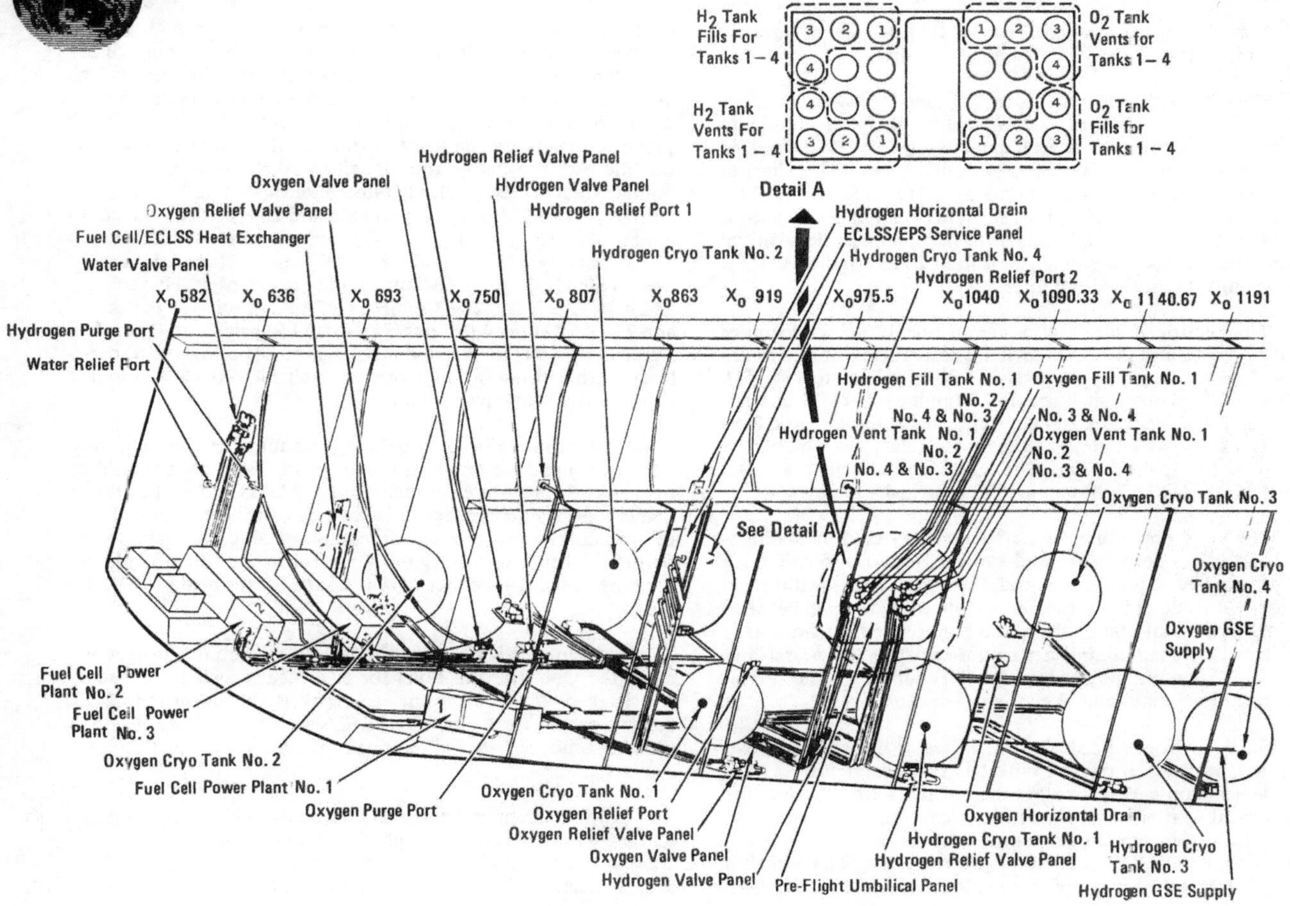

Space Shuttle Spacecraft Systems

Power Reactant Storage and Distribution System

Space Shuttle Spacecraft Systems

five tank sets can be installed. The five tank sets are all located in the mid fuselage under the payload bay area.

The oxygen tanks are identical and consists of inner pressure vessels of Inconel 718 and outer shells of aluminum 2219. The inner vessel is 849 millimeters (33.435 inches) in diameter and the outer shell is 934 millimeters (36.8 inches) in diameter. Each tank has a volume of 0.317 cubic meter (11.2 cubic feet) and stores 354 kilograms (781 pounds). Dry weight of each tank is 91 kilograms (201 pounds). The initial temperature of the stored oxygen is minus 176° (minus 185°F). Maximum fill time is 45 minutes.

The hydrogen tanks also are identical. Both the inner pressure vessel and the outer shell are constructed of aluminum 2219. The inner vessel's diameter is 1,054.5 millimeters (41.516 inches) and the outer shell is 1,155.7 millimeters (45.5 inches). The volume of each tank is 0.605 cubic meter (21.39 cubic feet) and each stores 41 kilograms (92 pounds). Each tank weighs 98 kilograms (216 pounds) dry. The initial storage temperature is minus 251°C (420°F). Maximum fill time is 45 minutes.

The inner pressure vessels are kept super-cold by minimizing conductive, convective, and radiant heat transfer. Twelve low-conductive supports suspend the inner vessel within the outer shell. Radiant heat transfer is reduced by a shield between the inner vessel and outer shell, and convective heat transfer is minimized by maintaining a vacuum between the vessel and shell. A vac-ion pump, powered by GSE during ground servicing operations, maintains the required vacuum level.

Each hydrogen tank has one heater probe with two elements, while each oxygen tank has two heater probes with two elements on each probe. As the reactants are depleted, the heaters add heat energy to maintain a constant pressure in the tank. The heaters operate in manual and automatic modes. The oxygen tank (O_2 TK) and hydrogen tank (H_2 TK) switches (AUTO, ON, OFF) for tanks 1-3 are located on Panel R1; switches for the O_2 and H_2 tank 4 heaters are on Panel A11. When a heater switch is positioned to AUTO, the heater is controlled by a tank heater controller. Each heater controller receives a signal from a tank pressure sensor. If pressure in a tank is equal to or below a specific pressure and the controller sends a "low pressure" signal to the heater logic and the heater is powered on, the pressure bands are: 10,340 to 10,650 mm of mercury (mmHg) (200 to 206 psia); H_2 tanks 3 and 4 — 11,219 to 11,529 mmHg (217 to 223 psia); O_2 tanks 1 and 2 — 41,618 to 42,239 mmHg (805 to 817 psia); and O_2 tanks 3 and 4 — 43,118 to 43,738 mmHg (834 to 846 psia). When H_2 tanks 1 and 2 pressure is 11,374 to 11,684 mmHg (220 to 226 psia), H_2 tanks 3 and 4 — 12,253 to 12,563 mmHg (237 to 243 psia), O_2 tanks 1 and 2 — 43,423 to 44,048 mmHg (840 to 852 psia), and O_2 tanks 3 and 4 — 44,927 to 45,548 mmHg (869 to 881 psia), the respective controller sends a "high pressure" signal to the heater logic and the heater involved is shut off.

A dual mode heater operation capability is available for pairs of O_2 and H_2 tanks. If the heaters of both Tanks 1 and 2 or Tanks 3 and 4 are selected to the AUTO mode, the tank heater logic is interconnected. In this case, the heater controllers of both tanks must send a low pressure signal to the heater logic before the heaters will turn on. Once the heaters are on, a high pressure signal from either tank will turn off the heaters in both tanks.

In the manual mode, the flight crew controls the heaters by using the ON/OFF positions for each heater switch on panel R1. High or low pressure in each tank is shown on the CRT display or the Cryo O_2, H_2 PRESS gauge on Panel 02; the specific tank is selected through a rotary switch (S2) on Panel 02.

During ascent and entry, O_2 heater switches for tank set 1 and 2 are in B AUTO and H_2 heater switches for tank set 1 and 2 are in A and B AUTO. When tank sets 3 and 4 are added, O_2 and H_2 heater switches are in A and B AUTO during ascent and

Space Shuttle Spacecraft Systems

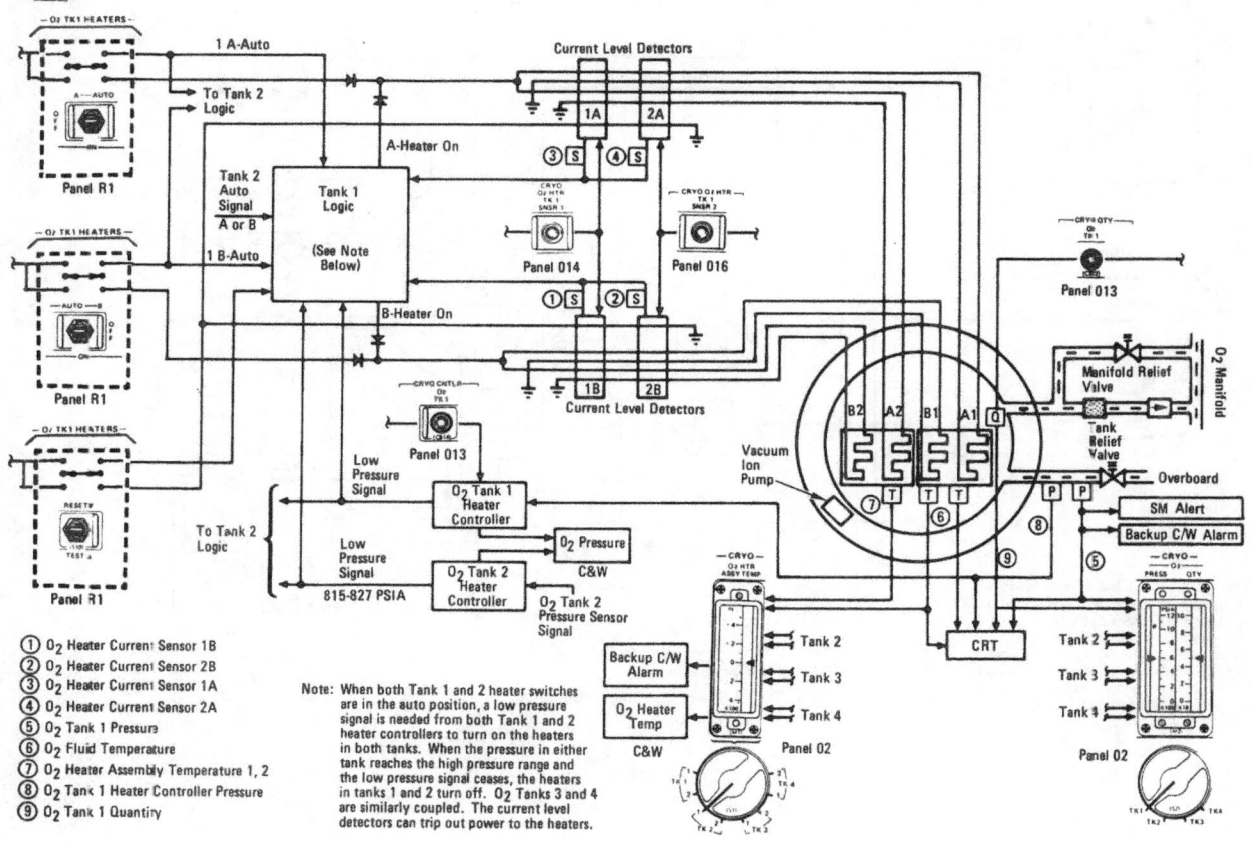

PRSD - O_2 Tank 1

Space Shuttle Spacecraft Systems

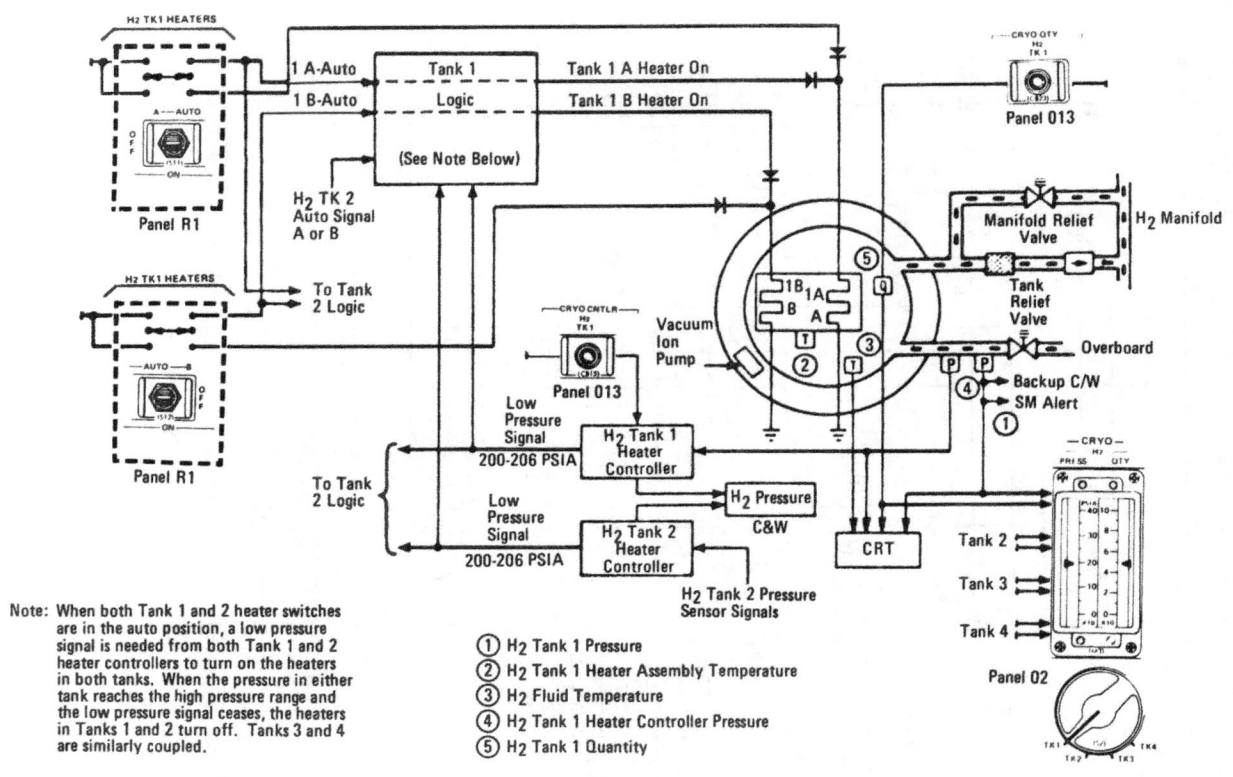

Note: When both Tank 1 and 2 heater switches are in the auto position, a low pressure signal is needed from both Tank 1 and 2 heater controllers to turn on the heaters in both tanks. When the pressure in either tank reaches the high pressure range and the low pressure signal ceases, the heaters in Tanks 1 and 2 turn off. Tanks 3 and 4 are similarly coupled.

① H_2 Tank 1 Pressure
② H_2 Tank 1 Heater Assembly Temperature
③ H_2 Fluid Temperature
④ H_2 Tank 1 Heater Controller Pressure
⑤ H_2 Tank 1 Quantity

PRSD - H_2 Tank 1

Space Shuttle Spacecraft Systems

entry. The position of the heater switches on orbit is dependent on the mission.

The cryo O_2 HTR ASSY TEMP meter on Panel O2, in conjunction with a rotary switch (S1) to select one of the two heaters in each tank, permits temperatures of the heater element to be displayed. The range of the display is from minus 425°F to plus 475°F. The temperature sensor in each heater also is hard-wired directly to the yellow O_2 HEATER TEMP C/W light on Panel F7. This light will illuminate if the temperature is at or above 349°F. A signal also is sent to the computers, where software checks the limit and, if at or above 349°F, the BACKUP C/W light on Panel F7 illuminates. This signal also is transmitted to the CRT and telemetry.

Two current level detectors are built into each O_2 tank heater to prevent electrical shorts. The second detector is redundant. Each detector is divided into A and B detectors, one measuring heater A current and the other measuring heater B current. The detectors are powered by circuit breakers on Panels 014, 015, 016, and ML86B and are identified as CRYO O_2 HTR TK 1, 2, 3, 4 Snsr (sensor) 1,2. Each detector measures the current in and out of a heater and if the current difference is 0.9 amps or greater for 1.5 milliseconds, a trip signal is sent to the heater logic to remove power from the heaters, regardless of the heater switch position. If one element of a heater causes a "trip-out," power to both elements are removed. An "O_2 TK HEATERS 1, 2, 3/TEST/RESET" switch on Panel R1 and on Panel A11 for TK-HEATER 4 can be used to reapply power to that heater when positioned to RESET. The TEST position will cause a 1.4 amp delta current to flow through all four detectors of a specified O_2 tank, causing them to trip-out. During on-orbit operations, the flight crew will be alerted to a current level detector "try-out" by an SM ALERT on Panel F7 and on the CRT.

Each O_2 and H_2 tank has a quantity sensor powered by a circuit breaker. These are identified on Panel 013 as CRYO QTY O_2 (or H_2) TK1 and TK2 and on Panel ML86B as CRYO QTY O_2 (or H_2) TK3 and TK4. Data from the quantity sensors is sent to Panel 02, where a rotary switch (S2) is used to select the tank for display on the CRYO O_2 (or H_2) QTY meters. The range of the meters is 0 to 100. The data also is sent to the CRT.

There are two tank pressure sensors for each O_2 and H_2 tank. One sensor transmits its data to the tank heater controllers and to the yellow O_2 (or H_2) PRESS C/W lights on Panel F7, which illuminate if O_2 tank pressure is below 27,945 mmHg (540 psia) or above 50,973 mmHg (985 psia) or if H_2 tank pressure is below 7,917 mmHg (153 psia) or above 15,204 mmHg (293.8 psia). The signal also is transmitted to the CRT and on Panel 02 (S2) used to select the tank for display on the CRYO O_2 (or H_2) PRESS meter. The data also goes to the SM ALERT, BACKUP C/W on Panel F7 and to telemetry. The range of the O_2 meter is 0 to 1,200 psia, and of the O_2 meter is 0 to 400 psia.

The O_2 and H_2 fluid temperature sensors transmit data to the CRT and telemetry.

Each tank set (one hydrogen and one oxygen tank) has a hydrogen/oxygen control box that contains the electrical logic for the hydrogen and oxygen heaters and controllers. The control box is located on coldplates in the mid body under the payload bay envelope.

The reactants from the tanks flow through two relief valve/filter package modules and valve modules, then to the fuel cells through a common manifold. Oxygen is supplied to the manifold from the tank at a pressure of 42,176 to 45,591 mmHg (815 to 881 psia) and hydrogen is supplied at a pressure of 10,350 to 12,575 mmHg (200 to 243 psia). The pressure of the reactants will be essentially the same at the fuel cell interface as it is in the tanks, since only a small pressure decrease occurs in the manifolds.

Space Shuttle Spacecraft Systems

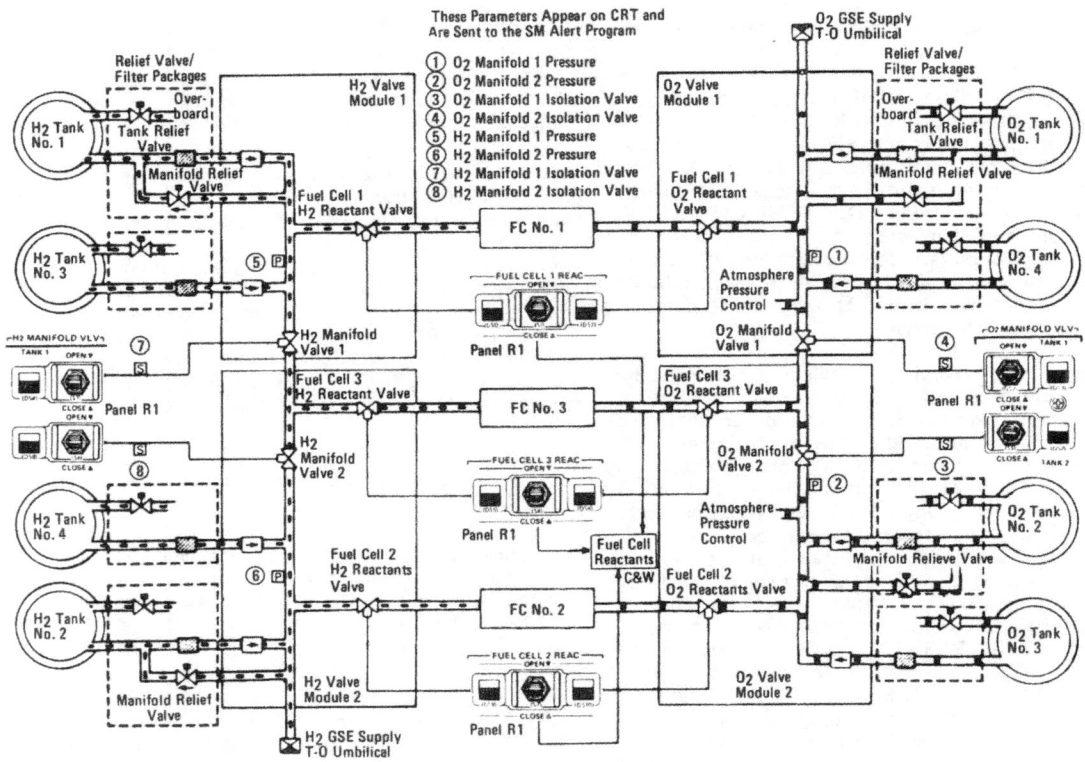

Cryo/Fuel Cell Interface

Space Shuttle Spacecraft Systems

After leaving the tank, the reactants first flow through a relief valve/filter package module which contains the tank relief valve and a 12-micron filter. The filter removes reactant impurities that could degrade the fuel cell. The valve relieves excessive pressure that builds up in the tank and a manifold valve relieves pressure in the manifold lines. The O_2 tank relief valve relieves at 52,008 mmHg (1,005 psia) and the H_2 tank relief valve relieves at 16,042 mmHg (310 psia).

The reactants then flow through the valve modules. There are four of these: H_2 valve modules 1 and 2 and O_2 valve modules 1 and 2. Each module contains a check valve for each tank line coming in, a manifold valve, and fuel cell reactant valves. The O_2 valve modules also contain the ECLSS atmospheric pressure control O_2 system 1 and 2 supply valves. All manifold valves and fuel cell reactant valves will normally be open during all mission phases. These valves are controlled by the flight crew through switches with talkbacks located on Panel R1. The check valve in each tank line prevents reactants from flowing from one tank to another in the event of a tank leak, which prevents a total loss of reactants. The fuel cell reactant valves allow reactants to flow into the fuel cells from the manifold or isolate the reactants from the fuel cell. The manifold valve isolates one reactant system but allows reactants from the remaining supply system to flow to two fuel cells.

The FUEL CELL REAC 1, 2, 3, OPEN/CLOSE switches and indicators are located on Panel R1. When the switch is positioned to OPEN, the respective H_2 and O_2 reactant valves are opened, allowing reactants to flow from the manifold into the fuel cell. The corresponding talkback indicators O_2, H_2 would indicate OP (open). When the switch is positioned to CLOSE, the corresponding H_2 and O_2 reactant valves are closed, isolating the reactants from the fuel cell and the corresponding talkback indicator would indicate CL (closed). Because it is critical to have reactants available at the fuel cells, the red FUEL CELL REAC light on Panel F7 would illuminate when any fuel cell reactant valve is closed, a C/W tone is sounded and the computer senses the closed valve and illuminates the BACKUP C/W light on Panel F7 and the SM ALERT and displays the data on the CRT. This alerts the flight crew that the fuel cell will be inoperative within approximately 20 seconds for an H_2 valve closure and 130 seconds for an O_2 valve closure.

The H_2 (or O_2) MANIFOLD TK1, 2 OPEN/CLOSE switches and indicators are located on Panel R1. When the switch is positioned to OPEN, the respective H_2 or O_2 manifold valve is opened, allowing reactant to flow from the manifold to the fuel cell reactant valve. The corresponding indicator would indicate OP. When the switch is positioned to CLOSE, the valve is closed, isolating reactant from the fuel cell reactant valve, and the indicator indicates CL. The closure of the O_2 and H_2 manifold valve isolates one reactant supply system to flow to two fuel cells and may also be used to isolate a malfunctioning fuel cell.

The manifold relief valves are a built-in safety device in the event a manifold valve and a fuel cell reactant valve are closed because of a malfunction. The reactants trapped in the manifold lines would be warmed up by the internal heat of the orbiter and overpressurize. The manifold relief valve will open at 15,007 mmHg (290 psi) for H_2 and 50,456 mmHg (975 psi) for O_2 to relieve pressure and allow the trapped reactants to flow back to their tanks. Both H_2 and O_2 manifold valves should never be closed at the same time because there is no provisions for relieving trapped reactants in the line between the two manifold valves. If both were closed, the pressure would build up very quickly and cause the line to rupture.

Two pressure sensors located in both the H_2 and O_2 valve modules transmit data to the CRT. This data is also sent to the systems management computer where its lower limit is checked and, if the H_2 and O_2 manifold pressures are below 7,762 mmHg (150 psia) and 10,350 mmHg (200 psia), respectively, an SM ALERT will occur.

Space Shuttle Spacecraft Systems

During preflight ground operations, the fuel cells are supplied by GSE reactants until approximately T minus 2 minutes, 35 seconds. The GSE supply pressure is 15,525 to 16,560 mmHg (300 to 320 psia) for H_2 and 51,750 to 52,785 mmHg (1,000 to 1,020 psia) for O_2, which is higher than the onboard orbiter power reactant storage distribution system. The GSE supply valves close automatically to transfer to internal reactants.

The fuel cells operate on only 453 kilograms (1,000 pounds) of hydrogen and oxygen and produces about 473 liters (125 gallons) of water in a three day mission. In a seven day mission, the fuel cells operate on 1,134 kilograms (2,500 pounds) of hydrogen and oxygen and produce about 1,135 liters (300 gallons) of water at 19 kilowatts including payload.

The three fuel cells operate as independent electrical power sources, each supplying its own isolated, simultaneously operating, 28-volt dc bus. The fuel cell consists of a power section, where the chemical reaction occurs, and an accessory section, which controls and monitors the power section performance. The power section, where hydrogen and oxygen are transformed into electrical power, water, and heat, consists of 62 small cells contained in three substacks. Manifolds run the length of these substacks and distribute hydrogen, oxygen, and coolant to the cells. The cells contain electrolyte consisting of potassium hydroxide (KOH) and water, an oxygen electrode (cathode) and a hydrogen electrode (anode).

Hydrogen is routed to the cell's hydrogen electrode, where it reacts with hydroxyl ions in the electrolyte. This electrochemical reaction produces electrons (electrical power), water, and heat with the electrons being routed through the orbiter's EPDC subsystem to perform electrical work. Oxygen is routed to the cell's oxygen electrode, where it reacts with the water and returning electrons to produce hydroxyl ions. The hydroxyl ions then migrate to the hydrogen electrode, where they enter into the hydrogen reaction at that electrode. The oxygen and hydrogen are reacted (consumed) in proportion to the orbiter's electrical power demand.

Excess water vapor is removed by an internal hydrogen system. Hydrogen and water vapor from the reactants exits the cell stack, is mixed with replenishing hydrogen from the storage and distribution system, and enters a condensor, where waste heat from the hydrogen and water vapor is transferred to the fuel cell coolant system. The resultant temperature decrease condenses some of the water vapor to water droplets. A centrifugal water separator extracts the liquid water and pressure-feeds it to potable tanks in the lower deck of the pressurized crew cabin. Water from the potable water storage tanks can be used for crew consumption and cooling of the Freon-21 coolant loops. The remaining circulating hydrogen is directed back to the fuel cell stack.

FUEL CELL POWERPLANTS

Each of the three fuel cell powerplants is reusable and restartable. The cells are located under the payload bay area in the forward portion of the orbiter mid fuselage.

The accessory section monitors the reactant flow, removes waste heat and water from the chemical reaction, and controls the temperature of the stack. The accessory section consists of the H_2 and O_2 flow system, the coolant loop, and the electrical control unit (ECU).

The fuel cell coolant system circulates a liquid fluorinated hydrocarbon (FC-40) and transfers the waste heat from the cell stack through the fuel cell heat exchanger of that fuel cell powerplant to the Freon-21 coolant loop system in the mid fuselage. Internal control of the circulating fluid maintains the cell stack at a normal operating temperature of approximately 93 °C (200 °F).

Space Shuttle Spacecraft Systems

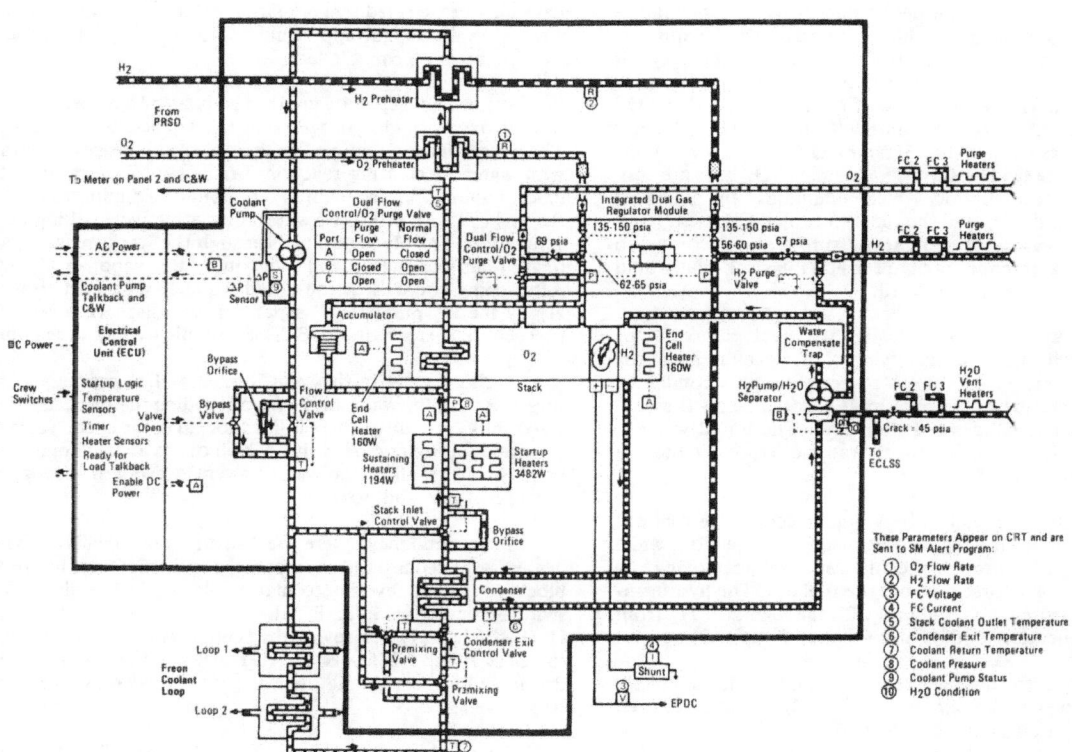

Fuel Cell 1 - Typical

Space Shuttle Spacecraft Systems

After the reactants enter the fuel cells, they flow through a preheater (where they are warmed) from cryogenic temperature to 4.4 °C (40 °F) or greater, a 5-micron filter, and a two-stage integrated dual gas module. The first stage of the regulator reduces both the H_2 and O_2 pressure to 6,986 to 7,762 mmHg (135 to 150 psia). The second stage of the regulator reduces the O_2 to a range of 3,208 to 3,363 mmHg (62 to 65 psia) and maintains the H_2 pressure at 232 to 310 mmHg (4.5 to 6.0 psia) differential below the O_2 pressure. The regulated O_2 lines are connected to the accumulator which maintains an equalized pressure between the O_2 and the fuel cell coolant. If the O_2 and H_2 pressure decreases, the coolant pressure is also decreased to prevent a large differential pressure inside the stack that would crack the thin walls of the fuel cell.

The incoming H_2 upon leaving the dual gas regulator module mixes with the hydrogen-water vapor exhaust from the fuel cell stack. This gas mixture is routed through a condensor where the temperature of the mixture is reduced below the boiling point of water, condensing the water. The liquid water is then separated from the hydrogen-water mixture by the H_2 pump/H_2O separator.

The H_2 pump circulates the H_2 gas back to the fuel cell stack, where some of the H_2 is consumed in the reaction, while the remainder flows through the fuel cell stack, removing the product water by evaporation from the fuel cell. The hydrogen-water vapor mixture then combines with the regulated H_2 from the dual gas generator module and the loop is started over.

The O_2 from the dual gas regulator module flows directly through two ports into a closed-end manifold in the fuel cell stack, achieving optimum oxygen distribution in the cells. All oxygen that flows into the stack is consumed, except during purge operations.

Reactant consumption is directly related to the current produced: if there are no internal or external loads on the fuel cell, no reactants will be used. Because of this direct proportion, leaks may be detected by comparing reactant consumption and current produced; an appreciable amount of excess reactants used indicates a probable leak.

Water and electricity are the products of the chemical reaction of oxygen and hydrogen that takes place in the fuel cells. The water must be removed or the stacks will become saturated with water, decreasing reaction efficiency. With an operating load of about 7 kilowatts, it takes about 7 to 8 minutes to flood the fuel cell with produced water, thus effectively halting power generation. The H_2 is pumped through the stack, reacting with oxygen and picking up and removing water vapor on the way. After being condensed, the liquid water is separated from the H_2 by the H_2 pump/H_2O separator and discharged from the fuel cell to be stored in the ECLSS potable water storage tanks.

A check valve in the water discharge line prevents a back pressure from the water tanks from flooding the fuel cells. If the water tanks are full or there is line blockage, the check valve will close and the water relief valve, which opens at 2,328 mmHg (45 psia), opens to allow the water to vent overboard through the H_2O relief line and nozzle.

For redundancy, there are two thermostatically activated heaters wrapped around the discharge and relief lines to prevent blockage caused by the formation of ice in the lines. Two switches on Panel R12, FUEL CELL H_2O LINE HTR and H_2O RELIEF HTR, provide the flight crew with the capability to select either AUTO A or AUTO B for the fuel cell water discharge line heaters and the water relief line and vent heaters, respectively.

Water line temperature is maintained by the thermostatically controlled heaters when in use between 54° and 68 °C (140 to 150 °F). The water relief valve temperature is maintained by the thermostatically controlled heaters when in use between 21° and 37 °C (70° to 100 °F). Temperature sensors

Space Shuttle Spacecraft Systems

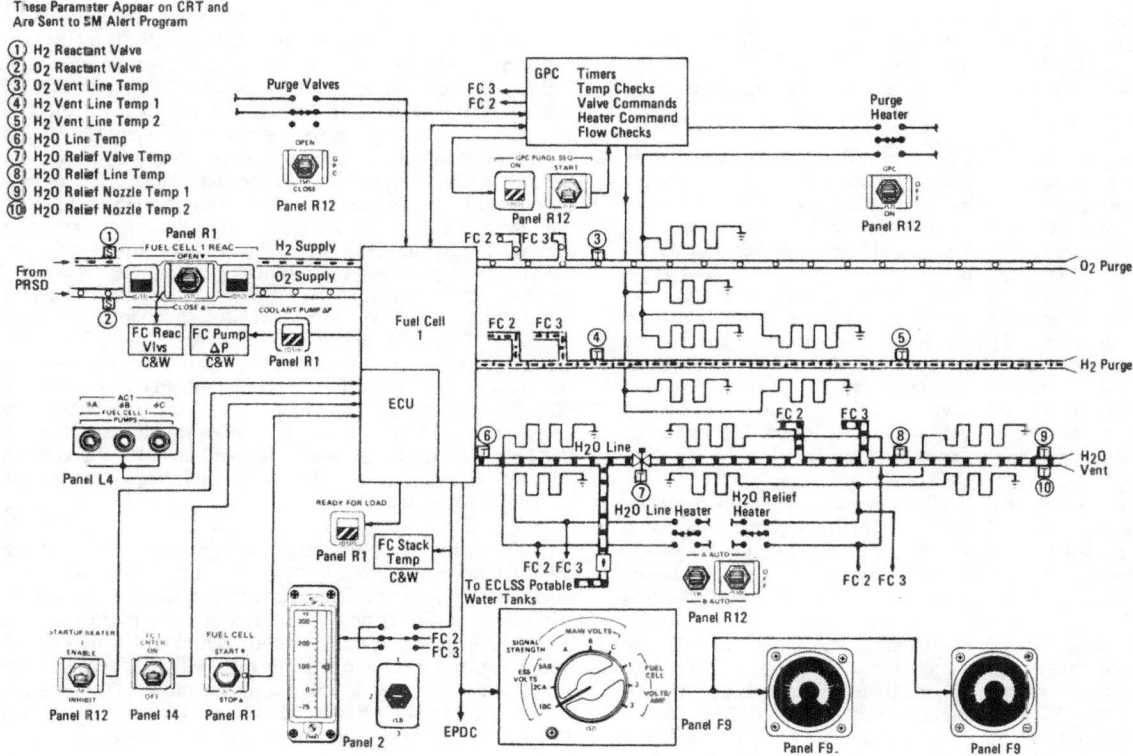

Fuel Cell 1 Typical - Displays and Controls

 **Space Shuttle Spacecraft Systems**

located on the fuel cell water discharge line, relief valve, relief line, and vent nozzle are displayed on the CRT.

If the KOH catalyst in the fuel cell migrates into the product water, a pH sensor located downstream of the H_2 pump/H_2O separator will sense the presence of the catalyst as a high pH and the crew is alerted with an SM ALERT and displayed on the CRT.

During normal fuel cell operation, the reactants are present in a closed-loop system and are 100-percent consumed in the production of electricity. Any inert gases or other contaminants will accumulate in and around the porous electrodes in the cells and reduce the reaction efficiency and electrical load support capability. Purging is therefore required at least every eight hours to cleanse the cells. When a purge is initiated by opening the purge valves, the O_2 and H_2 become open-loop systems with increased flows allowing the reactants to circulate through the stack, pick, up the contaminants, and blow them out overboard through the purge lines and vents. Electrical power is produced throughout the purge sequence, although no more than 8 kilowatts should be required from a fuel cell being purged because of the increased reactant flow.

The purge can be controlled manually or automatically by use of the switches on Panel R12. The basic purge sequence is the same regardless of the mode selected. First, the purge line heaters are activated to prevent any condensate from freezing and thus block the lines. The purge line heater switch is located on Panel R12 and identified as FUEL CELL PURGE HEATER. The switch positions are GPC, OFF, and ON; the purge line heaters are commanded on or off automatically if in the GPC position or manually by the flight crew in the ON and OFF positions. When minimum temperatures of 20 °C (69 °F) for the O_2 line and 26° and 4 °C (79° and 40 °F) for the H_2 1 and 2 lines are indicated by line temperature sensors and verified by the flight crew or GPC within 27 minutes, the purge sequence begins.

To initiate the automatic GPC mode, the FUEL CELL GPC PURGE SEQ switch on Panel R12 must be held in the START position until a talkback indicator turns GRAY (approximately 3 seconds). The automatic purge sequence will not be on if the indicator shows a BARBERPOLE. The GPC turns the purge line heaters on and monitors the temperature. If the temperatures are not up to minimum after 27 minutes, the GPC will issue an SM ALERT and display the data on the CRT. When the proper temperatures have been attained, the GPC will open for two minutes and then close the H_2 and O_2 purge valves for Fuel Cells 1, 2, and 3 in that order. The GPC checks the status of the purge valves by comparing the H_2 and O_2 flow rates with the fuel cell current output. Flow rates should increase 0.2 kilograms per hour (0.6 pounds per hour) for H_2 and 0.9 kilograms per hour (2.2 pound per hour) for O_2 when the purge valves are open. If the GPC determines that a valve has failed closed, it will issue an SM ALERT and display the data on the CRT. If the GPC determines that a valve has failed open, it will terminate the purge sequence, but will leave the heaters on and issue an SM ALERT with the data displayed on the CRT. Fifty-five minutes after the Fuel Cell 3 purge valves are closed (to ensure that the purge lines have been totally evacuated), the GPC turns off the purge line heaters and the GPC PURGE SEQ indicator shows a BARBERPOLE, signifying the end of the automatic purge sequence.

The manual fuel cell purge would be initiated by the flight crew placing FUEL CELL PURGE HEATER switch to ON when the proper temperatures have been attained. The ON position will open the H_2 and O_2 purge valves for Fuel Cells 1, 2, 3 in that order for two minutes and then close them. The CRT would display the H_2 and O_2 flow rates with the fuel cell current at the same values as in the automatic mode. The purge line heaters are turned OFF after 55 minutes from the last fuel cell purge by the flight crew ending the purge sequence.

In order to cool the fuel cell stack during its operation, distribute heat during fuel cell starting, and warm the cryogenic

Space Shuttle Spacecraft Systems

reactants entering the stack, the fuel cell circulates a coolant—fluorinated hydrocarbon (FC-40)—throughout the fuel cell. Fuel Cells 1, 2, and 3 are identical in respect to the coolant loop in the fuel cell and its interface with the ECLSS coolant loops.

At the coolant entrance to the fuel cell, the temperature of the coolant returning from the ECLSS Freon-40 loops is sensed before it passed through a 75-micron filter. After the filter, two temperature-controlled mixing valves allow some of the hot coolant to mix with the cool returning coolant to prevent the coolant flowing into the condensor exist control valve from undergoing repeated oscillations. The condensor exit control valve adjusts the flow of the coolant through the condensor to maintain the H_2/H_2O vapor exiting the condensor at a temperature between 64° and 67°C (148° and 153°F).

The stack inlet control valve maintains the temperature of the coolant entering the stack between 80° and 86°C (177° and 187°F). The accumulator is the interface with the O_2 cryogenic reactant to maintain an equalized pressure between the O_2 and the coolant (the O_2 and H_2 pressures are equalized at the dual gas regulator) to preclude a high pressure differential in the stack which could rupture the thin walls of the fuel cells. The pressure in the coolant loop is sensed before the coolant enters the stack.

The coolant is circulated through the fuel cell stack to absorb the waste heat from the H_2/O_2 reaction occurring in the individual cells. After leaving the stack, the temperature of the coolant is sensed with the data transmitted to the GPC, the FUEL CELL STACK TEMP meter through the Fuel Cell 1, 2, 3 switch located below the meter on Panel 2, and to the CRT display. The FUEL STACK TEMP yellow caution/warning light on Panel F7 and the BACKUP C/W and SM ALERT lights will illuminate if fuel cell and stack temperatures exceed certain limits: below 79°C (175°F) or above 100°C (212°F) for Fuel Cell 1, below 85°C (185°F) or above 105°C (222°F) for Fuel Cell 2, and below 90°C (195°F) or above 114°C (238°F) for Fuel Cell 3. The hot coolant from the stack flows through the O_2 and H_2 preheaters, where it warms the cryogenic reactants before they enter the stack.

The coolant pump utilizes three-phase ac power to circulate the coolant through the loop. The differential pressure (ΔP) sensor senses a pressure differential across the pump to determine the status of the pump. The FUEL CELL PUMP caution/warning light on Panel F7 will illuminate if Fuel Cell 1, 2, or 3 coolant pump pressure is below 2,587 or above 3,881 mmHg (50 and 75 psia). The SM ALERT light also will illuminate and a fault message will be sent to the CRT. If the coolant pump for Fuel Cell 1, 2, or 3 is OFF, the BACKUP C/W light illuminates and a fault message is sent to the CRT. Downstream of the pump is the temperature-actuated flow control valve. It adjusts the coolant flow to maintain the fuel cell coolant exit temperature between 92° and 112°C (196° and 235°F). The stack inlet control valve and flow control valve have bypass orifices to allow coolant flow through the coolant pump and to maintain some coolant flow through the condensor for water condensation, even when the valves are fully closed due to the requirements of thermal conditioning.

The coolant (that which is not made to bypass) exits the fuel cells to the ECLSS Freon-40 coolant loop fuel cell heat exchanger, where it transfers its excess heat to the Freon to be dissipated through the Freon-21 coolant loop system in the mid-fuselage.

In addition to thermal conditioning by means of the coolant loop, the fuel cell has internal startup, sustaining, and end cell heaters. The 3,482-watt startup heater is used only during startup to warm the fuel cell to its operational level. The two 160-watt end cell heaters keep the cells at the end of the fuel cell stack at the same temperature as the cells in the middle of the

 **Space Shuttle Spacecraft Systems**

stack. The 1,194-watt sustaining heaters normally are used during low power periods to maintain the fuel cells at their operational temperature.

The electrical control unit (ECU) is the brain of the fuel cell. The module is located in the fuel cell and contains the startup logic, heater thermostats, and 30-second timer; it is the interface with the controls and displays for fuel cell startup, operation, and shutdown. The ECU controls the supply of ac power to the coolant loops, H_2 pump/H_2O separator, the pH sensor, and the dc power supplied to the flow control bypass valve (open only during startup), the internal startup, sustaining, end cell heaters. The ECU also controls the status of the FUEL CELL 1, 2, 3 READY FOR LOAD and COOLANT PUMP Δ P talkback indicators on Panel R1.

The nine fuel cell circuit breakers which connect the three-phase ac power to the three fuel cells are located on Panel L4 and the fuel cell ECU receives its power from an essential bus through the FC (flight critical) CNTLR switch on Panel O14.

The FUEL CELL START/STOP switch on Panel R1 is used to initiate the start sequence or stop the fuel cell operation. When this switch is held in its momentary start position, the ECU connects the three-phase ac power to the coolant pump and H_2 pump/H_2O separator (allowing the coolant and the hydrogen-water vapor to circulate through these loops), and connects the dc power to the internal startup, sustaining, and end cell heaters and the flow control bypass valve. The switch must be held in the START position until the COOLANT PUMP Δ P talkback shows GRAY in approximately three to four seconds, which indicates that the coolant pump is functioning properly by creating a differential pressure across the pump. When the COOLANT PUMP Δ P indicates talkback BARBERPOLE, it indicates the coolant pump is not running.

The READY FOR LOAD talkback will show GRAY after the 30-second timer times out and the STACK OUT TEMP is above 86°C (187°F) (which can be monitored on Panel 2 in conjunction with the 1, 2, 3 switch located beneath the FUEL CELL STACK OUT TEMP meter); this indicates that the fuel cell is up to the proper operating condition and is ready for loads to be attached to it. It should not take longer than 15 minutes for the fuel cell to warm up and become fully operational, the actual time depending on the fuel cell initial temperature. The READY FOR LOAD indicator remains GRAY until the FUEL CELL START/STOP switch is placed to STOP, the FC CNTLR switch is placed to OFF, or the ESS (essential) BUS power is lost to the ECU.

The STARTUP HEATER ENABLE/INHIBIT switch on Panel R12 provides the crew control of the OFF/ON status of the startup heaters during fuel cell startup. The INHIBIT position allows the startup heaters to remain off and would be used only when starting an already warm fuel cell.

The Fuel Cell 1, 2, or 3 dc voltage and current (amps) can be monitored on the DC VOLTS and DC AMPS meters on Panel F9, using the FUEL CELLS VOLTS/AMP rotary switch to select a specific fuel cell.

The fuel cells will be on, but in standby, when the crew boards and the vehicle is powered by the ground support equipment (GSE). Just before liftoff (T minus 3 minutes and 30 seconds), the GSE is powered off and the fuel cells take over the vehicle electrical loads. Indication of the switchover can be noted on the CRT display and the DC AMPS meter. The fuel cell current will increase to approximately 220 amps, the O_2 and H_2 flow increases to approximately 1.8 and 0.2 kilogram per hour (4 and 0.6 lb/hour), respectively, and the fuel cell stack temperature increases slightly.

Space Shuttle Spacecraft Systems

Fuel cell standby consists of removing the electrical loads but continuing operation of the fuel cell pumps, controls, instrumentation, and valves, with the electrical power being supplied by the remaining fuel cells. A small amount of reactants are used to generate power for the fuel cell internal heaters.

Fuel cell shutdown, after standby, consists of stopping the coolant pump and H_2 pump/H_2O separator by positioning that fuel cell START/STOP switch on Panel R1 to the STOP position. If the environment in the fuel cell compartment beneath the payload bay is lower than 4 °C (40 °F), the fuel cell should be left in standby instead of shutdown to prevent it from freezing.

Each fuel cell powerplant is 355 millimeters (14 inches) high, 381 millimeters (15 inches) wide, and 1,016 millimeters (40 inches) long and weighs 91 kilograms (201 pounds).

The voltage and current of each is 2 kilowatts at 32.5 volts dc, 61.5 amps, and 12 kilowatts at 27.5 volts dc, 436 amps. Each fuel cell is capable of supplying 12 kilowatts peak, 7 kilowatts coverage of power. The three fuel cells are capable of a maximum continuous output of 21,000 watts with 15 minute peaks of 36,000 watts. The average power consumption of the orbiter is expected to be approximately 14,000 watts, or 14 kilowatts per day, leaving 7 kilowatts average available for payloads. Each fuel cell will be serviced between flights and reused until each accumulated 2,000 hours of on-line service.

ELECTRICAL POWER DISTRIBUTION AND CONTROL (EPDC)

The dc power generated by the three fuel cells is routed to main buses and essential buses located in the distribution and controller assemblies.

The three main dc buses (MNA, MNB, and MNC) are the primary source of dc power for all orbiter electrical equipment.

Each fuel cell is dedicated to a specific main bus: FC-1 to MNA, corresponding main bus by power contractors, which are dc-motor-driven remote switches rated at 500 amps and are located in the distribution assembly (DA). The DA, through fuses, powers the corresponding buses in the mid power controller (MPC) assembly; the forward power controller (FPC) assembly; the forward load controller (FLC) assembly; the aft power controller (APC) assembly and the aft load controller (ALC) assembly.

The power controller asemblies (PCA's) and load controller assemblies (LCA's) use remote switching devices to

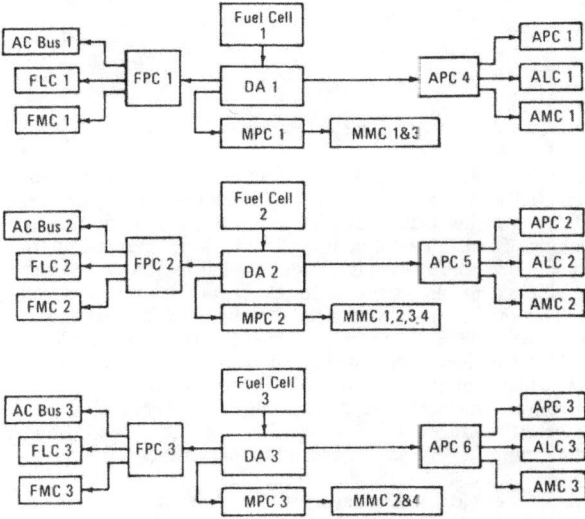

DC Power Distribution Block Diagram

Space Shuttle Spacecraft Systems

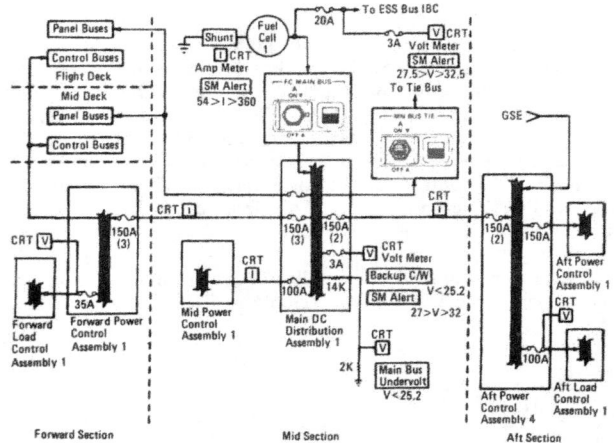

Main Bus A Distribution

distribute the loads. The PCA's contain remote power controllers (RPC's), which are solid-state, remote switching devices used for applying loads up to 20 amps and relays for remote switching. The LCA's contain hybrid circuit devices, which are solid-state, remote switching devices used as logic switches and as remotely controlled switches for electrical loads of 5 amps or less. The mid power controllers contain RPC's relays, and hybrid circuit drivers. The remote switching devices permit location of major electrical distribution buses close to the major loads, which eliminates the need for heavy electrical feeders to and from the crew cabin flight deck display and control panels. This greatly reduces spacecraft wiring, and thus weight, and permits more flexible electrical load management.

All the No. 1 distribution assemblies and controller assemblies go with FC-1 (fuel cell)/MNA, all No. 2 assemblies with FC-2/MNB, and all No. 3 assemblies with FC-3/MNC. The aft Power Controllers 4, 5, and 6 go with FC-1/MNA, FC-2/MNB, and FC-3/MNC, respectively. The FC/MAIN BUS A switch on Panel R1 is used to connect fuel cell 1 to MNA Distribution and Controller Assemblies 1, thus MNA buses throughout the vehicle are powered up through fuses. The indicator located adjacent to the FC/MAIN BUS A switch on Panel R1 indicates ON when Main Bus A is connected to Fuel Cell 1 and OFF when it is not connected. The FC/MAIN BUS B and FC/MAIN BUS C switches and indicators on Panel R1 are essentially the same, except for the respective bus.

MN BUS TIE A switch on Panel R1 can be used to tie MNA to MNB or MNC. The adjacent indicator would indicate ON when MNA is connected to the tie bus and OFF when it is not connected. The MN BUS TIE B and MN BUS TIE C switches and indicators on Panel R1 are essentially the same.

The Main Bus A, B, or C voltages can be displayed on the DC VOLTS meter on Panel F9 through the MAIN VOLTS A, B, or C rotary swtich. The MAIN BUS UNDERVOLTS red caution and warning light on Panel F7 will illuminate if Main Bus A, B, or C voltage is at 25.2 vdc, informing the crew that the minimum equipment operating voltage limit of 24 vdc is being violated. A backup caution/warning light will also illuminate at 25.2 vdc. An SM ALERT light will illuminate at a voltage of 27.0 vdc or less, alerting the flight crew to the possibility of a future low-voltage problem. A fault message also is transmitted to the CRT.

Depending on the criticality of electrical equipment, some electrical loads may receive power from two or three sources for redundancy only, and not for total power consumption, and for transient sensitive equipment.

Three essential buses, ESS 1BC, ESS 2CA, ESS 3AB, supply control power to the flight deck crew display and control switches which are necessary to restore power to a failed main

Space Shuttle Spacecraft Systems

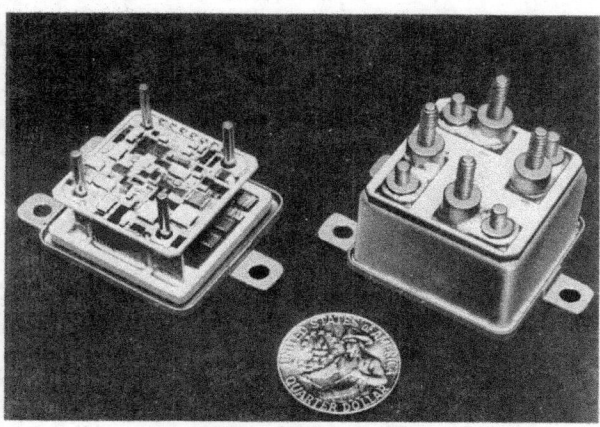

Remote Power Controller

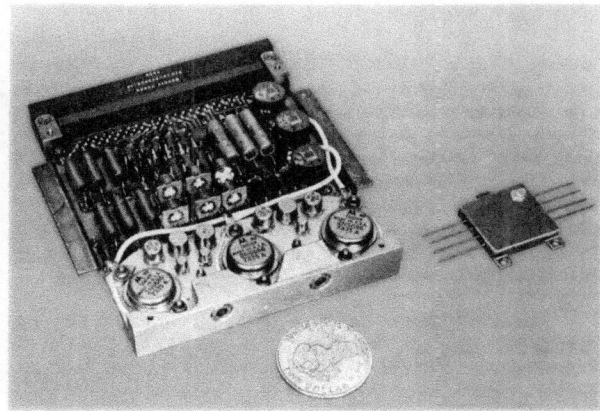

Hybrid Load Controller

dc or ac bus and to essential non-EPS electrical loads and switches.

The nominal fuel cell voltage is 27.5 to 32.5 volts dc, and the nominal main bus voltage range is 27 to 32 volts dc, which correspond to 12 and 2 kilowatt loads, respectively.

Each essential bus receives power from three redundant sources: a fuel cell and two main buses. ESS essential 1 BC bus receives power through the ESS BUS SOURCE FC1 switch on Panel R1 and Main Bus B and C through remote power controller's (RPC's) which are controlled by the ESS BUS SOURCE MN B/C switch, indicating Fuel Cell 1 and MN B and MN C are powering that essential bus. ESS 2 CA receives power through the ESS BUS SOURCE FC 2 switch and Main Bus C switch and Main Bus C and A through the ESS BUS SOURCE FC 3 switch and Main Bus A and B through RPC's which are controlled by the ESS BUS SOURCE A/B switch. Essential buses in the DA are distributed through fuses to the controller assemblies and to the flight deck and mid-deck display and control panels.

The ESS bus voltage can be monitored on the "DC Volts" meter on Panel F9 through the "ESS Volts 1 BC, 2 CA, or 3 AB" rotary switch. An "SM Alert" will illuminate to inform the flight crew if the essential bus voltage is less than 25 vdc. A fault message also is transmitted to the CRT.

Main buses located in the forward power controller's (FPC's) are connected through RPC's and diodes to nine control buses. The RPC's are powered continuously unless the CONTROL BUS PWR switch MNA, MNB, or MNC on Panel

Space Shuttle Spacecraft Systems

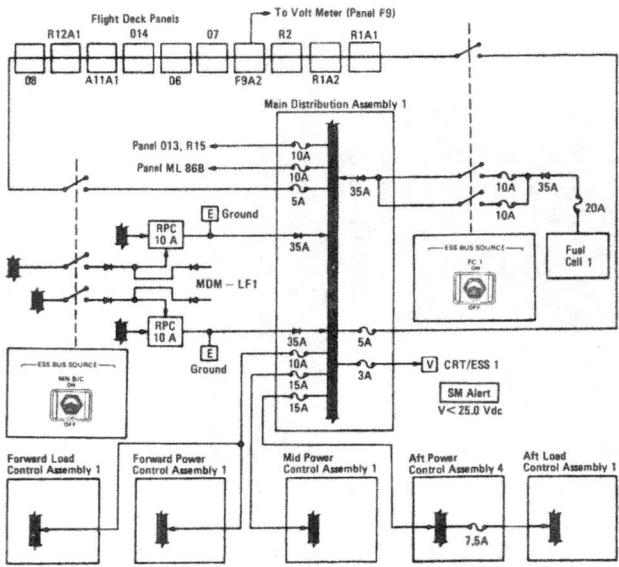

Essential Bus 1 BC Distribution

R1 is placed to the momentary RESET position, which turns the RPC power off and resets the RPC if it has been tripped out. Each control bus receives power from two main buses for system reliability. The control buses are CNTL AB 1, 2, and 3; CNTL BC 1, 2, and 3, and CNTL CA 1, 2, and 3. CNTL bus AB 1 receives power from MNA and B; the one indicates the number of the bus and not a fuel cell. An SM ALERT is illuminated if the control bus voltage is less than 24.5 vdc. A fault message is sent to the CRT. Mission Control Center (MCC) in Houston can monitor the status of each RPC. The control buses are located only on the flight deck and mid-deck areas behind the displays and controls. They are required and used only for control power to the display and control panel switches.

Until T minus 3 minutes and 30 seconds, power to the orbiter is load shared with the fuel cells and GSE, even though the fuel cells are on and capable of supplying power. Main bus power is suppied through the T-0 umbilicals, MNA through the left side umbilical, and MNB and C through the right side umbilical to APC (aft power controller), 4, 5, and 6. From APC 4, 5, and 6 the GSE power is directed to the DA, where the power is distributed throughout the vehicle. The power for the preflight test bus also is supplied through the T-0 umbilical. These test buses are scattered throughout the orbiter and are used for GSE-related functions, although they also power up the essential buses in the APC's, when on GSE. As in the main bus distribtion, essential bus power from the APC's is directed to the DA's and then distributed throughout the vehicle. T minus 3 minutes, 30 seconds, the ground turns off the GSE power to the main buses and the fuel cells automatically pick up the loads. At T-0, the T-0 umbilical is disconnected with the preflight Test Bus wires live.

The EPDC can provide 50 kwh to electrical energy to the payloads (based on a seven-day mission) when all three fuel cells are operating, the payload bay doors are fully opened, and a two-panel radiator kit supplements the baseline six-panel unit. The primary electrical interface is located at X_O 645 in the forward payload bay and is the primary payload (PRI PL) bus. This bus is capable of transferring 7 kw of average power at 27.2 vdc from Fuel Cell 3 through the positioning of the PAYLOAD PRI FC 3 switch on Panel R1. As a backup to Fuel Cell 3, the primary electrical interface can transfer 5 kw of average power at a minimum of 26.8 vdc from either MNB or MNC buses through the positioning of the PAYLOAD PRI MNB or PRI MNC switches on Panel R1. A talkback indicator adjacent to each of these switches on Panel R1 indicates ON when that bus is connected to the PRI PL bus and OFF when not connected.

Space Shuttle Spacecraft Systems

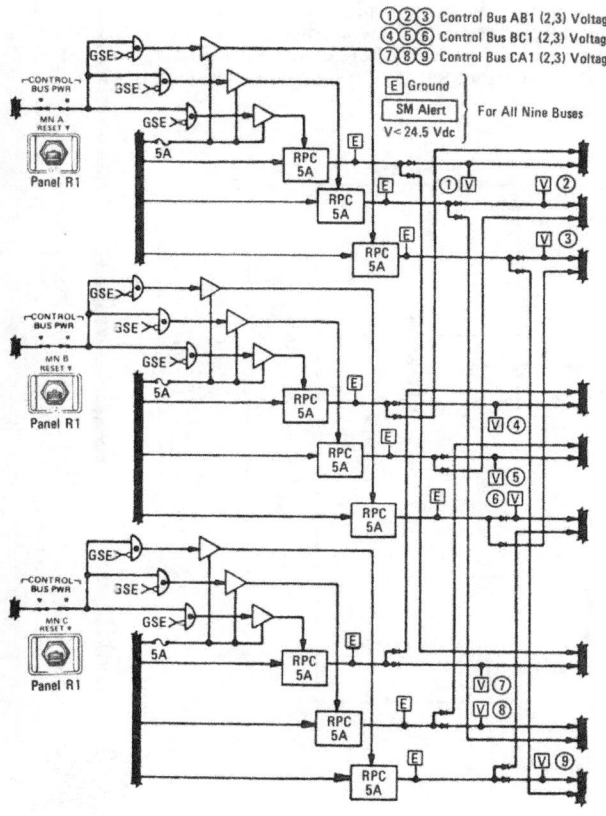

Control Buses

Orbiter and GSE Power Interface (MNA Only)

A backup to the PRI PL bus can be used by installation of an orbiter furnished kit. The backup (SEC PL) bus can transfer 5 kw of average power directly from MNB bus. If the SEC PL bus is used, the ability to connect MNB to the PRI PL bus is lost. If payload equipment requiring electrical power from two separate orbiter sources is required, this backup interface would be used.

There are two additional primary buses in the aft section of the payload bay at X_O 1307, referred to as aft payload

# Space Shuttle Spacecraft Systems

(AFT PL) bus MNB and MNC. These can power the AFT PL bus MNB and MNC when the PAYLOAD AFT MNB or AFT MNC switch is positioned to ON, on Panel R1. Each bus can transfer 1.5 kw of average power and a peak of 2 kw at a minimum of 25.2 vdc. They are used by payloads in the aft section of the payload bay.

Two other sources — auxiliary payloads A and B (AUX PLA and AUX PLB) are located adjacent to the primary and secondary buses in the mid payload bay area at X_O 645. They can supply a total of 0.4 kw average power from MNA and MNB buses through positioning of the PAYLOAD AUX switch to ON, on Panel R1. The two auxiliary payload buses

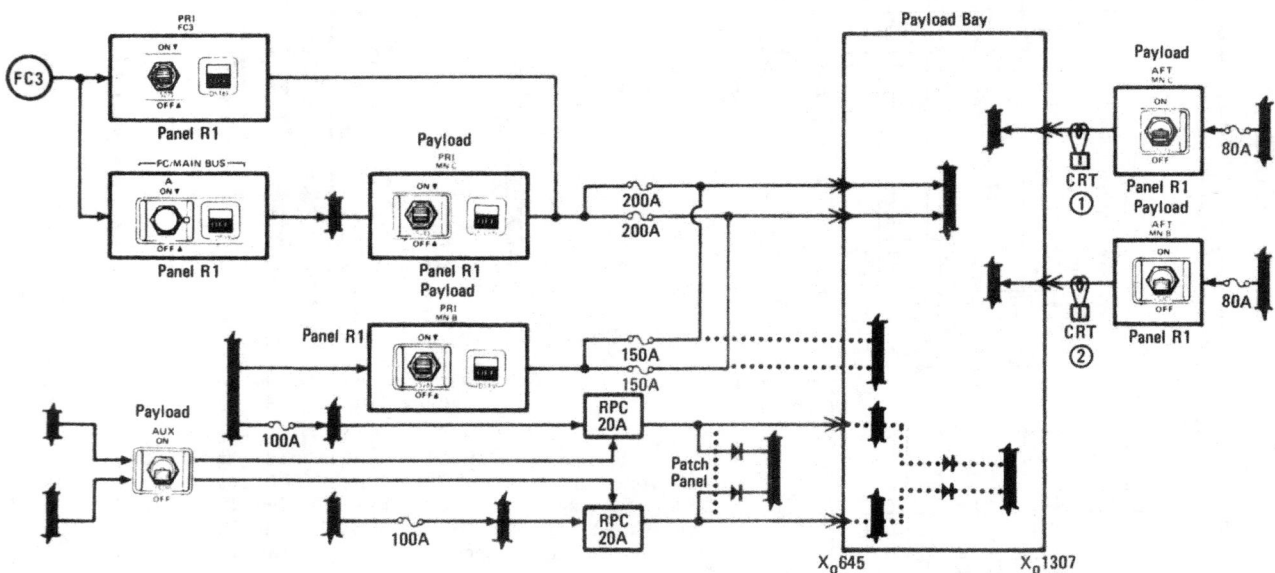

Orbiter/Payload Power Interface Schematic

Space Shuttle Spacecraft Systems

may be dioded together to form one bus, AUX PL A/B, redundantly powered by two sources. The auxiliary buses are planned to be used for emergency equipment and functions only.

Two or more feeders to the payload may be used simultaneously, but two orbiter power sources may not be tied directly within the payload. Any payload equipment requiring electrical power from two separate orbiter sources will ensure isolation of these power sources so that no single failure in a load, or succession or propagation of failures in a load, will cause an out-of-limits condition to exist on the orbiter system equipment on more than one bus.

The PAYLOAD CABIN switch on Panel R1 provides MNA or MNB power to patch panels located under the on-orbit, payload specialist, and mission specialist stations located at the aft flight deck. These patch panels supply power to the payload-related equipment located on panels at these stations. Two three-phase circuit breakers, AC 2 CABIN PL 3φ and AC 3 CABIN PL 3φ, on Panel MA73C provide ac power to the patch panels.

Alternating-current power is generated and made available to using systems through three ac buses: AC 1, AC 2, and AC 3. The inverters convert dc to ac and the ac bus distribution assemblies contain the ac buses and the ac bus sensors. The ac power is distributed to the flight and mid-deck display and control panels and in the payload bay deployment mechanisms, in payloads, and in the aft main engine controllers and payload bay doors. The inverters are located in the forward avionics bays of the crew compartment.

The three phase inverters for AC 1 receive power from MNA, AC 2 from MNB, and AC 3 from MNC.

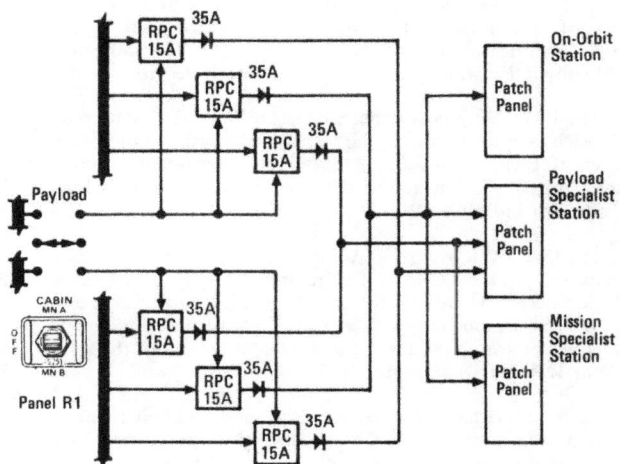

Note: The purpose of patch panels is to supply power to payload-related equipment located on panels at these stations.

Payload Cabin Switch

Each bus consists of three separate phase buses connected in a three-phase array. Static inverters, one for each phase, are rated at 750 va (volt-amps) per phase and have an output voltage of 117 plus three minus one volts at 400 plus or minus 2 Hertz. The inverter efficiency is approximately 76.5 percent.

The three-phase inverters for AC 1 receive power only from MNA, AC 2 from MNB, and AC 3 from MNC. The INV

Space Shuttle Spacecraft Systems

PWR switches 1, 2, 3 on Panel R1 are used to apply main bus power to the inputs of each phase inverter. An indicator adjacent to each switch shows the status of the switch; all three inverters of INV 1 have MNA bus power and all must be on before the indicator shows ON, as an example. The indicator would show OFF when bus power is not connected to INV 1.

The INV/AC Bus switches 1, 2, 3 on Panel R1 are used to apply each inverter output to the respective ac bus. An indicator adjacent to each switch shows its status, all three inverters of inverter 1 must be connected to their respective ac buses before the indicator indicates ON.

The INV PWR and INV/AC BUS switches must have control power from the AC CONTR circuit breakers on Panel R1 in order to operate. Once ac power has been established, these circuit breakers are opened to prevent any inadvertent disconnection, either by switch failure or accidental movement of the INV PWR or INV/AC BUS switches.

Each ac bus has a sensor, a switch, and a circuit breaker for flight crew control. The AC 1, 2, 3 SNSR circuit breakers are located on Panel O13 and apply essential bus power to its respective AC BUS SNSRS 1, 2, 3 switch on Panel R1 and operational power to the respective INV/AC BUS switch indicator. The AC BUS SNSR 1, 2, 3 switch selects the mode of operation of the ac bus sensor: AUTO TRIP, MONITOR, or OFF. The ac bus sensor monitors each ac phase bus for over or under voltage, and each phase inverter for an overload signal. The overvoltage limits are bus voltages greater than 123 to 127 vac for 50 to 90 milliseconds. The undervoltage limits are bus voltages less than 102 to 108 vac for 6.5 to 8.5 milliseconds. The overload is when any ac phase current is greater than 14.5 amps for 10 to 20 seconds or is greater than 17.3 to 21.1 amps for 4 to 6 seconds.

When the respective AC BUS SNSR switch is positioned to the AUTO TRIP position and an overload or overvoltage condition exists, the ac bus sensor will illuminate the respective yellow AC VOLTAGE or AC OVERLOAD caution and warning light on Panel F7 and trip out (disconnect) the inverter from its respective phase bus for the bus/inverter causing the problem. There is only one AC VOLTAGE and one AC OVERLOAD caution and warning light; as a result, all nine inverters/ac phase buses can illuminate the lights. The AC VOLTS meter and rotary switch (AC 1, ϕA, ϕB, ϕC; AC 2, ϕA, ϕB, ϕC; AC 3, ϕA, ϕB, ϕC) on Panel F9 or the CRT display would be used to determine which inverter or phase bus caused illumination of the light. The phase bus causing the problem would show 0 volts. Because of the various three-phase motors throughout the vehicle, there will be a small induced voltage on the phase bus if there is only one phase which has loss of power. Before power can be restored to the tripped bus, the trip signal to the INV/AC BUS switch must be removed by positioning the AC BUS SNSR switch to OFF, then back to the AUTO TRIP position, which extinguishes the caution and warning light. The INV/AC BUS switch is then positioned to ON, restoring power to the failed bus. If the problem is still present, the sequence will be repeated.

If an undervoltage exists, the yellow AC VOLTAGE caution and warning light on Panel F7 will illuminate but the inverter will not be tripped out from its phase bus.

When the AC BUS SNSR 1, 2, 3 switches are in the MONITOR position, the ac bus sensor will monitor for an overload, overvoltage, and undervoltage and illuminate the applicable caution and warning light but will not trip out the phase bus/inverter causing the problem..

When the AC BUS SNSR switches are OFF the ac bus sensors are non-operational and all caution and warning and trip out capabilities are inhibited.

A backup caution and warning light will illuminate for overload or over/undervoltage conditions. The SM ALERT will

Space Shuttle Spacecraft Systems

occur for over/undervoltage conditions. A fault message also is sent to the CRT.

There are 10 motor controller assemblies (MCA's) used on the orbiter; three are in the forward area, four are in the mid body area, and three are in the aft area. Their only function is to supply ac power to the ac motors used for vent doors, air data doors, star tracker doors, payload bay doors, payload bay latches, and reaction control system/orbital maneuvering system (RCS/OMS) motor-actuated valves. The MCA's contain main buses, ac buses, and hybrid relays, which are the remote switching devices for switching the ac power to electrical loads. The main buses are used only to supply control or logic power to the hybrid relays so that ac power can be switched on/off. If a main bus is lost, the hybrid relays using that main bus will not operate.

The three forward motor controller (FMC) assemblies (FMC-1, FMC-2, and FMC-3) correspond to MNA/AC 1,

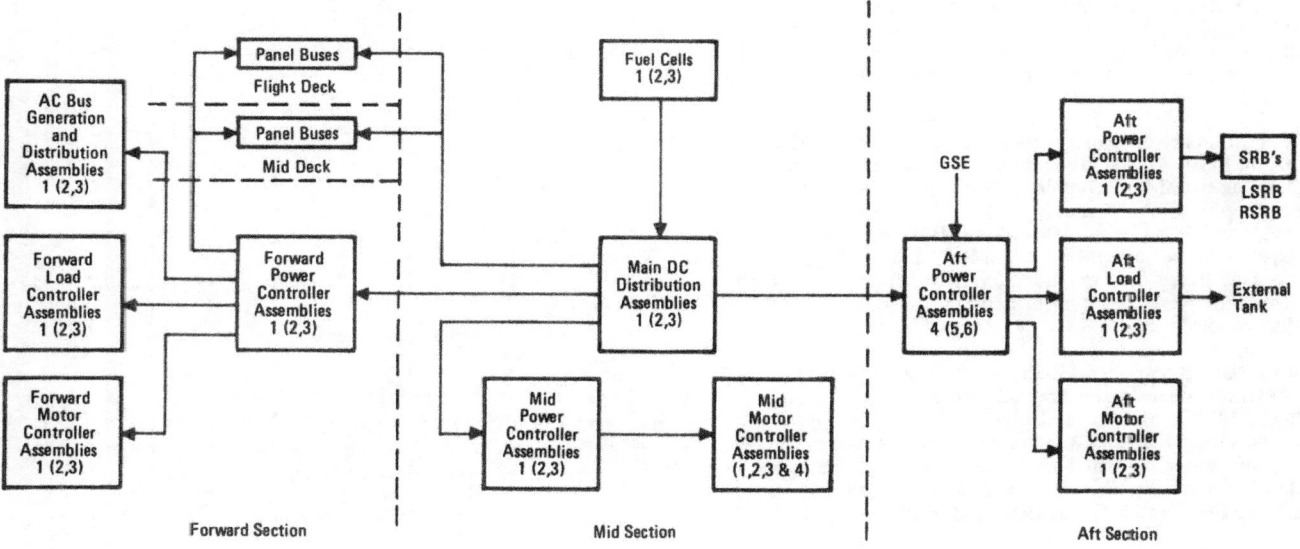

Electrical Power Distribution Block Diagram

Space Shuttle Spacecraft Systems

MNB/AC 2, and MNC/AC 3, respectively. Each FMC contains a main bus, an ac bus, and an RCS ac bus. The main bus supplies control or logic power to the relays associated with both the ac bus and RCS ac bus. The ac bus supplies power to the forward left and right vent doors, the star tracker Y and Z doors, and the air data left and right doors. The RCS ac bus supplies power to the forward RCS manifold and tank isolation valves.

The aft motor controller assemblies (AMC-1, AMC-2, and AMC-3) correspond to MNA/AC 1, MNB/AC 2, and MNC/AC 3, respectively. Each AMC assembly contains a main bus and its corresponding ac bus and a main RCS/OMS bus and its corresponding RCS/OMS ac bus. Both main buses are used for control or logic power for the hybrid relays. The ac bus is used by the aft RCS/OMS manifold and tank isolation and crossfeed valves.

The mid motor controller assemblies (MMC-1, MMC-2, MMC-3, and MMC-4) contains two main dc buses and two corresponding buses. MMC-1 contains MN bus A and B and their corresponding buses, AC 1 and 2. MMC-2 contains MNB and C and AC 2 and AC 3 buses. MMC-3 and MMC-4 contain the same respective buses as MMC-1 and MMC-2. Loads for the main buses/ac buses are vent doors, payload bay door and latches, radiator panel deployment actuator and latches, and payload retention latches.

The electrical components in the mid body are mounted on coldplates and cooled by the Freon-21 system coolant loops. The PCA's, LCA's, MCA's, and inverters located in the forward avionics bays 1, 2, and 3 are mounted on coldplates and cooled by the water coolant loops. The inverter distribution assemblies in forward avionics bays 1, 2, and 3 are air-cooled. The LCA's, PCA's, and MCA's located in the aft avionics bays

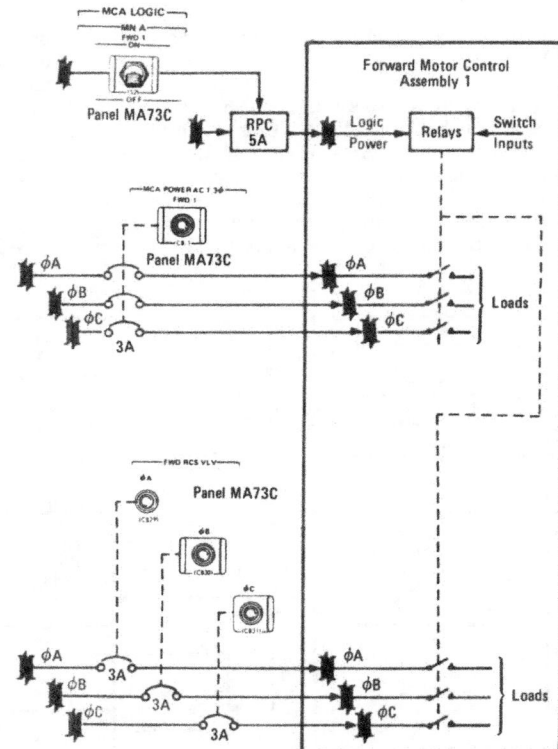

Forward Motor Control Assembly

239

Space Shuttle Spacecraft Systems

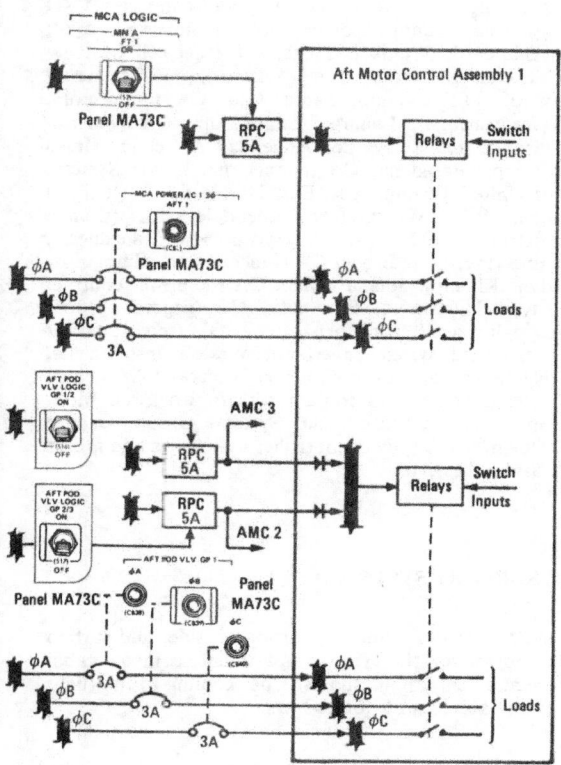

Aft Motor Control Assembly

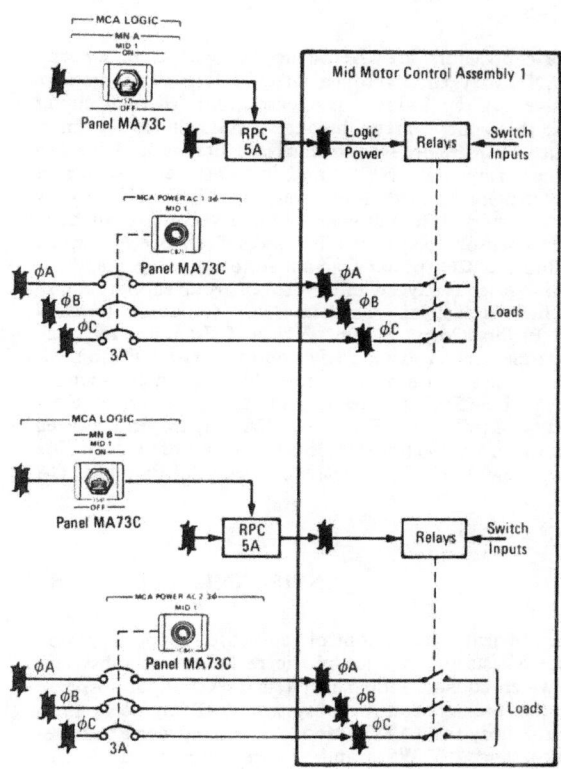

Mid Motor Control Assembly

Space Shuttle Spacecraft Systems

are mounted on coldplates and cooled by the Freon-21 system coolant loops.

The contractors are: Aerodyne Controls Corp., Farmingdale, NY (oxygen, hydrogen check valve, water pressure relief valve); Aiken Industries, Jackson, MI (thermal circuit breakers; three-phase circuit breakers); AIL Cutler Hammer, Milwaukee, WI (remote control circuit breaker); American Aerospace Farmingdale, NY (ac and dc current sensors, current level detector); Applied Resources, Fairfield, NJ (rotary switch); Rockwell International Autonetics Group, Anaheim CA (ac bus sensor, load controller assemblies); Beech Aircraft Corp., Boulder, CO (power reactant storage hydrogen and oxygen tanks, gaseous oxygen and hydrogen group support equipment); Bell Industries, Gardena, CA (modular terminal boards); Bendix Corp., Sidney, NY, and Franklin, IN (high density connectors); Bussman Division of McGraw Edison, St. Louis, MO (fuses, fuse holders, fuse dc limiter high current); Brunswick-Circle Seal, Anaheim, CA (water check valve); Consolidated Controls, El Segundo, CA (hydrogen, oxygen solenoid valve, undirectional/bidirectional shutoff valve); Cox and Co., New York, NY (heaters); Deutsch, Banning, CA (general-purpose connector); Fairchild Stratos, Manhattan Beach, CA (cryogenic fluid and gas supply disconnects); G/H Technology Co., Santa Monica, CA (connector cryo); Hamilton Standard, Windsor Locks, CT (fuel cell heat exchanger); Haveg Industries, Inc., Winooski, VT (general-purpose wire); ITT Cannon, Santa Ana, CA (connectors, bulkhead feedthrough); Labarge, Santa Ana, CA (general-purpose wire); Leach Relay, Los Angeles, CA (relay); Malco Microdot Corp., Pasadena, CA (connector); Power Systems Division of United Technologies, East Hartford, CT (fuel cell powerplants); R.V. Weatherford, Glendale, CA (shunt); Statham Instruments, Oxnard, CA (cryo pressure transducer); Tayco Engineering, Long Beach, CA (fuel cell water dump nozzle); Teledyne Kinetics, Solana Beach, CA (dc power contractor); Teledyne Thermatics, Elm City, NC (general-purpose wire); Westinghouse Electric Corp., Lima, OH (remote power controller electrical system inverters); Weston Instruments, Newark, NJ (electrical indicator meter) Brunswick-Wintec El Segundo, CA (reactant and coolant filters); Rockwell International Space Transportation and Systems Group, Orbiter Division, Downey, CA (power controller assemblies and motor controller assemblies).

ENVIRONMENTAL CONTROL AND LIFE SUPPORT SYSTEM

The environmental control and life support system (ECLSS) consists of an atmospheric revitalization subsystem (ARS), water coolant loop subsystem (WCLS), atmosphere revitalization pressure control subsystem (ARPCS), active thermal control subsystem (ATCS), food, water, and waste management subsystem (FWWS), and airlock support subsystem. These subsystems interact to provide a habitable environment in the crew compartment for the crew and passengers.

The ARS provides humidity, carbon dioxide, and carbon monoxide control for the crew compartment, controls cabin temperature and ventilation, and provides cooling to the flight deck avionics, cabin, and avionics bays.

The ARPCS controls cabin pressure and oxygen partial pressure, nitrogen pressurization of the potable and waste water

Space Shuttle Spacecraft Systems

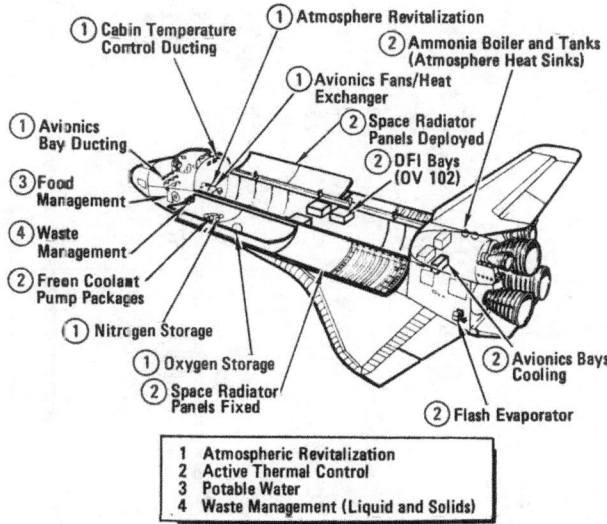

Environmental Control and Life Support System

tanks, and storage of nitrogen and emergency oxygen consumables.

The WCLS collects heat from the cabin atmosphere and electronics and transfers it to the ATCS, which rejects it to space via water and Freon coolant loops. The ATCS also removes and rejects waste heat from the fuel cells, payload, and mid-body and aft-located electronic units, while providing heating of the hydraulic systems when needed.

Potable water is produced by the three fuel cells aboard the spacecraft and stored in tanks for crew consumption and personal hygiene; during certain phases of the mission it is also used for cooling of the Freon coolant loops. Waste water collected from the cabin heat exchanger is stored in tanks along with crew members' waste water. Solid waste remains in the waste management system until the orbiter is serviced on the ground.

The orbiter crew compartment provides a life-sustaining environment for a crew of four and has accommodations for four passengers. The crew cabin volume with the airlock inside the mid deck of the crew compartment is 66 cubic meters (2,325 cubic feet). If the airlock is outside in the payload bay area, the volume is 74 cubic meters (2,625 cubic feet).

ATMOSPHERIC REVITALIZATION CONTROL SUBSYSTEM

The ARS maintains a habitable environment for the crew and passengers and a conditioned environment for the electronic avionics equipment located inside the crew cabin. The ARS consists of the water coolant loops (WCL), and the cabin air loops, and pressure control.

CABIN PRESSURE. The cabin is pressurized to 760 millimeters of mercury (mmHg) ±103 mmHg (14.7 ±0.2 psia) and maintained at an average 80-percent nitrogen and 20-percent oxygen mixture by the ARS.

Oxygen partial pressure is maintained between 152 and 178 mmHg (2.95 and 3.45 psi), with sufficient nitrogen added to achieve the cabin total pressure of 760 plus or minus 10 mmHg (14.7 ±0.2 psia).

Oxygen is obtained from two sources: the primary and secondary power reactant super-critical cryogenic oxygen storage supply system. The supercritical cryogenic oxygen supply system is located in the lower portion of the mid fuselage

Space Shuttle Spacecraft Systems

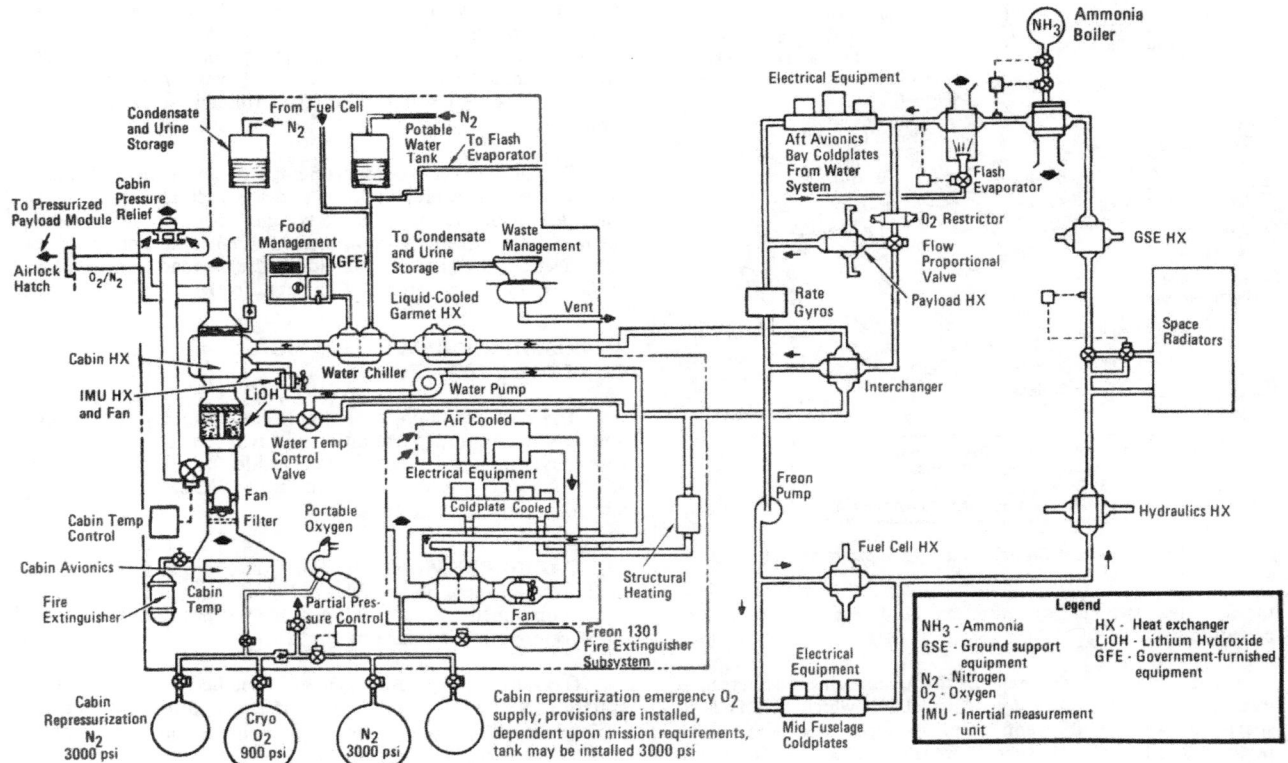

ECLSS Schematic

Space Shuttle Spacecraft Systems

and is also utilized by the three power reactant fuel cells. The emergency gaseous oxygen supply is located in the lower forward portion of the mid fuselage and is installed as a kit which provides the capability of flying a mission with or without it.

Nitrogen is obtained from the primary or secondary gaseous nitrogen storage supply system located in the lower forward portion of the mid fuselage. For normal orbital operations, one oxygen and nitrogen system is used. For launch and entry, both the primary and secondary systems will be used.

The heart of the ARS is a nitrogen/oxygen control panel and a supply panel, oxygen partial pressure sensor, and crew compartment positive and negative pressure relief valves.

Primary and secondary nitrogen are provided to the control panel from the N_2 storage tanks via the supply panel. The N_2/O_2 control panel selects and regulates primary or secondary O_2 and N_2. There is also a cross-over between the primary and secondary systems. The control panel can regulate the emergency gaseous oxygen supply, if installed and supply oxygen for airlock support. The primary and secondary systems are used together during airlock repressurization.

The gaseous nitrogen primary and secondary storage tanks No. 1 and 2 and O_2/N_2 supply panel are located in the lower forward portion of the mid fuselage. The N_2 storage tanks are serviced to a nominal pressure of 153,240 mmHg (2,964 psia) at 26°C (80°F). The emergency oxygen supply tank (if installed) in the lower forward portion of the mid fuselage is serviced to a nominal pressure of 126,150 mmHg (2,440 psia) at 26°C (80°F) and stores 29.7 kilograms (67.6 pounds) of gaseous oxygen to provide high flow along with gaseous nitrogen. This emergency supply maintains the cabin at 414 mmHg (8 psi) and oxygen partial pressure at 103.5 mmHg (2 psia), if installed.

The O_2 and N_2 systems provide makeup gas for oxygen consumption by the crew and passengers. Four crew members use an average of 0.79 kilogram (1.76 pounds) of O_2 per person per day. Up to 3.49 kilograms (7.7 pounds) of nitrogen and 4 kilograms (9 pounds) of oxygen are expected to be used per day for normal loss of crew cabin air to space and metabolic usage. The O_2 and N_2 also provides for repressurization of the airlock and pressurization of the potable and waste water tanks. The tanks are pressurized to 879 mmHg (17 psi).

The supercritical cryogenic oxygen primary and secondary storage supply systems are controlled individually by ATM (atmospheric) PRESS (pressure) CONTROL O_2 (oxygen) SYS 1 and 2 SUPPLY switches on flight deck and display control Panel L2 (when switch and display nomenclature is printed in all caps—e.g. it indicates that it is the exact way it appears on the display and control panel). When a switch is positioned to OPEN, oxygen supply system is directed to a heat exchanger in Freon-21 coolant loop system No. 1 (for system No. 1 oxygen) and coolant loop system No. 2 (for system No. 2 oxygen) which warms the supercritical cryogenic oxygen supply to the O_2 regulator of that system. When the switch is closed, the oxygen storage supply system is isolated from the O_2 regulator. An indicator above the switches will show when a valve is open and closed.

The oxygen supply system is directed to an O_2 REG (regulator) INLET manual valve when the respective ATM PRESS CONTROL O_2 SYS SUPPLY valve is open. The manual O_2 REG INLET valve of that system, when open, directs that O_2 supply to its O_2 regulator. The O_2 regulator in each system reduces the oxygen supply source pressure to 5,175 millimeters of mercury (mmHg) 100 psi with a minimum flow rate capability of 34 kilograms (75 lbs) per hour. Each regulator is a two-stage regulator with the second stage functioning as a relief valve when the differential pressure across the second stage is 12,678 mmHg (215 psi). The relief pressure is vented into the crew module cabin. The regulated pressure from system No. 1 and system No. 2 is routed to its respective 14.7 psi CABIN REG (regulator) INLET manual valve, 8 psi EMERG

# Space Shuttle Spacecraft Systems

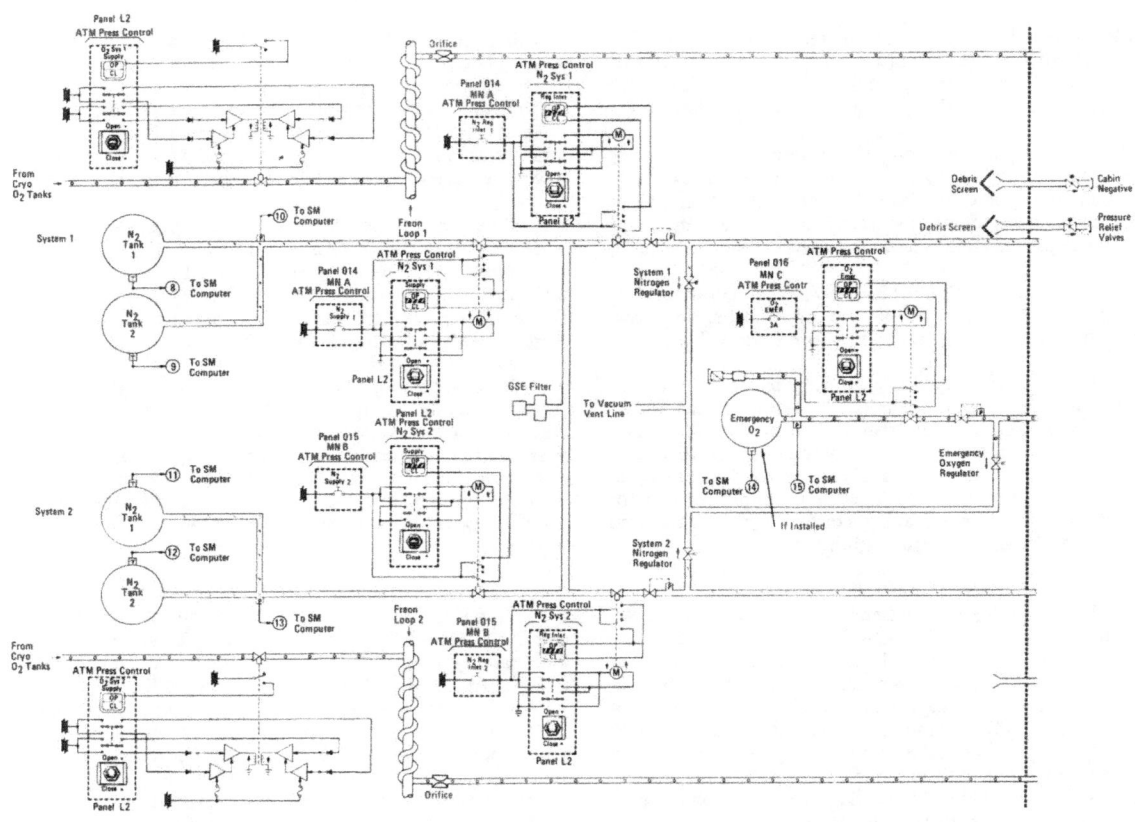

Atmospheric Revitalization System Pressurization (Sheet 1 of 3)

Space Shuttle Spacecraft Systems

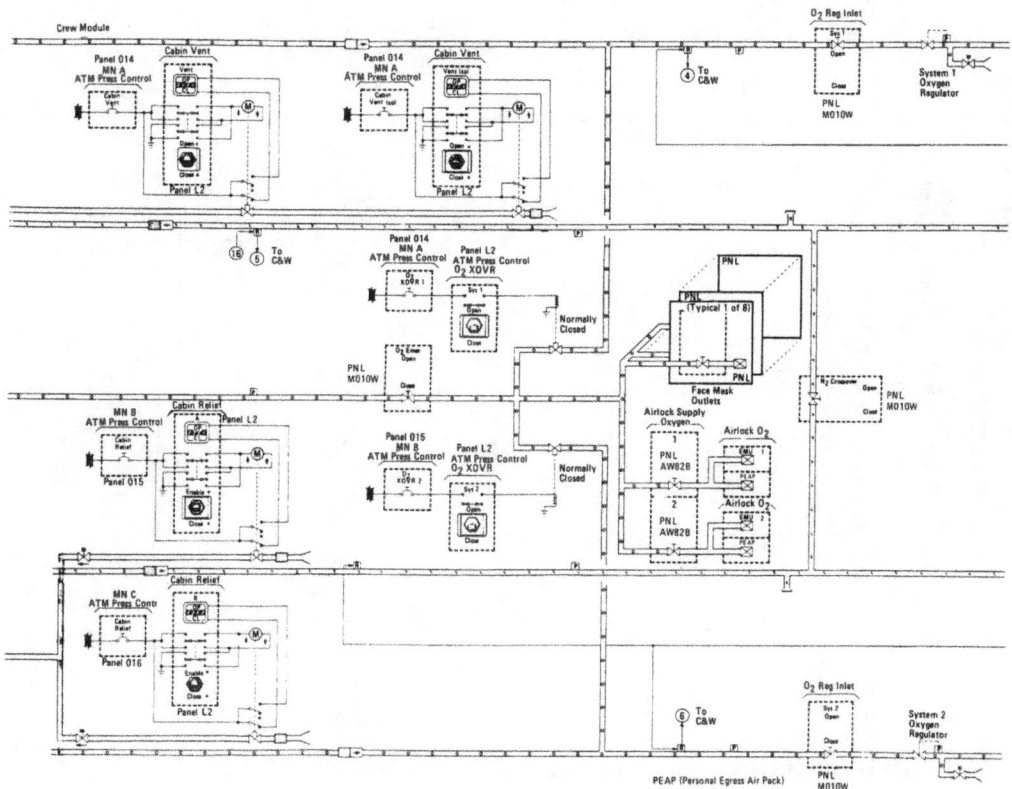

Atmospheric Revitalization System Pressurization (Sheet 2 of 3)

Space Shuttle Spacecraft Systems

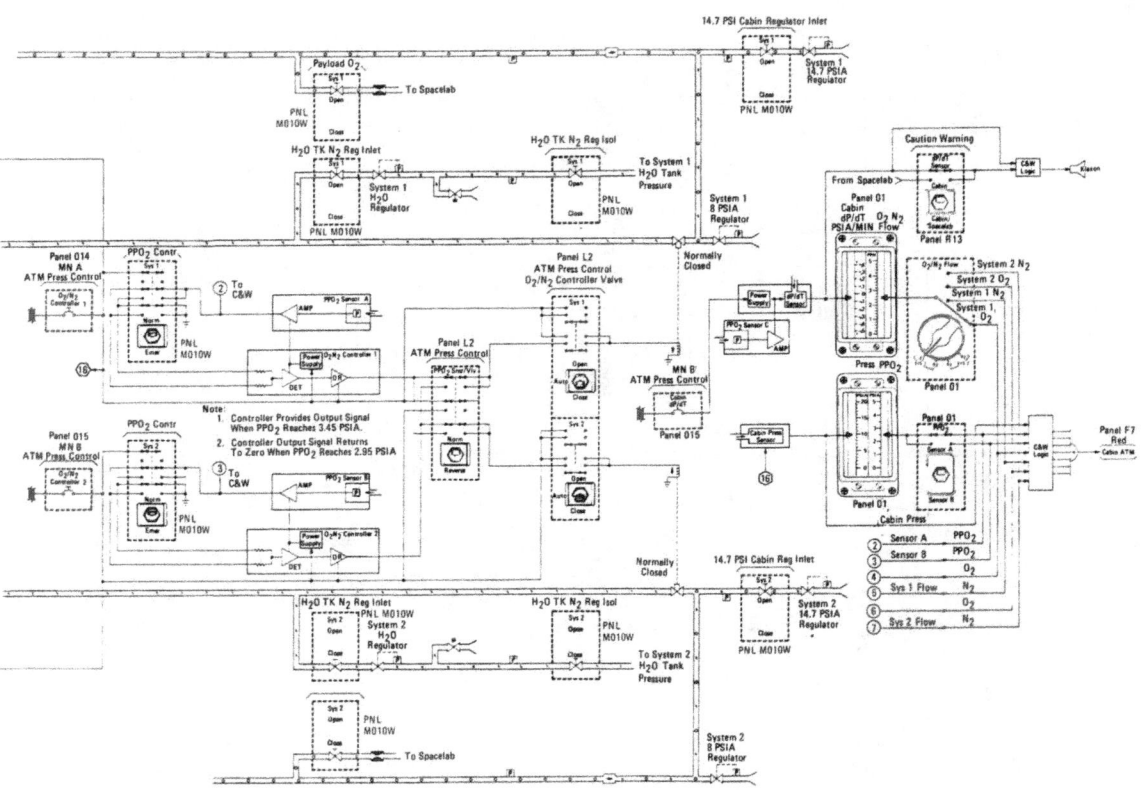

Atmospheric Revitalization System Pressurization (Sheet 3 of 3)

Space Shuttle Spacecraft Systems

(emergency) REG (regulator), the PAYLOAD O_2 manual valve and the XOVER (crossover) O_2 valve. The check valve in each O_2 supply line between the O_2 regulator and the 14.7 CABIN REG INLET prevents O_2 reverse flow to the 5,175 mmHg (100 psi) regulator.

The two valves in the crossover manifold are controlled by ATM PRESS CONTROL O_2 PRESS SYS 1 and SYS 2 XOVER switches on Panel L2. When a switch is opened, that oxygen supply system is directed to the AIRLOCK SUPPLY OXYGEN 1 and 2 manual valves, AIRLOCK O_2 1 and 2 EMU (extravehicular mobility unit), PEAP (personal egress air pack), and the eight face mask outlets. If both ATM PRESS CONTROL O_2 PRESS SYS 1 and SYS 2 XOVER valves are open, oxygen supply systems No. 1 and 2 are interconnected. When the respective ATM PRESS CONTROL O_2 PRESS switch is closed, that oxygen supply system is isolated from the crossover manifold.

The emergency O_2 tank (if installed) is made of a filament wound Kevlar fiber with an Inconel liner. The O_2 tank is serviced to a nominal 126,150 mmHg (2,440 psia) at 26 °C (80 °F) and stores 29.7 kilograms (67.6 lb) of gaseous oxygen. The emergency oxygen supply system is controlled by the ATM PRESS CONTROL O_2 EMER switch on Panel L2. The emergency O_2 supply system is directed to its regulator when the ATM PRESS CONTROL O_2 EMER switch is open and isolated from the regulator when the switch is closed. An indicator above the switch indicates OP (open) when the valve is open, CL (close) when the valve is closed, and BARBERPOLE when the motor operated valve is in transit.

The emergency O_2 regulator reduces emergency O_2 supply pressure to 15,525 mmHg (300 psi) when the O_2 emergency valve is open. The two-stage regulator has a relief valve. When the differential pressure across the relief valve is 64,687 mmHg (1,250 psi), the valve will operate. The relief pressure is vented overboard.

The emergency O_2 regulated pressure is directed to an O_2 EMER manual valve. When the manual valve is open, the emergency O_2 supply is directed into the O_2 crossover manifold. The emergency O_2 supply is normally isolated except for ascent and entry.

The check valve between the Freon-21 coolant loop and crossover valve in the primary and secondary supercritical oxygen supply system prevents O_2 flow from one supply source to the other when the crossover valves are open.

The primary and secondary gaseous nitrogen supply tanks are identical to the emergency gaseous oxygen supply tank except that a titanium liner is used. Each gaseous nitrogen (N_2) tank is serviced to a nominal pressure of 153,240 mmHg (2,964 psia) at 26 °C (80 °F) with a volume of 134,086 cubic centimeters (8,181 cubic inches). The two N_2 tanks in each system are manifolded together. The primary and secondary nitrogen supply systems are controlled by ATM PRESS CONTROL N_2 SYS 1 and 2 SUPPLY switches on Panel L2. When a switch is opened, that nitrogen supply system is directed to an ATM PRESS CONTROL N_2 SYS REG (regulator) INLET valve. When the switch is closed, the nitrogen supply system is isolated from the N_2 SYS REG INLET valve. An indicator above each switch indicates CL (close) when the valve is closed, OP (open) when the valve is open and BARBERPOLE when the motor operated valve is in transit.

Each nitrogen inlet valve is controlled by its respective ATM PRESS CONTROL N_2 SYS REG INLET 1 and 2 switch on Panel L2. When the individual switch is open, that valve permits that N_2 source pressure to that system N_2 regulator, providing the respective N_2 supply valve is open. When the individual switch is closed, the N_2 supply pressure is isolated from that system N_2 regulator. An indicator above each switch indicates CL (close) when the valve is closed, OP (open) when the valve is open and BARBERPOLE when the motor operated valve is in transit.

Space Shuttle Spacecraft Systems

The N_2 regulators in the primary and secondary supply systems reduce pressure to 10,350 mmHg (200 psi). Each N_2 regulator is a two-stage regulator with the second stage functioning as a relief valve. The second stage relieves pressure overboard at 14,237 mmHg (245 psi).

The regulated N_2 pressure of each N_2 system is supplied to the N_2 crossover valve, the H_2O (water) tank regulator inlet valve, and the O_2/N_2 controller valve in each N_2 system.

The N_2 XOVER (crossover) manual valve connects both regulated N_2 systems when the valve is open. When it is closed, the N_2 regulated supply systems are isolated from each other. A check valve between the N_2 regulator and N_2 crossover valve in each N_2 regulated supply line prevents reverse flow of N_2 when the crossover valve is open.

The H_2O TK (tank) N_2 REG INLET valve in each N_2 regulated supply system permits N_2 to flow into the H_2O regulator when that valve is manually opened. The closed position isolates the N_2 regulated supply from the H_2O regulator.

The H_2O tank regulator of each system reduces the 10,351 mmHg (201 psi) supply pressure to 879 mmHg (17 psi). Each H_2O tank regulator is a two-stage regulator. The second stage relieves pressure into the crew module cabin at a differential pressure of 1,034 mmHg (20 psi).

Two partial pressure oxygen (PPO_2) sensors are located in the crew module mid deck cabin air supply duct. The PPO_2 sensors in the mid deck cabin air supply duct are sensor's A and B and provide inputs to the PPO_2 CONTR (controller) SYS 1 and 2 controller and switch, respectively.

When a PPO_2 CNTR switch is positioned to NORM (normal), the corresponding PPO_2 controller and PPO_2 sensor, in conjunction with the ATM PRESS CONTROL PPO_2 SNSR (sensor)/VLV switch on Panel L2 in the NORM position, provide electrical power to the corresponding ATM PRESS CONTROL O_2/N_2 CNTLR VLV switch on Panel L2 for SYS 1 and SYS 2. When the O_2/N_2 CNTRL VLV switch is set on AUTO, electrical power automatically controls the nitrogen valve in the corresponding regulated nitrogen supply. If O_2 is required in the crew module cabin, the nitrogen valve is automatically closed, the 14.7 psi cabin regulator opens (providing the 14.7 psi CABIN REG INLET manual valve on that system is open), the 10,350 mmHg (200 psi) nitrogen supply in the manifold drops below 5,175 mmHg (100 psi), and oxygen flows through the check valve and cabin regulator into the crew module cabin. When sufficient oxygen is present in the cabin as determined by the PPO_2 sensor, the nitrogen valve is opened and 10,350 mmHg (200 psi) nitrogen enters the manifold. The nitrogen closes the oxygen check valve and flows through the 14.7 psi cabin regulator into the crew module cabin. Oxygen partial pressure is maintained at 178 mmHg (3.45 psi). The OPEN and CLOSE positions of the N_2/O_2 CNTLR VLV SYS 1 and SYS 2 switch on Panel L2 permit the crew to control the nitrogen valve in each system manually. The REVERSE position of the PPO_2 SNSR/VLV switch on Panel L2 allows Controller B to SYS 1 and Controller A to SYS 2.

If the 14.7 psi CABIN REG INLET manual valves of SYS 1 and SYS 2 are closed, the crew module cabin pressure will decrease to 8 psi. The PPO_2 CONTR SYS 1 and SYS 2 switches on Panel L2 are positioned to EMER (emergency) for the corresponding nitrogen system which selects the 113.8 mmHg (2.2 psi) oxygen partial pressure. The corresponding PPO_2 sensor and controller, through the corresponding PPO_2 CONTR switch and the PPO_2 SNSR/VLV switch on NORM, provide electrical inputs to the corresponding O_2/N_2 CNTRL VLV switch. The electrical output from the applicable O_2/N_2 CNTLR VLV switch controls the nitrogen valve in that supply system in the same manner as in the 14.7 psi mode except that the crew module cabin partial pressure is maintained at 113.8

Space Shuttle Spacecraft Systems

mmHg (2.2 psi). In this mode, the crew members would use supplemental oxygen from the PEAP's.

The O_2 system No. 1 and 2 and N_2 system No. 1 and 2 flows are monitored and sent to the O_2/N_2 FLOW rotary switch on Panel 01. The rotary switch permits SYS 1 O_2 or N_2 or SYS 2 O_2 or N_2 flow to be monitored on the O_2/N_2 FLOW meter on Panel 01 in PPH (pounds per hour).

PPO_2 Sensors A and B monitor the oxygen partial pressure and transmit the signal to the PPO_2 SENSOR select switch on Panel 01. When the PPO_2 SENSOR switch is positioned to SENSOR A, oxygen partial pressure from Sensor A is monitored on the PPO_2 meter on Panel 01 in psia. If the switch is set on SENSOR B, oxygen partial pressure from Sensor B is monitored.

The cabin pressure sensor transmits directly to the CABIN PRESS meter on Panel 01 and is monitored in psia.

The RED CABIN ATM caution and warning light on Panel F7 would illuminate from any of the following monitored parameters:

Cabin pressure below 14.0 psia or above 15.4 psia
PPO_2 below 2.8 psia or above 3.6 psia
O_2 flow rate above 5 lbs/hr
N_2 flow rate above 5 lbs/hr

A klaxon will sound in the crew module cabin if the dP/dT (which stands for change in pressure versus change in time) is greater than 0.05 psi per minute.

The temperature and pressure of the primary and secondary nitrogen and emergency oxygen tanks are monitored and transmitted to the SM (system management) computer. This information is used to compute O_2 and N_2 quantities.

If the crew module cabin pressure is lower than the pressure outside the cabin, the negative pressure relief valves will open in a 10 to 36 mmHg (0.2 to 0.7 psi) range differential permitting flow into the crew module cabin. The maximum flow rate at 25 mmHg (0.5 psi) differential is 0 to 296 kilograms (654 lbs) per hour.

The crew module cabin vent and vent isolation valves provide the capability of venting the cabin to ambient following the prelaunch cabin pressure integrity test. These two valves are in series, thus both valves must be open to vent the cabin. The CABIN VENT VALVE ISOL (isolation) switch on Panel L2 opens and closes the cabin vent isolation valve. An indicator above the switch indicates OP (open) when the valve is open, CL (close) when the valve is closed, and BARBERPOLE when the valve is in transit. The CABIN VENT VALVE switch on Panel L2 opens and closes the cabin vent valve. An indicator above the switch indicates in the same manner as in the cabin vent isolation valve. When both these valves are open, the maximum flow 10 mmHg (0.2 psi) differential is 408 kilograms (900 lbs) per hour.

The two parallel cabin relief valves provide protection against overpressurization of the crew module cabin above 828 mmHg (16.0 psi) differential. Each cabin relief valve has a backup, motor-operated isolation valve. CABIN RELIEF switch A controls cabin relief valve A and CABIN RELIEF switch B control cabin relief valve B. When a switch is positioned ENABLE, the corresponding cabin relief valve is enabled and when the switch is positioned to CLOSE, the valve is disabled. An indicator above each switch shows whether the valve is open or closed. Each cabin relief valve will flow a maximum of 68 kilograms (150 lbs) per hour.

Approximately one hour and 26 minutes before launch, the crew module cabin is pressurized by ground support equipment to approximately 864 mmHg (16.7 psi) for a leak check.

Space Shuttle Spacecraft Systems

The CABIN VENT ISOL and VENT valves are then opened and the crew module is vented down to 786 mmHg (15.2 psi) or lower. The CABIN VENT ISOL and VENT valves are then closed.

CABIN AIR

Cabin air cools the cabin avionics electronic units, the crew, and passengers. Some avionics are cooled by air loops within the avionics bays. These loops are not pressure isolated from the crew cabin, although each avionics bay contains a closeout cover to minimize air interchange and thermal losses to the cabin environment: therefore, equipment contained in these air loops meets outgassing and flammability requirements to minimize toxicity levels resulting from outgassing materials. Low-toxicity materials also are used in the crew cabin habitable areas.

The crew module cabin contains five air loops: the cabin, three avionics bays and inertial measurement unit cooling loop. The crew module cabin atmosphere is drawn through the cabin through a 300-micron filter by one of two cabin fans located downstream of the filter.

Each of the cabin fans is controlled individually by the CABIN/FAN switch on Panel L1. CABIN FAN A switch turns cabin fan A on and off. CABIN FAN B switch controls cabin fan B. Normally only one fan is used at a time.

The cabin air is then ducted to the two lithium hydroxide (LiOH) canister where carbon dioxide (CO_2) is removed and activated charcoal removes odors and trace contaminants. CO_2 is maintained at 7.6 mmHg (0.147 psia) maximum. The two LiOH-activated charcoal canisters are replaced alternately every 12 hours (for four crew members) via an access door in the mid deck floor.

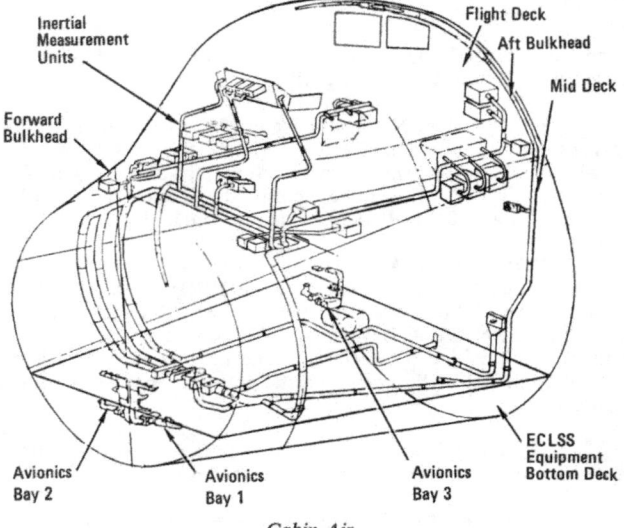

Cabin Air

The cabin atmosphere is then ducted to the crew module cabin heat exchanger where the cabin air is cooled by the water coolant loops. Humidity condensation is removed from the cabin heat exchanger by a fan separator which draws air and water from the cabin heat exchanger, separates the air and water, routes the water into waste water tanks, and ducts the air through its exhaust into the cabin. The two separator fans are individually controlled by the HUMIDITY SEP (separator) switch on Panel L1. HUMIDITY SEP switch A controls separator fan A and HUMIDITY SEP switch B operates separator fan B. The fan separators separate the air and water by centrifugal force and remove up to 118 kilograms (4 lbs) of water per hour. Only one fan separator is used at a time.

Space Shuttle Spacecraft Systems

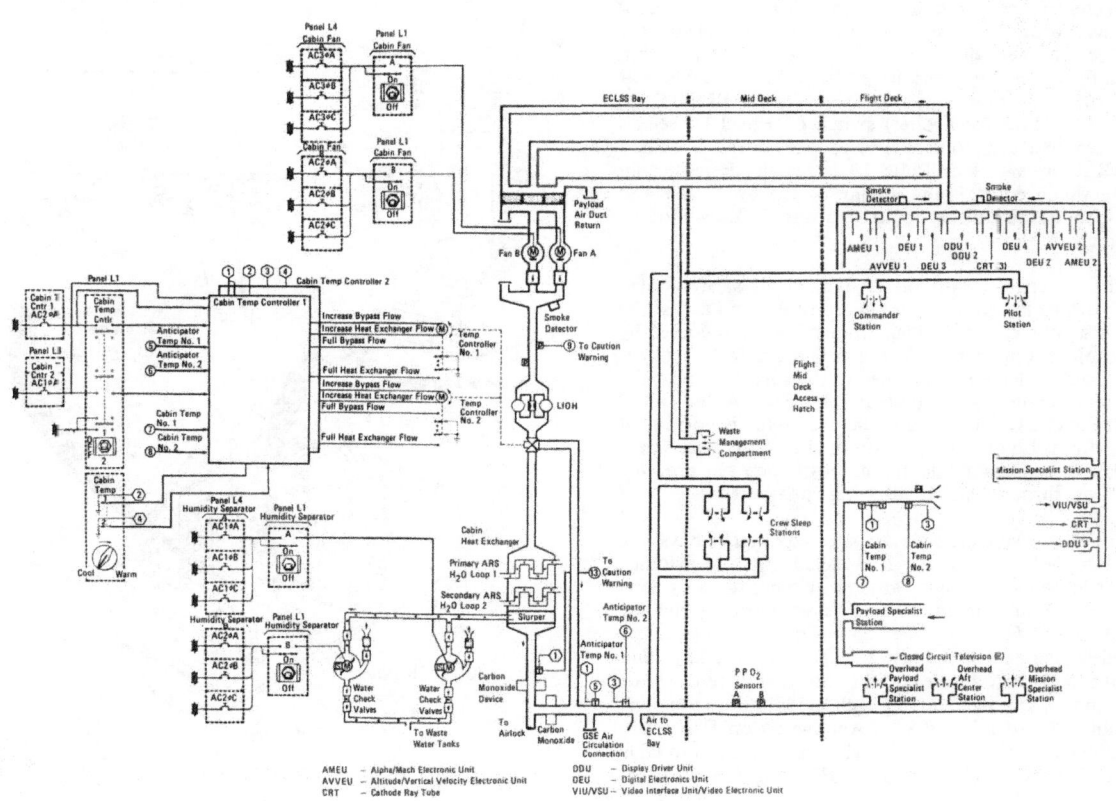

Crew Module Cabin Air

 **Space Shuttle Spacecraft Systems**

A small portion of the revitalized/conditioned air from the cabin heat exchanger is ducted to the carbon monoxide removal unit, which converts carbon monoxide into CO_2. A bypass duct carries cabin air around the heat exchanger and it mixes with the revitalized/conditioned air to control the crew module cabin return air at the selected temperature of between 18 to 26 plus or minus 1 °C (65 to 80 plus or minus 2 °F). The CABIN TEMP (temperature) CNTLR (controller) switch on Panel L1 selects the cabin temperature controller and the CABIN TEMP SELECTOR rotary switch on Panel L1 selects the desired cabin temperature which controls the mixing of the bypass air and revitalized/conditioned air before its return to the crew module cabin ducting.

The three inertial measurement units (IMU's) are cooled by cabin air drawn through the 300-micron filter and across the three IMU's by one of three parallel fans. The air is cooled by the water coolant loops which flow through the IMU heat exchanger and the cooled air is returned to the cabin. Each of the IMU fans is controlled by individual IMU FAN A, B, and C switches on Panel L1. When the applicable switch is positioned to ON, that fan is on and when positioned to OFF, that fan is off. A check valve installed at the outlet of each fan prevents reverse air flow through the standby fan chambers.

Each of the three electronic avionics equipment bays has identical air-cooled systems. Air is directed into the avionics bays at floor level and is drawn through avionics units by connectors at the back of each unit. The air then returns to the fan package inlet and 300-micron filter upstream of two fans. The air is cooled by the heat exchanger for each avionics bay. The water coolant loops cool the air and the cooled air is returned to the avionics bays. The two fans in each avionics bay are controlled by an AV (avionics) BAY FAN switch on Panel L1. When the applicable AV BAY FAN switch is positioned to ON, that fan is on and when positioned to OFF, that fan is off. A

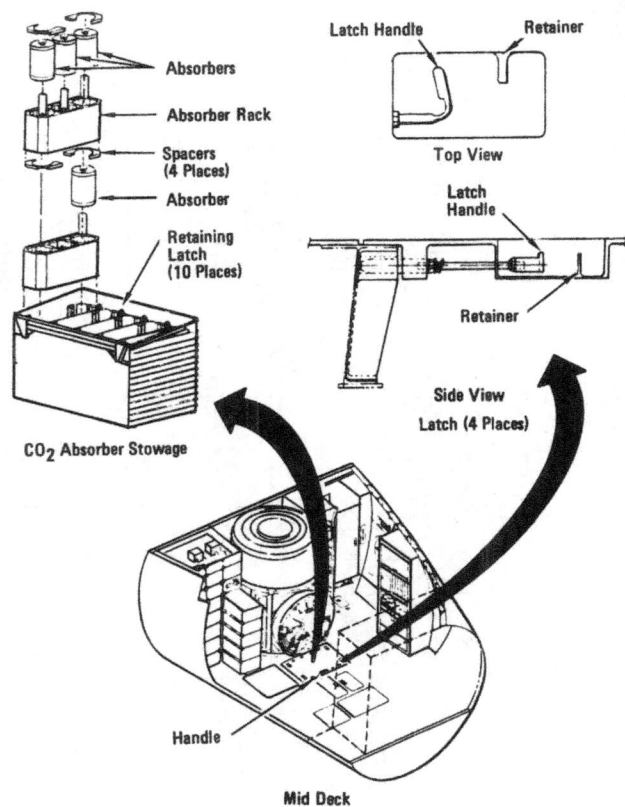

Carbon Dioxide Absorbers

Space Shuttle Spacecraft Systems

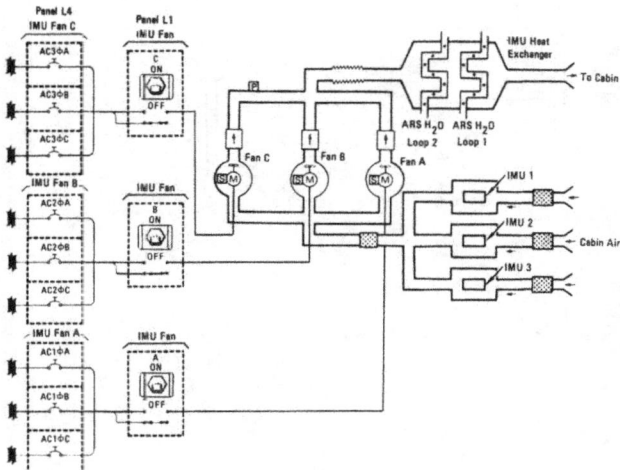

IMU — Inertial Measurement Unit

check valve in the outlet of each fan prevents a reverse flow in the standby fan.

The air outlet temperature in each avionics bay and the cabin heat exchanger is monitored and sent to the AIR TEMP (temperature) rotary switch on Panel 01. The rotary switch positioned to AV BAY 1, 2 or 3 position or CAB (Cabin) HX (heat exchanger) out position permits that temperature to be displayed on the AIR TEMP meter on Panel 01.

The air outlet temperature in each avionics bay and the cabin heat exchanger provides inputs to the YELLOW AV BAY CABIN AIR caution and warning light on Panel F7. The light would illuminate if any of the avionics bay outlet temperatures are above 57°C (135°F), if the cabin heat exchanger outlet temperature is above 18°C (65°F), or if the cabin fan delta pressure is below 5 mmHg (0.1 psi) or above 15 mmHg (0.3 psi).

If the payload bay contains the Spacelab pressurized module, a kit is installed to provide ducting from the crew cabin into the tunnel from the crew compartment mid deck to the Spacelab.

The fan separators, cabin heat exchanger, avionics heat exchangers and inertial measurement unit heat exchanger, waste water tanks, LiOH filters, carbon monoxide unit, and waste and potable water tanks are located beneath the mid deck crew compartment floor.

WATER COOLANT LOOP SUBSYSTEM

The WCLS provides thermal conditioning of the crew cabin by collecting heat through air-to-water heat exchangers and transferring the heat from the water coolant loops to the Freon coolant loops.

There are two complete and separate water (H_2O) coolant loops that flow side by side and have the capability of operating at the same time. The only difference between the H_2O Loop No. 1 and 2, is that Loop No. 1 has two H_2O pumps and Loop No. 2 has one pump.

Some of the electronic units in the avionics bays are mounted on coldplates with H_2O flowing through the coldplates. The heat generated by the electronic unit is transferred to the coldplate and into the H_2O which carries the heat away from the electronic unit. Coldplates mounted on the shelves in

Space Shuttle Spacecraft Systems

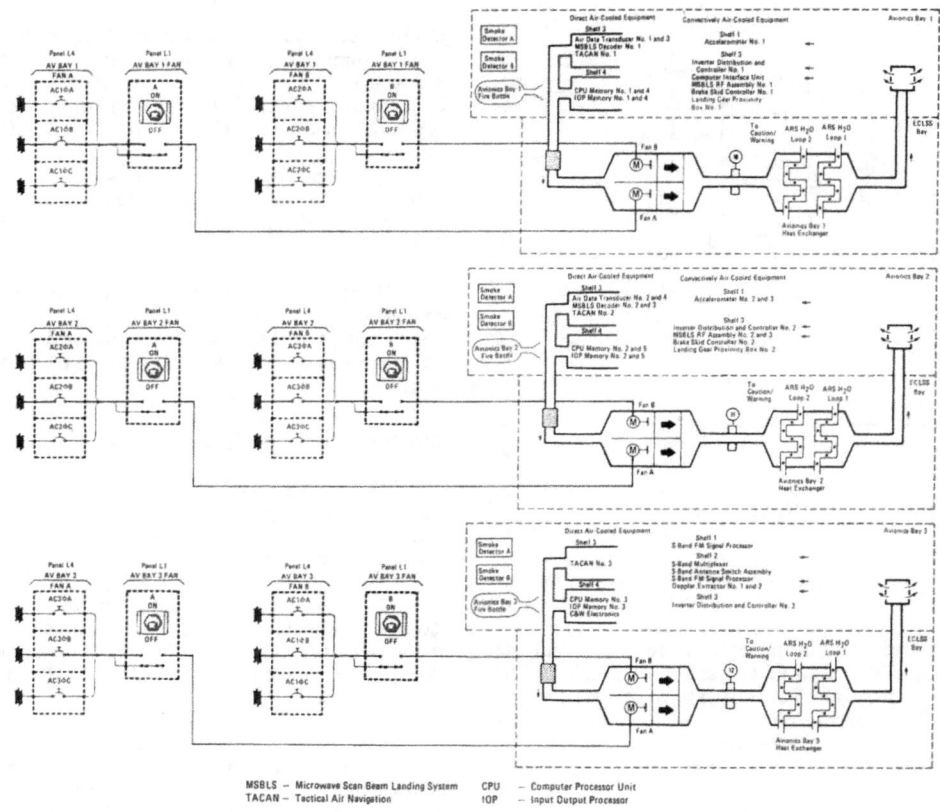

Avionics Bay Cooling

 **Space Shuttle Spacecraft Systems**

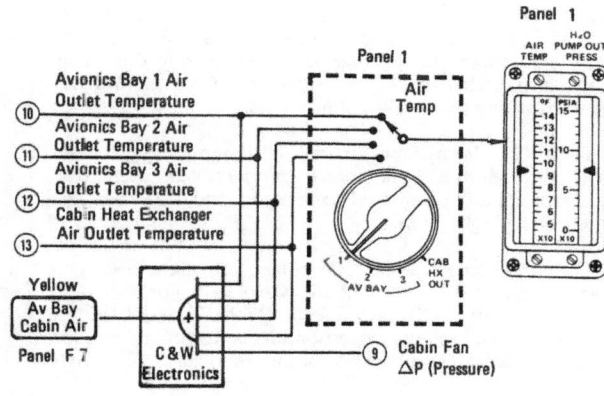

Avionics Bay and Cabin Heat Exchanger Temperature Monitoring and Caution/Warning

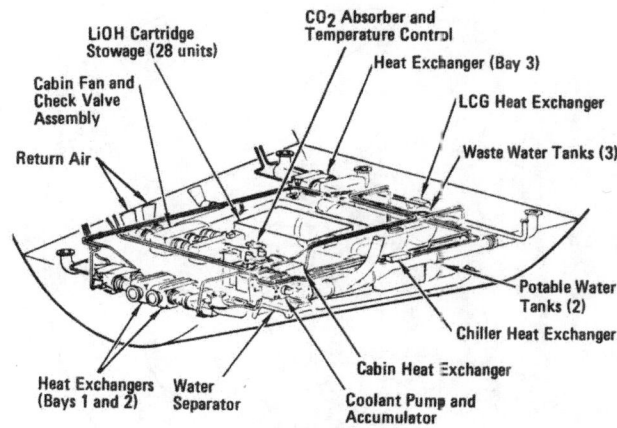

Crew Cabin Bottom Deck ECLSS

the avionics bays are connected in a series — parallel arrangement with respect to the H_2O flow.

The H_2O pumps in Loop No. 1 are controlled by the H_2O pump Loop 1 A and B switch on Panel L1 in conjunction with the Loop 1 GPC (general purpose computer) OFF and ON switch on Panel L1. The H_2O PUMP LOOP 1 switch positioned to GPC permits the GPC to energize relays, which cycles H_2O pump A or B for six minutes over a period determined by the GPC and the position of the H_2O PUMP LOOP 1 A or B position. The ON position of the Loop No. 1 switch energizes the relays and allows H_2O pump A or B to operate as determined by the position of the H_2O PUMP LOOP 1 A or B switch. The OFF position of the Loop No. 1 switch de-energizes the relays, which prohibits operation of the Loop No. 1 H_2O pump A and B. A ball-type check valve downstream of the H_2O pumps prevents reverse flow through the standby pump.

The H_2O pump in Loop No. 2 is controlled by the Loop No. 2 GPC ON, OFF switch on Panel L1. When the switch is in the GPC position, the Loop NO. 2 H_2O pump is controlled by the GPC as in the case of the H_2O pump in Loop No. 1.

Normally Water Loop No. 2 will be in operation for launch and entry and during on-orbit operations. Water Loop No. 1 pump B is in operation under GPC control during on-orbit operations.

H_2O Loops No. 1 and 2 flow side by side through the same areas. Downstream of the H_2O pump in each loop, the H_2O flow splits three ways. One leg goes through the Avionics Bay

Space Shuttle Spacecraft Systems

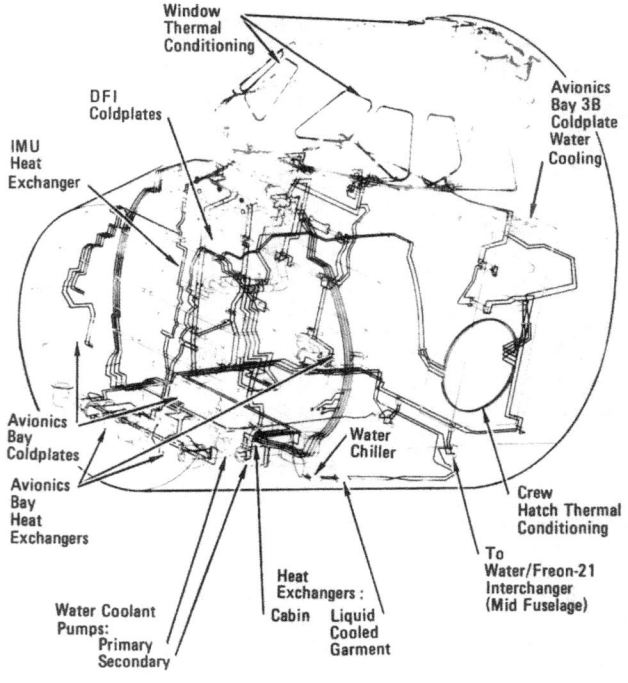

Water Coolant

No. 1 heat exchanger and coldplates and provides thermal conditioning of the crew ingress/egress hatch. Another leg goes through the Avionics Bay No. 2 heat exchanger and coldplates and provides thermal conditioning of the flight deck cabin windows. The third leg goes through the flight deck MDM (multiplexer/demultiplexer) coldplates with a predetermined amount of H_2O bypassing these coldplates and splits again into two parallel paths for the Avionics Bay No. 3 and 3B coldplates. All of these paths come together again and flow to the Freon-21/H_2O heat exchanger where excess heat from the WCL is transferred to the Freon-21 coolant loops.

The WCL loop flows from the Freon-21/H_2O interchanger through the liquid-cooled garment heat exchanger, H_2O water chiller, cabin heat exchanger, and IMU heat exchanger to the H_2O pump inlet.

The controller for each H_2O coolant loop is enabled by its respective loop BYPASS MODE and MAN switch on Panel L1. The AUTO position of the LOOP 1 BYPASS and LOOP 2 BYPASS switch allows the corresponding controller to position the bypass valve of that H_2O loop automatically. The MAN (manual) position of the LOOP 1 BYPASS and LOOP 2 BYPASS switch disables the automatic control of the bypass valve of that H_2O loop and enables the corresponding LOOP 1 BYPASS and LOOP 2 PYPASS MAN INCR DECR switch. The crew would position the LOOP 1 BYPASS and LOOP 2 BYPASS MAN switch to INCR (increase) or to DECR (decrease) to control that bypass valve in that H_2O coolant loop manually. The bypass valve is adjusted manually prior to launch to provide 408 to 453 kilograms (900 to 1,000 pounds per hour) through the interchanger. The control system remains in the manual mode for the entire flight.

The accumulator in each H_2O coolant loop provides a positive pressure on the H_2O pump of the corresponding H_2O loop in addition to providing for thermal expansion capability in that H_2O loop. Each accumulator is pressurized with gaseous nitrogen at 983 to 1,811 mmHg (19 to 35 psi).

The pressure at the outlet of the H_2O pump in ech coolant loop is monitored and sent to the H_2O PUMP OUT PRESS

Space Shuttle Spacecraft Systems

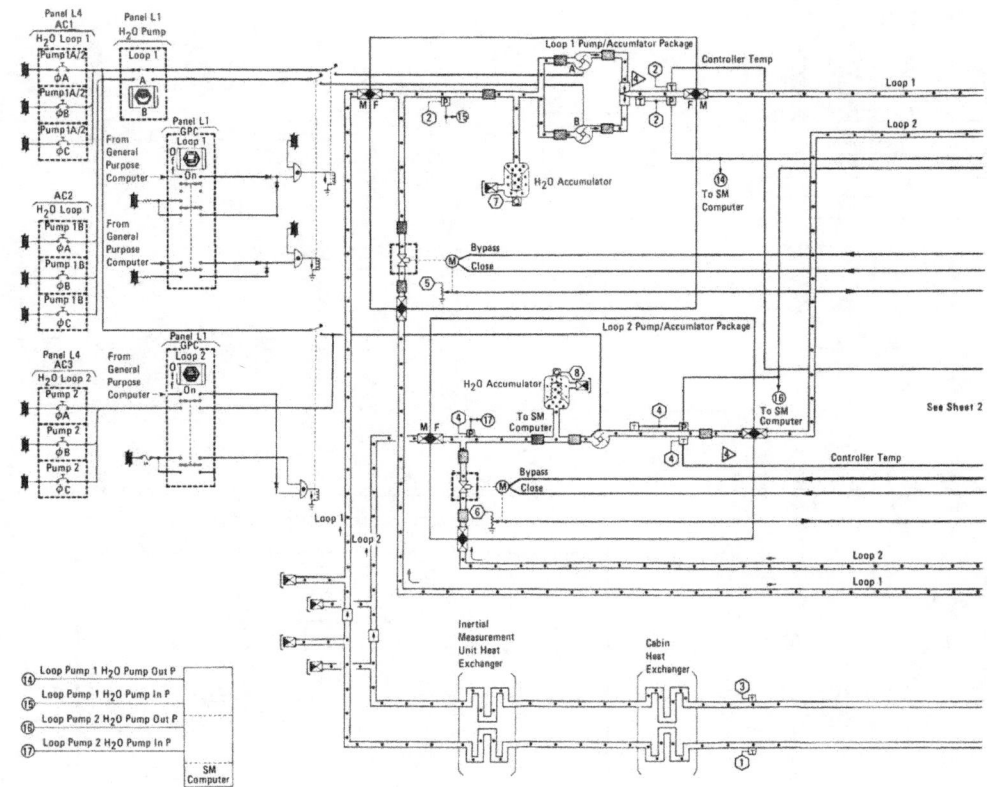

Water Coolant Loops (Sheet 1 of 3)

Space Shuttle Spacecraft Systems

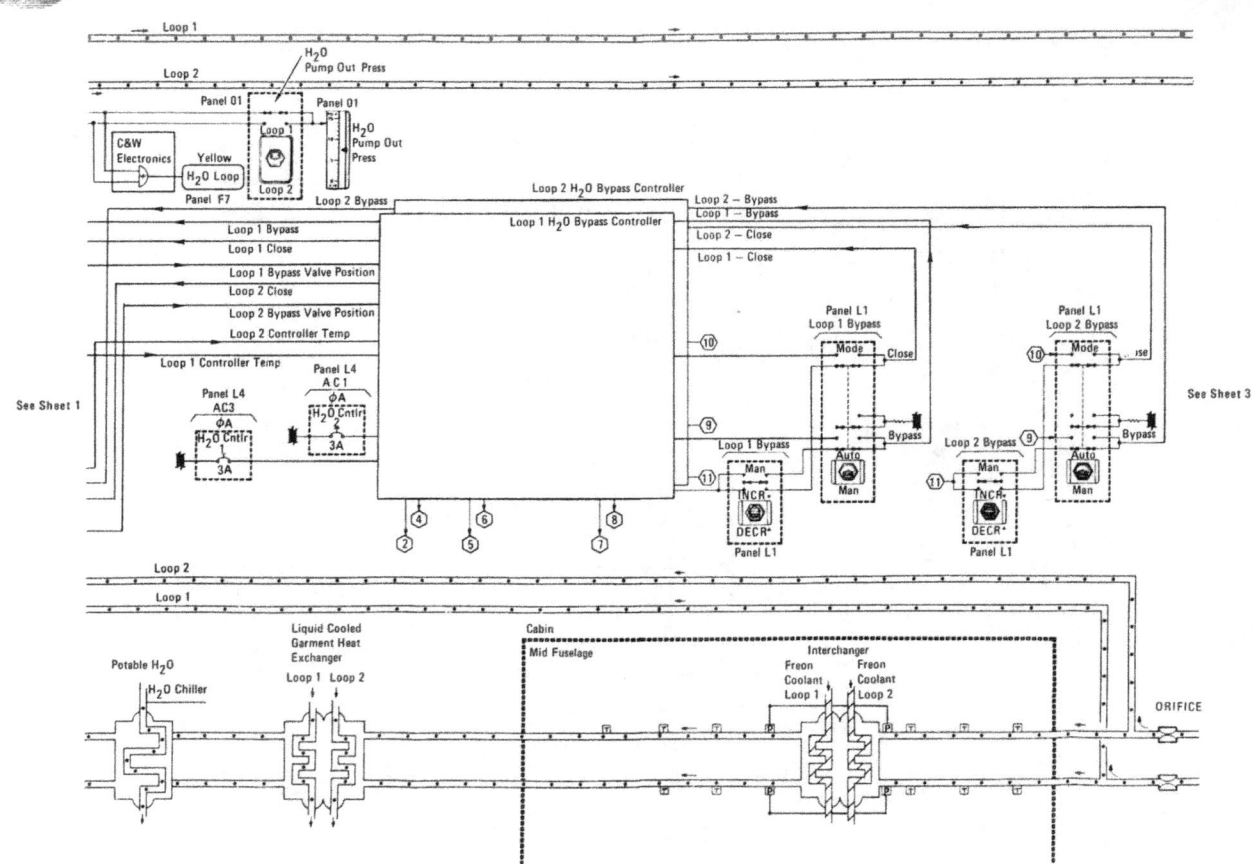

Water Coolant Loops (Sheet 2 of 3)

 **Space Shuttle Spacecraft Systems**

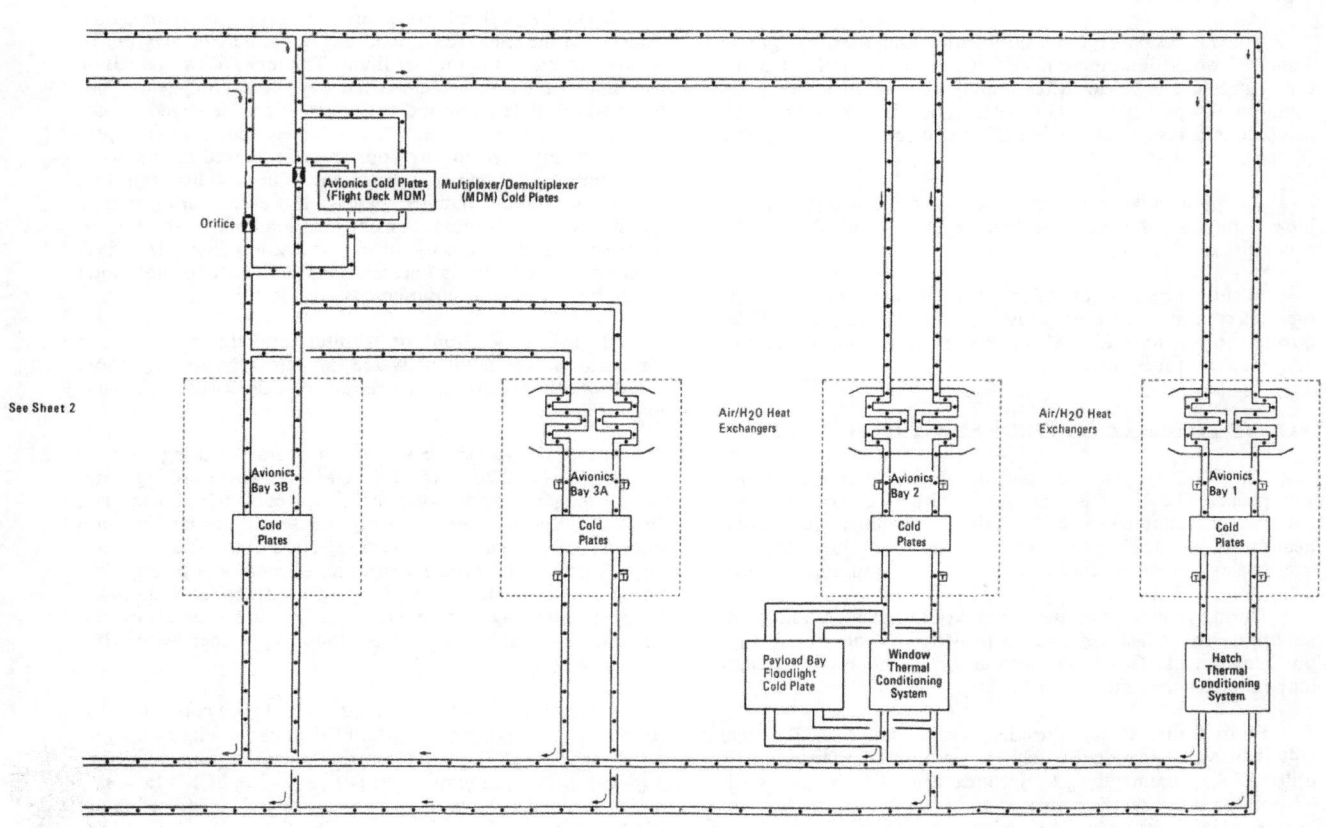

Water Coolant Loops (Sheet 3 of 3)

Space Shuttle Spacecraft Systems

switch on Panel 01. When the switch is on LOOP 1 or LOOP 2, that H2O coolant loop pressure can be monitored on the H2O PUMP OUT PRESS meter on Panel 01 in psia.

The YELLOW H2O LOOP caution and warning light on Panel F7 would illuminate if H2O Coolant Loop No. 1 pump outlet pressure is below 2,328 mmHg (45 psi) or above 4,114 mmHg (79.5 psi) or if H2O Coolant Loop No. 2 pump outlet pressure is below 2,328 mmHg (45 psi) or above 4,191 mmHg (81 psi).

The pump inlet and outlet pressure of each H2O coolant loop is monitored and transmitted to the SM GPC for CRT capabilities.

In summary, with use of the crew module cabin structural thermal capacity, the crew cabin will not exceed 32°C (90°F) during entry or until after flight crew egress, assumed to be 15 minutes after touchdown.

ACTIVE THERMAL CONTROL SUBSYSTEM

The ATCS provides orbiter heat rejection during all mission phases. The ATCS is composed of two Freon coolant loops (FCL's), coldplate networks for avionics cooling, liquid/liquid heat exchangers for orbiter systems cooling, and three heat sink subsystems (radiators, flash evaporator, and ammonia boiler).

During ground operations (checkout, pelaunch, and postlanding), orbiter heat rejection is provided by the ground support equipment (GSE) heat exchanger in the Freon coolant loops through ground system cooling.

From liftoff to an altitude greater than 42,672 meters (140,000 feet) — approximately 125 seconds — thermal lag is utilized. Approximately 125 seconds after liftoff, the flash evaporator subsystem is activated and provides orbiter heat rejection of the Freon coolant loops via water boiling. Flash evaporator operation continues until the payload bay doors are opened on orbit.

When the payload bay doors are opened, radiator panels attached to the forward payload bay doors may or may not be deployed depndent upon that flight. The forward two panels on each side of the orbiter if deployed away from the payload bay doors will radiate from both sides and if not deployed radiate only from one side. The aft radiator panels on the forward portion of the aft payload bay doors remain affixed to the doors and radiate only from the upper surface. On-orbit heat rejection is provided by the radiator panels; however, during orbital operations where combinations of heat load and spacecraft attitude exceed the capacity of the radiator panels, the flash evaporator subsystem is automatically activated to meet total system heat rejection requirements.

At the conclusion of orbital operations the flash evaporator subsystem is activated, and the payload bay doors closed with the radiator panels retracted, if deployed in preparation for entry.

The flash evaporator subsystem operates during entry to an altitude of 36,576 meters (120,000 feet) where boiling water can no longer provide adequate Freon coolant temperatures. Through the remainder of the entry phase and postlanding until ground cooling is connected, heat rejection of the Freon coolant loops is provided by the evaporation of ammonia through the use of the ammonia boiler. When ground cooling is initiated during postlanding, the ammonia boilers are shut down and heat rejection of the Freon coolant loops is provided by the GSE heat exchanger.

There are two complete and identical Freon coolant loops; Loop No. 1 and Loop No. 2. Both Freon coolant loops operate at the same time. There are two Freon coolant pumps in each loop with only one pump active per loop. The FCL's flow side by side except for the radiator panels. Freon-21 is utilized in the FCL's.

Space Shuttle Spacecraft Systems

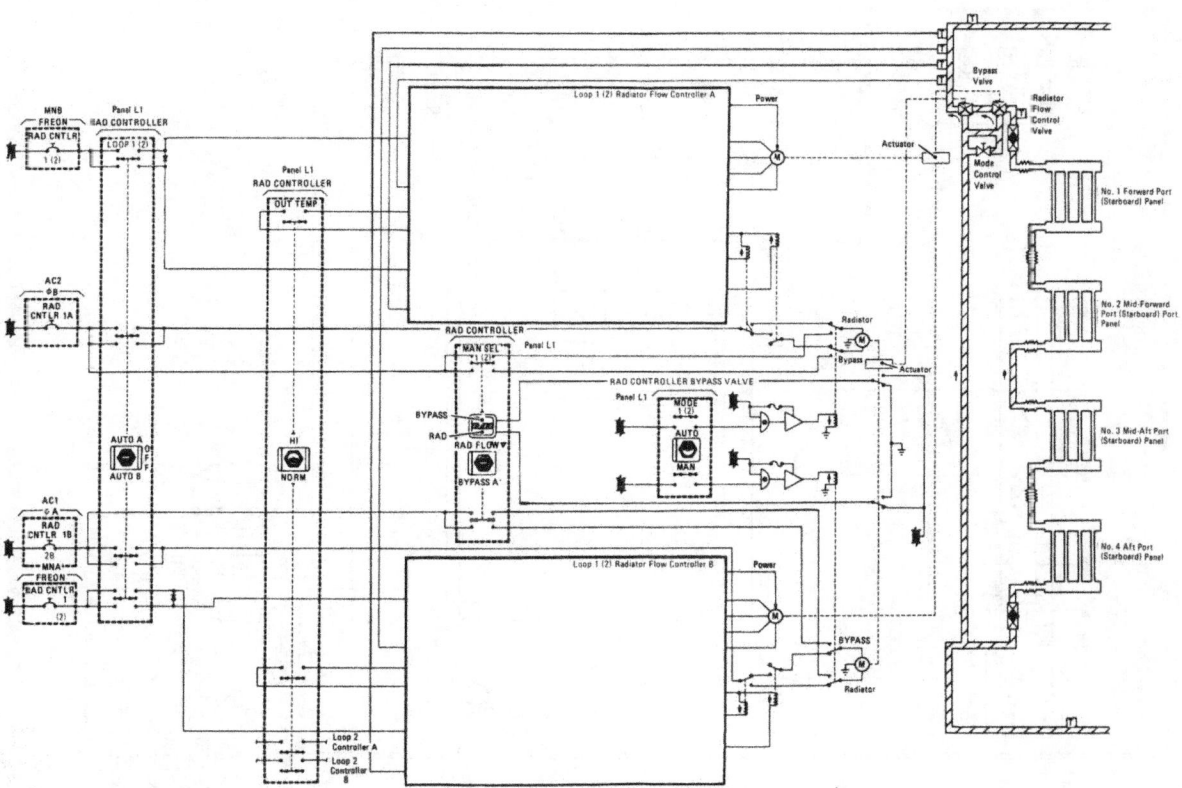

Active Thermal Control Freon 21 Coolant Loop (Sheet 1 of 3)

Space Shuttle Spacecraft Systems

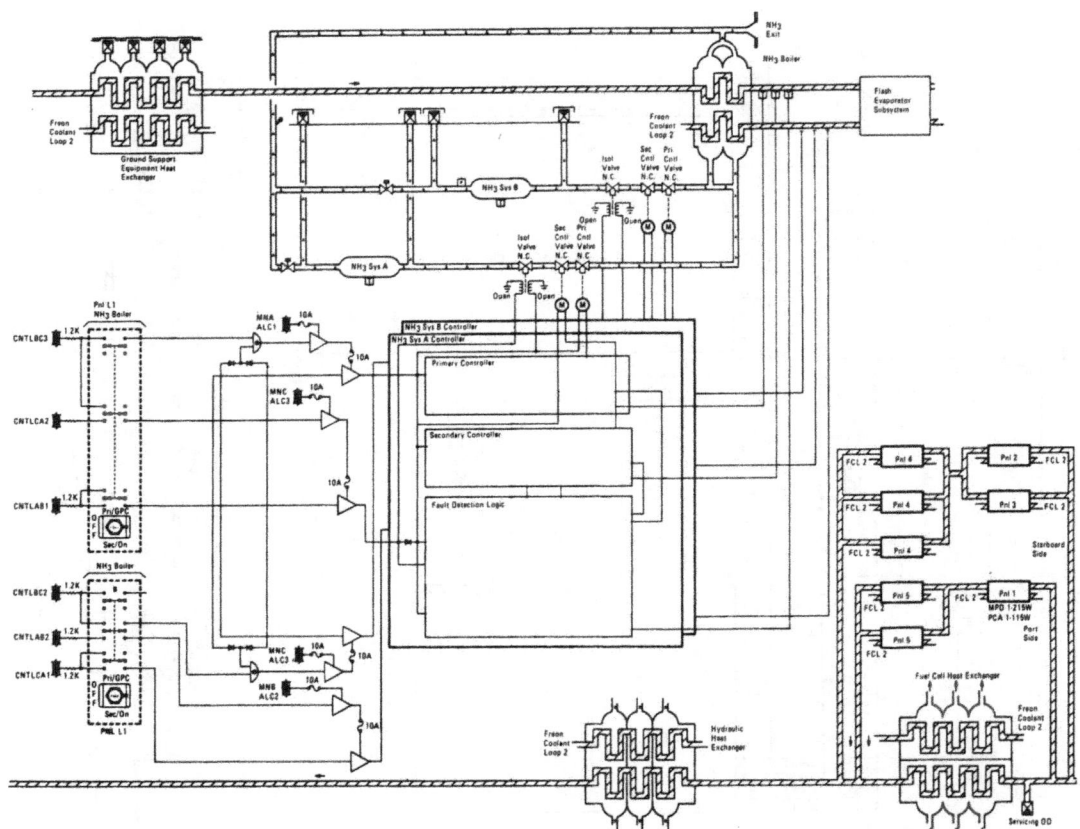

Active Thermal Control Freon 21 Coolant Loop (Sheet 2 of 3)

Space Shuttle Spacecraft Systems

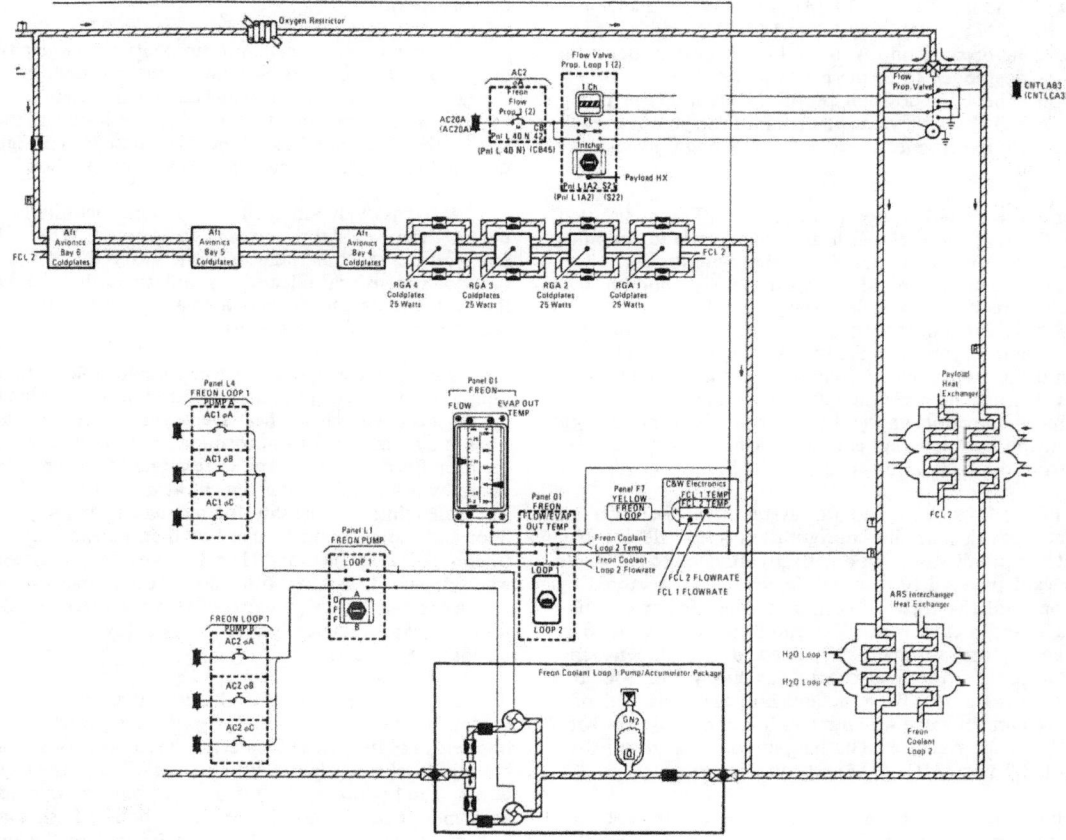

Active Thermal Control Freon 21 Coolant Loop (Sheet 3 of 3)

Space Shuttle Spacecraft Systems

The Freon pumps in each Freon coolant loop are controlled by individual FREON PUMP switches on Panel L1. When the FREON PUMP LOOP 1 or LOOP 2 switch is positioned to A, the Freon pump A in the Freon coolant loop is in operation. If positioned to B, the Freon pump B in that loop is in operation. The OFF position prohibits either Freon pump operation. A ball-check valve downstream of the pumps in each Freon coolant loop prevents a reverse flow through the standby pump.

When a Freon coolant pump is operating, Freon is routed in parallel through the fuel cell heat exchangers and the midbody coldplate network to cool electronics avionics units. The Freon coolant from the midbody coldplate network and fuel cell heat exchanger reunites in a series flow path before entering the hydraulics heat exchanger, which extracts energy from the Freon-21 coolant loop to heat the hydraulic systems fluid loops during on-orbit hydraulic circulation thermal conditioning operations. During the prelaunch and boost phase of the mission and during the atmospheric flight portion of entry through touchdown, the hydraulic system heat exchanger transfers excess heat from the hydraulic systems to the Freon-21 loops.

The FCL's flow to the radiator system, which consists of three radiator panels (baseline configuration — two deployable, if required by that mission, and one fixed) attached to the inside of the forward payload bay doors (deployable, if required by that mission) and the forward section of the aft payload bay doors (fixed) and a flow control assembly for each loop. The radiator panels are normally bypassed during ascent and descent. On-orbit, the flow control assembly controls the temperature of the loop (mixed radiator outlet) through use of a variable flow control valve which mixes hot bypassed flow with cold flow from the radiator. The temperature is controlled to either 3 °C (38 °F) or 13 °C (57 °F) setpoint temperature.

The radiator panels on each side of the orbiter are configured to flow in series while flow within each panel is parallel through a bank of tubes connected by an inlet and outlet collector manifold.

To increase heat rejection capability for large payloads requiring 29,000 Btu (British thermal units) per hour of heat rejection an additional radiator panel can be kitted into the network by attaching a fixed radiator panel to the inside of the aft section of the aft payload bay doors. The baseline configuration is designed for payloads rejecting 21,500 Btu per hour.

A bypass valve in each FCL system permits Freon-21 to bypass the radiators except on-orbit. When Freon-21 temperatures at the radiator outlet exceed 5 °C (41 °F), the radiator system heat rejection capability has been exceeded and the flash evaporators are activated automatically to produce the required Freon-21 temperature.

During boost and entry, each deployable radiator (if required by that mission) panel is secured to the payload bay door by six motor-driven latches. Deployment for on-orbit operations is by a motor-driven, torque-tube-lever arrangement. The aft four fixed radiator panels are attached to the payload bay doors by a ball joint arrangement at a maximum of 12 places. The ball joints compensate for movement of the payload bay door and radiator panel caused by thermal expansion and contraction of each member. The forward four radiator panels, when deployed, expose both sides of the radiator panels to increase the heat rejection capability of the Freon-21 loops. The four forward radiator panels are deployed 35.5° from the payload bay doors.

The ATCS pumps, interchanger, fuel cell heat exchanger, payload heat exchanger, flow proportioning valve modules, and mid body coldplates are located in the lower forward portion of the mid fuselage. The radiators are attached to the payload bay doors. The hydraulic systems' heat exchanger, ground support equipment heat exchanger, ammonia boiler, flash evaporator, and aft avionic bay coldplates are located in the orbiter aft

Space Shuttle Spacecraft Systems

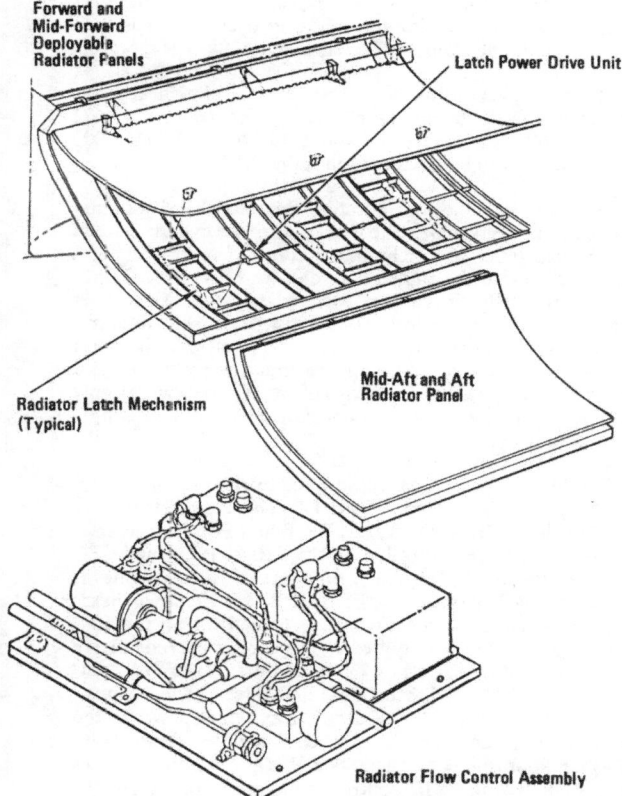

Radiators and Radiator Flow Control Assembly

fuselage. The radiator flow control assemblies are located in the lower aft portion of the mid fuselage.

The radiator panels are constructed of an aluminum honeycomb face sheet 3,200 millimeters (126 inches) wide and 8,128 millimeters (320 inches) long. The forward deployable radiator panels are two sided with a core thickness of 22 millimeters (0.9 inch). They have longitudinal tubes bonded to the internal side of both facesheets. The forward deployable panels contain 68 tubes each, with a tube spacing of 48 millimeters (1.90 inches). Each tube has an inside diameter of 3.32 millimeter (0.131 inch). Each side of the forward deployable radiator panels has a coating bonded by an adhesive to the facesheet consisting of silver backed Teflon tape for proper emissivity properties. The aft fixed panels are one sided with a core thickness of 12.7 millimeters (0.5 inch) with tubes only on the exposed side of the panel and a coating bonded by an adhesive to the exposed facesheet. The aft panels contain 26 longitudinal tubes with a tube spacing of 125 millimeters (4.96 inches), and each tube has an inside diameter of 4.57 millimeter (0.18 inch). The additional thickness of the forward radiator panels is required to meet deflection requirements when the orbiter is exposed to ascent acceleration.

The radiator panels on the left-hand side (port) facing forward are connected in series with Freon-21 coolant Loop 1. The panels on the right-hand-side (starboard) facing forward are connected in series with Freon-21 coolant Loop 2.

The RAD (radiator) CONTROLLER LOOP 1 and LOOP 2 switch on Panel L1 enables Loops No. 1 and 2 controllers A and B. When the switch for a loop is positioned to AUTO A, radiator controller A automatically controls the radiator flow control valve in that loop, which maintains the desired radiator mixed outlet temperature as determined by the RAD CONTROLLER OUT TEMP switch on Panel 1A. When the switch

Space Shuttle Spacecraft Systems

is turned to AUTO B, controller B is enabled and automatically controls the radiator control valve in the corresponding loop as it did in the case of AUTO A.

The RAD CONTROLLER OUT (outlet) TEMP switch on Panel L1 enables the selected controller A or B in Loop No. 1 or 2 to control the radiator outlet temperature of that loop. The radiator outlet temperatures in Loops No. 1 and 2 are automatically controlled at 3 °C (38 °F) when the switch is on NORM (normal) and at 13 °C (57 °F) when it is on HI. The flash evaporator is activated automatically when the radiator outlet temperature exceeds 50 °C (41 °F).

The RAD CONTROLLER MODE 1 and MODE 2 switch on Panel L1 permits automatic control of radiator flow control valve or manual control of the radiator bypass valve. When in the AUTO position for Loop No. 1 or 2, the respective radiator flow control valve automatically controls the radiator outlet temperature. When in the MAN position, the automatic control of the radiator flow control valve in the corresponding loop is inhibited and the bypass valve is controlled by the RAD CONTROLLER MAN (manual) SEL (select) 1 or 2 switch on Panel L1. If the MAN SEL switch is positioned to the RAD FLOW, the bypass valve permits the Freon-21 to flow through the radiators. If positioned to BYPASS position for that loop, the bypass valve permits the Freon-21 in that loop to bypass the radiators.

The indicator located above the RAD CONTROLLER MAN SEL 1 and 2 switch on Panel L1 indicates the position of the bypass valve in that loop. The indicator indicates BYPASS when the bypass valve is in the bypass position, BARBERPOLE when the motor operated valve in that loop is in transit from the bypass or radiator flow position, and indicates RAD when the bypass valve in that loop is in the radiator flow position.

Freon-21 from the radiator flow control valve assembly is routed to the ground support equipment (GSE) heat exchanger, which is used during ground operations (checkout, prelaunch and postlanding) for orbiter heat rejection of the Freon-21 coolant loops. Freon-21 from the GSE heat exchanger is then directed through the ammonia boiler, then the flash evaporator.

The flash evaporator is used to reject orbiter heat loads from the Freon-21 coolant loops during ascent above 42,672 meters (140,000 feet) and entry above 36,576 meters (120,000 feet) altitude and to supplement the radiators in orbit.

There are two flash evaporators (high-load and topping) contained in one envelope. The evaporators are cylindrical and have a finned inner core. The hot Freon-21 from the coolant loops flows around the finned core and water is sprayed onto the core from the nozzles in each evaporator. The water vaporizes, cooling the Freon-21. In the low-pressure areas above 36,576 meters (120,000 feet), water vaporizes quickly. The water changing from liquid to vapor removes approximately 1,000 Btu per hour per 2.2 kilograms (1 pound) of water. The water supply is obtained from the potable water storage tanks and supply systems A and B.

The flash evaporators are controlled by the FLASH EVAP CONTROLLER switches on Panel L1. The evaporators have three controllers: PRI (primary) A, PRIB and SEC (secondary). These controllers are controlled by the PRIA, PRIB and SEC switches on Panel L1. Normally, only one of these switches is used at a time. When one of the PRIA, PRIB or SEC switches is positioned to GPC, that controller is turned on by the BFS (backup flight system) computer as the orbiter ascends above 42,672 meters (140,000 feet) and is turned off by the BFS computer during entry at 36,576 meters (120,000 feet). The ON position of the PRIA, PRIB, or SEC switch provides power to the flash evaporator controller directly. OFF removes power from the flash evaporator controller. The PRIA controller utilizes water system A; the PRIB controller utilizes water system B. The SEC controller uses water system A if the SEC switch on Panel L1 is in SPLY A or B if the SEC switch on Panel L1 is in

Space Shuttle Spacecraft Systems

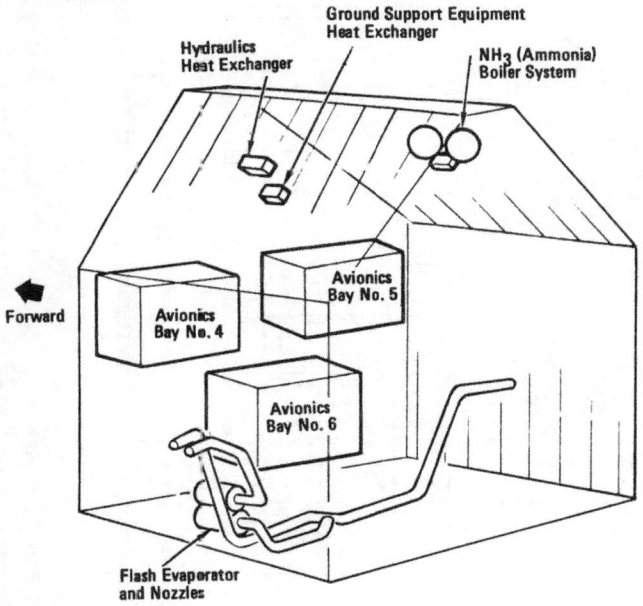

Aft Fuselage ECLSS Components

SPLY B and the HI-LOAD EVAP switch is in the ENABLE position.

The PRIA and B controllers control evaporator outlet Freon-21 loop temperatures at 3°C (39°F) and the SEC controller controls evaporator outlet Freon-21 loop temperatures at 16°C (62°F).

The applicable flash evaporator controller pulses water into the evaporators, cooling the Freon-21. The steam generated in the topping evaporator is ejected through two sonic nozzles at opposing sides on each side of the aft end of the orbiter, reducing payload water vapor pollutants on orbit and minimizing vent thrust effects on the orbiter guidance navigation and control system. The hi-load evaporator is used in conjunction with the topping evaporator during ascent and entry when higher Freon-21 coolant loop temperatures impose a greater heat load which requires a higher heat rejection. The HI-LOAD EVAP ENABLE switch on Panel L1 must be in the ENABLE position for hi-load evaporator operation. After leaving the high-load evaporator, the Freon-21 would also flow through the topping evaporator for additional cooling. The steam generated by the hi-load evaporator is ejected through a single sonic nozzle on the left-hand (port) side aft end of the orbiter facing forward. The hi-load evaporator would not normally be used on orbit because it has a propulsive vent and might pollute the payload.

The topping evaporator can be used to dump excess potable water from the storage tanks. In this mode, the radiator flow control valve assembly has an alternate control temperature of 13.8°C (57°F), which is used during forced water dumping.

Heaters are employed on the topping and hi-load steam ducts of the flash evaporator to prevent freezing. The HI-LOAD DUCT HTR (heater) switch on Panel L1 positioned to A provides electrical power to the thermostatically controlled A heaters on the hi-load evaporator steam duct and steam duct exhaust. The B position provides electrical power to the thermostatically controlled B heaters on the hi-load evaporator steam duct and steam duct exhaust. The A/B position provides electrical power to both the A and B heaters. The C position provides electrical power to the thermostatically controlled C heaters on the hi-load evaporator steam duct and steam duct exhaust. The OFF position removes electrical power from the heaters.

# Space Shuttle Spacecraft Systems

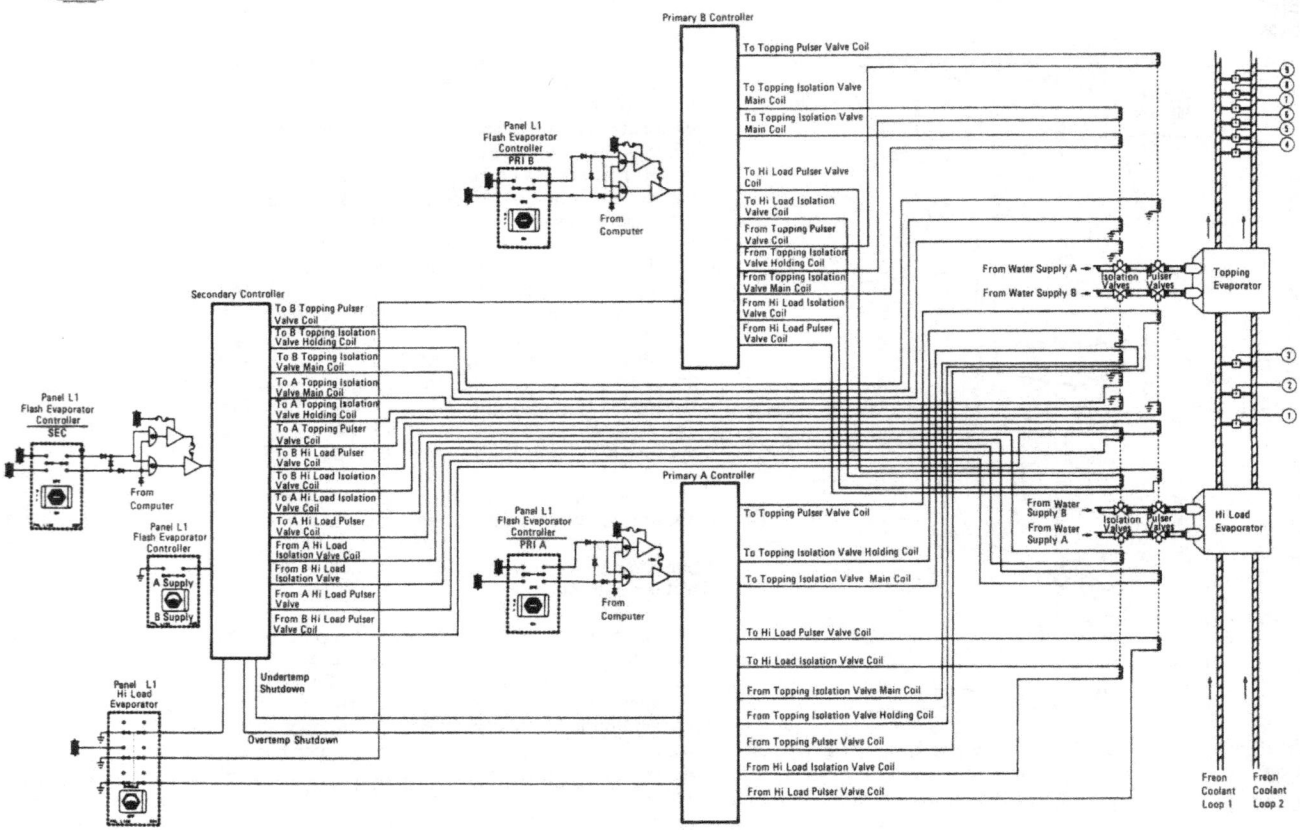

Flash Evaporator

Space Shuttle Spacecraft Systems

The TOPPING EVAPORATOR HEATER DUCT switch on Panel L1 positioned to A provides electrical power to the thermostatically controlled A heaters on the topping evaporator. The B position provides electrical power to the thermostatically controlled B heaters. The A/B position provides electrical power to both the A and B heaters. The C position provides electrical power to the thermostatically controlled C heaters. The OFF position removes electrical power from the heaters.

The TOPPING EVAPORATOR HEATER L (left) and R (right) NOZZLE switches on Panel L1 provide electrical power to the topping evaporator left and right nozzles. The L and R AUTO A position provides electrical power to the left and right A nozzle heaters to maintain nozzle temperatures between 4° and 21 °C (40° and 70 °F). The L and R AUTO B position provides electrical power to the left and right B nozzle heaters.

The ammonia boilers are used below 36,576 meters (120,000 ft) during entry. There are two individual storage and control systems that feed a boiler containing common ammonia passages and individual Freon-21 coolant loop passages. This provides a safe return from orbit for any combination of failures in both the Freon-21 coolant loops and ammonia boilers. The ammonia boilers are enabled by the NH3 CONTROLLER A and B switches on Panel L1. The NH3 CONTROLLER A and B switches are positioned to PRI (primary)/GPC before entry. As the orbiter descends through 36,576 meters (120,000 ft), the BFS commands controller A and B on. The ammonia (NH3) boiler is a shell-and-tube type with a single pass on the ammonia side and two passes from each Freon-21 coolant loop. The NH3 flows in the ammonia tubes and the Freon-21 coolant loops flow over the tubes, cooling the Freon-21. Freon temperature is maintained at 1.1 °C (34 °F) by regulating the flow of NH3 through the boiler. At each Freon-21 coolant loop outlet of the ammonia boiler, Freon-21 temperature is monitored by three temperature sensors. One sensor is associated with the primary NH3 flow control valve of that loop and another with the secondary NH3 flow control valve. The third sensor automatically switches control from the primary to secondary system in the event of low Freon-21 coolant loop outlet temperature. The NH3 boiler exhaust is vented overboard in the aft section of the orbiter adjacent to the bottom right side of the vertical tail. The boiler continues to operate and provide heat rejection until a ground cooling cart is connected to the GSE heat exchanger after touchdown.

The NH3 CONTROLLER A and B switch positioned to SEC (secondary) ON provides electrical power directly to the NH3 controllers and boiler, which operates the NH3 boiler. The OFF position removes electrical power from the NH3 controllers and boiler. The two complete ammonia boiler systems each have an ammonia storage tank. The capacity of each tank is 29 kilograms (64.7 lbs). There is about 20 kilograms (46 lbs) usable for each system. Each ammonia tank is pressurized with helium at an operating pressure between 4,295 and 28,462 mmHg (83 to 550 psi). There are three valves in the plumbing leading to each ammonia boiler. The isolation valve is opened by the primary or secondary controller. The next two control valves in line are modulating valves whose position is dependent on the amount of current to the control motor. The full closed position of these two valves inhibits about 75 percent of the flow. The fail mode of the control valves is open. If the fault circuitry detects the Freon-21 outlet temperature below minus 0.4 °C (31.25 °F) for greater than 10 seconds, an automatic switchover occurs in the controller for the NH3 system and secondary controller takes over. A relief valve provides over pressurization protection for each ammonia tank.

Freon Coolant Loops No. 1 and 2 are routed from the flash evaporator in series, then into parallel paths. A portion of the Freon-21 loop is directed to the aft avionics bays in the aft fuselage of the orbiter, where some of the electronic avionics units in Avionics Bays 6, 5 and 4 are mounted on coldplates which transfer the heat generated from the electronic avionics units to the Freon-21, which carries the generated heat away.

Space Shuttle Spacecraft Systems

The Freon-21 coolant loops also flow through the coldplates of Rate Gyro Assemblies 1, 2, 3 and 4.

The remaining parallel path downstream of the flash evaporator heats oxygen by means of a heat exchanger to 4.4°C (40°F) prior to entering the cabin. The source of the oxygen is the cryogenic storage and distribution system. This path branches in parallel at the flow proportioning valve in each Freon-21 loop.

The FLOW PROP (proportioning) VLV (valve) switch on Panel L1 for each coolant loop controls the flow of Freon-21 to the payload heat exchanger or the water/Freon-21 interchanger. The INTCHGR (interchanger) position of the LOOP 1 and LOOP 2 switches controls the respective flow proportioning valve to allow maximum Freon-21 flow through the water/Freon-21 interchanger. The PAYLOAD HX (heat exchanger) position of the LOOP 1 and LOOP 2 switch controls the respective flow proportioning valve to allow maximum Freon-21 flow through the payload heat exchanger. The indicator above the respective LOOP 1 and LOOP 2 switch on Panel 01 indicates ICH when the water/Freon interchanger position, BARBERPOLE when that valve is in transit from ICH or PL, and indicates PL (payload) when that valve is in the payload heat exchanger position.

The parallel paths from the water/Freon-21 interchanger and payload heat exchanger are reunited with the parallel path from the aft avionics bay and rate gyro assemblies. The coolant loop then returns to its Freon-21 coolant pump.

The accumulator in each Freon-21 coolant loop is a metal, bellows-type pressurized with gaseous nitrogen. The accumulator provides for thermal expansion and keeps a positive suction on the coolant pumps.

The Freon-21 coolant loop temperature and flow rate are monitored on Panel 01. When the FREON FLOW EVAP OUT TEMP switch is positioned to LOOP 1 or LOOP 2, the respective Freon-21 coolant loop evaporator outlet temperature is monitored on the FREON EVAP OUT TEMP in degrees Fahrenheit and the Freon flow is monitored on the FREON FLOW in PPH (pounds per hour). This information is also transmitted to the RED FREON LOOP caution and warning light on Panel F7. The RED FREON LOOP light would illuminate if the Freon-21 Coolant Loop No. 1 or 2 evaporator outlet temperature fell below 0°C (32°F) or rose above 15.6°C (60°F) or if the Freon-21 flow rate is below 544 kilograms (1,200 lbs) per hour.

FOOD, WATER AND WASTE MANAGEMENT

The food, water, and waste management (FWW) subsystem provides the basic life support functions for the flight crew.

Each potable water tank has a usable capacity of 74 kilograms (165 pounds), is 901 millimeters (35.5 inches) in length and 393 millimeters (15.5 inches) in diameter, and weighs 17.9 kilograms (39.5 pounds) dry.

Potable water is generated by the three fuel cells at a maximum of 11.34 kilograms (25 lbs) per hour. The hydrogen-enriched water from the fuel cell passes through a hydrogen (H_2) separator, where 95 percent of the excess hydrogen is removed. The H_2 separator consists of a matrix of silver palladium tubes which have an affinity for H_2. The H_2 is dumped overboard through a vacuum vent.

The water from the H_2 separator is directed to the water storage system, which can consist of a total of four tanks. The tanks are identified as A, B, C and D. Each tank has a solenoid inlet and outlet valve. Water for Tank A passes through a microbial check valve. The microbial check valve adds approximately 3-5 parts per million iodine to the water. The water from the microbial check valve is directed to Tank A and the galley (if

Space Shuttle Spacecraft Systems

the galley is installed). The crew can select cooled or ambient water. Cooling is accomplished by passing through the water chiller, where heat is rejected to the water coolant loop.

When the Tank A inlet valve is closed or Tank A is full, the water is directed through a 77 mmHg (1.5 psi) relief valve which routes the water to Tank B.

When the Tank B inlet valve is closed or Tank B is full, the water is directed through another 77 mmHg (1.5 psi) relief valve to Tanks C and D. The inlet and outlet valves for each tank can be opened or closed selectively to use water; however, the Tank A outlet valve fill always remains closed since the water has been treated by passage through the microbial filter for crew consumption.

The controls for the water supply system are located on Panels R12 and ML31C. Tanks A, B and C are controlled from Panel R12, Tank D from Panel ML31C.

Tanks A, B and C have their own SUPPLY H_2O INLET and OUTLET switch on Panel R12. When the SUPPLY H_2O INLET TK A, B, or C switch is positioned to OPEN, the inlet valve for that tank allows water into the tank. If positioned to CLOSE, the inlet valve isolates the water inlet from that tank. An indicator located above the respective switch on Panel R12 indicates OP (open) when the corresponding valve is open, BARBERPOLE when that valve is in transit and CL (close) when that valve is closed. Tank D has its own SUPPLY H_2O TK INLET switch and indicator on Panel ML31C and operates in the same manner as the ones on Panel R12.

A SUPPLY H_2O GALLEY SPLY (supply) VLV (valve) switch on panel R12 permits or isolates water from Tank A to the galley. The switch has OPEN and CLOSE positions and an indicator above the switch shows whether the valve is open or closed or BARBERPOLE when that valve is in transit.

When the valve is open, water is supplied to an Apollo water dispenser and water gun at ambient and chilled temperatures for drinking and food reconstitution. The ambient water temperature range is 18 to 35 °C (65 to 95 °F) and the chilled water temperature range is 6 to 13 °C (43 to 55 °F) for a payload in lieu of galley arrangement. In the galley arrangement the water supply to the galley is directed to a hot water heater, which provides hot water at a temperature range between 68 and 73 °C (155 to 165 °F). Chilled water is supplied at the galley at a temperature range of 7 and 12 °C (45 to 55 °F).

Tank A is used for crew consumption. To prevent contamination the Tank A OUTLET valve will remain closed. Tank B is used for flash evaporator cooling on-orbit. Tanks A and B may also be dumped overboard as necessary to provide space for water storage. Tanks C and D are saved full of water for contingency purposes.

Each of the water tanks is pressurized from the nitrogen pressure supply system at a pressure of 905 mmHg (17.5 psi) to force the water from the water storage tanks for flash evaporator use. The H_2O ALTERNATE PRESS switch on Panel L1 provides the capability of referencing the water tank pressurization system to ambient cabin pressure if the N_2 system fails. If the switch is positioned to OPEN, cabin atmosphere pressure is supplied to the water tanks for pressurization. The CLOSE position isolates the cabin atmosphere pressure from the water tank pressurization supply system.

From the water supply tanks, two evaporator feed lines referred to as SYSTEM A and SYSTEM B are routed to the flash evaporators in the aft fuselage. A crossover valve between the two supply systems is controlled by a SUPPLY H_2O CROSSOVER VLV swich on Panel R12. If the switch is positioned to OPEN, the crossover valve allows all six water tanks for the flash evaporator or overboard dumping. When the switch is positioned to CLOSE, the crossover valve is closed,

Space Shuttle Spacecraft Systems

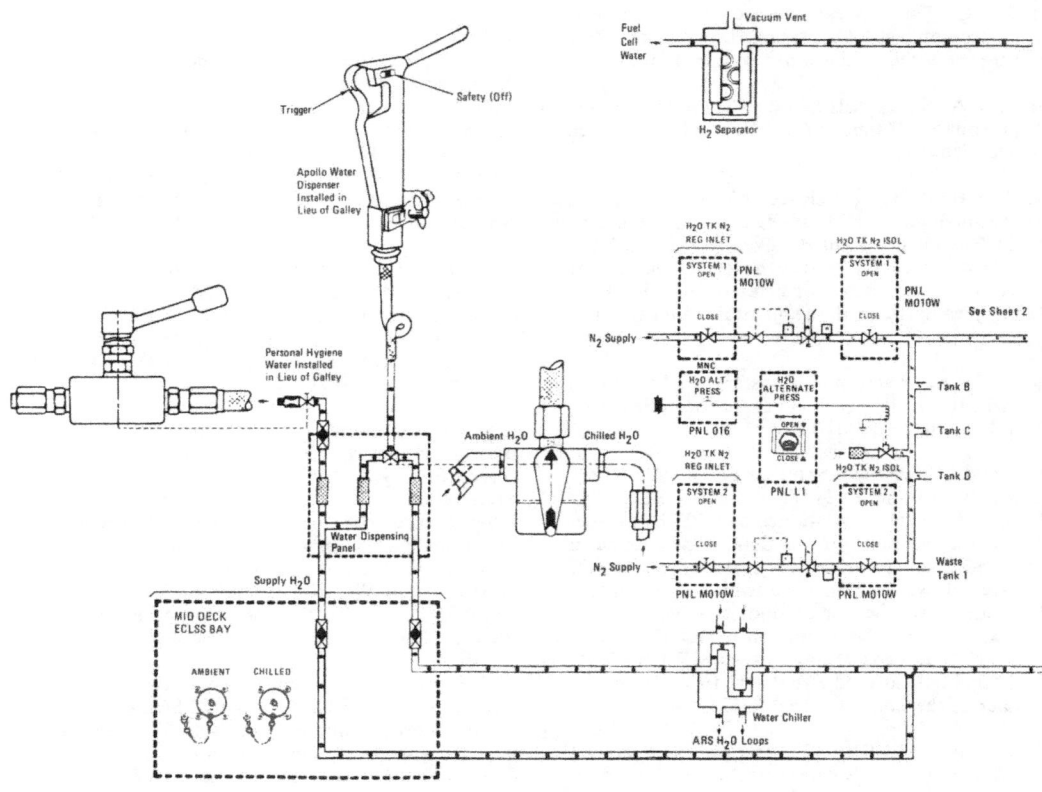

Potable Water (Sheet 1 of 3)

Space Shuttle Spacecraft Systems

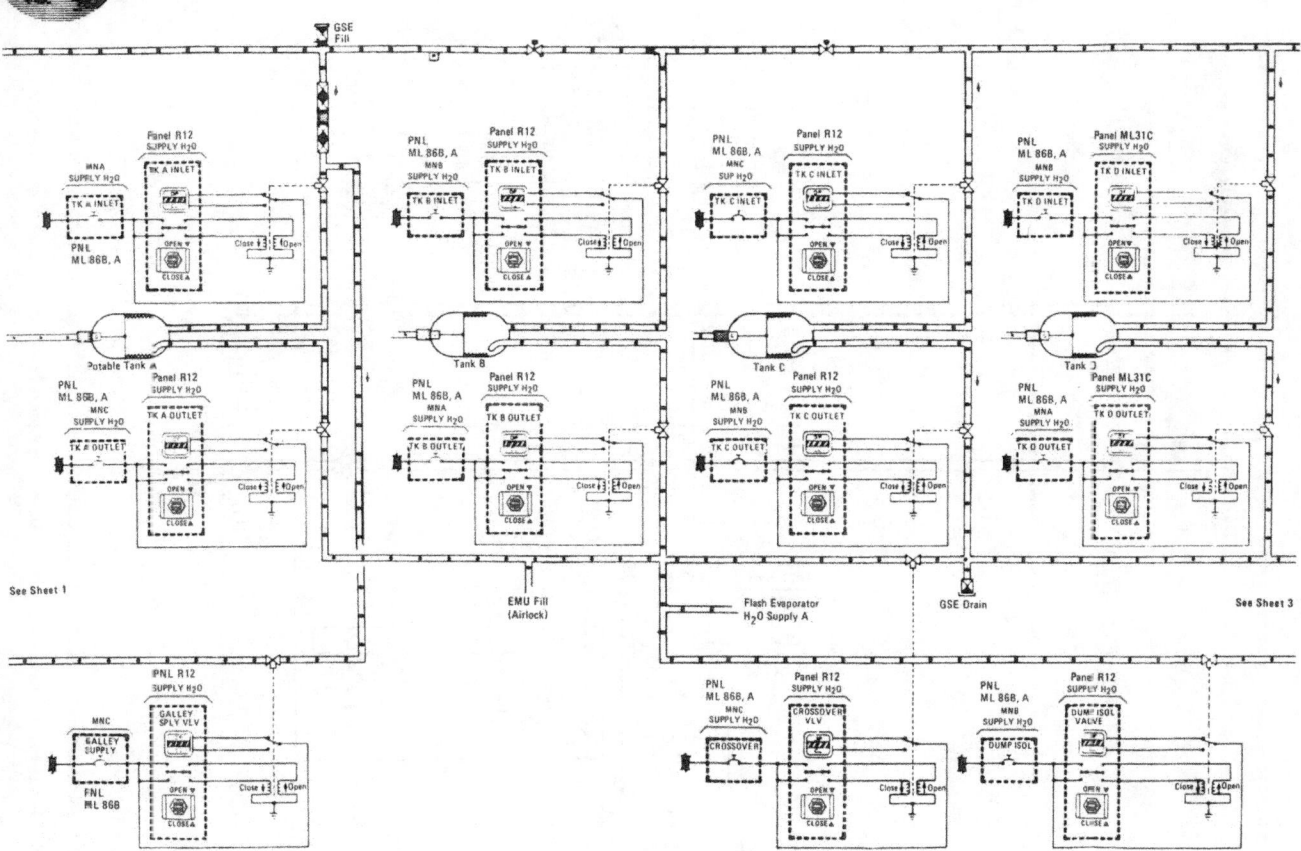

Potable Water (Sheet 2 of 3)

 **Space Shuttle Spacecraft Systems**

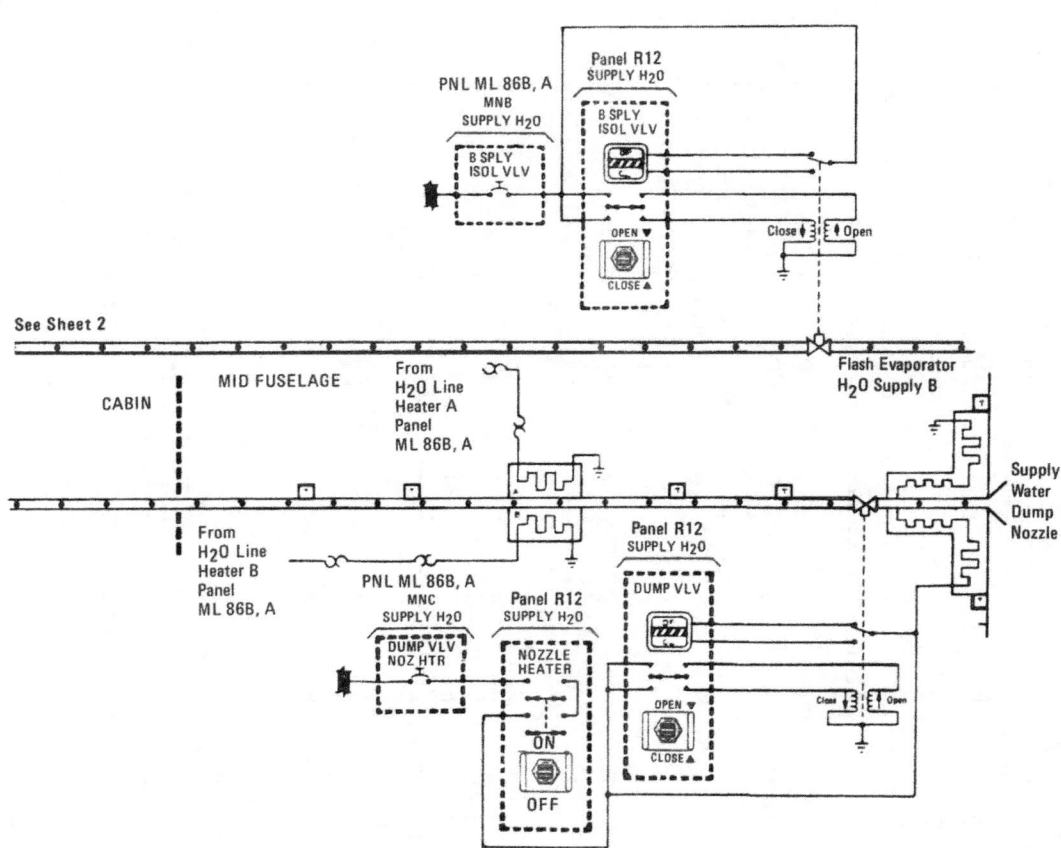

Potable Water (Sheet 3 of 3)

Space Shuttle Spacecraft Systems

and Tanks C and D cannot be used for the flash evaporator A water supply. But by opening the B supply isolation valve, these tanks can flow to B water supply and, thus, the flash evaporator. An indicator above the switch indicates OP (open) when the valve is open, CL (close) when the valve is closed, and BARBERPOLE when the valve is in transit.

The water supply system B to the flash evaporator has an additional supply isolation valve. This valve is controlled by the SUPPLY H_2O B SPLY (supply) ISOL VLV switch on Panel R12. An indicator above the switch on Panel R12 indicates the same as in the previous paragraph.

Water from Tank A, when full, and Tank B can also be dumped overboard. The overboard dump consists of a dump isolation valve in the crew module cabin and a dump valve in the mid fuselage. Both valves are closed unless performing a dump. The SUPPLY H_2O DUMP ISOL VLV switch on Panel R12 opens and closes the dump isolation valve in the crew module cabin. The SUPPLY H_2O DUMP VLV switch on Panel R12 controls the dump valve in the mid fuselage. Indicators above the switches indicate OP (open) when the valves are open, CL (close) when the valves are closed, and BARBERPOLE when the valves are in transit.

The water dump nozzle has a heater to prevent freezing. The heater is controlled by the DUMP VALVE ENABLE/NOZZLE HEATER switch on Panel R12. The nozzle heater is powered when the switch is positioned to ON.

There are thermostatically controlled line heaters upstream of the water dump nozzle on the line. The heaters are powered by circuit breakers H_2O LINE HTR A and B on Panel ML86B.

The system A and B water feedlines to the flash evaporator are approximately 30 meters (100 ft) long. Redundant heaters installed along the length of the water lines are controlled by the FLASH EVAP FEEDLINE HTR switches on Panel L2. The switch enables the thermostatically controlled heaters on H_2O system A and B. Heater circuit one or two on PR1 then A or B may be selected. The OFF position of either switch removes electrical power from the heaters.

The orbiter is equipped with food and facilities for food stowage and preparation and dining to provide each crew member with three meals plus snacks per normal day in orbit; two meals on launch day; one meal on reentry day; and an additional 96 hours of contingency food. The food supply and food preparation facilities are furnished by the government and are designed to accommodate variations in the number of crew and duration of flight, ranging from a crew of two for one day to a crew of seven for 30 days.

In a payload in lieu of galley, the food preparation system is limited. It consists of the water dispenser, food warmer, food trays, food (meal menu and pantry), and food system accessories.

The food warmer is a small, portable, thermostatically controlled unit that can warm meals for two crew members simultaneously. The food trays serve as a dining surface with restraints for food items and provide each crew member with associated dining accessories. The food consists of individually packaged items of dehydrated, thermo-stabilized, irradiated, intermediate moisture, natural form and beverage form. Meal accessories include salt, pepper, sauces, etc., as well as candy, gum, vitamin tablets, wipes, utensils, thermal pads, drinking containers, and germicidal tablets.

For the galley arrangement flights, the food preparation system consists of the galley, food trays, work/dining table, food, and food system accessories.

The galley is a multi-purpose facility that provides food preparation facilities, stowage of meal accessories, food trays and oven inserts, and food and volume for seven crew members.

 **Space Shuttle Spacecraft Systems**

OMITTED

Space Shuttle Spacecraft Systems

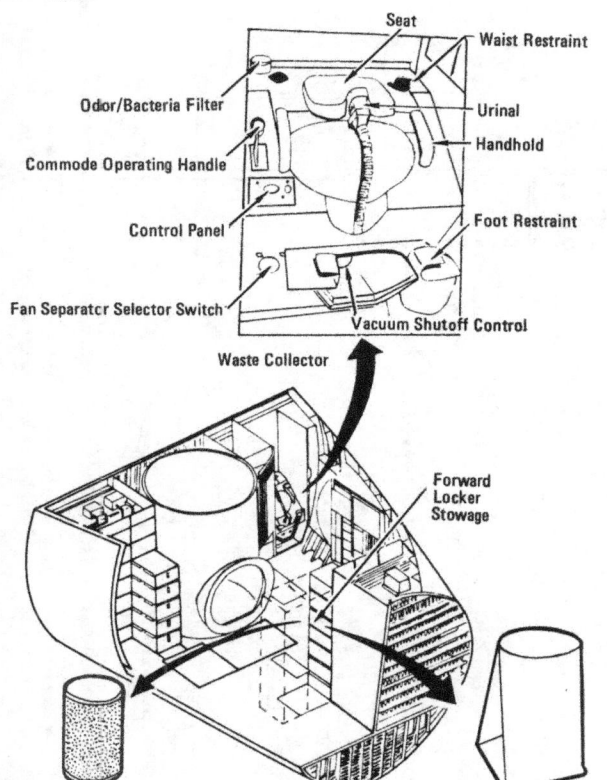

Location of Waste Collection in Mid Deck

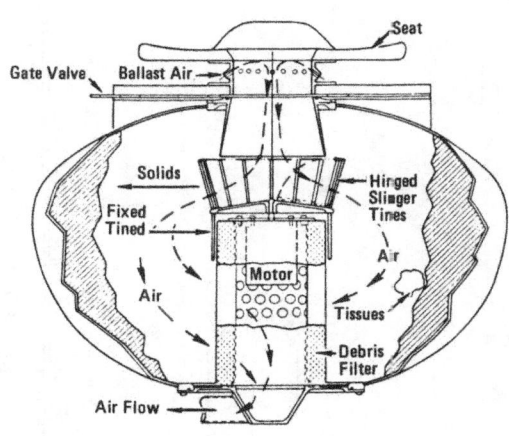

COMMODE OPERATION

Pulling up gate valve control activates slinger motor. As slinger reaches operational speed (1500 rpm), hinged slinger tines unfold outward. Pushing gate valve control forward opens gate valve for commode use. Feces enters commode through seat opening, drawn in by ballast air flow. Slinger tines shred feces and deposit it in thin layer on commode walls. Tissues move up over slinger tines and settle at bottom of collector. Ballast air passes through debris filter and hydrophobic filter to fan separators.

For emesis disposal, fecal/emesis selector switch on WCS control panel is moved to emesis position, rotational speed of slinger tines is slowed and tines do not unfold. This allows unobstructed passage of fecal/vomitus bag into commode.

Waste Collector

 **Space Shuttle Spacecraft Systems**

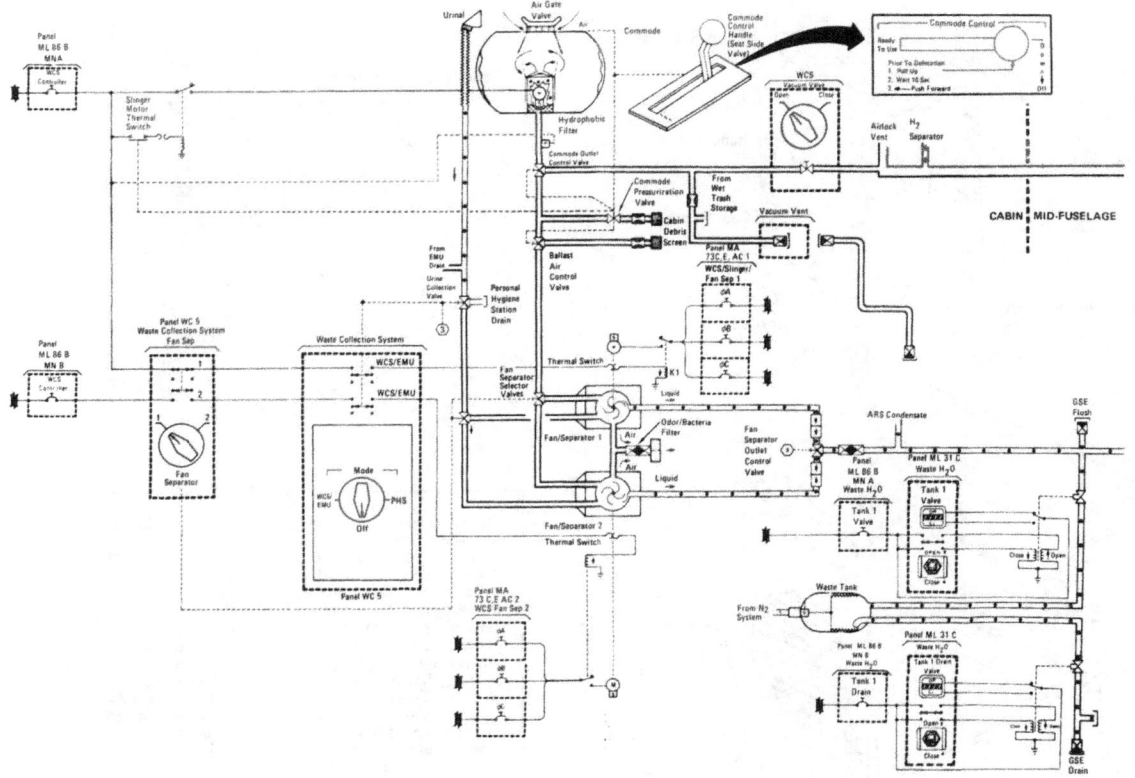

Waste Collection (Sheet 1 of 2)

 **Space Shuttle Spacecraft Systems**

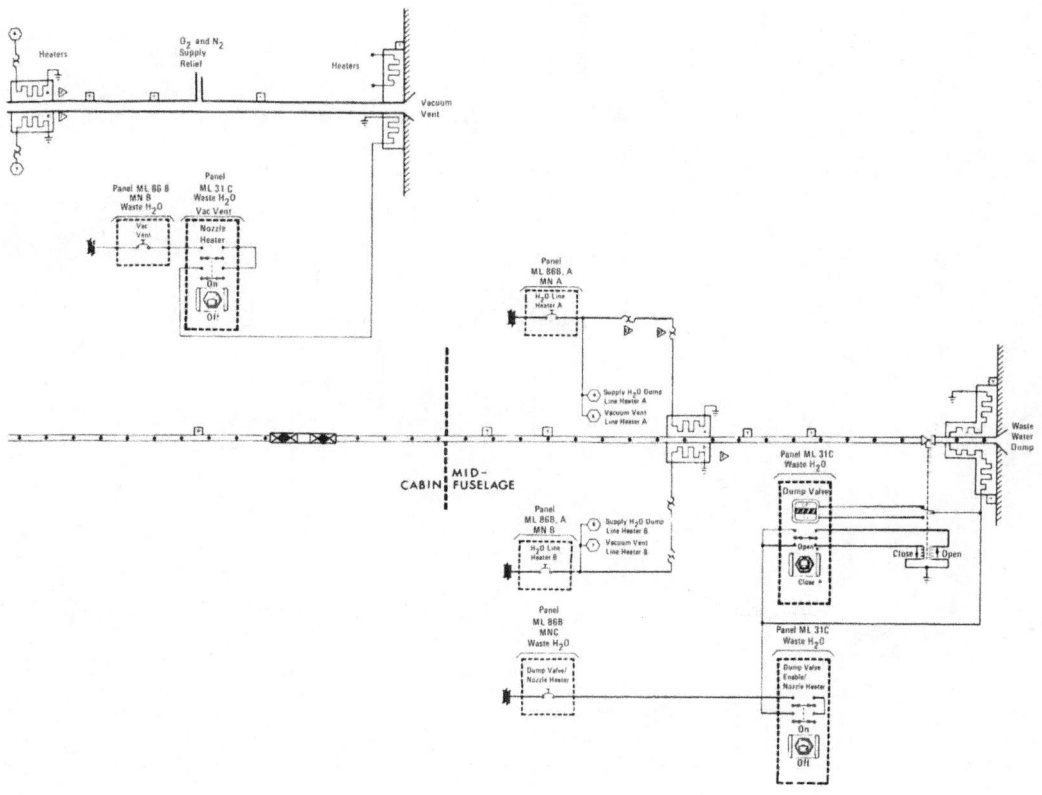

Waste Collection (Sheet 2 of 2)

 **Space Shuttle Spacecraft Systems**

OMITTED

Space Shuttle Spacecraft Systems

The commode has a storage capacity equivalent to 210 crew-member days of vacuum-dried feces and toilet tissue. Each crew member-day usage results in 0.12 kilogram (0.27 pound) of fecal and paper waste, including 0.09 kilogram (0.2 pound) of moisture. The commode can accommodate up to four usages per hour.

Heaters are installed on the vacuum vent line and are thermostatically controlled. Heaters are also installed on the vacuum vent nozzle and are enabled by the WASTE H_2O VAC VENT NOZZLE HEATER switch on panel ML31C when positioned to ON. The WCS waste gases are vented overboard through the vacuum vent line and nozzle.

Personal hygiene accommodations for the crew include a personal hygiene station, personal hygiene kits, pressure-packaged personal hygiene agents, towel dispenser, and tissue dispenser.

In a payload in lieu of galley, the personal hygiene station is located in the mid deck of the crew cabin and provides ambient temperature water with no drain. In the galley arrangement, the personal hygiene station is on the aft side of the galley in the mid deck of the crew cabin and provides ambient and hot water plus a drain to the urinal assembly.

Personal hygiene kits provide for brushing teeth, hair care, shaving, nail care, etc. Pressure-packaged personal hygiene agents are for cleaning hands, face, and body. A seven-day supply of towels is provided for each crew member; additional towel dispensers are provided for each crew member for each additional seven days. Tissues are provided to supply each crew member for seven days and dispensers are added for each seven days added to a mission.

Two privacy curtains are attached to the waste collection compartment door. One is attached to the top of the door and interfaces with the edge of the inter-deck access. The other is attached to the door and interfaces with the galley. The deployed curtain isolates the waste collection compartment and galley personal hygiene station from the rest of the orbiter mid deck.

The waste water tank receives waste water from the ARS crew compartment humidity separator and the waste collection system urine separator.

The waste water tank usable capacity is 74 kilograms (165 pounds). It is 901 millimeters (35.5 inches) in length and 381 millimeters (15 inches) in diameter and weighs 17.9 kilograms (39.5 pounds) dry.

The waste water tank is pressurized with the same gaseous nitrogen source supply as the potable water tanks. The waste

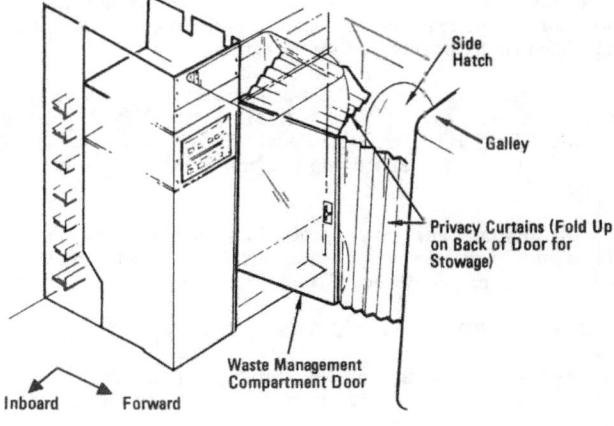

Privacy Curtains

Space Shuttle Spacecraft Systems

water tank has an inlet and outlet valve. The outlet valve is opened only for ground servicing and controlled by the WASTE H2O TK/DRAIN VLV switch on panel ML31C. The valve is opened when the switch is positioned to OPEN and closed when OP (open) when the valve is open, CL (close) when the valve is closed, and BARBERPOLE when the valve is in transit.

The inlet valve permits waste water from the cabin heat exchanger humidity separator and waste fan/separators to enter the waste tank. The inlet valve is controlled by the WASTE H2O TANK 1 VLV switch on Panel ML31C. The switch opens and closes the valve. An indicator above the switch indicates the position of the valve in the same manner as the previous paragraph.

In order for waste water to be dumped overboard, the waste water dump valve must be opened. This valve is controlled by the WASTE H2O DUMP VLV switch on Panel ML31C. The WASTE H2O DUMP ENABLE/NOZ HTR switch must be on before the dump valve can be activated. An indicator above the switch indicates the position of the valve in the same manner as the previous paragraph.

A heater installed on the waste water dump nozzle is turned on and off by the WASTE H2O DUMP VLV ENABLE/NOZ HTR switch on Panel ML31C. Heaters are installed on the waste water dump line and are thermostatically controlled.

Waste water tanks may be added in parallel if a mission requires additional stowage of waste water. The potable water tanks have a similar overboard dump capability.

AIRLOCK SUPPORT SUBSYSTEM

An airlock is located in the mid deck. The airlock and airlock hatches permit extravehicular activities (EVA) flight crew members to transfer from the mid deck crew compartment into the payload bay without depressurizing the orbit crew cabin.

Normally, two extravehicular mobility units (EMU's) are stowed in the mid-deck. The EMU's are an integrated space suit assembly and life support system which provides the capability for the flight crew members to leave the pressurized orbiter crew cabin and work outside the cabin in space.

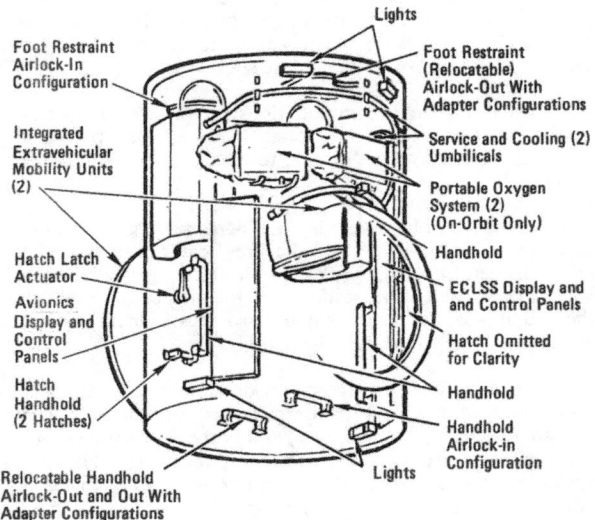

Airlock

Space Shuttle Spacecraft Systems

The airlock is normally located inside the mid-deck of the spacecraft's pressurized crew cabin. It has an inside diameter of 1,600 millimeters (63 inches), is 2,108 millimeters (83 inches) long, and has two 1,016 millimeter (40 inch) diameter D-shaped openings, 914 millimeter (36 inches) across, plus two pressure sealing hatches and a complement of airlock support systems. The airlock volume is 4.24 cubic meters (150 cubic feet).

The airlock is sized to accommodate two fully suited flight crew members simultaneously. The airlock support provides airlock depressurization and repressurization, EVA equipment recharge, liquid cooled garment water cooling, EVA equipment checkout, donning and communications. All EVA gear, checkout panel, and recharge stations are located against the internal walls of the airlock.

The airlock hatches are mounted on the airlock. The inner hatch is mounted on the exterior of the airlock (orbiter crew cabin mid-deck side) and opens in the mid-deck. The inner hatch isolates the airlock from the orbiter crew cabin. The outer hatch is mounted in the interior of the airlock and opens in the airlock. The outer hatch isolates the airlock from the unpressurized payload bay when closed and permits the EVA crew members to exit from the airlock to the payload bay when open.

Airlock repressurization is controllable from inside the orbiter crew cabin mid-deck and from inside the airlock. It is performed by equalizing the airlock and cabin pressure with airlock hatch-mounted equalization valves mounted on the inner hatch. Depressurization of the airlock is controlled from inside the airlock. The airlock is depressurized by venting the airlock pressure overboard. The two D-shaped airlock hatches are installed to open toward the primary pressure source, the orbiter crew cabin, to achieve pressure assist sealing when closed.

Each hatch has six interconnected latches with a gearbox/actuator, a window, a hinge mechanism and hold-open

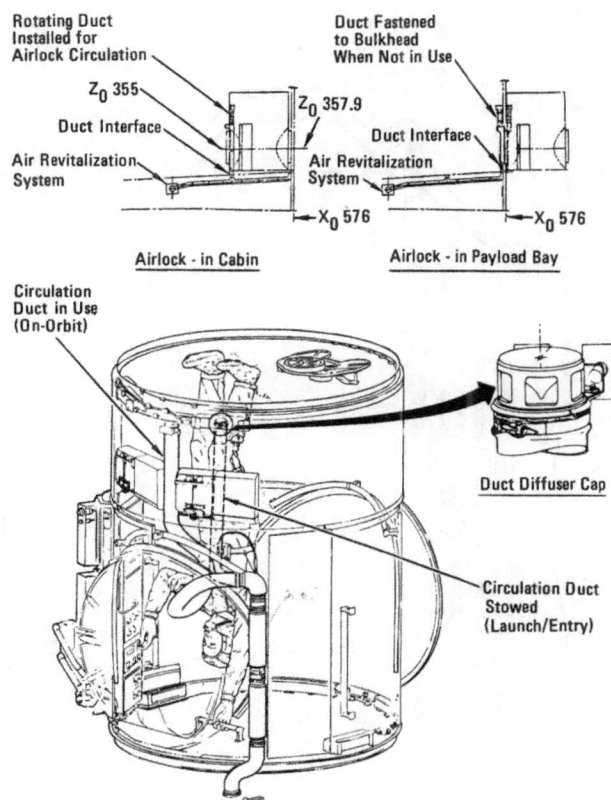

Airlock in Mid Deck Without Tunnel Adapter

Space Shuttle Spacecraft Systems

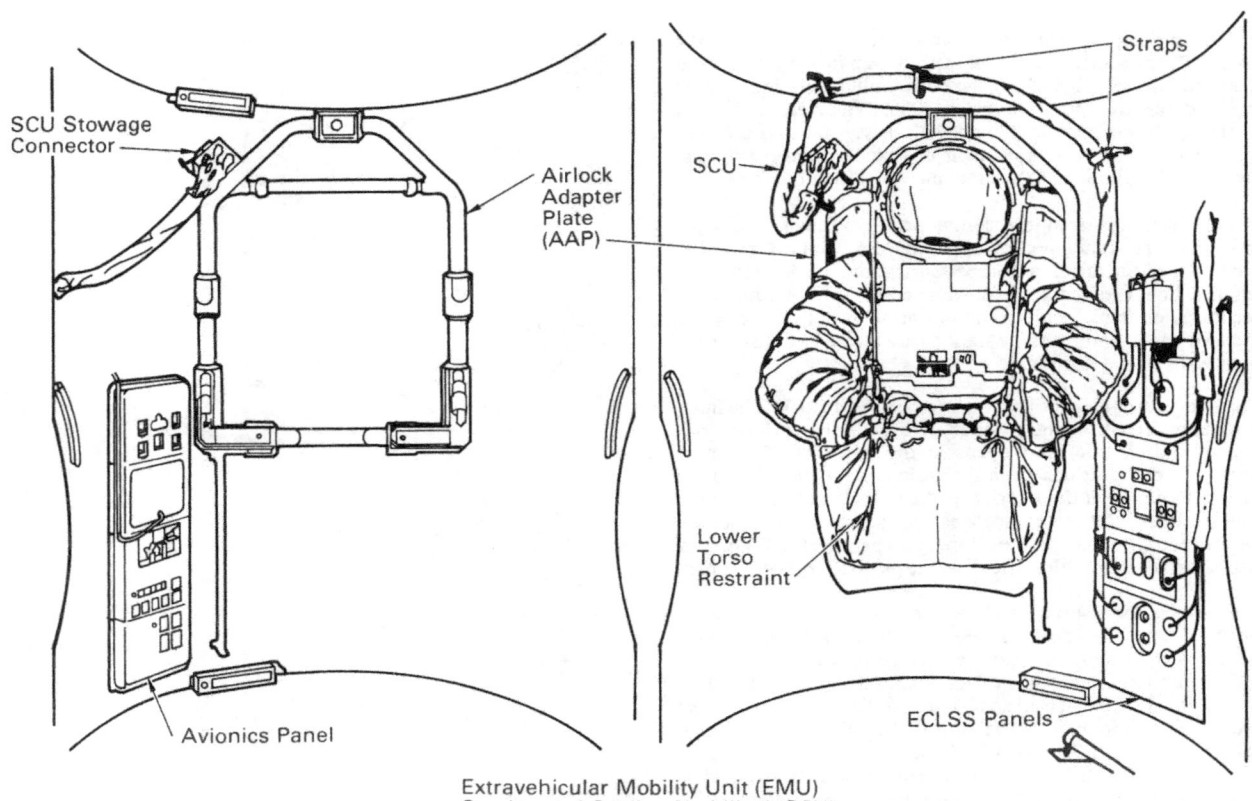

Extravehicular Mobility Unit (EMU)
Service and Cooling Umbilical (SCU)

Airlock Stowage Provisions

Space Shuttle Spacecraft Systems

Extravehicular Mobility Unit (EMU)

device, a differential pressure gage on each side, and two equalization valves.

The window in each airlock hatch is 101 millimeters (4 inches) in diameter. The window is used for crew observation from the cabin/airlock and the airlock/payload bay. The dual window panes are made of polycarbonate plastic and mounted directly to the hatch using bolts fastened through the panes. Each hatch window has dual pressure seals with seal grooves located in the hatch.

Each airlock hatch has dual pressure seals to maintain pressure integrity for the airlock. One seal is mounted on the airlock hatch and the other on the airlock structure. A leak check quick disconnect is installed between the hatch and the airlock pressure seals to verify hatch pressure integrity prior to the flight.

The gearbox with latch mechanisms on each hatch allows the flight crew to open and/or close the hatch during transfers and EVA operation. The gearbox and the latches are mounted on the low pressure side of each hatch, with a gearbox handle installed on both sides to permit operation from either side of the hatch.

Three of the six latches on each hatch are double acting. They have cam surfaces which force the sealing surfaces apart when the latches are opened, thereby acting as crew assist devices. The latches are interconnected with "push-pull" rods and an idler bellcrank installed between the rods for pivoting the rods. Self-aligning dual rotating bearings are used on the rods for attachment to the bellcranks and the latches. The gearbox and hatch open support struts are also connected to the latching system, using the same rod/bellcrank and bearing system. To latch or unlatch the hatch, a rotation of 440 degrees on the gearbox handle is required.

Space Shuttle Spacecraft Systems

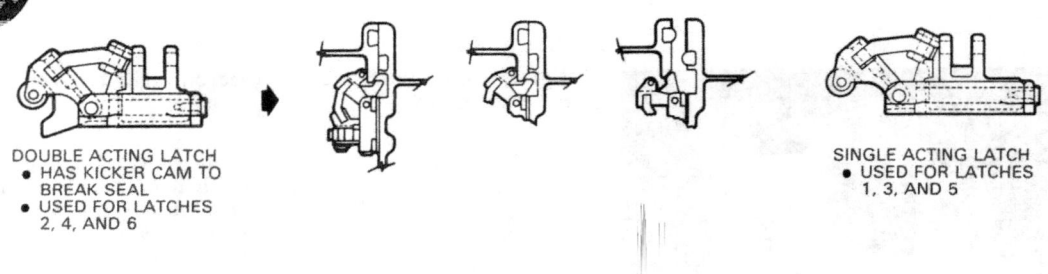

DOUBLE ACTING LATCH
- HAS KICKER CAM TO BREAK SEAL
- USED FOR LATCHES 2, 4, AND 6

SINGLE ACTING LATCH
- USED FOR LATCHES 1, 3, AND 5

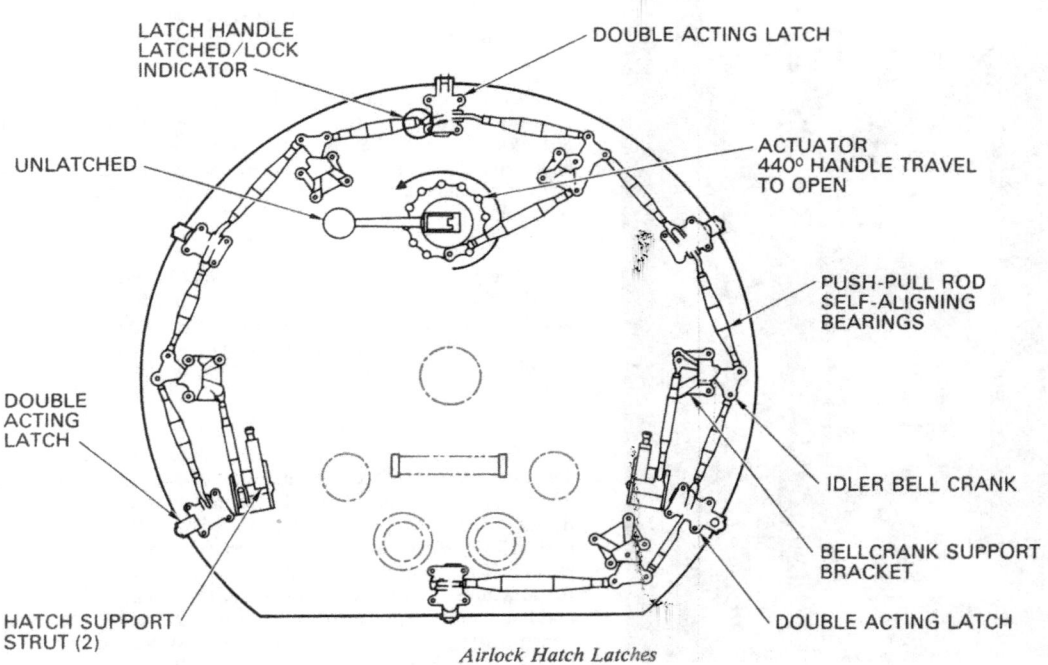

Airlock Hatch Latches

Space Shuttle Spacecraft Systems

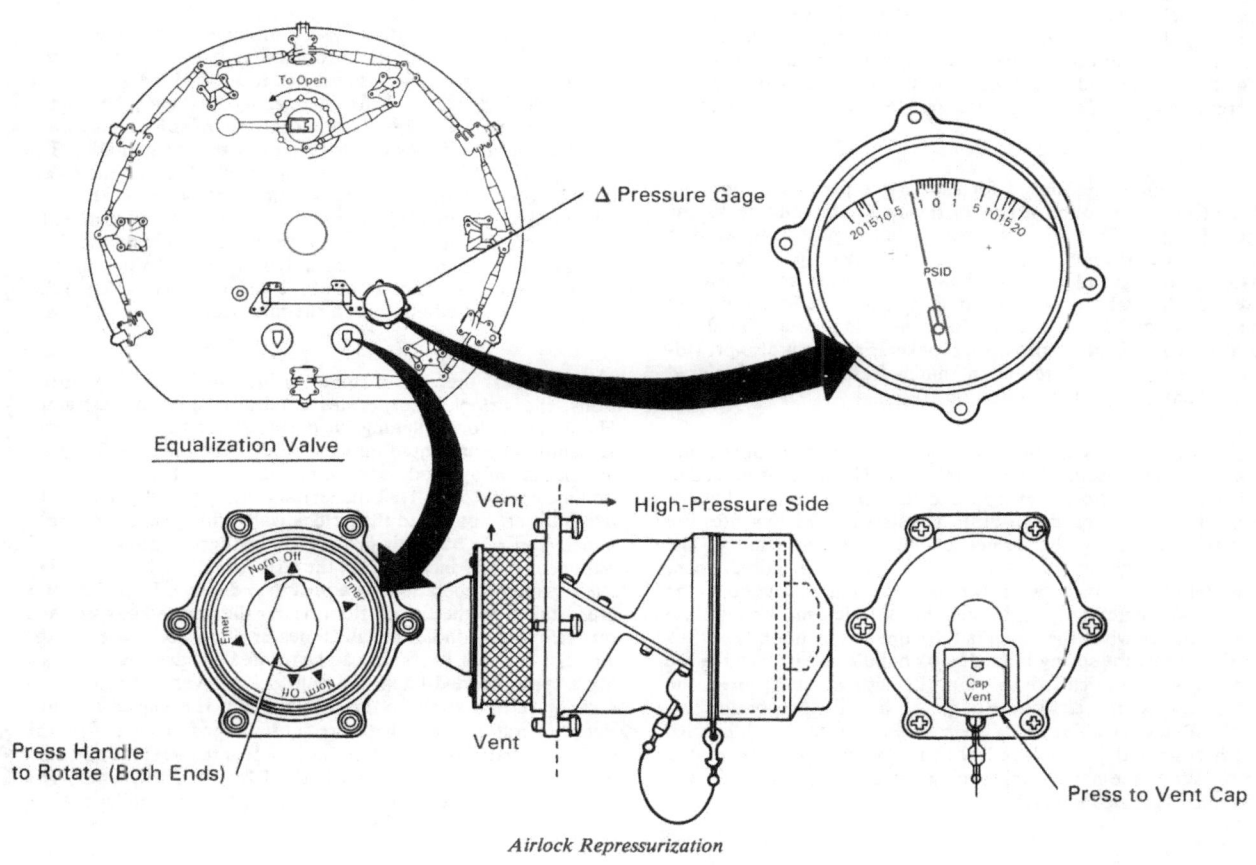

Airlock Repressurization

# Space Shuttle Spacecraft Systems

The hatch actuator/gearbox is used to provide the mechanical advantage to open/close the latches. The hatch actuator lock lever requires a force of 35 to 44 Newtons (8 to 10 pounds) through an angle of 180 degrees to unlatch the actuator. A rotation of 440 degrees minimum with a force of 133 Newtons (30 pounds) maximum applied to the actuator handle is required to operate the latches to their fully unlatched positions.

The hinge mechanism for each hatch permits a minimum opening sweep into the airlock or the crew cabin mid-deck. The inner hatch (airlock to crew cabin) is pulled/pushed forward to the crew cabin approximately 152 millimeters (6 inches). The hatch pivots up and to the starboard (right) side. Positive locks are provided to hold the hatch in both an intermediate and a full open position. To release the lock, a spring-loaded handle is provided on the latch hold-open bracket. Friction is also provided in the linkage to prevent the hatch from moving if relased during any part of the swing.

The outer hatch (in airlock to payload bay) opens and closes to the contour of the airlock wall. The hatch is hinged to be first pulled into the airlock and then pulled forward at the bottom and rotated down until it rests with the low pressure (outer) side facing the airlock ceiling (mid-deck floor). The linkage mechanism guides the hatch from the closed/open, open/closed position with friction restraint throughout the stroke. The hatch has a hold-open hook which snaps into place over a flange when the hatch is fully open. The hook is relased by depressing the spring-loaded hook handle and by pushing the hatch toward the closed position. To support and protect the hatch against the airlock ceiling, the hatch incorporates two deployable struts. The struts are connected to the hatch linkage mechanism and are deployed when the hatch linkage is rotated open. When the hatch latches are rotated closed, the struts are retracted against the hatch.

The airlock hatches can be removed in-flight from the hinge mechanism via pip pins, if required.

Airlock air circulation system provides conditioned air to the airlock during non-EVA operation periods. The airlock revitalization system duct is attached to the outside airlock wall at launch. Upon airlock hatch opening in-flight, the duct is rotated by the flight crew through the cabin/airlock hatch and installed into the airlock and held in place by a strap holder. The duct has a removable air diffuser cap installed on the end of the flexible duct which can adjust the airflow from 0 to 97 kilograms per hour (216 pounds per hour). The duct must be rotated out of the airlock prior to closing the cabin/airlock hatch for airlock depressurization. During the EVA preparation period, the duct is rotated out of the airlock and can be used as supplemental air circulation in the mid-deck.

To assist the crew member in pre- and post-EVA operations, the airlock incorporates handrails and foot restraints. Handrails are located alongside the avionics and ECLSS panels. A handhold is mounted on each side of the hatches. They are aluminum alloy and oval configurations 19.05 by 33.52 millimeters (0.75 by 1.32 inches) and are painted yellow. The handrails are bonded to the airlock walls with an epoxyphenolic adhesive. Each handrail provides a handgrip clearance of 57 millimeters (2.25 inches) from the airlock wall to the handrail to allow gripping operations in a pressurized glove. Foot restraints are installed on the airlock floor nearer the payload bay side and the ceiling handhold installed nearer the cabin side of the airlock. The foot restraints can be rotated 360 degrees by releasing a spring-loaded latch and will lock in every 90 degrees. A rotation release knob on the foot restraint is designed for shirt sleeve operation, and therefore must be postioned before the suit is donned. The foot restraint is bolted to the floor and cannot be removed in flight and is sized for the EMU boot. The crew member ingresses by first inserting the foot under the toe

Space Shuttle Spacecraft Systems

bar and then the heel is pressed down by rotating the heel from inboard to outboard until the heel on the boot is captured.

There are four floodlights in the airlock. The lights are controlled by switches in the airlock on panel AW18A; light 2 can also be controlled by a switch on mid-deck Panel M013Q, allowing illumination of the airlock prior to entry. Lights 1, 3, and 4 are powered by buses MNA, B, and C respectively and light 2 is powered by ESS1BC. The circuit breakers are on Panel ML86B.

If the airlock is relocated to the payload bay from the mid deck, it will function in the same manner as in the mid deck. Insulation would be installed on the airlock exterior for protection from the extreme temperatures of space.

In preparation for an EVA, the flight crew member will first don a liquid cooled and ventilation garment (LCVG). It is similar to "long-john" underwear into which have been woven many feet of flexible tubing that circulates cooling water. The liquid cooled and ventilation garment is worn under the pressure and gas garment to maintain desired body temperature.

A urine collection device (UCD) is worn for collection of urine in the suit. It stores approximately 0.9 liter (approximately one quart) of urine. It consists of adapter tubing, storage bag and disconnect hardware for emptying after an EVA into the orbiter waste water system.

The airlock provides stowage for two Extravehicular Mobility Units (EMU's) and two service and cooling umbilicals (SCU's) and various miscellaneous support equipment.

Both EMU's are mounted on the airlock walls by means of an airlock adapter plate (AAP).

The prime contractor to NASA for the space suit/life support system is United Technologies' Hamilton Standard Division in Windsor Locks, CT. Hamilton Standard is program systems manager for the space suit/life support system in addition to designer and builder. Hamilton Standard's major subcontrctor is ILC Dover of Frederica, DE, which fabricates the space suit.

The EMU's provide the necessities for life support, such as oxygen, carbon dioxide removal, a pressurized enclosure, temperature control and meteoroid protection during EVA.

The EMU space suit comes in various sizes so that prior to launch, flight crew members can pick their suits "off the rack." Components are designed to fit male and female from the 5th to the 95th percentiles of body size.

The life support system is self contained and contains seven hours of expendables such as oxygen, battery power for electrical power, water for cooling, and lithium hydroxide for carbon dioxide removal and a 30 minute emergency life support system during an EVA.

The airlock adapter plate in the airlock also provides a fixed position for the EMU's to assist the crew member during donning, doffing, checkout and servicing. Each EMU weighs approximately 102 kilograms (225 pounds) and the overall storage envelope is 660 x 711 x 1,016 millimeters (26 x 28 x 40 inches). For launch and entry, the lower torso restraint, a cloth bag attached to the airlock adapter plate (AAP) with straps, is used to hold the lower torso and arms securely in place.

To don the EMU, the crew member enters the airlock and dons the lower torso assembly which has boots attached. The lower torso consists of the pants, boots and the hip, knee and ankle joints. The hard, upper torso assembly includes the life support backpack and provides the structural mounting interface for most of the EMU including helmet, arms, lower torso, portable life support system, display and control module and electrical harness. The arm assembly contains the shoulder joint

Space Shuttle Spacecraft Systems

and upper arm bearings that permit shoulder mobility as well as the elbow joint and wrist bearing. The gloves contain the wrist disconnect, wrist joint and insulation padding for palms and fingers. The helmet consists of a clear polycarbonate bubble neck disconnect and ventilation pad. An EVA visor assembly is attached externally to the helmet which contains visors which are manually adjusted to shield the crew member's eyes. The upper and lower torsos are connected with a waist ring.

In addition, the portable life support system consists of an EMU electrical harness that provides bioinstrumentation and communications connections; a display and control module that is chest mounted which contains all external fluid and electrical interfaces and controls and displays; the portable life support subsystem referred to as the "backpack" which contains the life support subsystem expendables and machinery; a secondary oxygen pack mounted on the base of the portable life support subsystem which contains a 30 minute emergency oxygen supply and a valve and a regulator assembly, and an in-suit drink bag that stores liquid in the hard upper torso which has a tube projecting up into the helmet to permit the crew member to drink while suited.

The orbiter provides electrical power, oxygen, liquid cooled ventilation garment cooling and water to the EMU's in the airlock via the SCU for EVA pre- and post-EVA operations.

The service and cooling umbilical (SCU) is launched with the orbiter end fittings permanently connected to the appropriate ECLSS panels and the EMU connected to the airlock adapter plate stowage connector. The SCU contains communication lines, electrical power, water and oxygen, recharge lines and drain lines. It allows all supplies (oxygen, water, electrical, and communication) to be transported from the airlock control panels to the EMU before and after EVA without using the EMU expendable supplies of water, oxygen and battery power that are scheduled for use in the EVA. The SCU also provides EMU recharge. The SCU umbilical is disconnected just

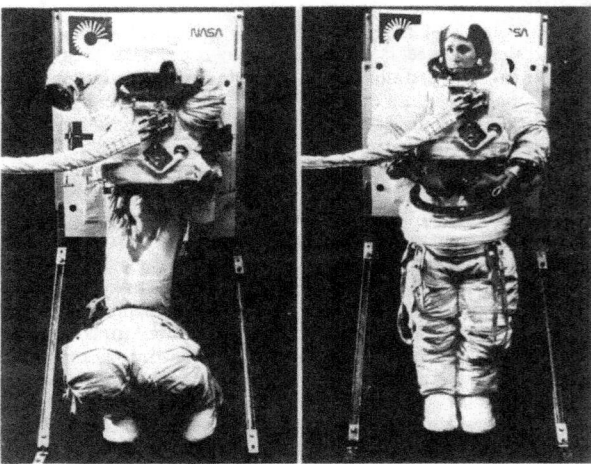

Extravehicular Mobility Unit (EMU)

before the crew member leaves the airlock on an EVA and upon return to the airlock after an EVA. Each SCU is 3,657 millimeters (144 inches) long and 88 millimeters (3.5 inches) in diameter and weighs 9.1 kilograms (20 pounds). Actual usable length after attachment to the control panel is approximately 2 meters (7 feet).

The airlock has two display and control panels. The airlock control panels are basically split to provide either ECLSS or avionics operations. The ECLSS panel provides the interface for the SCU waste and potable water, liquid cooled ventilation garment cooling water, EMU hardline communication, EMU power and oxygen supply. The avionics panel includes the airlock lighting, the airlock audio system, and the EMU power and battery recharge controls. The avionics panel is

Space Shuttle Spacecraft Systems

located on the starboard (right) side of the cabin airlock hatch and the ECLSS panel on the port (left) side. The airlock panels are designated AW18H, AW18D, and AW18A on the port side and AW82H, AW82D, and AW82B on the starboard side. The ECLSS panel is divided into EMU1 functions on the starboard side and EMU2 functions on the port side.

Airlock communications are provided with the orbiter audio system at airlock panel AW82D where connectors for the headset interface unit (HIU's) and the EMU's are located at airlock panel AW18D which is the airlock audio terminal (ATU). The HIU's are inserted in the crew-member communications carrier unit (CCU1 and CCU2) connectors on airlock panel AW82D. The CCU's are also known as the "Snoopy Cap" which fits over the crew member's head and snaps into place with a chin guard. It contains a microphone and headphones for two-way communications and receiving caution and warning tone. The adjacent two-position switches labeled CCU1 and CCU2 POWER enable transmit functions only, as reception is normal as soon as the HIU's are plugged in. The EMU1 and EMU2 connectors on the same panel to which the service and cooling umbilical (SCU) is connected include contacts for EMU hard-line communications with the orbiter prior to EVA. Panel AW18D contains displays and controls used to select access to and control volume of various audio signals. Control of the airlock audio functions can be transferred to the mid-deck ATU's panel M042F, by placing the CONTROL knob to MIDDECK position.

During EVA, the Extravehicular Communicator (EVC) is part of the same UHF system which is used for air-to-air and air-to-ground voice communications between the orbiter and landing site control tower and the orbiter and chase aircraft. The EVC provides full duplex (simultaneous transmission and reception) communications between the orbiter and the two EVA crew members and continuous data reception of electrocardiogram signals from each crew member by the orbiter and orbiter processing and relay of electrocardiogram signals to the ground. The UHF airlock antenna in the forward portion of the payload bay provides the UHF-EVA capability.

Panel AW18H in the airlock provides 17 plus or minus 0.5 vdc at five amperes at both EMU electrical connector panels, Panel AW82D, in EVA prep. Bus MNA or B can be selected on the BUS SELECT switch and then the MODE switch is positioned to POWER. The BUS SELECT switch provides a signal to a remote power controller (RPC) which applies 28 vdc from the selected bus to the power/battery recharger. The MODE switch in the POWER position makes the power available at the SCU connector and also closes a circuit that provides a battery feedback voltage charger control which inhibits EMU power when any discontinuity is sensed in the SCU/EMU circuitry. The MODE switch in the POWER position also applies power through the SCU for the EMU microphone amplifiers for hardline communication. When the SCU umbilical is disconnected for EVA, the EMU operates on its self contained battery power. For post-EVA, when the SCU is reconnected to the EMU, selecting a bus and the CHARGE position on the MODE switch charges the portable life support system battery at 1.55 plus or minus 0.05 amps. When the battery reaches 21.8 plus or minus 0.1 vdc and/or the charging circuit exceeds 1.55 plus or minus 0.05 amps, a solenoid controlled switch internal to the battery charger removes power to the charging circuitry. The EMU silver zinc battery provides all electrical power used by the portable life support system during EVA and is filled with electrolyte and charged prior to flight.

Cooling for the flight crew members before and after the EVA is provided by the liquid cooled garment circulation system via the SCU and LCG (liquid cooled garment) SUPPLY AND RETURN connections on panel AW82B. These connections are routed to the orbiter liquid cooled garment heat exchanger which transfers the collected heat to the orbiter Freon-21 coolant loops. The nominal loop flow of 113 kilograms per hour (250 pounds per hour) is provided by the EMU/portable life support system water loop pump. The

Space Shuttle Spacecraft Systems

system circulates chilled water at 10 degrees Celsius (50 °F) maximum to the liquid cooled ventilation garment inlet and provides a heat removal capability of 2,000 Btu (British Thermal Units) per hour per crew member. When the SCU is disconnected the portable life support system provides the cooling. Upon return from the EVA, the portable life support system is reconnected to the SCU and the crew member cooling is provided as it was in the EVA prep.

With the suit connected to the SCU, oxygen at 15,525 to 46,575 mmHg (300 to 900 psia) is supplied through airlock panel AW82B from the orbiter oxygen system when the OXYGEN valve is in the OPEN position on the airlock panel. This provides the suited crew member with breathing oxygen, preventing depletion of the portble life support system oxygen tanks prior to the EVA. Prior to the crew member sealing the helmet, an oxygen purge adapter hose is connected to the airlock panel to flush nitrogen out of the suit.

The crew member will prebreathe pure oxygen in the EMU for approximately three and one-half hours prior to the EVA. This is necessary to remove nitrogen from their blood before working in the pure oxygen environment of the EMU due to the orbiter pressurized crew cabin mixed gas atmosphere of 20 percent oxygen and 80 percent nitrogen at a pressure of 760 plus or minus 10 mmHg (14.7 plus or minus 0.2 psia). Without prebreathing, bends occur when an individual fails to reduce nitrogen levels in the blood prior to working in a pressure condition that can result in nitrogen coming out of solution in the form of bubbles in the bloodstream. This condition results in pain in the body joints, possibly because of restricted blood flow to connective tissues or the extra pressure caused by bubbles in the blood at joint area. During prebreathe, the suit is at 77 mmHg (1-1/2 psia).

When the SCU is disconnected, the portable life support system provides oxygen for the suit. When the EVA is completed and the SCU is reconnected, the orbiter oxygen supply begins recharging the portable life support system, providing the OXYGEN valve on panel AW82B is OPEN. Full oxygen recharge takes approximately one hour (allowing for thermal expansion during recharge) and the tank pressure is monitored on the EMU display and control panel as well as on the airlock oxygen pressure readout.

Each EMU is pressurized to 207 mmHg (4.0 psid) differential. They are designed for a 15 year life with cleaning and drying between flights.

The EMU WATER SUPPLY and WASTE valves are opened during the EVA prep by switches on Panel AW82D. This provides the EMU, via the SCU, access to both the orbiter potable water and waste water systems. The support provided to the EMU portable life support system is further controlled by the EMU display and control panel. Potable water — supplied from the orbiter at 828 plus or minus 25 mmHg (16 plus or minus 0.5 psi), 45 to 58 kilograms per hour (100 to 300 pounds per hour), and 4 to 37 °C (40 to 100 °F) — allowed to flow to the feedwater reservoir in the EMU which provides pressure which would "top-off" any tank not completely filled. Waste water, condensate, developed in the portable life support system is allowed to flow to the orbiter waste water system via the SCU whenever the regulator connected at the bacteria filters (airlock ends of the SCU) detects upstream pressure in excess of 828 plus or minus 25 mmHg (16 plus or minus 0.5 psi).

When the SCU is disconnected from the EMU, the portable life support system assumes this function. When the SCU is reconnected to the EMU upon completion of the EVA, the same functions as in pre-EVA are performed except that the water supply is allowed to continue until the portable life support system water tanks are filled, which takes approximately 30 minutes.

In preparation for the EVA from the airlock, the airlock hatch to the orbiter crew cabin is closed and depressurization of the airlock begins.

Space Shuttle Spacecraft Systems

Airlock depressuriation is accomplished by a three position valve located on the ECLSS (Environmental Control Life Support System) panel AW82A in the airlock. The airlock depressurization valve is covered with a pressure/dust cap. Prior to removing the cap from the valve, it is necessary to vent the area between the cap and valve by pushing the vent valve on the cap. In-flight storage of the pressure/dust cap is adjacent to the valve. The airlock depressurization valve is connected to a 50 millimeter (2 inch) inside diameter stainless steel overboard vacuum line. The AIRLOCK DEPRESS valve controls the rate of depressurization by varying the valve diameter size. Depressurization is accomplished in two stages. The CLOSED position prevents any airflow from escaping to the overboard vent system.

When the crew members have completed the prebreathe in the EMU's for 3.5 hours, the airlock is depressurized from 760 mmHg (14.7 psia) to 258 mmHg (5 psia) by position labeled "5" on the AIRLOCK DEPRESS valve which opens the depressurization valve and allows the pressure in the airlock to decrease. Pressure during depressurization can be monitored by the delta pressure gage on either airlock hatch. A delta pressure gage is installed on each side of both airlock hatches.

At this time the flight crew performs an EMU suit leak check, electrical power is transferred from the umbilicals to the EMU batteries, the umbilicals are disconnected and the suit oxygen packs are brought on line.

The second stage of airlock depressurization is accomplished by positioning the AIR LOCK DEPRESS valve closed to "0" which increases the valve diameter and allows the pressure in the airlock to decrease from 258 mmHg (5 psia) to 0 mmHg (0 psia). The depressurization of the airlock is accomplished within eight minutes at rates no more than 5.1 mmHg (0.1 psia) per second. The suit sublimators are activated for cooling, EMU system checks are performed and the airlock/payload bay hatch can be opened. The hatch is capable of opening against a 10 mmHg (0.2 psia) differential maximum.

Hardware provisions are installed in the orbiter payload bay for use by the crew member during the EVA.

Handrails and tether points are located on the payload bulkheads, forward bulkhead station X_O 576 and aft bulkhead station X_O 1307, and along the sill longeron on both sides of the bay to provide translation and stabilization capability for the EVA crew member. The handrails are designed to withstand a load of 90.72 kilograms (200 pounds), 127.01 kilograms (280 pounds) maximum in any direction. Tether attach points are designed to sustain a load of 260.37 kilograms (574 pounds), 364.69 kilograms (804 pounds) maximum, in any direction.

The handrails have a cross section of 33 by 19 millimeters (1.32 x 0.75 inches). They are made of aluminum alloy tubing and are painted yellow. The end braces and side struts of the handrails are constructed to titanium. An aluminum alloy end support standoff functions as the terminal of the handrail. Each end support standoff incorporates a 25.4 millimeter (one inch) diameter tether point.

A 7.62 meter (25 foot) crew member safety tether is attached to each crew member at all times during an EVA.

The tether consists of a reel case with an integral "D" ring, a reel with a light takeup spring, a cable and a locking hook. The safety tether hook is locked onto the slidewire before launch and the cable is routed and clipped along the port (left) and starboard (right) handrails to a position just above the airlock/payload bay hatch. After opening the airlock hatch and before egress, the crew member attaches a waist tether to the "D" ring of the safety tether to be used. The other end of the

# Space Shuttle Spacecraft Systems

waist tether is hooked to a ring on the EMU waist bearing. The crew member may select either the port or the starboard safety tether. With the selector on the tether in the locked position, the cable will not retract or reel out. Moving the selector to the unlocked position allows the cable to reel out and the retract feature to take up slack. The cable is designed for a maximum load of 398 kilograms (878 pounds). The routing of the tethers follows the handrails, allowing the crew member to deploy and restow his tether during translation.

The two slidewires, approximately 14.11 meters (46.3 feet) long, are located in the longeron sill area on each side of the payload bay. They start approximately 2.83 meters (9.3 feet) aft of the forward bulkhead and extend approximately 14.11 meters (46.3 feet) down the payload bay. The slidewires withstand a tether load of 260.37 kilograms (574 pounds) with a safety factor of 1.4 or 364.49 kilograms (804 pounds) maximum.

The airlock/cabin hatch has two pressure equalization valves which can be operated from both sides of the hatch for repressurizing the airlock volume. Each valve has three positions, CLOSED, NORM (Normal), and EMERG (Emergency) and is protected by a debris pressure cap on the intake (high-pressure) side of the valve, which on the outer hatch must be vented for removal. The caps are tethered to the valves and also have small Velcro spots which allow temporary stowage on the hatch. The exit side of the valve contains an air diffuser to provide uniform flow out of the valve.

Through the use of the equalization valve/valves in the various positions, the airlock can be repressurized in a normal mode to 760 mmHg (1,417 psia) within five minutes at rates no more than 5.1 mmHg (0.1 psia) per second. If both equalization valves are positioned to EMERG, the airlock can be repressurized to 760 mmHg (14.7 psia) within 20 plus or minus five seconds at a rate no more than 51.75 mmHg (1.0 psia) per second. The hatch is capable of opening against a 10 mmHg (0.2 psia) differential maximum.

The airlock is initially pressurized to 258 mmHg (5 psia) and the umbilicals are connected and electrical power is transferred back to umbilical power. The airlock is then pressurized to equalize with the cabin pressure, followed by EMU doffing and the crew members' recharge of the EMU's.

The orbiter provides accommodations for three two-flight-crew member EVA's of six-hour duration per flight at no weight or volume cost to the payload. Two of the EVA's are for payload support and the third is reserved for orbiter contingency. Additional EVA's can be considered with consumables charged to payloads.

For Spacelab missions, the airlock remains in the crew compartment mid-deck and a tunnel adapter is installed in the payload bay which mates with the airlock and the Spacelab tunnel.

The airlock, tunnel adapter, hatches, tunnel extension and tunnel permits the flight crew members to transfer from the spacecraft pressurized mid-deck crew compartment into Spacelab in a pressurized shirt sleeve environment.

In addition, the airlock, tunnel adapter and hatches permit the EVA flight crew members to transfer from the airlock/tunnel adapter in the space suit assembly into the payload bay without depressurizing the spacecraft crew cabin and Spacelab.

The tunnel adapter is located in the payload bay and is attached to the airlock at orbiter station X_O 576 and attached to the tunnel extension at X_O 660, thus the Spacelab tunnel and Spacelab. The tunnel adapter has an inside diameter of 1,600 millimeters (63 inches) at the widest section and tapers in the cone area at each end, to two 1,016 millimeter (40 inch) diameter D-shaped openings, 914 millimeters (36 inches) across. A 1,016 millimeter (40 inch) diameter D-shaped opening, 914 millimeters (36 inches) across is located at the top of the tunnel adapter. Two pressure sealing hatches are located in the tunnel

Space Shuttle Spacecraft Systems

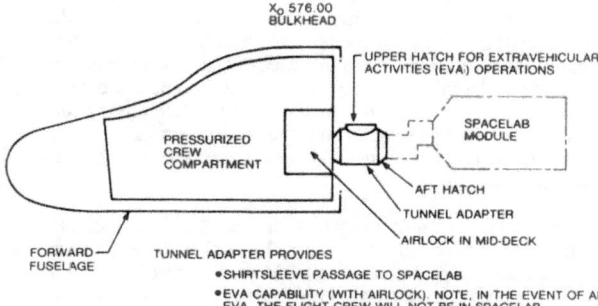

Airlock/Tunnel Adapter

adapter, one at the upper area of the tunnel adapter and one at the aft end of the tunnel adapter. The tunnel adapter is constructed of 2219 aluminum and is a welded structure with 60 by 60 millimeter (2.4 by 2.4 inch) exposed structural ribs on exterior surface and an external waffle skin stiffening.

The hatch located on the mid-deck side of the airlock is mounted on the exterior of the airlock and opens into the mid-deck. This hatch isolates the airlock from the spacecraft crew cabin. The hatch located in the tunnel adapter aft end isolates the tunnel adapter/airlock from the tunnel extension tunnel and Spacelab. This hatch opens into the tunnel adapter. The hatch located in the tunnel adapter at the upper D-shaped opening isolates the airlock/tunnel adapter from the unpressurized payload bay when closed and permits the EVA crew members to exit from the airlock/tunnel adapter to the payload bay when open. This hatch opens into the tunnel adapter.

Airlock repressurization is controllable from inside the orbiter crew cabin mid-deck and from inside the airlock. It is performed by equalizing the airlock and cabin pressure with airlock

Tunnel Adapter

hatch-mounted equalization valves mounted on the inner hatch. Depressurization of the airlock is controlled from inside the airlock. The airlock is depressurized by venting the airlock pressure overboard. The airlock hatch is installed to open toward the primary pressure source, the orbiter crew cabin, to achieve pressure assist sealing when closed. The two hatches in the tunnel adapter are also installed to open toward the primary pressure source, the orbiter crew cabin, to achieve pressure assist sealing when closed.

Each hatch has six interconnected latches (with the exception of the aft hatch which has 17) with a gearbox/actuator, a window, a hinge mechanism and hold-open device, a differential pressure gage on each side, and two equalization valves.

Space Shuttle Spacecraft Systems

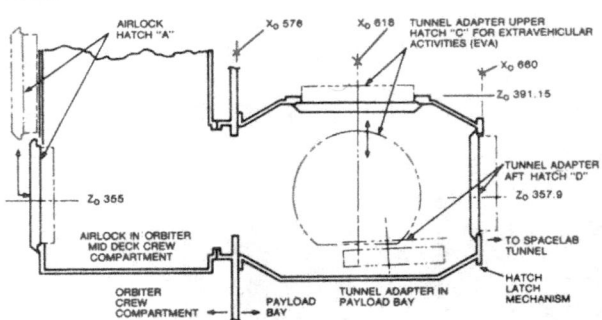

Airlock and Tunnel Adapter Hatch Mechanical Systems

The window in each hatch is 101 millimeters (4 inches) in diameter. The window is used for crew observation from the cabin/airlock, tunnel adapter to tunnel, and tunnel adapter to payload bay. The dual window panes are made of polycarbonate plastic and mounted directly to the hatch using bolts fastened through the panes. Each hatch window has dual pressure seals with seal grooves located in the hatch.

Each hatch has dual pressure seals to maintain pressure integrity. One seal is mounted on the hatch and the other on the structure. A leak check quick disconnect is installed between the hatch and the pressure seals to verify hatch pressure integrity prior to flight.

The gearbox with latch mechanisms on each hatch allows the flight crew to open and/or close the hatch during transfers and EVA operation. The gearbox and the latches are mounted on the low pressure side of each hatch, with a gearbox handle installed on both sides to permit operation from either side of the hatch.

Three of the six latches on each hatch are double acting (with the exception of the aft hatch which has two). They have cam surfaces which force the sealing surfaces apart when the latches are opened, thereby acting as crew assist devices. The latches are interconnected with "push-pull" rods and an idler bellcrank installed between the rods for pivoting the rods. Self-aligning dual rotating bearings are used on the rods for attachment to the bellcranks and the latches. The gearbox and hatch open support struts are also connected to the latching system, using the same rod/bellcrank and bearing system. To latch or unlatch the hatch, a rotation of 440 degrees on the gearbox handle is required.

The hatch actuator/gearbox is used to provide the mechanical advantage to open/close the latches. The hatch actuator lock lever requires a force of 35 to 44 Newtons (8 to 10 pounds) through an angle of 180 degrees to unlatch the actuator. A rotation of 440 degrees minimum with a force of 133 Newtons (30 pounds) maximum applied to the actuator handle is required to operate the latches to their fully unlatched positions.

The hinge mechanism for each hatch permits a minimum opening sweep into the tunnel adapter or the spacecraft crew cabin mid-deck. The airlock crew cabin hatch in the mid-deck is pulled/pushed forward to the mid-deck approximately 152 millimeters (6 inches). The hatch pivots up and to the starboard (right) side. Positive locks are provided to hold the latch in both an intermediate and a full open position. To release the lock, a spring-loaded handle is provided on the latch hold-open bracket. Friction is also provided in the linkage to prevent the hatch from moving if released during any part of the swing.

The aft hatch is hinged to be first pulled into the tunnel adapter and then pulled forward at the bottom. The top of the hatch is rotated towards the tunnel and downward until the hatch rests with the Spacelab side facing the tunnel adapter floor. The linkage mechanism guides the hatch from the closed/open, open/close position with friction restraint throughout the stroke. The hatch is held in the open position by straps and Velcro.

Space Shuttle Spacecraft Systems

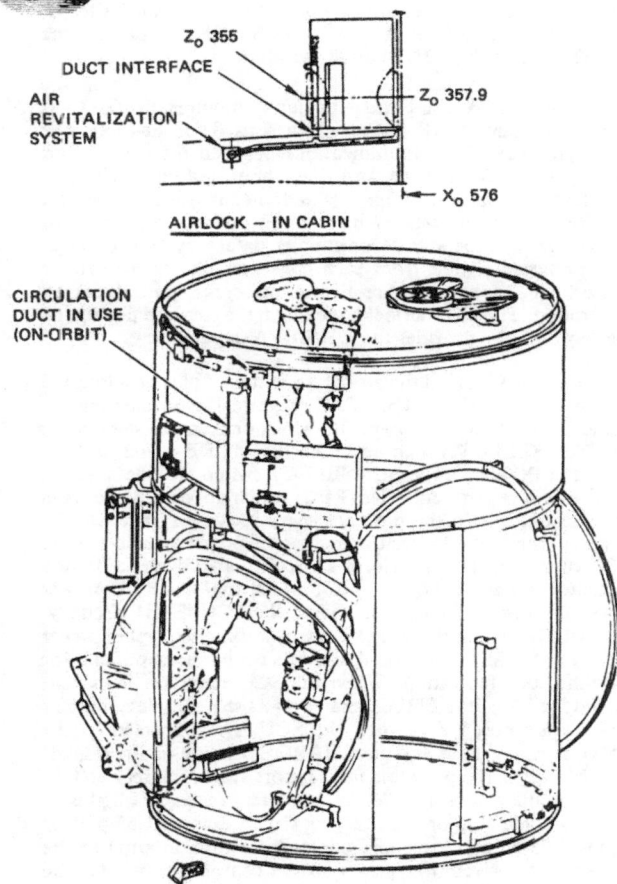

Airlock With Tunnel Adapter For Spacelab

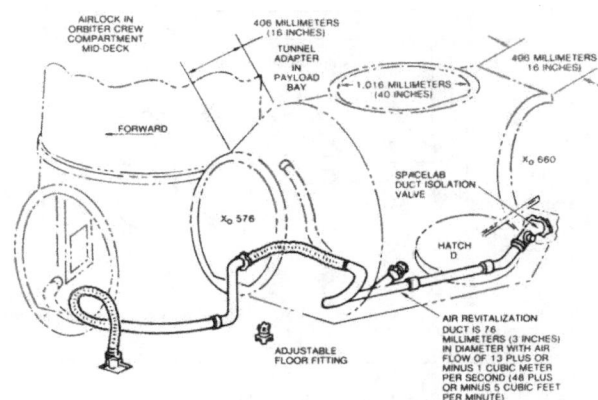

Environmental Control Life Support System (ECLSS) Air Circulation Duct Routing

The upper (EVA) hatch in the tunnel adapter opens and closes to the port (left) wall of the tunnel adapter. The hatch is hinged to be first pulled into the tunnel adapter and then pulled forward at the hinge area and rotated down until it rests against the port wall of the tunnel adapter. The linkage mechanism guides the hatch from the closed/open, open/closed position with friction restraint throughout the stroke. The hatch is held in the open position by straps and Velcro. The hatches can be removed in-flight from the hinge mechanism via pip pins, if required.

The spacecraft environmental control life support system (ECLSS) provides conditioned air to the airlock, tunnel adapter, and tunnel during non-EVA operation periods. Upon airlock hatch opening in-flight, the duct is attached to the spacecraft ECLSS. The duct must be disconnected from the spacecraft ECLSS prior to closing the airlock hatch.

Space Shuttle Spacecraft Systems

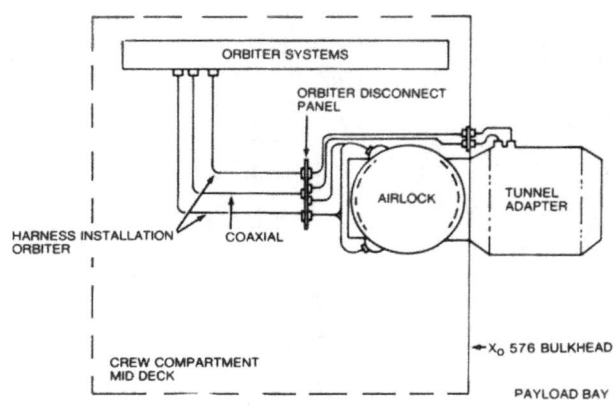

Airlock/Tunnel Adapter Harness Installation and Coaxial Antenna

Airlock communications are provided with the orbiter audio system at airlock panel AW82D where connectors for the headset interface units (HIU's) and the EMU's are located at airlock panel AW18D which is the airlock audio terminal (ATU). The HIU's are inserted in the crew-member communications carrier unit (CCU1 and CCU2) connectors on airlock panel AW82D. The CCU's are also known as the "Snoopy Cap" which fits over the crew member's head and snaps into place with a chin guard. It contains a microphone and headphones for two-way communications and receiving caution and warning tone. The adjacent two-position switches labeled CCU1 and CCU2 POWER enable transmit functions only, as reception is normal as soon as the HIU's are plugged in. The EMU1 and EMU2 connectors on the same panel to which the service and cooling umbilical (SCU) is connected include contacts for EMU hard-line communications with the orbiter prior to EVA. Panel AW18D contains displays and controls used to select access to and control volume of various audio signals. Control of the airlock audio functions can be transferred to the mid-deck ATU's panel M042F, by placing the CONTROL knob to MIDDECK position.

During EVA, the Extravehicular Communicator (EVC) is part of the same UHF system which is used for air-to-air and air-to-ground voice communications between the orbiter and landing site control tower and the orbiter and chase aircraft. The EVC provides full duplex (simultaneous transmission and reception) communications between the orbiter and the two EVA crew members and continuous data reception of electrocardiogram signals from each crew member by the orbiter and orbiter processing and relay of electrocardiogram signals to the ground. The UHF airlock antenna in the forward portion of the payload bay provides the UHF-EVA capability.

Panel AW18H in the airlock provides 17 plus or minus 0.5 vdc at five amperes at both EMU electrical connector panels, panel AW82D, in EVA prep. Bus MNA or B can be selected on the BUS SELECT switch and then the MODE switch is positioned to POWER. The BUS SELECT switch provides a signal to a remote power controller (RPC) which applies 28 vdc from the selected bus to the power/battery recharger. The MODE switch in the POWER position makes the power available at the SCU connector and also closes a circuit that provides a battery feedback voltage charger control which inhibits EMU power when any discontinuity is sensed in the SCU/EMU circuitry. The MODE switch in the POWER position also applies power through the SCU for the EMU microphone amplifiers for hardline communication. When the SCU umbilical is disconnected for EVA, the EMU operates on its self contained battery power. For post-EVA, when the SCU is reconnected to the EMU, selecting a bus and the CHARGE position on the MODE switch charges the portable life support system battery at 1.55 plus or minus 0.05 amps. When the battery reaches 21.8 plus or minus 0.1 vdc and/or the charging circuit exceeds 1.55 plus or minus 0.05 amps, a solenoid controlled switch internal to the battery charger removes power to the charging circuitry. The

Space Shuttle Spacecraft Systems

EMU silver zinc battery provides all electrical power used by the portable life support system during EVA and is filled with electrolyte and charged prior to flight.

Cooling for the flight crew members before and after the EVA is provided by the liquid cooled garment circulation system via the SCU and LCG (liquid cooled garment) SUPPLY AND RETURN connections on panel AW82B. These connections are routed to the orbiter liquid cooled garment heat exchanger which transfers the collected heat to the orbiter Freon-21 coolant loops. The nominal loop flow of 113 kilograms per hour (250 pounds per hour) is provided by the EMU/portable life support system water loop pump. The system circulates chilled water at 10 degrees Celsius (50 °F) maximum to the liquid cooled ventilation garment inlet and provides a heat removal capability of 2,000 Btu (British Thermal units) per hour per crew member. When the SCU is disconnected the portable life support system provides the cooling. Upon return from the EVA, the portable life support system is reconnected to the SCU and the crew member cooling is provided as it was in the EVA prep.

With the suit connected to the SCU, oxygen at 15,525 to 46,575 mmHg (300 to 900 psia) is supplied through airlock panel AW82B from the orbiter oxygen system when the OXYGEN valve is in the OPEN position on the airlock panel. This provides the suited crew member with breathing oxygen, preventing depletion of the portable life support system oxygen tanks prior to the EVA. Prior to the crew member sealing the helmet, an oxygen purge adapter hose is connected to the airlock panel to flush nitrogen out of the suit.

The crew member will prebreathe pure oxygen in the EMU for approximately 3 and one-half hours prior to the EVA. This is necessary to remove nitrogen from their blood before working in the pure oxygen environment of the EMU due to the orbiter pressurized crew cabin mixed gas atmosphere of 20 percent oxygen and 80 percent nitrogen at a pressure of 760 plus or minus 10 mmHg (14.7 plus or minus 0.2 psia). Without prebreathing, bends occur when an individual fails to reduce nitrogen levels in the blood prior to working in a pressure condition that can result in nitrogen coming out of solution in the form of bubbles in the bloodstream. This condition results in pain in the body joints, possibly because of restricted blood flow to connective tissues or the extra pressure caused by bubbles in the blood at joint area. During prebreathe, the suit is at 77 mmHg (1-1/2 psig).

When the SCU is disconnected, the portable life support system provides oxygen for the suit. When the EVA is completed and the SCU is reconnected, the orbiter oxygen supply begins recharging the portable life support sytem, providing the OXYGEN valve on panel AW82B is OPEN. Full oxygen recharge takes approximately one hour (allowing for thermal expansion during recharge) and the tank pressure is monitored on the EMU display and control panel as well as on the airlock oxygen pressure readout.

Each EMU is pressurized to 207 mmHg (4.0 psid) differential. They are designed for a 15 year life with cleaning and drying between flights.

The EMU WATER SUPPLY and WASTE valves are opened during the EVA prep by switches on panel AW82D. This provides the EMU, via the SCU, access to both the orbiter potable water and waste water systems. The support provided to the EMU portable life support system is further controlled by the EMU display and control panel. Potable water — supplied from the orbiter at 828 plus or minus 25 mmHg (16 plus or minus 0.5 psi), 45 to 58 kilograms per hour (100 to 300 pounds per hour), and 4 to 37 °C (40 to 100 °F) — allowed to flow to the feedwater reservoir in the EMU which provides pressure which would "top-off" any tank not completely filled. Waste water, condensate, developed in the portable life support system is allowed to flow to the orbiter waste water system via the SCU whenever the regulator connected at the bacteria filters (airlock

 **Space Shuttle Spacecraft Systems**

end of the SCU) detects upstream pressure in excess of 828 plus or minus 25 mmHg (16 plus or minus 0.5 psi).

When the SCU is disconnected from the EMU, the portable life support system assumes this function. When the SCU is reconnected to the EMU upon completion of the EVA, the same functions as in pre-EVA are performed except that the water supply is allowed to continue until the portable life support system water tanks are filled, which takes approximately 30 minutes.

In preparation for the EVA from the airlock, all hatches are closed and depressurization of the airlock begins.

Airlock/tunnel adapter depressurization is accomplished by a three position valve located on the ECLSS (Environmental Control Life Support System) panel AW82A in the airlock. The airlock depressurization valve is covered with a pressure/dust cap. Prior to removing the cap from the valve, it is necessary to vent the area between the cap and valve by pushing the vent valve on the cap. In-flight storage of the pressure/dust cap is adjacent to the valve. The airlock depressurization valve is connected to a 50 millimeter (2 inch) inside diameter stainless steel overboard vacuum line. The AIRLOCK DEPRESS valve controls the rate of depressurization by varying the valve diameter size. Depressurization is accomplished in two stages. The CLOSED position prevents any airflow from escaping to the overboard vent system.

When the crew members have completed the prebreathe in the EMU's for 3.5 hours, the airlock/tunnel adapter is depressurized from 760 mmHg (14.7 psia) to 258 mmHg (5 psia) by position labeled "5" on the AIRLOCK DEPRESS valve which opens the depressurization valve and allows the pressure in the airlock to decrease until the flight crew closes the valve at 258 mmHg (5 psia). Pressure during depressurization can be monitored by the delta pressure gage on the airlock hatch. A delta pressure gage is installed on each side of the hatches.

At this time the flight crew performs an EMU suit leak check, electrical power is transferred from the umbilicals to the EMU batteries, the umbilicals are disconnected and the suit oxygen packs are brought on line.

The second stage of airlock depressurization is accomplished by positioning the AIRLOCK DEPRESS valve closed to "0" which increases the valve diameter and allows the pressure in the airlock to decrease from 258 mmHg (5 pisa) to 0 mmHg (0 psia). The depressurization of airlocks/tunnel adapter is accomplished within 18 minutes at rates no more than 5.1 mmHg (0.1 psia) per second during normal operations. The suit sublimators are activated for cooling, EMU system checks are performed and the airlock/payload bay hatch can be opened. The hatch is capable of opening against a 10 mmHg (0.2 psia) differential maximum.

Hardware provisions are installed in the orbiter payload bay, and on the tunnel adapter, tunnel and Spacelab for use by the crew member during the EVA.

Handrails and tether points are located on the payload bulkheads, forward bulkhead station X_O 576 and aft bulkhead station X_O 1307, and along the sill longeron on both sides of the bay to provide translation and stabilization capability for the EVA crew member. The handrails are designed to withstand a load of 90.72 kilograms (200 pounds), 127.01 kilograms (280 pounds) maximum in any direction. Tether attach points are designed to sustain a load of 260.37 kilograms (574 pounds), 364.69 kilograms (804 pounds) maximum, in any direction.

The handrails have a cross section of 33 by 19 millimeters (1.32 by 0.75 inches). They are made of aluminum alloy tubing and are painted yellow. The end braces and side struts of the handrails are constructed of titanium. An aluminum alloy end support standoff functions as the terminal of the handrail. Each end support standoff incorporates a 25.4 millimeter (one inch) diameter tether point.

Space Shuttle Spacecraft Systems

A 7.62 meter (25 foot) crew member safety tether is attached to each crew member at all times during an EVA.

The tether consists of a reel case with an integral "D" ring, a reel with a light takeup spring, a cable and a locking hook. The safety tether hook is locked onto the slidewire before launch and the cable is routed and clipped along the port (left) and starboard (right) handrails to a position just above the airlock/payload bay hatch. After opening the airlock hatch and before egress, the crew member attaches a waist tether to the "D" ring of the safety tether to be used. The other end of the waist tether is hooked to a ring on the EMU waist bearing. The crew member may select either the port or the starboard safety tether. With the selector on the tether in the locked position, the cable will not retract or reel out. Moving the selector to the unlocked position allows the cable to reel out and the retract feature to take up slack. The cable is designed for a maximum load of 398 kilograms (878 pounds). The routing of the tethers follows the handrails, allowing the crew member to deploy and restow his tether during translation.

The two slidewires, approximately 14.11 meters (46.3 feet) long, are located in the longeron sill area on each side of the payload bay. They start approximately 2.83 meters (9.3 feet) aft of the forward bulkhead and extend approximately 14.11 meters (46.3 feet) down the payload bay. The slidewires withstand a tether load of 260.37 kilograms (574 pounds) with a safety factor of 1.4 or 364.49 kilograms (804 pounds) maximum.

The airlock/cabin hatch has two pressure equalization valves which can be operated from both sides of the hatch for repressurizing the airlock volume. Each valve has three positions, CLOSED, NORM (Normal), and EMERG (Emergency) and is protected by a debris pressure cap on the intake (high-pressure) side of the valve, which on the other two hatches must be vented for removal. The caps are tethered to the valves and also have small Velcro spots which allow temporary stowage on the hatch. The exit side of the valve contains an air diffuser to provide uniform flow out of the valve.

Through the use of the equalization valve/valves in the various positions, the airlock can be repressurized in a normal mode to 760 mmHg (14.7 psia) within 13 minutes at rates no more than 5.1 mmHg (0.1 psia) per second during normal operations. If both equalization valves are positioned to EMERG, the airlock/tunnel adapter can be repressurized to 760 mmHg (14.7 psia) in 65 plus or minus 5 seconds at rates no more than 51.75 mmHg (1.0 psi) per second. The hatch is capable of opening against a 10 mmHg (0.2 psia) differential maximum.

The airlock is initially pressurized to 258 mmHg (5 psia) and the umbilicals are connected and electrical power is transferred back to umbilical power. The airlock is then pressurized to equalize with the cabin pressure, followed by EMU doffing and the crew members' recharge of the EMU's.

The orbiter provides accommodations for three two-flight-crew member EVA's of six-hour duration per flight at no weight or volume cost to the payload. Two of the EVA's are for payload support and the third is reserved for orbiter contingency. Addtional EVA's can be considered with consumables charged to payloads.

PERSONAL EGRESS AIR PACKS

The portable oxygen system used in the spacecraft are Personal Egress Air Packs (PEAP's).

The PEAP's are designed to be used with the Launch Entry Helmet (LEH) that will be worn by each flight crew member. The LEH's are used during launch and entry and may be used for EVA prebreathing to denitrogenize the flight crew member's circulatory system.

Space Shuttle Spacecraft Systems

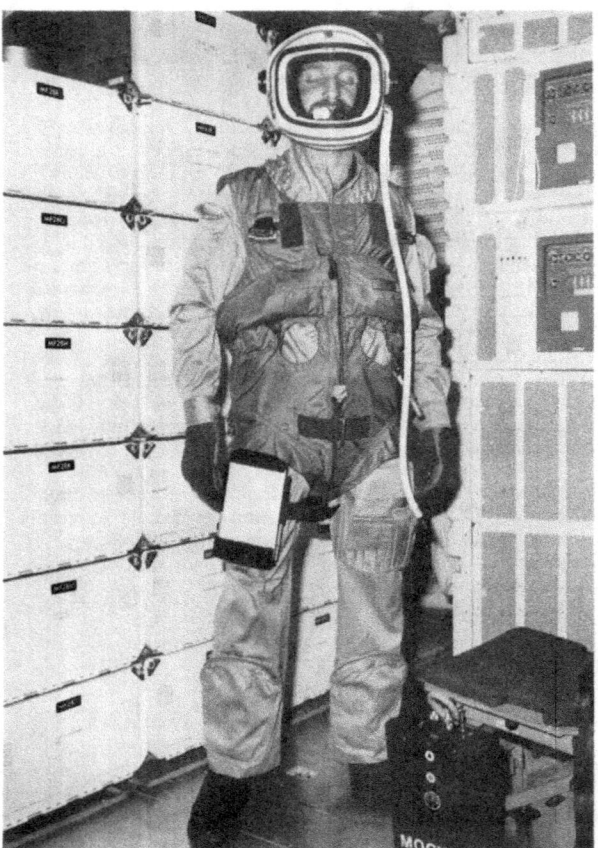

Integrated Harness Restraint, Launch and Entry Helmet, and Personal Egress Air Pack

The PEAP located at each flight crew member's seat is supplied with oxygen from the atmospheric revitalization system (ARS). Two oxygen regulators are installed to reduce the oxygen pressure to 5,175 millimeters of mercury (mmHg) (100 psi) for each PEAP. Oxygen is supplied from the regulator to each PEAP through quick disconnect flexible hoses. Each PEAP will supply that crew member with an approximate six minute walk-around capability when disconnected from the oxygen supply system. There are 10 oxygen system locations available: two in the airlock, four on the mid deck ceiling, and four on the aft center console.

Before the EVA's, the crew member must be denitrogenized to prevent the bends when EVA's are begun in the 212 mmHg (4.1-psi) EMU suit.

ANTIGRAVITY SUIT

The AGS prevents pooling of body fluids and aids in maintaining circulating blood volume during entry. Pooling of body fluids can occur when high "g" loads are imposed on the body. It is particularly noticeable after crew members have had more than three days of zero "g" activity.

The bladders in the AGS receive oxygen from the power reactant storage distribution cryogenic system, or the ECLSS ARS emergency oxygen system, if installed. When the bladders of the AGS are activated, pressure is applied to the crew member's lower extremities and to the abdomen to prevent the pooling of body fluids.

The AGS weighs 2.2 kilograms (5 pounds).

Contractors involved with the ECLSS are Hamilton Standard Division of United Technologies Corp., Windsor Locks, CT (atmospheric revitalization, Freon-21 coolant loops, heat exchangers, cabin fan assembly, debris trap, CO_2 absorber, humidity control heat exchanger, avionics fan, accumulators,

Space Shuttle Spacecraft Systems

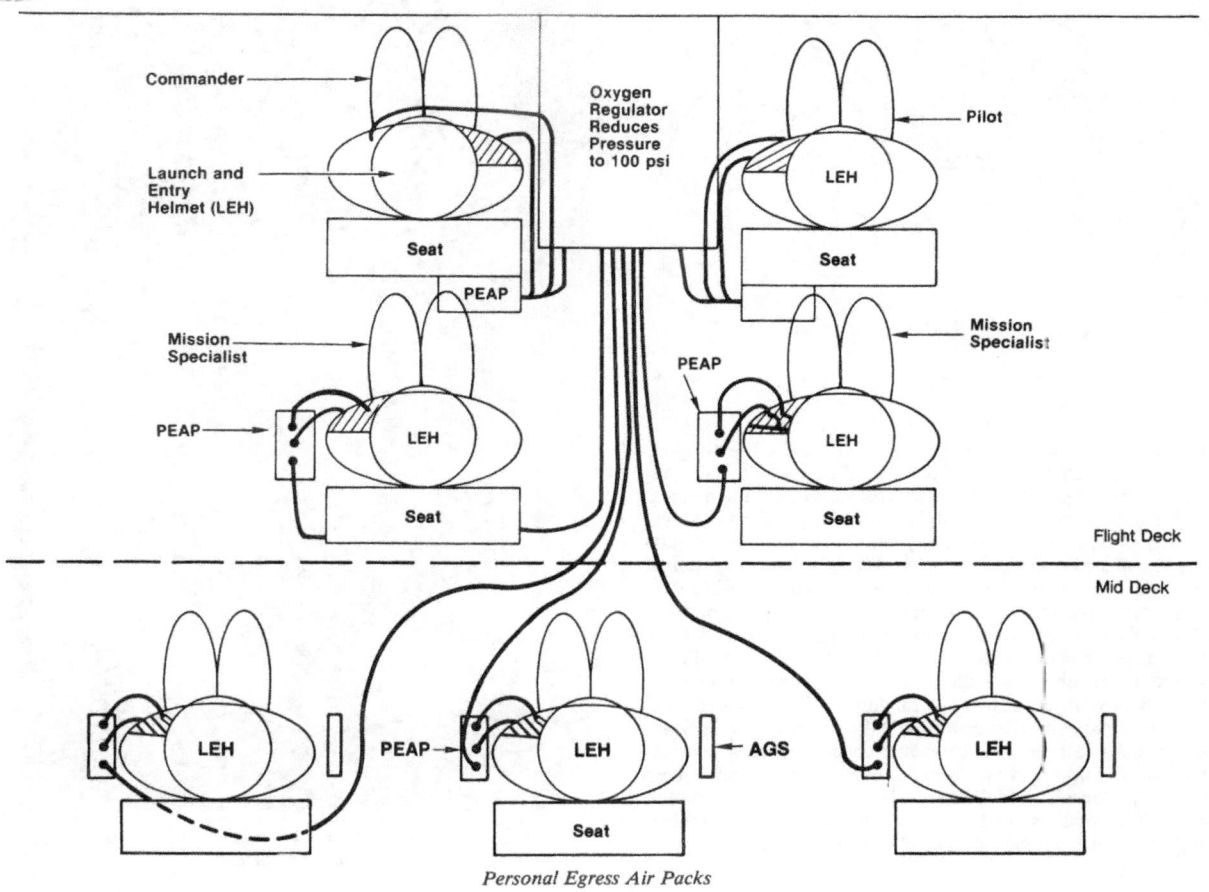

Personal Egress Air Packs

Space Shuttle Spacecraft Systems

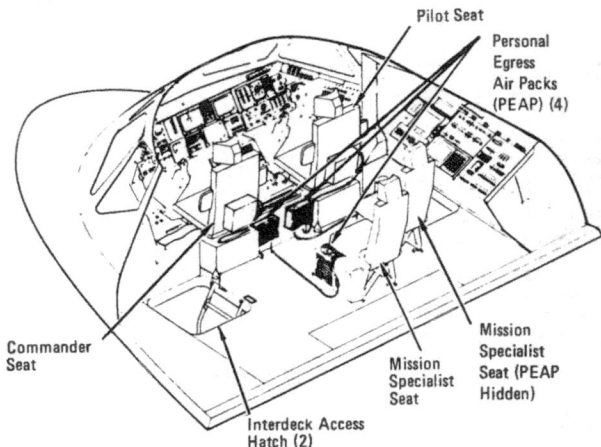

Personal Egress Air Packs

flash evaporators, water management panel EVA life support system and EMU's); Carlton Controls, East Aurora, NY (atmospheric revitalization pressure control subsystem and airlock support components); Aerodyne Controls Corp., Farmingdale, NY (water pressure relief valve, oxygen check valve); Aeroquip Corp., Marman, Los Angeles, CA (couplings, clamps, retaining straps, and flexible air duct); AiResearch Manufacturing Co. Garrett Corp., Torrance, CA (ground coolant unit); Anemostat Products, Scranton, PA (cabin air diffuser); Arrowhead Products Division of Federal Mogul, Los Alamitos, CA (couplings, flex air duct, flexible connector, connector drain system convoluted bellows); Brunswick, Lincoln, NE (atmospheric revitalization oxygen, nitrogen tanks); Brunswick, Circle Seal, Anaheim, CA (water relief valve, water check valve); Brunswick Wintec, El Segundo, CA (water relief valve, water check valve, water filter); Consolidated Controls, El Segundo, CA (unidirectional/bidirectional shutoff valve, water solenoid latching valve); Cox and Co., New York, NY (water relief valve, vent

Personal Egress Air Pack Launch Entry Helmet Anti' "G" Suit for Entry

Space Shuttle Spacecraft Systems

nozzle and port heater, water boiler steam vent line heater); Dynamic Corp., Scranton, PA (cabin diffuser); Fairchild Stratos, Manhattan Beach, CA (ammonia boiler); General Electric, Valley Forge, PA (waste collector); Metal Bellows Co., Chatsworth, CA (potable and waste water tanks, flex metal tubes); RDF Corp., Hudson, NH (temperature sensor/transducer); Symetrics, Canoga Park, CA (Freon fluid disconnects, water boiler quick disconnects); Seaton Wilson, Inc., subsidiary of Systron-Donner, Burbank, CA (water and coolant system quick disconnects); Tavis Corp., Mariposa, CA (Freon flow meter); Tayco Engineering, Long Beach, CA (urine, waste water, O_2, N_2 waste dump); Titeflex Division, Springfield, MA (water coolant flex line); Vacco Industries, El Monte, CA (potable water inline pressure relief valve); Vought Corp., Dallas, TX (radiators and flow control assembly).

AUXILIARY POWER UNIT

The auxiliary power unit (APU) is a hydrazine-fueled turbine-driven power unit and generates the mechanical shaft power to a pump that produces pressure for the orbiter's hydraulic system. There are three separate APU's, three hydraulic pumps, and three hydraulic systems, all located in the aft section of the spacecraft.

The APU's and their fuel systems are isolated from each other. The three independent hydraulic systems are connected to the main engine thrust vector control (TVC) switching valve actuators and the aerosurface actuator switching valves. Thus a single system failure will not affect full operational performance. If two systems fail, the third system will provide sufficient hydraulic power to operate all actuators at a reduced rate.

The three APU's/hydraulic pumps operate during launch and boost. They provide the three main engines with propellant valve control, thrust vector control by hydraulically gimbaling the three main engines, and control of the orbiter hydraulic actuators of the aerosurface elevons for aerodynamic elevon load relief during boost.

The APU's and pumps are restarted prior to deorbit and operate continuously through landing and rollout for hydraulic positioning of orbiter aerosurfaces (elevons, rudder/speed brake, body flap) during the atmospheric flight portion of entry and to provide hydraulics for nose and main gear deployment, nose gear wheel steering, and main landing gear braking.

Except for on-orbit checkout, the APU/hydraulic pumps are dormant during the orbital flight phase.

The three APU's are located in the orbiter aft fuselage. Each system consists of a fuel tank, a fuel feed system, an APU and controller, an exhaust duct, lube oil cooling system, and fuel/lube oil vents and drains. Redundant electrical heater systems and insulation thermally control the system above 7 °C (45 °F) to prevent fuel freezing and provide required oil viscosity. The insulation is used to minimize the electrical heater size and to control high surface temperatures to safe limits on the turbine and exhaust ducts.

Each APU fuel system provides fuel to its respective fuel pump and control valves, then to the gas generator. Gas generator catalytic action decomposes the fuel and the resultant hot gas drives the APU two-stage turbine. The APU turbine assembly provides mechanical power to the APU gearbox which drives the APU fuel pump, hydraulic pump, and oil lube pump. The APU lube oil system is circulated through a lube oil heat exchanger in the APU/hydraulic water spray boiler to cool the lube oil system. The turbine exhaust of each APU flows over the exterior of the gas generator exterior, cooling it, and then is

 **Space Shuttle Spacecraft Systems**

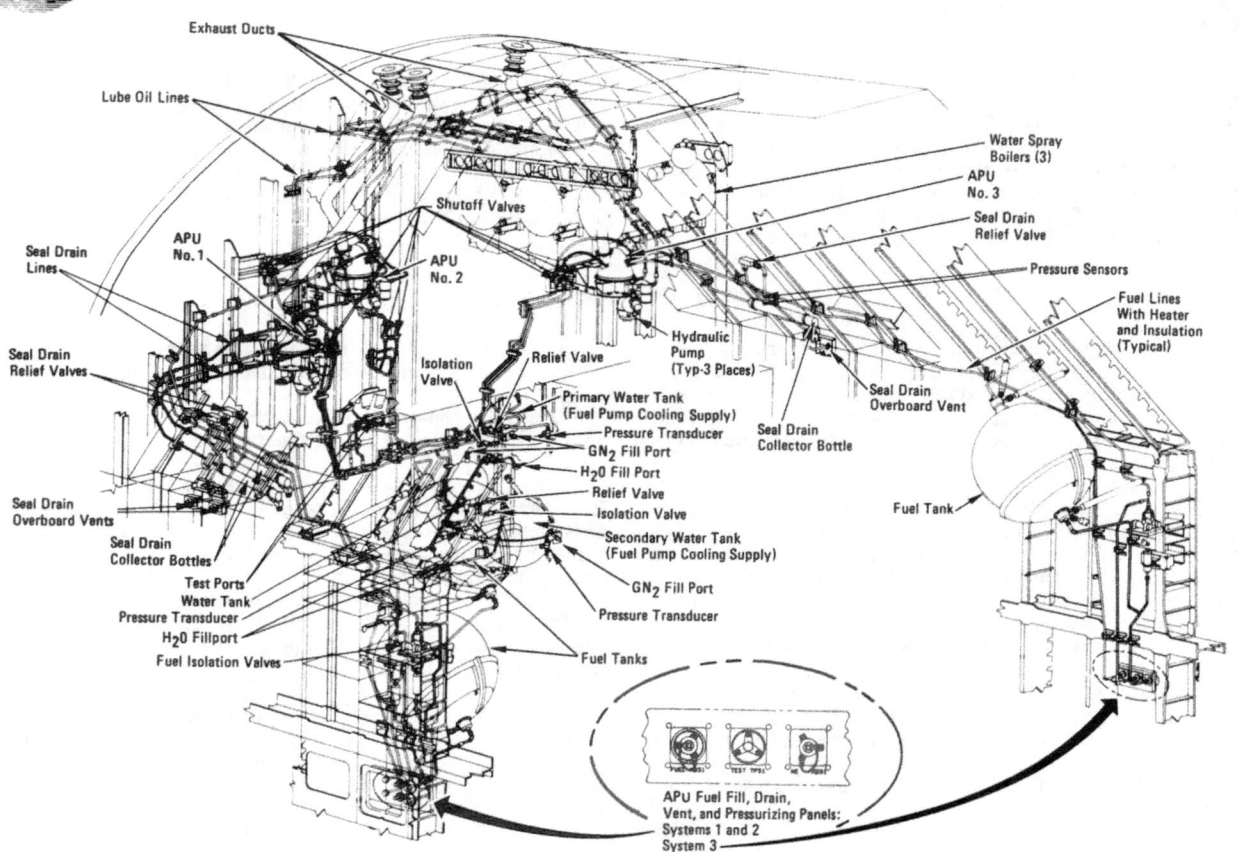

Auxiliary Power Unit/Water Boilers

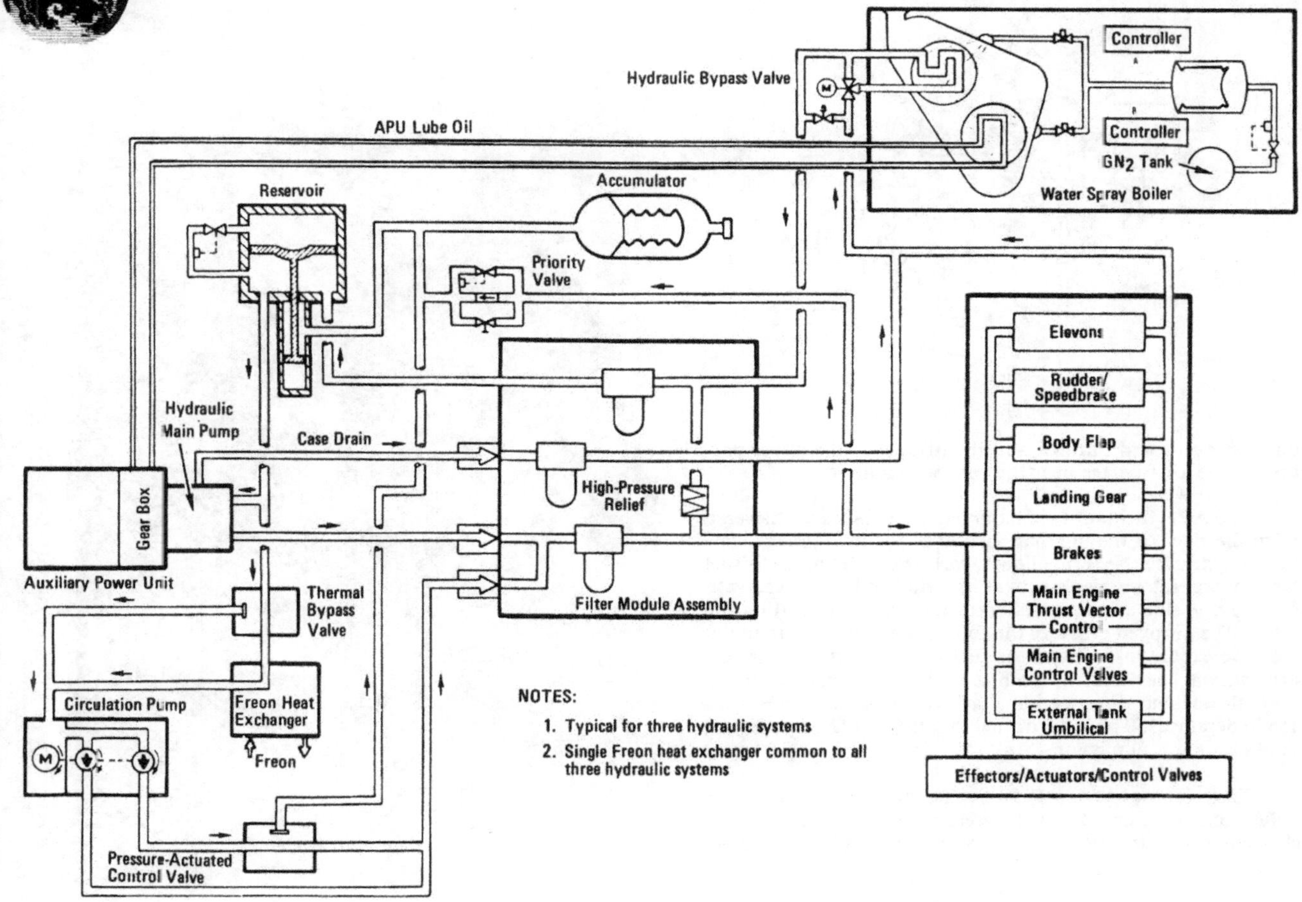

Space Shuttle Spacecraft Systems

APU, Water Spray Boiler, and Hydraulic System Schematic

Space Shuttle Spacecraft Systems

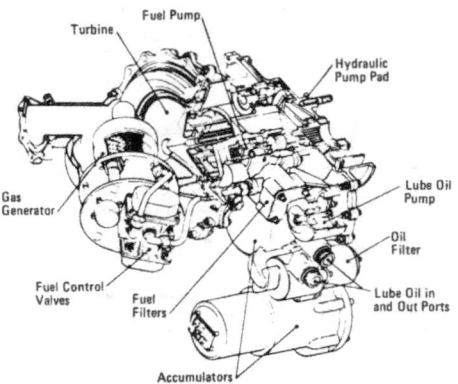

Auxiliary Power Unit

directed overboard through an exhaust duct at the upper portion of the aft fuselage near the vertical stabilizer.

The APU fuel tanks are mounted on supports cantilevered from the sides of the internal portion of the aft fuselage. The fuel is hydrazine (N_2H_4), a liquid, earth-storable fuel. The fuel tank incorporates a diaphragm at its center and is serviced with fuel on one side and the pressurant (gaseous nitrogen) on the other. The nitrogen provides the force acting on the diaphragm (positive expulsion) to expel the fuel from the tank to the fuel distribution lines and maintain a positive fuel supply to the APU throughout its operation. Each fuel tank is serviced with 158 kilograms (350 pounds) of fuel to satisfy APU operating requirements for all missions.

Each of the three APU's is 508 millimeters wide, 553 millimeters high, and 457 millimeters deep (20 x 21.80 x 18 inches). Each of the three APU controllers is 152 millimeters

Auxiliary Power Unit

Space Shuttle Spacecraft Systems

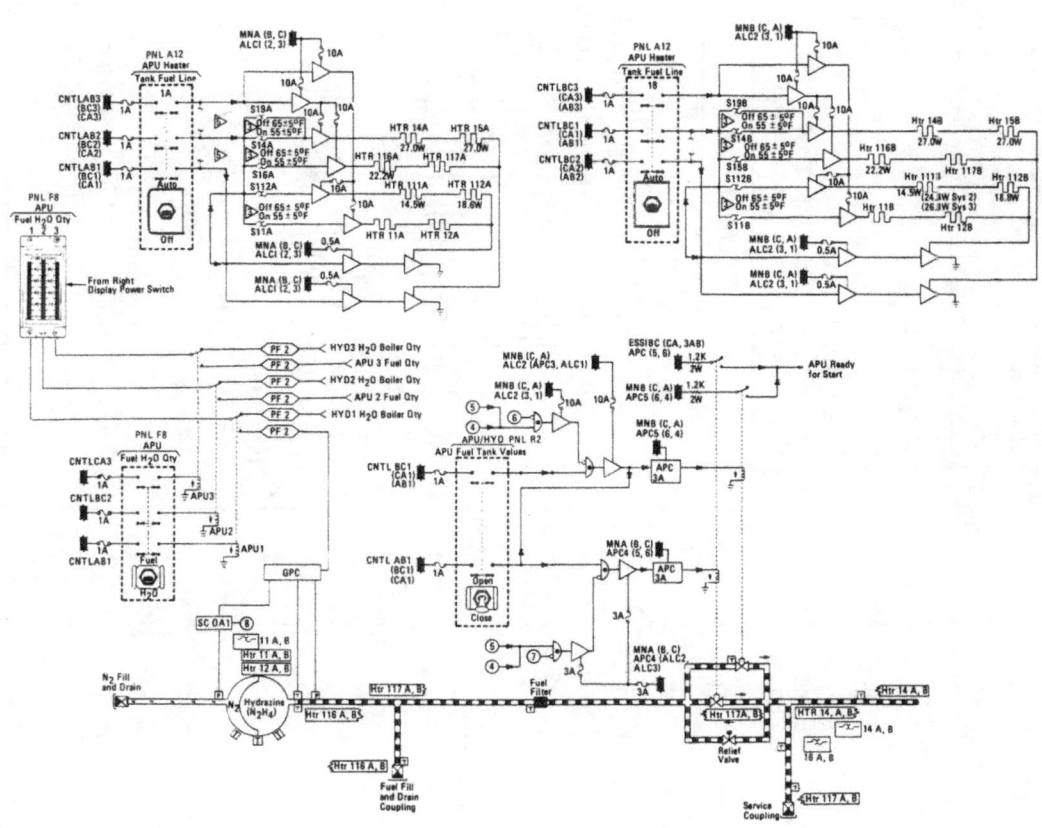

APU Heater System and APU Fuel System

Space Shuttle Spacecraft Systems

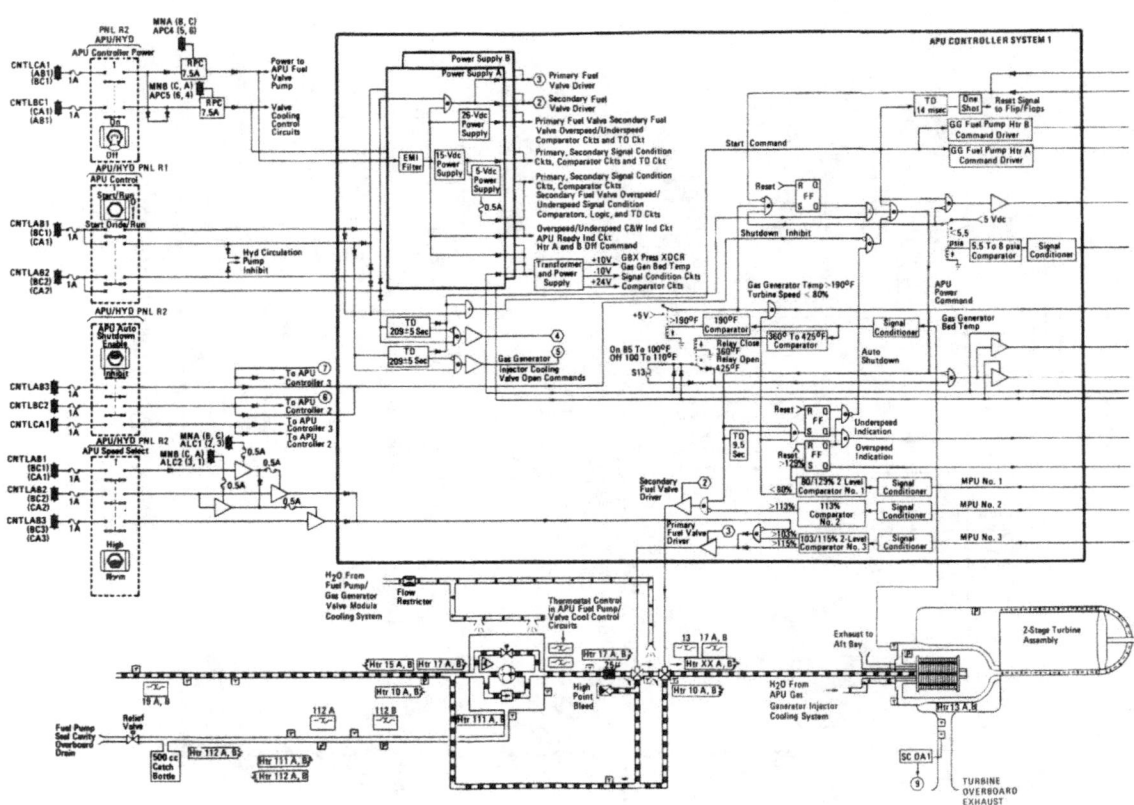

APU Controller System 1 and APU Fuel System

Space Shuttle Spacecraft Systems

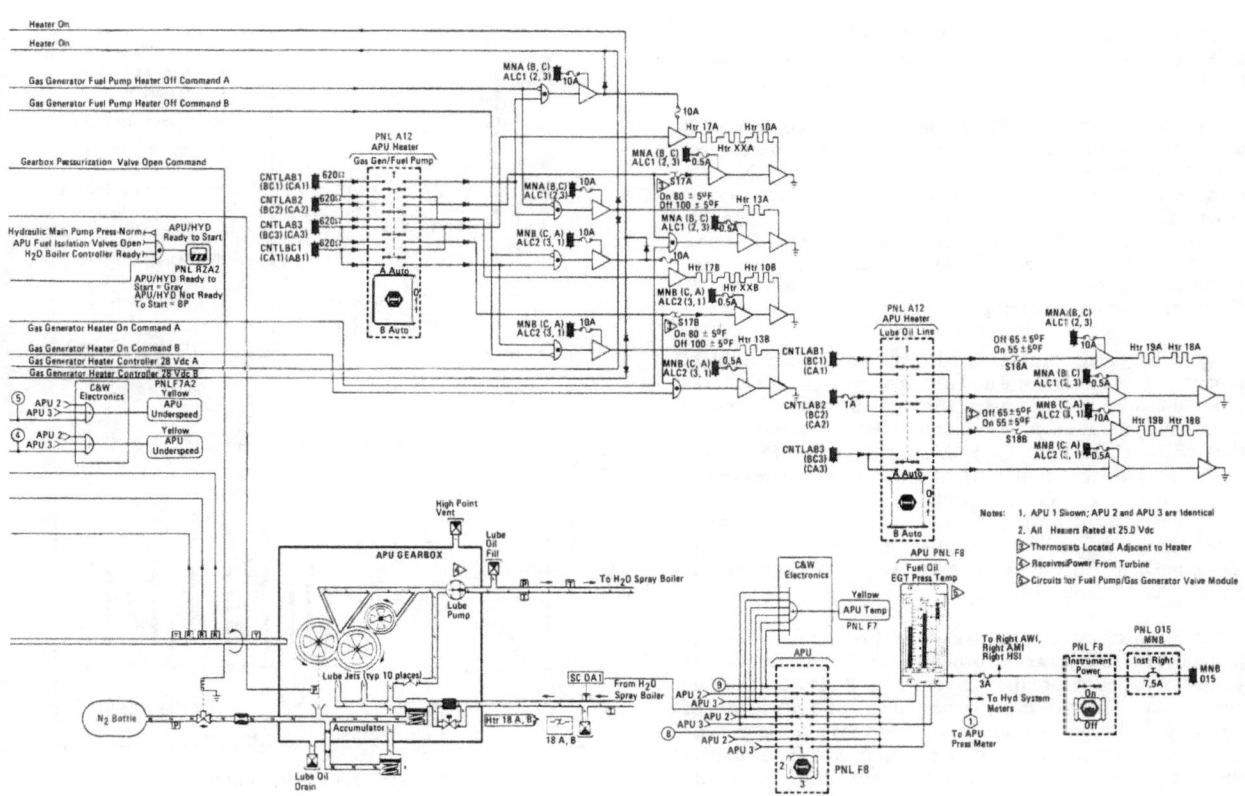

APU Heater System and APU Gearbox

Space Shuttle Spacecraft Systems

wide, 193 millimeters high, and 482 millimeters long (6 x 7-1/2 x 19 inches).

The rated horsepower of each APU is 135. The weight of each is approximately 39 kilograms (88 pounds), with its controller weighing approximately 6.8 kilograms (15 pounds).

The fuel tanks are 711-millimeter (28-inch) diameter spheres. Fuel tanks No. 1 and No. 2 are located on the left (minus Y) side of the orbiter's aft fuselage, and tank No. 3 is located on the right (plus Y) side. The fuel tanks are serviced through fill and drain couplings bolted to the corresponding APU servicing panel which is located on the side of the fuselage. The nitrogen service connections are located on the same panel. The initial gaseous nitrogen pressure is 17,077 mmHg (millimeters of mercury) (330 psi).

All APU system controls and displays are located on flight deck panels. The temperature and nitrogen pressure in each fuel tank is monitored and processed through the orbiter's system management general-purpose computer (GPC) and transmitted to the APU FUEL/H$_2$O QTY display on Panel F8. If the display select switch is in the FUEL position, the quantities in APU fuel tanks 1, 2, and 3 are displayed simultaneously on the display meter in percent (%). (When switch and display nomenclature is printed in all caps—e.g., APU FUEL/H$_2$O QTY—it indicates that it is the exact way it appears on the display and control panel.)

The fuel distribution system supplies the fuel from the fuel tank to the APU. Filters are incorporated in the distribution line to remove particles before the fuel arrives at the APU. The fuel distribution line branches into two parallel paths downstream of the fuel tank and filter. An isolation valve is installed in each of these paths, which converge into a single path immediately downstream of the valves. These valves are installed to isolate

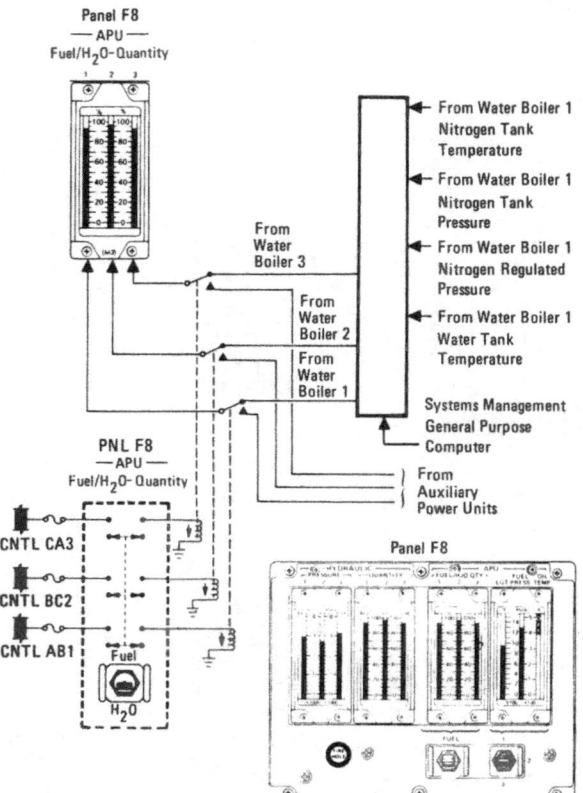

APU and Hydraulic Indicator

Space Shuttle Spacecraft Systems

the fuel tank supply from the APU. The valves are controlled by the applicable system APU FUEL TK VLV switch located on Panel R2. The fuel tank isolation valves are open for APU operation and closed when the APU is not in operation. The valve internal design contains a reverse relief feature to allow relief of pressure trapped in the line when both APU fuel tank valves are closed. The relief feature will function if the downstream pressure increases to a pressure range of 20,700 to 25,875 mmHg (400 to 500 psi).

The fuel/nitrogen fill, drain/vent, and test point couplings permit servicing and ground checkout of the fuel distribution system. The lube oil system couplings permit servicing and checkout of the lube oil system.

The APU CONTROL POWER switch for each APU also is located on Panel R2. The switch applies or removes 28 Vdc (volts direct current) power to individual APU controllers. Each APU controller provides the checkout logic prior to starting the APU, detects malfunctions, and controls the APU turbine speed, gearbox pressurization, and fuel pump/gas generator heaters when the APU is not operating. In the OFF position, power is removed from the controller.

The APU HEATER GAS GEN/FUEL PUMP switch for each APU is located on Panel A12. When the switch is positioned to AUTO A or AUTO B, electrical power is provided to the fuel pump and gas generator heaters for each APU. The fuel pump temperature is controlled automatically by thermostats which prevent the fuel from freezing. The thermostats are set to cycle on/off between 26 and 38 °C (80-100 °F). The gas generator heaters are on when the switch is in AUTO A or AUTO B prior to the start of the APU. When the APU is initiated, the APU controller automatically turns off both gas generator and fuel pump heaters. The OFF position removes power from heater circuits.

The APU/HYD READY TO START indicator for each APU is located on Panel R2. The indicator shows gray when that APU is ready to start; that is, when the APU gas generator temperature is above 87 °C (190 °F), turbine speed less than 80%, APU gearbox pressure above 284 mmHg (5.5 psi), H_2O boiler controller ready, APU fuel isolation valves open, and hydraulic main pump depressurized. When the APU is initiated, and the turbine speed is greater than 80 percent of normal speed, the indicator will show a barberpole pattern.

The APU CONTROL START/RUN switch for each APU is located on Panel R2. When the switch is in the START/RUN position, the APU controller is activated, and the fuel pump and gas generator heaters are turned off. The APU is then started by the APU controller. The OFF position removes the start signal from the APU controller.

To start, the APU controller opens the normally closed secondary valve shutoff of the gas generator valve module and fuel flows through the fuel pump bootstrap start bypass circuit (the primary fuel control valve of the gas generator valve module was already in the open position) to the APU gas generator. Gas generator catalytic action decomposes the fuel, creates a hot gas, and feeds the hot gas exhaust product to the APU turbine. The APU two-stage turbine assembly provides mechanical power to the APU gearbox and drives the APU fuel pump. The APU turbine must come up to speed in 9.5 seconds or the APU will automatically shut down. Because of the gearbox driving the APU fuel pump, the pump increased the fuel pressure at its outlet and sustains pressurized fuel to the gas generator valve module and gas generator.

The APU SPEED SELECT switch for each APU (Panel R2) selects the speed at which each APU controller operates its APU. The NORM position controls APU speed at 74,160 rpm, 103 plus or minus 8%. The HIGH position controls the APU speed at 81,360 rpm, 113 plus or minus 8%, with a second backup at 82,800 rpm, 115 plus or minus 8%.

When the upper APU turbine speed is reached, the normally open port of the primary fuel control valve of the gas

Space Shuttle Spacecraft Systems

generator valve module assembly closes, and the normally closed port of the control valve opens, allowing the fuel to bypass the fuel pump inlet. A relief valve provides pressure relief for the APU fuel pump outlet pressure. The APU fuel pump operates at a nominal speed of 3,918 rpm and provides a nominal pressure of 73,485 mmHg (1,420 psi).

The APU speed is controlled in the following manner. When the lower speed limit is reached, the primary fuel control valve opens, allowing fuel to the gas generator; it closes when the upper speed limit is reached, shutting off fuel to the gas generator.

The frequency and duration of primary fuel control valve cycling is a function of the hydraulic load on the APU as well as hydraulic system dynamics.

The secondary shutoff valve of the gas generator valve module assembly provides a backup in the event of a malfunction of the primary control valve.

The APU OVERSPEED yellow caution and warning light, on Panel F7, will illuminate if APU 1, 2, or 3 turbine speed is above 92,880 rpm, 129 plus or minus 1%. The APU UNDERSPEED yellow light will illuminate if the turbine speed is less than 57,600 rpm, 80 plus or minus 3%, normal speed.

In the event of an APU overspeed or underspeed shutdown, the HYD PRESS yellow light on Panel F7 also would illuminate for the corresponding hydraulic system, since the APU driving the hydraulic pump for the system would be off; the yellow HYD PRESS light illuminates when any hydraulic system drops below 144,900 mmHg (2,800 psi).

The INHIBIT position of the APU AUTO SHUTDOWN switch on Panel R2 bypasses the automatic shutdown sequence from the APU controllers and the 9.5-second speed time delay if the APU CONTROL switch is in the START/RUN position.

The START ORIDE/RUN position of each APU CONTROL switch on Panel R2 will override the APU pre-start conditions (gas generator temperature above 87 °C [190 °F], turbine speed less than 80%, and gearbox pressure above 284 mmHg [5.5 psi]) to permit a start of the respective APU, if one or more of the prestart conditions are not met. This switch also activates the APU gas generator active cooling system which provides the capability to restart an APU. The restart is inhibited for 209 seconds after the switch is positioned, during which time the gas generator is cooled by flowing water through its cooling passages.

The APU turbine provides the mechanical drive to the APU two-stage gearbox to drive the hydraulic pump at a nominal speed of 3,918 rpm. In addition to driving the hydraulic pump, the output shaft drives the APU fuel pump and the lube oil pump.

Each APU oil lube system uses the power train gears as scavenge pumps to supply lube oil to the inlet of the lube pump. The lube oil pump increases the lube pressure to a nominal 3,984 mmHg (77 psi), directs the lube oil through the APU/hydraulic water spray boiler to cool it and returns the lube oil to the APU gearbox accumulators.

The accumulators allow thermal expansion of the lube oil, accommodate gas initially trapped in the external lube circuit, regulate lube pressure, and act as a zero-gravity, all-attitude lube reservoir. The nitrogen pressurization system for each gearbox is activated when the gearbox case pressure is below 232 mmHg (4.5 psi) plus or minus 77 mmHg (1.5 psi) by the corresponding APU controller, assuring that gearcase pressure is sufficiently above the requirements for proper scavenging and lube pump operation.

Each APU turbine exhaust flows over the exterior of the gas generator, cooling it, and then is vented overboard through its own independent exhaust duct at the upper portion of the aft

Space Shuttle Spacecraft Systems

fuselage near the vertical stabilizer. Overtemperature detection of each APU exhaust is monitored on display and control Panel F8.

Each APU turbine exhaust gas temperature, lube oil temperature, and fuel pressure is routed to Panel F8. The APU SELECT switch position 1, 2, or 3 permits the respective APU exhaust gas temperature, fuel pressure, and lube oil temperature to be displayed on the APU EGT (exhaust gas temperature), FUEL PRESS, and OIL TEMP meter.

The APU TEMP yellow caution and warning light on Panel F7 will illuminate if APU 1, 2, or 3 exhaust gas temperature is above 682 °C (1,260 °F) or if APU 1, 2 or 3 lube oil temperature is above 143 °C (290 °F).

The APU HEATER TANK FUEL LINE switches on Panel A12 for each APU provide the operation of the thermostatic controlled heaters located on the respective APU fuel tank fuel lines. The thermostats maintain the temperature between a nominal 12 and 18 °C (55 to 65 °F).

The heaters for the gas generator and the fuel pump and gas generator valve module water system are also controlled by the APU HEATER TANK FUEL LINE switches on Panel A12. Thermostats for the water systems maintain the temperature between 26 and 35 °C (80 to 90 °F).

The APU HEATER TANK FUEL LINE switches for each APU are divided into an A and B system switch for each APU. The A switch controls the A heaters and the thermostats provide automatic control. The B switch controls the B heaters and the thermostats provide automatic control. The OFF position of each switch removes power from the respective heater circuits.

The APU lube oil lines on each APU have a heater system. These heaters are controlled by a heater switch for each APU on Panel A12. When the HEATERS LUBE OIL LINE switch on APU is in the A AUTO position, the A lube oil heaters are powered and controlled automatically by thermostats between 12 and 18 °C (55-65 °F). The B AUTO heater switch position provides the same capability to the B heater system. The OFF position of each switch removes power from the heater circuits.

The APU fuel pump and gas generator valve module is cooled by water spray following APU shutdown on completion of the ascent boost phase and after orbital checkout. The water spray cooling prevents hydrazine decomposition in the APU fuel pump and valve due to heat soakback in each APU. The cooling system consists of a primary (A) and secondary (B) independent water supply to the three APU's. Each water system consists of a 419-millimeter (16.5-inch) diameter tank, a 6.35-millimeter (0.25-inch) diameter line to each APU, heaters, and control valves. Each water tank is loaded with 9.5 plus or minus 0.2 kilograms (21 plus or minus one pound) of water. Each tank is pressurized with nitrogen between 2,587 and 3,053

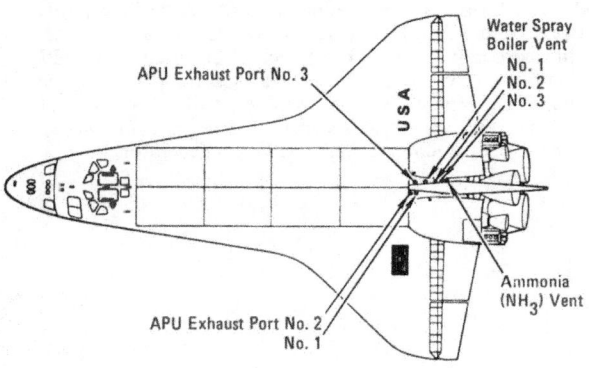

APU Exhaust and Water Boiler Vent

Space Shuttle Spacecraft Systems

mmHg (50 to 59 psi). The pressure acts on a diaphragm in each tank to expel the water into the lines to the control valves.

When the APU's are shut down, the APU FUEL PUMP/VLV COOL switch A or B on Panel R2 is positioned to AUTO. With the A switch on AUTO, the 71-76°C (160 to 170°F) thermostats on each APU, through the timer in the water controller for each APU, open control Valve A on each APU to permit water to spray onto the valve module and fuel pump for 1.25 seconds, then close Valve A for 4 seconds, etc. The cooling system is activated for two hours and 45 minutes after APU shutdown. The B switch controls Valve B in the same manner as in the A case. Nitrogen pressurization in each water tank is referred to as a blowdown system (pressure decay continues until the water is expelled from each tank). The water is exhausted into the aft fuselage compartment.

The gas generator active cooling is used only where the normal cooldown time of 180 minutes is not available. This provides cooling of the gas generator injector in each APU. The injector is cooled by circulation of water through the injector. One water tank services all three APU's. The system consists of the one water tank and 29-millimeter (1.18-inch) diameter line to each APU, heaters, and a control valve. The water tank is a 238-millimeter (9.4-inch) diameter tank loaded with 2.72 plus or minus 0.2 kilograms (6 plus or minus 0.5 pounds) of water. The tank is pressurized with nitrogen at a nominal pressure of 4,398 mmHg (85 psi) to expel the water into the lines to the control valve.

When the APU CONTROL switch on Panel R2 for each APU is positioned to START ORIDE and the temperature sensor for each APU injector is above 212°C (415°F), each APU controller opens its water control valve for 209 plus or minus 5 seconds and directs water into the gas generator injector to cool it. When the timer in the APU controller has timed out, the control valve is closed. It should be noted that the APU CONTROL switch on panel R2 for each APU will not activate the water cooling system in the START position. The water from the gas generator is exhausted into the aft fuselage.

The APU's are designed for 50 hours of maintenance-free operation. The current life of the gas generator in each APU is 20 hours, however, because of degradation of the catalyst; thus, the gas generator life is 60 hours with two refurbishments of the catalyst bed.

The contractors involved with the APU system are Sundstrand Corp., Rockford, IL (APU and APU controller); Consolidated Controls, El Segundo, CA (APU fuel isolation valve); Pressure Systems Inc., Los Angeles, CA (APU fuel tank); SSP Products, Inc., Burbank, CA (APU exhaust duct assembly); Sundstrand Data Control, Redmond, WA (APU heater thermostat); Cox and Co., New York, NY (APU fuel tank, fuel and lube line heaters); Brunswick Wintec, El Segundo, CA (APU fuel line filter); J.C. Carter Co., Costa Mesa, CA (APU servicing coupling); Wright Components Inc., Clifton Springs, NJ (fuel pump seal cavity drain catch, relief valve); Rocket Research Corp., Redmond, WA (APU gas generator).

Space Shuttle Spacecraft Systems

WATER SPRAY BOILER

The water spray boiler (WSB) maintains auxiliary power unit (APU) lube oil and hydraulic fluid temperatures below 129°C (265°F) and 108°C (228°F) respectively during APU/hydraulic system operation. Each of the three hydraulic APU/hydraulic systems has its own WSB. The WSB uses a gaseous nitrogen (GN_2) system for positive expulsion of water into the boiler section for cooling and dumps the generated steam overboard. There are two redundant electronic controllers installed on each WSB.

Each WSB consists of a boiler, expendable water supply, regulated gas pressurization system for water expulsion, separate water feed valves for independent control of water to cool the APU lube oil and hydraulic fluid, and APU lube oil cooling and hydraulic fluid bypass valves. Redundant electric controllers provide complete automatic operation.

Each WSB provides an expendable heat sink for the orbiter hydraulic system's fluid and APU lube oil system on the

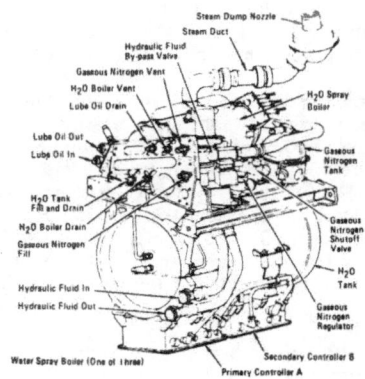
Water Spray Unit

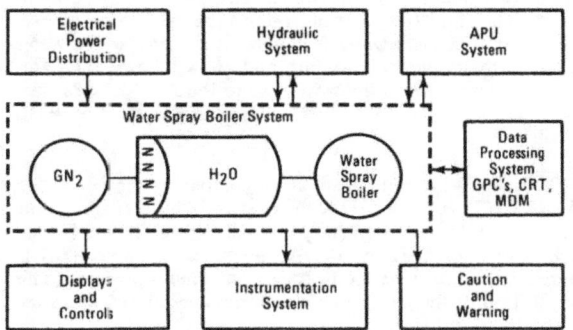
Water Spray Boiler System

Water Spray Boilers in Aft Fuselage

 **Space Shuttle Spacecraft Systems**

ground, during the boost phase, orbital checkout, deorbit, and entry through landing and rollout. The core of each WSB is the stainless-steel, crimped-tube bundle oil cooler. The oil cooler of each WSB consists of a hydraulic fluid section and an APU lube oil section. The hydraulic fluid and APU lube oil flow through their respective sections of the tube bundle oil cooler.

The GN_2 pressure for each WSB is contained in its respective 152-millimeter (6-inch) spherical pressure vessel. The pressure vessel contains 0.34 kilograms (0.76 pounds) of GN_2 at a nominal pressure of 126,787 mmHg (millimeters of mercury) (245 psi) at 37 °C (70 °F). The GN_2 storage system of each WSB is directed to its respective water storage tank. Each storage vessel contains sufficient GN_2 to expel all the water from the tank and to allow for relief valve venting during ascent and leakage of one cubic centimeter per minute over a thirty-day period.

The GN_2 shutoff valve between the GN_2 pressure vessel and H_2O storage tank of each WSB permits isolation of high pressure GN_2 supply. Each GN_2 valve is controlled by its respective boiler N_2 supply 1, 2, or 3 switch on the flight deck crew display and control panel R2. The GN_2 shutoff valve is latched in the open or close position and consists of two independent solenoid coils which allow valve control from either the primary or secondary controller. The respective switch is positioned to control each valve.

A single-stage regulator is installed between the GN_2 pressure shutoff valve and the H_2O storage tank. The GN_2 regulator for each WSB regulates the high pressure GN_2 between 1,267 and 1,345 mmHg (24.5 to 26 psi) to the H_2O storage tank.

A relief valve is incorporated internally to each GN_2 regulator to prevent the H_2O storage tank pressure from exceeding 1,707 mmHg (33 psi) due to heat soak-back when operating or in the event of a failed open GN_2 regulator. The GN_2 relief valve opens between 1,552 and 1,707 mmHg (30 to 33 psi).

The H_2O supply for each WSB is stored in a positive displacement aluminum water tank containing a welded metal bellows separating the stored water inside the bellows from the GN_2 expulsion gas. The 63 kilograms (140 pounds) of H_2O in each WSB provide total heat rejection.

Non-redundant pressure and temperature sensors located downstream of the GN_2 pressure vessel and on the H_2O tank for each WSB transmit the pressures and temperatures via the A controller to the Systems Management General Purpose Computer (SM-GPC). The SM-GPC computes the pressure/volume/temperature and transmits the H_2O tank quantity to the flight deck crew display and control panel F8 for each WSB. The APU FUEL/H_2O QTY switch on panel F8 is positioned to H_2O which allows the H_2O quantity of each WSB to be displayed on the APU FUEL/H_2O QTY 1, 2, or 3 meter. Thus H_2O quantity is only available when the A controller is powered.

Downstream of the H_2O storage tank, the feedwater lines to each water boiler split into two parallel lines: one line goes to the hydraulic fluid flow section, one to the APU lube oil section. A hydraulic fluid water feed valve is installed in the water line to the hydraulic fluid section and an APU lube oil water feed valve is installed in the water line to the APU lube oil section. Each valve is controlled independently by the WSB controller.

The two WSB controllers are regulated by the respective BOILER CNTLR PWR/HTR switches 1, 2, and 3 on the flight deck crew display and control panel R2. When the applicable switch is positioned to A, the A controller for that WSB is powered; or, if positioned to B, the B controller is powered. The OFF position of the applicable switch removes electrical power from both controllers.

Space Shuttle Spacecraft Systems

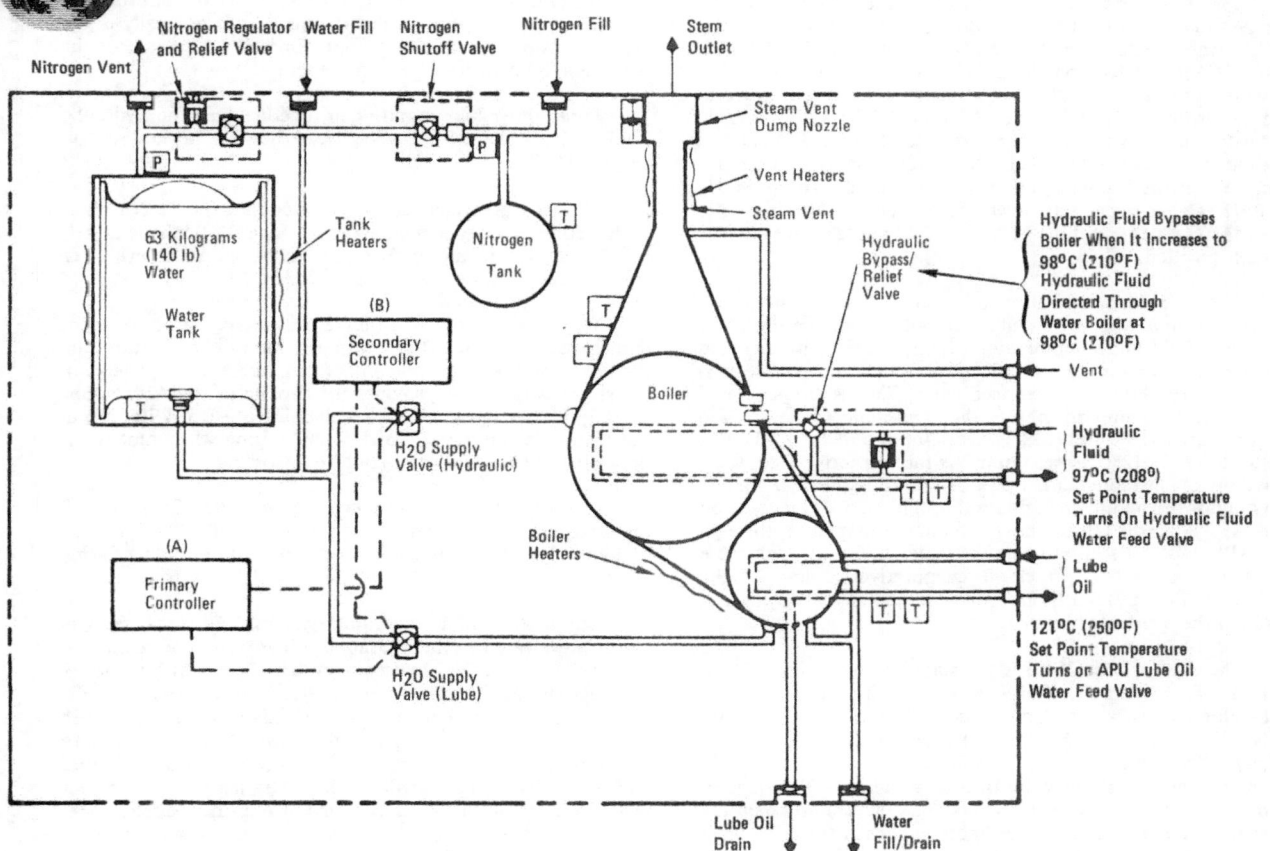

Water Spray Boiler (One of Three)

Space Shuttle Spacecraft Systems

The BOILER CNTLR switches 1, 2, and 3 enable (provide the automatic control functions) the specific controller A or B which was selected for that WSB by the BOILER CNTLR PWR/HTR switch on panel R2. When the applicable controller A or B is enabled for that WSB, a ready signal is transmitted to the corresponding APU/HYD READY TO START talkback indicator (along with other prerequisites from the APU and hydraulic system) on the flight deck crew display and control panel R2, if the following additional conditions are met: GN_2 shutoff valve is open, steam vent nozzle temperature is greater than 54 °C (130 °F), and the hydraulic fluid bypass valve is in the correct position.

The enabled controller of that WSB monitors the hydraulic fluid and APU lube oil outlet temperature. The hydraulic fluid outlet temperature controls the hydraulic fluid water feed valve and the APU lube oil outlet temperature controls the APU lube oil water feed valve. Signals are generated based on a comparison of the hydraulic system fluid temperature to its 97 °C (208 °F) set point and of the APU lube oil to its 121 °C (250 °F) set point. As the respective feed water valve opens, instantaneous flows of 6.8 kilograms per minute (15 pounds per minute) maximum through the hydraulic section and 4.5 kilograms (10 pounds) per minute maximum through the APU lube oil section enter the water boiler through these respective spray bars to effect evaporative cooling of the hydraulic fluid and APU lube oil with the steam vented out through the overboard steam vent.

The core of each WSB is the stainless steel crimped tube bundle. The hydraulic fluid section is divided into three 431-millimeter (17-inch) long passes of crimped tubes (first pass—234 tubes, second pass—224 tubes, and third pass—214 tubes). The tubes are 3.18 millimeters (0.125 inches) in diameter, with a wall thickness of 0.25 millimeters (0.010 inch). Crimps located every six millimeters (0.24 inches) break up the internal boundary layer and promote enhanced turbulent heat transfer. The APU lube oil section is comprised of two passes with 103 crimped tubes in its first pass and 81 smooth tubes in the second pass. Although the second pass is primarily a low pressure drop return section, approximately 15 percent of the APU lube oil heat transfer occurs there.

Three spray bars manifolded together feed the hydraulic fluid section while two spray bars feed the APU lube oil section of each WSB.

The separate water feed valves modulate the water flow to each section of the tube bundle core of each WSB independently in 200-millisecond pulses that vary from one pulse every ten seconds to one pulse every 0.250 seconds.

When the orbiter is in the vertical launch attitude on the launch pad with the APU/hydraulic pump combination and the WSB's in operation, the APU tube bundle is completely immersed in water. This provides for cooling of the APU's lube oil. Liquid level sensors in the spray boiler of each WSB prevent the hydraulic water feed valve from pulsing when that water boiler is full to avoid water spillage or loss.

It is noted that during prelaunch, the APU's lube oil cooling is required within six to seven minutes after APU start. The hydraulic systems fluid probably will not require cooling during ascent.

Due to the unique hydraulic system fluid flows, control valves are located in the hydraulic system fluid line section of each WSB. Normally, hydraulic system fluid flows up to 79 liters per minute (21 gallons per minute); however, the hydraulic system experiences one- to two-second flow spikes up to 238 liters per minute (63 gallons per minute). If these spikes were to pass through the boiler, pressure drop would increase nine-fold and the boiler would flow-limit the hydraulic system. To prevent this, a relief function is provided by a spring-loaded poppet valve which opens when the hydraulic fluid pressure drop exceeds 2,484 mmHg (48 psi) and is capable of flowing 162 liters

Space Shuttle Spacecraft Systems

per minute (43 gallons per minute) at a differential pressure of 2,587 mmHg (50 psi) across the boiler. A temperature-controlled diverter valve allows the hydraulic fluid to bypass the boiler when the hydraulic fluid has decreased to 87 °C (190 °F). At 98 °C (210 °F), the controller commands the diverter valve to direct the fluid through the boiler. When the hydraulic fluid cools to 87 °C (190 °F), the controller again commands the diverter valve to bypass the fluid around the boiler.

Two hours before APU/hydraulic checkout in orbit, and before deorbit, the controllers for each WSB are enabled to allow the steam vent heater to preheat the steam vent nozzle and WSB surface heaters for the water boiler surfaces. When the APU/hydraulic combination is started and the hydraulic fluid and APU lube oil flow commences and the fluid temperatures rise, spraying is initiated as required.

During entry into the atmosphere, whenever the boiling-point of water is greater than 87 °C (190 °F), the liquid level sensor is enabled and the boiler operates in a flooded mode.

There are three 863 x 787 x 482 millimeter (34 x 31 x 19 inch) WSB's located in the aft fuselage of the orbiter, one for each independent hydraulic system/APU combination. The dry weight of each WSB is 82 kilograms (181 pounds). The WSB's require electrical power and control from the orbiter as its only outside source.

The WSB's are mounted in the orbiter aft fuselage between X_O 1340 and 1400 and on the Z_O plane 488 and minus 15, plus 15 in the Y_O plane. Insulation blankets cover each WSB.

The contractor for the WSB's is Hamilton Standard Division, United Technologies Inc., Windsor Locks, CT.

HYDRAULIC SYSTEM

The hydraulic system consists of three independent hydraulic systems. Each of the three auxiliary power units provides mechanical shaft power to drive a hydraulic pump, and each of the three hydraulic pumps provides the hydraulic pressure for the respective hydraulic system.

Each hydraulic system provides hydraulic pressure for operation of actuators to control orbiter aerosurfaces (elevons, rudder/speed brake, and body flap), the three main engine gimbals (thrust vector control), main engine valves, external tank umbilical retraction, landing gear deployment, main landing gear brakes and anti-skid control, and nose gear steering.

The hydraulic systems are capable of operation when exposed to forces or conditions caused by acceleration, deceleration, normal "g", zero "g", hard vacuum, and extreme low temperatures encountered in orbit. The systems are active during liftoff, ascent, and orbital insertion for main engine thrust vector control and propellant valve control, in addition to elevon load relief during ascent.

In orbit, the hydraulic system's fluids are circulated periodically by electric-motor-driven circulation pumps to absorb heat from the Freon-21 hydraulic heat exchanger and distribute it to all areas of the hydraulic systems.

The hydraulic system is restarted for deorbit, entry, and descent for actuation of the aerodynamic control surfaces (elevons, rudder/speed brake, and body flap) during atmospheric flight, for deploying the main and nose landing gear, for actuation of the main landing gear brakes and anti-skid system, and for actuation of nose landing gear steering.

Space Shuttle Spacecraft Systems

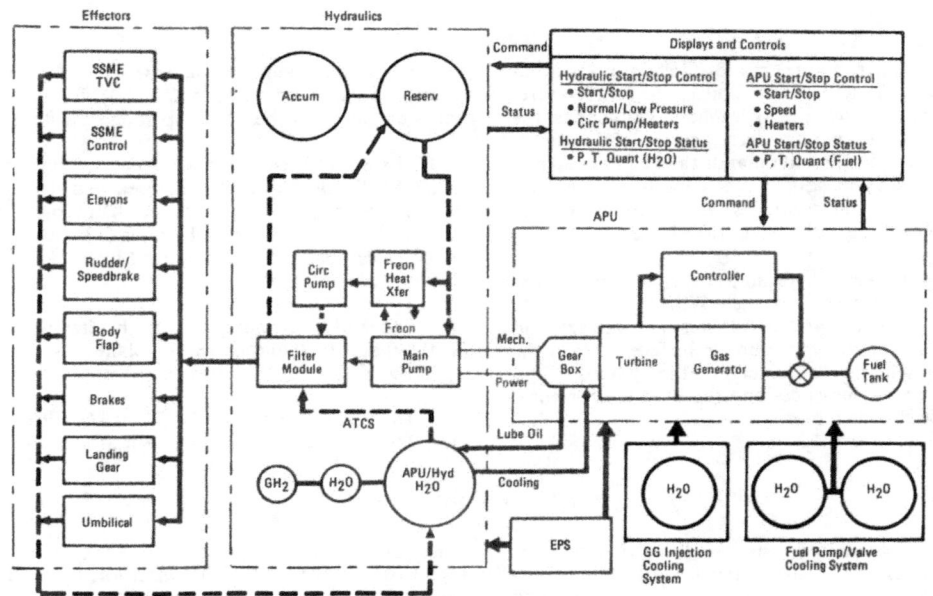

APU/Hydraulic System

The hydraulic systems are designated 1, 2, and 3. Each consists of the components necessary to generate, distribute, control, monitor, and utilize hydraulic pressure and thermally condition the hydraulic fluid.

The hydraulic pump for each hydraulic system is a variable displacement-type pump. Each hydraulic pump has an electrically operated depressurization valve. The depressurization valve for each hydraulic pump is controlled by its respective HYD MAIN PUMP PRESS switches 1, 2, or 3 located on the flight deck display and control panel R2. When the applicable HYD MAIN PUMP PRESS switch is positioned to LOW, the respective depressurization valve is energized, which reduces the hydraulic pump discharge pressure from its normal 150,075 to 160,425 millimeters of mercury (mmHg) (2,900 to 3,100 psi) output to 25,875 to 51,570 mmHg (500 to 1,000 psi), thereby reducing the auxiliary power unit (APU) torque requirements during the start of the auxiliary power unit. (When display and

323

Space Shuttle Spacecraft Systems

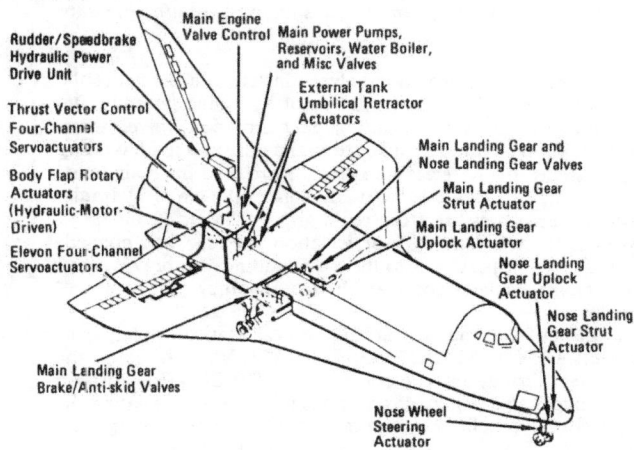

Hydraulic Systems

Hydraulic System Pumps

control panel nomenclature is printed in all caps—e.g., HYD MAIN PUMP PRESS—it indicates that it is the exact way it appears on the panel.)

Prior to the start of the APU, the applicable hydraulic system APU/HYD READY TO START talkback indicator on the flight deck display and control panel R2 should be gray. In order for the applicable talkback indicator to indicate gray, the respective hydraulic system HYD MAIN PUMP PRESS switch must be in LOW, the respective BOILER CNTLR/PWR/HTR, BOILER CNTLR, and BOILER H_2 SUPPLY switches must be in ON position on panel R2 and the boiler ready signal must be present which consists of four parameters: boiler stream above 54 °C (130 °F), N_2 (nitrogen) valve open, bypass valve powered, and boiler enabled.

When the applicable auxiliary power unit has been started, the respective HYD MAIN PUMP PRESS switch is positioned from LOW to NORM. This de-energizes the respective depressurization valve, allowing that hydraulic pump to increase its outlet pressure from 25,875 to 51,750 mmHg (500 to 1,000 psi) to 150,075 to 160,425 mmHg (2,900 to 3,100 psi). Each hydraulic pump is a variable displacement pump which provides 0 to 238 liters (0 to 63 gallons) per minute at 150,250 mmHg (3,000 psi) nominal, with the APU at normal speed; and 263 liters (69.9 gallons) per minute at 155,250 mmHg (3,000 psi) nominal, with the APU at high speed.

Each hydraulic system has a filter module assembly which filters the hydraulic fluid entering and leaving the hydraulic pump in addition to a filter in the hydraulic pump case drain and a filter in the return line to the reservoir.

A high-pressure relief valve contained within the filter module for each hydraulic system relieves the hydraulic pump supply line pressure into the return line in the event the supply line pressure exceeds 199,237 mmHg (3,850 psi) differential.

Space Shuttle Spacecraft Systems

A pressure sensor in the filter module for each hydraulic system monitors the hydraulic system source pressure and displays the pressure on the flight deck display and control HYDRAULIC PRESSURE 1, 2, and 3 meters on panel F8. The same hydraulic pressure sensor for each system also provides an input to the YELLOW HYD PRESS caution and warning light on the flight deck display and control panel F7, if the hydraulic pressure of systems 1, 2, or 3 is below 142,830 mmHg (2,760 psi). The RED BACKUP caution and warning light on Panel 7 will illuminate if the hydraulic pressure of systems 1, 2, or 3 is at 142,830 mmHg (2,760 psi).

The aerosurfaces (elevons, rudder/speed brake and body flap) are powered by hydraulic pressure and the movement of the applicable aerosurface is accomplished mechanically. Each of the aerosurface actuation units have switching valves which interface with all three hydraulic systems with the exception of the body flap. One of the hydraulic systems is the normal supply system and the other two are standby systems No. 1 and No. 2 in the case of the elevon actuators. All three systems are utilized for the body flap and rudder/speed brake actuation units, thus a loss of a hydraulic system results in a reduced response rate of actuation for the body flap. The aerosurfaces are controlled by the guidance, navigation, and control (GN&C) system in respect to command functions.

The orbiter nose and main landing gear deployment is accomplished by hydraulic pressure. Hydraulic system No. 1 pressure is the only system utilized to hydraulically actuate the landing gear uplock hooks for release of each landing gear. A pyrotechnic initiator provides the emergency backup of the No. 1 hydraulic system. The pyrotechnic initiator on each landing gear mechanically actuates the landing gear uplock hooks for release of each landing gear. Nose gear steering is accomplished with hydraulic system pressure from the No. 1 hydraulic system only. The left and right main landing brakes provide a backup for nose gear steering. This is accomplished by alternately applying the left and right main landing gear brakes which will castor the nose wheel for steering.

The orbiter main landing gear brakes utilize No. 1 and No. 2 hydraulic systems as the primary source of hydraulic pressure, and No. 3 hydraulic system as the standby hydraulic pressure source.

A landing gear isolation valve is installed in the hydraulic supply line to each of the landing gear hydraulic systems. The No. 1 hydraulic system landing gear isolation valve, when opened, allows hydraulic pressure to the nose and main landing gear systems, the nose gear steering system, and the main landing gear brakes; and, when closed, isolates the No. 1 hydraulic system source pressure from these areas. The No. 2 or No. 3 hydraulic system landing gear isolation valves, when opened, allow access, respectively, to the main landing gear brakes, and, when closed, isolates access to the main landing gear brakes.

Each of the landing gear isolation valves is controlled by its respective HYDRAULICS LG HYD ISOL VLV switch on the flight deck display and control panel R4. The CLOSED position of the applicable switch closes that isolation valve, isolating the respective hydraulic source pressure from that landing gear system. A talkback indicator located above the respective switch on panel R4 would indicate CL (closed). The landing gear isolation valves are closed during prelaunch, boost, and entry. The OPEN position of the respective switch opens the respective isolation valve, allowing that hydraulic system source pressure to its landing gear system. The respective talkback indicator would indicate OP (open). The landing gear isolation valves are opened on-orbit to permit thermal conditioning of the landing gear hydraulic fluid and system. The GPC (general purpose computer) position permits the GPC to open landing gear isolation valve system 3 for 15 minutes, system 2 for 10 minutes, and system 1 for 5 minutes prior to landing to permit thermal conditioning of the landing gear hydraulic fluid and system.

Hydraulic system No. 1 landing gear retract/circulation valve is controlled by the HYDRAULICS LG RET/CIRC VLV switch on the flight deck display and control panel R4. The retract/circulation valve is closed normally during prelaunch,

Space Shuttle Spacecraft Systems

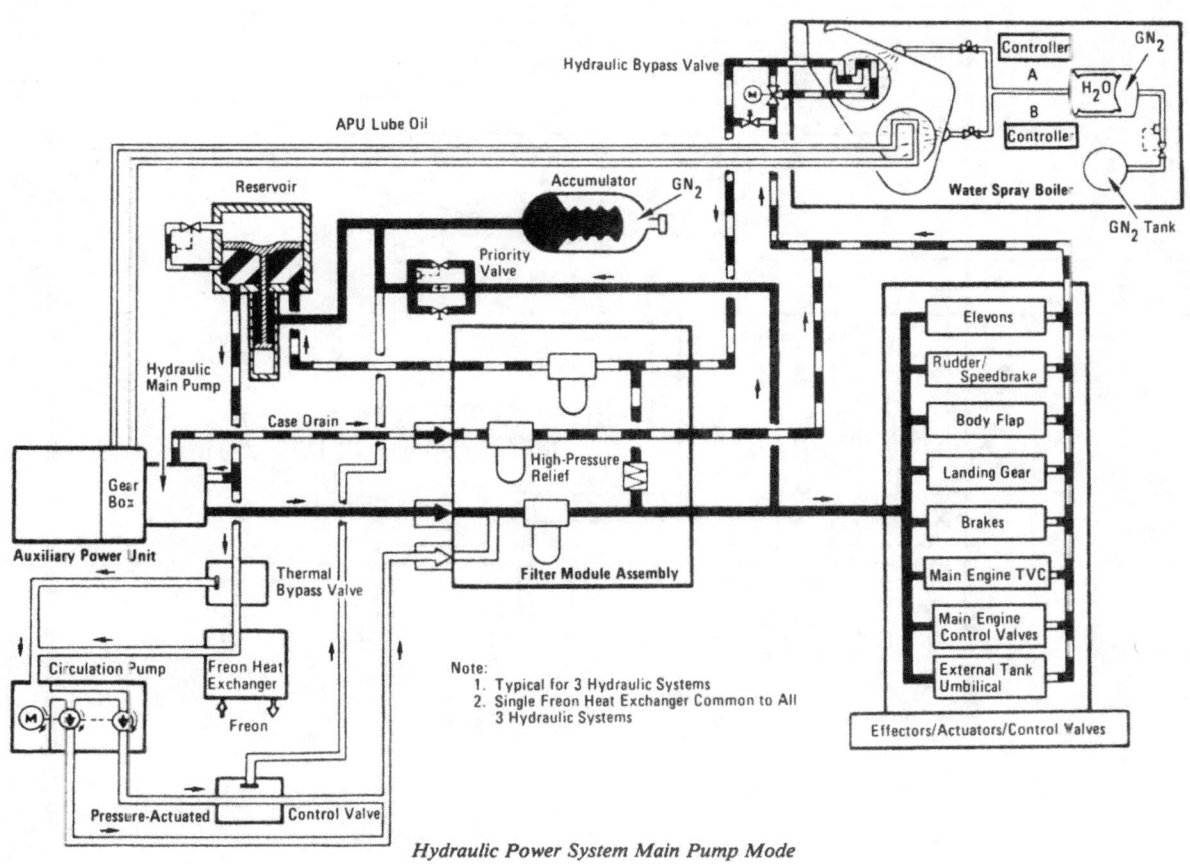

Hydraulic Power System Main Pump Mode

 **Space Shuttle Spacecraft Systems**

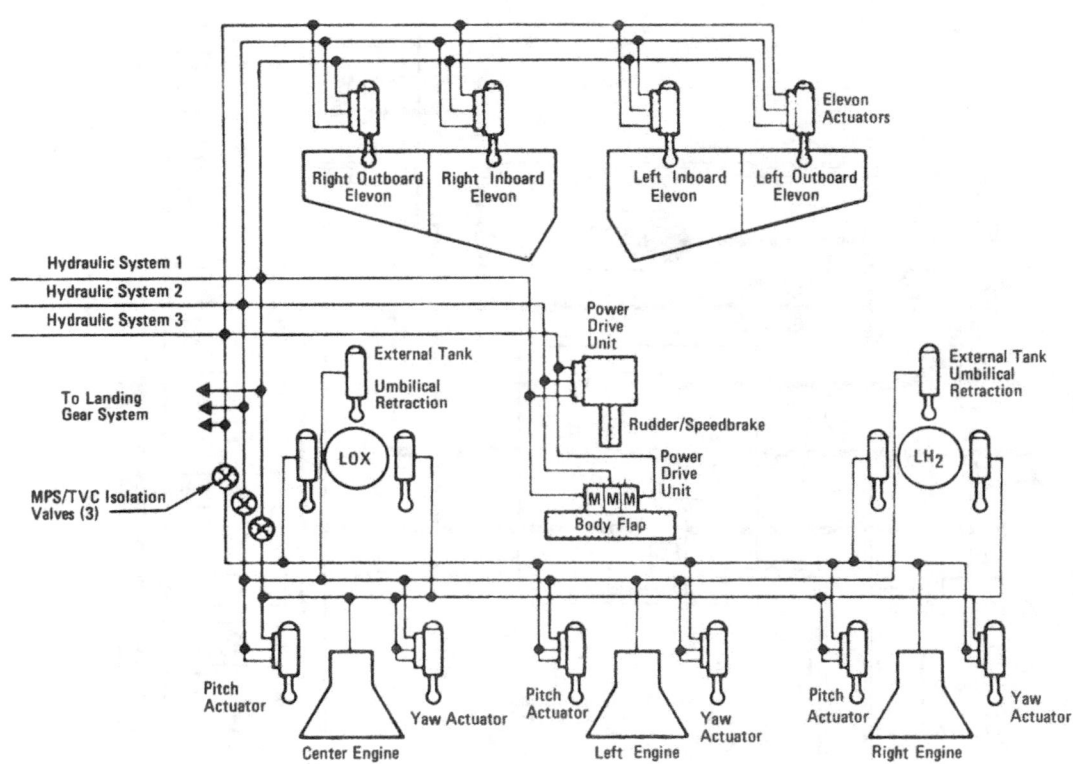

Hydraulic Actuator System — Landing Gear and Brake

Space Shuttle Spacecraft Systems

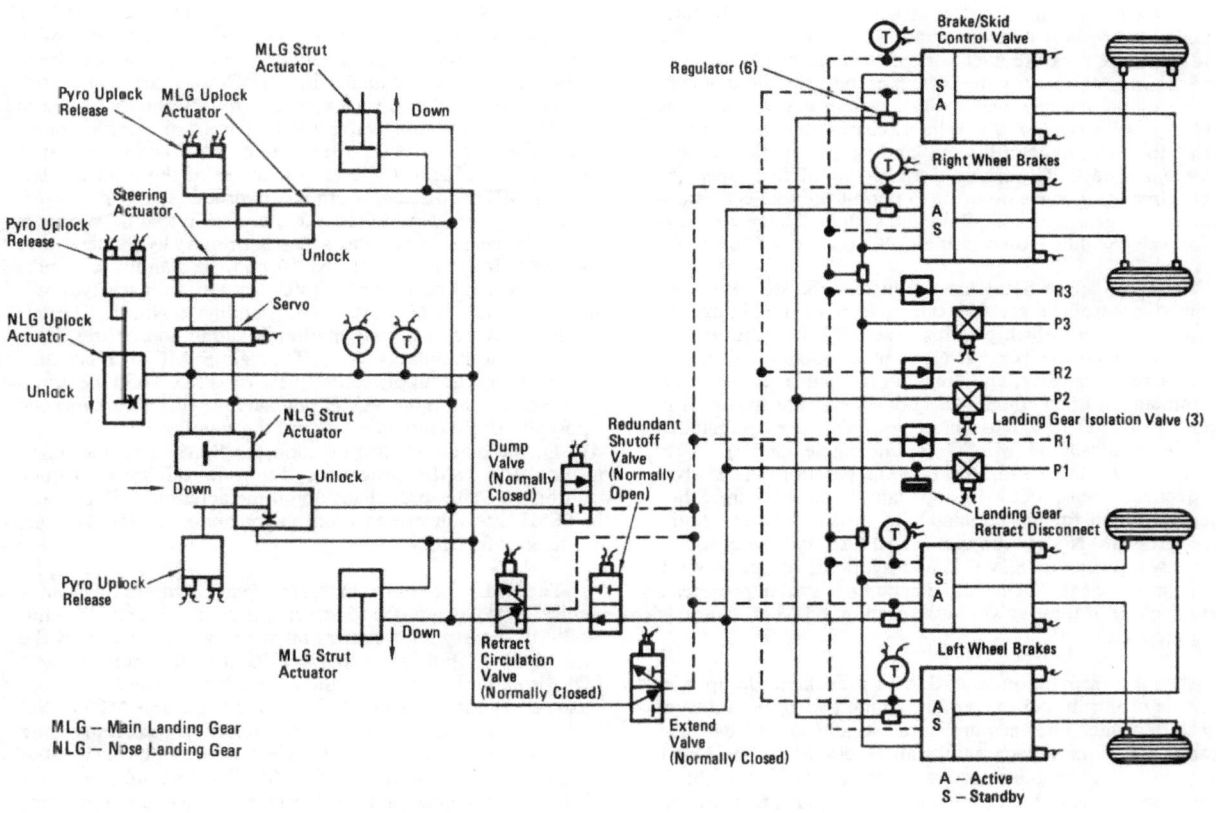

Hydraulic Actuator System — Landing Gear and Brake

Space Shuttle Spacecraft Systems

boost, and entry. When the LG RET/CIRC VLV is positioned to OPEN, the No. 1 hydraulic system landing gear retract/circulation valve is energized open which permits system source pressure to circulate through the retract lines through the nose and main landing gear uplock actuators and return through the extend lines (the pressure inlet port of the landing gear control valve is open to the return line) which permits the thermal conditioning of the landing gear hydraulic fluid and system. The CLOSE position closes the retract/circulation valve which prevents the thermal conditioning landing gear hydraulic fluid and system. The GPC position permits the GPC to open the retract/circulation valve on-orbit to permit thermal conditioning of landing gear hydraulic fluid and system. The retract/circulation valve will be closed in the GPC mode prior to entry.

The three Space Shuttle main engines (SSME) and their associated controllers provide the control of the individual hydraulic actuators which positions each SSME preburner oxidizer valve, main oxidizer valve, chamber coolant valve, fuel preburner oxidizer valve, and the main fuel valve. These valves are commanded open for SSME ignition and are sustained in the open position through ascent. These valves are commanded closed hydraulically at MECO (main engine cutoff). After SSME shutdown and ET (external tank) separation, these valves are sequenced open for SSME propellant dump and purge, then sequenced closed for the remainder of the mission. Hydraulic source pressure No. 1 supplies SSME 1; hydraulic source pressure No. 2 supplies SSME 2; and hydraulic source pressure No. 3 supplies SSME 3. The SSME pneumatic system provides a backup closure of these valves in the event of a loss of hydraulic source pressure.

After ET separation and SSME propellant dump and purge, the orbiter liquid oxygen and liquid hydrogen umbilical at the ET/orbiter interface are retracted and locked by three hydraulic actuators at each umbilical. Hydraulic system No. 1 source pressure is supplied to one actuator at each umbilical; hydraulic system No. 2 source pressure is supplied to a second actuator at each umbilical; and hydraulic system No. 3 source pressure is supplied to a third actuator at each umbilical.

Each SSME is provided with thrust vector control by a pitch and yaw actuator which is controlled by the ascent thrust vector control system. Each actuator is powered hydraulically which mechanically gimbals the SSME for start and launch position and for thrust vector control during the ascent. Each actuator has a switching valve which allows a primary hydraulic source pressure to power the actuator. A standby hydraulic system is available in the event of failure of the primary. The center SSME pitch actuator primary hydraulic source pressure is hydraulic system No. 1 and the standby hydraulic supply is No. 1. The center SSME yaw actuator primary hydraulic source pressure is hydraulic system No. 3 and the standby hydraulic supply is No. 1. The left SSME pitch actuator primary hydraulic source pressure is No. 2 and the standby is No. 1. The left SSME yaw actuator primary hydraulic supply source pressure is No. 1 and the standby is No. 2. The right SSME pitch actuator primary hydraulic supply source pressure is No. 3 and the standby is No. 2. The right SSME yaw actuator primary hydraulic supply source pressure is No. 2 and the standby is No. 3. After MECO, the actuators will position the SSME's to the dump position for SSME propellant dump to minimize attitude disturbance. After propellant dump the actuators will position the SSME's to the stowed position for minimum aerodynamic interference for entry.

The hydraulic source pressure of each hydraulic system is supplied or isolated to the SSME engine hydraulic actuators and the SSME thrust vector control pitch and yaw actuators by the MPS (main propulsion system)/TVC (thrust vector control) ISOL (isolation) VLV 1, 2, and 3 switches on the flight deck display and control panel R4. When the applicable MPS/TVC ISOL VLV switch is positioned to OPEN, the corresponding hydraulic source pressure is supplied to the aforementioned areas; and when positioned to CLOSE, that hydraulic system is isolated. A talkback indicator located above the respective

Space Shuttle Spacecraft Systems

switch indicates OP when that valve is open or indicates CL when that valve is closed. The MPS/TVC ISOL VLV 1, 2, and 3 are open during prelaunch and ascent and are closed after main propulsion dump, and remain closed except to reposition the SSME's prior to deorbit, if required.

The return line of each hydraulic system is directed to its respective WSB. There is one WSB for each hydraulic system. The WSB provides the expendable heat sink for each orbiter hydraulic system and each of the APU lube oil systems during prelaunch, boost phase, on-orbit checkout, de-orbit, and entry through rollout and landing.

Due to the unique hydraulic system fluid flows, hydraulic fluid control valves are located in the return line of the hydraulic system to the WSB. Normally, the hydraulic system fluid flows up to 79 liters per minute (21 gallons per minute), however, the hydraulic system experiences one- to two-second flow spikes up to 238 liters per minute (63 gallons per minute). If these spikes were to pass through the WSB, pressure drop would increase nine-fold and the WSB would flow-limit the hydraulic system. To prevent this, a relief function is provided by a spring-loaded poppet valve which opens when the hydraulic fluid pressure exceeds 2,484 mmHg (48 psi) and is capable of flowing 162 liters per minute (43 gallons per minute) at 2,587 mmHg (50 psi) differential across the WSB. A temperature controller diverter valve allows the hydraulic fluid to bypass the boiler when the hydraulic fluid has increased to 98 °C (210 °F). At 98 °C (210 °F), the controller commands the diverter valve to direct the hydraulic fluid through the WSB. When the hydraulic fluid cools to 87 °C (190 °F), the controller again commands the diverter valve to bypass the fluid around the WSB.

A reservoir in each hydraulic system return line provides positive pressure fluid to the main hydraulic pump and the circulation pump in that hydraulic system. Within each hydraulic reservoir, there is a differential area piston (40 to 1 area ratio between the reservoir side and accumulator side) which is actuated by that hydraulic system pressure and provides the pressurized fluid to the main hydraulic pump and circulation pump of that system. The reservoir of each hydraulic system also provides for volumetric expansion and contraction. The capacity of the MIL-H-83282 hydraulic fluid (synthetic hydrocarbon, reduces fire hazard) in each reservoir is 30,321 cubic centimeters (1,850 cubic inches). The quantity of each reservoir is monitored on the flight deck display and control panel F8 as HYDRAULIC QUANTITY 1, 2, and 3 in percentages. The quantity in each reservoir is 30 liters (8 gallons). A pressure relief valve in each reservoir protects that reservoir from overpressurization and relieves at 6,210 mmHg (120 psi) differential.

A hydraulic accumulator in each hydraulic system dampens pressure impulses and minimizes the main hydraulic pump and circulation pump pressure ripples. Each accumulator is a piston type and precharged with gaseous nitrogen (GN_2) at 87,975 to 90,562 mmHg (1,650 to 1,750 psi). The GN_2 capacity of each accumulator is 243 cubic centimeters (96 cubic inches). The hydraulic fluid volume is 129 cubic centimeters (51 cubic inches).

A circulation pump in each hydraulic system is utilized to provide the thermal conditioning of the hydraulic fluid in that system. The circulation pump in each hydraulic system is electrically driven.

A pressure sensing priority valve in each hydraulic system automatically closes and traps the accumulated hydraulic pressure permitting the reservoir to supply a positive hydraulic pressure from the reservoir to the main hydraulic pump and circulation pump in altitude start conditions. A dump valve is provided for each accumulator and is utilized to depressurize the respective hydraulic reservoir for ground operations only.

Each circulation pump is controlled by its respective HYD CIRC PUMP POWER switches 1, 2, and 3 on the flight deck

 # Space Shuttle Spacecraft Systems

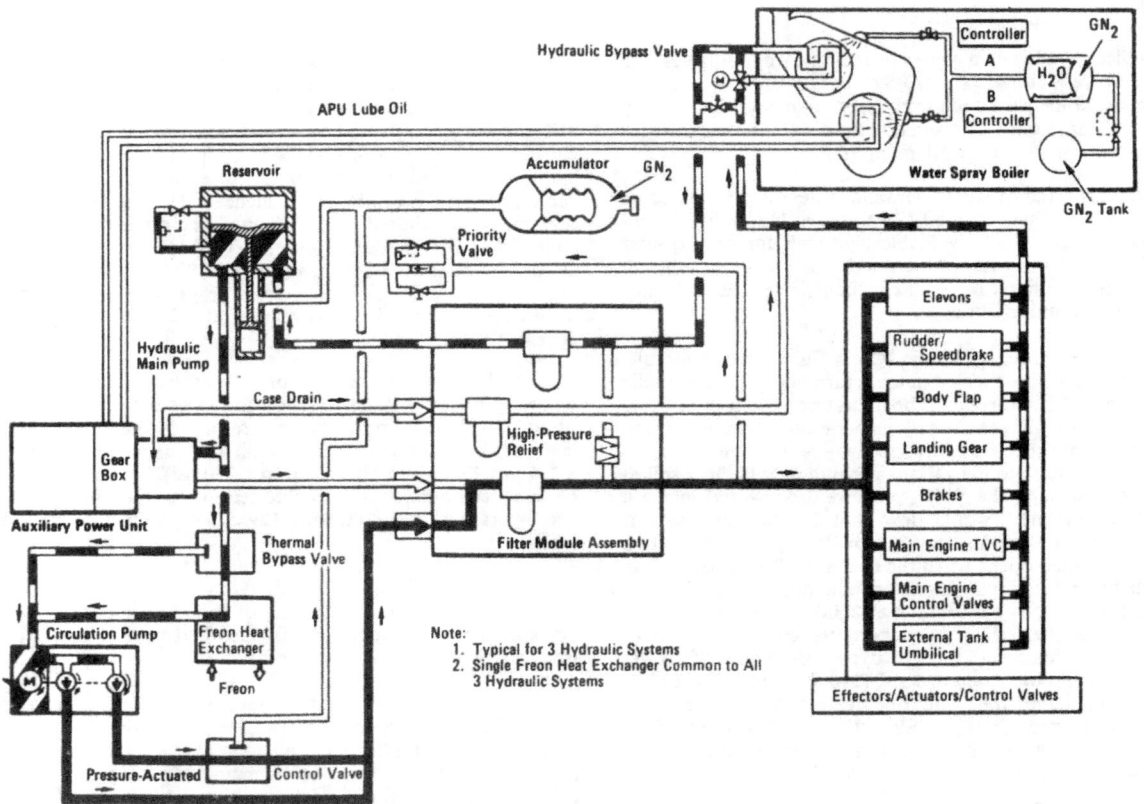

Hydraulic Power System Circulation Mode

Space Shuttle Spacecraft Systems

display and control panel A12. The HYD CIRC PUMP POWER switches 1, 2, and 3 on panel A12 provide MNA, if in MNA; or MNB, if in MNB, to the respective APU/HYD CIRC PUMPS 1, 2, and 3 switches on panel R2.

The ON position of the HYD CIRC PUMP 1, 2, and 3 switches on panel R2 provide electrical power to its respective hydraulic system circulation pump providing that the APU START/RUN switch is not in the START/RUN or START/ORIDE position. It is noted that all three circulation pumps operate in prelaunch, prior to APU start and at postlanding. The GPC position of the HYD CIRC PUMP 1, 2, and 3 switches provide GPC automatic control of the respective circulation pump providing the APU/START/RUN switch is not in START/RUN or START/ORIDE position. In the GPC automatic mode, there are two modes: one is thermostat-controlled and the other is timer-controlled. In the thermostat-controlled mode, the first hydraulic system temperature sensor to reach below minus 17°C (0°F) is the first system to be circulated. The GPC commands that circulation pump off when all temperature sensors in that hydraulic system being circulated is minus 6°C (20°F) or when that circulation pump has been on for a good period in excess of a specified minimum run time and a second hydraulic system requires circulation. The GPC timer control consists of one temperature sensor in each hydraulic system which has excessively low and high temperature limits, which demand continuous circulation of that hydraulic system and the circulation pump run time for each pump is set to a desired value along with the delay time between pump runs and will continue to operate and cycle the three circulation pumps until the temperature high and low limits are readjusted to normal set points. The OFF position removes electrical power from the circulation pump. At postlanding, the APU/hydraulic system is shutdown. The circulation pumps are activated to circulate the heated hydraulic fluid through the WSB for cooling.

The circulation pump in each hydraulic system consists of a high-pressure and low-pressure, two-stage gear pump driven by a 28-Vdc induction electric motor with a self-contained inverter. Protection against excessive electronic component temperature is provided by directing the inlet fluid flow around these components and through the electric motor before entering the pumps. The low-pressure stage is rated at 10.9 liters per minute (2.9 gpm) at 181,125 mmHg (350 psi). The circulation pumps in each hydraulic system maintains the desired hydraulic fluid temperatures in prelaunch prior to APU start and provides orbital thermal control of the hydraulic fluid by transferring heat from the active thermal control system Freon-21 coolant loop/hydraulic heat exchanger to that hydraulic system. After postlanding/rollout, the circulation pump in each hydraulic system provides thermal conditioning of the hydraulic fluid after APU shutdown through the WSB to limit hydraulic fluid temperature rise due to heat soakback. In the event of pressure loss in the accumulator due to leakage on-orbit, the high-pressure stage pump delivers 0.3 liters per minute (0.1 gpm) at a discharge pressure up to 129,375 mmHg (2,500 psi) to repressurize the accumulator. This function is regulated by a pressure-activated control valve.

Insulation and electrical heaters are used on those portions of the hydraulic systems which are not adequately thermal conditioned by the individual hydraulic system circulation pump due to stagnant hydraulic fluid areas.

Redundant electrical heaters are installed on the body flap differential gearbox, rudder/speed brake mixer gearbox, the four elevon actuators, the aft fuselage body flap A and B seal cavity drain line, and rudder/speed brake cavity drain line.

The heaters are enabled by the HYDRAULIC HEATER switches on the flight deck display and control panel A12. There are HYDRAULIC HEATER switches A and B for the RUDDER/SPD BK, BODY FLAP, ELEVON, and AFT FUSELAGE. The AUTO A and B position of each of these switches permits the corresponding electrical Bus A or B to power the redundant heaters at each location. The thermostats

Space Shuttle Spacecraft Systems

in each electrical heater system cycle the heaters on/off automatically. The OFF position of the applicable switch removes electrical power from that heater system.

The contractors involved with the hydraulic systems are Arkwin Industries, Westbury, NY (hydraulic reservoir, filter, and control valves); Purolator Inc., Newbury Park, CA (hydraulic filter module); Parker-Hannifin Corp., Irvine, CA (hydraulic accumulator); Abex Corp., Aerospace Division, Oxnard, CA (hydraulic pump); Crissair Inc., El Segundo, CA (hydraulic check valve and flow restrictor); Hi-Temp Insulation Inc., Camarillo, CA (hydraulic blanket insulation); Bertea Corp., Irvine, CA (external tank umbilical retractor actuator, main and nose landing gear uplock actuator); Lear Siegler, Elyria, OH (hydraulic disconnect); Moog Inc., East Aurora, NY (main engine gimbal servo actuators and elevon servo actuators); Pneu Devices, Goleta, CA (hydraulic thermal control shutoff valve and electric-motor-driven circulation pump); Pneu Draulics, Montclair, CA (priority valve hydraulic reservoir); Resistoflex, Roseland, NJ (hydraulic system line connectors); Sterer Engineering and Manufacturing, Los Angeles, CA, (hydraulic solenoid shutoff valve, main engines and landing gear, nose landing gear steering and damping system, 3-way solenoid operating valve landing gear uplock and control valves); Sundstrand, Rockford, IL (rudder/speed brake actuation unit and body flap actuation unit); Symetrics, Canoga Park, CA (hydraulic quick disconnects); Titeflex Division, Springfield, MA (hydraulic system hose); Whittaker Corp., North Hollywood, CA (hydraulic accumulator dump valve); Wright Components, Inc., Clifton Springs, NJ (hydraulic latching solenoid valve); Hamilton Standard Division of United Technologies Corp., Windsor Locks, CT (water boiler hydraulic thermal unit and ground support equipment hydraulic cart).

LANDING GEAR SYSTEM

The landing gear system is a conventional aircraft tricycle configuration with steerable nose gear and main left and right landing gears.

The landing gear system includes a shock strut assembly constructed of high-strength, stress-corrosion-resistant steel alloys, aluminum alloys, stainless steel, and aluminum bronze. Each gear is made up of two wheel and tire assemblies. The nose landing gear is steerable, and each of the two main landing gears has a brake assembly with antiskid protection.

The shock strut assembly of each gear is a pneudraulic shock absorber containing gaseous nitrogen and hydraulic fluid. Because the orbiter will be under zero-g conditions during space flight, a floating diaphragm separates the gaseous nitrogen from the hydraulic fluid to maintain absorption integrity.

Cadmium-titanium plating and urethane paint are applied to landing gear strut surfaces for space flight protection. The wheel halves are forged aluminum, primed, and painted with two coats of urethane paint.

The landing gears, during the spacecraft's approach, are extended by the orbiter's commander or pilot, who first depresses a landing gear system "Arm" pushbutton and then a "DN" pushbutton. The gear is fully extended within ten seconds.

Initiation of the system hydraulically releases uplock hooks, which permit the landing gear to free-fall into the extended position. Springs and hydraulic actuators assist the free fall.

Space Shuttle Spacecraft Systems

Pyrotechnic actuators may also unlock the uplock hooks if the hydraulic system malfunctions. When fully extended, the gear is locked by spring-loaded bungees.

Doors on the nose and main right and left gears are operated through mechanical linkage attached to the door and fuselage and powered by strut camming action during gear extension. As the uplock hook is released and the gear begins its descent, the doors open in sequence. The gear strut actuators include an oil snubber to control the rate of extension and prevent damage to the downlink linkages.

The landing gears may not be retracted in flight. Retraction is a ground operation. For retraction, each gear is hydraulically moved forward and up. The mechanical linkage closes the doors, and the uplock hook is engaged to retain the gears in the housing. The nose gear is housed in the forward fuselage, and two doors cover the area. The right and left main gears are retracted into the wings and have one large door covering the area.

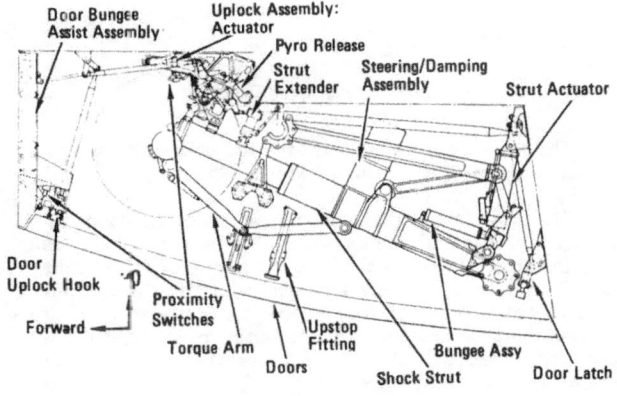

Nose Landing Gear Stowed

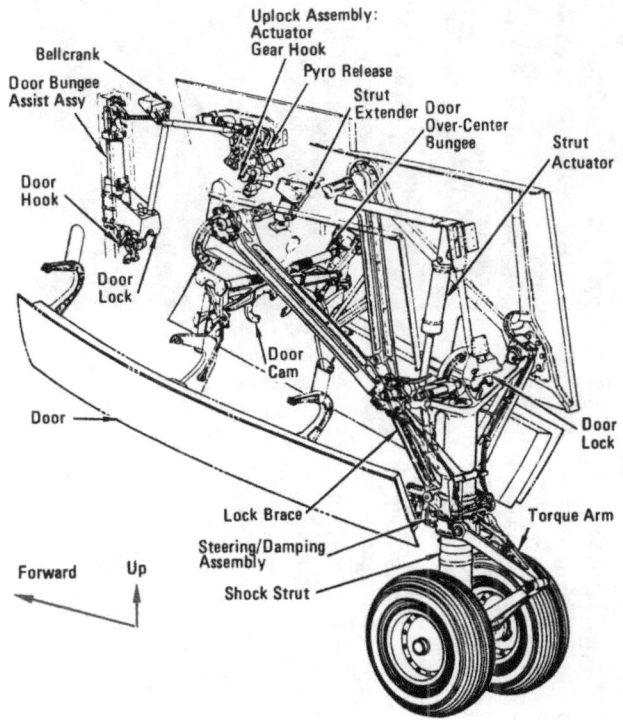

Nose Landing Gear Deployed

Each landing gear door has high-temperature, reusable surface insulation (HRSI) tiles bonded to the outer surface and a thermal barrier to protect the landing gear from the high-temperature thermal loads encountered during entry.

 **Space Shuttle Spacecraft Systems**

Nose Landing Gear

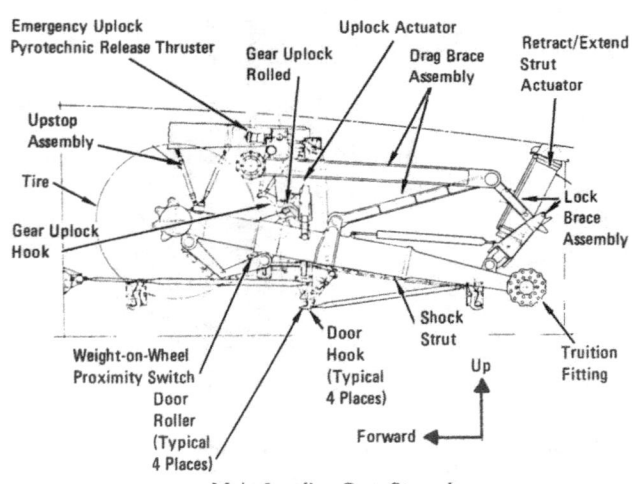
Main Landing Gear Stowed

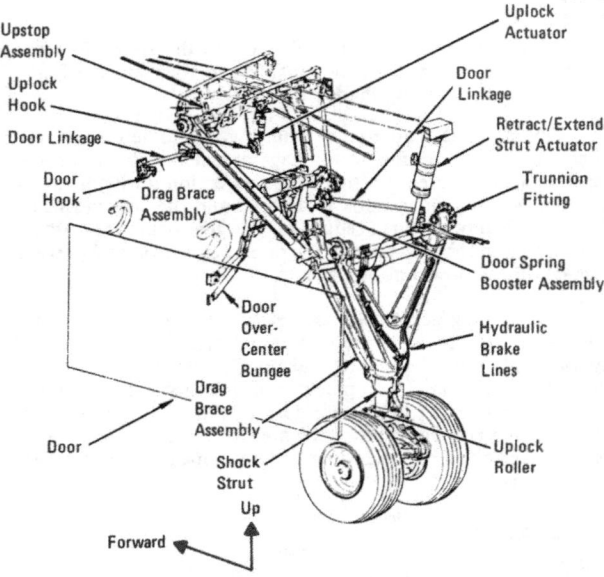
Main Landing Gear Deployed

The gears are deployed only after the spacecraft has an indicated airspeed of less than 300 knots (345 mph).

The nose and main landing gear are deployed with hydraulic pressure. Hydraulic system No. 1 is the only system used to actuate the uplock hooks for release of each landing gear. Pyrotechnic initiators serve as backup to the hydraulic system. Hydraulic system No. 1 also is the only system used to position the nose gear for steering. During rollout, the orbiter also may be steered by differential main gear brake application, which casters the nose wheel for steering.

335

Space Shuttle Spacecraft Systems

Main Landing Gear

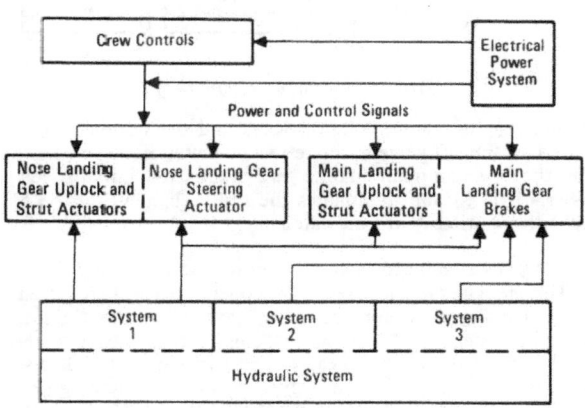

Landing/Deceleration Interfaces

The orbiter main landing gear brakes use hydraulic systems No. 1 and 2 as the primary sources and No. 3 as a standby source.

An isolation valve is installed in the hydraulic supply source line to each landing gear hydraulic system. When the system No. 1 isolation valve is opened, pressure is applied to the nose and main landing gear deployment system, the nose gear steering system, and the main landing gear brakes. The hydraulic system No. 2 or 3 landing gear isolation valves, when opened, allow pressure to be applied to the main landing gear brakes.

The landing gear isolation valves are controlled by HYDRAULICS LG HYD ISOL VLV switches on flight deck display and control Panel R4. (When display and control panel nomenclature is printed in all caps—e.g., HYDRAULICS LG HYD ISOL VLV—it indicates that it is the exact way it appears on the panel.) When these switches are closed, the hydraulic sources are isolated from the landing gear system; when open, they are connected to the systems. An indicator located above each switch shows whether the valve is closed (CL) or open (OP). The isolation valves normally are closed during prelaunch, boost, and entry. The hydraulic system No. 1 isolation valve is opened on orbit to permit thermal conditioning of the landing gear deployment hydraulic systems; it is closed before entry. The GPC (general purpose computer) position of the isolation valve switches enables the computer to open the landing gear isolation valves 15, 10, and 5 minutes, for system No. 3, 2, and 1, respectively, before landing for thermal conditioning of the landing gear hydraulic brake systems.

The HYDRAULICS LG RET/CIRC VLV switch on Panel R4 controls the landing gear retract/circulation valve in hydraulic system No. 1. The retract/circulation valve normally

Space Shuttle Spacecraft Systems

is closed during prelaunch, boost, and entry. When it is open, hydraulic system No. 1 source pressure circulates through the retract lines to the nose and main landing gear uplock actuators for thermal conditioning of the system No. 1 landing gear hydraulic fluid. The pressure enters through an orifice in the piston head and returns through the extend lines (the pressure inlet port of the landing gear control valve is closed; however, the outlet port of the landing gear control valve is open to the system No. 1 return line). The hydraulic fluid flow to the landing gear strut actuators dead ends at the actuators' retract ports. The GPC position of the switch permits the computer to automatically (temperature controlled) open and close the retract/circulation valve to allow on-orbit thermal conditioning of the landing gear hydraulic system number one fluid. The retract/circulation valve is closed in the GPC mode when the landing gear system is ARMED for deployment. Deployment controls for the nose and main landing gear are located on the flight deck display and control panel. The commander (CDR) and pilot (PLT) each has landing gear indicators and control stations. The commander's indicator and controls are on panel F6; and pilot's indicators and controls are on panel F8. At each station is a NOSE, LEFT, and RIGHT landing gear indicator and an ARM and DN (down) pushbutton switch/light indicator with guard covers.

The landing gear position indicators indicate UP when the gear is up and locked, BARBER POLE when the gear is deploying or retracting, and DN when the gear is down and locked.

Landing gear deployment is initiated when the CDR or PLT depresses the guarded ARM pushbotton switch/light indicator, then the guarded DN pushbutton switch/light indicator. This is accomplished at least 15 seconds prior to predicted touchdown and at a speed no greater than 300 knots (345 mph).

Depressing the ARM pushbutton switch/light indicator energizes latching relays that close the hydraulic system No. 1

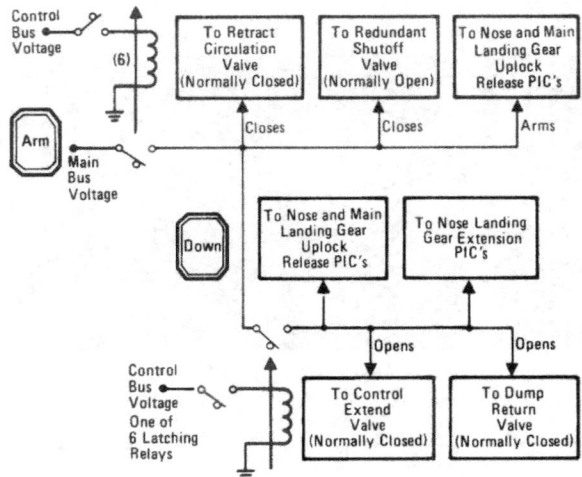

Arm/Down Circuit (Simplified)

landing gear retract/circulation valve and the normally open redundant shutoff valve to the retract/circulation valve. It also arms the nose and main landing gear pyrotechnic initiator controller's (PIC's) and illuminates the yellow light in the ARM pushbutton switch/light indicator.

The DN pushbutton switch/light indicator is then depressed. This energizes latching relays that open the hydraulic system No. 1 landing gear control valve, permitting the fluid in hydraulic system No. 1 to flow through the inlet port to the outlet port to the landing gear extend hydraulic lines. It also opens the landing gear dump valve, allowing the landing gear retract line fluid to flow into the hydraulic system No. 1 return line and illuminating the green light in the DN pushbutton switch/light indicator.

Space Shuttle Spacecraft Systems

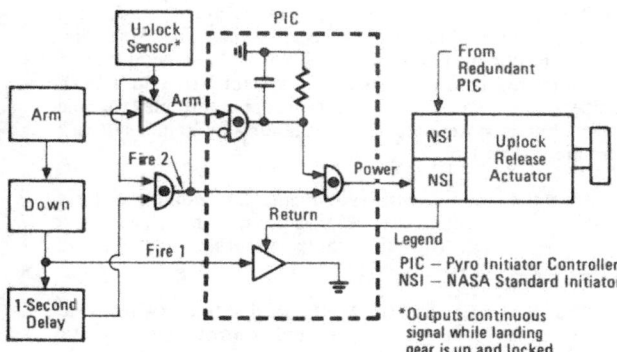

Uplock Release and Pyro Actuator Circuit (Simplified)

Hydraulic system No. 1 source pressure is routed to the nose and main landing gear uplock actuators, which releases the nose and main landing gear uplock hooks and door uplock hooks. As the uplock hooks are released, the gear begins its descent, and mechanical linkage attached to the doors and fuselage is powered by landing gear strut camming action during the gear extension, which opens the landing gear doors. There are two landing gear doors for the nose gear and one door for each main landing gear. The landing gear free-falls into the extended position, and the strut actuators aid in the deployment. The hydraulic strut actuator incorporates a hydraulic fluid flow-through orifice (snubber) to control the rate of landing gear extension and thereby prevent damage to the gear's down-lock linkages.

If hydraulic system No. 1 fails to release the landing gear—in one second after the DN pushbutton is depressed—the nose, left, and right main landing gear uplock sensors (proximity switches) will provide inputs to the PIC's for initiation of the redundant NASA Standard Initiators (NSI's) (nose, left, and right main landing gear pyrotechnic backup release system). They release the same uplock hooks as the hydraulic system. The nose landing gear, in addition, has a PIC and redundant NSI's that initiate a pyrotechnic power assist thruster strut extender two seconds after the DN pushbutton is depressed.

The landing gear drag brace overcenter lock and spring-loaded bungee lock the nose and main landing gear in the down position.

The uplock and downlock sensors (proximity switches) on each landing gear control the respective LANDING GEAR NOSE, LEFT, and RIGHT indicators.

The LDG GR/ARM/RN RESET switch positioned to RESET on the flight deck display and control panel A12 unlatches the relays that were latched during landing gear deployment by the LANDING GEAR ARM and DN pushbutton light/switch indicators. This is primarily a ground function, which will be performed only during landing gear deactivation.

The RESET position also will extinguish the yellow light in the ARM pushbutton switch/light indicator and the green light in the DN pushbutton switch/light indicator. In addition, the hydraulic system No. 1 landing gear dump valve is closed, the retract/circulation valve is opened only if the switch is in the OPEN position, and its redundant shutoff valve opened (deenergized) deenergizes the landing gear PIC circuits, and the landing gear control valve closes off the source pressure to the landing gear.

The nose landing gear tires are 32 by 8.8 inches and will withstand a burst pressure of not less than 3.2 times the normal inflation pressure of 15,515 mmHg (300 psi). The inflation agent is gaseous nitrogen. The maximum allowable load per nose landing gear tire is 20,412 kilograms (45,000 pounds) and rated at 225 knots (258 mph) landing speed.

Space Shuttle Spacecraft Systems

The nose landing gear shock strut has a 558-millimeter (22-inch) stroke. The maximum allowable derotation rate is approximately 9.4 degrees per second or 3.3 meters per second (11 feet per second), vertical sink rate.

The main landing gear tires are 44.5 by 16.21 inches. The normal inflation pressure is 16,301 mmHg (315 psi). The inflation agent is gaseous nitrogen. The maximum allowable load per main landing gear tire is 55,792 kilograms (123,000 pounds). (If the orbiter touches down with a 60/40 percent load distribution on a strut's two tires, with one tire supporting the maximum load of 55,792 kilograms (123,000 pounds), then the other tire can support a load of only 37,381 kilograms (82,410 pounds). Therefore, the maximum tire load on a strut is 93,173 kilograms (205,410 pounds) with a 60/40-percent tire load distribution. The tires are rated at 225 knots (258 mph).

The main landing gear shock strut stroke is 406 millimeters (16 inches). The allowable main gear sink rate for a 85,276-kilogram (188,000-pound) orbiter is 2.9 meters per second (9.6 feet per second) and 1.8 meters per second (6 feet per second) with a 102,513-kilogram (226,000-pound) orbiter. With a 20-knot (23-mph) crosswind, the maximum allowable gear sink rate for a 85,276-kilogram (188,000-pound) orbiter is 1.8 meters per second (6 feet per second) and approximately 1.5 meters per second (5 feet per second) with a 102,513-kilogram (226,000-pound) orbiter.

The landing gear tires have a life of five nominal landings.

MAIN LANDING GEAR BRAKES

Each of the orbiter's four main landing gear wheels have electro-hydraulic disc brakes and an anti-skid system.

Each main landing gear wheel has a disc brake assembly consisting of nine discs, four rotors, and five stators. The carbon-lined beryllium rotors are splined to the inside of the wheel and rotate with the wheel. The carbon-lined beryllium stators are splined to the outside of the axle assembly and do not rotate with the wheel. When the brakes are applied, eight hydraulic actuators in the brake assembly press the discs together, providing brake torque. Four of these actuators are manifolded together from hydraulic system No. 1 in a brake chamber. The remaining four actuators are manifolded together from hydraulic system No. 2. The standby hydraulic system is hydraulic system No. 3.

As in the landing gear deployment, the landing gear isolation valve in hydraulic system No. 1, 2, and 3 must be open to allow the applicable hydraulic source pressure to the main landing gear brakes.

The BRAKES MNA, MNB, and MNC switches are located on the flight deck display and control panel O14, O15, and O16 and allow electrical power to the brake/anti-skid control boxes A and B. The ANTI SKID switch located on panel L2 provides electrical power for enabling the anti-skid portion of

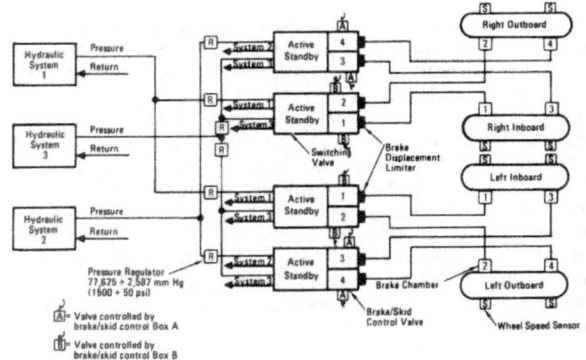

Wheel Brake Subsystem

Space Shuttle Spacecraft Systems

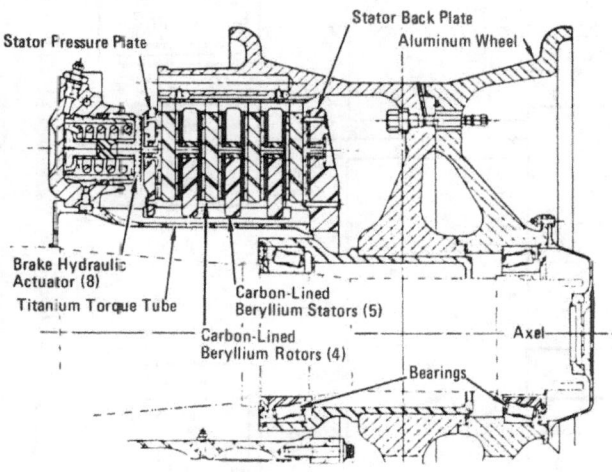

Wheel Brake Assembly

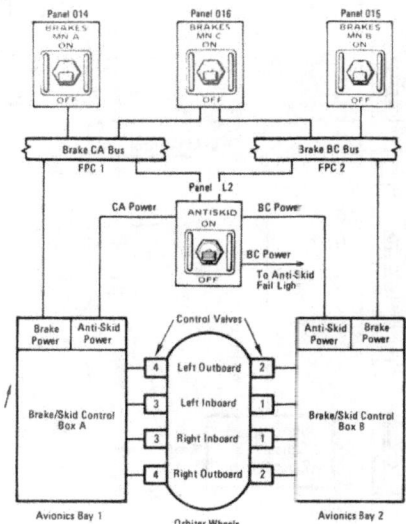

Brake/Skid Control Power

the braking system boxes A and B. The BRAKES MNA, MNB and MNC switches are positioned to ON to supply electrical power to the brake boxes A and B and to OFF to remove electrical power. The ANTI-SKID switch is positioned to ON to enable the anti-skid system and to OFF to disable the system.

When weight is sensed on the main landing gear, the brake/anti-skid boxes A and B are enabled, permitting the main landing gear brakes to become operational.

The main landing gear brakes controlled by the CDR or PLT brake pedals located on the rudder pedal assemblies at the CDR and PLT station. The pedals are adjustable by a handle. The braking commands are accomplished by the CDR or PLT initiating toe pressure on the top of the rudder pedal assembly.

Each brake pedal (left and right) has four linear variable differential transducers (LVDT's). The left pedal transducer unit will output four separate braking signals through the brake/skid control boxes for braking control of the two left main wheels. The right pedal transducer unit does likewise for the two right main wheels. When toe pressure is applied to the brake pedal, the transducers transmit electrical signals of 0 to 5 vdc to the brake/anti-skid control boxes. The pedal with the greatest toe pressure becomes the controlling pedal through electronic OR circuits. The electrical signal is proportional to the toe pressure. The electrical output energizes the main landing gear brake coils proportionally to brake pedal deflection allowing the desired hydraulic volume to be directed to the main

Space Shuttle Spacecraft Systems

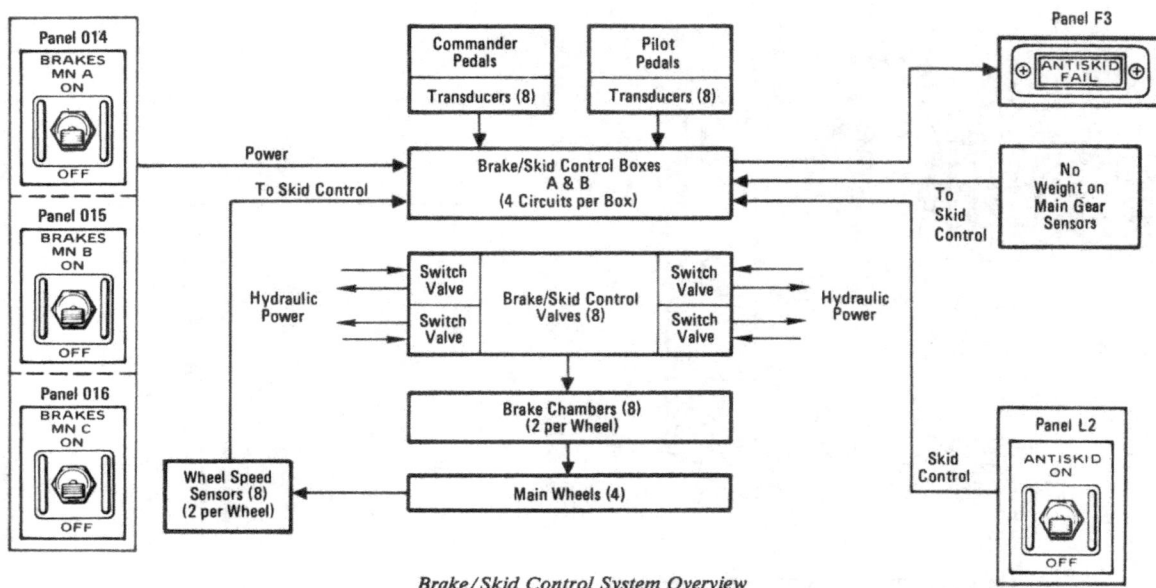

Brake/Skid Control System Overview

landing gear brakes for braking action. The brake system bungee at each brake pedal provides the braking artificial feel to the crew member.

Each of the three hydraulic systems' source pressure of 155,250 millimeters of mercury (mmHg) (3,000 psi) is reduced by a regulator in each of the brake hydraulic systems to 77,625 mmHg (1,500 psi). Hydraulic system No. 1 and 2 are the normal hydraulic supply to the brake system, with system No. 3 as the standby. Switching valves in each brake system provide automatic switching if either of the active hydraulic systems is lost.

The anti-skid portion of the brake system provides optimum braking by preventing tire skid or wheel lock, and subsequent tire damage.

Each main landing gear wheel has two speed sensors that supply wheel rotational velocity information to the anti-skid monitoring circuits. If wheel velocity is approximately 30 percent below the average of all four wheels, the anti-skid control

Space Shuttle Spacecraft Systems

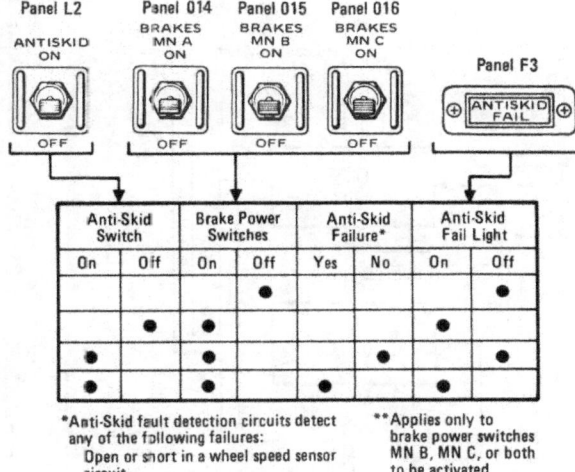

Anti-Skid Fail Light Status

circuits consider that wheel is locked or in a skid condition. The anti-skid coil in that brake/skid control valve is then energized, reversing polarity of that brake coil and inhibiting the hydraulic pressure to the brakes of that wheel until that wheel's velocity is increased again. The anti-skid control system also considers a flat tire. If that wheel's velocity is approximately 30 percent below the average velocity of all four wheels for at least two seconds, the hydraulic pressure on the adjacent wheel is limited to 41,400 plus or minus 5,175 mmHg (800 plus or minus 100 psi) to prevent brake and/or tire damage.

Anti-skid control is disabled between 10 to 15 knots (11 to 17 mph) to prevent loss of manual braking during parking and maneuvering. The anti-skid system control circuits contain fault detection logic. The ANTI-SKID yellow caution and warning light located on the flight deck display and control panel F4 will illuminate if the anti-skid fault detection circuit detects an open or short in a wheel speed sensor, open or short in a anti-skid control valve servo coil, or a failure in an anti-skid control circuit. A failure of the aforementioned items will only deactivate the failed circuit, not the total anti-skid control. If the BRAKE POWER switches are ON and the ANTI-SKID switch is OFF, the ANTI-SKID caution/warning light will illuminate.

With one brake chamber on each wheel providing hydraulic pressure, the orbiter crew can preform a normal-energy stop; however, the breaking distance will be approximately 30 to 40 percent longer.

In a normal-energy stop with an 85,276-kilogram (188,000-pound) orbiter, beginning with initial braking at 153 knots (176 mph), 871 meters (2,860 feet) is required to stop the orbiter with 77,625 mmHg (1,500 psi) of hydraulic pressure. An emergency stop with a 102,513 kilogram (227,000-pound) orbiter beginning with initial braking at 178 knots (204 mph), requires 1,330 meters (4,364 feet) to stop the orbiter with 77,625 mmHg (1,500 psi) of hydraulic pressure.

The brakes have five normal energy stop landings. One emergency stop landing will require brake refurbishment.

NOSE WHEEL STEERING

The orbiter nose wheel is steerable upon landing. The nose wheel is electro-hydraulically steerable automatically through use of the general purpose computer (GPC) and flight control system AUTO mode. The nose wheel may also be steered by use

 **Space Shuttle Spacecraft Systems**

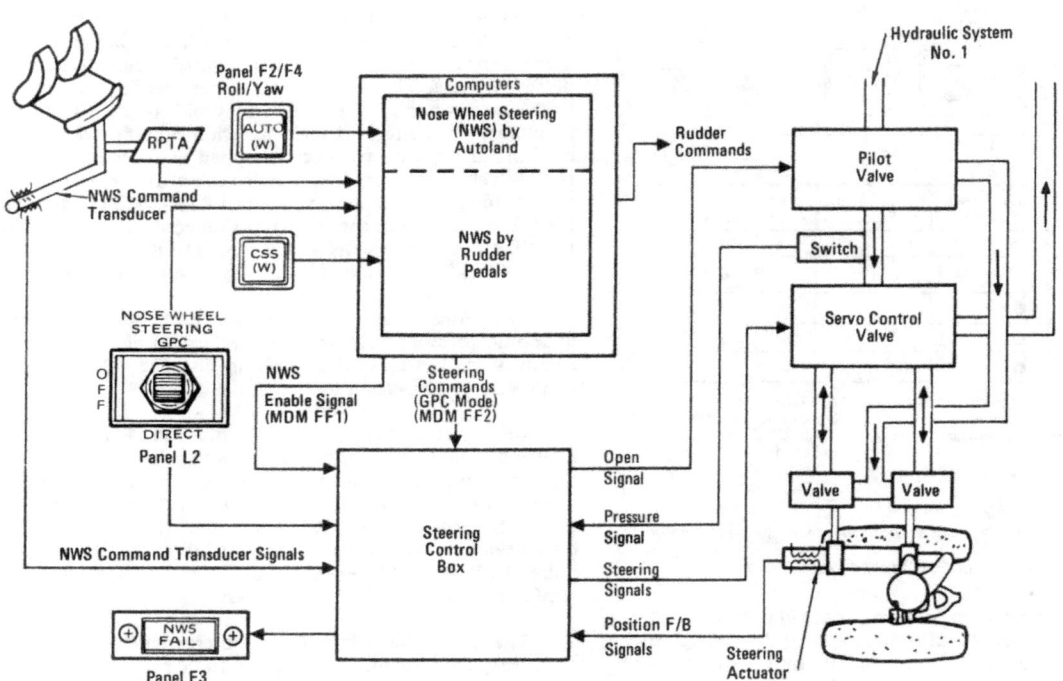

Nose Wheel Steering

Space Shuttle Spacecraft Systems

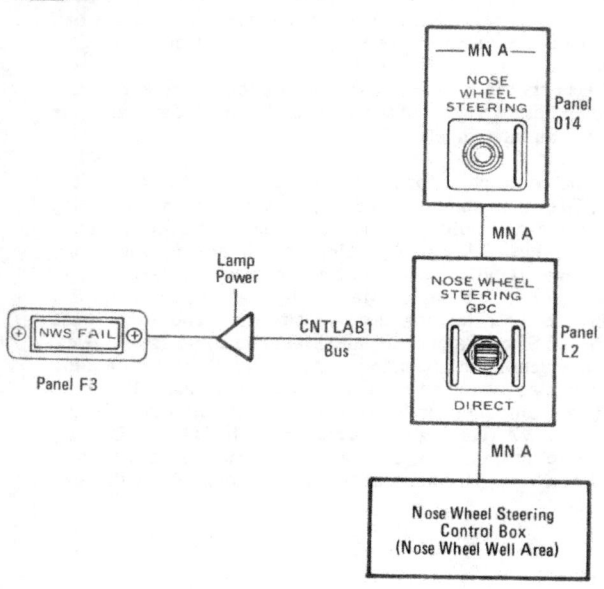

Nose Wheel Steering Power

of the CDR or PLT rudder pedals in the control stick steering (CSS) mode.

Only hydraulic system No. 1 supplies hydraulic system pressure for steering the nose wheel in either of the aforementioned modes. If system No. 1 is inoperative, the CDR or PLT can apply the left and right main landing gear brakes, alternately, which will caster the nosewheel and provide nose wheel steering.

The NOSE WHEEL STEERING switch on the flight deck display and control panel L2 enables the nose wheel steering solenoid control valves, which directs hydraulic system No. 1 pressure to the nose wheel steering servo valve when the NOSE WHEEL STEERING switch is in the GPC position. The GPC position, in conjunction with the flight control system ROLL/YAW AUTO Pushbutton switch/light indicator on panel F2 or F4, allows the AUTOLAND computer mode to provide steering commands to the nose wheel servo actuator when weight on the nose gear (main and nose gear proximity switches plus vehicle attitude) is sensed and the hydraulic pressure switch detects that the hydraulic pressure is at least 69,862 mmHg (1,350 psi). Position feedback is transmitted from the nose wheel steering actuator for position verification and fault detection. It is noted that if the hydraulic pressure falls to 51,750 plus or minus 7,762 mmHg (1,000 plus or minus 150 psi), nose wheel steering is disabled. When nose wheel steering is not activated, the nose wheel steering damping actuator prevents nose wheel shimmy and allows the nose wheel to free castor through plus or minus 10 degrees of rotation when differential braking is used.

When the NOSE GEAR STEERING switch is positioned to DIRECT, the nose wheel steering solenoid control valves are enabled, which directs hydraulic system No. 1 pressure to the nose wheel steering servo actuator. The DIRECT position, in conjunction with the flight control system ROLL/YAW CSS pushbutton switch/light indicator on panels F2 and F4 allows the CDR or PLT rudder pedals to provide steering commands directly from the steering command transducer on the rudder pedals to the servo actuator, providing hydraulic pressure to the steering actuator when weight on the nose is sensed. Position feedback is transmitted from the nose wheel steering actuator for position verification and fault detection.

When the flight control system ROLL/YAW AUTO pushbutton switch/light indicator is depressed, it illuminates a

Space Shuttle Spacecraft Systems

white light. When the ROLL/YAW CSS Pushbutton switch/light indicator is depressed, it illuminates a white light. Only ROLL/YAW AUTO or ROLL/YAW CSS can be selected at a time, not both simultaneously.

When the NOSE WHEEL STEERING switch is in GPC or DIRECT and the nose wheel steering system fault detection logic detects loss of hydraulic pressure, open or short in the servo control valve circuitry, open or short in the position feedback, rate position error, open or short in the command transducer, broken linkage or loss of electrical power, the NWS (nose wheel steering) FAIL yellow caution and warning light on Panel F3 will illuminate, and the nose wheel reverts automatically to free castor.

If nose wheel steering has failed, the NOSE WHEEL STEERING switch is positioned to OFF, extinguishing the NWS FAIL yellow light. Nose wheel steering can then be accomplished by differential braking. In differential braking, the CDR or PLT applies toe pressure to the rudder pedals, alternately to the left and right main landing gear brakes; this will castor the nose wheel for nose wheel steering. When the NOSE WHEEL STEERING switch is in the OFF position, hydraulic system No. 1 pressure to the nose wheel steering system is off, allowing the nose wheel to be in the free castor mode.

Heaters can be installed on the landing gear brake system for temperature control of the brake hydraulic system if they are required on later flights.

The contractors for the landing gear are B.F. Goodrich, Troy, OH (main and nose landing gear wheel and main landing gear brake assembly and the nose/main gear tires); Bertea Corp., Irvine, CA (main landing gear hydraulic uplock actuator, main landing gear strut actuator and nose landing gear uplock actuator); Menasco Manufacturing CO., Burbank, CA (main and nose landing gear shock struts and drag brace assembly); Sterer Engineering and Manufacturing, Los Angeles, CA (nose gear steering/damping and solenoid-operated landing gear uplock control valves); Crane Co., Hydro Aire, Burbank, CA (main landing gear brake antiskid system); Eldec Corp., Lynwood, WA (landing gear proximity switch); OEA, Denver, CO (nose gear uplock release pyro thruster); Scott Inc., Downers Grove, IL (main landing gear uplock release thruster actuator).

AVIONICS SYSTEMS

The Space Shuttle avionics system controls, or assists in controlling, most of the Shuttle systems. Its functions include automatic determination of the vehicle status and operational readiness, implementation sequencing and control for the external tank and solid rocket boosters during launch and ascent, performance monitoring, digital data processing, communications and tracking, payload and system management, and guidance, navigation, and control, as well as electrical power distribution for the orbiter, external tank, and solid rocket boosters.

Automatic vehicle flight control can be used for every phase of the mission except docking, which is a manual operation by the flight crew. Manual control—referred to as the control stick steering (CSS) mode—also is available at all times as a flight crew option.

The avionics equipment is arranged to facilitate checkout, access, and replacement with minimal disturbance to the other subsystems. Almost all electrical and electronics equipment is

Space Shuttle Spacecraft Systems

installed in three areas of the orbiter: the flight deck, the forward and aft avionics equipment bays in the mid deck of the orbiter crew compartment, and the aft avionics equipment bays in the orbiter aft fuselage. The flight deck of the orbiter crew compartment is the center of avionics activity, both in flight and on the ground, except during hazardous servicing operations before flight. Before launch, the orbiter avionics system is linked to ground support equipment through umbilical connections.

The Space Shuttle avionics system consists of more than 300 major electronic "black boxes" located throughout the vehicle, connected by more than 300 miles of electrical wiring. The black boxes are connected to a set of five computers through common party lines, called data busses. The black boxes offer dual or triple redundancy for every function.

The avionics are designed to withstand multiple failures through redundant hardware and software (computer programs) managed by the complex of five computers; this is called a fail-operational/fail-safe capability. Fail-operational/fail-safe capability is provided by a combination of hardware and software redundancy. Fail-operational performance means that after a first failure in a system, redundancy management allows the vehicle to continue on its mission. Fail-safe means after a second failure, the vehicle still is capable of returning to a landing site safely.

The status of the individual avionics components is checked by a system monitoring computer program. The status of critical vehicle functions such as the payload door position, external tank and solid rocket booster separation mechanisms, and excessive temperatures for certain areas are monitored continuously and displayed to the crew.

DATA PROCESSING SYSTEM

The orbiter relies on computerized control and monitoring for successful performance. The data processing system (DPS) through the use of various hardware components and its self-contained computer programming (software) provides this monitoring and control.

The data processing system consists of five general-purpose computers for computation and control; two magnetic-tape mass memories for large-volume bulk storage: time-shared, serial digital data busses (essentially party lines) to accommodate the data traffic between the computers and other orbiter systems; 19 multiplexer/demultiplexer units to convert and format data at various systems; three engine interface units to command the orbiter main engines; and four multi-function television (cathode ray tube—CRT) display systems so the crew can monitor and control the vehicle and payload systems.

The software stored in and executed by the orbiter computers is the most sophisticated and complex set of programs ever developed for aerospace use. The programs are written to

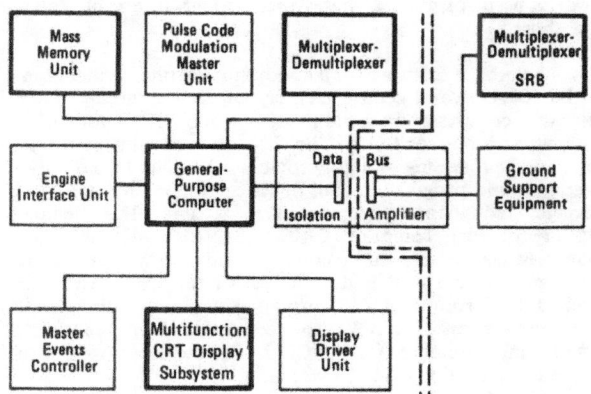

Data Processing System

Space Shuttle Spacecraft Systems

accommodate almost every aspect of the Space Shuttle vehicle operations including vehicle checkout at the Rockwell Palmdale (CA) assembly facility, pre-launch and final countdown, turnaround activity at the Kennedy Space Center, and control during ascent, on-orbit, entry, and landing and abort or other contingency mission phases.

In-flight programs monitor the status of vehicle systems; provide consumable computations; control the opening and closing of the payload bay doors; operate the remote manipulator system; perform fault detection and annunciation, provide for payload monitoring, commanding, control, and data acquisition; provide antenna pointing for the various communication systems; and provide primary and backup guidance, navigation, and control for ascent, on-orbit, entry, landing, and abort.

These computer programs are written so they can be executed by a single computer or by all computers executing an identical program in the same time frame. The multicomputer mode is used for critical phases such as launch, ascent, entry, and abort.

The orbiter software for a major mission phase must fit into the 106,496-word central memory of each computer. Each computer consists of a central processor unit (CPU) and an input/output processor (IOP). The CPU performs the arithmetic and logical processing of data, provides control and handling of interrupts and program control of its corresponding IOP, and manages redundant systems such as sensors. The memory capacity of each computer CPU is 81,920 words. All data transmissions between the computers and vehicle systems are performed by the IOP under control of the CPU. The IOP receives data from the CPU, formats it, and relays commands to vehicle systems. The IOP also receives data from the vehicle systems and formats it for the CPU. The memory capacity of each computer IOP is 24,576 words.

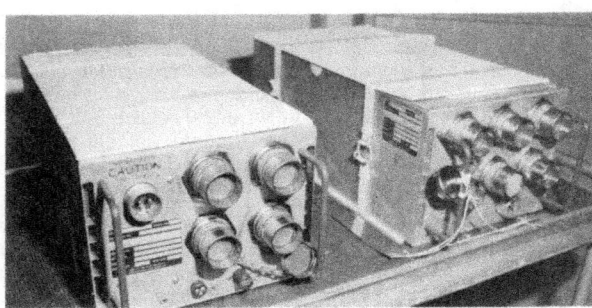

General-Process Computer (IOP and CPU)

To accomplish all of the computing functions for all mission phases, approximately 400,000 words of computer memory are required. To fit the software needed into the computer memory space available, computer programs have been subdivided into nine memory groups corresponding to functions executed during specific flight and checkout phases. As an example, one memory group accommodates final countdown, ascent, and aborts; another on-orbit operations; and another the entry and landing computations. Different memory groups support checkout and ground turnaround operations and system management functions. Thus, in addition to central memory stored in the computers themselves, 34,000,000 bytes of information can be stored in two mass memories.

The orbiter computers are loaded with different memory groups from magnetic tapes containing the desired program. In this way all the software needed can be stored in mass memory units (magnetic tape machine) and loaded into the computers only when actually needed. Critical programs and data are loaded in both mas memories and protected from erasure. Normally, one mass memory is activated for use and the other is held in

Space Shuttle Spacecraft Systems

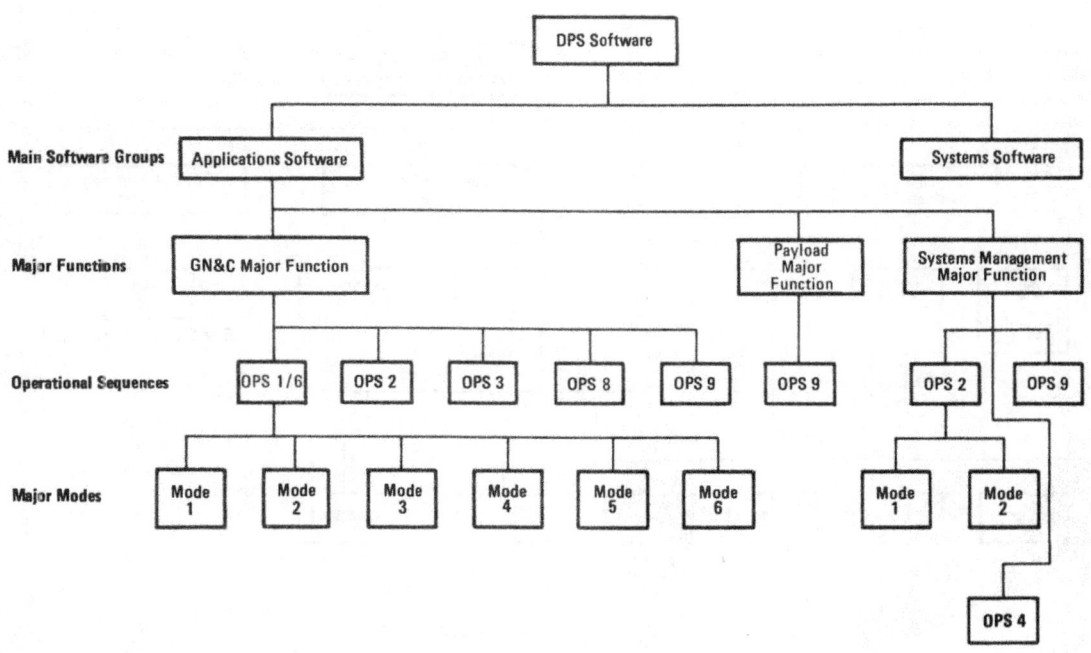

Data Processing Software

 **Space Shuttle Spacecraft Systems**

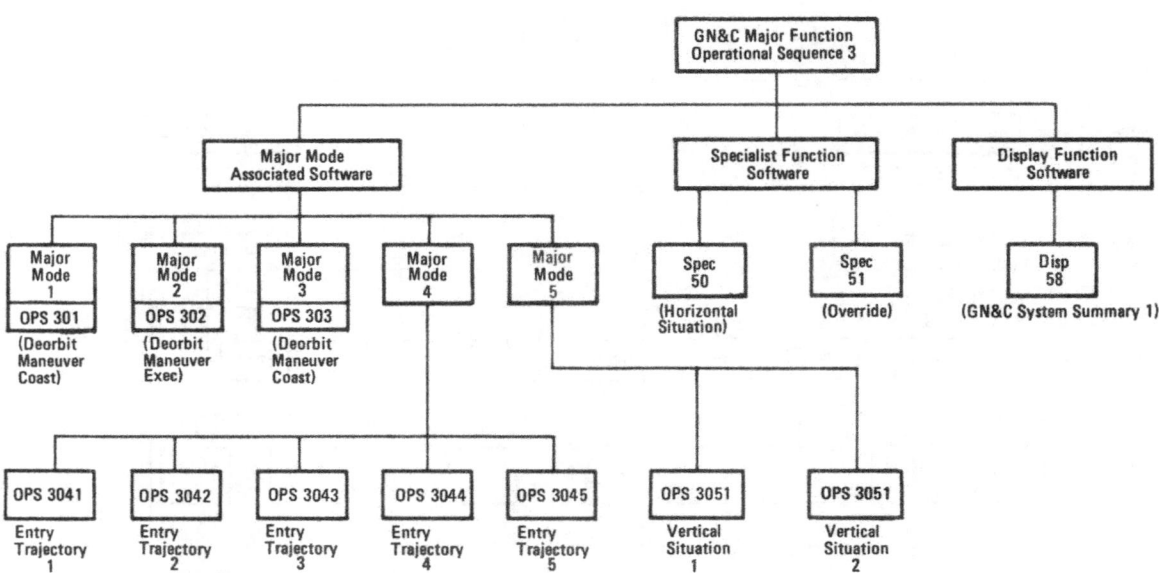

Operational Sequence Associated Software

Space Shuttle Spacecraft Systems

reserve. However, it is possible to use both simultaneously on separate data busses or communicating with separate computers. The data stored in the mass memories include prelaunch and preflight test routines, fault isolation diagnostic test programs, cathode ray tube (CRT) display formats, overlay program segments to be loaded during specific mission phases, and duplicate copies of resident on-line programs for initial loading, reloading, or reconfiguration of the computers. The mass memories are an advanced form of data storage and fill the gap between slow access drives of high storage capacity and discs or drums with fast access but relatively low storage capacity. In contrast to disc or drum memories, the mass memories consume power only when active.

The DPS software is divided into two major groups, called system software group and applications processing software.

At the top is the system software group, which consists of three sets of programs: the flight computer operating program (the executive), which controls the processors, monitors key system parameters, allocates computer resources, provides for orderly program interrupts for higher priority activities, and updates computer memory; the user interface programs, which provide instructions for processing crew commands or requests; and the system control program, which initializes each computer and arranges for the multicomputer operation during flight-critical periods. The system software group programs tell the computers how to perform and how to communicate with other equipment.

The second level of memory groups is the applications processing software. This group contains (1) specific software programs for guidance, navigation, and control which are required for launch, ascent flight to orbit, maneuvering on orbit, entry and landing on a runway; (2) systems management programs which contain instructions for loading memories in the main engine computers and for checking the instrumentation system in addition to aiding in vehicle subsystem checkout and in ascertaining that crew displays and controls perform properly and update the inertial measuring unit state vectors; (3) payload processing programs which contain instructions for control and monitoring of orbiter payload systems which can be revised depending on the nature of the payload; and (4) vehicle checkout programs which are required to handle data management, performance monitoring, and special processing and display and control processing.

The two software program groups are combined to form a memory configuration for a specific mission phase. The software programs are written in HAL/S (high-order assembly language/Shuttle) especially developed for use in real-time space applications. These programs are grouped by function and partitioned into memory configuration. When requested, memory is reconfigured from mass memory so operating sequences for the needed function can be overlaid into the main computer memory.

The highest level of the applications software is the OPS (operational sequences) which is required to perform part of a mission phase. Each OPS is a set of unique software required to perform phase-oriented tasks. An OPS can be further subdivided into groups called major modes, each representing a portion of the OPS mission phase. As an example, the launch phase (OPS-1) is subdivided into six major modes.

Each major mode has with it an associated CRT display which provides the flight crew with information concerning the current portion of the mission phase. The display function of OPS software presents a fixed format of data and configuration status on a CRT, which is not subject to flight crew manipulation and is used only to provide the flight crew with information.

The specialist (SPEC) function of the OPS software is a block of software associated with one or more OPS which has an associated CRT display and enables the flight crew to

Space Shuttle Spacecraft Systems

monitor and manipulate the vehicle systems via item keyboard entries.

The multifunction CRT display system provides the flight crew with the ability to interface with and control the onboard software, observe vehicle system data, and monitor error or fault messages. The system is composed of three components: display electronics units (DEU), keyboard units (KBU), and display units (DU), which include the CRTs.

The flight crew has two keyboards, on the left and right sides of the flight deck display and control center console. There are three DU-CRTs on the flight deck forward display and control console. Each CRT is 127 x 177 millimeters (5 x 7 inches). There is also one keyboard and one DU-CRT at the aft side station flight deck display and control console. The three DU-CRTs at the forward console are connected to each of the forward center console keyboards. The DU-CRT at the aft station is connected to the keyboard at that station.

The DU uses a magnetic-deflected electrostatic-focused CRT. When supplied with deflection signals and video input, the CRT will display alphanumeric and graphic information. Characters can be flashed and the CRT brightness varied for individual characters. The CRT has a single-color (green) phosphor.

The four DEUs provide storage of display data, the computer/keyboard unit and computer/display unit interface display generation, updating, and refreshing, keyboard entry error checking, and keyboard entry echoing to the DUs.

The three keyboard units provide the crew with a controlling interface for software operations and management. Each keyboard has 32 momentary-contact (pushbutton) function and numeric keys. Using these keys, the flight crew can ask the computer more than 1,000 questions about the flight and condition of the vehicle. The DUs provide the flight crew almost im-

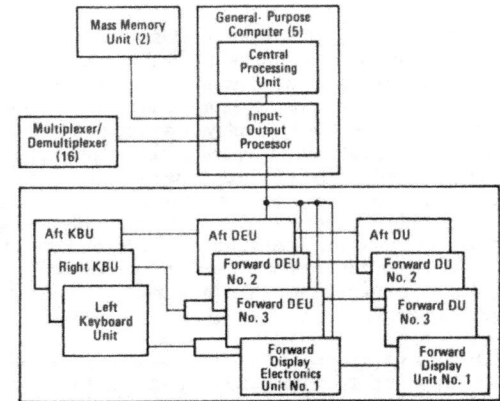

Data Processing System Hardware

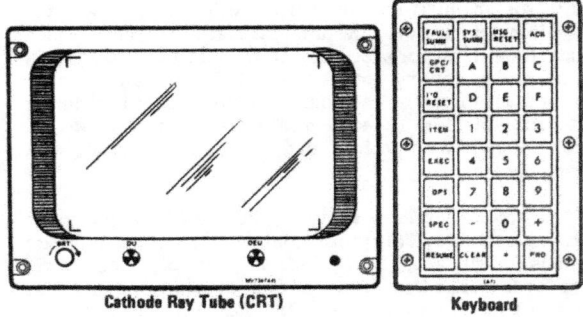

Cathode Ray Tube and Keyboard

Space Shuttle Spacecraft Systems

mediate response to the inquiries through display graphs, trajectory plots, and prediction about flight progress. The flight crew controls the Space Shuttle system operation through the use of the keyboards. In conjunction with the DUs, the flight crew can alter the system configuration, change data or instructions in the computer main memory, change memory configurations corresponding to different mission phases, respond to error messages and alarms, request special programs to perform specific tasks, run through operational sequences for each mission phase, and request specific displays.

The input-output processor (IOP) of each computer has 24 independent processors, each of which controls one of the 24 data busses used to transmit digital data between the computers and vehicle systems, and secondary channels between the telemetry system and units that collect instrumentation data.

Cathode Ray Tube Display

The data transfer technique uses time-division data multiplexing with pulse-code modulation. In this system, data channels are multiplexed together, one after the other, and information is coded on any given channel by a series of binary pulses corresponding to discrete information. The information transmission word length is 28 bits. The first three bits provide synchronization and indicate whether the information is commands or data. The next five bits identify the destination or source of the information. For command words, 19 bits identify the data transfer or operations to be performed; for data words, 16 of these 19 bits contain the data and 3 bits define the word validity. The last bit of each word format is for an odd parity error test. The 24 data busses are connected to each IOP via multiplexer interface adapters (MIAs) which receive, convert, and validate the serial data in response to discrete signals calling for available data to be transmitted or received. The data busses are organized into seven groups: intercomputer communication, display keyboard, flight critical, mass memory, payload launch/boost, and instrumentation.

Interface adaptation between the data bus network and most vehicle systems is accomplished by multiplexer/demultiplexers (MDM). The MDMs are used in numerous remote locations in the vehicle to handle the functions of serial data time multiplexing/demultiplexing associated with the digital data busses and for signal conditioning. The MDMs act as translators and put information on or take it off the data busses. There are 19 MDMs on the orbiter and two on each SRB.

The MDMs receive from the vehicle systems hundreds of analog signals which can be minus 5 to plus 5 vdc, 28-vdc discrete signals, and serial or digital words, and converts these into a digital-serial output (which is a digitized representation of the signals and data). The digital-serial outputs are transmitted via the data busses to the computers and to the pulse code modulation (PCM) master unit.

 **Space Shuttle Spacecraft Systems**

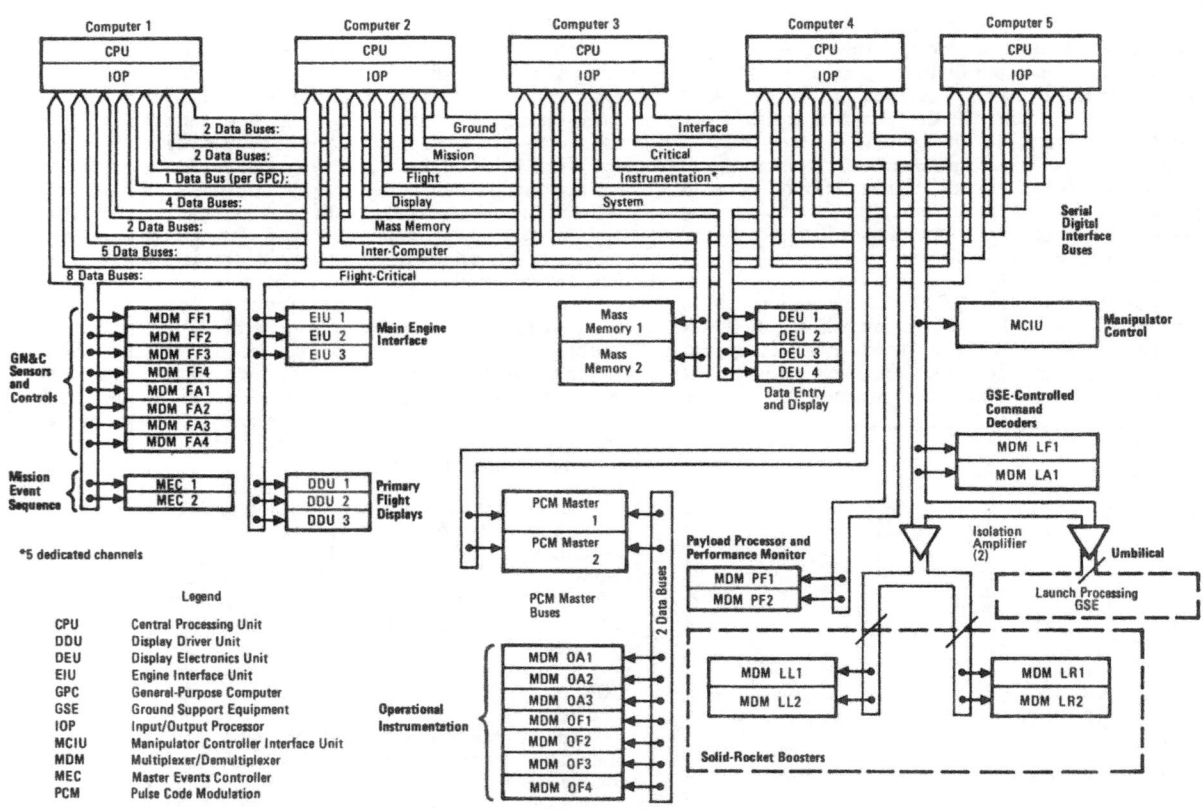

Data Processing System

Space Shuttle Spacecraft Systems

The technique of transferring serial-digital computer data via the data busses to the MDMs and then to vehicle systems is called multiplexing.

Each computer sends serial-digital downlist data through four instrumentation busses to the pulse code modulation master unit, where it is mixed with instrumentation and payload data and transmitted to ground downlink telemetry.

The PCM also formats the vehicles operational instrumentation into serial digital for transmittal to telemetry.

The four instrumenmtation busses also transmit non-flight-critical orbiter system data from the PCM master unit to each computer for display on the flight crew CRTs. The PCM master unit contains a programmable read-only memory (PROM) for accessing subsystem data, a random-access memory (RAM) in which to store system data, and a memory in which data from the computers are stored for incorporation in the downlink telemetry.

The MDMs controlled by either the computers or pulse code modulation master unit are a demand response system via the data busses. A command from either can order the applicable MDM to collect data and transmit the data back to the controlling hardware.

Uplink software provides for the ground to send commands and data to the orbiter via the S-band transponder or Ku-band communications link. The orbiter uplink software interfaces with network signal processors to the computers via one of the flight-critical MDMs.

All data is time-tagged by three master timing units (MTU), which provide a Greenwich Mean Time (GMT) base as well as mission elapsed time and event time. The system software in each computer selects the GMT from one of the MTUs or from the computer's own internal clock with frequent updates from the MTUs as a function of timekeeping redundancy management. The timekeeping software can be controlled by the flight crew via manipulation of the specialist function CRT display. The MTUs also supply synchronizing signals to other electronic circuits.

All computer communications with the DU-CRT display system are transmitted and received over the display keyboard busses.

The guidance, navigation, and control system is composed of the four orbiter computers and other major components which make up the primary flight control system. The computers use a program called the digital autopilot to control the vehicle through launch, ascent, on-orbit, deorbit, entry, and landing. The guidance, navigation, and control system provides automatic or manual (control stick steering) control of the vehicle in all flight phases. During launch most of the computer commands are directed to gimbal the main engines and solid rocket boosters. To circularize the orbit, in orbit, and for deorbit, the computer directs the orbital maneuvering system. At external tank separation, in orbit, and during a portion of entry, vehicle attitude control commands are directed to the reaction control system. In atmospheric flight the computers direct the orbiter aerodynamic flight control surfaces.

During critical mission phases (launch and entry), four of the computers are assigned to perform GN&C tasks, operating as a cooperative redundant set. One computer acts as a commander of a given data bus in the flight control scheme and initiates all bus transactions. The noncommander computers on the same bus listen to all incoming data that the commander requests. Thus, each response to a request by any computer is heard by all performing the same redundant operations and verified for consistently identical output. The computer redundancy management software module centers around the concept

Space Shuttle Spacecraft Systems

that each computer compares its outputs with the other computers in the set. If the comparison disagrees, this disagreement is displayed to the flight crew as a CRT message; however, processing continues.

Each of the computers operating in a redundant set operates in synchronized steps and cross-checks results of processing about 440 times per second. If a computer operating in a redundant set fails to meet the synchronization requirements for redundant set operations, it would be removed from the set. Each computer performs about 325,000 operations per second during critical phases of the mission.

As an example, Computers 1 through 4 are operating in a redundant set when Computer 1 fails to stay synchronized with the other three. Computer 1 software recognizes the disagreement and also recognizes that it has been voted failed by the other three computers. Computer 1, therefore, sets a self-fail vote and does not vote the other three computers failed.

All intercomputer communications other than synchronization are transmitted and received over four specific buses. Cross-strapping of the four buses to the four computers allows each computer access to the status of the data received or transmitted by the other computers, making possible the verification of identical results among the four computers. The four computers are loaded with the same software programs. Each bus is assigned to one of the four computers in the command mode and the remaining computers operate in the listen mode for the bus. Each computer has the ability to receive data with the other three computers, pass data to the others, request data from the others, and perform any other tasks required to operate the redundant set. No vehicle systems are connected to these buses.

The flight-critical buses are directed into groups of four to be compatible with the grouping of the four computers. Commands to flight deck crew flight control system (dedicated) displays and the forward GN&C system as well as the data from the forward GN&C sensors are transmitted and received over one group of buses (flight-critical FC-1 through FC-4). The data commands to the aft GN&C system, as well as the data from the aft GN&C components, are transmitted and received over FC-5 through FC-8 buses. Each bus in a group is assigned to a separate computer operating in a command mode. The computer in the command mode issues data requests and commands to the applicable vehicle systems over its assigned FC (dedicated) bus. The remaining three buses in each group are assigned to the remaining computers to operate in the listen mode. A computer operating in the listen mode can only receive data. Thus, if Computer 1 operates in the command mode on bus FC-1, it listens on the three remaining buses. In this manner, each computer commands on one bus of a flight-critical group and listens on the remaining three.

Each flight-critical bus in a group of four is commanded by a different computer. There are multiple units of each GN&C hardware item (sensors, controllers, flight control effector) and each unit is wired to a different MDM and flight-critical bus. The MDM and bus can be assigned to another computer. The flight computer operating system in systems software in each of the redundant set computers activates a GN&C executive program and issues commands to the bus and MDM to provide a set of input data. Each MDM receives the command from the computer assigned to it, acquires the requested data from the GN&C hardware wired to it, and sends the data to all four computers.

When the sets of GN&C hardware data arrive at the computers via the MDMs and data buses, the data is generally not in the proper format, units, or form for use by flight control, guidance, or navigation. A subsystem operating program for each type of hardware processes the data to make it usable by GN&C software. These programs contain the software necessary for hardware operation, activation, self-testing, and moding. The level of redundancy varies from two to four

Space Shuttle Spacecraft Systems

depending on the particular unit. The software which processes data from the redundant GN&C hardware is called redundancy management. This performs two functions: selects, from redundant sets of hardware data, one set of data for use by flight control, guidance, and navigation; and detects data which is out of tolerance, identifies the faulty unit, and announces the failure to the flight crew and to the data collection software.

In the case of four hardware units, the redundancy management software utilizes three and holds the fourth in reserve and utilizes a middle value select until one of the three is bad, then uses the fourth. If one of the remaining three is lost, it would downmode to two and use the average of the two. If one of the remaining two were lost it would downmode to one and pass the only data it receives.

The three engine interface units between the computers and three main engine controllers accept computer main engine commands, reformat then, and transfer them to each main engine controller. In return, the engine interface units accept data from the main engine controller, reformat it, and transfer it to computers and operational instrumentation. Main engine functions such as ignition, gimbaling, throttling, and shutdown are controlled by the main engine controller internally through inputs from the guidance equations which are computed in the orbiter computers.

During non-critical flight periods in orbit, only one or two computers are used for GN&C tasks and another for payload operations and system management. The remaining three can be used either for payload management or deactivated on standby.

The fifth onboard computer is used as a GN&C backup in the flights; however, it also provides unique functions such as system non-critical function monitoring and payload command and monitoring. The fifth computer has a separate independent software design and coding activity to protect against generic software failures in the primary computer set.

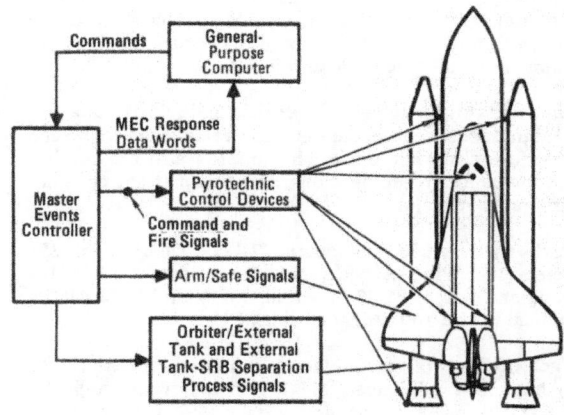

Master Events Controller

The payload in the orbiter may have up to five safety-critical status parameters hardwired, so that these parameters and others can be recorded as a part of the orbiter's system management which is transmitted and received over the two payload buses. To accommodate the various forms of payload data, the payload data interleaver integrates payload data into the orbiter avionics so it can be transmitted to ground telemetry.

The two master events controllers under computer control provide signals for arming and safing pyrotechnics, and for command and fire signals for pyrotechnics in separation processes.

Data bus isolation amplifiers are the interfacing device among GSE, the solid rocket booster MDMs, and the orbiter launch data bus. They transmit or receive multiplexed data in any direction. The amplifiers enable multiplexed communications over the longer data bus cables which connect the orbiter

Space Shuttle Spacecraft Systems

and GSE. The receiving section of the amplifiers detects low-level coded signals, discriminates against noise, and decodes the signal to standard digital data at very low bit error rate; the transmit section of the amplifiers then re-encodes the data and retransmits it at full amplitude and low noise.

Data bus couplers couple the vehicle multiplexed data and control signals between the data bus and cable studs connected to the various electronic units. The couplers also provide impedance matching on the data bus, line termination, dc isolation, and noise rejection.

Each CPU is 193 millimeters (7-1/2 inches) high, 257 millimeters (10-1/8 inches) wide, and 497 millimeters (19-1/2 inches) long and weighs 25.85 kilograms (57 pounds). The IOPs are identical in size and weight to the CPUs.

Each of the two mass memories is 193 millimeters (7-1/2 inches) high, 294 millimeters (11-1/2 inches) wide, and 381 millimeters (15 inches) long and weighs 9.97 kilograms (22 pounds). The MDMs are 330 by 254 by 177 millimeters (13 by 10 by 7 inches) and weigh 16.64 kilograms (36.7 pounds) each. The data bus isolation amplifiers are each 177 by 152 by 127 millimeters (7 by 6 by 5 inches) and weigh 3.4 kilograms (7-1/2 pounds). Each data bus coupler is 16.39 cubic centimeters (one cubic inch) in size and weighs less than 28 grams (one ounce).

The five CPUs and IOPs are located in the crew compartment mid deck avionics bays 1, 2, and 3 and are cooled by fans. The mass memories are located in the crew compartment mid deck avionics bays 2 and 3; each MM is mounted on a coldplate and cooled by a water loop.

The forward flight-critical MDMs are located in the crew compartment mid deck avionics bays 1, 2, and 3. They also are mounted on coldplates and cooled by a water loop. The aft flight-critical MDMs are located in the aft fuselage avionics bays, are mounted on coldplates, and cooled by Freon-21 coolant loops.

COMMUNICATIONS

There are two communications systems used in communicating between the orbiter and the ground. One system is referred to as the S-band system and the other is referred to as the Ku-band system.

The existing worldwide S-band system of ground stations provide up to approximately 20 percent of a satellite's or a spacecraft's orbit, limited to brief periods when the satellite or spacecraft are within the line of sight of a given tracking station. Each tracking station in the network can handle at most two satellites or spacecraft at one time and most stations can handle but one. Moreover, much of the equipment at the ground stations is almost 20 years old and inadequate to meet the demands of Space Shuttle and today's advanced spacecraft.

With the advent of the Tracking and Data Relay Satellite System (TDRSS), when fully operational, the TDRS system can provide continuous global coverage of earth orbiting satellites above 1,200 kilometers (750 miles) up to an altitude of about 5,000 kilometers (3,100 miles). At lower altitudes there will be brief periods when satellites or spacecraft over the Indian Ocean near the equator will be out of view. The TDRS operational system will be able to provide almost full time coverage not only for the Space Shuttle spacecraft but up to 26 other near earth-orbiting satellites or spacecraft simultaneously. When TDRSS is fully operational, ground stations of the worldwide Spaceflight Tracking and Data Network (STDN) will be closed or consolidated in savings in personnel, operating, and maintenance costs with the exception of Merritt Island, FL, Ponce de Leon, FL, and Bermuda which will remain open to support the launch of the Space Transportation System and the landing of the Space Shuttle spacecraft when landing at Kennedy Space Center, FL.

It is noted, that deep space probes and earth orbiting satellites above approximately 5,000 kilometers (3,100 miles) will use the three ground stations of the Deep Space Network

 **Space Shuttle Spacecraft Systems**

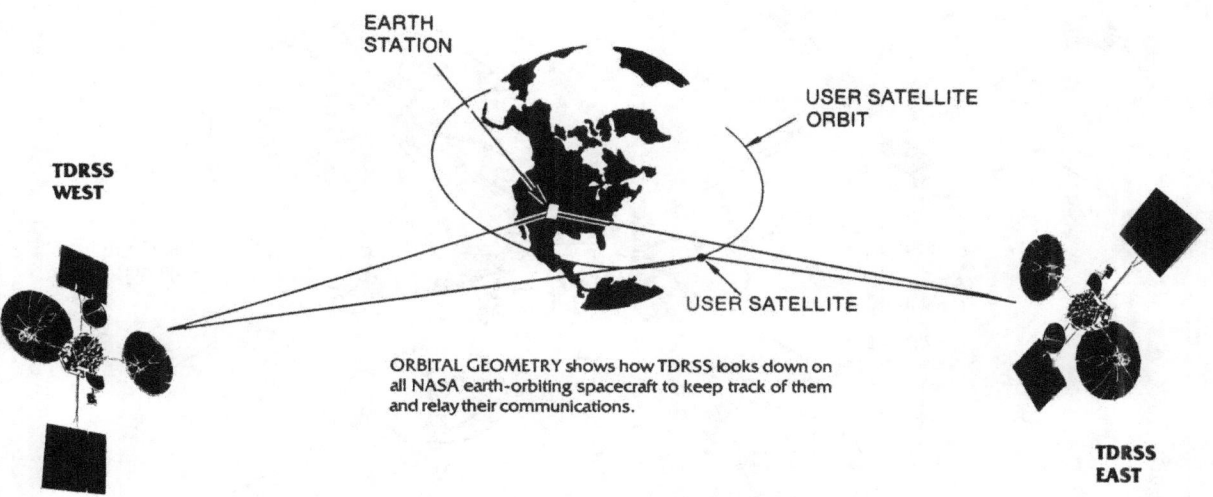

Tracking and Data Relay Satellite System

Space Shuttle Spacecraft Systems

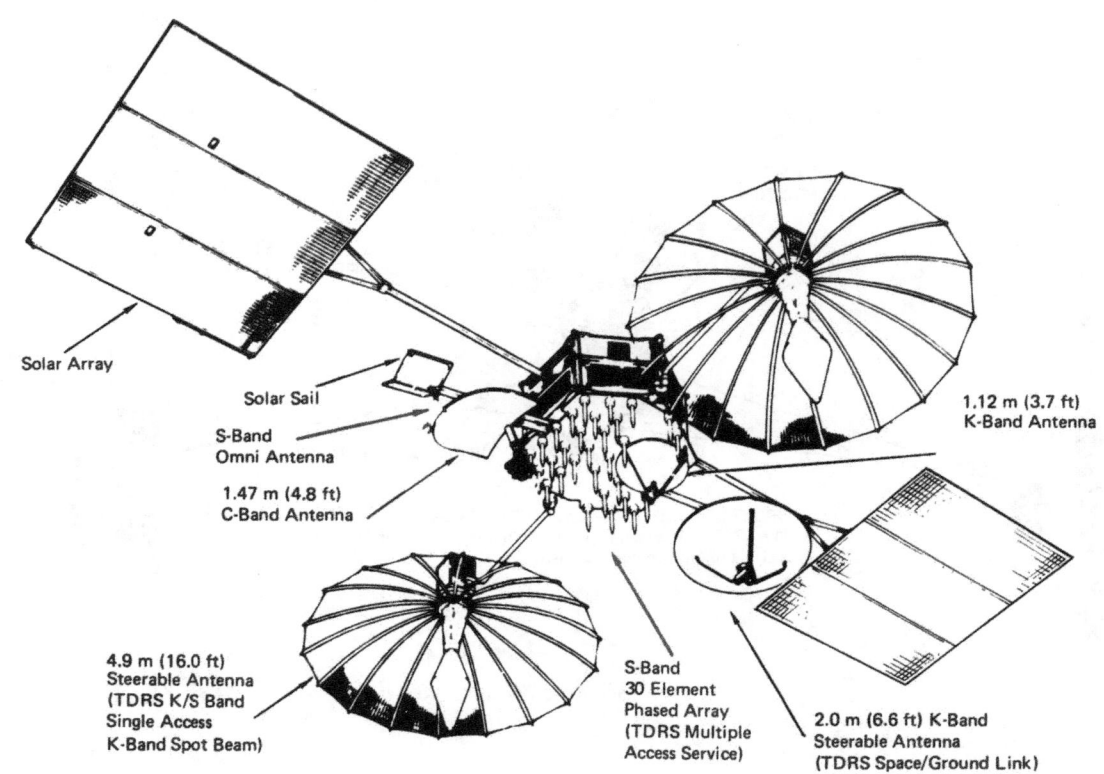

TDRS Satellite

Space Shuttle Spacecraft Systems

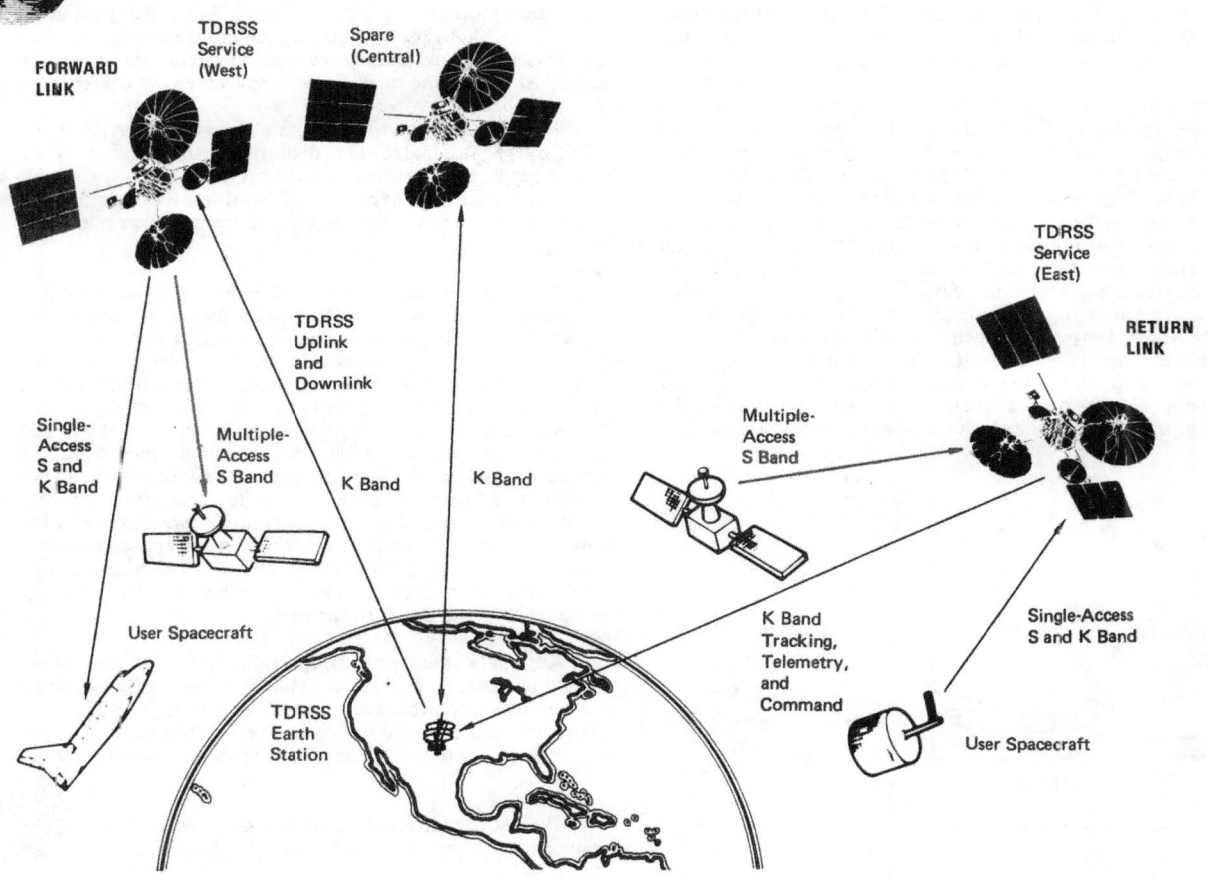

Linking Three Identical and Interchangeable Satellites With Earth Station

Space Shuttle Spacecraft Systems

(DSN) operated for NASA by the Jet Propulsion Laboratory, Pasadena, CA. The STDN stations that are co-located with the three DSN stations, Goldstone, CA, Madrid, Spain, and Orroral, Australia, will be consolidated with the DSN.

The fully operational TDRS system will consist of a satellite over the equator at 41 degrees West longitude over the Atlantic Ocean (which is now on station) and is referred to as TDRS-East (TDRS-A). Next, TDRS-B will be carried into earth orbit aboard the Space Shuttle and deployed from the Space Shuttle spacecraft and positioned over the Pacific Ocean at the equator southeast of Hawaii at 171 degrees West longitude and will be referred to as TDRS-West. TDRS-C will be the next one to be carried aboard the Space Shuttle spacecraft and deployed and positioned at the central station as a backup just west of South America over the Pacific Ocean at 79 degrees West longitude. The TDRS satellites are positioned at geosynchronous orbit above the equator at an altitude of 35,880 kilometers (22,300 statute miles). At this altitude, because the speed of the TDRS satellite is the same as the rotational speed of earth, they remain "fixed" in orbit over one location. The eventual positioning of two TDRS satellites will be 130 degrees apart at geosynchronous orbit instead of the usual 180 degrees spacing. This 130 degree spacing reduces the ground station to one instead of two, if the satellites were spaced at 180 degrees.

The TDRS system will serve as a radio data relay, carrying voice, television, analog, and digital data signals. The TDRS system offers three frequency band services: S-band, C-band, and high capacity Ku-band. The C-band transponders operate at 4-6 gigahertz and the Ku-band TDRS transponders operate at 12-14 gigahertz.

The highly automated TDRS system ground station is located at NASA's White Sands Test Facility, New Mexico, and is owned and managed by Spacecom, which NASA also leases. The ground station provides a location at a longitude with a clear line-of-sight to the TDRS satellites and at a location where rain conditions are very remote, as rain can interfere with the K-band uplink and downlink channels. It is one of the largest and most complex communication terminals ever built. All satellite or spacecraft transmissions are relayed by the TDRS satellites and funneled through the White Sands ground station. A NASA sophisticated operational control system located adjacent to the ground terminal at White Sands and at Goddard Space Flight Center (GSFC), MD, enables NASA to schedule TDRSS support of each user and to distribute the user's data directly from White Sands to the user.

Automatic data processing equipment at White Sands ground terminal (WSGT) aids in making user satellite tracking measurements, controls and communications. Equipment in the TDRS and in the ground station, collects system status data for transmission along with user satellite at spacecraft data to NASA.

The data acquired by the TDRS satellites is relayed to the single centrally treated ground terminal at NASA's White Sands

TDRSS System Elements

Space Shuttle Spacecraft Systems

Test Facility in New Mexico. From White Sands, the raw data is sent directly by domestic communications satellite (DOMSAT) to NASA control centers at Johnson Space Center (JSC) Houston, TX, for Space Shuttle operations and the Goddard Space Flight Center (GSFC), MD, which schedules TDRSS operations and controls a large number of satellites. To increase system reliability and availability, there is no signal processing done aboard the TDRS satellites, they will act as repeaters, relaying signals to and from the ground station or to and from user satellites or spacecraft. No user signal processing is done onboard the TDRS satellites.

The liftoff and ascent phase of a Space Shuttle mission launched from the Kennedy Space Center (KSC), FL, will use the Space Shuttle spacecraft S-band system to transmit/receive through the Merritt Island (MILA), FL, Ponce de Leon (PDL), FL, and Bermuda (BDA) STDN tracking stations and Mission Control Center—Houston (MCC-H), in a high data rate mode.

Upon leaving the line of sight tracking station at BDA, the Space Shuttle spacecraft S-band system will transmit/receive through the TDRS satellite and the TDRS system, thus the White Sands facility and MCC-H, but in a low data rate mode. The low data rate mode must be used because of limited power since the Space Shuttle spacecraft S-band system does not have a high enough signal gain to handle the high data rate mode.

When the Space Shuttle spacecraft is on orbit and its payload bay doors are opened, the Space Shuttle spacecraft Ku-band antenna stowed in the forward portion of the payload bay starboard (right) side, will be deployed. One drawback of the Ku-band system, is its narrow pencil beam, which makes it difficult for the antennas on TDRS to lock on to the signal. The Space Shuttle spacecraft will use the S-band system to lock the Ku-band antenna into position first, because the S-band system has a larger beam width. Once the S-band signal has locked the Ku-band antenna into position, the Ku-band signal is turned on. The Ku-band system provides a much higher gain signal with a smaller antenna than the S-band system. With communications acquisition, if the TDRS satellite is not detected within the first eight degrees of spiral conical scan, the search is automatically expanded to 20 degrees. The entire TDRS search requires approximately three minutes. The scanning stops when an increase in the received signal is sensed. Thus, the Ku-band antenna is gimbaled which permits it to acquire TDRS. The Space Shuttle spacecraft Ku-band antenna is a 914 millimeter (36 inch) density antenna. The Space Shuttle spacecraft Ku-band system and antenna will then transmit/receive through the TDRS satellite in view, thus the White Sands facility ground station and MCC-H in a high data rate mode.

There are times when in view of a TDRS satellite, that transmission/receiving on the Ku-system will be interrupted due to the Space Shuttle spacecraft blocking its Ku-band antenna view to the TDRS satellite because of a Space Shuttle spacecraft attitude requirement or when certain payloads aboard the Space Shuttle spacecraft cannot allow Ku-band radiation to be hit by the main beam of the Space Shuttle spacecraft Ku-band antenna. The main beam of the Space Shuttle spacecraft Ku-band antenna produces 340 volts per meter at the antenna, but decreases in distance, such as 200 volts per meter, 20 meters (65 feet) away from its antenna. Dependent upon the payload, a program can be instituted into the Space Shuttle spacecraft Ku-band antenna control system, which would limit the azimuth and elevation angle, which would inhibit its Ku-band antenna from directing its beam into the area of that onboard payload. This is referred to as an obscuration zone. This program would be instituted from MCC-H. In other cases, such as deployment of a satellite from the Space Shuttle spacecraft payload bay, the Space-Shuttle spacecraft Ku-band system would be turned off during deployment and turned on after deployment.

In preparation of entry, the Space Shuttle spacecraft payload bay doors must be closed for entry, thus its Ku-band antenna must be stowed. It is noted, if the Space Shuttle spacecraft antenna cannot be stowed, provisions are incorporated to

Space Shuttle Spacecraft Systems

jettison the assembly from the spacecraft so the spacecraft payload bay doors can be closed for entry. The Space Shuttle spacecraft will transmit/receive through the S-band system and the TDRS satellite in view and the TDRS system, thus the White Sands facility ground station and MCC-H in the low data rate mode until reaching communications blackout in entry. There remains a question at the time of the printing of the booklet as to whether data can be received from the spacecraft through TDRSS during the blackout period. It is noted that in the event of a landing at Edwards Air Force Base, CA, the Dryden Flight Research Facility at Edwards Air Force Base has only a receive capability, whereas the Goldstone, California Deep Space Network station has both a transmit/receive capability. After communications blackout in entry, the Space Shuttle spacecraft will again operate in S-band through the TDRS system in the low data rate mode to as low a view as possible until reaching the S-band landing site ground station, which would then transmit/receive in the high data rate mode on S-band.

S-BAND SYSTEM. The Space Shuttle spacecraft S-band system operates in the S-band portion of the radio frequency spectrum of 1,700 to 2,300 mHz.

The S-band forward link (previously referred to as uplink) is phase modulated (PM) on a center carrier frequency of either 2106.4 mHz (primary) or 2041.9 mHz (secondary) for NASA and is operative through STDN or TDRS. The two S-band forward link frequencies prevents interference when two Space Shuttle spacecraft are in operation at the same time, as one Space Shuttle spacecraft could select the high frequency and the other could select the low frequency.

The S-band return link (previously referred to as downlink) is PM on a center carrier frequency of 2287.5 mHz (primary) or 2217.5 mHz (secondary) for NASA and is operative through STDN or TDRS. The two S-band return link frequencies also prevents interference when two Space Shuttle spacecraft are in operation at the same time. Thus, one Space Shuttle

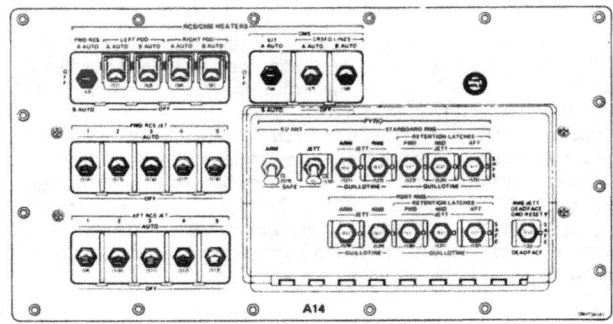

spacecraft would operate on the forward link high frequency of 2106.4 mHz (primary) and a return link high frequency of 2287.5 mHz (primary) through STDN or TDRS and the other Space Shuttle spacecraft would operate on the forward link low frequency of 2041.9 mHz (secondary) and a return link low frequency of 2217.5 mHz secondary through STDN or TDRS.

The Department of Defense S-band forward link is PM on a center carrier frequency of either 1831.8 mHz (primary) or 1775.5 mHz secondary from the the Air Force Satellite Control Facility (SCF) through its own ground stations and is not operative through TDRS as the S-band power amplifiers are not powered in the DOD SCF mode, thus cannot operate through TDRS. The two S-band forward link frequencies prevents interference when two Space Shuttle spacecraft are in operation at the same time, as one Space Shuttle spacecraft could select the high frequency and the other could select the low frequency.

The Department of Defense S-band return is PM on a center carrier frequency of 2287.5 mHz (primary) or 2217.5 mHz (secondary) through the Air Force ground stations to the Air Force SCF and is not operative through TDRS as the S-band power amplifiers are not powered in the DOD SCF

Space Shuttle Spacecraft Systems

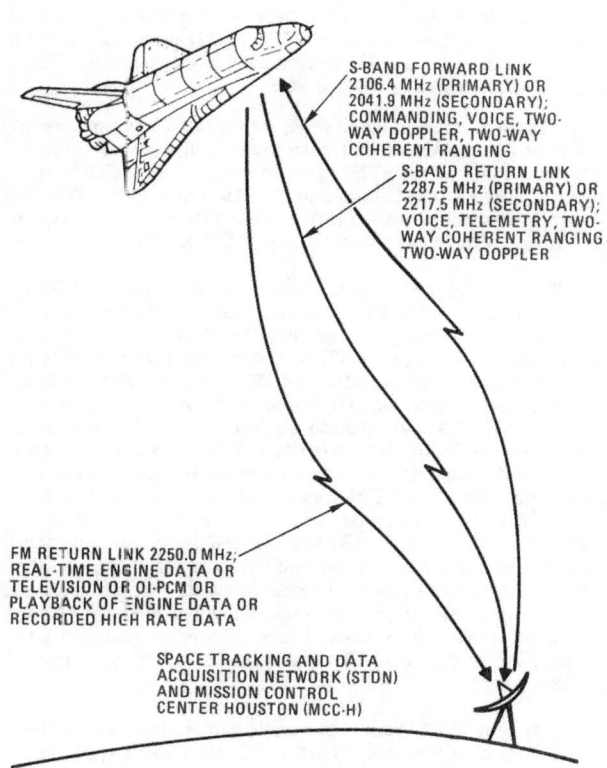

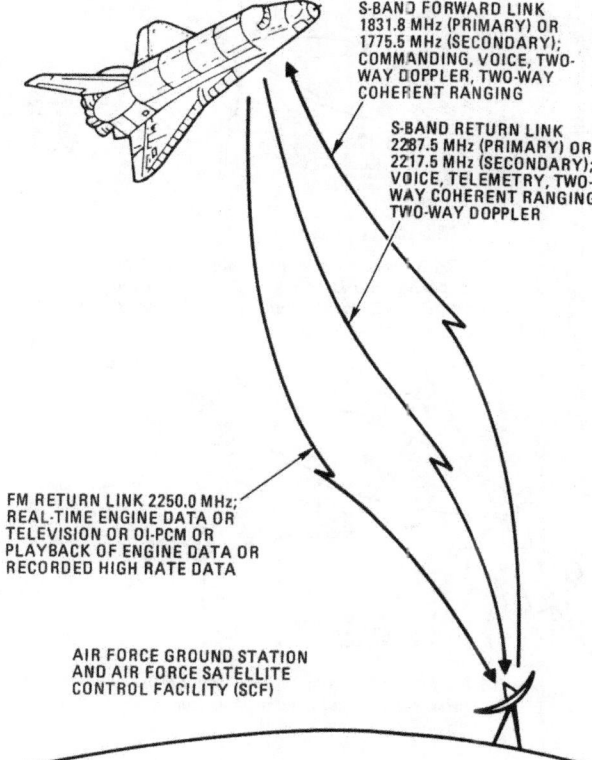

NASA S-Band Communications With Space Tracking and Data Acquisition Network

Department of Defense S-Band Communications With Air Force Ground Station and Air Force Satellite Control Facility

Space Shuttle Spacecraft Systems

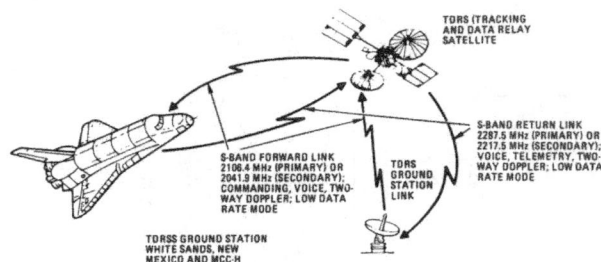

NASA S-Band Communications With TDRSS (Tracking and Data Relay Satellite System) and Mission Control Center Houston (MCC-H)

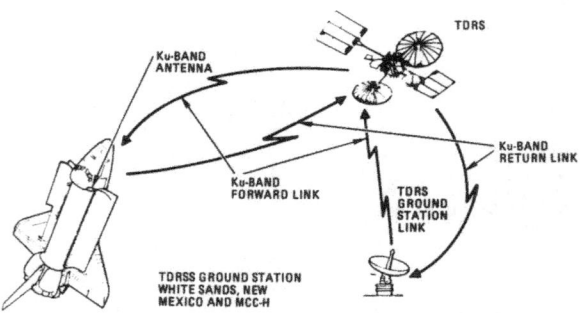

NASA Ku-Band Communication With TDRSS and Mission Control Center Houston

mode, thus cannot operate through TDRS. The two S-band return link frequencies also prevents interference when two Space Shuttle spacecraft are in operation at the same time. Thus, one Space Shuttle spacecraft would operate on the forward link high frequency of 1831.8 mHz (primary) and a return link of 2287.5 mHz (primary) from the Air Force SCF and its ground stations and the other Space Shuttle spacecraft would operate on the forward link of 1775.5 mHz (secondary) and a return link of 2217.5 mHz (secondary) through the Air Force ground stations and Air Force SCF.

A frequency modulated (FM) return link (previously referred to as downlink) can be transmitted simultaneously with the PM return link to the STDN ground station and MCC-H or to the Air Force ground station to the Air Force SCF. The FM return link is not operative through the TDRS. The FM return link is on a center carrier frequency of 2250.0 mHz.

The S-band PM forward link originates from MCC-H through the NASA STDN ground stations used for launch, liftoff, ascent or landing or through the White Sands facility ground station through the TDRS system and TDRS satellite to the Space Shuttle spacecraft. The DOD S-band PM forward link originates from the Air Force Satellite Control Facility (SCF) through its own ground stations to the Space Shuttle spacecraft. NASA or DOD Air Force SCF has a choice of two forward link frequencies for data transfer, but not both at the same time. The S-band PM forward link transfers a high data rate of 72 kbps (kilo-bits-per second), consisting of two air to ground voice channels at 32 kbps each, and one command channel at 8 kbps, two way doppler and two-way tone ranging. The S-band PM forward link transfers a low data rate of 32 kbps, consisting of one air to ground voice channel at 24 kbps and one command channel of 8 kbps, two-way doppler and two-way ranging. Note: the two-way ranging is not operative through TDRS.

The S-band PM return link can originate from one of two S-band PM transponders aboard the Space Shuttle spacecraft. Each transponder can return link on a frequency 2287.5 mHz (primary) or 2217.5 mHz (secondary), but not both at the same time. The S-band return link from the Space Shuttle spacecraft will transmit the data through the NASA STDN ground stations used for launch, liftoff, ascent, or landing or through the TDRS

Space Shuttle Spacecraft Systems

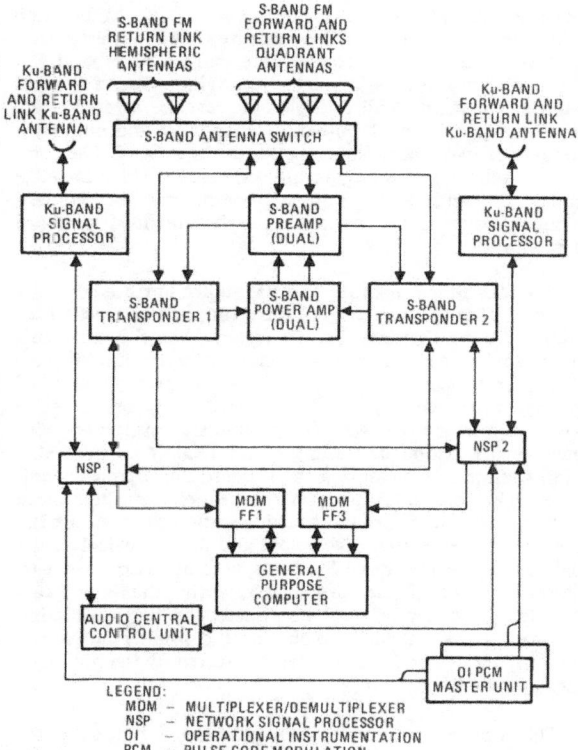

Ku-Band and S-Band System

satellite and TDRS system through the White Sands facility ground station to MCC-H. The DOD S-band return link from the Space Shuttle spacecraft will transmit the data through the Air Force ground stations to the Air Force SCF. The S-band return high data rate of 192 kbps consists of two air-to-ground voice channels at 32 kbps each, and one telemetry link of 128 kbps and two way doppler and two-way ranging. It is noted, that the two-way doppler and two-way ranging would only be operative when in view of the NASA STDN ground stations at launch, liftoff, ascent or landing or in view of the DOD Air Force SCF ground stations. The two way ranging will not be operative with the TDRSS. The S-band return link low data rate consists of one air-to-ground voice channel at 32 kbps and one telemetry link at 64 kbps and two-way doppler and two-way ranging. As noted the two-way doppler and two-way ranging would be utilized in the same manner as in the high data rate mode.

Four quadrant S-band antennas are located on the forward fuselage center skin of the Space Shuttle spacecraft approximately 90 degrees apart. The antennas are covered with a reusable thermal protection system. In the Space Shuttle spacecraft on the flight deck looking out the forward windows, the quadrant antennas are to the upper right, lower right, lower left, and upper left. The four S-band PM quadrant antennas are the radiating elements for transmitting of the S-band PM return link and for receiving the S-band PM forward link. Each quad antenna has a forward and aft pattern which is selected either by the flight crew or by onboard computer control or by ground control. As an example; lower right aft or lower right forward effectively, provides eight antennas.

The Space Shuttle spacecraft S-band antenna switch assembly provides the signal switching among the two S-band transponders and any one of the four quadrant antennas. The proper quadrant forward or aft pattern antenna to be used is selected automatically under onboard computer control, by ground command from MCC-H, or manually by the flight crew

Space Shuttle Spacecraft Systems

utilizing the displays and controls on the spacecraft flight deck. In the automatic mode, the onboard computer selects the proper quadrant antenna to be used whenever an S-band transponder is active. The antenna selection is based upon the computed line of sight to the NASA STDN ground station used for launch, liftoff, ascent or landing or the TDRS satellite in view. The antenna selection is based upon the computed line of site to the DOD Air Force SCF ground station. The antenna switching commands are sent to the switch assembly via the payload multiplexer/demultiplexers (MDM's).

Two identical S-band PM transponders are located in the Space Shuttle spacecraft. Only one transponder operates at a given time. The other transponder is a redundant backup. The S-band PM transponder receives the forward link and transmits the return link. The selected transponder transfers the forward link commands and voice to the network signal processor (NSP), receives the return link telemetry and voice from the NSP.

The selected transponders also provides a coherent turnaround of the PM forward and PM return two-way doppler and two-way tone ranging signals. The two-way doppler and two-way ranging signals are operative when in view of the NASA STDN ground stations at launch, liftoff, ascent or landing or the DOD Air Force SCF ground stations. The two-way doppler is operative through the TDRS, but the two-way ranging is not operative through TDRS.

The two onboard network signal processors (NSP's) receive commands, forward link, and transmit, return link, telemetry data to the selected S-band transponder. Only one NSP operates at a time, the other providing a redundant backup. The selected NSP receives either one or two analog voice channels from the onboard Audio Central Control Unit (ACCU), depending on whether one (in low data rate mode) or both (in high data rate mode) of the air-to-ground channels are being used, and converts these analog voices to digital voice signals, time-division-multiplexes them with the Pulse Code Modulation (PCM) telemetry from the Pulse Code Modulation Master Unit (PCMMU) and sends the composite signal to the S-band PM transponder for transmission on the return link. On the forward S-band PM link, the NSP does just the reverse. It receives the composite signal from the S-band transponder and outputs it either one or two analog voice signals to the ACCU. The composite forward link also has ground commands, which the NSP decodes and sends to the onboard computers. The onboard computers route the commands to the intended onboard systems.

The S-band PM forward and PM return link are normally coherent so that the frequency of the return link is directly proportional to the forward link frequency. The S-band transponder provides a coherent turnaround of the forward link carrier frequency necessary for the two-way doppler (change of frequency with velocity) data. This would only be operative when in view of the NASA STDN ground stations in view with launch, liftoff, ascent or landing or the DOD Air Force SCF ground stations. By measuring the forward link frequency and knowing what return link frequency to expect from the Space Shuttle spacecraft, the ground tracking station can measure the double doppler shift that takes place and used to calculate the radial velocity of the Space Shuttle spacecraft with respect to the ground station. These links are PM so that the S-band carrier center frequency will not be affected by the modulating wave. It would be impossible to obtain valid doppler data of the S-band carrier center frequency were affected by the modulating technique.

The S-band transponder also provides a subcarrier for two-way tone ranging. This would only be operative when in view of the NASA STDN ground stations in view with launch,

Space Shuttle Spacecraft Systems

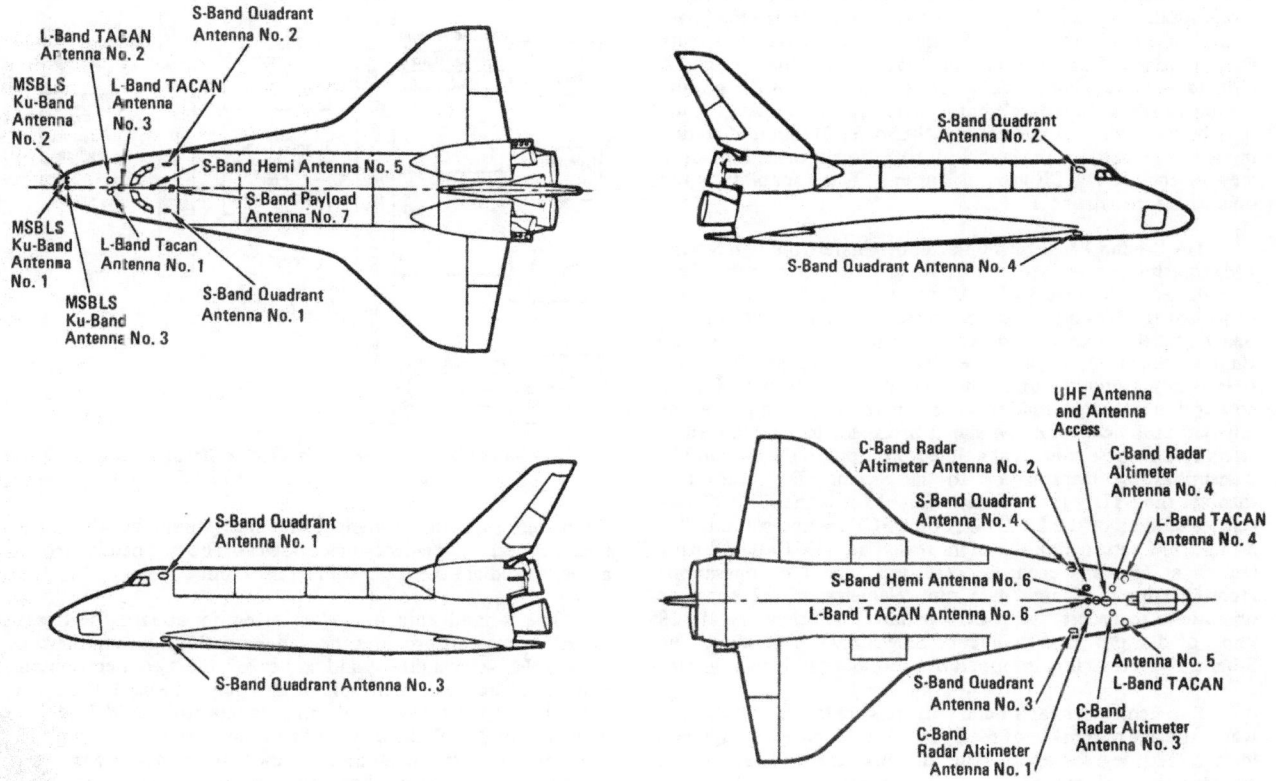

Orbiter Antennas

Space Shuttle Spacecraft Systems

liftoff, ascent or landing or in view of the DOD Air Force. SCF ground stations. This capability is not operative through the TDRSS. The ground station will forward link ranging tones at 1.7 mHz and compute vehicle slant range from the time delay in receiving the return links 1.7 mHz tones to determine the Space Shuttle spacecraft range. The Space Shuttle spacecraft azimuth is determined from the ground station antenna angles. A C-band skin tracking mode is also provided from the ground station to track the Space Shuttle spacecraft and again, would only be used when in view of the NASA STDN ground station in view, associated with, launch, liftoff, ascent or landing or in view of the DOD SCF ground stations. This capability is not operative through the TDRSS.

The S-band FM return link can originate from two S-band FM transmitters aboard the Space Shuttle spacecraft. Both transmitters are tuned to 2250.0 mHz. The S-band FM signal processor receives inputs and processes data from five onboard Space Shuttle spacecraft sources: the three Space Shuttle Main Engine (SSME's), engine interface units (EIU's); the video (television) switching unit; the operational recorders for recorder dump; the payload recorder for recorder dump; and the payload umbilical. The FM signal processor is commanded to select one of these sources at a time for output to the S-band FM transmitter for transmission to the S-band FM return link through the STDN ground station used for launch, liftoff, ascent or landing or the DOD Air Force SCF ground station. The S-band FM return link transfers real time SSME data during launch at 60 kbps each; or real time video; or operations recorders dump of high data rate telemetry at 192 kbps; or operations recorders dump of low data rate telemetry at 128 kbps, or dump of one channel of SSME data at 60 kbps. The S-band FM return link is inoperative through the TDRS system.

Two hemispherical S-band antennas are located on the forward fuselage outer skin of the Space Shuttle spacecraft approximately 180 degrees apart. The antennas are covered with a reusable thermal protection system. In the spacecraft on the

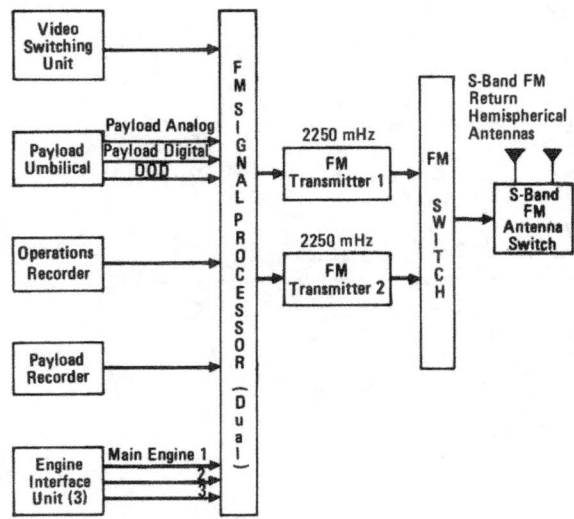

S-Band (FM) Frequency Modulation Signal Processor (Not Operative Through TDRS)

flight deck, the hemispherical antennas would be above the head (upper) and below the feet (lower). The two hemispherical antennas radiate the S-band FM return link.

The S-band antenna switch assembly aboard the Space Shuttle spacecraft provides the signal switching among the two S-band FM transmitters and either of the two hemispheric antennas. The proper hemispheric antenna to be used is selected automatically under onboard computer control, by MCC-H, or manually by the flight crew utilizing displays and controls on the flight deck. In the automatic mode, the onboard computer selects the proper hemispheric antenna to be used whenever an

Space Shuttle Spacecraft Systems

S-band FM transmitter is active. The antenna selection is based upon the computed line of sight to the NASA STDN ground station used for launch, liftoff, ascent or landing or DOD Air Force SCF ground.

The basic difference between the Space Shuttle spacecraft quadrant and hemispherical antennas is that the hemispheric antennas have a larger beam width, while the quadrant antennas have a higher antenna gain. The hemispheric antennas are so-called because there are two of them, one on top of the orbiter and one on the bottom. The quadrant antennas are so-called because there are four of them, two on each side of the orbiter, one on the upper half and one of the lower half of each side, which provides nearly total coverage in all directions.

An S-band payload antenna is located on the forward fuselage outer skin of the Space Shuttle spacecraft, just aft of the upper hemispherical antenna. The payload antenna is covered with a reusable thermal protection system. This antenna is used as the radiating element for S-band transmitting/receiving to and from the Space Shuttle spacecraft to detached payloads through forward and return links.

KU-BAND SYSTEM. The Ku-band system operates in the Ku-band portion of the radio frequency spectrum between 15,250 and 17,250 mHz.

When the Ku-band antenna is deployed aboard the Space Shuttle spacecraft and handover from the S-band system to the Ku-band system occurs the Space Shuttle spacecraft onboard NSP operates with the Ku-band signal processor rather than the S-band transponder. The data stream is then directed through the Ku-band signal processor and Ku-band antenna to the TDRS satellite in view to the TDRS system White Sands facility ground station to MCC-H on the return link and reverse the process for the forward link. If the forward link of Ku-band is lost, the system will fail safe to S-band.

The Ku-band system forward link consists of a mode one and two through the TDRS satellite in view. Mode one consists of 72 kbps data (two air-to-ground voice at 32 kbps each and 8 kbps command) and 128 kbps text and graphics (used in place of the teleprinter) and 16 kbps synchronization. Mode two consists of 72 kbps operational data (two air-to-ground voice at 32 kbps each and 8 kbps command).

The Ku-band system return link consists of channel one, mode one and two; plus one channel two, mode one and two; and one channel three. Channel one, mode one and two consists of 192 kbps operational data (128 kbps operational data telemetry/payload interleaver plus two air-to-ground voice at 32 kbps each); plus one of channel two, mode one and two selection of four; 1) payload digital data from 16 kbps to 2 mbps (mega): or 2) payload digital data from 16 kbps to 2 mbps: or 3) operations recorder playback from 60 kbps to 1,024 kbps: or 4) payload recorder playback from 25.5 kbps to 1,024 kbps; plus one of the following from channel three; mode one attached payload digital data (real-time or playback) from 2 mbps to 50 mbps; or mode two, television (color or black/white) composite video; or mode two, real time attached payload digital data or payload analog data.

AUDIO SYSTEM. The audio system interfaces with the caution and warning (C/W) system for reception of C/W (tone) signals, with UHF (ultra high frequency), S-band and Ku-band systems for transmission and reception of external signals (air-to-air and air-to-ground) and with the three TACAN (Tactical Air Navigation) sets for receiver selection and signal monitoring. The NSP supplies S-band and Ku-band signals to the audio system for transmission and/or reception at various flight crew stations in the spacecraft. The attached payload (Spacelab), payload bay, launch umbilical and the operations recorders all have electrical interface with the audio system for transmission and/or reception of signals.

 **Space Shuttle Spacecraft Systems**

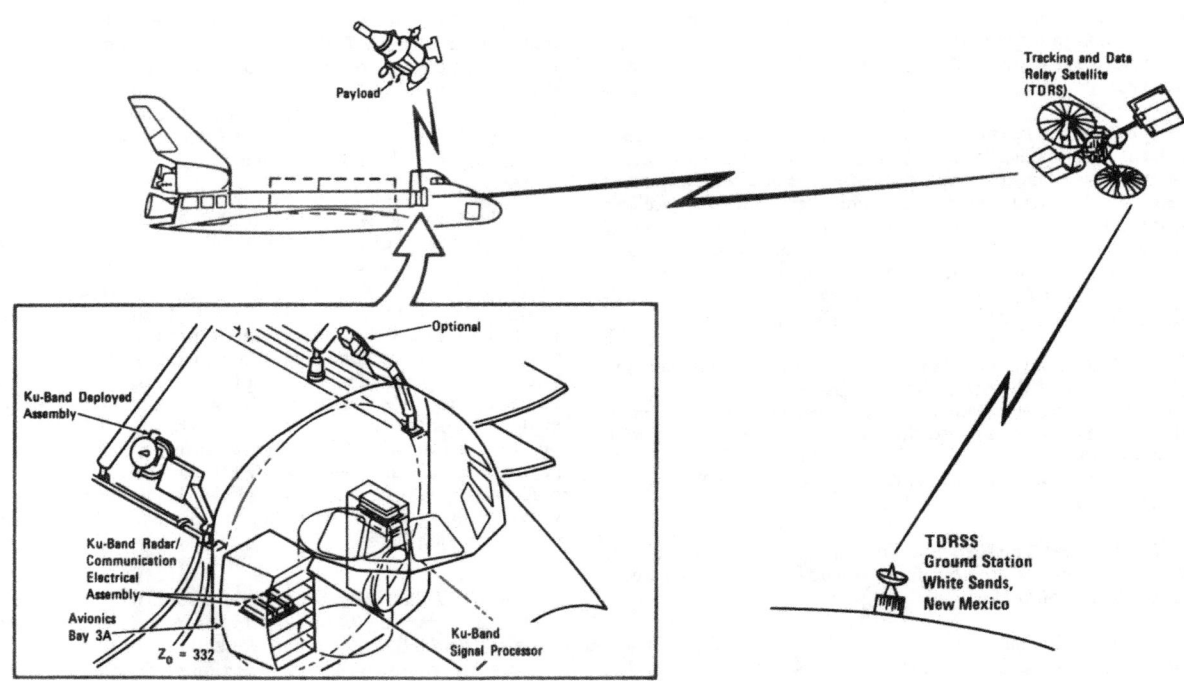

Ku-Band Antenna

Space Shuttle Spacecraft Systems

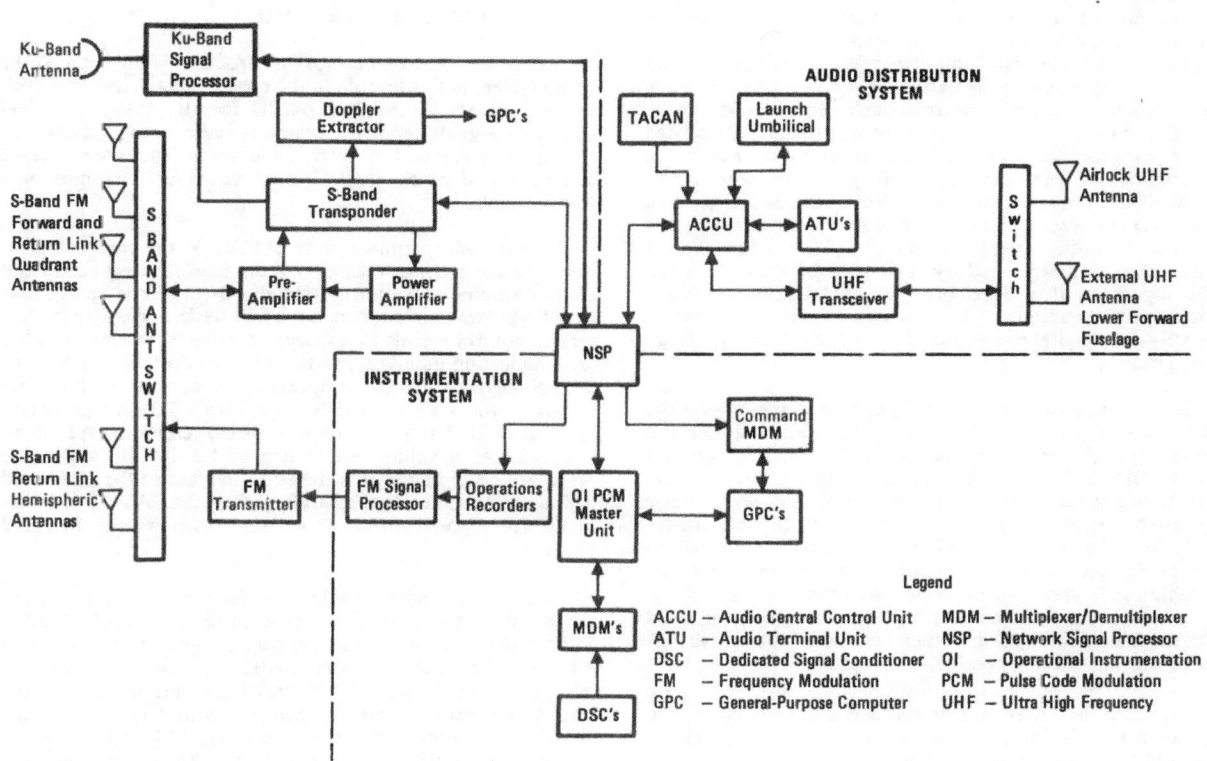

Communications System

Space Shuttle Spacecraft Systems

The audio distribution system (ADS) is a digital system which greatly reduces the number of wires necessary to carry electronic signals between system components. Electronic impulses can be given identifying characteristics and sorted into groups of signals. The coded impulses called "bits" of information and are generated by the particular position of each switch on the various spacecraft control panels. Several bits can be reduced in number by a multiplexer to a particular identifying impulse. Bit groups from several sources are reduced so that a large number of signals can be sent along a single wire. Up to 128 bits of information can be encoded by the audio terminal unit (ADU) into a serial data word and sent along one wire to a decoder in the audio central control unit (ACCU) where the bits are identified and separated by their original characteristics. Audio signals are then distributed by the ACCU's to the appropriate ADS components. There are no digital voice signals in the ADS—only digital enable signals. All ADS voice signals are analog (audio).

During launch and entry, the flight crew members wear the launch and entry helmets (LEH) as a head protection device that provides the flight crew members with face protection, light and sound attenuation and communication capabilities. The enclosed environment of the helmet attenuates the severe noise levels encountered at launch and allows intelligible air-to-ground communications. To provide communication capabilities, the helmet contains a microphone, earphone and connector and cable for interface with the headset interface unit (HIU). The microphone is built into the helmet and can be positioned to suit the individual flight crew member. The earphones are integrated into the helmet's padded ear muffs.

On orbit, the LEH's are stowed and a wireless crew communication unit (WCCU) is used with communication cables and the HIU's are used as a backup.

For extravehicular activities (EVA's), the communications carrier assembly (CCA) is a headset integrated into a skull cap, often referred to as a "Snoopy Hat."

The audio central control unit (ACCU) is the heart of the audio system and is located in the crew compartment mid deck forward avionics bay. The ACCU identifies, switches, and distributes signals between the various audio distribution system (ADS) components. Both digital and audio signals are received and processed by the ACCU, but the ACCU transmits only audio signals.

Eight audio terminal units (ATU's) in the crew compartment are audio control panels and are used to select access and control volume to various audio signals. There is an ATU at each flight crew station, two in the mid-deck, and one in the airlock to control signals to headsets or helmets via crew member communication umbilical /EMU. The mid deck has an ATU to control signals through a speaker microphone unit (SMU) located at this station. Signals to or from ATU's are processed by the ACCU. The VOX (voice activated) sensitivity, ATU control and master volume are controlled by circuitry within the ATU. All other knobs or switches on the face of the ATU send digital enable signals to the ACCU (except the "AUD" function of the ATU power switch which supplies power to the ATU circuits).

Four ATU's have a redundancy feature in which control of a particular panel may be switched to another ATU. The left commanders (CDR), may be switched to the right, pilots (PLT) ATU and the right PLT ATU control may be switched to the left CDR. The CDR, PLT control knobs are located on the respective panel. The mission specialist (MS) ATU control may be switched to the payload specialist (PS) ATU, and the airlock control may be switched to the mid-deck. The control knob for

Space Shuttle Spacecraft Systems

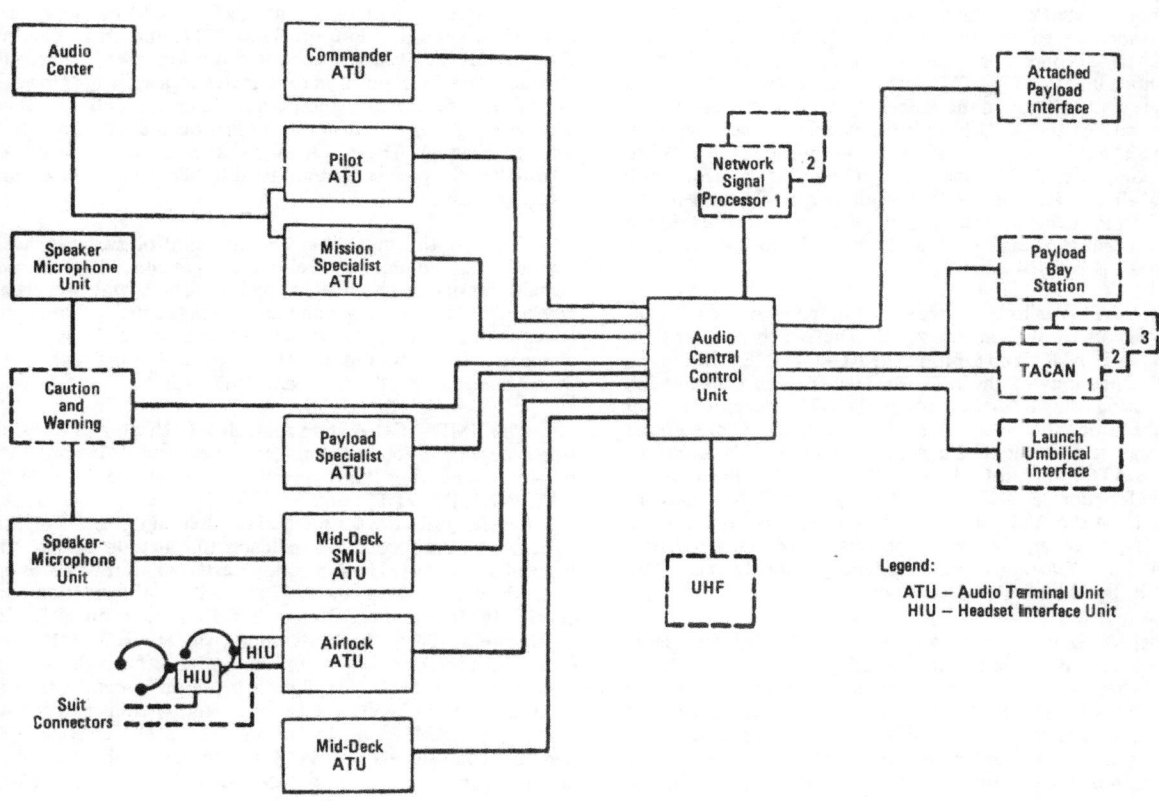

Audio Distribution System

Space Shuttle Spacecraft Systems

the MS ATU is located below the CRT 4 keyboard which is below and left of the MS ATU. Control of these latter two ATU's is not reversible as with the CDR and PLT ATU's. Airlock crew member communications unit (CCU)/EMU 1 control is switched to the mid deck CCU ATU, and airlock CCU/EMU 2 control is switched to the mid deck speaker microphone unit (SMU) ATU. Both functions are switched with the single control knob in the airlock ATU. In the NORM position, control of the ATU is with the panel to which the knob belongs. The other position of the knob indicates the ATU to which control can be transferred. The ATU control knob changes all ATU functions to the alternate ATU except the master volume control. Redundancy protection is for use in the event of a failure or malfunction of any of the four ATU's that have an ATU control knob.

Each ATU has its own three-position power switch to control all signals to or from the ATU. The switch positions are AUD/TONE, AUD, and OFF. In the AUD/TONE position, all available functions of the ATU are armed, and transmission and receptions may be made through the ATU, depending upon the position of other switches on the ATU. C/W tone digital enable signals are sent to the ACCU to allow C/W audio to reach the ATU, thus to CCU or SMU. The AUD position has the same functions as AUD/TONE except that C/W signals are blocked from the ATU. The OFF position shuts off power to the ATU power supply, which powers the ATU amplifiers. Siren (AP) and Klaxon (fire) C/W signals go directly to an SMU even with the SMU and/or ATU power off.

Each ATU has a two position, spring-loaded-off paging switch which must be held in the PAGE position to activate the circuit. When activated, the switch enables the ATU to transmit to all other ATU's, the EVA transceiver, the attached payload circuit (Spacelab). Any number of stations may simultaneously use the paging circuit, and the circuit may be used regardless of the position of the various individual channel control switches.

On all ATU's, the two air-to-ground (AG) channels, the air-to-air channel (A/A), and the two intercom (ICOM A and B) channels have individual three-position control switches for selecting access to particular channels for transmission or reception. The switch positions are T/R, RCV, and OFF. The T/R position permits transmission or reception over the selected channel. The RCV position deactivates transmission capability on the selected channel, and permits reception only of signals. The OFF position deactivates transmission and reception on the selected channel. These control switches do not turn on any transmitter or receiver, but are used to allow access to a transmitter or receiver.

Each of the individual channel control switches has a thumb-wheel volume control which permits adjustment of signal intensity on the related channel. The thumb wheels are labeled from zero (lowest volume) to nine (highest volume), and cover a range of approximately 27 decibels in 3 decibel increments. There is also a volume control thumb-wheel for TACAN signals on the CDR and PLT ATU's.

The XMIT/ICOM knob controls the selection of transmission capability in four combinations of external/intercom transmissions. The four knob positions are labeled PTT/HOT, PTT/VOX, PTT/PTT, and VOX/VOX. In each case, the first set of initials indicates the method of external transmission activation, and the second set indicates the method of intercom transmission. PTT/HOT position, external transmissions are made through push-to-talk (PTT) activation of a rotation hand control (RHC) at the CDR or PLT station, or an HIU (any CCU station) ATU, or an SMU (mid deck) and HOT MIC is activated and the intercom is continuously live from the selected station (HIU or SMU XMIT must be keyed to enable external transmission). PTT/VOX position, external transmissions are made by the XMIT function of an HIU, SMU, or RHC PTT and intercom signals are VOX (voice activated). PTT/PTT position provides access to external and intercom channels

Space Shuttle Spacecraft Systems

through the PTT of an RHC, an HIU, or an SMU and the RHC PTT will activate any external and intercom channels selected. VOX/VOX provides access to external and intercom channels and is voice activated.

VOX sensitivity regulates the loudness of the signal required to activate the VOX feature. The MAX setting requires a higher decibel level to activate the circuit than the MIN setting.

Master speaker volume control of all incoming signals to earphones or speaker is adjusted with the MASTER VOL knob. The master volume control acts in series with the volume control wheels for the individual channels, A/G 1 and 2, A/A, ICOM A and B, and TACAN. Master volume knobs are located on the CDR, PLT, and airlock CCU ATU's, and on the mid deck SMU ATU. There are two master volume knobs on the airlock ATU labeled 1 and 2 and control volume to the respective CCU/EMU outlets in the airlock. The knobs are labeled from 1 (minimum volume) to 9 (maximum volume).

The mid deck ATU and the MID DECK SPEAKER AUDIO panel, controls operation of the SMU located at that station. In addition to the features of the other ATU's, the speaker microphone unit (SMU), ATU has a three position power switch labeled OFF, SPKR, and SPKR/MIC. In the OFF position, no signals go through the ATU. In the SPKR position, the SMU operates as a speaker only. In the SPKR/MIC position, the SMU can be used as either a speaker or a microphone. The SMU is located in the mid deck ceiling. The power switch for the SMU is located on the ATU. Signals to or from the SMU are selected on the ATU. A three position, spring-loaded off switch on the face of the SMU operates in conjunction with the PTT function of the ATU. The positions are: XMIT, for access to external transmissions; ICOM, for internal communications; and the unlabeled off (center) position blocks outgoing PTT signals from the SMU. The XMIT position sends signals over selected intercom circuits and to any external transmitters selected by the SMU ATU. The ICOM position excludes signals to external transmitters and allows signals to be sent over the selected ICOM channel(s) (A and/or B). Keying the ICOM or XMIT switch will override the speaker, except for C/W emergency signals. In the VOX mode, the first signal to activate the circuit, either MIC or SPEAKER, will have priority, except for emergency C/W signals. The MIC KEY light is a two-position adjustable intensity light that operates in conjunction with SMU transmissions. The intensity of the top half of the light (MIC) is adjustable by the MIC LEVEL knob; the brighter the light, the louder the signal. The bottom half of the light (KEY) will illuminate when a PTT function is selected and the circuit is keyed. Siren (ΔP) and Klaxon (fire) C/W tones go directly to the speakers, even if the speaker power switch is off.

The audio center panel is located between the two aft viewing windows at the aft flight station and has three functions; UHF control, electrical interface capability with external vehicles and the payload bay, and operations recorders selection. All switches on the audio center panel send digital impulses to the ACCU to enable the selected function provisions for communicating with Spacelab and the payload bay are controlled by the audio center panel. Sets of ON-OFF toggle switches labeled SPACELAB and PL BAY OUTLETS electrically connect the particular function to the ADS. The SPACELAB has seven switches to enable the following functions: A/A, A/G 1, A/G 2, ICOM A, ICOM B, PAGE and TONE (C/W). The PL BAY OUTLETS sub-panel has two ON-OFF switches, one for ICOM A and one for ICOM B, to enable the respective intercom function.

Two rotary knobs labeled VOICE RECORD SELECT, control various audio signals to be sent to the operational recorder via the NSP. A/G 1, A/G 2, A/A, ICOM A, or ICOM B audio can be sent to either recorder. Any two signals may be recorded at the same time, one on channel 1 and the other on channel 2. Either, or both channels may be turned off. Signals to the operational recorders cannot be monitored by the flight crew.

Space Shuttle Spacecraft Systems

The CCU/EMU outlets and power switches provide electrical connection to headset/helmet cables and have an ON-OFF toggle switch at each outlet to control electrical power to the respective cables. The CDR and PLT stations each have three position CCU switches labeled CCU, OFF, and SUIT. The CCU position permits power flow to a WCCU or HIU. The OFF position blocks MIC power from either a WCCU, HIU or LEH helmet. The SUIT position permits power to reach the LEH microphone. All other CCU's are two-position switches labeled CCU and OFF and function the same as the CDR and PLT switches. The EMU switches are located in the airlock below the EMU outlets on the panel labeled POWER/BATTERY CHARGER. The individual three-position switches are labeled EMU 1, EMU 2, and BUS SELECT. Each switch provides the capability of selecting MNA or MNB DC power for the EMU, or turning power on/off to the EMU.

The WCCU (wireless crew communications unit) consists of a wall unit and a leg unit to be worn by a flight crew member in the crew compartment when in orbit. The wall unit connects to a CCU outlet and remains attached to the crew compartment wall by Velcro until stowed for entry. Each wall and leg unit transmits on a unique pair of UHF frequencies, therefore, wall and leg units must be used together. Each set is identified by a letter on the set (e.g. A). The wall unit is further identified by enclosing the letter in a box (e.g. leg unit A works with wall unit A). Each unit is stowed with its cabling attached. The wall unit has a 584 millimeter (23 inch) cable to interface with the CCU outlet, and the leg unit has a 558 millimeter (22 inch) cable attached to a mini-headset.

The only assembly necessary when the WCCU is unstowed is to insert and tighten the flexible antenna into the bottom of each wall and leg unit. The wall unit receives power from the CCU outlet, therefore the on/off/volume knob is not used and the battery pack is empty. The MASTER VOL control is set to full volume. All other switches are set as required; typically, the individual communication loop VOL thumbwheels will be set at 2, if all communication loops are used and correspondingly higher as fewer communication loops are used. The leg unit is stowed with battery pack installed and is attached to the flight crew members leg with a wrap around elastic strap. The rotary on/off/volume knob (unlabeled) is rotated clockwise past the on/off detent and the volume set as desired. Battery changeout is accomplished by depressing the battery pack latch pushbutton lever (unlabeled) and sliding the battery pack off the unit. Sliding the new battery pack into the unit causes both the electrical connector and mechanical connector to latch.

The mini headset secures to the back of the ear via a hook type device. The mini headset has an adjustable microphone, a molded ear plug that serves as the earphone and a cable and connector interface with the WCCU. The corresponding ATU is configured for the desired communication loop. The mini-headset cable and connector can interface also with the HIU.

The headset interface unit (HIU) is a portable microphone switch which is considered part of a CCU. The HIU provides volume control and PTT capabilities to the CCA "Snoopy Helmet" used for EMU/EVA, the LEH's, and as a backup the WCCU mini-headset. The HIU has a clip for attaching to a flight crew member's flight suit. The unit has a rocker-type three-position switch and a volume control knob. The switch positions are XMIT, ICOM, and an unlabeled, spring-loaded off, center position. The XMIT position allows access to intercom and external circuits, while the ICOM position is for intercom only. The volume control knob acts in series with the volume controls on the associated ATU.

The ACCU contains circuitry that activates signals from the launch umbilical connection ICOM A and B channels. Any crew station ATU can then be configured to transmit and/or receive ICOM signals from the ground via the umbilical. Only ICOM signals are processed through the umbilical.

Space Shuttle Spacecraft Systems

The CDR and PLT ATU are the only ATU's at which access to TACAN signals can be controlled. There are two TACAN switches. The two position ON-OFF switch either allows reception of TACAN signals or blocks incoming TACAN signals (OFF). The other switch which is a three-position switch labeled 1, 2, and 3 allows selection of the particular TACAN set to be monitored. There are no transmission capabilities over the TACAN channels. The TACAN is a polar coordinate system which provides distance and bearing information to a TACAN station selected. This system operates in a band of frequencies from 962 to 1,213 mHz. The TACAN ground station identification signals of the TACAN ground stations call letters in Morse code are repeated every 40 seconds.

UHF SYSTEM. Control of UHF transmission is accomplished through the UHF control knob on the CDR's overhead panel. Three two position toggle switches, labeled XMIT FREQ, ANTENNA, and SQUELCH, are on the same panel. The XMIT frequency switch selects one of the two UHF frequencies, 296.8 mHz primary or 259.7 mHz secondary, for external transmission. The antenna switch selects the UHF antenna on the lower forward fuselage external skin of the orbiter or the airlock antenna located in the forward payload bay. The airlock antenna is used only for air-to-air communications during EVA. The UHF antenna on the lower forward fuselage of the orbiter is covered with a reusable thermal protection system. The SQUELCH switch permits ON selection or OFF deselection of UHF squelch. A five-position rotary knob on the UHF control panel activates power to the UHF transceiver and selects any of the following modes of UHF transmission. EVA transmissions made on one frequency selected by "Xmit Freq" switch and reception is on the other frequency; OFF—removes all electrical power; SIMPLEX—transmission and reception are both made on the frequency selected by the "Xmit Freq" switch; SIMPLEX + G Rcv—same as SIMPLEX except that reception of the UHF guard (emergency) frequency (243.0 mHz) is also possible; G T/R—transmission and reception are both made on the UHF guard (emergency) frequency. Access to transmission and reception of UHF signals is controlled by three two-position toggle switches located on the bottom of the audio center panel at the aft station. The switches are labeled T/R for transmission/reception, OFF for blocking UHF signals to or from the UHF transceiver. There is also a switch for both of the A/G channels and the A/A channel. All three of the UHF frequencies (296.8 mHz, 259.7 mHz, and 243.0 mHz) are pre-set in the UHF transmitter and cannot be altered by the flight crew.

The UHF system is used during the orbital phase of the flight for EVA activities. The EVA astronaut UHF communications to and from the orbiter are through the UHF airlock antenna. The two existing UHF frequencies of 296.8 mHz and 259.7 mHz are used plus an additional UHF frequency of 279.0 mHz is added to the EVA EMU's backpack. The 279.0 mHz frequency can transmit/receive only between the two EVA astronauts and the orbiter, not ground stations.

One EVA astronaut would operate in Mode A, transmitting data and voice to the orbiter on 259.7 mHz, transmit voice to the other EVA astronaut on 259.7 mHz, receive voice from the orbiter on 296.8 mHz and receive voice from the other EVA astronaut on 279.0 mHz. The remaining EVA astronaut would operate in Mode B, transmitting data and voice to the orbiter on 279.0 mHz, transmit voice to the other EVA astronaut on 279.0 mHz, receive voice from the orbiter on 296.8 mHz and receive voice from the other EVA astronaut on 259.7 mHz. The orbiter would then operate through a switch in the orbiter via the UHF EVA relay mode by retransmission over air-to-ground (A/G) through the orbiter S-band system to the STDN ground station or through the orbiter S-band system through TDRS or through the orbiter Ku-band system through TDRS or vice-versa. It is noted, as a backup procedure, when over an UHF ground station only, the EVA astronauts, orbiter and ground would switch to the 259.7 UHF frequency, simplex.

The UHF system may also be used after entry during the approach and landing phase of the mission. The air-to-ground

Space Shuttle Spacecraft Systems

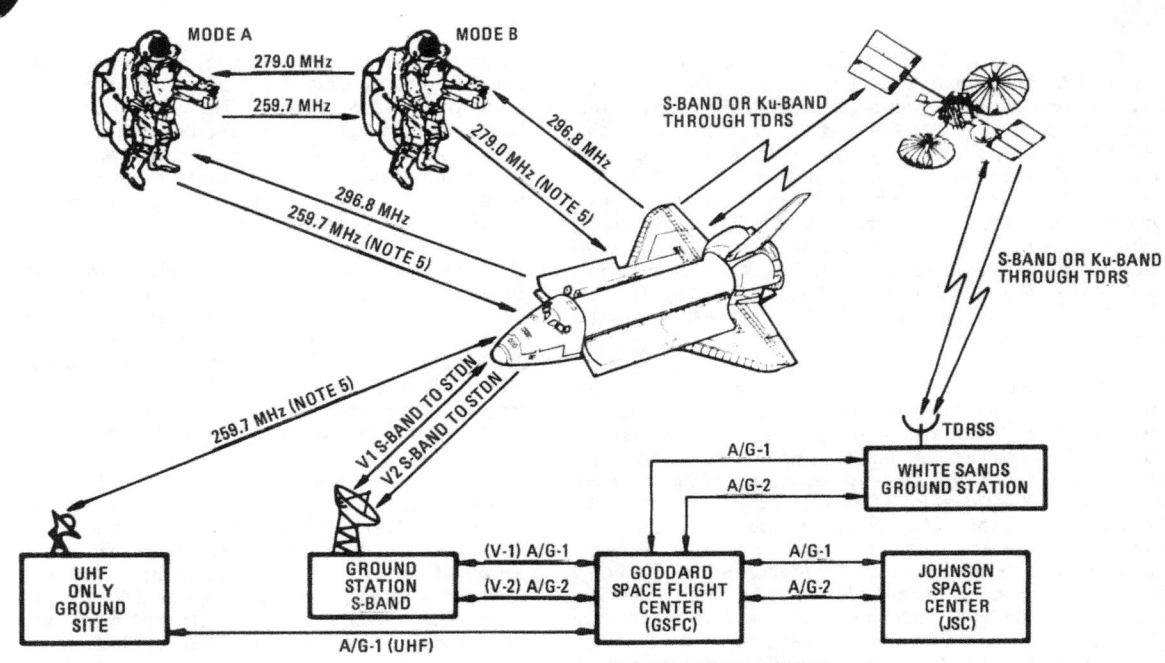

NOTE:
1. CAPSULE COMMUNICATOR (CAPCOM) UPLINK AIR-TO-GROUND (A/G)-1 ONLY. (S-BAND GROUND SITES)
2. HOUSTON COMMUNICATION MONITOR A/G-2
3. GROUND SITE MONITOR UHF
4. BACKUP A/G LINES ARE DIRECT TO JSC
5. IF TWO-WAY COMM IS NEEDED OVER UHF ONLY GROUND SITE, EVA, ORBITER, AND GROUND WILL SWITCH TO 259.7 SIMPLEX

EVA OPERATIONAL MODES

HL	A	B	BU
TWO-WAY HARDLINE COMM BETWEEN ORBITER AND EXTRAVEHICULAR MOBILITY UNIT (EMU) VIA SCU (SERVICE AND COOLING UMBILICAL)	XMIT 259.7 DATA AND VOICE / RCV 296.8 VOICE / RCV 279.0	XMIT 279.0 DATA AND VOICE / RCV 296.8 VOICE / RCV 259.7	XMIT/RCV 259.7 / RCV 295.8

Extravehicular (EVA) Air-to-Ground Configuration

Space Shuttle Spacecraft Systems

voice communications would take place between the Space Shuttle spacecraft and the landing site control tower and chase planes (if used).

INSTRUMENTATION. The instrumentation system consists of transducers, signal conditioners, pulse code modulation encoding equipment, operational recorders, timing equipment, and onboard checkout equipment. The instrumentation system is made up of an operational instrumentation (OI) system. The OI system interfaces with other avionics systems, external tank, solid rocket boosters, and ground support equipment.

The OI system senses and acquires, conditions, digitizes, formats, and distributes data for display, telemetry, recording, and checkout. It provides for PCM recording, voice recording, and master timing for onboard systems. The equipment consists of two pulse code modulation master units (PCMMU's), two operational recorders, one payload recorder, one master timing unit, and various multiplexer/demultiplexers (MDM's), signal conditioners, and sensors.

The dedicated signal conditioners (DSC's) provide inputs to the OI from such transducer signals as frequency, voltage, current, pressure, temperature (variable resistance and thermocouple), and displacement (potentiometer), 28- and 5-volt dc discretes. The signal conditioners convert their input signals to an analog signal of 0 to 5 volts dc or to a 28- or 5-volt dc discrete output signal.

The output signals of the OI DSC's are directed to the flight deck crew displays, C/W system, and a corresponding MDM. The MDMs convert the analog signal to serial digital data (a digitized representation of the applied voltage). The MDM's send this serial digital data to a PCMMU upon request through the OI data buses. When the MDM is addressed by the PCMMU, the MDM will select, digitize, and send the requested data to the PCMMU in serial digital form.

The OI PCMMU receives digital data from the OI MDM's in addition to computer downlist data from the onboard computers, and combines them to form the PCM telemetry for the S-band return link or Ku-band system. The PCMMU controls the data received from the MDM's; downlist data from the computers is under the control of the flight software. All data received by the PCMMU is stored in memory and periodically updated. The PCMMU also sends data to the onboard computers on request.

The OI PCMMU has two formatter memories: programmable read only and random access (RAM). The former is programmed only before launch; the latter is reprogrammed several times during flight. The PCMMU will use the format memories to downlink data from the computers and data from the OI MDM's into PCM telemetry data streams. These data streams are sent to the network signal processor, which then sends the data to the operations recorders and to the S-band transponder or Ku-band transmitter for transmission to the ground.

Only one of the redundant OI PCMMU's and network signal processor's operates at a time. The ones used are controlled by the crew through the flight deck display and control panel.

It is noted that the primary port of an MDM operates with PCMMU 1 and the secondary port of an MDM operates with PCMMU 2.

A payload data interleaver will accept data simultaneously from up to five attached and one detached payload, interleave it, and send it to the PCMMU. The input data is in serial digital data streams. Data temporarily stored in the PCMMU memory can be accessed by the PCMMU telemetry formatter and by the onboard computers. The payload data interleaver is programmed on board from mass memory via the computers to select specific data from each payload PCM signal and store it within its buffer memory locations.

Space Shuttle Spacecraft Systems

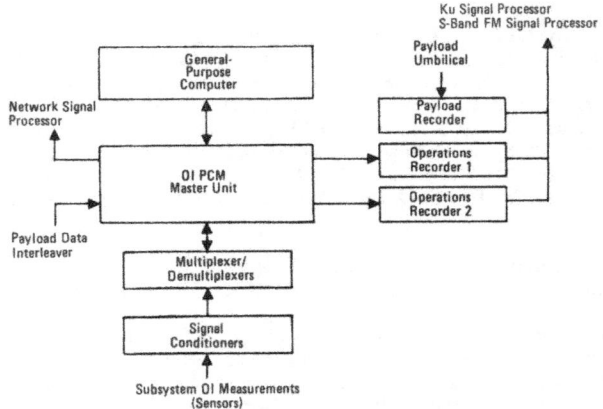

Instrumentation

Area	Characteristics
Signal Conditioning	• 13 DSC's conditioning approximately 1200 channels • 6 WBSC to MDM • Approximately 45 high-level transducers • Distributes data to PCM data system, panel displays, C-W, flight-critical MDM's
PCM Data System	• Acquires data through 7 OI MDM's approximately 2800 measurements • Receives and provides data to 5 computers • Accepts payload data through PDI or payload data bus (Spacelab) • Provides output data in 64 and 128-kbps formats • Provides synchronization to PDI and NSP • Provides data to NSP and T-0 umbilical
Payload Data Recorders	• Receives data directly from payload or PLSP • Provides data to GPC's and PCM • Stores 3 channels of engine data at 60-kbps rate • Stores interleaved voice and data at 96, 128, or 192-kbps rate • Stores payload data • Provides for playback of recorded data during and after mission
Master Timing	• Provides time reference to computers, OI PCM, display panel, and payload • Provides synchronization to instrumentation and other subsystems

OI System

The OI PCMMU's receive a synchronization clock signal from the master timing unit. If this signal is not present, the PCMMU provides its own timing and continues to provide timing signals to the payload data interleaver and network signal processor.

There are many different PCM telemetry formats that control the measurement groupings the PCMMU assembles into the data streams for different mission phases. Those to be used for each mission are stored in the onboard computer's mass memory. When the ground or flight crew want to change the format, a command is sent to the computers, which will then load the desired format from mass memory into the formatter RAM of the PCMMU.

The five onboard computers are capable of data processing. They perform programmed computations and then prepare the data for telemetry transmission by means of a downlist. Four of the computers are assigned to the primary flight control system in the orbiter and provide primary downlist data. The fifth computer is assigned to the backup flight control system and to systems management and payload management and provides downlist data. The computer downlist data consists of a table of values compiled in the computer. The format can be changed by ground command or by the flight crew. The downlist data is sent from the computers to the PCMMU, where it is

Space Shuttle Spacecraft Systems

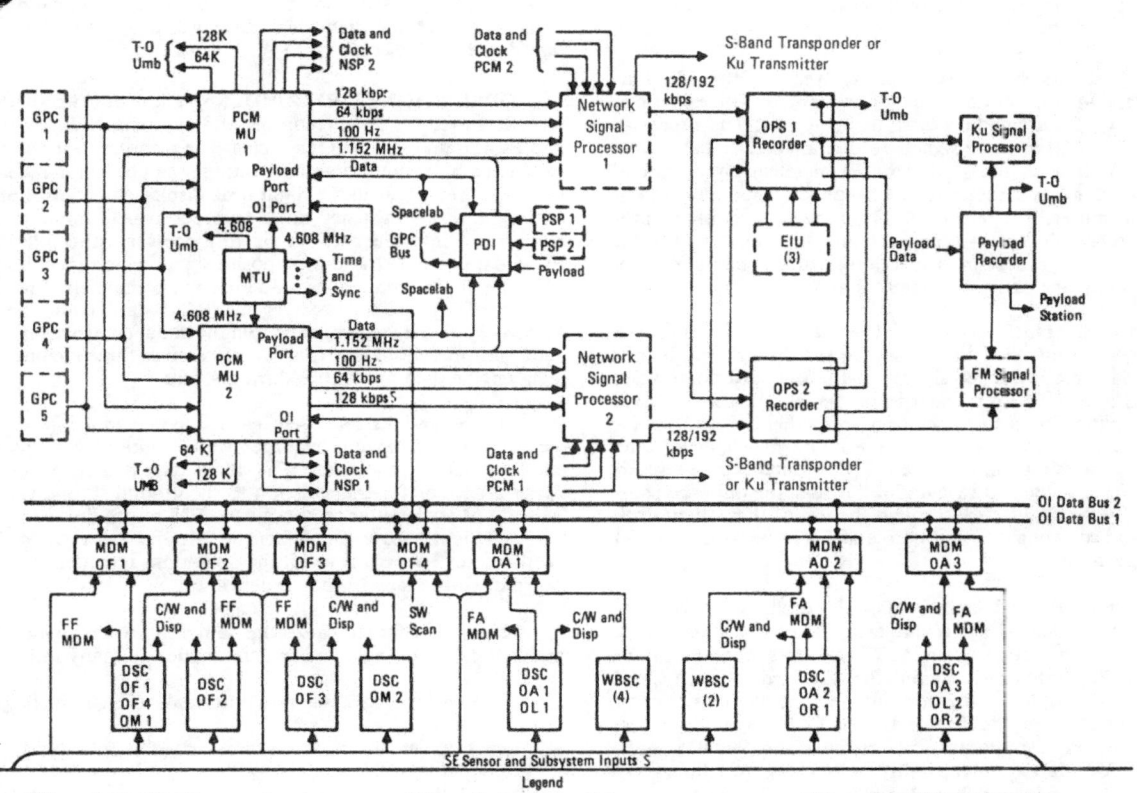

OI System

Space Shuttle Spacecraft Systems

combined with operational instrumentation data to form the PCM downlink data stream.

The network signal processor receives the PCMMU telemetry data streams and also one or two analog voice channels from the spacecraft audio central control unit. The processor converts the analog voice signals to digital voice signals, time-division multiplexes them with the PCM telemetry data, and sends the composite signal to the S-band transponder or Ku-band transmitter. The return rate from the network signal processor can be switched by ground command at any point in the data cycle. The processor also outputs to the operations recorders. Only one processor operates at a time.

MASTER TIMING UNIT. The master timing unit is a stable crystal-controlled timing source for the orbiter. It provides serial time reference signals to the onboard computers, PCMMU's, and various time display panels. It also provides synchronization to instrumentation and other systems. It includes separate time accumulators for Greenwich mean time (GMT) and mission elapsed time (MET), which can be reset or updated from the ground via uplink through the onboard computer or by the flight crew through the use of their flight deck display and control panel keyboard and CRT (cathode ray tube) time displays.

The signal flows from the 4.608-mHz oscillators to the output of the GMT and MET accumulators. The three independent GMT and three independent MET counters operate simultaneously. Separate time accumulators are used for each GMT and MET clock, and they accumulate time in days, hours, minutes, seconds, and milliseconds. The GMT capability is 366 days, 23 hours, 59 minutes, 59 seconds, and 999.875 milliseconds. For MET, the capability is 365 days, 23 hours, 59 minutes, 59 seconds, and 999.875 milliseconds. Both can be updated and reset by ground equipment before flight or from the onboard controls by the flight crew. During flight, the GMT and MET accumulators are updated at a predetermined time by uplink and onboard computer or by voice command and entered through the flight deck display and control panel keyboard and CRT display.

OPERATIONAL RECORDERS. There are two of these recorders used for serial recording and dumping of digital voice and PCM data from the OI systems. The recorders normally are controlled by ground command, but they can be commanded by the flight crew through the flight deck display and control panel keyboard or by switches on a recorder panel. Input to the recorders is from the network signal processor, either 128-kbps PCM data or a 192-kbps composite signal which includes the 128-kbps PCM data and two 32-kbps voice channels. The network signal processor receives the PCM data from the OI PCMMU and the voice signals from the audio control center. In addition, operations recorder No. 1 receives three channels of main engine data at 60 kbps during ascent.

The operations recorders can be commanded to dump recorded data from one recorder while continuing to record real-time data on the other. The dump data is sent to the FM signal processor for transmission to the ground station via the S-band FM transmitter on the S-band FM return link or to the Ku-band signal processor. When the ground has verified that the data they received is valid, the operations recorders can use that part of the tape to record new data.

A single recorder can store and reproduce digital and analog data both singly and in combination at many rates.

The single recorder function is normally used in the flights.

Recorder functions can be summarized as follows:

- Data In, Recorder No. 1:

  - Accepts three parallel channels of engine data at 60 kbps during ascent

Space Shuttle Spacecraft Systems

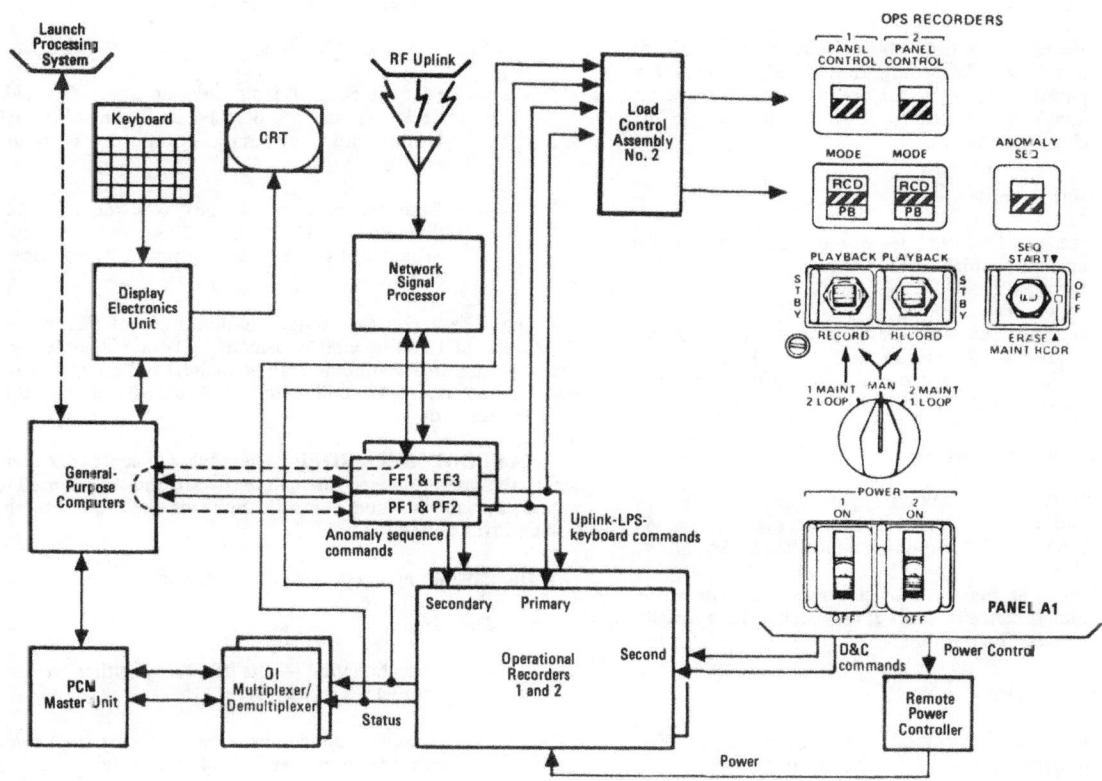

Operational Recorders

Space Shuttle Spacecraft Systems

- Accepts 128/192 kbps of interleaved PCM data and voice which serially sequences from Track 4 to Track 14.

- Accepts real-time data from network signal processor. Recording time is 32 minutes for parallel record and 5.8 hours for serial record on Tracks 4 through 14 at 381 millimeters (15 inches) per second.

• Data In, Recorder No. 2:

- Accepts 128/192 kbps of interleaved PCM voice and data which serially sequences from Track 1 through Track 14.

- Accepts real-time data from network signal processor. Recording time is 7.5 hours at 381 millimeters (15 inches) per second for serial record on 14 tracks.

• Data Out, Recorder No. 1:

- In flight payback of engine interface unit data and network signal processor digital data via S-band FM transponder or Ku-band transmitter.

- In-flight playback of anomaly PCM data for maintenance recording; playback of data serially to GSE T-O umbilical.

• Data Out, Recorder No. 2:

- In-flight playback of digital data via S-band FM transponder or Ku-band transmitter.

- In-flight playback of anomaly PCM data for maintenance recording; playback of data serially to GSE T-O umbilical.

• Recorder Control:

- Manual control from mission specialist flight deck aft station display and control panel. Uplink and onboard computer keyboard control.

- Recorder speeds of 190, 381, 609, and 3,048 millimeters (7.5, 15, 24, and 120 inches) per second provided by hardware programs plug direct command.

The tape recorders contain a minimum of 731 meters (2,400 feet) of 12 millimeters (0.5-inch) by 1-mil magnetic tape. They operate at 609 millimeters (24 inches) per second in the record mode, and 3,048 millimeters (120 inches) per second in the playback mode.

PAYLOAD RECORDER. The payload recorder is identical to the operations recorders in hardware and will be used to record payload data and dump in flight via the S-band or the Ku-band transmitter.

The payload and recorder capabilities are:

• Data In:

- Accepts digital inputs of 64 kbps either serial or parallel up to 14 tracks.

- Accepts analog data from 1.9 kHz to 2 mHz either serial or parallel up to 14 tracks.

 **Space Shuttle Spacecraft Systems**

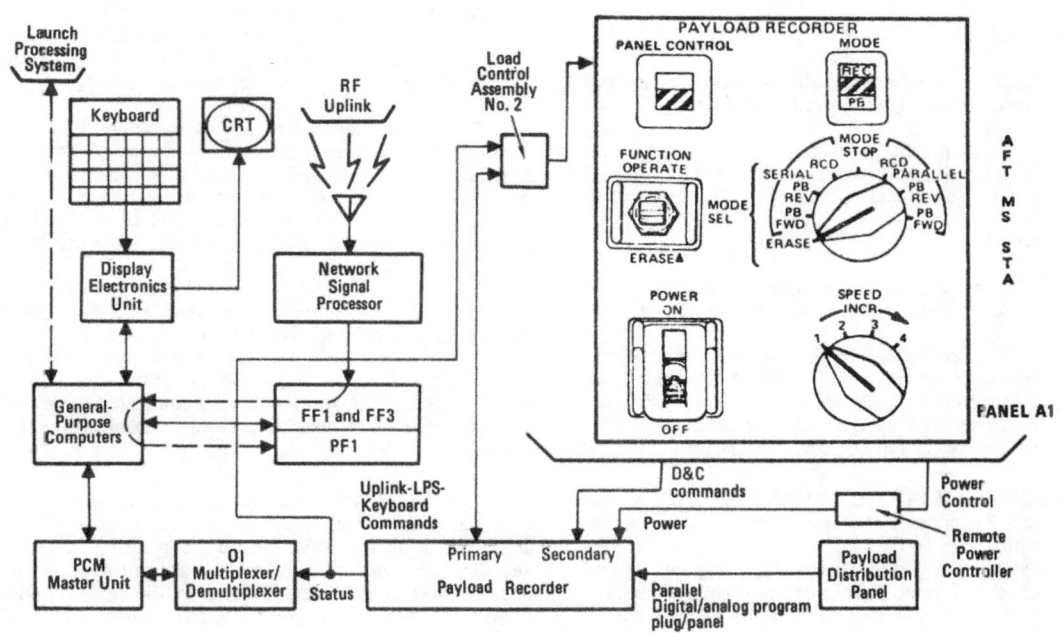

Payload Recorder

Space Shuttle Spacecraft Systems

- Serial/parallel track programming is determined by premission payload distribution panel wiring.

- Record time is 32 minutes for parallel record or 7.5 hours for serial record at 381 millimeters (15 inches) per second.

• Data Out:

- In-flight playback of digital data via S-band transponder or Ku-band transmitter.

- In-flight playback (analog/digital) data to payload distribution panel; playback of data to GSE via T-O umbilical.

• Control

- Manual control from mission specialist flight deck aft station display and control panel.

- Uplink and computer keyboard by computers. Recorder speeds of 381, 762, 1,524 and 3,048 millimeters (15, 30, 60, and 120 inches) per second provided by hardware program plug. Recorder hardwired for continuous run modes.

MODULAR AUXILIARY DATA SYSTEM. The Modular Auxiliary Data System (MADS) is an onboard instrumentation system that measures and records selected pressure, temperature, strain, vibration, and event data to support payloads and experiments and to determine orbiter environments during the flights of the orbiter. MADS supplements the operational instrumentation (OI) that exists in the orbiter. The MADS equipment conditions, digitizes, and stores data from selected sensors and experiments.

MADS collects detailed data during ascent, orbit, and entry to define the vehicle response to the flight environment, permit correlation of data from one flight to another, and enable comparison of one orbiter's flight data to the flight data of another orbiter.

All of the MADS equipment installed in the orbiter are structurally mounted and environmentally compatible with the orbiter and mission requirements. Due to its location, the MADS will not intrude into the payload envelope.

The MADS consists of a pulse code modulation (PCM) multiplexer, a frequency division multiplexer (FDM), a power distribution assembly (PDA), and appropriate signal conditioners mounted on shelf 8 beneath the payload bay liner of the mid-fuselage. The MADS also consists of a MADS control module (MCM) and a MADS recorder that are mounted below the mid deck floor.

MADS will record approximately 246 measurements throughout the orbiter. These measurements are from the orbiter airframe and skin and the orbital maneuvering system/reaction control system (OMS/RCS) left hand pod only. Measurements of MADS components are connected to existing operational instrumentation for real time monitoring of MADS status.

The MADS interfaces with the orbiter through the orbiter electrical distribution system and the inputs to the operational instrumentation for MADS status monitoring. Coaxial cables and wire harnesses from the sensors are routed through the orbiter payload bay harness bundles to the signal conditioners, PCM, and FDM, attached to mid-fuselage shelf 8. After the signal conditioners and the multiplexers have processed the data, four outputs of the FDM and one output of the PCM is routed forward to the MCM, which will then record them on

Space Shuttle Spacecraft Systems

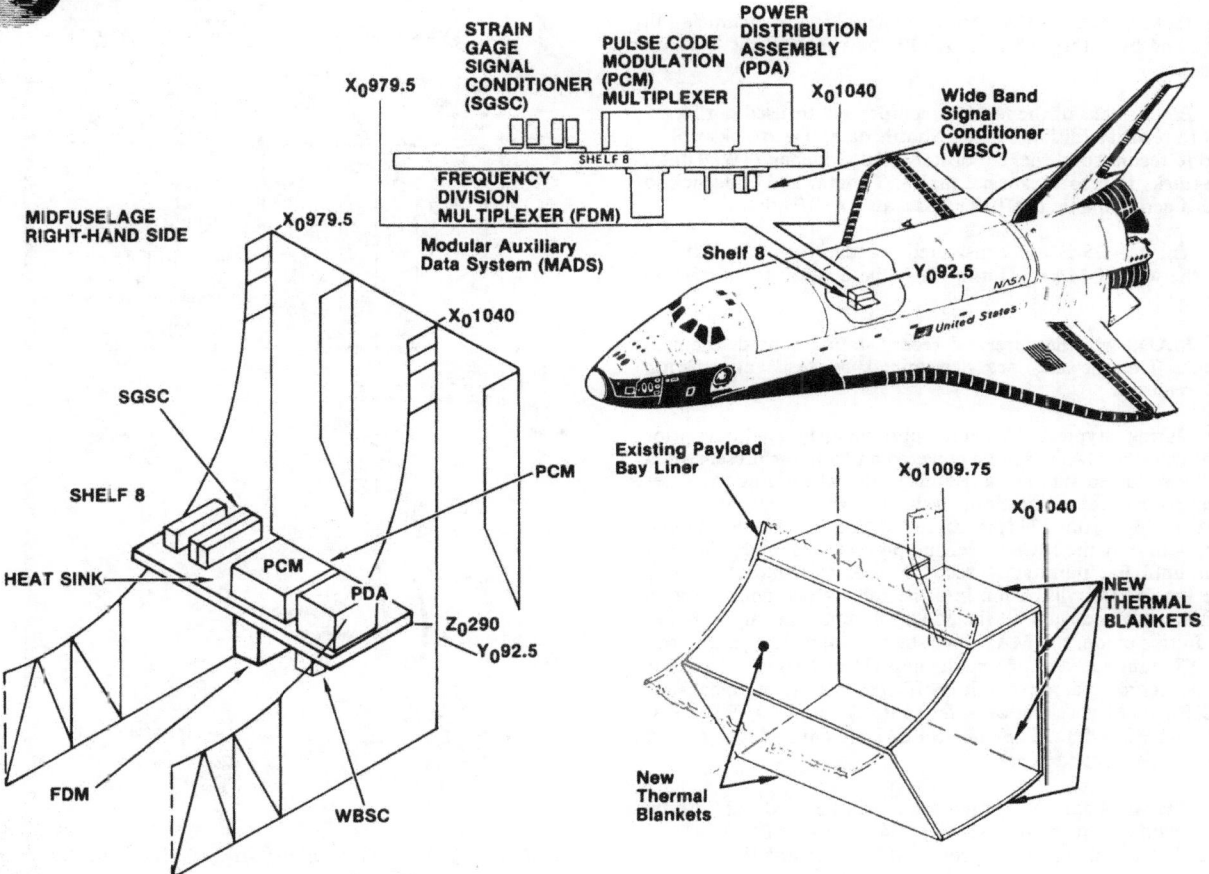

Modular Auxiliary Data System (MADS) Mid-Fuselage

Space Shuttle Spacecraft Systems

five tracks of the MADS recorder. The same five channels will be routed back through the X_O 1307 bulkhead to the T-0 umbilical.

Eight tracks of the MADS recorder will be used during ascent to record additional Space Shuttle data. Two tracks will be used to record solid rocket booster (SRB) wideband (WB) data, five tracks to record external tank (ET) data, and one track to record aerodynamic coefficient package (ACIP) data.

The MADS is not considered mandatory for launch nor will the loss of MADS during flight be a cause for a mission abort.

MADS will measure and record data for predetermined events. These events are determined by test and mission requirements.

During a typical mission at approximately five hours prior to launch, the MADS will be powered on from the preset switch configuration to supply a prelaunch manual calibration. After completion of the calibration, all switches will be returned to the preset configuration. This leaves the MADS in the standby position, with only the MCM receiving power. This mode will continue until five minutes 30 seconds prior to launch, at which time the MADS will be put into the full system mode through uplink commands and all the MADS components are powered on. In this mode, the MADS will be recording at a continuous (CONT) tape speed of 381 millimeters (15 inches) per second. It will be recording aerodynamic coefficient identification package (ACIP), flight acceleration safety cutoff (FASCO), ET, SRB, WB, and PCM data. The MADS PCM will have a bit rate of 64 kilo-bits-per second (kbps).

The wideband (WB) only mode will be used only during the prelaunch automatic (AUTO) and manual (MAN) calibrations. In this mode, the recorder will be recording the AC and

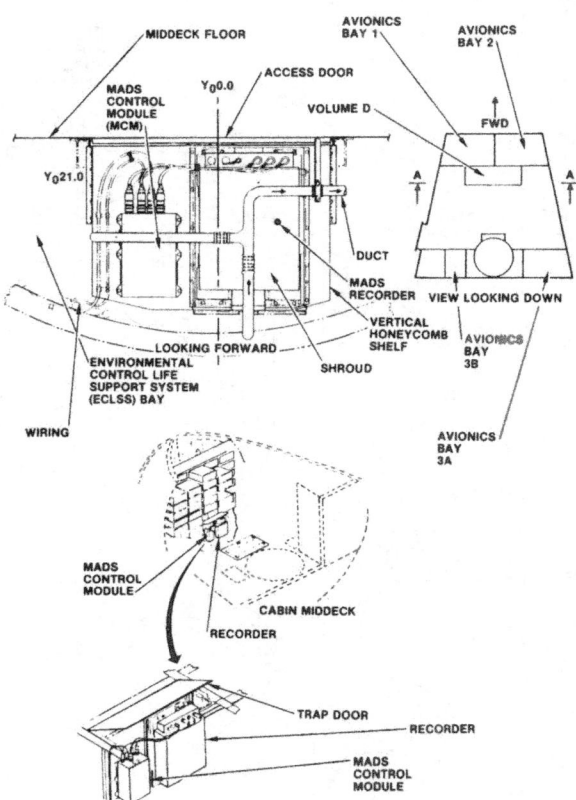

Modular Auxiliary Data System (MADS) Crew Compartment

Space Shuttle Spacecraft Systems

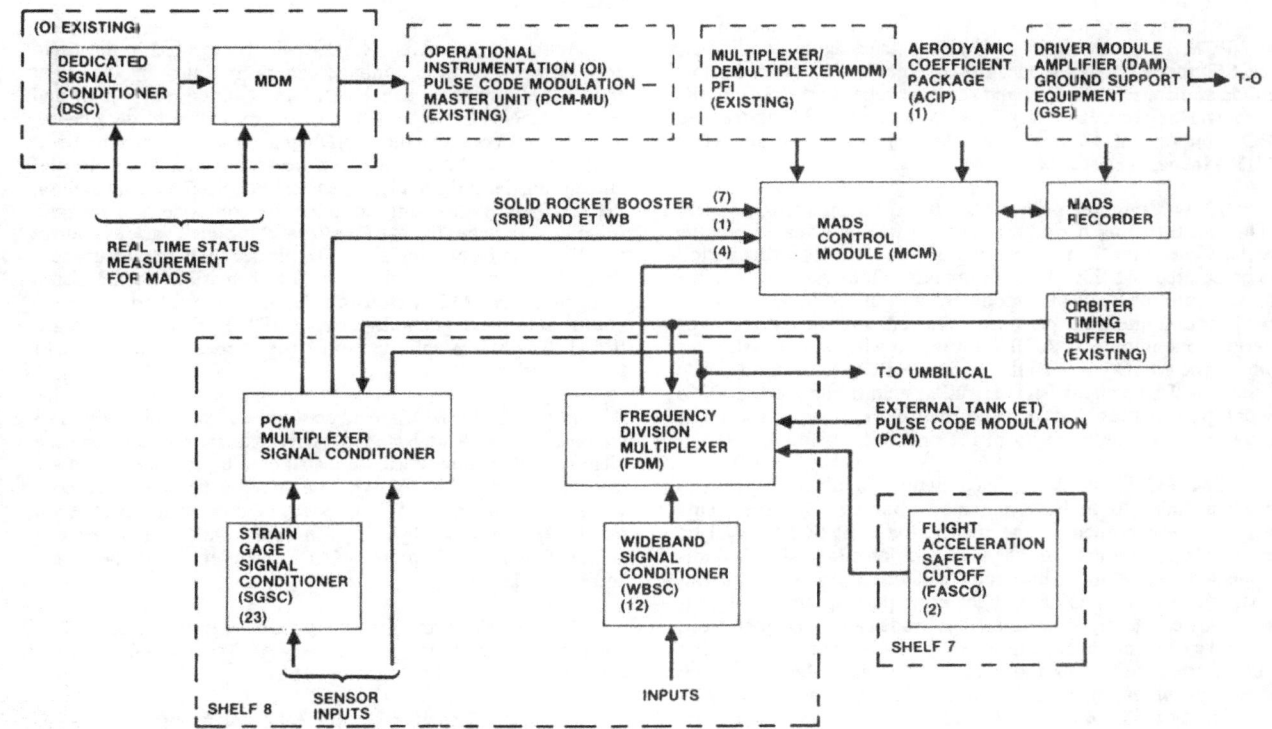

Modular Auxiliary Data System (MADS) Block Diagram

Space Shuttle Spacecraft Systems

DC current calibration levels provided by the FDM. Each manual calibration level will be recorded for 10 seconds at a tape speed of 381 millimeters (15 inches) per second in the continuous mode.

At 12 minutes after launch MADS will be commanded into the PCM snapshot (S/S) with strain gage signal conditioner (SGSC) mode. In this mode, the recorder will be in the sample mode and conserves power and recorder tape. In this S/S mode, data will be recorded every 10 seconds every 10 minutes at a PCM bit rate of 32 kbps and a tape speed of 95 millimeters (3-3/4 inches) per second.

At two minutes prior to the orbital maneuvering system (OMS)-2 thrusting period, commands will be given to put the MADS back into the full system mode until the thrusting period is completed. At this time, commands will be given to put the MADS into the PCM only mode, which will continue during the orbit until a quiescent period is achieved. During the quiescent period, one minute of ACIP calibration will be required, after which the MADS will continue in the PCM only mode. The system will be switched to the full system mode for the OMS separation thrusting periods and then be returned to the PCM only mode for the majority of the on-orbit mission.

The PCM with strain gage signal conditioners (SGSC) mode is similar to the PCM only mode, but strain measurements will also be recorded during this period. The SGSC's will be cycled along with the other MADS components to signal conditioners to warm up. This mode will occur between two full system modes to minimize flight crew participation and conserve power and recorder tape. This mode can be initiated from the full system mode or returned to the full system mode by one uplink command. This mode can be put into the PCM only mode by commanding the SGSC off, which is done manually by positioning switch 4 on panel A7A2 in the OFF position. This mode is used on orbit.

At two minutes before the deorbit thrusting period, the MADS will be put into the full system mode for one hour to record descent (entry) data. At the conclusion of the one hour period, the MADS will be powered down for the entire postlanding period.

With the use of the MADS switches located in the flight crew compartment, commands can be initiated by the flight crew. These switches are located on two panels, C3A5 and A7A2. Panel C3A5 is located on the forward flight deck center console and contains the MADS master power switch (S14). This switch will be used to turn power on or off during prelaunch, postlanding and emergencies. Panel A7A2 is located on the aft flight station and contains the component power and functional switches for MADS. From this panel, various control functions can be accomplished. To reduce flight crew participation, all commands should be uplink if possible from Mission Control Center (MCC) Houston (H) and transmitted to the onboard multiplexer/demultiplexer (MDM), Payload Forward (PF)-1. The MDM will then route the commands to the MCM for processing.

Power for the MADS will be supplied from the orbiter's 28 vdc main buses A and B. The ACIP experiment is a separate identity, but its power will be distributed by the MADS power distribution assembly (PDA). The ACIP experiment will consume power when the WB is powered on, using switch 5 on panel A7A5. The 64 kbps of PCM data from the ACIP experiment will be recorded on the MADS recorder during the ascent and entry phases.

The flight acceleration safety cutoff located on shelf 7 in the mid fuselage, directly above the MADS shelf 8, interfaces 12 vibration measurements with the MADS.

The MADS shelf 8 components will be protected from overheating by a passive thermal control system that will be

Space Shuttle Spacecraft Systems

used to constrain maximum temperatures. The MADS installation is thermally isolated from the orbiter structure by 1.2 millimeter (0.049 inches), thin wall titanium struts. The installation is also enclosed from the orbiter environment by a 38 millimeter (1.5 inch) bulk insulation enclosure.

Each measurement uses either a thermocouple, resistance thermometer, radiometers, vibration sensor, strain gage, or pressure transducer.

The MADS recorder is a Bell and Howell 28-track wideband modular airborne recording system (MARS) similar to the *Columbia* development flight instrumentation (DFI) missions and orbiter experiments (OEX recorders). The recorder is capable of simultaneously recording, and subsequently reproducing, 28 tracks of digital biphase L data or any combination of wideband analog and digital biphase L data equal to 28 tracks.

All 28 tracks can be output simultaneously with adequate levels to drive the input circuitry of the driver amplifier module (DAM) which is part of the MADS equipment that is not installed in the orbiter. It is support equipment that will be carried on and used for dumping the data recorder during the checkout or postlanding.

The total weight of the MADS is 290 kilograms (641 pounds).

KU-BAND RENDEZVOUS RADAR. The orbiter Ku-band system includes a rendezvous radar which is used to skin-track satellites or payloads that are in orbit. This makes it easier for the orbiter to rendezvous with any satellite or payload in orbit. For large payloads that will be carried into orbit, one section at a time, the orbiter will rendezvous with that payload that is already in orbit to add on the next section. The Ku-band antenna gimbaling permits it to radar search for space hardware. The Ku-band system is first given the general location of the space hardware for orbiter computers. The antenna then makes a spiral scan of the area to pinpoint the target.

Radar search for space hardware may use a wide spiral scan up to 60 degrees. Objects may be detected by reflecting the radar beam off the surface of the target (passive mode) or by using the radar to trigger a transponder beacon on the target (active mode).

TEXT AND GRAPHICS. The text and graphics is government furnished equipment (GFE). It is basically a hard copy machine that operates via telemetry and provides the capability to transmit text, materials, maps, schematics, and photographs to the orbiter via the Ku-band system.

EQUIPMENT LOCATION. Instrumentation equipment, except for sensors and selected dedicated signal conditioners, are located in the forward and aft avionics bays. Sensors and dedicated signal conditioners are located throughout the orbiter in areas selected on the basis of accessibility, minimum harness requirements, and functional requirements. Effective use of remote data acquisition techniques was considered for optimizing equipment location. The factors which were considered in determining equipment location were weight, power, physical size, redundancy, and wire density and length to each compartment and interconnect wiring.

The abbreviation OA refers to operational, aft, OF to operational forward, OL to operational left, OR to operational right, OM to operational mid.

TELEVISION. The orbiter closed circuit television (CCTV) system includes both interior and exterior cameras. In this description, the television (TV) system is limited to the discussion of the crew cabin operations. The payload bay and remote manipulator system television is discussed in the payload deployment and retrieval section of this booklet.

Space Shuttle Spacecraft Systems

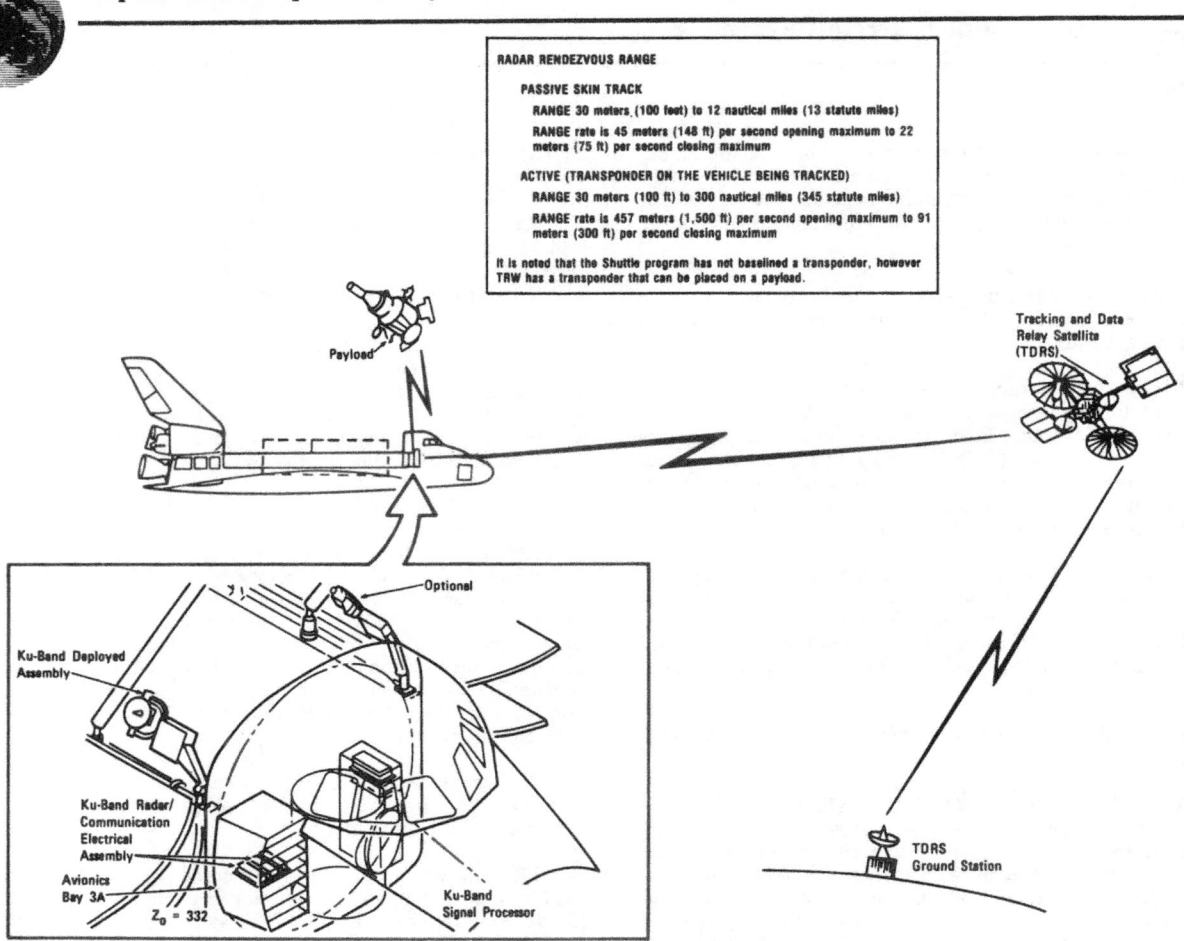

Ku-Band Radar Communication System

Space Shuttle Spacecraft Systems

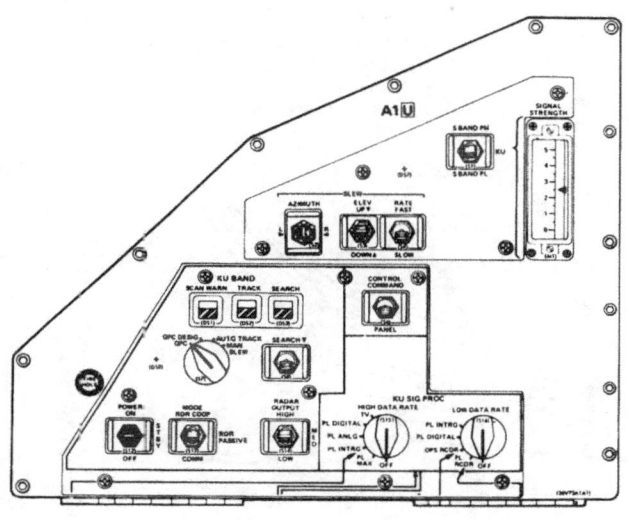
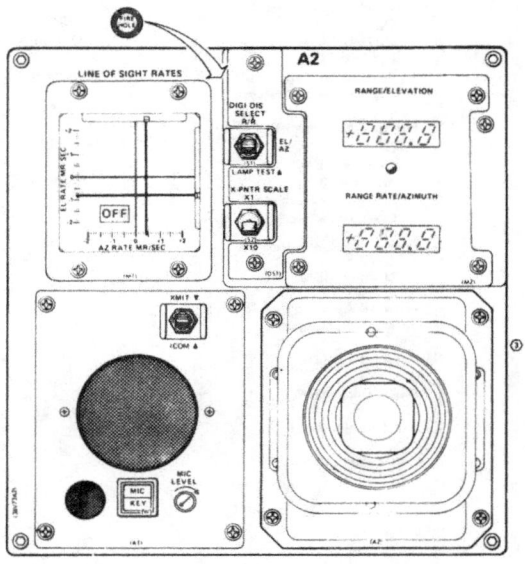

The TV system is used to document a wide variety of on-orbit activities and events inside the crew cabin. These activities and events include crew activities and operations, experiment observations, inspections and data retrieval, hardware observations, and crew compartment areas and configurations.

The system consists of TV camera, TV lens, TV wide angle lens, TV viewfinder monitor, TV cable 3 meter (10 feet) and 6 meter (20 feet), TV viewfinder monitor cable, TV console monitor, video tape recorder, video tape recorder cassette, and TV system displays and controls.

All TV operations involving forward link commands and data return links are via the S-Band or Ku-band system.

During launch and entry, the TV system components are stowed in lockers, with the exception of the video tape recorder (VTR) and monitors 1 and 2 which are installed on panels L12 and A3, respectively.

Space Shuttle Spacecraft Systems

① **Forward Avionics Bay 1**
 DSC
 OI Data MDM
 PCM MU 1
 Payload recorder
 Payload data interleaver

② **Forward Avionics Bay 2**
 DSC
 OI data MDM
 PCM MU 2
 Operations Recorder 1
 Operations Recorder 2

③ **Forward Avionics Bay 3A**
 DSC
 OI data MDM

④ **Forward Avionics Bay 3B**
 Master timing unit

⑤ **Forward RCS DSC**

⑥ **Flight Deck MDM**

⑦ **Aft Avionics Bay 4**
 DSC
 OI data MDM

⑧ **Aft Avionics Bay 5**
 DSC
 OI data MDM

⑨ **Aft Avionics Bay 6**
 DSC
 OI data MDM

⑩ **OMS DSC's (4)**

⑪ **Fuel cell DSC's (2)**

Legend
DSC — Dedicated signal conditioner
MDM — Multiplexer/demultiplexer
MU — Master unit
SCA — Signal conditioner assembly

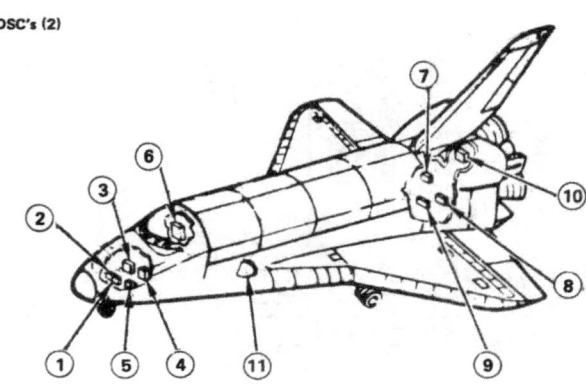

Instrumentation Equipment Location

The crew is required to deploy, set up, and stow the TV system. Control of the system is available from the ground through forward commands and onboard from panel A7 and the camera assembly. The ground and crew can control the system power, camera and lens functions, and input/output video configuration selection. Only the crew can determine control mode (ground and/or onboard) and assign the input to monitors 1 and 2. The ground can affect the monitor 1 (2) display if a split screen display from a multiplexer is the input to the monitor. In this case, the ground can change the camera inputs assigned to the multiplexer. Also, the ground can simultaneously return link a camera selected for monitor 1 (2) and adjust the camera settings (zoom, etc.) and affect the monitor 1 (2) display. VTR record and playback operations are controlled only by the crew.

Space Shuttle Spacecraft Systems

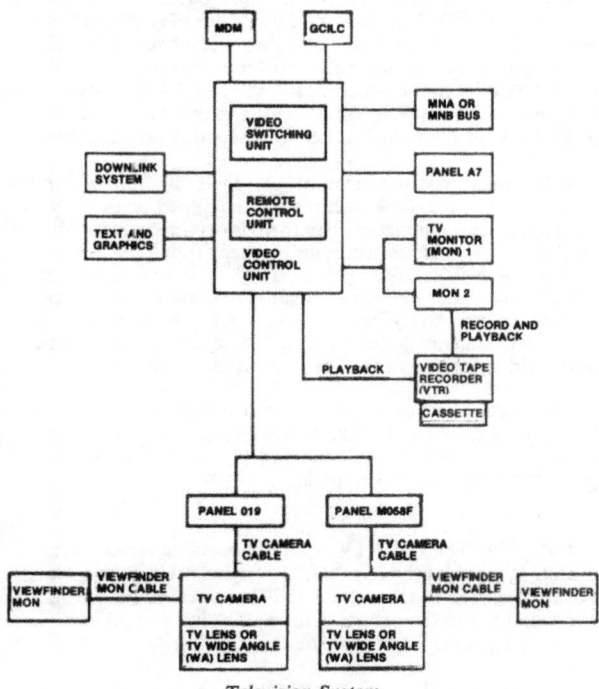

Television System

All interior cameras are equipped with a color lens. Monitor 1 (2) and the TV viewfinder monitor provide a monochrome (black and white) output, regardless of the lens assembly utilized, that allows the crew to adjust and view the video scenes. When viewing interior camera scenes, a flicker will be noticed.

The flicker is caused by the rotating color wheel used to generate the field-sequential-color signal that allows color outputs on the ground. The color conversion process on the ground eliminates the flicker before the color signal is distributed.

The camera is a basic monochrome camera that converts light images into a composite video signal (picture plus sync) and can produce either a black and white or color image depending on the lens assembly used.

The TV lens provided with the TV system is a color lens. The lens is an F/1.4 zoom lens. The lens is equipped with a six-segment three-color rotating filter wheel that produces sequential red, green, and blue color fields. The lens is equipped with motorized lens control functions for varying the zoom, iris, and focus. These lens functions are controlled manually via lens switches, remotely via panel A7, or via ground command.

The viewfinder monitor is a small portable monochrome monitor which enables the crew to view camera video output for picture quality and scene verification when monitors 1 and 2 are not accessible or available.

The TV cable provides an interface between the camera and the two TV system input stations located in the crew cabin at Panels 019 and M058F, respectively. The cable provides the camera with 28 vdc power, camera/lens commands along with the sync signal, and video including camera/lens data back to the control unit for distribution.

The viewfinder monitor cable provides an interface between the camera and the monitor. The cable provides the monitor with 28 vdc power and camera video input. The cable is 2.74 meters (9 feet) in length.

The two TV console monitors fixed in the aft flight deck crew station on Panel A3 are identical and are arranged one

 ## Space Shuttle Spacecraft Systems

over the other. The top one is referred to as monitor 1 and the bottom as monitor 2. The monitors are used primarily to provide the crew with a means of viewing payload bay/remote manipulator system camera outputs for picture quality, scene verification, and to conduct camera operations. These monitors can also be used to display cabin TV output when it is more convenient than using the viewfinder monitor. Monitor 2 is also the source of video for the video tape recorder. Each monitor is a monochrome monitor and has the capability of displaying any of the camera inputs available and includes a split screen image from any two cameras. This feature is generated by the control unit and selected via Panel A7 commands. The video source may be selected from the Panel A7 selection, the downlink signal, and a direct video tape record video playback, monitor 2 only. The monitors are capable of displaying a set of crosshairs for camera pointing alignment and superimposing alphanumeric camera data on the screen. The data consists of a camera ID (identification) number, camera pan and tilt angles, and camera temperatures if any camera is in an overtemperature condition. The monitors are equipped with displays and controls to directly control most monitor operations; however, the assignment of an input source to the monitor is controlled by Panel A7.

The video tape recorder is a modified, off-the-shelf video recorder. The video tape recorder has the capability of also recording audio with the video through the crew communications umbilical connector on the video tape recorder housing. This connector only interfaces the video tape recorder and does not allow the crewman to access the vehicle communications system. The video tape recorder utilizes 30 minute tape cassettes which can be played back for downlink via a combination of crew and ground operation. Only recorded video, not audio, is downlinked. All actual video tape recorder operations must be preformed by the crew including everything from tape changeout to video tape recorder activation. The video tape recorder is configured to receive its video input via monitor 2 (to record video on the video tape recorder, the desired camera output must be displayed on monitor 2 which is connected directly to the video tape recorder). For onboard playback, the recorded video can be reviewed on monitor 2 by positioning the monitor 2 source switch to DIRECT and initiating a playback mode. For downlink purposes, the video tape recorder is connected to the payload 1 (PL 1) input which is a Panel A7 VIDEO INPUT choice—PL 1. The switch panel located above the video tape recorder includes a circuit breaker for the video tape recorder.

Video tape recorder uses off-the-shelf 30 minute tape cassettes. The video tape recorder is equipped with a NO VIDEO light that illuminates when there is no video source present at the recorder. When recording, a check should be made to verify that the NO VIDEO light is off. There is also an end-of-tape (EOT) light that indicates when the cassette is out of tape. When this occurs, the video tape recorder will automatically stop. Tape changeout is just a matter of ejecting the cassette from the video tape recorder like any cassette recorder.

Viewfinder monitor is 203 x 107 x 91 millimeters (8 x 4.25 x 3.60 inches) and weighs 1.81 kilogram (4 pounds). TV console monitor is 317 x 254 x 177 millimeters (12.5 x 10 x 7 inches) and weights 9.52 kilograms (21 pounds).

EMU-TV. The EMU-TV is a fully portable remote television unit. It provides the capability to transmit black-and-white television pictures to the orbiter CCTV system from virtually any location about the orbiter exterior. Orbiter reception is via either of the S-band FM hemispheric antennas.

The assembly fits over the EMU helmet and light assembly. It is battery powered (28 vdc) and transmits video at 1775.7 mHz to a video receiver/processing unit installed in the orbiter's mid-deck. When the receiver is connected to the TV input station at panel M058F by a standard 6 meter (20 foot) TV power cable, the IVA (intravehicular) flight crew members can view real-time EVA video on either monitor by selecting the mid deck

 ## Space Shuttle Spacecraft Systems

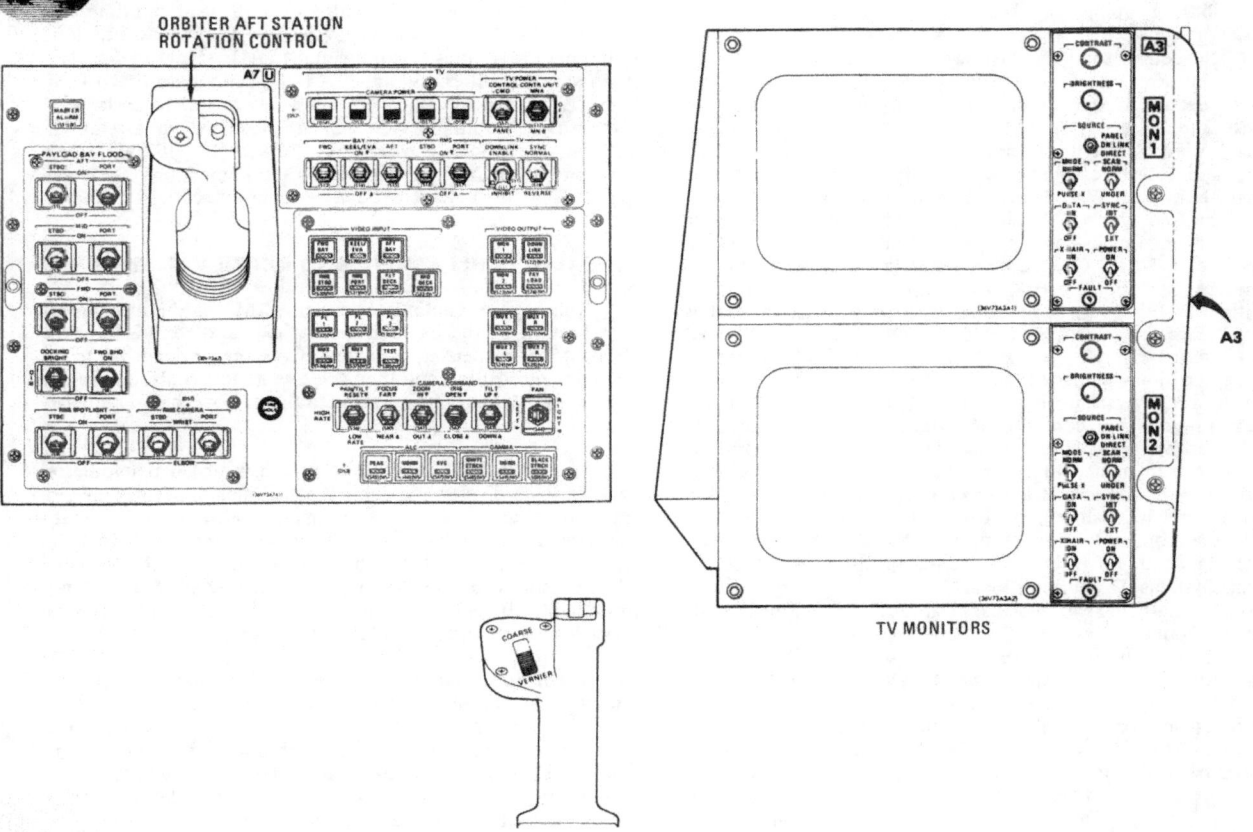

TV MONITORS

RMS Rotation Control

Space Shuttle Spacecraft Systems

camera input on Panel A7. EVA video can also be selected for return link or taping.

The contractors involved with the instrumentation and communication systems are Aydin, Vector Division, Newton, PA (wideband frequency division multiplexing); Communications Components, Costa Mesa, CA (UHF antenna); Conrac Corp., West Caldwell, NJ (mission timer event timer, ground command interface logic box, FM signal processor); Eldec Corp., Lynwood, WA (dedicated signal conditioner); Endevco, San Juan Capistrano, CA (piezoelectric accelerometer, acoustic pickup piezoelectric—acoustic and vibration); Gulton Industries, Costa Mesa, CA (accelerometer linear flow frequency, vibration, acoustic); Harris Corp., Electronic Systems Division, Melbourne, FL (pulse code modulation master unit); Harris Corp., Electronic Systems Division, Baltimore, MD (payload data interleaver); Hughes, El Segundo, Ca (Ku-band radar, communication system deployable antenna and electrical assembly); K-West, Westminsters, CA (wideband signal conditioner, strain gauge signal conditioner); Magnavox, Ft. Wayne, IN (UHF receiver-transmitter mount, UHF receiver-transmitter); Micro Measurements, Romulus, MI (strain gauge); RDF Corp., Hudson, MH (sensors, transducers); Rosemount, Inc., Eden Prairie, MN (transducer, sensors); Radio Corp. of America, Astro-Electronics Division, Princeton, NJ (closed circuit television); Spectran, La Habra, CA (sensors); Sperry Rand Corp., Flight Systems Division, Phoenix, AZ (multiplexer/demultiplexer); Stratham Instruments, Oxnard, CA (transducers); Systron-Donner, Concord, CA (accelerometer); Teledynamics Division of Ambac Industries, Fort Washington, PA (S-band transmitter, FM transmitter, S-band transceiver); Telephonics Division, Instruments Systems Corp., Huntington, NY (orbiter audio distribution system); TRW Systems, Electronic Systems Division, Redondo Beach, CA (S-band payload interrogator, S-band network equipment, network signal processor, payload signal processor); Transco Products, Venice, CA (S-band switch); Watkins Johnson, Palo Alto, CA (C-band radar altimeter antenna, L-band TACAN, S-band quad antenna, S-band hemiantenna, S-band payload antenna, S-band power amplifier); Wavecom, Northridge, CA (S-band multiplexer); Westinghouse Electric Corp., Systems Development Division, Baltimore, MD (master timing unit), AIL, Huntington, NY (S-band preamplifier assembly), AVCO, Wilmington, MA (Ku-band MSBLS antenna, Ku-band waveguide); Teledyne-Microwave, Mountain View, CA (S-band switch assembly); Teledyne-Electronics, Newberry Park, CA (S-band FM transmitter); RCA, Government Communications Systems, Camden, NJ (extravehicular activity communications system).

PAYLOAD DEPLOYMENT AND RETRIEVAL SYSTEM

The remote manipulator system (RMS) is the mechanical arm portion of the payload deployment and retrieval system (PDRS) that maneuvers a payload from the payload bay to its deployment position and then releases it. It can also grapple a free-flying payload, maneuver it to the payload bay, and berth it.

The basic RMS configuration consists of a manipulator arm, an RMS display and control panel (including rotation and translation hand controls), and a manipulator controller interface unit which interfaces with the orbiter computer. Most missions will require only one manipulator arm, which normally will be installed on the port (left) side longeron of the orbiter payload bay. It also can be installed on the starboard (right) side if needed. A two-arm installation also can be used.

When both manipulator arms are installed, they can be operated only one at a time, since only a single software package (computer programs) and a single set of display and control panel hardware are provided. The fifth onboard computer controls the RMS. The RMS takes up 32 percent of the CPU (computer processor unit) in the one computer for RMS operation and 30 percent for manual augmented operation. Wiring is in the aft station for both RMS's.

Space Shuttle Spacecraft Systems

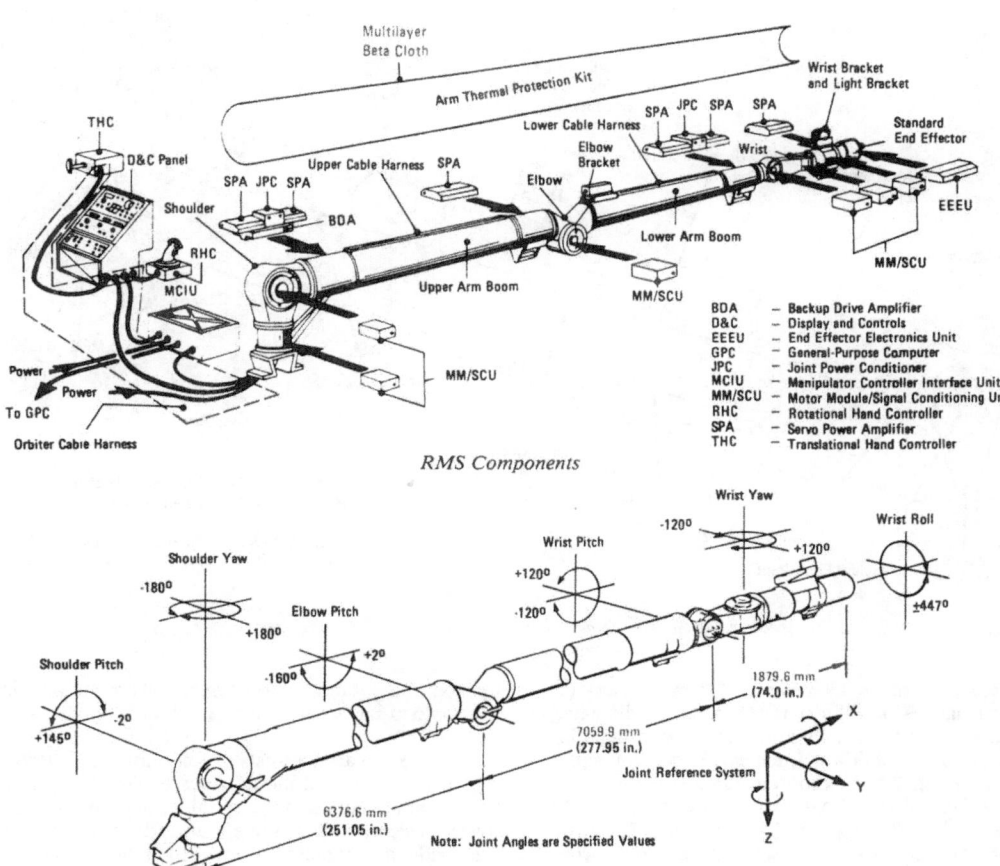

RMS Components

BDA — Backup Drive Amplifier
D&C — Display and Controls
EEEU — End Effector Electronics Unit
GPC — General-Purpose Computer
JPC — Joint Power Conditioner
MCIU — Manipulator Controller Interface Unit
MM/SCU — Motor Module/Signal Conditioning Unit
RHC — Rotational Hand Controller
SPA — Servo Power Amplifier
THC — Translational Hand Controller

Mechanical Arm — Stowed Position and Movement Configruation

 **Space Shuttle Spacecraft Systems**

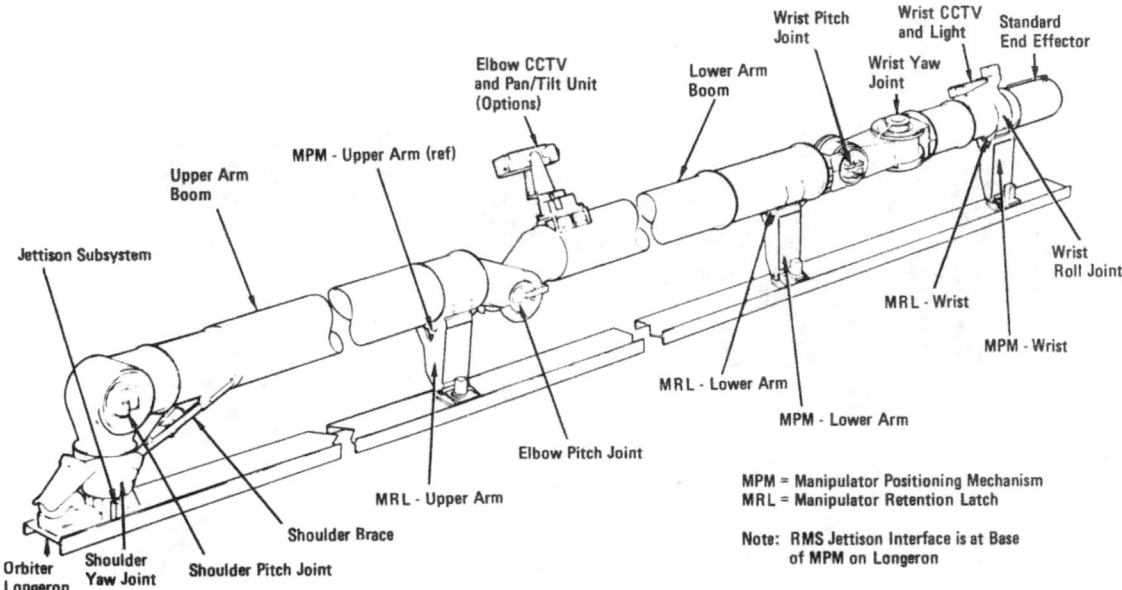

Mechanical Arm — Stowed Position and General Arrangement

The manipulator arm is 15 meters, 76.2 millimeters (50 feet, 3 inches) in length, 381 millimeters (15 inches) in diameter, and has six degrees of freedom. In conjunction with handling aids, it can remove and install a 4.5-meter (15-foot diameter), 18-meter (60-foot) long, 29,484-kilogram (65,000-pound) payload. The arm weight is 410 kilograms (905 pounds) and the total system weight is 450 kilograms (994 pounds). The RMS will rotate 31.36 degrees towards the payload bay doors when opened and rotates 31.36 degrees towards the payload bay so the payload bay doors can be closed.

The RMS arm consists of joint housings, electronics housing, arm booms, and shoulder brace. There are two booms: the upper, which connects the shoulder and elbow joints, and the lower, which connects the elbow and wrist joints. The booms are made of graphite/epoxy, 330 millimeters (13 inches) in

Space Shuttle Spacecraft Systems

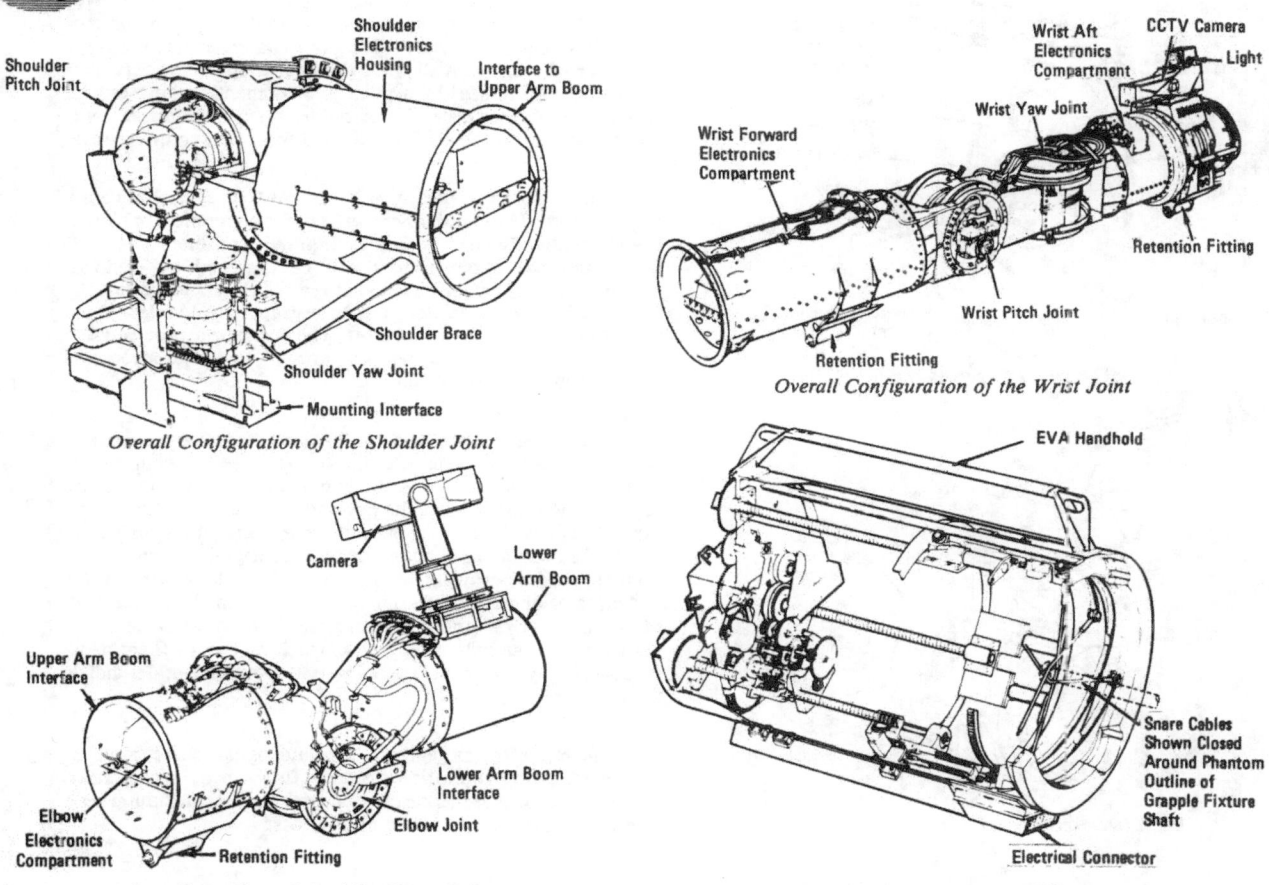

Overall Configuration of the Shoulder Joint

Overall Configuration of the Wrist Joint

Overall Configuration of the Elbow Joint

Standard Snare Type End Effector

Space Shuttle Spacecraft Systems

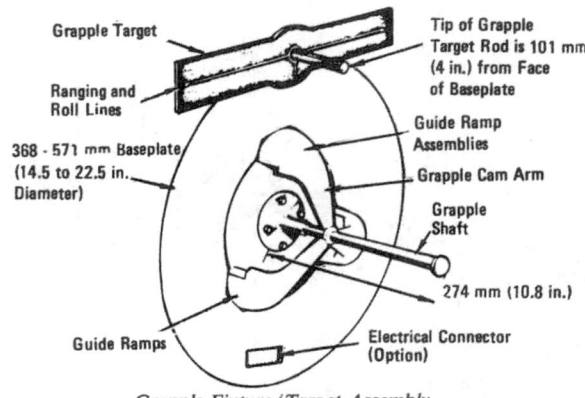

Grapple Fixture/Target Assembly

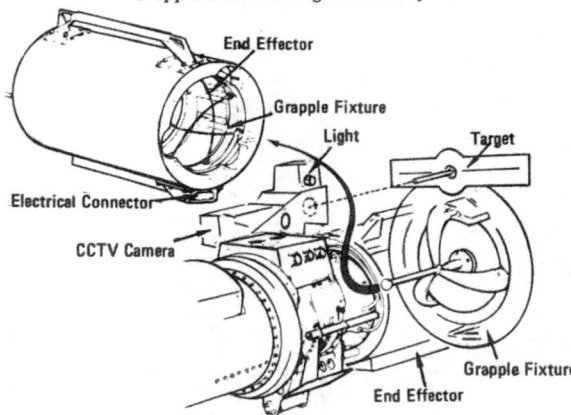

End Effector/Grapple Fixture Interface

diameter, by 5 meters (17 feet) and 6 meters (20 feet) respectively, attached by metallic joints. The composite weight in one arm is 42 kilograms (93 pounds). The joint and electronic housings are made of aluminum alloy. A shoulder brace, used only during launch, minimizes high pitch axis moment loading on the shoulder pitch gear train. The shoulder brace is unlatched by a switch located on the aft flight deck display and control panel.

The RMS can operate with standard or special-purpose end effectors. The standard end effector can grapple a payload, keep it rigidly attached as long as required, and then release it. Special-purpose end effectors will be designed by payload developers. They can be installed instead of the standard end effector during ground turnaround or be grappled and released by the standard end effector in orbit. The special-purpose end effector will receive electrical power through a connector located in the standard end effector.

The standard end effector has two functions: capture/release and rigidize/derigidize. Capture/release is accomplished by rotating an inner cage assembly containing three wire snares to open and close around the payload-mounted standard grapple fixture. A switch on the back of the RMS rotation hand control (RHC) commands capture or release. Rigidize/derigidize is accomplished by drawing the snare assembly into the rear of the end effector or moving the snares forward toward the open end of the effector. In the automatic mode, rigidization is automatic; when manually operated, a switch on the aft flight deck station display and control panel is used to rigidize or derigidize the effector.

The end effector generates six data signals corresponding to the following indications: snares fully open, snares fully closed, payload present, carriage fully extended, maximum tension level crossed, and zero tension crossed.

# Space Shuttle Spacecraft Systems

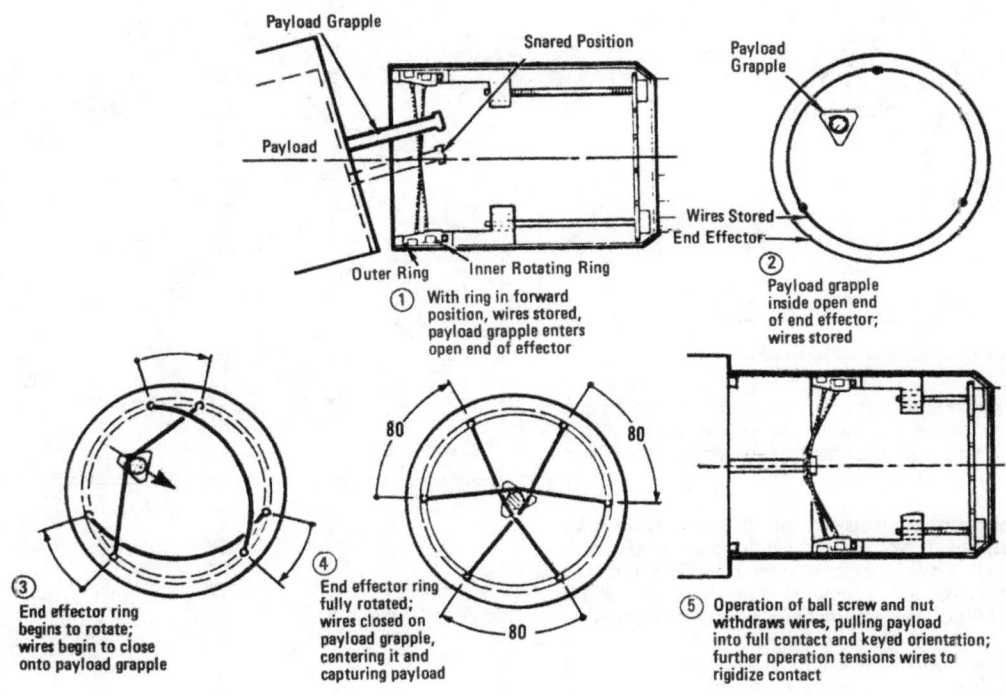

Snare Capture and Rigidization Sequence

Space Shuttle Spacecraft Systems

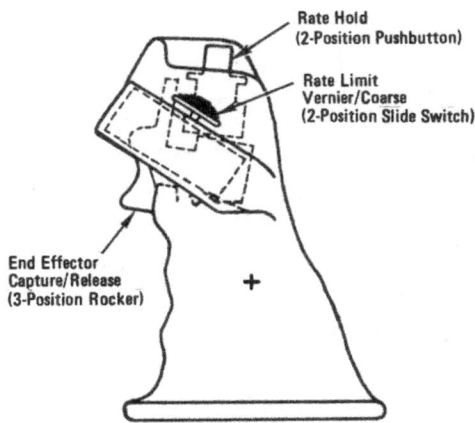

RMS Rotation Hand Control Switches

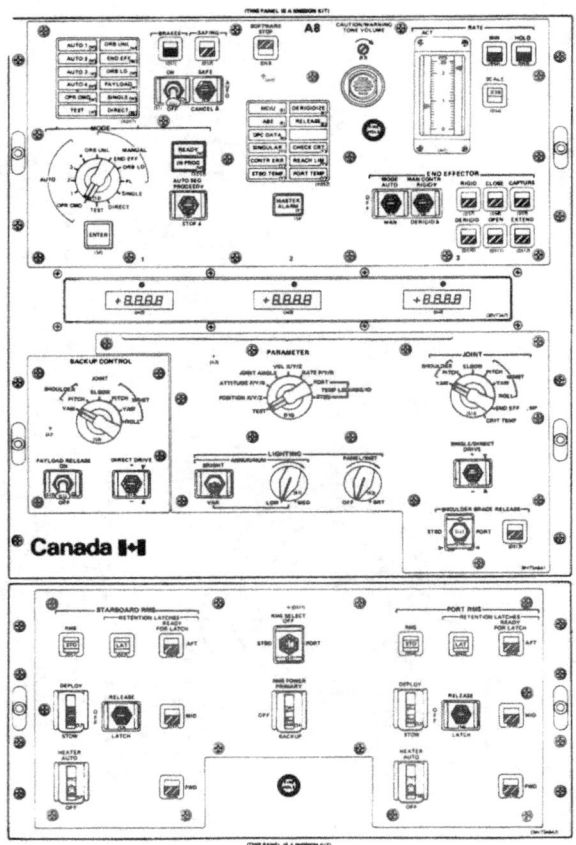

Display and Control Panel A8

The arm has provisions for a closed-circuit TV camera and a viewing light on the wrist section, as well as closed-circuit TV camera and a pan and a tilt unit at the elbow lower arm transition.

The RMS operator controls arm position and attitude by viewing it through the aft or overhead windows at the aft flight deck station, as well as by using closed-circuit TV from both the arm and payload-bay-mounted cameras. Two closed-circuit TV monitors at the aft flight deck station have split-screen capability.

The RMS has both passive and active thermal control systems. The passive system consists of multilayer insulation blankets and thermal coatings. The active system consists of 26 heaters on each arm that supply 520 watts of power at 28 vdc. The heater system uses redundant buses on each arm, so if a failure occurs on one, the other is capable of supplying full

Space Shuttle Spacecraft Systems

heater power. The heaters operate automatically to maintain the temperature within the joints above −25°C (−14°F). Heater circuits are individually switched off as the corresponding temperature reaches 0°C (32°F). Twelve temperature thermisters per arm monitor the temperatures, which can be displayed at the aft flight deck station.

Every joint of the arm is driven electromechanically. The joint drive train consists of a dc drive motor providing joint actuation, an output gear train that controls output speeds from the motor input, an optical encoder on the gearbox output shaft, and a mechanical brake on the motor output shaft.

The end effector drive train consists of a dc drive motor, a brake and clutch associated with the snare system, brake and clutch associated with the rigidization carriage and a differential unit. A spring mechanism is used for backup release.

The joint motor tachometers are the prime means of motion sensing, augmented by optical encoders. Tachometer data is supplied to control algorithms, which convert input drive commands to an output rate demand resolved for each joint of the arm. The algorithms output this rate demand within limits defined according to arm and individual joint loading conditions present at the time of computation. The algorithms supply the rate demand to control either end effector speed or position. The maximum attainable commanded velocity for the end effector and individual joints is limited by arm loading conditions, as is the maximum torque that can be applied to an individual joint under certain conditions. The aspect of arm control is provided by end effector velocity, joint rate, and motor current limiting within the software system under normal operating conditions. Joint velocity is limited during software-supported control modes by specifying a rate limit for each joint by the software system. Current limiting by the computer occurs during capture/rigidization operations. When the capture command is detected, the software commands zero current to all joint ser-

Aft Flight Deck RMS Crew Station/Crew Interface

vos, except for the wrist roll joint servo; thus, for a short period, there is a "limp" arm, except for the wrist roll joint. This is to allow for constrained motion adjustment during deployment.

Normal braking is accomplished by motor deceleration, while the joint brakes are used for emergency or driving contingency operations only. Backdriving occurs when the payload or moving arm transmits kinetic energy into the drive train.

Space Shuttle Spacecraft Systems

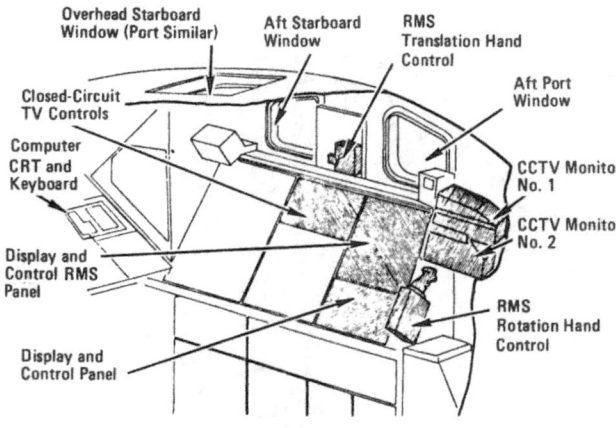

RMS Controls and Displays

The RMS can be operated in any one of five different modes: automatic, manual augmented, manual single-joint drive, direct drive, and manual backup drive.

The normal loaded arm movement rate is up to 0.06 meters per second (0.2 feet per second) and the unloaded arm movement rate is up to 0.60 meters per second (2 feet per second), no payload for the latter. Rate of movement can be controlled within 0.009 meters per second (0.03 feet per second) and 0.09 degrees per second.

The manual augmented mode is used to grapple a payload, maneuver it into or out of the payload retention fittings or handling aids, and grapple or stow a special-purpose end effector in orbit. The manual augmented mode enables the operator to direct the end-point of the arm using two 3-degree-of-freedom hand controllers to control end effector translation and rotation rate. The control algorithms process the hand controller signals into a rate demand to each joint of the arm. The operator can carry out manual augmented control of the arm using any four coordinate sytems: orbiter, end effector, payload, or orbiter loaded.

When the manual orbiter mode is selected, rate commands through the aft flight deck station RMS translation hand control (THC) result in motions at the tip of the end effector which are parallel to the orbiter-referenced coordinate frame and compatible with the up/down, left/right, in/out direction of the THC. Commands from the aft flight deck station RHC result in rotation at the tip of the end effector, which are also about the orbiter-referenced coordinate frame.

The manual end effector mode is to maintain compatibility at all times between rate commands at the THC and RHC and the instantaneous orientation of the end effector. The end effector mode is used primarily for grappling operations in conjunction with a wrist-mounted CCTV camera which is oriented with the end effector coordinates and rolls with the end effector. The CCTV scene presented on the television monitor has viewing axes which are oriented with the end effector coordinate frame. This results in compatible motion between the rate commands applied at the hand controllers and movement of the background image presented on the television monitor. Up/down, left/right, in/out motions of the THC results in the same direction of motion of the end effector as seen on the television monitor, except that the background in the scene will move in the opposite direction. Therefore, the operator must remember to use a "fly to" control strategy and apply commands to the THC and RHC that are toward the target area in the television scene.

The manual orbiter loaded mode is to enable the operator to translate and rotate a payload about the orbiter axis with the point of resolution of the resolved rate algorithm being at a predetermined point within the payload, normally the center of

Space Shuttle Spacecraft Systems

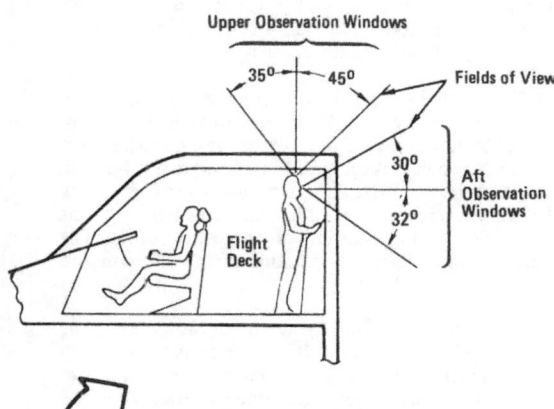

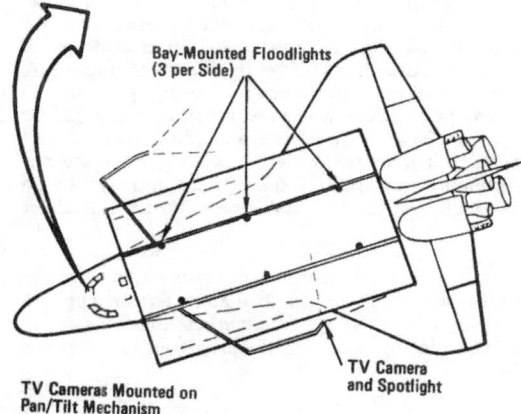

Payload Bay Television Cameras and Floodlights

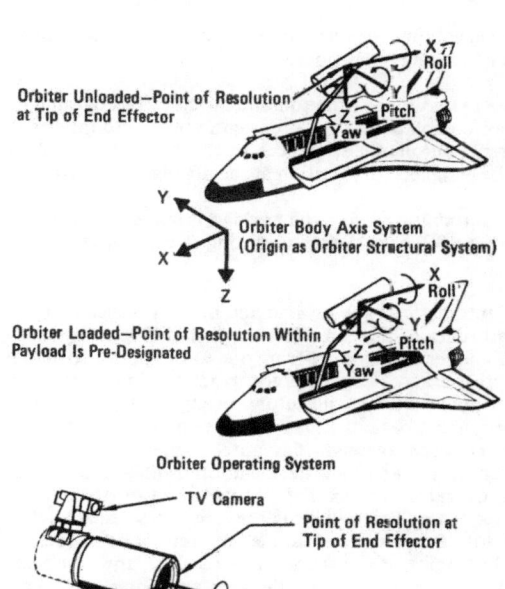

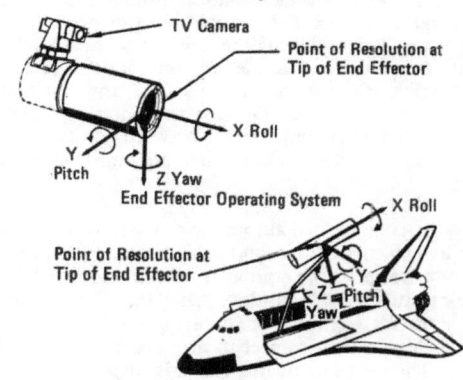

Control Coordinate Operating Systems

Space Shuttle Spacecraft Systems

geometry. This allows for pure rotations of the payload, which is useful for berthing operations.

There are two types of automatic modes, preprogrammed and operator commanded. The preprogrammed auto mode can store up to 20 automatic sequences in the computer, four of which can be assigned for selection at the aft flight deck station.

In the automatic modes, the payload is maneuvered to different locations for data taking according to a preprogrammed sequence.

Each automatic sequence is made up of a series of positions and attitudes of the end effector which define a trajectory of motion. The series may have from one to 199 points to define the trajectories. Pauses may be preprogrammed into the trajectory at any point. These will automatically cause the arm to come to rest, from which it may be able to proceed with the automatic sequence through the auto sequence "Proceed/Stop" switch on the aft flight deck station display and control panel. The operator can use the "Stop" position to halt the automatic sequence. This will bring the arm to rest, the switch is positioned to "Proceed" to resume the automatic sequence. When the last point in the sequence is reached, the computer will terminate the movement of the arm and enter a position hold mode. The speed of the end effector between points in a sequence is governed by the individual joint rate limits set in the RMS software.

The operator-commanded automatic mode moves the end effector from its present position and orientation to a new one defined by the operator to the computer via the keyboard and RMS cathode ray tube (CRT) display. After the data is keyed in, the RMS software verifies that the acquired position and orientation are "legal" with respect to arm configuration and reach envelope. The outcome of this check is displayed on the CRT. After the check, a "Ready" light will be displayed and the operator can execute the automatic sequence by placing the automatic sequence switch to "Proceed." The end effector will move in a straight line to the required position and orientation and then enter the hold mode. The operator can stop and start the sequence through the automatic sequence switch.

The single-joint drive control mode enables the operator to move the arm on a joint-by-joint basis with full computer support, thereby enabling full use of joint drive characteristics on a joint-by-joint basis. The operator supplies a fixed drive signal to the control algorithms via a toggle switch at the aft flight deck station. The algorithms supply joint rate demands to the selected joint while holding position on the other joints. The single-joint drive mode is used to stow and unstow the arm and drive it out of joint travel limits.

Direct-drive control is a contingency mode. It bypasses the manipulator control interface unit (MCIU), computer, and data buses to send a direct command to the motor drive amplifier (MDA) via hardwires. The direct-drive mode is used when the MCIU or computer has a problem that necessitates arm control by the direct drive mode to maneuver the loaded arm to a safe payload release position or to maneuver the unloaded arm to the storage position. The operator must place the brake on and select direct drive on the mode select switch. Since this is a contingency mode, full joint performance characteristics are not available. Computer-supported displays may or may not be available, depending on the fault that necessitated use of direct drive.

Backup drive control is a contingency mode used when the prime channel drive modes are not available. The backup is a degraded joint-by-joint drive system. It meets the fail-safe requirement of the RMS by using only the drive train of the prime channel.

Safing and braking are the two methods available for bringing the arm to rest. Safing can be accomplished by the operator from the aft flight deck station or by the MCIU in

Space Shuttle Spacecraft Systems

receipt of certain failure indications. Operator-initiated safing is sent on hardwires to the input latches, setting them to zero and thus resulting in zero current to each joint independent of computer commands.

The RMS has a built-in test capability to detect and display critical failures. It monitors the arm based electronics (ABE), display and controls, and the MCIU software checks in the computer monitor computations. Failures are displayed on the aft flight deck station panel and on the CRT and also are available for downlinking through orbiter telemetry.

All of the major systems of the ABE are monitored by built-in test equipment. The MCIU checks the integrity of the communications link between itself and the ABE, display and control, and the orbiter computer. It also monitors end effector functions, thermistor circuit operation, and its own internal consistency. The computer checks cover an overall check of each joint's behavior through the consistency check, encoder data validity, and end effector behavior, as well as the proximity of the arm to reach limits, soft stops, and singularities.

The caution/warning annunciators are located on the aft flight deck station display panel. There are six caution annunciators (port temperature, starboard temperature, reach limit, singularity, control error, and check CRT) and five warning annunciators (release, derigidize, ABE, GPC data, and MCIU). A "Master Alarm" light and an audio signal attract the flight crew member's attention whenever a fault condition is detected.

A jettisoning system is installed within the Rockwell-provided manipulator positioning mechanism in the event the RMS cannot be stowed. Three floodlights are installed on each side of the payload bay. A portion of the orbiter closed circuit television (CCTV) system supports the payload deployment retrieval operations. The payload deployment retrieval operator uses the payload bay TV cameras, the remote manipulator arm cameras, the TV monitors, and the TV controls and displays to assist in

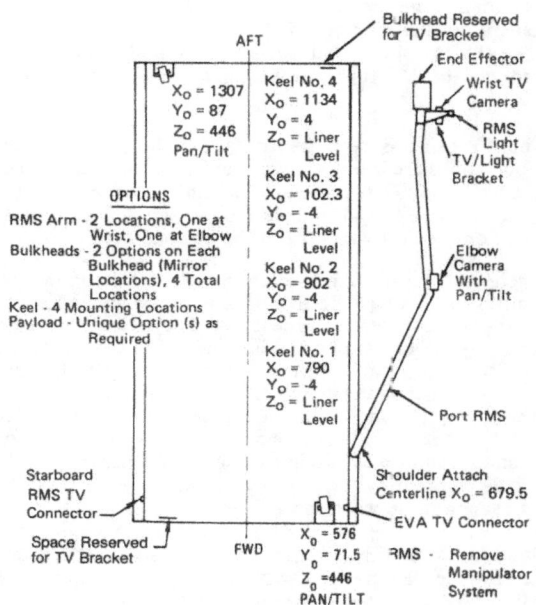

Remote Manipulator System

all phases of the payload deployment retrieval system operations. There are five TV cameras available and they can be positioned in the following locations, depending upon mission needs: arm wrist, arm elbow, forward port bulkhead, forward starboard bulkhead, aft port bulkhead, aft starboard bulkhead and keel (one of four predetermined positions).

The wrist TV camera is mounted on the roll joint of the arm; the elbow TV camera is mounted on the lower arm boom

Space Shuttle Spacecraft Systems

next to the elbow joint. The payload bay bulkhead TV camera brackets are attached to the aft and forward bulkheads. The keel camera bracket is mounted to the bottom of the payload bay. The TV monitors and the displays and controls are mounted on the aft flight deck display and control panel station.

The TV cameras used for payload deployment and retrieval operations are identical and, therefore, interchangeable. They are black and white cameras. The cameras have a pan/tilt unit, which provides plus or minus 170° in pan and tilt, except when used on the arm's wrist or in the payload keel.

There are two black and white monitors. The monitors' electronic crosshairs have both vertical and horizontal components at the electrical center of the image. They are used to align the cameras with targets and sighting aids. The crosshairs are also used to align overlays with the monitor image. Alphanumerics are available on the monitors. The pan and tilt angles are displayed in degrees and tenths of degrees when the monitors display full scene images. The alphanumerics can be turned off. Each monitor can display two images simultaneously. The right or left half of the monitor will display the center half of the selected camera scene when the split screen mode is used.

Spar Aerospace Limited, Toronto, Canada, is the prime contractor to the National Research Council for development of the RMS for NASA. CAE Electronics Ltd, Montreal is responsible for the displays and controls in the orbiter. RCA Ltd, Montreal is responsible for the electronic interfaces, provides servo amplifiers and power conditioners. Dilworth, Secord, Meagher and Assoc. Ltd (DSMA), Toronto is responsible for the end effector.

PAYLOAD RETENTION MECHANISMS

Nondeployable payloads are retained by passive retention devices, whereas, deployable payloads are secured by motor-driven, active retention devices.

Payloads are secured in the orbiter payload bay by means of the payload retention system or are equipped with their own unique retention systems.

The orbiter payload retention system provides three-axis support for up to five payloads per flight. After the initial orbiter development flights, the payload bay will be modified to accommodate attach fittings for five payloads.

The payload retention mechanisms secure the payloads during all mission phases and provides for installation and removal of the payloads when the orbiter is either horizontal or vertical.

Attachment points in the payload bay are in 99-millimeter (3.933-inch) increments along the left- and right-side longerons and along the bottom centerline of the bay. Of the potential 172 attach points on the longerons, 48 are unavailable because of the proximity of spacecraft hardware. The remaining 124 may be used for carrier/payload attachment; of these, 116 may be used for deployable payloads. Along the centerline keel, 89 attach points are available, 75 of which may be used for deployable payloads. There are 13 longeron bridges per side and 12 keel bridges available per flight. Only the bridges required for a particular flight are flown. The bridges are not interchangeable because of main frame spacing, varying load capability, and subframe attachments.

The longeron bridge fittings are attached to the payload bay frame at the longeron level and at the side of the bay. Keel bridge fittings are attached to the payload bay frame at the bottom of the payload bay.

The payload trunnions are the interfacing portion of the payload with the orbiter retention system. The trunnions that interface with the longeron are 82 millimeters (3.25 inches) in diameter and 177.8 or 222.2 millimeters (7 or 8.75 inches) long, depending upon where they are positioned along the payload

Space Shuttle Spacecraft Systems

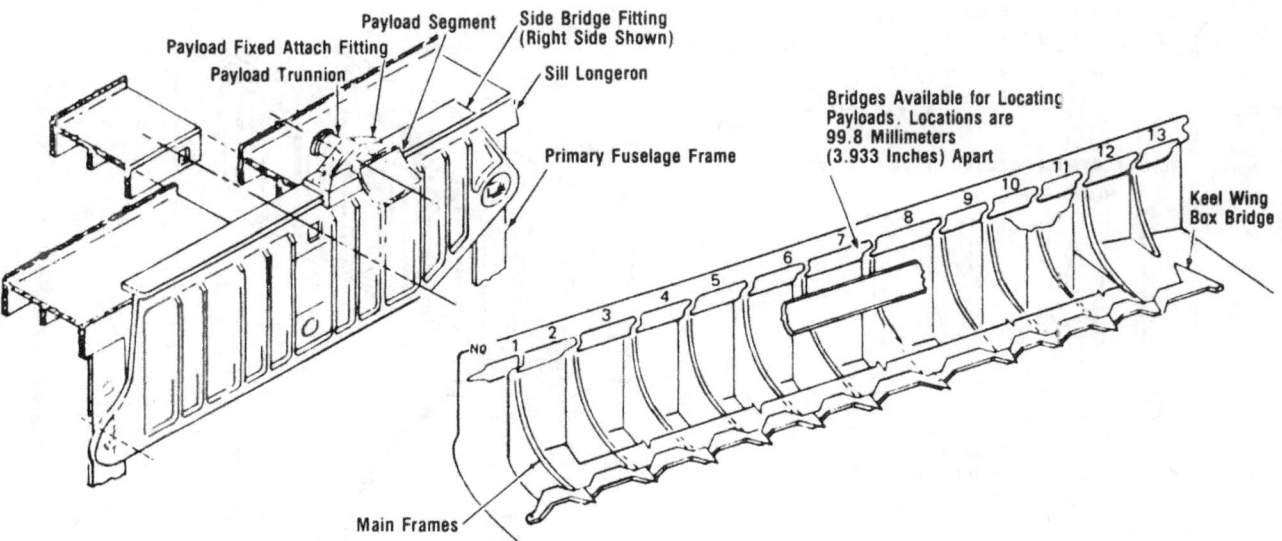

Payload Retention

bay. The keel trunnions are 76.2 millimeters (3 inches) in diameter and vary in length from 101.6 to 292.1 millimeters (4 to 11.5 inches), depending upon where they fit in the payload bay.

The orbiter/payload attachments are the trunnion/bearing/journal type. The longeron and keel attach fitting have a split, self-aligning bearing for nonrelease-type payloads in which the hinged half is bolted closed. For on-orbit deployment and retrieval payloads, the hinged half fitting releases or secures the payload by latches that are driven by dual redundant electric motors.

Payload guides and scuff plates are used to assist in deploying and berthing payloads in the payload bay. The payload is constrained in the X direction by guides and in the Y direction by scuff plates and guides. The guides are mounted to the inboard side of the payload latches and interface with the payload trunnions and scuff plates. The scuff plates are attached to the payload trunnions and interface with the payload guides.

The guides are V shaped with one part of the V being 50.8 millimeters (2 inches) taller than the other part. Parts are available to make either the forward or aft guide, the tallest.

Space Shuttle Spacecraft Systems

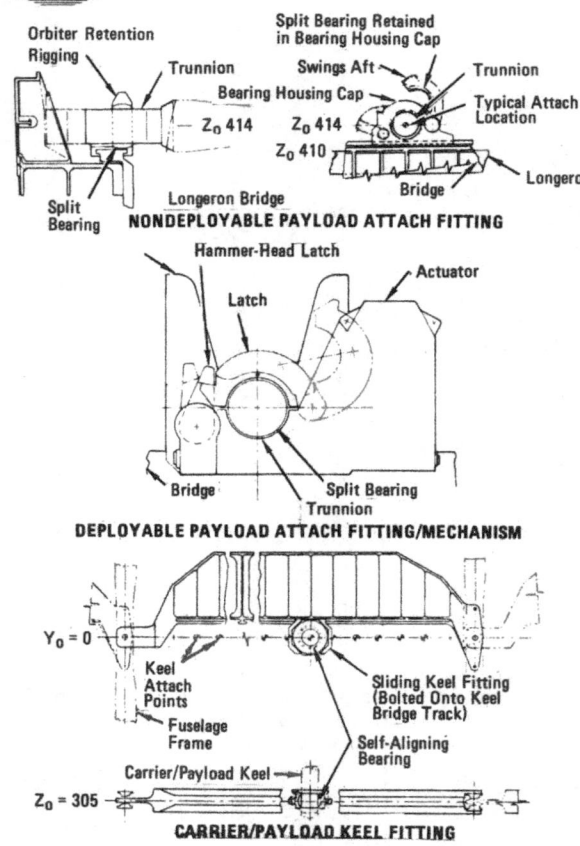

Standard Attach Fittings for Payloads

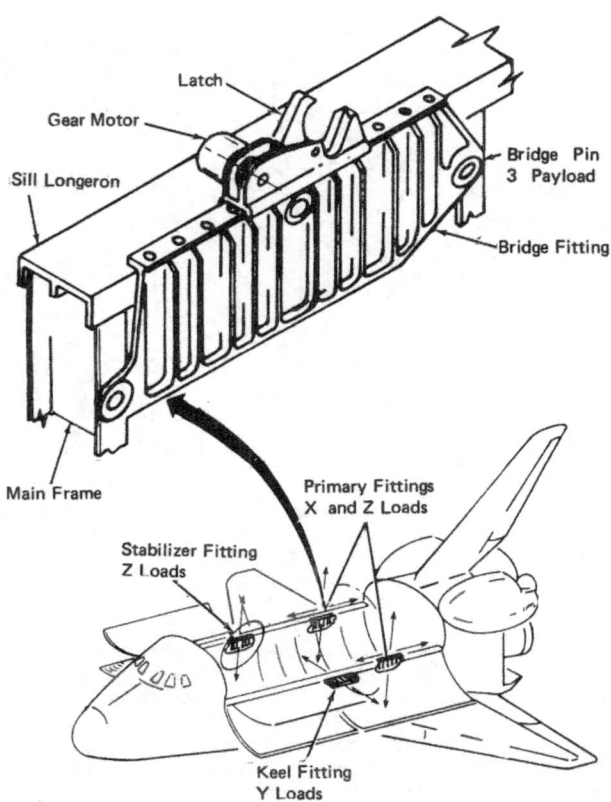

Active Payload Retention System

Space Shuttle Spacecraft Systems

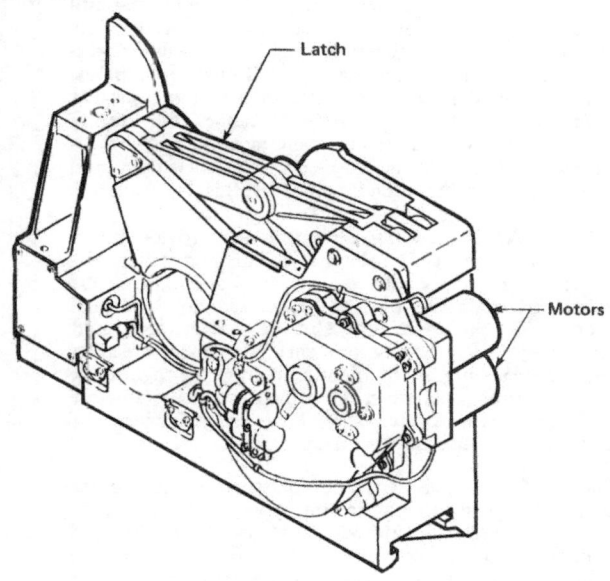

Payload Retention Latch

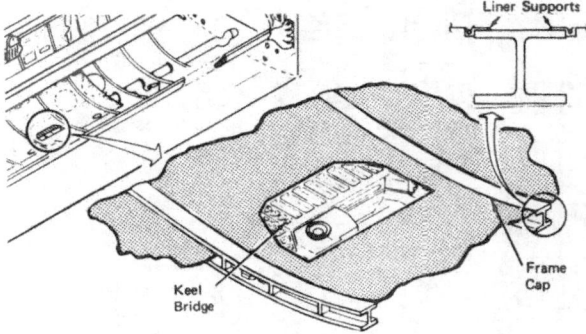

Active Keel Fitting

This difference enables the operator monitoring the berthing or deployment operations through the aft bulkhead TV cameras to better determine when the payload trunnion has entered the guide. The top of the tallest portion of the guide is 609.6 millimeters (24 inches) above the centerline of the payload trunnion when it is all the way down in the guide. The top of the guide has a 228.6-millimeter (9-inch) opening. These guides are mounted to the 203.2-millimeter (8-inch) guides that are a part of the longeron payload retention latches.

The payload scuff plates are mounted to the payload trunnions or the payload structure. There are normally three or four longeron latches and a keel latch for on-orbit deployment and retrieval of payloads. These latches are controlled by dual redundant electric motors with either or both motors releasing or latching the mechanism. The operating time of the latch is 30 seconds with both motors operating or 60 seconds with one motor operating. The latch/release switches on the aft flight deck display and control panel station control the latches. Each longeron latch has two microswitches sensing the ready-to-latch condition. Only one is required to control the ready-to-latch talkback indicator on the aft flight deck display and control panel station. Each longeron latch also has two microswitches to indicate latch and two to indicate release. Only one of each is required to control the latch or release talkback indicator on the aft flight deck display and control panel station. The keel latch also has two microswitches that sense when the keel latch is closed with the trunnion in it. Only one of the switches is required to operate the talkback indicator on the aft flight deck

Space Shuttle Spacecraft Systems

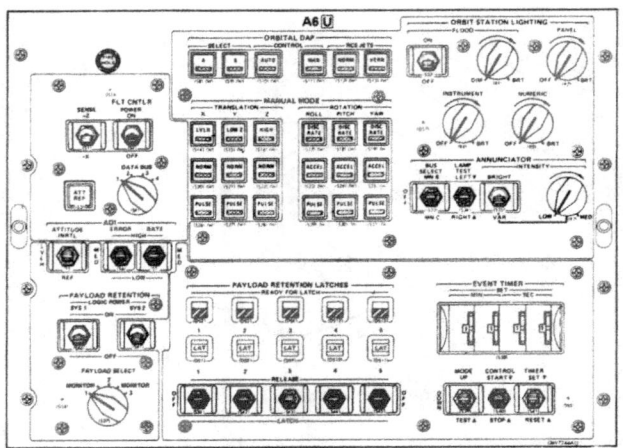

display and control panel station. The keel latch also has two microswitches that verify if the latch is closed or open, with only one required to control the talkback indicator on the aft flight station display and control panel station.

It is noted that the keel latch centers the payload in the yaw direction in the payload bay; therefore the keel latch must be closed before the longeron latch is closed. The keel latch can float plus or minus 69 millimeters (plus or minus 2.75 inches) in the X direction.

NAVIGATION AIDS

The navigation system used during entry consists of the inertial measurement units (IMU's) and navigation aids: TACAN (tactical air navigation), MSBLS (microwave scan beam landing system), air data system and radar altimeter. The three IMU's maintain an inertial reference and provide velocity changes until MSBLS is acquired. Navigation-derived air data is needed during entry as inputs to the guidance, flight control and flight-crew-dedicated displays. Such data is collected from below Mach 3 through landing. The navigation-derived data is used as a backup to TACAN, which supplies range and bearing measurements and is available beginning at an altitude of approximately 44,196 meters (145,000 feet). TACAN is used until MSBLS acquisition or until landing if MSBLS is not available.

AIR DATA SYSTEM. The air data system provides information on movement of the orbiter in the air mass (flight environment) during entry. There are two air data probe assemblies, each consisting of a probe, an actuator, and dual drive motors. The probes are stowed during ascent, orbit, and the initial entry heat load environment. The probes are then deployed independently by the flight crew from the flight deck display and control panel when the orbiter's velocity is below Mach 3. The probes are located on the lower left and lower right side of the forward fuselage nose area.

The air data system senses air pressures related to the spacecraft movement through the atmosphere for updating the navigation state vector in altitude, guidance in steering and speed brake command calculations, flight control for control law computations and for display on the commander's and pilot's AMI's (alpha Mach indicators) and the commander's and pilot's AVVI's (altitude vertical velocity indicators). The AMI's display essential flight parameters relative to the spacecraft travel in the air mass such as angle of attack (alpha), acceleration, Mach/velocity, and knots equivalent airspeed. The AVVI's display such essential flight parameters as radar altitude, barometric altitude, altitude rate, and altitude acceleration. Prior to deployment of the air data system probe, the AVVI's would receive their inputs from the navigation attitude processor. The AMI's would receive their inputs from the navigation attitude processor and IMU's prior to air data

Space Shuttle Spacecraft Systems

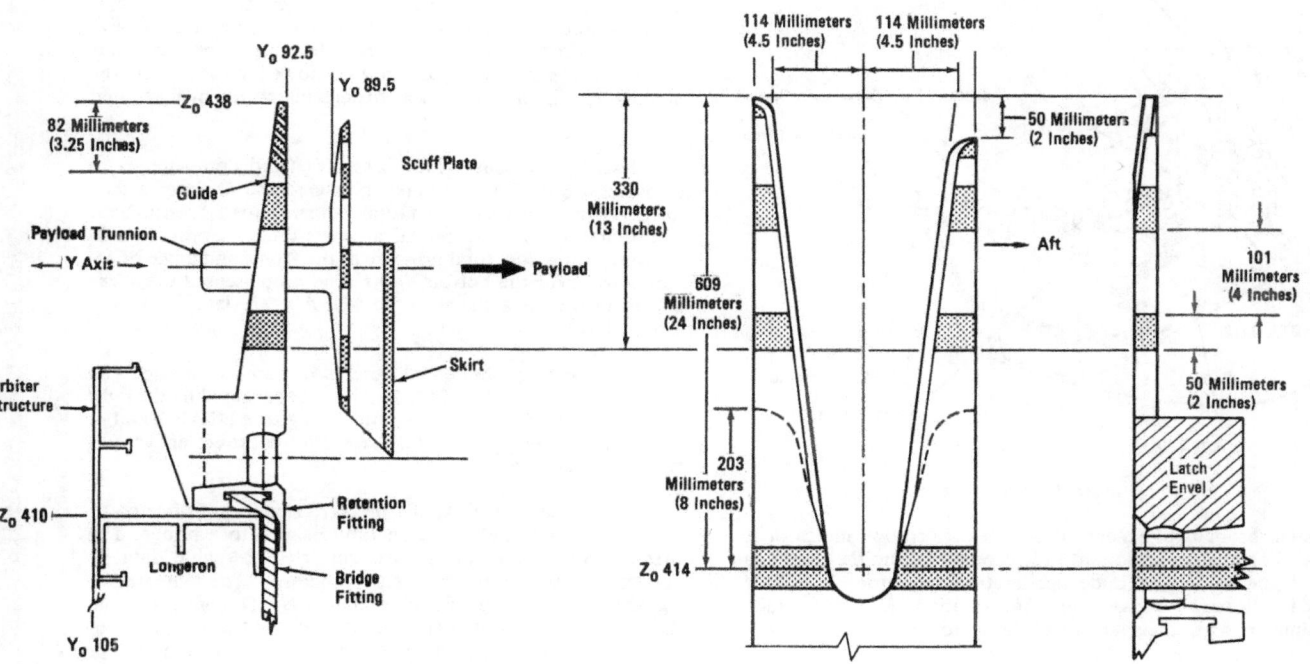

Orbiter Payload Guide and Trunnion/Scuff Plate (Nominal)   *Orbiter Active Latch Guide*

 **Space Shuttle Spacecraft Systems**

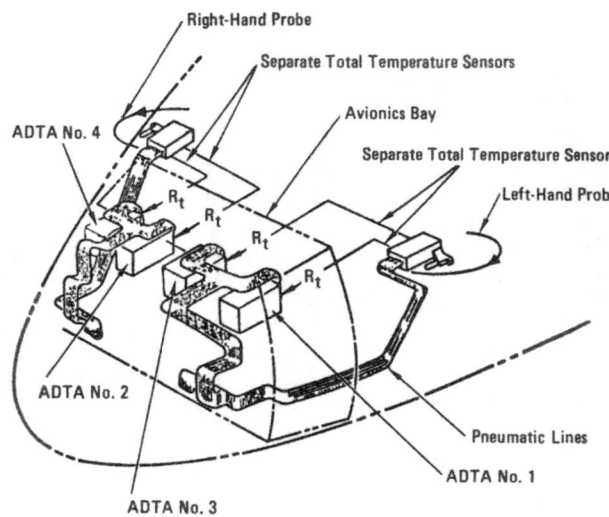

Air Data System

system probe deployment. The AMI acceleration indicator remains on the navigation attitude processor from the IMU's, as does the AVVI's altitude acceleration indicator. The AVVI radar altitude receive their information from the radar altimeters when the spacecraft is down to 5,000 feet in altitude.

Each probe is independently deployed by an actuator consisting of two motors connected by a differential and appropriate gearing and limit switches. The flight crew controls the deployment from switches on the flight deck display and control panel. The motors, through mechanical gear reductions drives the probe to the deployed position. When the probe is fully deployed, limit switches in the probe assembly remove power from the electrical motors. If deicing of the probe is required, heaters can be turned on. Deployment time is 15 seconds for two-motor operation and 30 seconds for single-motor operation. The probe mechanism (except for the probe itself) has thermal protection covering in the stowed position. In the deployed position, mechanical fittings match the orbiter mold line.

Each probe senses four pressures (static pressure, total pressure, angle of attack upper pressure, and angle of attack lower pressure) as well as two total temperature (T_t) resistances. The four pressures are sensed at ports on each probe: static pressure at the side, total pressure at the front, and angle of attack lower near the bottom front. The probe-sensed pressures and temperatures are sent to the ADTA's (air data transducer assemblies).

The left probe-sensed pressures are connected by pneumatic tubing to ADTA's 1 and 3. Those sensed by the right probe are connected by pneumatic tubing to ADTA's 2 and 4. Temperatures and sensed pressure from the probes are sent to the same ADTA's.

Within each ADTA, the pressure signals are directed to four transducers and the temperature signal to a bridge. The pressure transducer analogs are converted to digital data by counters controlled by the digital processor. The temperature signal is converted by an analog/digital (A/D) converter. The digital processor corrects errors and linearizes the pressure data and converts the temperature bridge data to temperatures in degrees centigrade. This data is sent to the digital output device, which converts the signals into serial digital format, and then to the onboard computers for updating the navigation state vector. The data also is sent to the commander and pilot AVVI's (altitude/vertical velocity indicators), AMI's (alpha/Mach indicator), and CRT (cathode ray tube).

Space Shuttle Spacecraft Systems

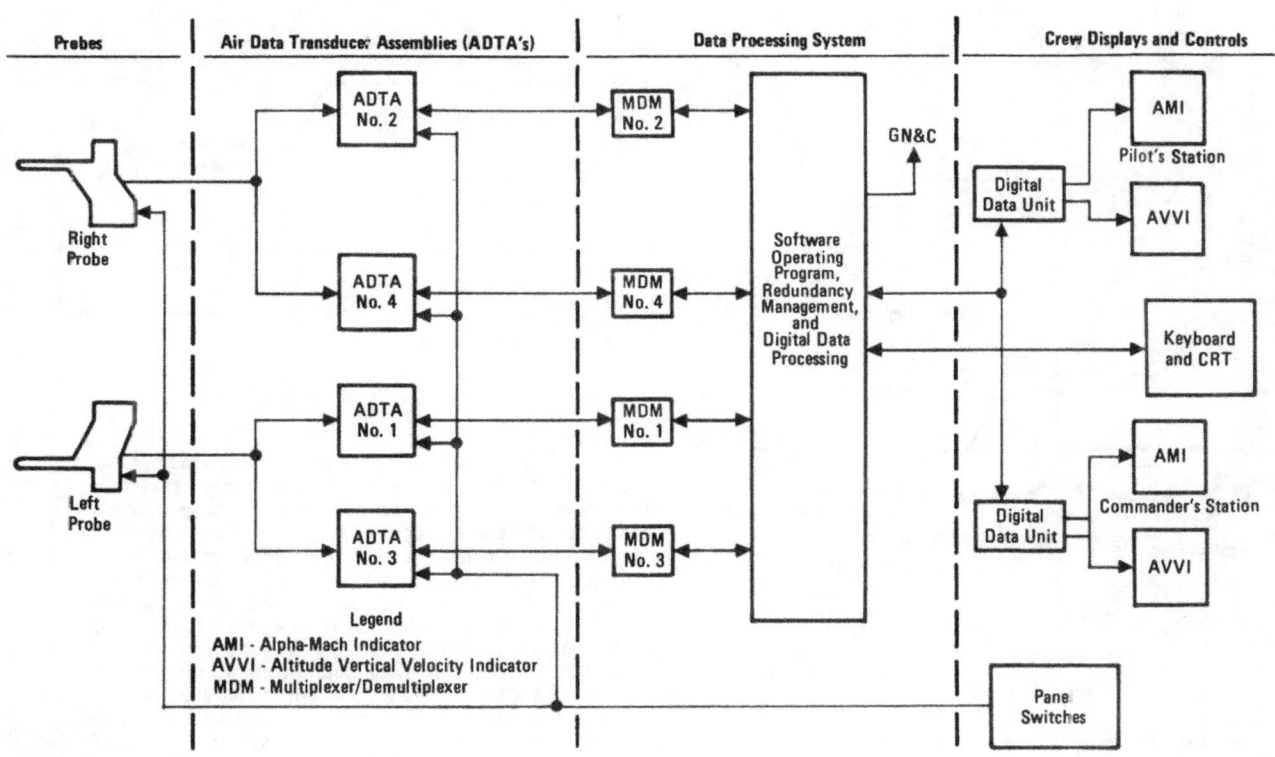

Air Data Interfaces

Space Shuttle Spacecraft Systems

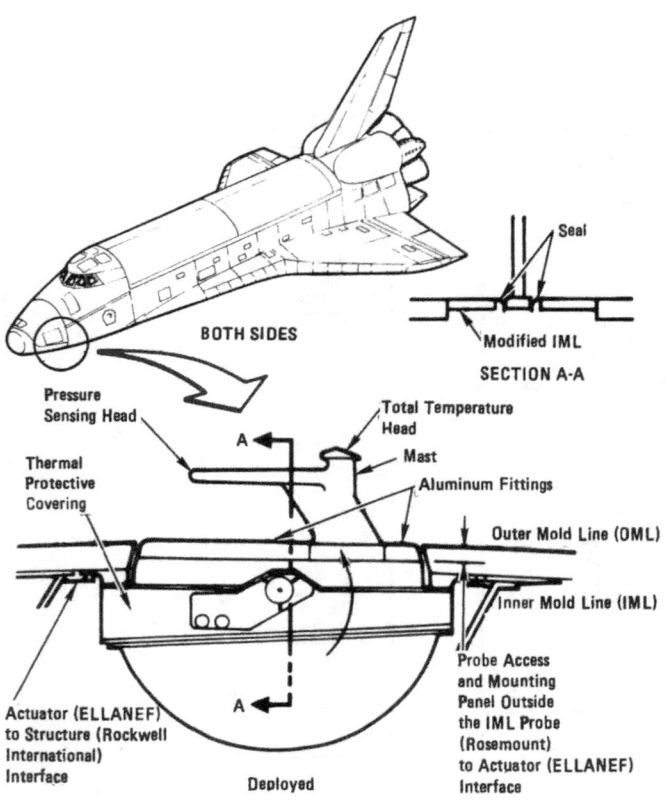

Air Data Sensor Subsystem General Arrangement

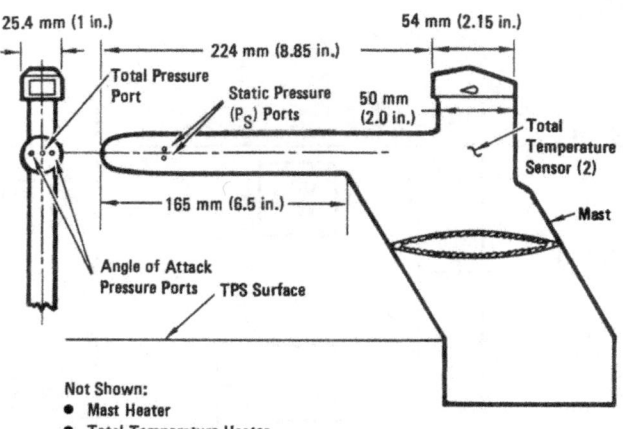

Air Data Probe Details

The four computers compare the pressure readings from the four ADTA's for error. If all the pressure readings compare within a specified value, one set of pressure readings from each probe is summed and averaged and sent to the software. If one or more pressure signals of a set of probe pressure readings fail, the failed set data flow from that ADTA to the averager is stopped and the software will receive data from the other ADTA of that probe. If both sets from a probe fail, the software operates on data from the two ADTA's connected to the other probe. The best total temperature from all four ADTA's is sent to the software. A fault detection would illuminate the "Air Data" red caution/warning light, backup C/W alarm light, and master alarm light and sound the audible tone and a

Space Shuttle Spacecraft Systems

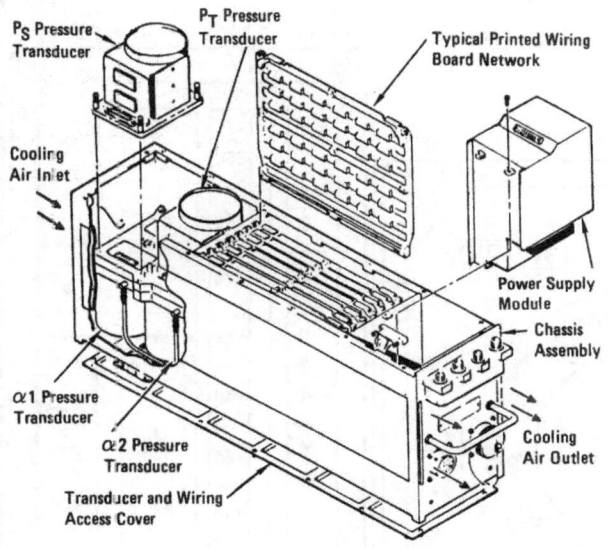

Air Data Transducer Assembly

fault message on the CRT. A communication fault will illuminate the "SM Alert". Prior to the air data probe deployment, the commander and pilot AVVI's and AMI's receive information from the navigation attitude processor.

The commander and pilot AVVI's and AMI's receive inertial information from the navigation attitude processor when their "Air Data" switches are in the "Navigation" position. When the air data probes are deployed, the commander and pilot can position their "Air Data" switches to the left or right air data probe for display.

The AVVI's display altitude acceleration ("Alt Accel"), altitude rate ("Rate"), navigation/air data system (NAV/ADS), and radar altitude ("Alt") information and the AMI's display angle of attack ("Alpha"), acceleration ("Accel"), Mach/velocity ("M/Vel"), and equivalent airspeed ("EAS").

All but the alpha indicators (a moving drum) and the altitude acceleration indicators (a moving pointer displayed against a fixed line) are moving tapes behind fixed lines. The angle of attack indicator reads from -18 to +60 degrees, the acceleration indicator from -50 to +100 fps^2 (feet per second-squared), the Mach/velocity indicator from Mach 0 to 4 and 4,000 to 27,000 fps; equivalent airspeed from 0 to 500 knots, altitude acceleration from -13.3 to +13.3 fps^2, altitude rate from -740 to +740 fps, altitude from -1,000 to 450,000 feet, then changes scale to +40 to +165 nautical miles (barometric altitude), and radar altitude from 0 to 5,000 feet.

Failure warning flags are provided for all four scales on the AVVI's and AMI's. The flags appear in the event of a malfunction in the indicator or in received data. In the event of power failure, all four flags appear.

The four ADTA's are located in the orbiter crew compartment mid-deck forward avionics bays and are convection cooled. Each is 123.6 millimeters (4.87 inches) high, 539.7 millimeters (21.25 inches) long, 110.9 millimeters (4.37 inches) wide, and weighs 8 kilograms (19.2 pounds).

TACAN (TACTICAL AIR NAVIGATION). TACAN provides orbiter position with respect to a ground-based TACAN station. The orbiter is equipped with three TACAN sets which operate in a redundant set mode. Each TACAN set has two antennas, one on the orbiter lower forward fuselage and one on the orbiter upper forward fuselage. The TACAN's are the airborne portion of the global navigation system for military

 **Space Shuttle Spacecraft Systems**

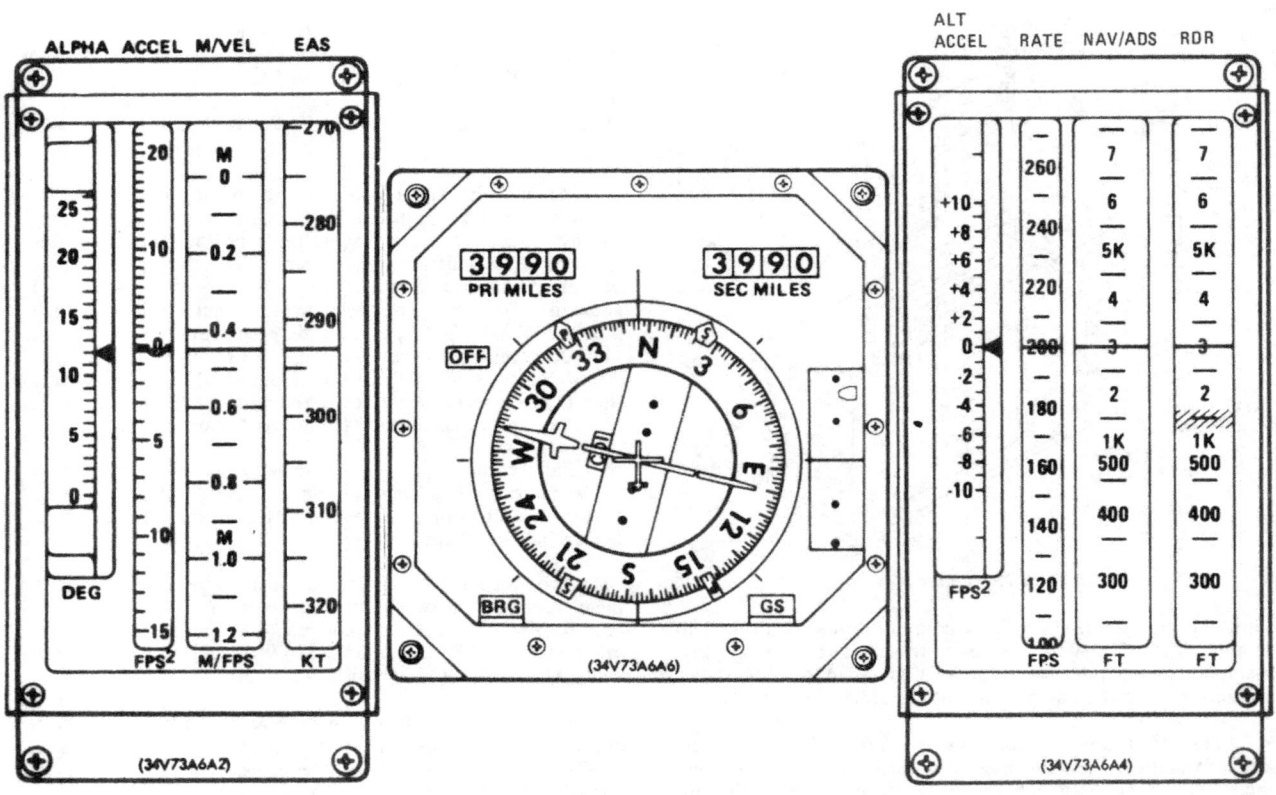

AMI (left), HSI (center), and AVVI (right)
Panel F6 and F8

Space Shuttle Spacecraft Systems

and civil aircraft operating at L-band (1-gigahertz) frequencies. The TACAN sets are used as an external navigation aid in the orbiter during the entry phase and RTLS (return-to-launch-site) abort.

The ground-based portion of the TACAN is a part of the global navigation network. Normally several ground stations will be used during entry after leaving L-band communications blackout and during the terminal area energy management (TAEM) phases. Each ground station has an assigned frequency (L band) and a three-letter Morse code identification to the orbiter audio system for the commander and pilot. The ground station transmits on one of 252 (126X and 126Y) pre-selected frequencies (channels) which correspond to the frequencies the onboard TACAN sets are capable of receiving. These frequencies are spaced at 63-megahertz intervals.

The ground beacon of the selected TACAN station constantly transmits a signal from which the onboard TACAN receiver units are capable of receiving through its antennas. Every 37.5 seconds (± 2.5), the selected TACAN ground station transmits a coded three-letter Morse code identification as part of the transmission. When the onboard TACAN discerns this code, it separates it from the range and bearing data and produces a signal which is used by the orbiter audio system to produce an audible signal allowing the commander and pilot to confirm TACAN lock-on and station selection.

Each onboard TACAN is controlled by a rotary switch on the flight deck display and control panel. Switch settings are "Off", "RCV" (receive), "T/R" (transmit/receive), and "GPC" (computer mode).

The onboard TACAN antennas are controlled by a "TACAN Antenna" switch. In "Auto," the computers control antenna selection automatically. Upper and lower antennas can be selected manually by the flight crew on the flight deck display and control panel.

In the GPC mode, the onboard computers control TACAN channel selection automatically. Ten TACAN ground stations are programmed into the software, divided into three geometric regions: The acquisition region (three stations), the navigation region (six stations), and the landing site region (one station).

During orbital operations, landing sites are grouped into "mini-table" and "maxi-table" programs. The maxi-table provides data sets that will support a broad range of trajectories for contingency deorbits and enables reselection of runway and navigation and data sets for those deorbits. The mini-table consists of three runways determined by the flight crew with one initialized as a primary runway. The mini-table is transferred from entry operations and becomes unchangeable. Entry guidance is targeted from one of the three runways as selected by the crew, initialized with the primary runway for the well-defined trajectory and nominal end of mission data sets. Since the TACAN units are placed in groups of ten and ten TACAN units from one group (primary) form the TACAN half of the mini-table, the secondary and alternate runways should be from the same group as the primary runway to assure TACAN coverage. The runways with MSBLS are acquired and operation is automatic, with the flight crew provided with necessary controls and displays to evaluate MSBLS performance and take over manually if required. The runways with MSBLS must be in the primary or secondary slot in the mini-table for the mini-table to copy the MSBLS data. The maxi-table is an I-loaded table of 18 runways data sets and MSBLS data for runways and 50 TACAN data sets. In orbital operations, the landing site function provides the capability to transfer data from the maxi-table to the mini-table.

The acquisition region is the area in which the onboard TACAN sets automatically start searching for a range lock-on of one of three ground stations, at approximately 44,196 meters (145,000 feet). After one TACAN acquires a range lock, the other two will lock on to the same ground station. When at least

Space Shuttle Spacecraft Systems

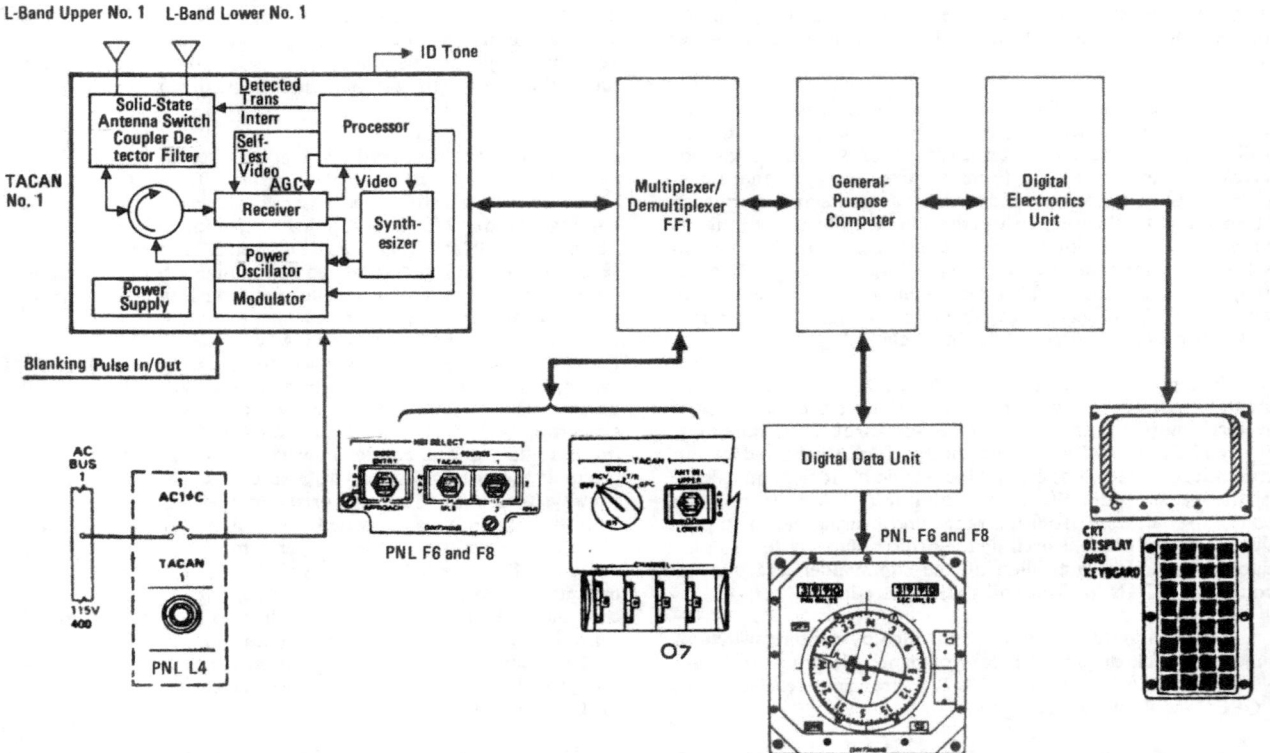

TACAN Receiver-Transmitter

Space Shuttle Spacecraft Systems

two TACAN sets lock on, TACAN range and bearing are used by navigation to update state vector until MSBLS selection and acquisition.

When the distance to the landing site is approximately 120 nautical miles (138 statute miles), the TACAN begins the navigation region of interrogating the six navigation stations. As the spacecraft progresses, the distance to the remaining stations is computed and the next nearest station is selected automatically when the spacecraft is closer to it than to the previous locked-on station. Only one station is interrogated when the distance to the landing site is less than approximately 20 nautical miles (23 statute miles). Again, the TACAN sets will automatically switch from the last locked-on navigation region station to begin searching for the landing site station. TACAN azimuth and range are provided on the CRT horizontal situation display. TACAN range and bearing cannot be used to produce a good estimate of the altitude position component, so navigation uses barometric altitude derived from the air data system probes, which are deployed by the flight crew at approximately Mach 3.

The onboard TACAN sets detect the phase angle between magnetic north and the position of the spacecraft with respect to the ground TACAN beacon to determine bearing. Periodically the onboard TACAN sets will emit an interrogation pulse which causes the selected TACAN ground station to respond with distance measuring equipment pulses. The slant range (orbiter to ground station) is computed by the onboard TACAN sets by measuring the elapsed time from interrogation to valid reply and subtracts known system delays. When approaching a ground TACAN station, the range will decrease. After the course has been selected, the onboard TACAN sets derive course deviation data.

The range and bearing data are used in the entry phase by navigation to update the state vector position components, area navigation for display on the flight deck display and control panel HSI's (horizontal situation indicators) after the data is transformed during TACAN operation, and for display of raw TACAN data on the CRT.

TACAN redundancy management consists of processing and mid-value selecting range and bearing data. The three TACAN sets are compared to determine if a significant difference is detected. When all three TACAN sets are good, redundancy management selects middle values of range and bearing. If one of the two parameters is out of tolerance, the remaining two will average that parameter. If a fault is verified, the "SM Alert" light is illuminated and a CRT fault message occurs for the applicable TACAN set.

Each TACAN set can be controlled manually by the flight crew on the flight deck display and control panel by its rotary switch and four thumbwheels. The thumbwheels are used to select the TACAN station for reception of bearing data when the rotary switch is on "Rcv" (receive). The "T/R" position allows both bearing and range data, with the thumbwheels used to select the station.

Since the TACAN does not provide altitude data, this is furnished by the onboard barometric altimeter to the onboard navigation.

The three TACAN sets are located in the orbiter crew compartment mid-deck avionics bays and are convection cooled. Each set is 193.5 millimeters (7.62 inches) in height, 193.5 millimeters (7.62 inches) in width, 318 millimeters (12.53 inches) in length, and weighs 13 kilograms (30 pounds).

HORIZONTAL SITUATION INDICATOR. Horizontal situation indicators (HSI) are located on the flight deck display and control panel at the commander and pilot stations. Each HSI displays a pictorial view of the spacecraft's location with

 **Space Shuttle Spacecraft Systems**

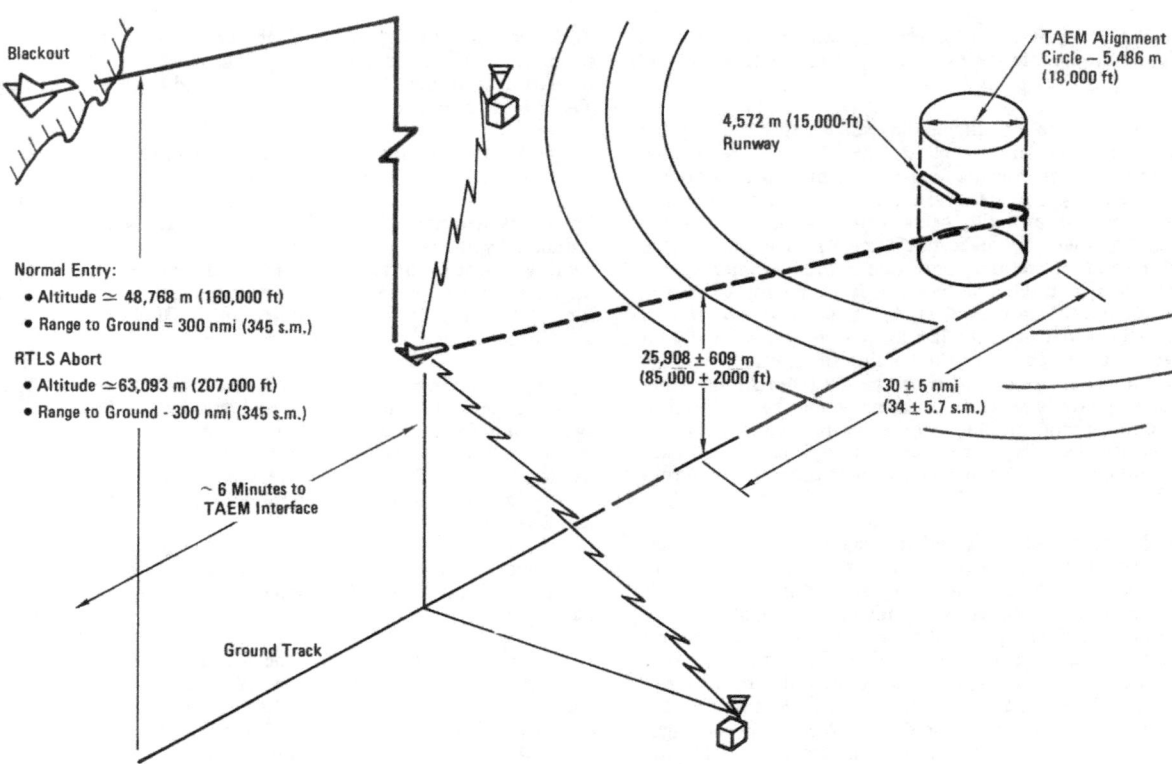

TACAN Approach Geometry

Space Shuttle Spacecraft Systems

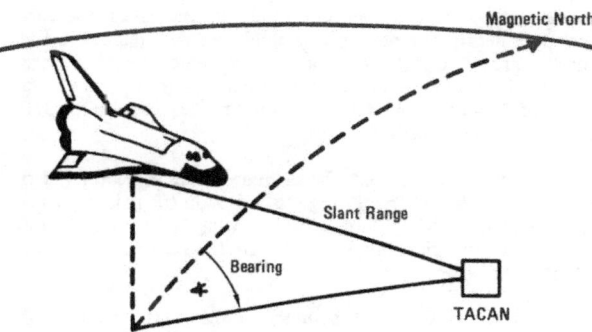

Tactical Air Navigation

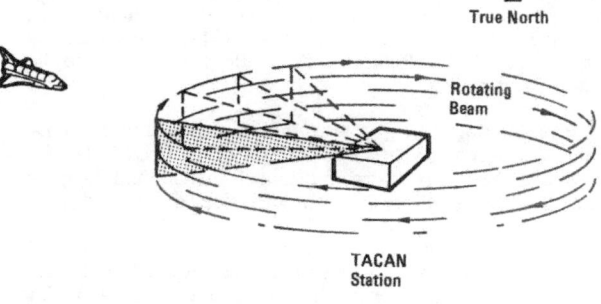

- TACAN station transmits RF burst when beam points to true north
- Aircraft receiver times interval from burst to beam impingement and derives bearing

TACAN Operation

respect to various navigation points during entry through rollout and during RTLS (return-to-launch-site) abort.

Three switches are associated with each HSI. One switch selects the mode ("Entry," "TAEM," or "Approach") and the other two select the source of data ("Nav," "TACAN," or "MLS" from source 1, 2, or 3). When in the "Nav" position, that HSI is supplied with data derived from navigation attitude processor. When in "TACAN," the HSI is supplied with data derived from TACAN 1, 2, or 3. When in "MLS," the HSI is supplied with data derived from MSBLS 1, 2, or 3.

Each HSI displays magnetic heading, runway magnetic course, course deviation, glide slope deviation, primary and secondary bearing, primary and secondary distance, and flags to indicate validity.

The magnetic heading (the angle between magnetic north and vehicle direction measured clockwise from magnetic north) is displayed by the compass card and can be read under the lubber line located at the top of the indicator dial. (A lubber line is a fixed line on a compass aligned to the longitudinal axis of the craft.) The compass card is positioned at zero degrees (N-north) when the heading input is zero. When the heading point is increased, the compass card rotates counterclockwise.

The course pointer points to the landing runway magnetic heading and rotates around the inside edge of the compass card. The scale reading on the compass card at the tip of the course pointer is the selected course. When the course input is set at zero (N), the course pointer is positioned at the lubber line regardless of the position of the compass card. The course pointer is driven relative to the HSI case and not relative to the HSI compass card. The software substracts the vehicle's magnetic heading from the course and this difference is represented by digital input to the HSI, when the digital input is increased, the course pointer rotates clockwise. There is no

Space Shuttle Spacecraft Systems

reversal of the pointer movement when it passes through 360 degrees.

The course deviation bar is zeroed when in "Entry" and indicates the displacement of the vehicle to the right or left of the extended runway centerline plus or minus 10 degrees when in TAEM and plus or minus 2.5 degrees when in Approach and Landing. When the source deviation input is zero, the deviation bar is aligned with the ends of the course pointer. An orbiter deviation off course to the left causes the deviation bar to deflect to the right (command to fly right).

The glide slope indicator shows deviation above or below the average TAEM glide path to Waypoint 1 (WP 1) for the nominal entry point (NEP) at the primary runway. In Entry, the glide slope pointer is stowed, and the flag is displayed. The indication is plus or minus 5,000 feet (1,524 meters) when in TAEM, and plus or minus 1,000 feet (304 meters) when in Approach and Landing. A deviation of the orbiter above the glide slope causes the pointer to deflect downward (to command fly down).

When in Entry, the primary bearing pointer (P) indicates the spherical bearing to WP 1 for the NEP at the primary landing runway and the secondary bearing pointer (S) indicates the spherical bearing to WP 1 for NEP at the secondary landing runway. When in TAEM, the primary bearing pointer indicates bearing to WP 1 on the selected heading alignment cylinder for NEP at the primary runway and the secondary bearing pointer indicates the bearing to the center of the selected heading alignment cylinder primary landing runway. When in Approach and Landing, both primary and secondary bearing pointers show the bearing to WP 2 at the primary runway. The pointers show bearings relative to the compass card when the card is positioned in accordance with the heading input data. When the bearing inputs are set to zero, the corresponding bearing pointer is positioned at the lubber line regardless of the position of the compass card. When the bearing inputs are increased, the corresponding bearing pointers rotate clockwise with respect to the compass card. The software subtracts the vehicle heading from these two bearings to assure appropriate indications. There is no reversal of the pointer movement when it passes through the 360 degrees in either direction.

When Entry is selected, the primary range indicator ("Pri Miles") shows the spherical range to WP 2 on the primary runway via WP 1 for NEP. The secondary range ("sec miles") indicator shows spherical range to WP 2 on the secondary runway via WP 1 for NEP.

When in TAEM, the primary range is the horizontal distance to WP 2 on the primary runway via WP 1 for NEP and the secondary range is the horizontal distance to the center of the selected heading alignment cylinder for the primary runway. When in Approach and Landing, both primary and secondary range indicators show horizontal distance to WP 2 on the primary runway. The range displays are on the magnetic wheels in the upper right and left corners of the HSI face. Each display ranges from 0 to 3,999 nautical miles. Both indicators use the same range data.

The bearing ("Brg") flag indicates that the heading, primary or secondary bearing, is invalid; the "CI" flag indicates that the course deviation display is invalid; and the glide slope ("GS") flag indicates that the glide slope deviation display is invalid.

Red and white diagonally striped flags are used to obscure the range indicators whenever their displays are invalid. An "Off" flag indicates that power is off or less than 18 volts.

MICROWAVE SCAN BEAM LANDING SYSTEM. The MSBLS or MLS (microwave landing system) is an airborne Ku-band receiver-transmitter landing and navigation aid with

Space Shuttle Spacecraft Systems

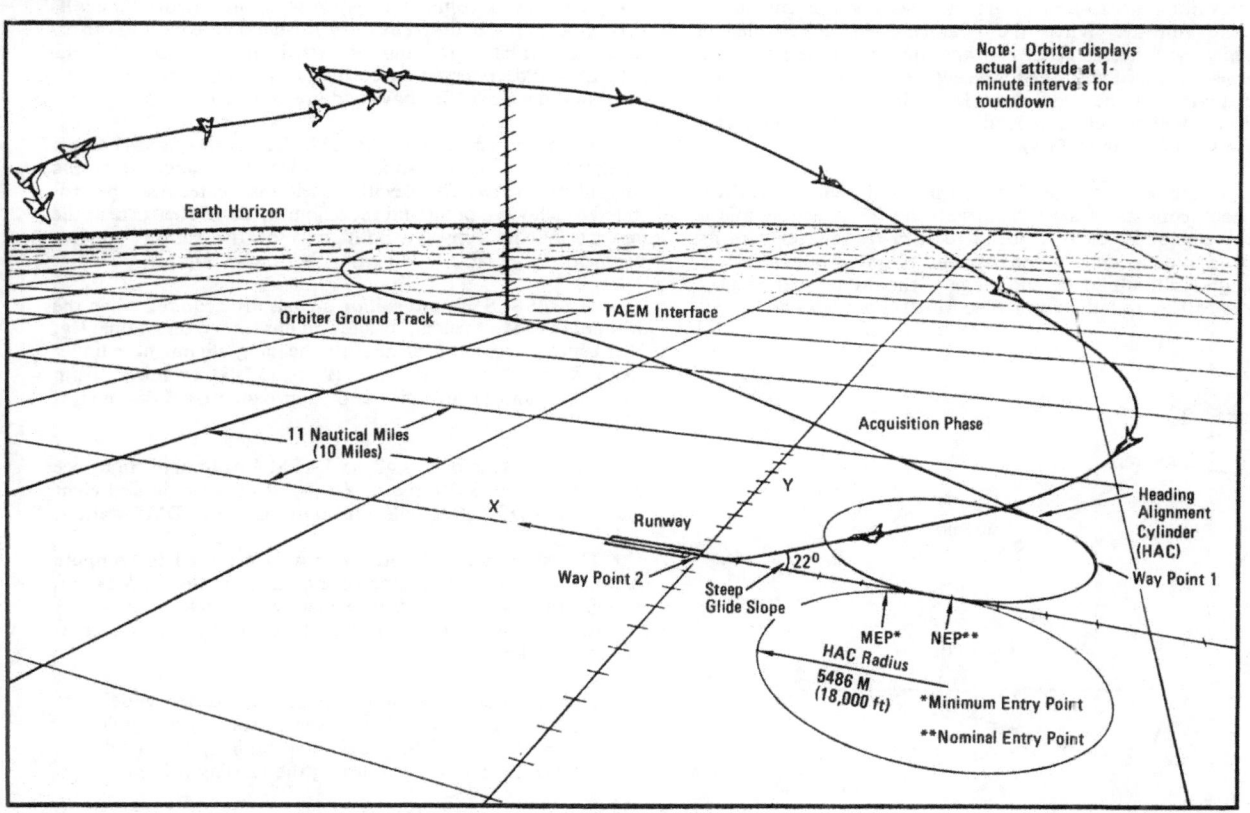

Entry Flight Profile

Space Shuttle Spacecraft Systems

decoding and computational capabilities. MSBLS provides orbiter position with respect to ground-based equipment placed near the runway. When the channel (specific frequency) associated with the targeted runway approach is selected, the airborne portion of the MSBLS receives elevation azimuth, and range data from the ground station. MSBLS is used during TAEM and approach and landing phases of the flight and return-to-launch-site (RTLS) abort.

The orbiter is equipped with three independent MSBLS sets. Each consists of a Ku-band antenna RF assembly and a decoder assembly. Each Ku-band receiver-transmitter with its decoder and data computation capabilities determines the elevation angle, azimuth angle, and range of the orbiter with respect to the MSBLS ground station. The MSBLS provides highly accurate three-dimensional position information to the orbiter to compute steering commands which maintain the orbiter on its proper flight trajectory. The three Ku-band antennas on the orbiter are located on the upper forward fuselage nose area. The MSBLS and decoder assembly are located in the crew compartment mid-deck avionics bays and are convection cooled.

The ground portion of the MSBLS consists of two shelters: an elevation shelter and an azimuth/DME (distance measuring equipment) shelter. The elevation shelter is located near the projected touchdown point and the azimuth/DME shelter near the far end of the runway. Both ends of the runway are instrumented to enable landing in either direction.

The MSBLS ground station signals are acquired when the orbiter is close to the landing site and has turned on its final leg. This usually occurs on or near the heading alignment cylinder about 8 to 12 nautical miles (9 to 13 statute miles) from touchdown and at an altitude of approximately 5,486 meters (18,000 feet).

Final tracking occurs at the TAEM-"Autoland" interface at approximately 3,048 meters (10,000 feet) altitude and eight nautical miles (9 statute miles) from the azimuth/DME station.

The MSBLS angle and range data are used to compute steering commands until the orbiter is over the runway approach threshold, at approximately 30-meter (100-foot) altitude. The autoland system may be overridden by the commander or pilot.

The HSI's show deviations from the selected glide slope and azimuth. When the orbiter is over the runway threshold, the radar altimeter is used to provide elevation (pitch) guidance. Azimuth/DME data is used during the landing rollout.

The three orbiter MSBLS sets operate on a common channel during the landing phase. The orbiter Ku-band antennas for

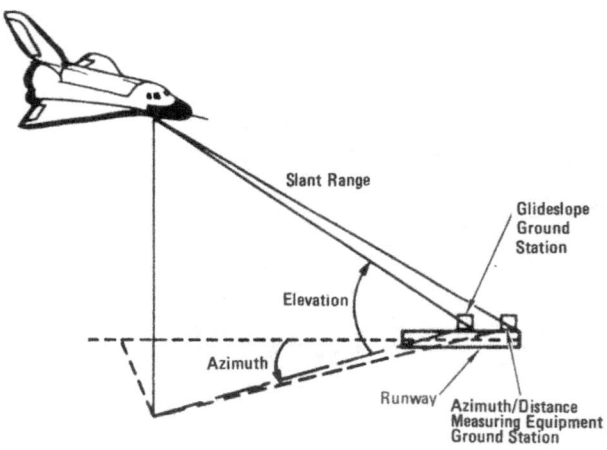

Microwave Landing System

Space Shuttle Spacecraft Systems

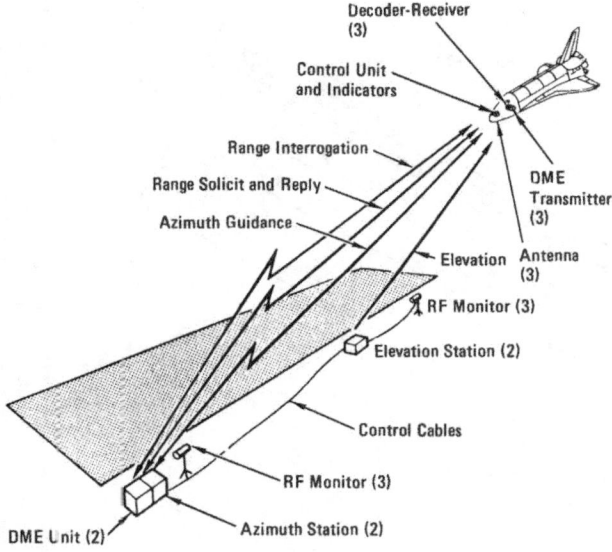

MSBLS Major Components and RF Links

each MSBLS are used for reception of angle data and transmission or reception of range data on a time-shared basis.

Each RF assembly routes the range, azimuth, and elevation information in RF form to its decoder assembly, which processes the information and converts it to digital words for transmission to the onboard computers via the MDM's for guidance and navigation.

Elevation, azimuth, and range data from the MSBLS are used by the GN&C system from acquisition until the runway approach threshold is reached. After that point, the azimuth and range data are used to control rollout. Altitude data is provided separately by the orbiter radar altimeter.

Each MSBL RF assembly transmits and receives precision slant range pulses. Range for DME is determined by measuring the time between the orbiter interrogation and the ground station reply.

Information on azimuth angle relative to the runway centerline is transmitted to the orbiter through a very narrow RF beam scanned 15 degrees to the right and left of the runway. This beam is modulated with many pulse pairs whose spacing varies with beam position as the beam scans the orbiter. An averaging operation is required to determine the center value. The orbiter can determine runway azimuth by measuring the received pulse pair spacing.

Absolute elevation angle referenced to horizontal information is received in the same manner as the azimuth, except the beam is scanned in a vertical direction, 3 degrees below and 27 degrees above the runway plane.

In addition, the position of the orbiter with respect to the runway is displayed on the HSI's. Elevation and azimuth are shown relative to a computer-derived glide slope on glide slope indicator and course deviation indicator needles. The range is displayed on the mileage indicators.

Since the azimuth/DME shelters are at the far ends of the runway, MSBLS can provide useful data until the orbiter is stopped. Azimuth data gives position in relation to the runway centerline and the DME gives distance from the orbiter to end of runway.

Each MSBLS set has channel select thumbwheels on the flight deck display and control panel which allow the flight crew

Space Shuttle Spacecraft Systems

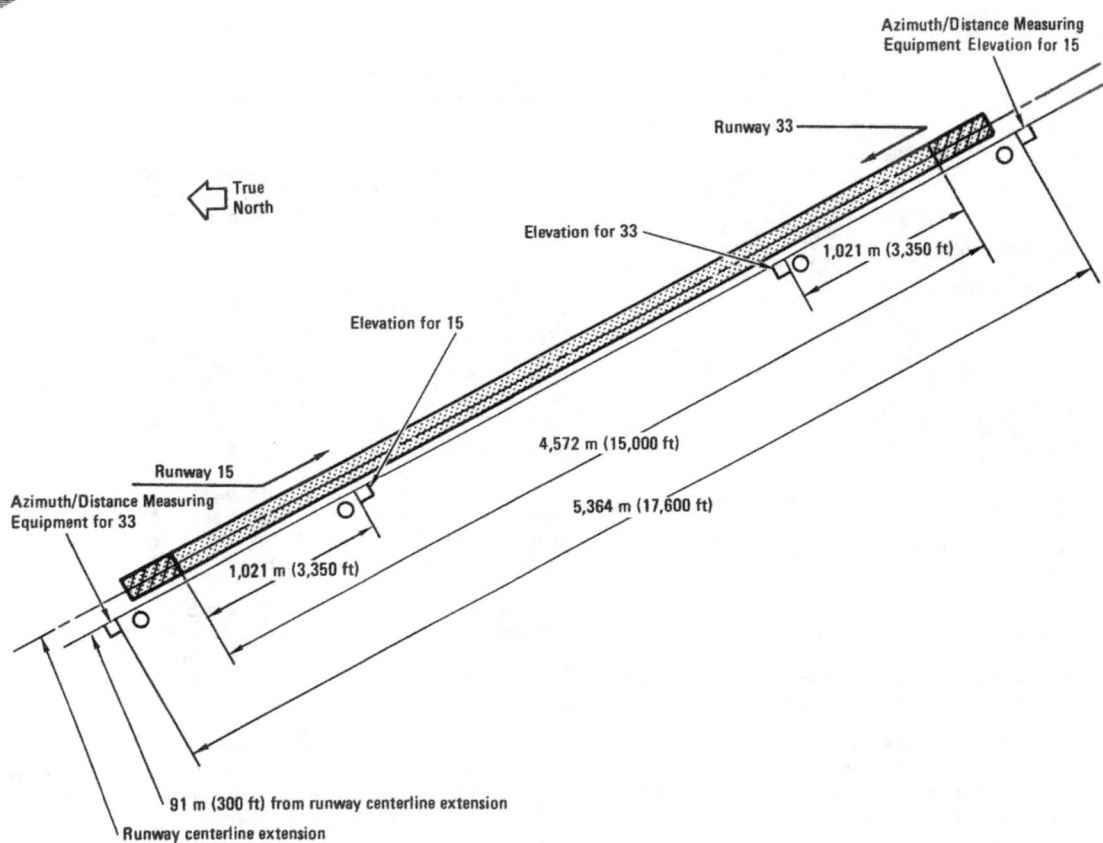

MSBLS Instrument Locations at KSC

Space Shuttle Spacecraft Systems

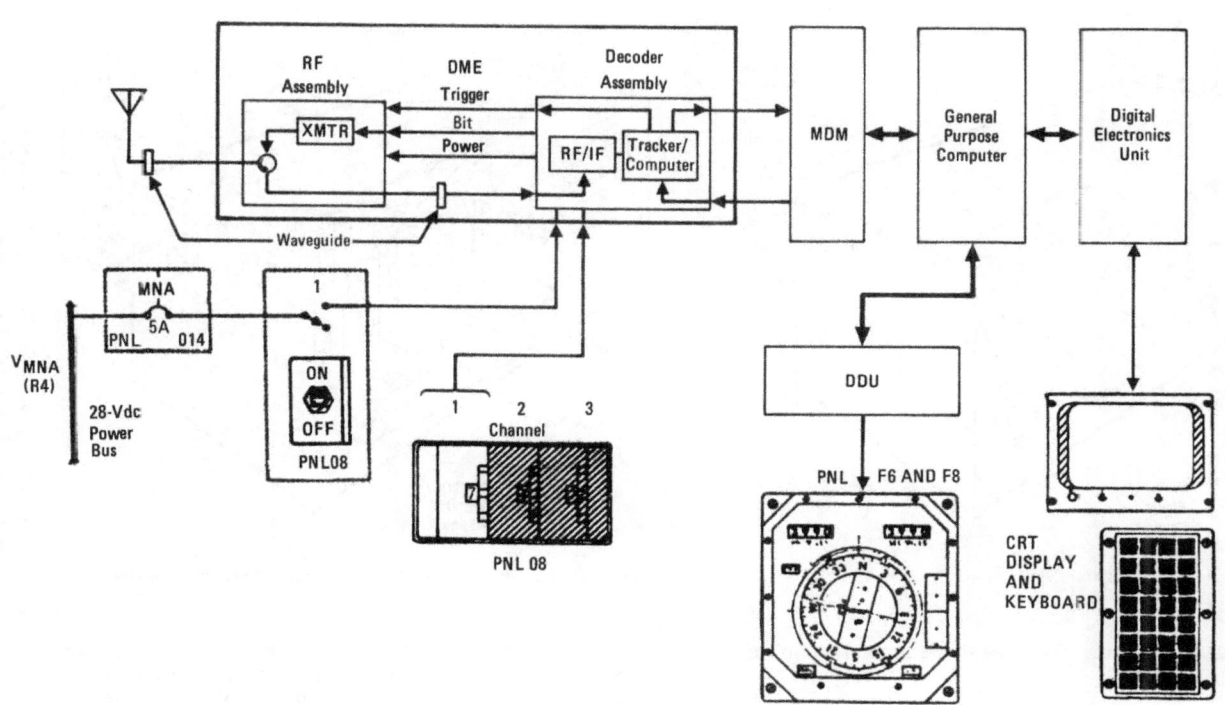

Microwave Landing System Interfaces

Space Shuttle Spacecraft Systems

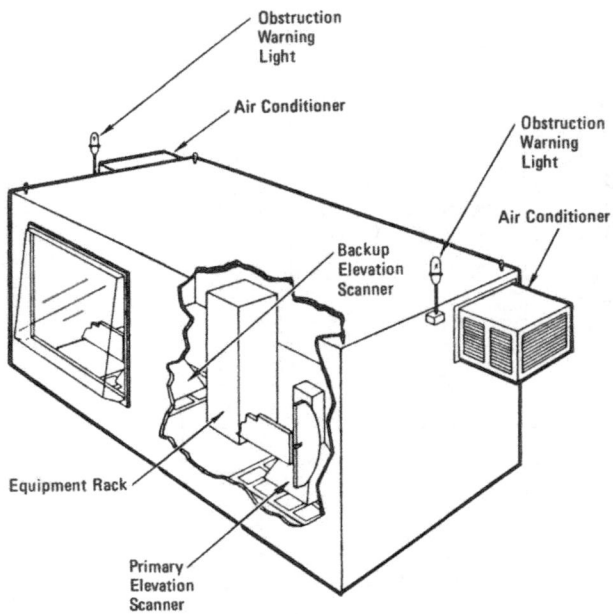

Elevation Shelter and Equipment

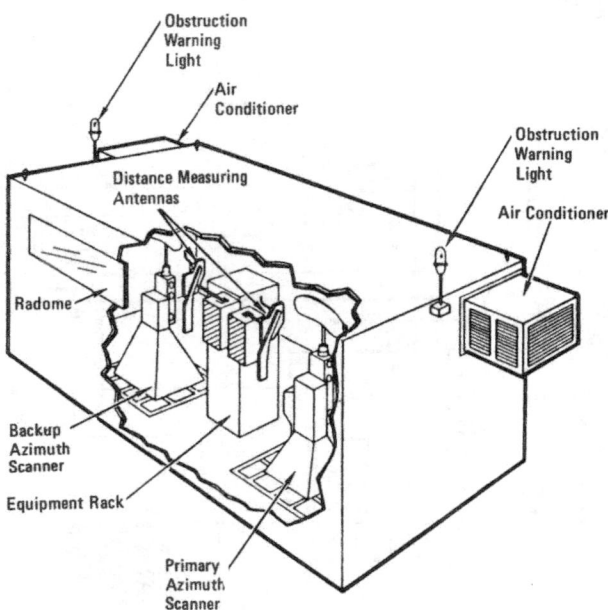

Azimuth Distance Measuring Equipment Shelter and Equipment

to manually select the channel for the ground station at the selected runway.

Redundancy management mid value selects azimuth and elevation angles for processing of Nav data. The three MSBLS sets are compared to determine if a significant difference is detected in one against the other.

When data from all three MSBLS sets are valid, redundancy management selects middle values of three ranges, azimuths, and three elevations. With two good MSBLS sets, the two ranges, azimuth, and elevation averaged. With only one good MSBLS set, its range, azimuth, and elevation are passed for display. When a fault is detected the "SM Alert" light is illuminated, and a CRT fault message shown.

Space Shuttle Spacecraft Systems

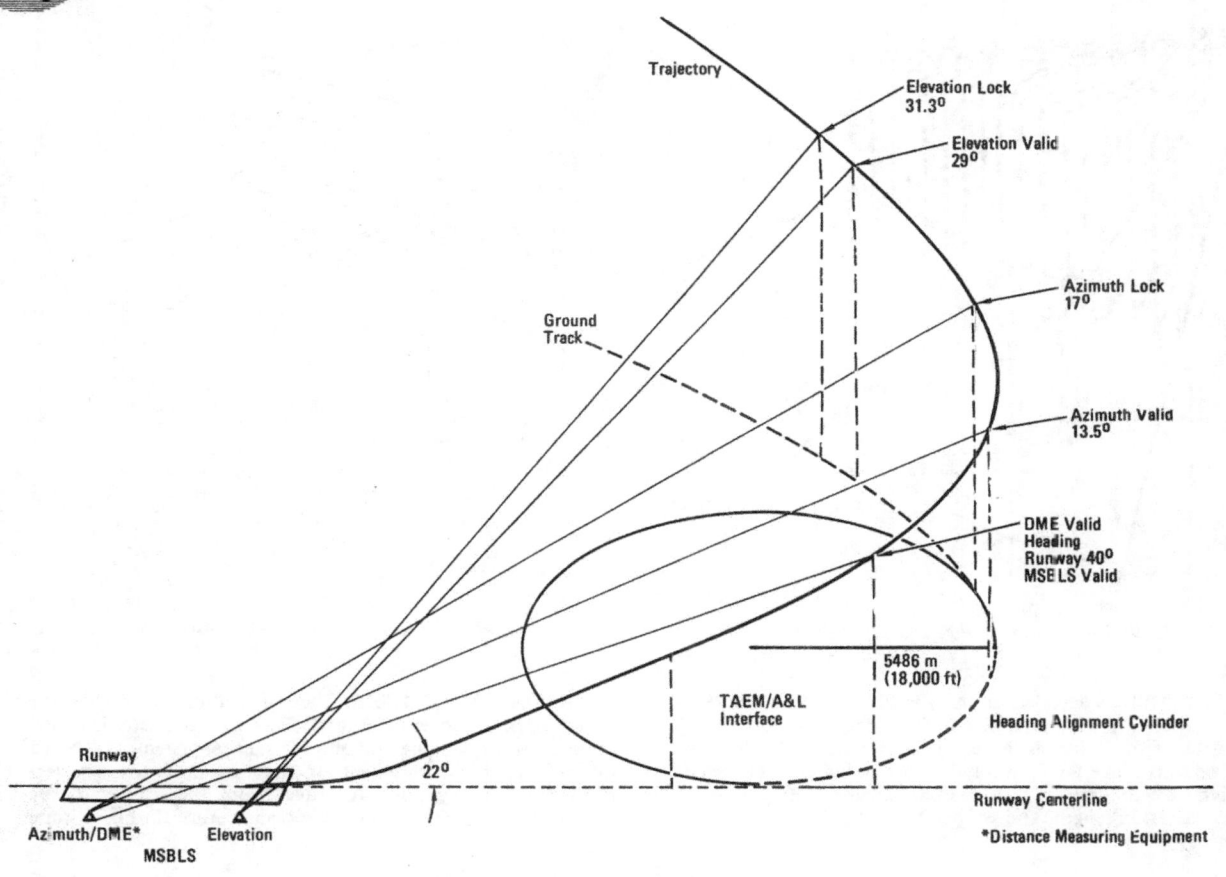

Microwave Scan Beam Landing System

Space Shuttle Spacecraft Systems

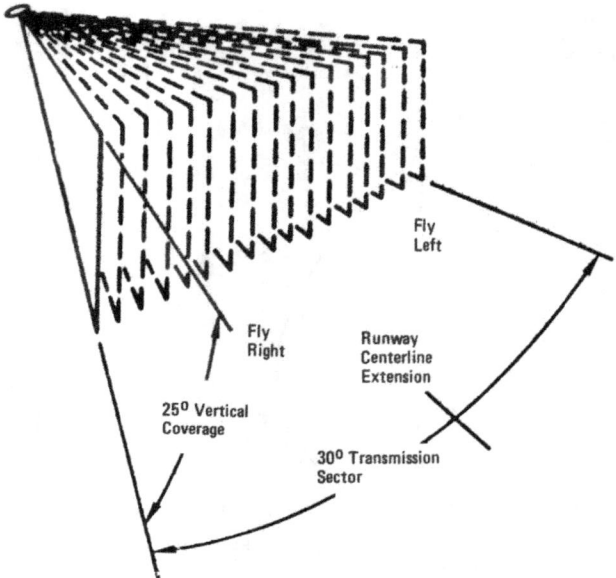

MSBLS Azimuth Coverage and Pulse Coding

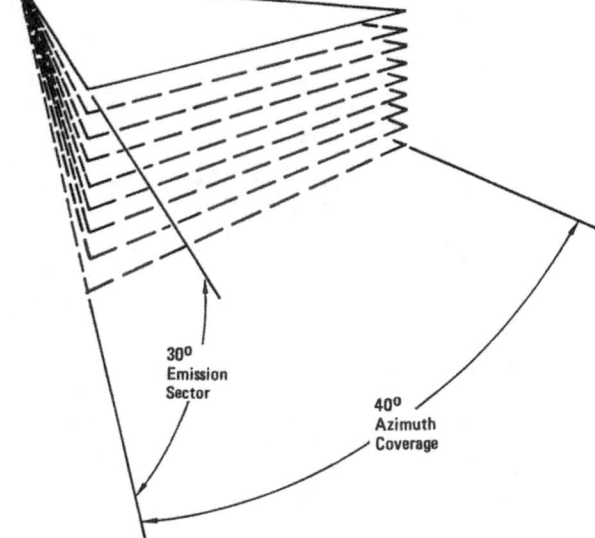

MSBLS Elevation Coverage and Pulse Coding

Each MSBLS decoder assembly is 209 millimeters (8.25 inches) in height, 127 millimeters (5 inches) in width, 410 millimeters (16.158 inches) in length, and weighs 7.9 kilograms (17.5 pounds). The RF assembly is 177.8 millimeters (7 inches) in height, 889 millimeters (3.50 inches) in width, 260 millimeters (10.25 inches) in length, and weighs 2.72 kilograms (6 pounds).

RADAR ALTIMETER. The radar altimeter provides the orbiter altitude above the ground. The radar altimeter is a low-altitude terrain tracking and altitude-sensing system. It is based on the precise measurement of time required for an electromagnetic energy pulse to travel from the orbiter to the nearest object on the ground below and return during altitude

Space Shuttle Spacecraft Systems

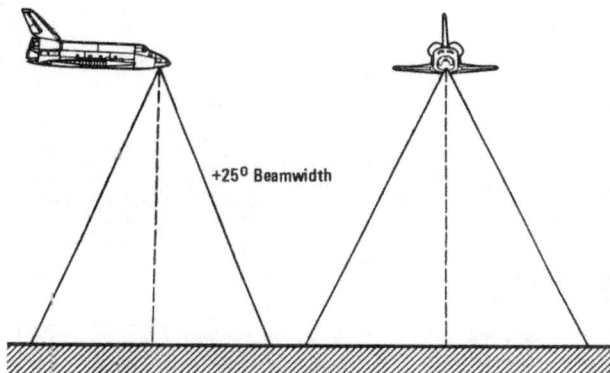

Radar Altimeter Antenna Beamwidth Coverage

rate changes up to 609 meters per second (2,000 fps). This enables tracking of mountain or cliff sides ahead or adjacent to the orbiter if these obstacles are nearer than the ground below, and provides warning of rapid changes in absolute altitude.

The radar altimeter is the primary sensor of the "Autoland" system and for touchdown guidance after the orbiter has crossed the runway threshold from an altitude of 30 meters (100 feet) down to touchdown. Before that point, radar altimeter data is displayed on the crew's AVVI's (altitude/vertical velocity indicators) up to 5,000 feet.

The orbiter contains two independent radar altimeter systems, each with a transmitting and receiving antenna. The systems are independent and can operate simultaneously without affecting each other. The four C-band antennas are located on the lower forward fuselage. The two receiver/transmitters are located in the mid-deck forward avionics bays and are convection cooled.

Altitude range signals are analog voltages proportional to the elapsed time required for return of the ground pulse, which is a function of height or distance to the nearest terrain. The onboard computers process the data for automatic flight control of orbiter altitude if in the autoland mode; otherwise, the data is used only for display on the AVVI's.

Since there are only two radar altimeters aboard the orbiter, the altitude data from the two units are averaged in the redundancy management for use in the autoland mode. The MSBLS elevation and range information then can be used down to 30 meters (100 feet).

Both commander and pilot stations have switches for selecting radar altimeter 1 or 2 for display on his AVVI. The display scale ranges from 0 to 5,000 feet and the altitude is displayed on moving tapes. Above 5,000 feet, the scale will be pegged. Below 5,000 feet, the indicator will show actual altitude, orbiter to ground. At 1,500 feet the RDR indicator changes scale. The "RA Off" flag will appear if there is a loss of power, loss of lock, data good-bad, or after three communication faults.

Each radar altimeter receiver-transmitter is 79 millimeters (3.13 inches) in height, 188.2 millimeter (7.41 inches) in length, 97 millimeters (3.83 inches) in width, and weighs 2 kilograms (4.5 pounds).

The contractors involved are: Eaton Corp., AIL Division, Farmingdale, NY (MSBLS); AiResearch Manufacturing Co., Garrett Corp., Torrance, CA (air data transducer assemblies); Hoffman Electronics Corp., Navigation Communication System Division, El Monte, CA (TACAN); Honeywell Inc., Minneapolis, MN (radar altimeter); Rosemount Inc., Eden Prairie, MN (air data sensor probes).

Space Shuttle Spacecraft Systems

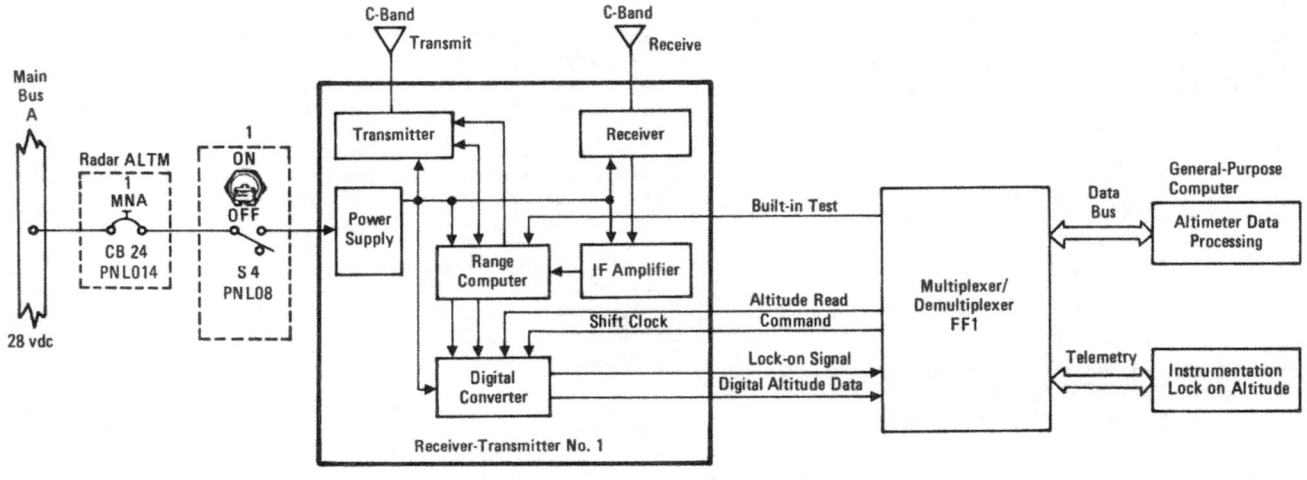

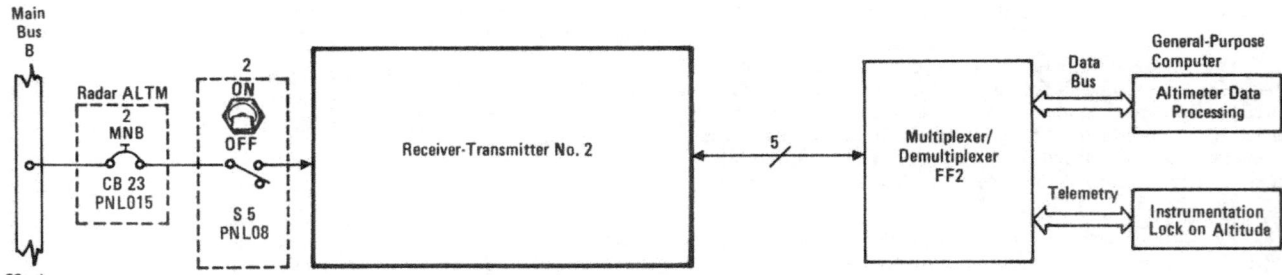

Radar Altimeter

Space Shuttle Spacecraft Systems

GUIDANCE, NAVIGATION, AND CONTROL

The guidance, navigation, and control (GN&C) system responds to software commands to provide vehicle control and also furnishes sensor and controller data to the GN&C software so that it can compute the commands.

The orbiter's five flight computers are organized into a redundant set of four which form the primary flight system, with the fifth computer used as the backup flight system in the initial development flight tests. The fifth computer operates independently of the other four.

These computers interface with various systems. The orbiter's flight forward and flight aft multiplexer/demultiplexers (MDM's) and data bus serve as the conduit for signals going to and from such units as the master timing unit, the sensors that provide velocity and attitude information, orbiter propulsion systems, aerodynamic surfaces, and displays and controls.

The GN&C system consists of two modes of operation: auto and manual (control stick steering). In the automatic mode, the primary flight system (PFS) essentially allows the computers to do all the flying, with the flight crew selecting the various operational sequences. The crew may control the vehicle in the control stick steering mode by using hand controls such as the rotation hand control (RHC), translation hand control (THC), speed brake/thrust controller (SBTC), and rudder pedals; however, the commands issued by the crew must pass through and be issued by the computers.

There are no direct mechanical linkages by which the flight crew can manipulate the various propulsion systems or aerodynamic aerosurfaces; thus, the orbiter is an entirely digitally controlled, fly-by-wire vehicle.

The multi-function cathode ray tube (CRT) display system presents GN&C and system status information to the flight crew and allows the crew extensive interaction with the computers via keyboard units on the flight deck display and control panel.

The mass memory units contain the primary and backup flight software to be loaded into the computers and communicate directly with the computers.

The computers also transmit pulse code modulation (PCM) telemetry data to the ground stations for monitoring of the internal state of the computer's status and operational information of all subsystems and to report progress of the flight.

The ground launch processing system (LPS) interfaces with the orbiter before launch via the launch umbilical, which provides direct access to the onboard computer's memory. In addition to performing vital ground launch sequencing tasks, the LPS, using a general memory read/write capability, can monitor and modify the onboard computer core (such as updating launch pad initial conditions).

The navigation system maintains an accurate estimate of the vehicle position and velocity, referred to as a state vector. From position and velocity, other parameters (acceleration, angle of attack) are calculated for use in guidance and for display to the crew. The current state vector is mathematically determined by integrating the equation of motion for coasting flight and by using the acceleration of the vehicle as sensed by the inertial measurement units (IMU's) for powered flight. The alignment of the IMU, and hence the accuracy of the resulting state vector, deteriorate as a function of time. Celestial navigation instruments (the star trackers and crew optical alignment sight) are used to keep the IMU's aligned while in orbit. The accuracy of the IMU-derived state vector is, however, insufficient

 **Space Shuttle Spacecraft Systems**

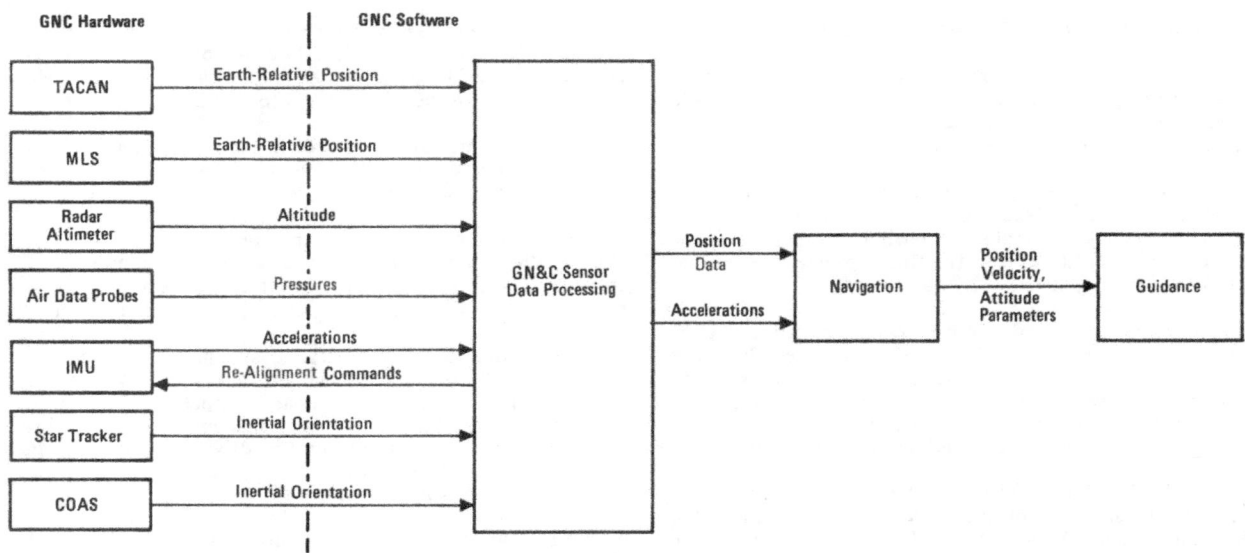

Navigation Interfaces

for either guidance or the flight crew to bring the spacecraft to a pinpoint landing. So, data from other navigation sensors — air data system, TACAN, microwave scan beam landing system (MSBLS), and radar altimeter — is blended into the state vector at different phases of entry to provide the necessary accuracy.

The guidance system computes and issues the propulsion system engine fire and gimbal commands (thrust vector control — TVC) for the main propulsion system engines, solid rocket boosters, and orbital maneuvering system (OMS) engines, the fire commands for the reaction control system (RCS) engines, and the orbiter aerodynamic surface deflection commands.

Flight control includes attitude processing and a digital autopilot which provides steering, thrust vector, attitude, and aerosurface control. Flight control receives commands from guidance software (automatic) or from the flight crew via hand

Space Shuttle Spacecraft Systems

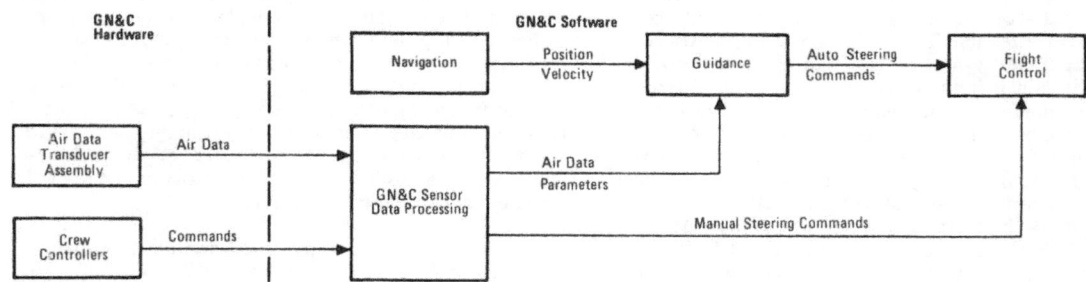

Guidance Interfaces

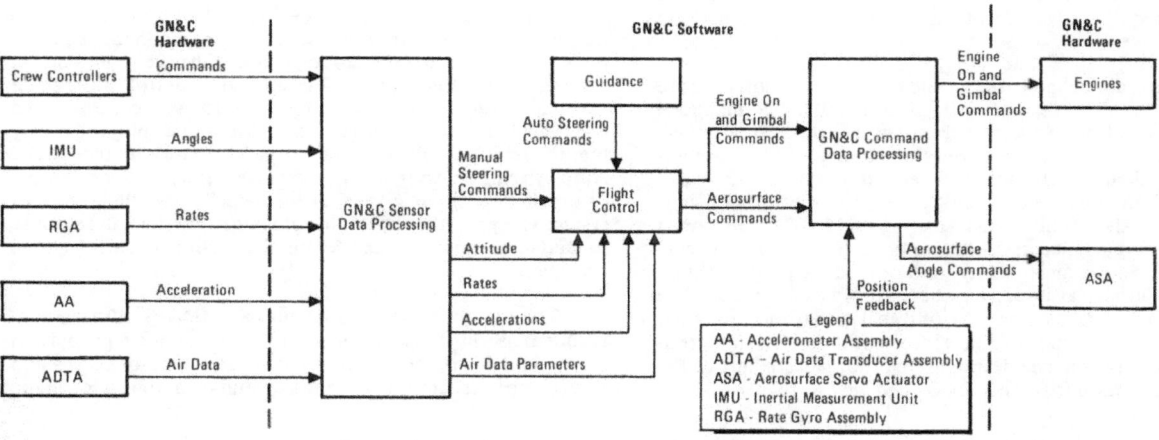

Flight Control Interfaces

Space Shuttle Spacecraft Systems

controllers (manual), which are in terms of vehicle dynamics (attitudes, rates, and accelerations) and processes them into commands for hardware (engine fire, engine gimbal, or aerosurface deflection). Flight control output commands are based on errors for stability augmentation. The errors are the difference between the commanded attitude, aerodynamic aerosurface position, body rate, or body acceleration and the actual attitude, position, rate, or acceleration. Actual attitude is derived from IMU angles, aerodynamic aerosurface position is provided by actuator feedback transducers through the aerosurface amplifiers (ASA's); body rates are sensed by the rate gyro assemblies (RGA's); and accelerations are sensed by the accelerometer assemblies (AA's). During atmospheric flight, flight control adjusts control sensitivity based on air data parameters derived from local pressure sensed by the air data system probes and performs body turn coordination using body attitude angles derived from IMU angles. Thus, the GN&C hardware required to support flight controls is a function of the mission phase.

Steering commands used by the flight control software are augmented by the guidance software or manually commanded using the hand controller or speed brake/thrust controller. When flight control software uses the steering commands computed by guidance software, it is termed automatic guidance; when the flight crew is controlling the vehicle via the hand controllers it is called control stick steering (CSS). The commands computed by guidance are those required to get from the current state (position and velocity) to a desired state (specified by target conditions, attitude, airspeed, runway centerline). The steering commands take the form of translational and rotational rates and accelerations. Guidance receives the current state from navigation software. The desired state or targets are part of the initialized software load and can be changed in flight.

Flight control software controls the vehicle via the selection of operational sequences/major modes (OPS/MM) in the onboard computers; this is accomplished via the keyboards at the flight deck commander and pilot stations. There is an operational sequence/major mode for prelaunch, first-stage ascent, second-stage ascent, orbit insertion OMS-1, orbit insertion OMS-2, orbit coast, maneuver execute, orbit checkout, deorbit maneuver, deorbit maneuver coast, entry guidance, and TAEM (terminal area energy management) guidance. In addition, the flight control system consists of two selectable modes — Auto and CSS — via pushbutton light indicators on the flight deck at the commander and pilot stations.

The flight control system two selectable modes are organized into two control channels: pitch and roll/yaw. The pushbutton will illuminate a white light to indicate the mode selected. Each pushbutton is a triple redundant, momentary contact, nonlatching switch.

In the Auto mode, the GN&C provides automatic control of the vehicle through selection of the operational sequence major mode via the onboard keyboard, thus the flight control system, "Auto Pitch" provides automatic control in the pitch axis and "Auto Roll/Yaw" provides automatic control in the roll/yaw axes. The flight crew monitors the displays to verify that the vehicle is following the correct trajectory. In the GN&C Auto system, the computers process the vehicle motion sensors to obtain the dynamic state of the vehicle. The flight control system processes the flight control laws (equations) in response to guidance commands and commands the flight control system effectors.

The computers have digital autopilot (DAP) software consisting of ascent thrust vector control (ATVC), main propulsion system (MPS) command processor, SRB processor, OMS and RCS processors, aerosurface control, and reconfiguration logic.

The DAP software generates acceleration profiles, trims, elevon load relief schedule, and scheduled gains as functions of

Space Shuttle Spacecraft Systems

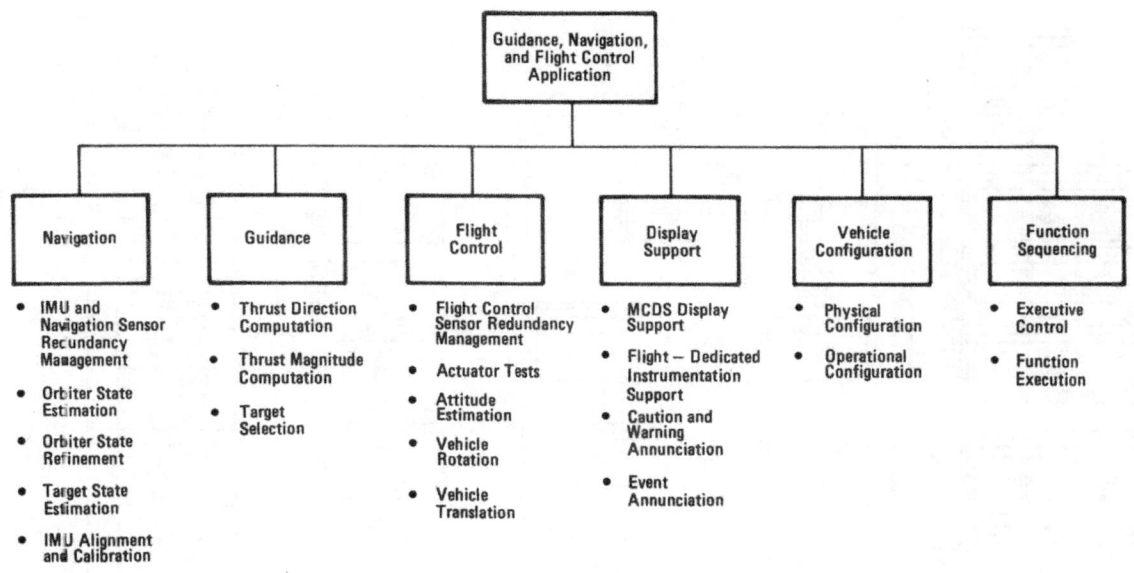

GN&C Structure

earth relative velocity magnitude, mission elapsed time, and vehicle mass. The DAP reconfiguration logic is the bookkeeper of the flight control software. In response to control mode changes, changes in failure status, and the occurrence of events, the reconfiguration logic generates the indicators that are needed by the DAP and guidance/steering for moding sequencing and initialization. The attitude processor and guidance/steering software interfaces with the DAP. The attitude processor provides the guidance/steering with current attitude information and flight control computes and flies out errors between the guidance — commanded and actual vehicle attitude.

LAUNCH. Prior to T minus 20 minutes, prelaunch functions are controlled by a ground computer network at the launch site. This is the launch processing system. The orbiter computers interact with the LPS from T minus 20 minutes

 **Space Shuttle Spacecraft Systems**

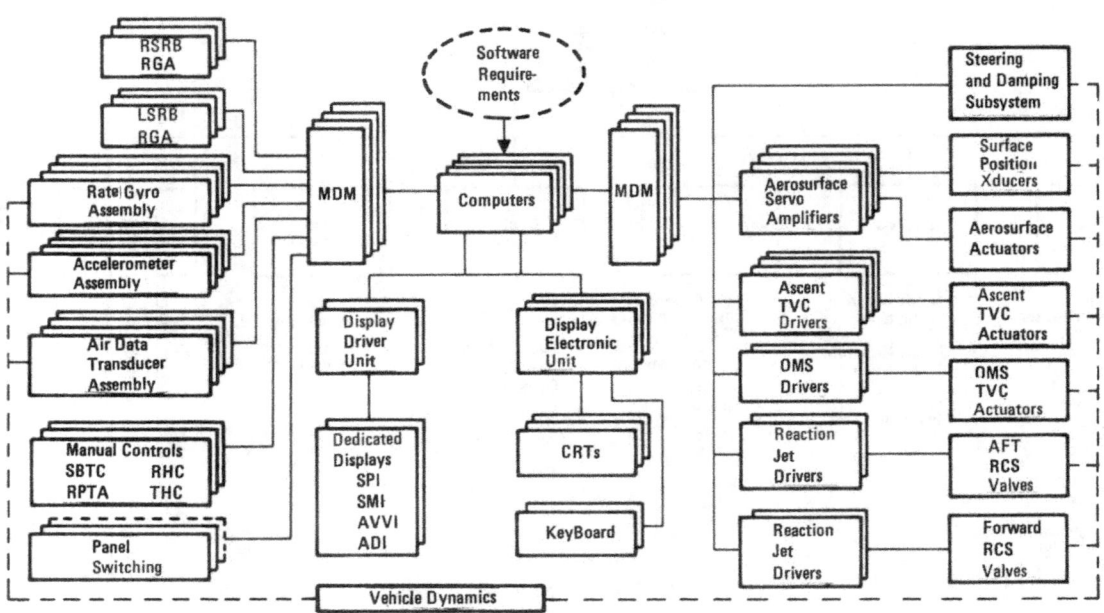

Flight Control System

Space Shuttle Spacecraft Systems

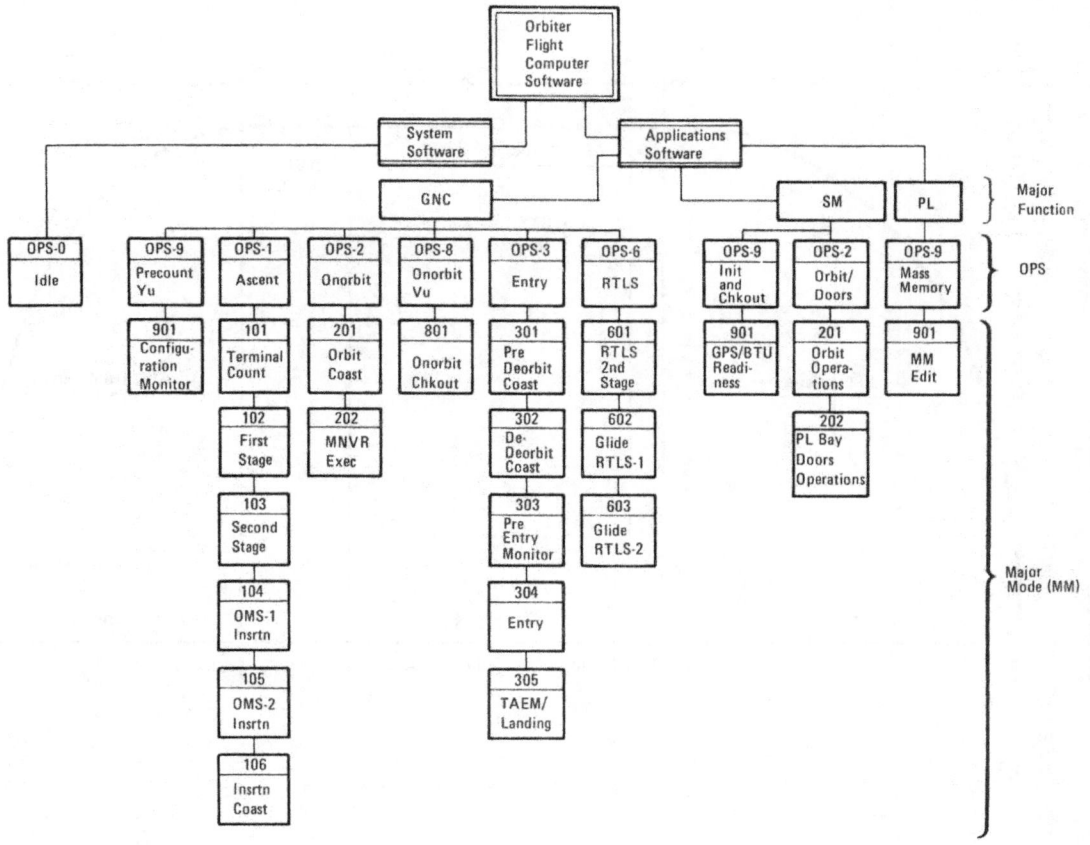

Orbiter Flight Control Software

Space Shuttle Spacecraft Systems

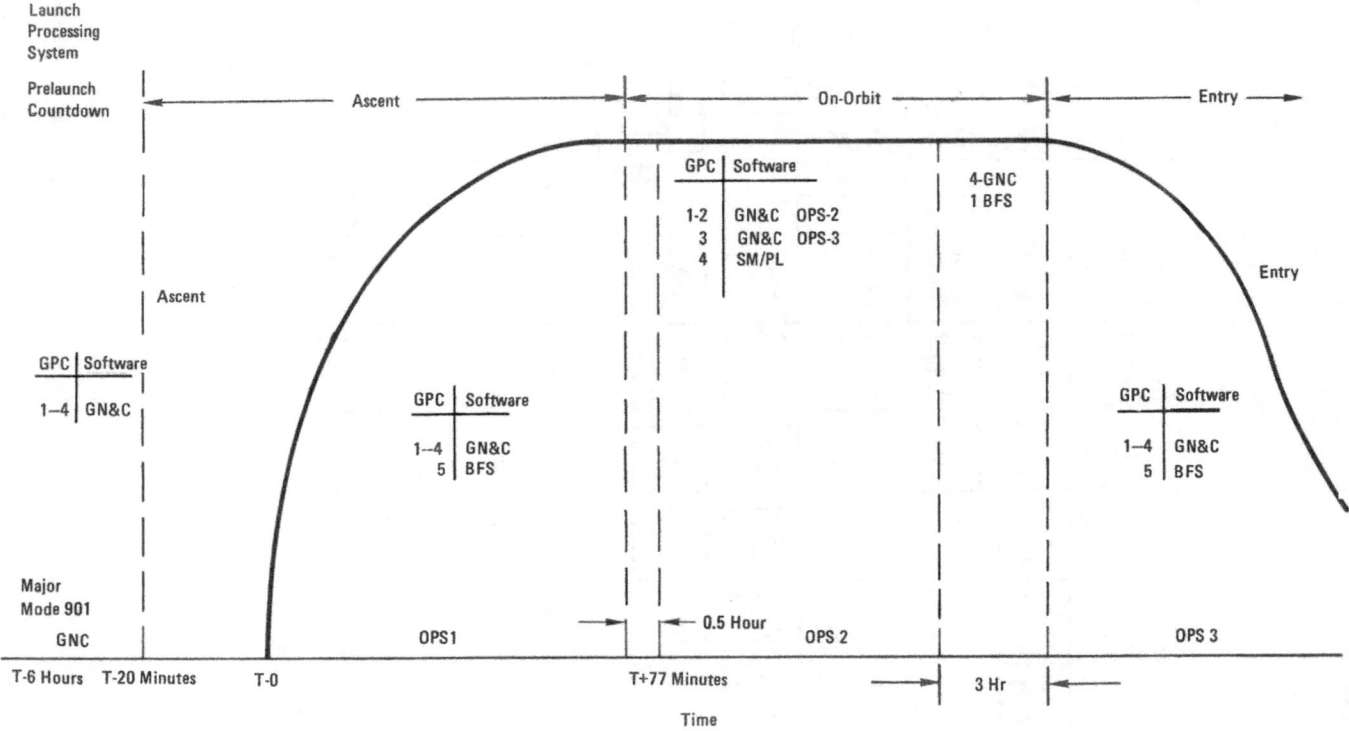

Orbiter Computer Configuration Nominal Mission Scenario

Space Shuttle Spacecraft Systems

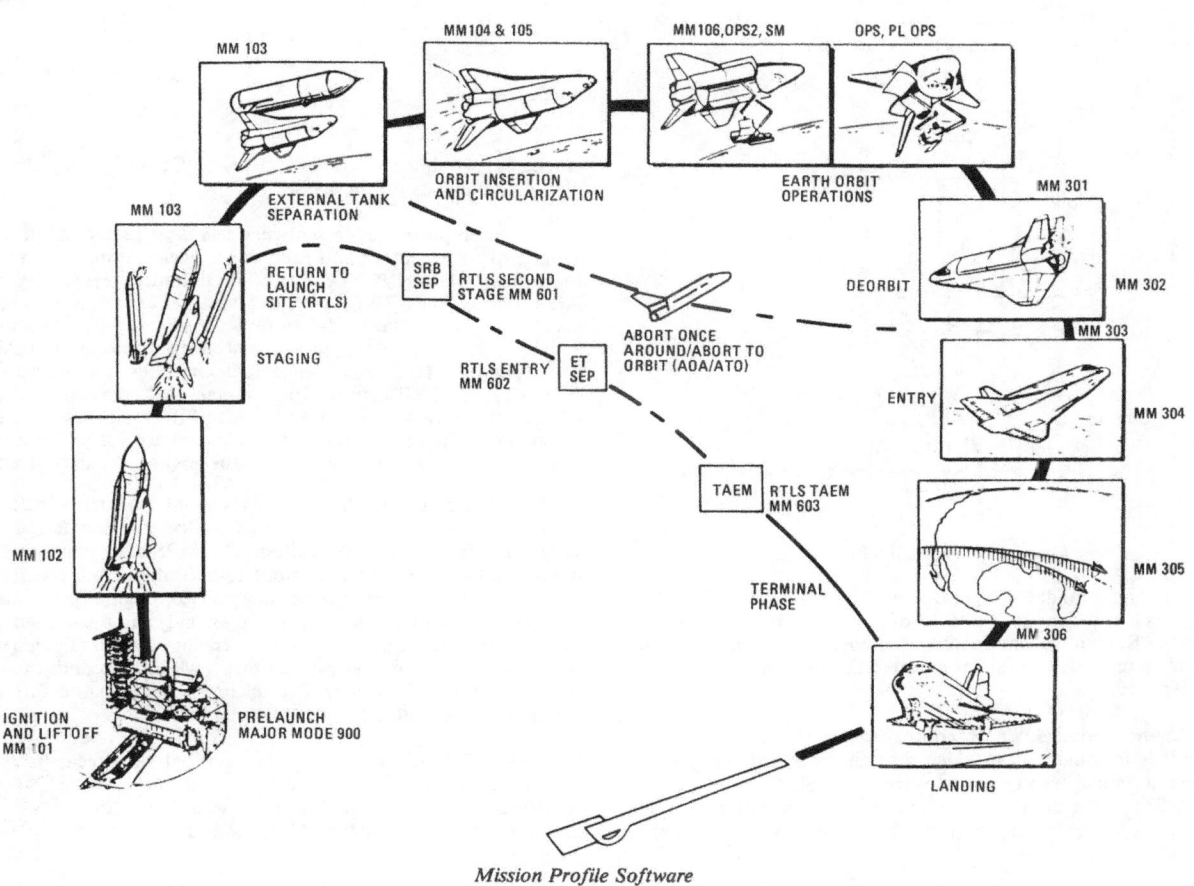

Mission Profile Software

 **Space Shuttle Spacecraft Systems**

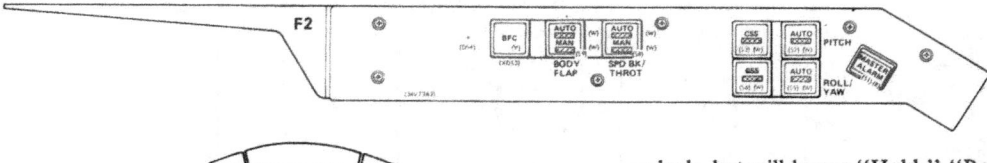

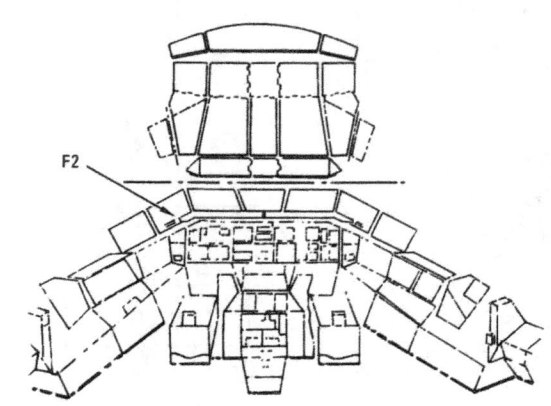

Flight Control System Pushbutton Indicators

(when they are loaded with software and formed into the four-computer redundant set) until liftoff. This interaction consists of the LPS sending commands to the computers, which follow out the commands and then respond to the LPS when the action is completed.

Launch countdown is controlled by the LPS until 31 seconds before launch, at which time the onboard computers' automatic launch sequence software is enabled by LPS command. From this point, the computers take control of the sequencing of events and perform functions by the onboard clock, but will honor "Hold," "Resume Count," and "Recycle" commands from the LPS.

The computer launch sequence sets flags to command the arming of SRB ignition and hold-down release system pyro initiator controllers (PIC's) and the T-0 umbilical release PIC's. After a delay, the SRB ignition PIC voltages are monitored for acceptable levels. The hold-down release system PIC's and the T-0 umbilical release system PIC's are monitored by the LPS. The computer launch sequence logic initiates a countdown "Hold" if the SRB ignition PIC voltages fall below an acceptable level at any time prior to issuance of the main engine start commands. If the SRB ignition PIC voltages are not acceptable, after the start commands are issued, the engines are shut down.

The computer launch sequence also controls certain critical main propulsion system valves and monitors the engine-ready indications from the main engines. The MPS start commands are issued by the onboard computers at T minus 6.6 seconds (staggered start — engine three, engine two, engine one — all approximately within one-fourth of a second) and the sequence monitors the thrust buildup of each engine. All three engines must reach the required 90 percent thrust within 3 seconds or an orderly shutdown is commanded up to 4.6 seconds and safing functions are initiated.

Normal thrust build-up to the required 90 percent thrust level will result in the engines being commanded to the liftoff position at T minus 3 seconds as well as the Fire 1 command being issued to arm the SRB's. At T minus 3 seconds, the vehicle

Space Shuttle Spacecraft Systems

base bending load modes are allowed to initialize (movement of approximately 650 millimeters [25-1/2 inches] measured at the top of the external tank - with movement towards the external tank).

At T-0, the two SRB's are ignited, under command of the four onboard computers, the four explosive bolts on each SRB are initiated (each bolt is 711 millimeters [28 inches] long and 88 millimeters [3.5 inches] in diameter), the two T-0 umbilicals (one on each side of the spacecraft) are retracted, the onboard master timing unit, event timer, and mission event timers are started, the three SSME's are at 104 percent, the ground launch sequence is terminated and modes the onboard computers.

The DAP ascent configuration logic, upon ascent initialization, places the main engine nozzles in the start position, places the SRB nozzles in the null position, and places the orbiter aerosurfaces in the launch position. Upon receipt of the command to prepare main engines for liftoff, the engines are commanded to the launch position.

There are no guidance functions performed during prelaunch. Also, there are no active flight control functions, except for the commands to gimbal the engines for test and launch positions and slewing of the aerosurfaces.

Navigation is initialized during prelaunch. At T minus eight seconds, the launch pad location (latitude, longitude, and attitude) of the navigation base is transformed to a position vector. Pad 39A at the Kennedy Space Center in Florida is at 28 degrees, 36 minutes, 30.32 seconds north latitude and 80 degrees, 36 minutes, 14.88 seconds west longitude. The Space Shuttle vehicle is oriented on Pad 39A with negative Z body axis (tail) pointed south. The launch azimuth is 64 degrees from north. The inertial measurement units are set inertial at T minus 20 minutes, but are biased for earth rotation until T minus 12 seconds, when the bias is removed.

ASCENT. Boost guidance commands an attitude hold for approximately the first 8 seconds after liftoff so that the launch pad tower is cleared, SRB nozzles are above the top of the launch tower lightning rod, at approximately 12 meters (41 feet). After the vehicle clears the launch umbilical tower, a pitch profile and a roll begins (heads down, wings level). By about T plus 20 seconds, the vehicle is at 180 degrees roll and 78 degrees pitch.

The vehicle flies upside down during the ascent phase. This orientation, together with trajectory shaping, establishes a trim angle of attack that is favorable for aerodynamic loads during the region of high dynamic pressure (max q) and yet results in a net positive load factor.

Thrust vector control is the hub of the wheel of flight control.

In the ascent phase, the four ascent thrust vector control (ATVC) drivers respond to commands from the guidance system. Thus, the TVC commands from guidance are transmitted to the ATVC drivers, which transmit electrical signals proportional to the commands to the servoactuators on each main engine and solid rocket booster.

Thrust vector control closes the acceleration and rate loops within the outer attitude loops to generate body axis attitude error rates which eventually are flown out by the main engines and SRB's. The main propulsion system processor of the DAP converts body axis and attitude error signals that are generated in TVC, into pitch and yaw engine bell deflection commands. The SRB processor of the DAP accomplishes the same functions as the MPS, except it is referred to as rock and tilt instead of pitch and yaw.

Each main engine and solid rocket booster servoactuator consists of four independent two-stage servoactuators which

Space Shuttle Spacecraft Systems

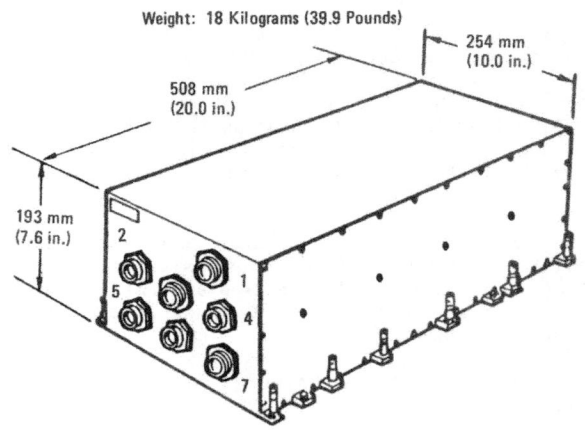

Ascent Thrust Vector Controller

receive signals from the ATVC's. All four two-stage servovalves in each servoactuator control one power spool in each actuator which positions the actuator ram and the respective SRB nozzle and main engine. The four two-stage servovalves in each actuator provide a force summed majority arrangement to position the power spool. With the four identical flight control system commands to the four two-stage servovalves, the actuator force sum action prevents a single erroneous command from affecting power ram motion. If the erroneous command persists for more than a predetermined time, differential pressure sensing activates a selector valve, isolating and removing the defective servovalve hydraulic pressure, permitting the three remaining channels and servovalves to control the actuator ram spool. A second failure would isolate and remove the defective servovalve hydraulic pressure in the same manner as the first failure, leaving only two remaining active channels. Each actuator is equipped with position transducers for position feedback to the TVC system. An FCS CHANNEL yellow caution/warning light on the flight deck display and control panel will illuminate if an SRB or main engine actuator channel fails.

The SRB pitch and yaw rate gyros are used exclusively during first stage, and control is switched to orbiter rate gyros when the SRB's are commanded to null in preparation for separation. Pitch and yaw axes and a combination of rate, attitude, and acceleration signals are blended to effect a common signal to both main engine and SRB TVC. In the roll axes, rate and attitude are summed to provide a common signal to both the main engine and SRB TVC.

Orbiter rate gyros are used by the flight control system during ascent, entry, and aborts as feedbacks to find rate errors which are used for stability augmentation and for display on the commander and pilot attitude director indicators (ADI's) rate needles. There are four orbiter rate gyro assemblies located in the mid fuselage under the payload bay. The rate gyros are referred to as rate gyro assemblies (RGA's) 1 and 2 and RGA's 3 and 4.

Each RGA contains three single-degree-of-freedom rate gyros positioned to sense rates about the roll, pitch, and yaw axes. All the RGA's contain identical gyros; the only difference is in the manner in which each gyro is positioned within the RGA. Each gyro has three axes. A motor forces the gyro to rotate about the spin axes. When the vehicle rotates about the gyro input axis, a torque results in a rotation about the output axis. An electrical voltage proportional to the angular deflection about the output axis and representing vehicle rate about the input axis, is generated and sent through the flight aft multiplexer/demultiplexers to the computers and RGA subsystem operating program (SOP). Selected rates are then sent to the ADI rate needles. The data is also sent to the display electronic unit CRT's.

Space Shuttle Spacecraft Systems

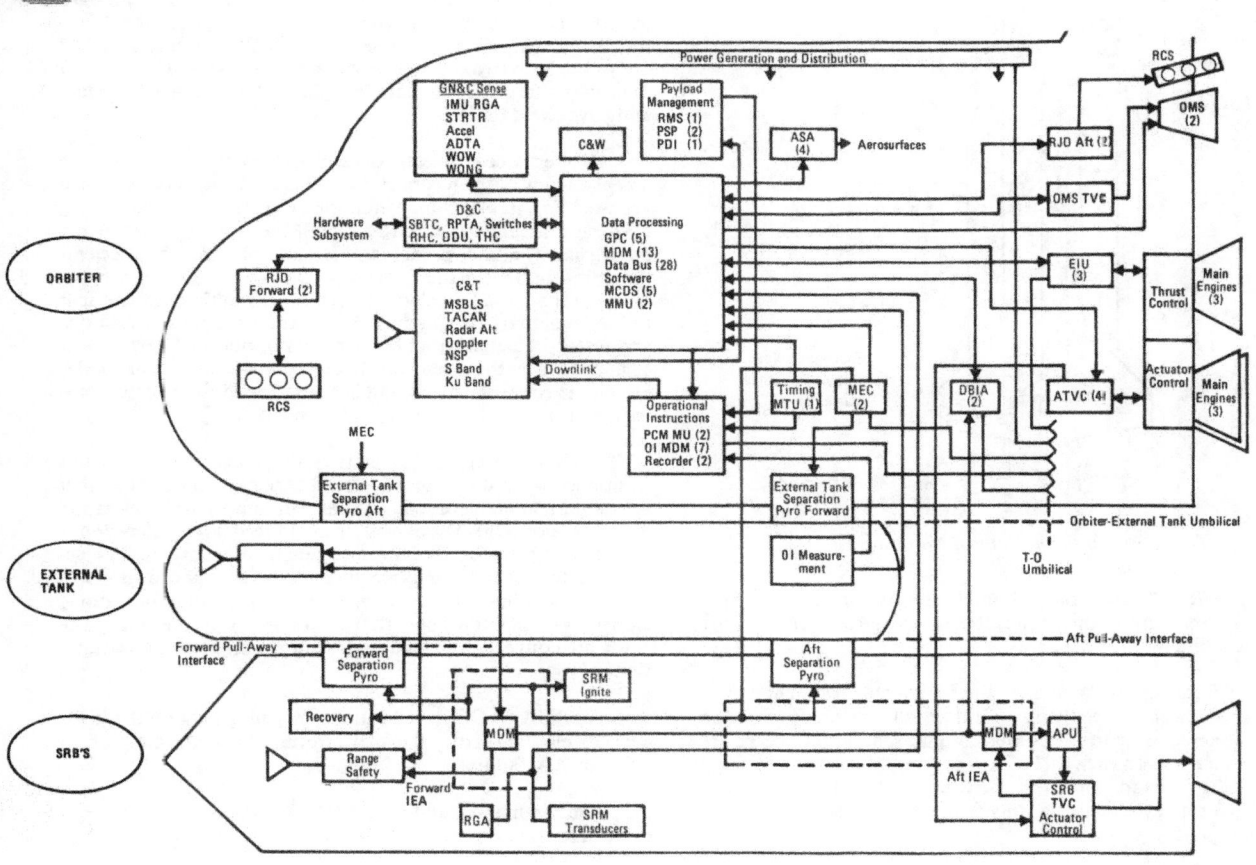

Avionics

Space Shuttle Spacecraft Systems

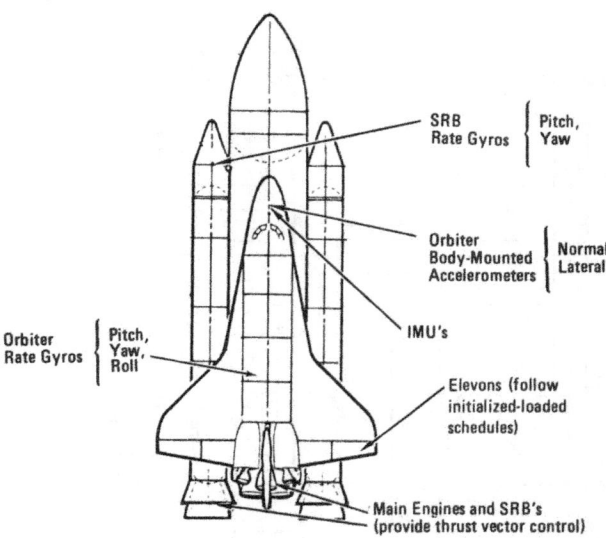

Ascent Flight Control Hardware

The RGA's are mounted on coldplates for cooling by the Freon-21 coolant loops. The RGA's require a five-minute warmup time.

SRB rate gyros are used by the flight control system during first-stage ascent as feedback to find rates from liftoff to two to three seconds prior to SRB-external tank separation. There are three rate gyros on each SRB. They are mounted on the forward ring in the forward skirt of the SRB-external tank attach point. The SRB rate gyros contain only two gyros, for sensing rates in the pitch and yaw axes.

SRB rate gyro data is sent through the orbiter forward MDM's to the computers. The flight control system DAP uses this data during first stage ascent. The SRB gyros are switched out of the loop two or three seconds before SRB separation and their yaw and pitch rate data is replaced by orbiter RGA pitch and yaw rate data.

Orbiter accelerometer assemblies (AA's) are used during ascent and entry for feedback to find acceleration errors, which are used for stability augmentation, elevon load relief during first-stage ascent, and during TAEM (terminal area energy management) and approach and landing phases. The accelerations transmitted to the forward MDM's are voltages proportional to the sensed acceleration. They are multiplexed and sent to the computers, where an AA operating program conditions and scales the signals from the selected normal and lateral AA's. The rates are then sent to flight control. The four body-mounted accelerometer assemblies are located in the crew compartment mid deck avionics bays 1 and 2.

Each assembly contains two single-axis accelerometers positioned so that one senses lateral acceleration and the other senses normal acceleration. Lateral (left and right) acceleration is sensed along the Y axis and normal (vertical) acceleration is sensed along the Z axis (yaw and pitch, respectively). The accelerometers use a pendulum light beam and photo diode to provide a reading of acceleration. Data is also sent to the display electronics unit CRT's and ADI error needles during entry. The AA's are convection cooled and require a five-minute warmup time.

An RGA/ACCEL red caution/warning light on the flight deck display and control panel informs the flight crew of an RGA or AA failure.

The commander and pilot ADI's display attitude, rate, and error data.

Space Shuttle Spacecraft Systems

ORBITER INERTIAL MEASUREMENT UNITS. The IMU's provide orbiter orientation in inertial space and accelerations along vertical axis. These contain two accelerometers to sense accelerations and two gyros which provide an inertially stabilized four-gimbal platform (called cluster) that can obtain vehicle attitude with respect to inertial space. There are three IMU's mounted on a navigation base inside the crew compartment forward of the flight deck display and control panel.

Each IMU gyro has two output axes. Angular motion about any axis perpendicular to the spin axis causes rotation about a third orthogonal axis, which contain components of rotation about the two output axes. The appropriate gimbals are torqued to null the gyro rotation. The gimbal torquing in response to vehicle motion results in an inertially stabilized platform. In addition to torquing in response to vehicle motion, the gyros may be pulse torqued (small angles) or slewed (large angles) by software which results in gimbal torquing to position or reposition the platform. Each gimbal has coarse (one-speed) and fine (eight-speed) resolvers. The resolver readouts are sent to the forward MDM's and then to the computers.

The two IMU accelerometers are a pendulous mass anchored to a case. Accceleration produces relative motion between the mass and the case. The motion is sensed and voltage is applied to counteract the motion. The voltage required is proportional to the acceleration. The voltage is electronically converted to pulses which are accumulated. The accumulated data pulses represent velocity and are sent to the computers via the forward MDM's. One accelerometer senses acceleration along two axes, X and Y, and the other senses delta velocity along a Z axis. These X, Y, and Z axes are not vehicle axes.

The three IMU's are initially aligned so that their cluster orientations are skewed relative to one another, which results in only one IMU being exposed to "gimbal flip" conditions at any one time. The gimbal sequence is outer roll, pitch, inner roll, and azimuth. The inner roll gimbal is redundant to the outer roll. By maintaining the pitch gimbal perpendicular to the yaw-azimuth gimbal and by mechanizing two roll gimbals, "gimbal lock" is avoided, but "gimbal flip" becomes a possibility with pure yaw motion at a pitch of 90 to 270 degrees.

The selected data from the IMU operating program is used by flight control for coordination and guidance for steering commands. Attitude angles are displayed on the ADI's and horizontal situation indicators (HSI's). The selected delta velocities are used by navigation to determine the three position and three velocity components of the state vector and by guidance and flight control for moding. IMU status is also displayed on the CRT's.

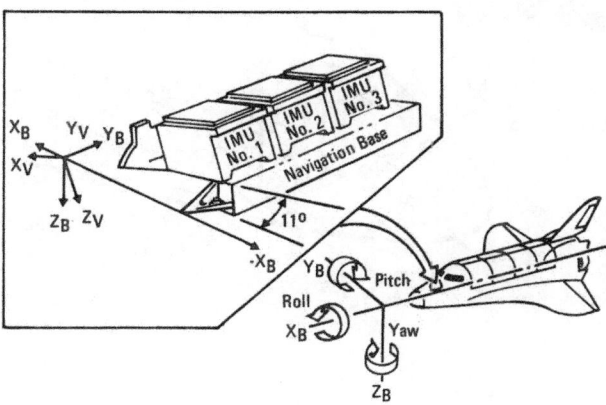

Orbiter IMU's

Space Shuttle Spacecraft Systems

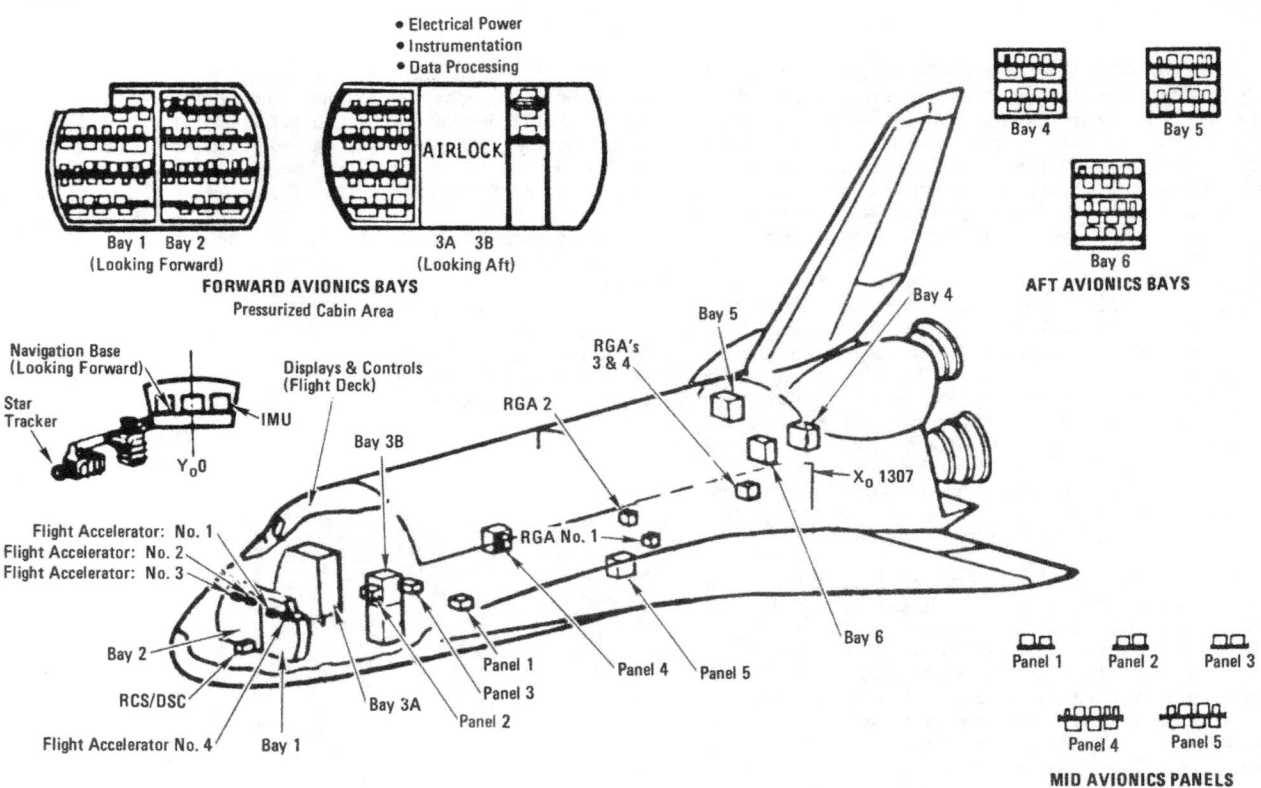

Equipment Installation

Space Shuttle Spacecraft Systems

The accuracy of the IMU deteriorates with time. If the errors are known, they can be physically or mathematically corrected. Software based on preflight calibrations is used to compensate for most of the inaccuracy. The star trackers and crew optical alignment sight (COAS) are used to determine additional inaccuracy and provide the means of aligning the IMU's to remove the errors, when in orbit.

Each IMU is cooled by cabin air drawn into and circulated within the casings by IMU fans. Critical temperatures are maintained by automatic heater circuits within each IMU. An IMU red caution/warning light on the flight deck display and control panel informs the flight crew of an IMU failure.

DEDICATED DISPLAY INSTRUMENTS. These are located on the orbiter crew compartment flight deck display and control panels. The dedicated display software performs the processing required to accept, prepare, and output guidance, navigation, and control data to the dedicated display indicators. The dedicated display indicators are the ADI's, HSI's, alpha-Mach indicators (AMI's), attitude vertical velocity indicators (AVVI's), and surface position indicators (SPI's). There are two sets of ADI's, HSI's, AMI's and AVVI's at the flight deck commander and pilot stations. There is only one SPI. Another ADI is located on the flight deck aft flight display and control panel.

The display driver unit (DDU) is an electronics unit that connects the computers and the primary flight displays. The DDU receives data signals from the computers and decodes them to drive the dedicated displays. The unit also provides dc and ac power for the ADI's and RHC's. The DDU contains logic for setting flags on the dedicated instruments for such items as data dropouts and failure to synchronize. The orbiter contains three DDU's, one at the commander's station, one at the pilot's station, and one at the aft station.

The three ADI's indicate the vehicle's roll, pitch, and yaw attitude via a gimbaled ball, attitude errors via three needles,

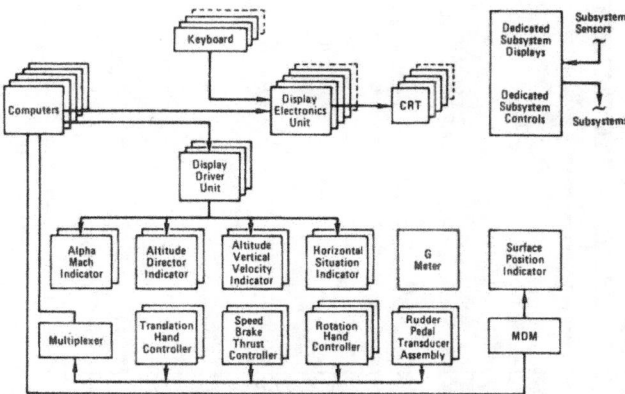

Display and Control System

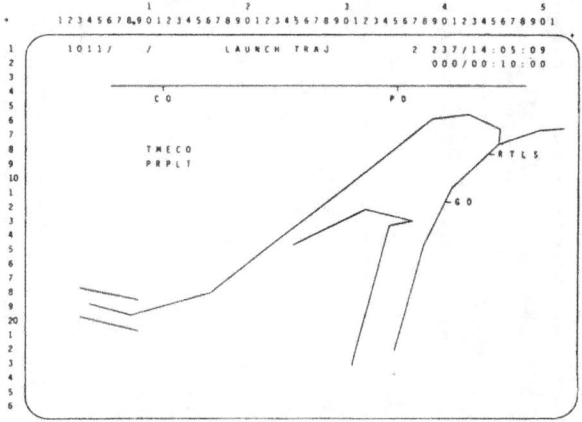

Launch Trajectory Typical Display

Space Shuttle Spacecraft Systems

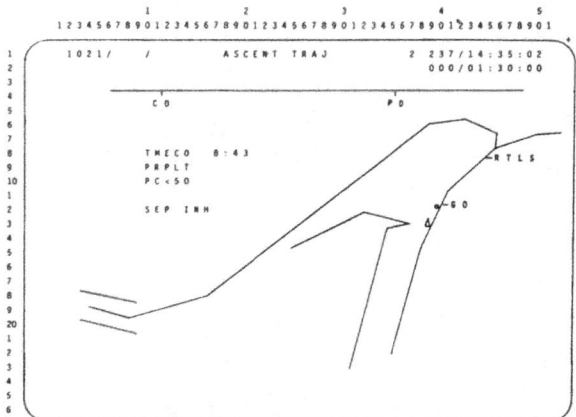

Ascent Trajectory Typical Display

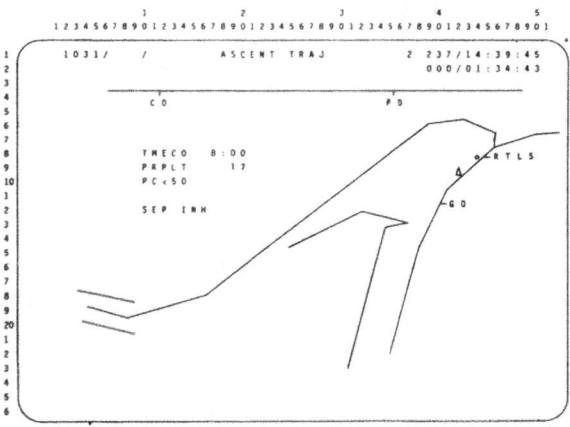

Ascent Trajectory Typical Display

and attitude rates via three pointers. The angles displayed on the three ADI's are generated in the attitude processor from IMU daa and indicate the orbiter's roll, pitch, and yaw via the gimbaled ball. The attitude error needles are superimposed on the top face of the ADI ball and provide attitude error information in roll, pitch, and yaw. The needles are driven in first- and second-stage ascent by attitude errors generated in guidance. The rate pointers are positioned on the indicator outside the ADI ball and provide rate information from the rate gyro assemblies. The ADI also provides an OFF flag to indicate when the attitude display may be invalid. The needles and pointers have a stored (out of view) position which is also used to indicate invalid conditions.

In first-stage ascent, the ADI's monitor the roll, pitch, and yaw of the vehicle and ensure that attitude rates remain within their limits. During second-stage ascent, the ADI's confirm that second-stage guidance has initiated the monitor hold for external tank separation.

FIRST-STAGE ASCENT. During first-stage ascent, it is sometimes necessary to relieve loads on elevon hinges that result from high shears. This is performed automatically by a load relief logic which drives the elevons to the appropriate positions to relieve the aerodynamic adverse hinge moment loads. This logic also holds the body flap and rudder/speed brake in place during ascent. The SPI displays the position of the aerosurfaces.

During first-stage ascent, the pre-stored guidance program works open-loop in that it selects an initialized-loaded roll, pitch, and yaw command based on vehicle relative velocity. An initialized-loaded table is accessed by software and commands the vehicle attitude needed to obtain the desired SRB separation attitude and orbit inclination.

Guidance also is responsible for issuing throttle commands to the main engines. These commands also are initialized-loaded. There are four initialized-loaded slots for throttle settings and they are referenced to relative velocity. The main

Space Shuttle Spacecraft Systems

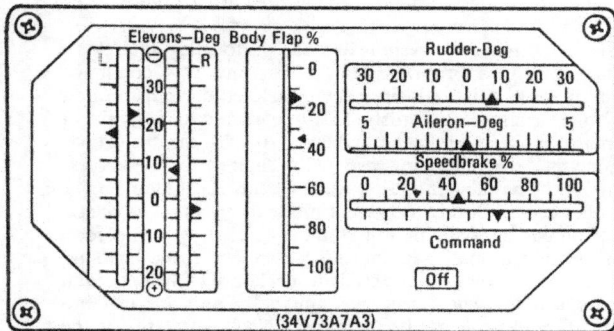

Attitude Director Indicator (ADI) - Displays 360° Maneuvers in Roll, Pitch, and Yaw
- Positive Roll Command – Ball Rotates Counterclockwise
- Positive Pitch Command – Ball Rotates From Top to Bottom
- Positive Yaw Command – Ball Rotates From Right to Left
- Error Needle Reflections to Positive Commands
  Roll – Left
  Pitch – Down
  Yaw – Left
- Rate Pointer Deflection to Positive Commands
  Roll – Right
  Pitch – Up
  Yaw – Right

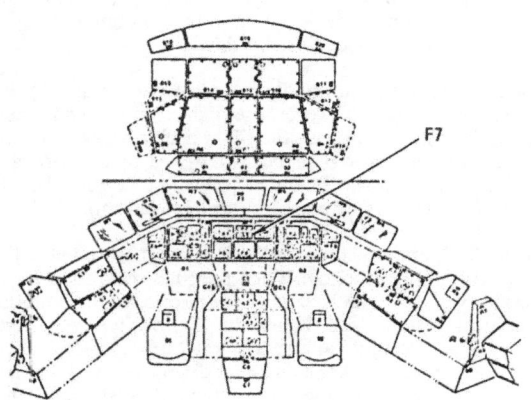

Surface Position Indicator

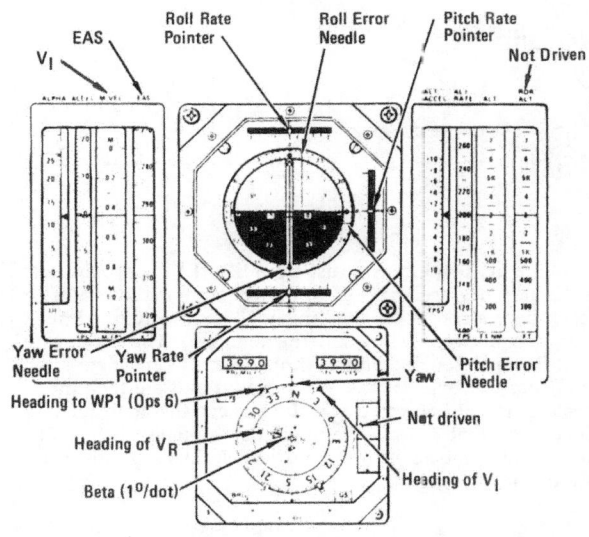

*Dedicatd Displays - First Stage Ascent
Panel F6 and F8*

Space Shuttle Spacecraft Systems

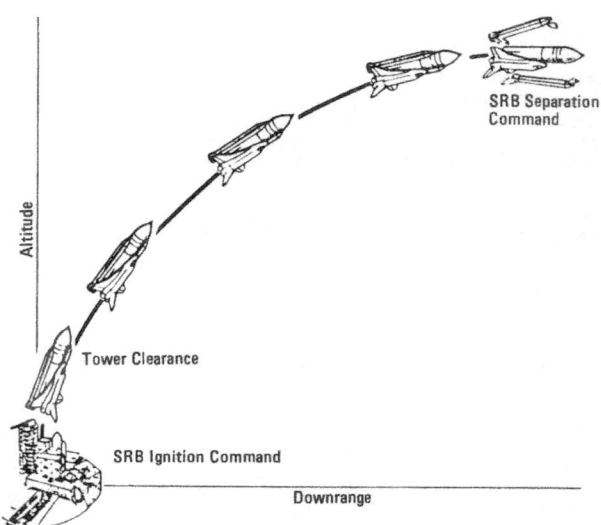

First Stage Ascent

engines are at 104-percent thrust from liftoff to approximately one minute; they are throttled to a predetermined lower thrust percent dependent upon that mission at 10-percent thrust/second to limit dynamic pressure (maximum q). The engines are held at the predetermined lower thrust percent dependent upon that mission for approximately 24 seconds and are then throttled back up to 104 percent at 10 percent thrust/second to maintain constant 3 g's in the initial development flights. The thrust profile reduces heating and vehicle loads during maximum dynamic pressure.

Navigation programs in first-stage ascent are responsible for providing accurate knowledge of the vehicle state vector through the use of IMU-sensed velocity changes and a mathematical model of the earth's gravitational forces. This function can be used to aid in driving the CRT's predictor.

Boost guidance is responsible for performing the "table lookup" of the appropriate roll, pitch, and yaw command depending on relative velocity. The table used depends on how many main engines are thrusting. If one engine fails, guidance automatically recognizes the failure and lofts the trajectory, commanding the remaining two main engines to 109-percent thrust for the remainder of first-stage ascent. Each main engine controller is used during the ascent phase to monitor the operation of an engine, issue inhibit commands to prevent a second engine from automatically shutting down when one has shut down, monitor the state of flight deck crew displays and switches from the switch processor, and issues appropriate commands. The controllers also monitor GN&C software for the proper time to check the main propulsion system liquid oxygen and liquid hydrogen low level sensors, monitor GN&C software for proper time of main engine cutoff (MECO), and issue the closure of the main propulsion system liquid oxygen and liquid hydrogen prevalves after engine shutdown.

SRB separation is normally performed automatically by the onboard computers; however, the flight crew can command separation through switches on the flight deck display and control panel. The automatic sequence is initiated by the software in the computers when the chamber pressure of both SRB's is below 2,587 millimeters of mercury (mmHg) (50 psi).

At SRB separation command, a three-axis attitude hold is commanded for four seconds. When the SRB separation command is received, the SRB nozzles are positioned to null and the flight control system is switched to the orbiter rate gyro assemblies. Four seconds after SRB separation, second-stage (main engine) guidance takes over. In addition, if a main engine shutdown is detected, the failed engine is positioned to null and trim changes are commanded on the remaining engines.

Space Shuttle Spacecraft Systems

SECOND-STAGE GUIDANCE. Second-stage guidance is completely changed from first stage. It is a closed-loop function that computes each cycle where the vehicle should be so that an initialized-loaded MECO target can be reached. Commands are generated to place the vehicle in a desired position with a desired velocity and with minimal fuel usage. The desired position and velocity are determined from a set of initialized-loaded conditions. The parameters defining MECO target are target-velocity, flight-path angle, target radius from earth center, and desired orbital plane. The MECO target conditions are selected primarily to ensure correct placement of the external tank splashdown.

The first phase of second-stage ascent extends from the SRB separation command until the earliest time at which an abort once around (AOA) is possible with a single engine failure. This assumption causes a lofted trajectory during the first phase so that MECO conditions can be reached with the use of only two engines.

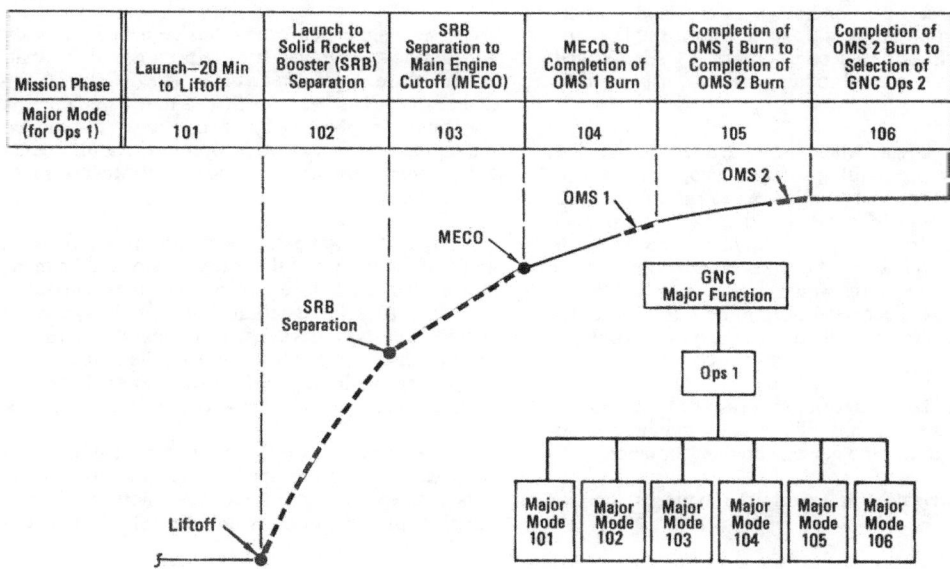

Ascent Major Modes

 **Space Shuttle Spacecraft Systems**

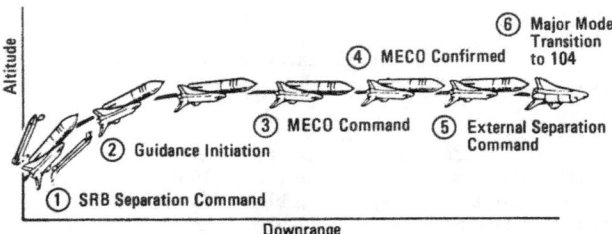

Second-Stage Ascent

The second phase of second-stage ascent begins at the earliest time at which an AOA is possible with a single engine failure and continues until an acceleration of 3 g's is reached. The remainder of the second phase assumes use of three engines.

The third phase begins when thrust acceleration reaches 3 g's. At this point, a 3-g limiting software program begins to adjust the throttle command so that acceleration does not exceed 3 g's. If the vehicle is sufficiently heavy, or if main engine performance is degraded, the 3-g limit may not be reached before MECO. In this case, the third phase would not be entered, and the second phase would continue until MECO. Also, if an engine failure occurred during the third phase, the acceleration would decrease, and the second phase would be re-entered.

The hardware utilized in second-stage ascent is the same as that used in first stage except that the body-mounted accelerometers are not used and the elevons are held in place.

Second-stage navigation is exactly the same as for first stage.

At MECO minus ten seconds, the MECO sequence begins and three seconds later the main engines are commanded to begin throttling at 10-percent thrust/second to 65 percent thrust. This is held for 6.7 seconds, and the main engines are then cut off.

External tank separation is performed automatically by the onboard computers; however, the flight crew can command it through the flight deck display and control panel. The automatic sequence is initiated by the GN&C moding, sequencing, and control when the main engine operations sequence has determined that all of the engines have shut down and the MECO CONFIRMED flag is set.

The main engine operations sequence is checked and if set and shutdown commands have been issued for that engine, its prevalves are closed after a delay. A MECO CONFIRMED flag is set after it has been confirmed that all engines have entered the shutdown phase. Also, a flag is set for the external tank separation sequence after the prevalves for all engines have been commanded closed. This is necessary because shutdown times may differ.

First, the computer operation program determines the mode of separation of if the separation is to be manually inhibited. It then arms the umbilical plate pyro initiator controllers, arms and fires the external tank tumble system after all the engine prevalves have been commanded closed, closes the feedline disconnect valves, gimbals the main engine nozzles to the proper position, deadfaces the external tank-orbiter interface, and unlatches and retracts the umbilical plates.

The sequence then arms the structural separation pyro controllers, performs some limit tests on certain body rates, and tests for feedline disconnect valve closure. If any test is not satisfied, the separation is inhibited and can occur only if the

 **Space Shuttle Spacecraft Systems**

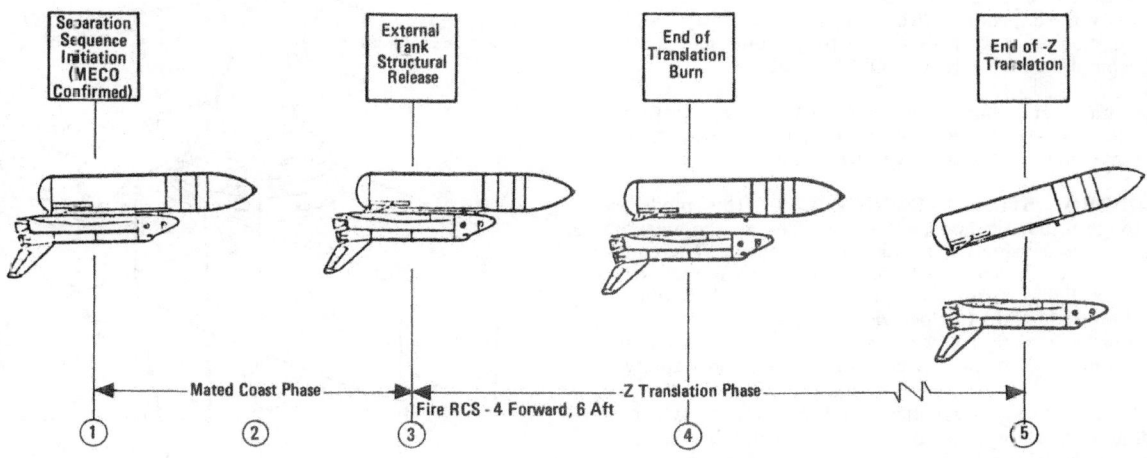

External Tank Separation

out-of-tolerance parameter comes back within tolerance or if the flight crew elects to continue the separation by overriding the inhibit. When either of these conditions is satisfied, the structural separation pyro initiators are fired.

After initiation of the orbiter-external tank separation sequence, there is an approximate 11-second mated coast and then the orbiter and external tank separate. The external tank tumble system is activated approximately two seconds after MECO to produce a tumble rate of 10 to 50 degrees per second after separation. The external tank is on a suborbital trajectory that results in an impact location in the Indian Ocean in KSC launches and in the South Pacific in Vandenberg AFB launches. External tank breakup nominally will occur during its entry into the earth's atmosphere at approximately 56,388 meters (185,000 feet) altitude.

At MECO, orbiter attitude commands (roll, pitch, and yaw) are frozen, and body rate damping is maintained during the coast period by the RCS. Just prior to external tank structural release, the RCS is inhibited, then reenabled immediately after external tank structural separation to an inertial attitude hold, followed by an RCS minus-Z translation maneuver. Four forward and six aft RCS jets are used to achieve a translation of 1.2 meters per second (four feet per second) vertically to ensure orbiter clearance from the arc to the rotating external tank. The orbiter continues to coast away from the external tank in the inertial altitude hold mode to obtain additional vertical

Space Shuttle Spacecraft Systems

clearance. When the orbiter has gained a velocity of 1.2 meters per second (four feet per second) vertically, the separation is flagged complete and the ground is responsible for issuing a go/no go for the impending OMS-1 thrusting sequence. OMS-1 time of ingition is set for MECO plus two minutes and is normally accomplished with the two OMS engines.

At launch and during the pre-MECO phase, the OMS engine nozzles are stowed to protect the nozzles from aerodynamic and thermal loads during ascent.

CONTROL STICK STEERING (CSS). This mode is similar to the automatic mode, except the flight crew commands three-axis motion using the rotation hand control. The computers process the commands from the RHC and motion sensors and the flight control system interprets the RHC motions (fore and aft, right and left, clockwise and counter-clockwise) as rate commands in pitch, roll, or yaw and process the flight control law (equations) to enhance control response and stability.

CSS pitch or roll/yaw modes are selected by depressing the applicable pushbutton indicator at the commander's or pilot's eyebrow/glareshield control and display panel. The pushbutton depressed will illuminate at both stations, downmoding that axis from automatic to CSS.

Three rotation hand controller's are provided, one each at the commander, and pilot stations and a third at the aft flight deck display and control panel. Each RHC moves about three axes. The RHC deflection produces a rotation in the same direction as the crew member's line of sight. Each RHC has a deadband of 0.25 degree in all three axes. To move the RHC beyond the deadband, an additional force is required. At an amount of deflection called the soft stop, a step increase in the force required for further deflection occurs. When a software detent position is exceeded, the RHC assumes manual control. If a malfunction occurs in the left (commander's) RHC or right

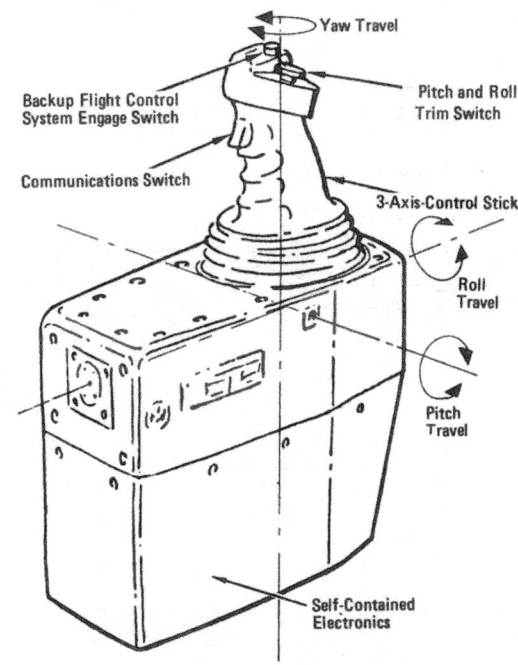

Rotation Hand Control

(pilot's) RHC, it illuminates a red left or right RHC caution/warning light on the display and control panel.

The aft RHC is used only in orbit, for rendezvous and docking with an unmanned vehicle. Software performs the necessary transformations on the aft RHC commands to make

Space Shuttle Spacecraft Systems

them common to the commander's and pilot's RHC in rotation of the same direction relative to the crew member's line of sight.

Each RHC contains nine transducers: three redundant transducers sense pitch deflection, three sense roll deflection, and three sense yaw deflection. The transducers produce an electrical signal proportional to the defelction of the RHC. The three transducers are called Channels 1, 2, and 3; the channel selected by redundancy management provides the command. Each channel is powered by a separate power supply in the associated display driver unit.

MANUAL THRUST VECTOR CONTROL. This capability is provided during first- and second-stage ascent. This is accomplished by substituting the inputs from the RHC for the automatic commands from guidance to gimbal the main engines and SRB's during first stage or the main engines during second stage. The DAP remains active to process the flight crew's input. This is referred to in this mission phase as manual thrust vector control (MTVC). MTVC is available at liftoff (SRB ignition command plus 0.365 second). One of the CSS pushbuttons at the glareshield/eyebrow panels must be depressed before MTVC is available.

Once MTVC is activated, the vehicle is in a rate-command/attitude-hold mode in all axes. When the RHC is in detent, with MTVC selected, the vehicle is in attitude hold. The DAP holds the vehicle in the attitude it had when the RHC was in detent. A rate command equal to zero replaces the rate command generated in guidance and control steering. The attitude error for the DAP is computed by integrating the rate gyro assemblies measured rates. However, there are limits on vehicle rates and attitude errors. If the limits are exceeded when attitude hold is requested by placing the RHC in detent, attitude hold will not be initiated until the rates and errors are within limits.

When the RHC is removed from detent, a rate command proportional to the amount of deflection replaces the rate com-

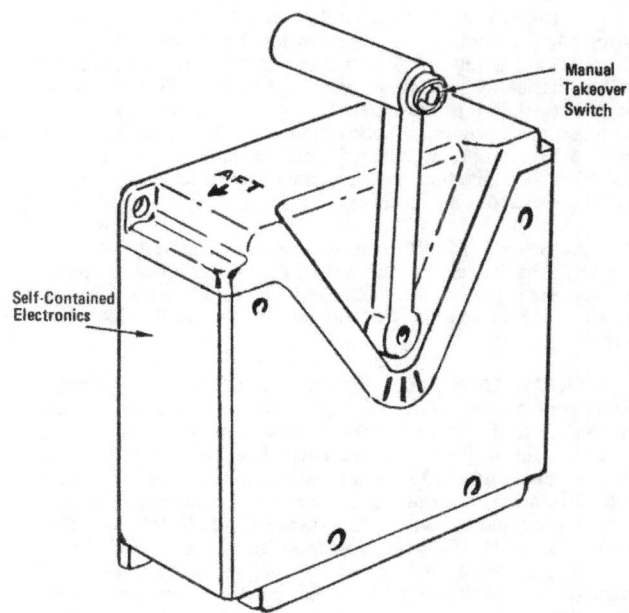

Speed Brake Thrust Control

mand previously generated. The attitude error is zeroed. The larger the deflection of the RHC, the larger the command. The flight control sytsem compares these commands with inputs from the rate gyro assemblies and accelerometer assemblies (what the vehicle is actually doing—motion sensors) and generates control signals to produce the desired rates. When the commander or pilot release the RHC, it returns to center and the vehicle will maintain in its present attitude (zero rates).

 **Space Shuttle Spacecraft Systems**

MANUAL THROTTLING. Throttling of the main engines also is possible through use of the speed brake/thrust control (SBTC) at the commanders or pilots station. Manual throttling is accomplished by depressing the takeover switch on the SBTC. The automatic thrust command is frozen until the MAN (manual) white light in the SPD BK/THROT pushbutton at either station is illuminated. This means the SBTC has matched the frozen automatic command. The takeover switch on the SBTC is then released and the pilot has control of throttling. This method prevents undesirable transients in throttle commands during takeover.

At the forward setting, thrust level is the highest. Rotating the SBTC back decreases the thrust. Each SBTC contains three transducers (Channels 1, 2, and 3) which produce a voltage proportional to deflection. Redundancy management selects the output.

TRANSITION DAP. The transition DAP flight control mode controls the orbiter in response to automatic or manual commands during insertion and deorbit. The effectors used to produce control forces and moments on the orbiter are the two OMS engines and the 38 primary RCS engines. At main engine cutoff, rotation and translation control of the spacecraft is provided by commands issued to the forward and aft RCS jets. The forward and aft RCS also provide attitude control and three-axis translation during external tank separation, on-orbit maneuvers and roll control for a single OMS engine operation. The OMS provides propulsive and three-axis control for orbit insertion, orbit circularization, orbit transfer, and rendezvous. Failure of a single OMS engine will not preclude a nominal orbit insertion.

Normally, the first OMS thrusting period raises the low elliptical orbit following external tank jettison and the second OMS thrusting places the spacecraft into a circular orbit as designated for that mission. For orbital maneuvers utilizing the OMS, any delta velocities greater than 1.8 meters per second (6 feet per second) utilize the two OMS engines.

Orbit insertion guidance, navigation, and flight control software, through the transition DAP, controls the external tank/orbiter separation maneuvering, OMS-1 and OMS-1 ignition attitude, OMS thrusting commands, OMS engine gimbaling for thrust vector control, and RCS thrusting commands in conjunction with the use of the ORBITAL DAP control panels.

Two ORBITAL DAP pushbuttons provide automatic and manual modes to the transition DAP program. The pushbuttons are illuminated by software commands in accordance with the mode selected and accepted by flight control software.

Navigation incorporates IMU delta velocities during powered flight and coasting flight to produce orbiter state (position and vector).

The transition DAP reconfiguration logic controls the moding, sequencing, and initialization of the control law modules and sets gains, deadbands, and rate limits. The steering processor is the interface between the guidance or manual steering commands and the transition DAP. The steering processor generates commands to the RCS processor, which generates the RCS jet commands required to produce the commanded spacecraft translation and rotation using attitude and rotational rate signals, or translation or rotation acceleration commands. The OMS processor generates OMS engine gimbal actuator thrust vector control commands to produce desired spacecraft/engine relationship for commanded thrust direction.

When in the automatic DAP mode, the external tank separation module initialized by the external tank separation sequencer compares Z delta velocities from the DAP attitude processor with an initialized-loaded desired Z delta velocity. Before this value is reached, the transition DAP and steering

Space Shuttle Spacecraft Systems

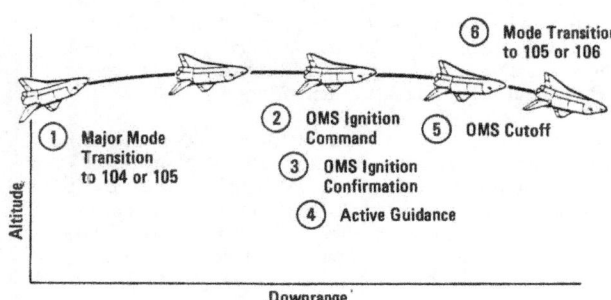

OMS-1/OMS-2 Sequence

processor sends commands to the RCS jet selection logic which uses a table look-up technique for the primary RCS jets and commands 10 RCS jets to fire in the plus Z direction. When the desired Z delta velocity is reached, the translation command is set to zero. Rotation commands are permitted during the external tank separation sequence.

The RCS jet selection module receives both RCS rotation and translation commands and, using the table look-up technique, outputs 75 discretes to the primary RCS jets to turn the 38 jets on or off.

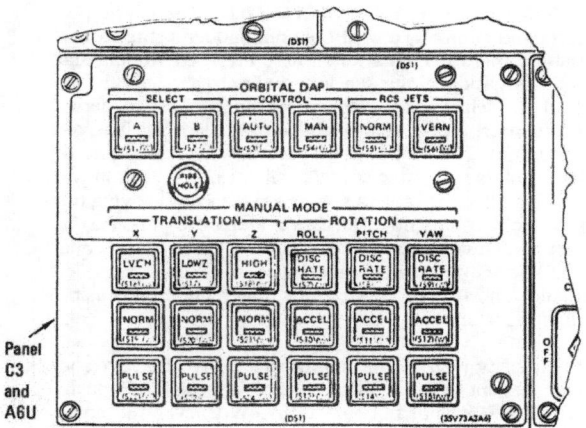

Orbital Digital Autopilot Pushbuttons

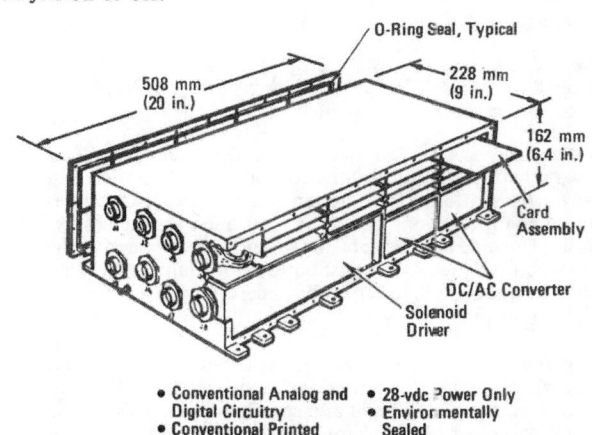

- Conventional Analog and Digital Circuitry
- Conventional Printed Circuit Cards
- 28-vdc Power Only
- Environmentally Sealed
- Cold-Plate-Cooled

	Weight, kg (lb)	Power (LRU Only), W	
		Normal Quiescent	Maximum
Aft Unit	15.0 (33.0)	27.4	62.6
Forward Unit	13.8 (30.4)	20.3	41.4

Reaction Jet Driver

Space Shuttle Spacecraft Systems

The RCS reaction jet driver forward and aft assemblies provide the turn-on/turn-off jet selection logic signals to the RCS jets. There is also a driver redundancy management program that permits only "good" RCS jets to be turned on. The RCS JET yellow caution/warning light on the flight deck display and control panel indicates a failed on, failed off, or leaking RCS jet.

The 19 RCS jets thrusting in the plus or minus Z direction provides Z translation and roll or pitch rotation control and are considered independent of other axes. The 12 RCS jets thrusting in the plus or minus Y direction provide Y translation, and yaw rotation only. The seven RCS jets thrusting in the plus or minus X direction provide X translation only.

Before the first OMS ignition, the spacecraft is maneuvered to the OMS ignition attitude by using the RHC and RCS jets; this reduces transient fuel losses.

ORBITAL DAP. The two orbital DAP control panels have two sets of pushbuttons, one set for manual rotation and one set for translation. There are nine pushbuttons for translation and nine for rotation. Those for translation are on axis-by-axis basis. The nine pushbuttons for rotation are for each of the three rotation body axes. The pushbuttons are illuminated in accordance with the mode selected and accepted by flight control software commands.

For the OMS thrusting period, the orbital state (position and vector) is produced by navigation incorporating IMU delta velocities during powered flight and coasting flight. This state is sent to guidance, which uses target inputs through the CRT to compute thrust direction commands and commanded attitude for flight control and thrusting parameters for CRT display. Flight control converts the commands into OMS engine gimbal angles for an automatic thrusting period. OMS thrust vector control for normal two-engine thrusting is entered by depressing the ORBITAL DAP AUTO pushbutton with both RHC's within software detents. OMS manual TVC for both OMS engines is entered by depressing the ORBITAL MAN DAP pushbutton or by moving the commander or pilot RHC out of any one of the three detents. The manual RHC rotation requests are converted into gimbal angles. OMS thrust in either case is applied through the spacecraft center of gravity.

Automatic thrust vector control for one OMS engine is identical to that for two OMS except that the RCS processor is responsible for roll control. Single-OMS engine thrust also is through the spacecraft's center of gravity, except when pitch or yaw rate commands are non-zero. If the left or right OMS engine fails, an OMS TVC red light on the flight deck display and control panel will illuminate.

Since an OMS cutoff is based on time rather than velocity, a velocity residual may exist following OMS cutoff. The residual is zeroed using the RCS via the THC.

TRANSLATION HAND CONTROLS. There are two translation hand controls, one at the commander's station and one at the aft flight deck station. These are used for manual control of translation along the longitudinal (X), lateral (Y), and vertical (Z) vehicle axes. The commander's THC is active during orbit insertion, in orbit, and during deorbit. The aft flight deck station THC is active only during the orbital phase. Each THC contains six three-contact switches, one each in the plus and minus directions for each of the three axes. Moving the THC to the right commands translation along the plus Y axis, closes three-switch contacts (referred to as Channels 1, 2, and 3), and the selected redundancy management channel provides the command. The aft THC is used when the flight crew is using the aft flight deck station.

In the transition DAP, the commander's THC is active and totally independent of the flight control orbital DAP pushbuttons or the RHC position or status. Whenever the commander's THC is out of detent, plus or minus X, Y, or Z,

Space Shuttle Spacecraft Systems

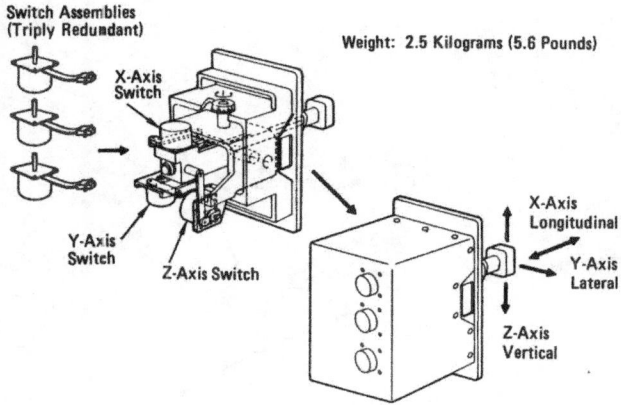

Translation Hand Control

translation acceleration commands are sent directly to the RCS jet selection logic for continuous RCS jet firings. Rotational commands may be sent simultaneously with translation commands within the limits of the RCS jet selection logic; if both plus X and minus Z translations are commanded simultaneously, plus X translation is given priority.

The attitude processor computes attitude from IMU gimbal angles and attitude errors from attitude and commanded attitude and sends these to the ADI's. The orbiter rate gyro assemblies provide body rates to flight control and ADI's.

OMS-2 is entered manually by the flight crew after OMS-1 and is similar to OMS-1 except for the different orbital parameters. Insertion is entered after the OMS-2 thrusting period. An orbital coast is achieved at this point.

ORBITAL OPERATIONS. Orbital guidance, navigation, and flight control software includes coast flight, rendezvous targeting, proximity operations, precision and rendezvous navigation, orbital maneuver guidance and flight control, and additional universal pointing and navigation display.

Navigation incorporates a drag model during coasting flight on IMU delta velocities during powered flight together with a gravity model to produce the orbiter state. This state is maintained primarily by navigation for entry; the only user in orbit is the local-vertical/local-horizontal function of universal pointing. For automatic rotations, attitude requests are input through the flight deck display and control panel CRT and attitude and rate information from flight control goes to guidance. Guidance sends attitude and rate commands to flight control and displays attitudes, errors, and rates on the CRT. Flight control converts these commands into RCS thrusting commands.

The orbital digital autopilot also can be operated in automatic or manual modes. AUTO or MAN pushbuttons are used to mode the RCS for automatic or manual rotation operation. The ORBITAL DAP SELECT A or B pushbuttons select the values the DAP will use from the DAP configuration parameter limits CRT display software loads. The values of attitude deadband, rate deadband, and vehicle change in rotation rate due to minimum impulse RCS thrusting are a function of DAP A or B selection and jet selection NORM (normal-primary) or VERN (vernier). Vernier selection prohibits all translation commands.

The jet selection logic uses a table look-up technique for the primary jets similar to the transition DAP. Unlike the transition DAP, the orbital jet selection logic checks for high or low rotation mode for both pitch and yaw for the forward and aft RCS jets. There are special tables for the high plus Z translation mode not used by transition DAP. Angular rate increments are

 **Space Shuttle Spacecraft Systems**

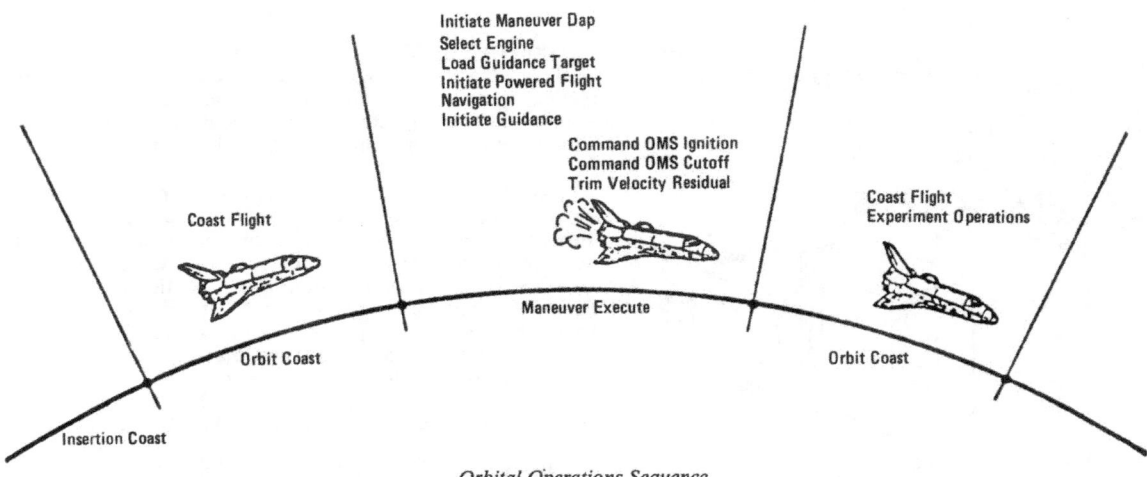

Orbital Operations Sequence

predicted for each group of primary jets and for each specific vernier jet. The primary jet predictions are based on one nominal inertia matrix and center of gravity. The vernier predictions are based on nominal or one of five different payload configuration with the manipulator arm extended) inertia matrixes. The rotation compensation logic uses a combination threshold set on the DAP specialist display and counter (or accumulator) in determining when to command RCS jets on for rotation compensation.

The AUTO pushbutton permits the automatic RCS rotation and movement of any RHC out of detent (any axis) downmodes the DAP from automatic to manual as will depressing the MAN pushbutton. All active RHC's must be in detent when an AUTO pushbutton is depressed in order to upmode to automatic RCS rotation mode. The control pushbuttons are illuminated by software commands in accordance with the mode selected and accepted by flight control software.

Automatic rotation commands are supplied by the universal pointing processor. The universal pointing processor via the operational sequence display provides: three-axis automatic maneuver, tracking local-vertical/local-horizontal about any body vector, rotating about any body vector at the DAP discrete rate, and stopping any of these options and commanding attitude hold. The parameters of these maneuvers are

Space Shuttle Spacecraft Systems

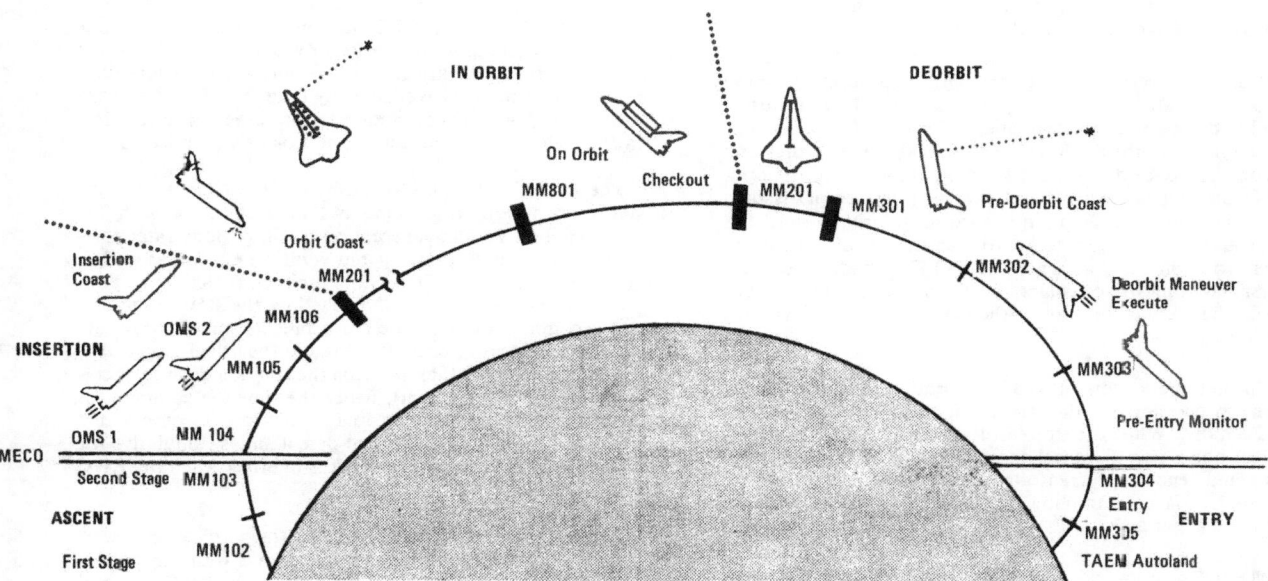

Orbit Insertion, Orbital Operations, Deorbit

displayed in current attitude, required attitude, attitude error, and body rates. Either total or DAP attitude errors may be selected for the display and the ADI error needles.

The automatic maneuver option calculates a commanded vehicle attitude and angular rate or to hold a vehicle attitude. The desired inertial commanded and rotation attitude is input into the operational sequence display in pitch, yaw, and roll. When the maneuver option is selected, universal pointing sends the required attitude increment and body rate to flight control, and flight control performs the maneuver when the DAP is in automatic.

The automatic rotation option calculates a rotation about a desired body axis. This option is used for passive thermal control, also known as barbecue. Pitch and yaw body components of the desired rotational axis are first input. The orbiter is maneuvered automatically or manually so that the rotational

Space Shuttle Spacecraft Systems

axis is oriented properly in inertial space. When the rotational option is selected, universal pointing will calculate the required body attitude and send it to flight control. Flight control performs the maneuver if the DAP is in automatic.

The local-vertical/local-horizontal automatic option calculates the attitude necessary to maintain LVLH with a desired body orientation. It calculates the attitude necessary to track the center of the earth with a given body vector. Pitch and yaw body components of the desired pointing vector are input first. Omicron (roll angle about the pointing vector) is then input. When the LVLH option is selected, the required LVLH attitude and the maneuver to get to that attitude are calculated and sent to flight control. When the LVLH attitude is reached, universal pointing will calculate an attitude and send it to flight control. Flight control performs the maneuver if the DAP is in automatic.

The automatic stop/attitude hold option cancels universal pointing processing of the automatic mauver, rotation, or LVLH options. When the stop/attitude hold option is selected, universal pointing will cancel the processing of the maneuver options and send the current attitude to flight control. When flight control is in automatic, attitude hold will be initiated about the current attitude.

Manual rotation commands are controlled by the three RHC's and either of the two orbital DAP control panels.

The rotation discrete rate pushbuttons select the "DAP-load" rates in effect (one for normal and one for vernier) designating specific rotational rates in the manual rotation mode. These body rates are commanded as long as the RHC is between the detent and the software softstop for the axis selected. Attitude hold is commanded when the RHC is in the detent, and the attitude reference used is the attitude sensed at the time the RHC reached the detent. Free drift exists on the affected axis between the softstop and the detent when the RHC is returning to the detent after having been past the softstop.

The ROTATION ACCEL pushbutton changes the mode to acceleration (continuous thrusting) when depressed and the RHC moved out of the affected axis detent in the manual rotation mode. This mode is available any time the RHC is past the software stop, regardless of mode selected. Free drift is in effect when the RHC is in the detent of the axis or axes selected.

The ROTATION PULSE pushbuttons enable a one-time rotation (open-loop) rate command in accordance with the value selected in the manual rotation moding. Body rate increments are set for both primary and vernier RCS jets as well as control acceleration available for each body axis for both primary and vernier RCS jets. Each time the RHC is out of detent, a command is generated at a predetermined body rate about a selected axis or axes. It is used in the calculation of the time required for the orbiter to reach the commanded rate increment (or rate increment sum), hence the time the command is active to the RCS jet selection logic. This rate is commanded each time the RHC is moved out of detent and is cumulative for multiple RHC commands. Free drift is in effect when the RHC is below the softstop.

Rotation options may be mixed as they are made axis by axis. Requests for changes in rotation options with the manual mode pushbuttons are not recognized when the RHC is out of detent.

Manual translation commands are controlled by the two THC's and either of the two ORBITAL DAP control panels.

There are two translation acceleration options: high and normal. The high mode commands all RCS jets in that axis to thrust, whereas in the normal mode, only selected RCS jets thrust. The moding is a function of the jet selection logic software and cannot be changed by the flight crew. Applicable RCS

Space Shuttle Spacecraft Systems

jets will thrust continuously as long as the THC is out of detent in one or more axes.

The TRANSLATION PULSE pushbuttons select a specific delta velocity in feet per second each time the THC is moved out of the detent for the axis or axes concerned in the manual translation mode. The commands are cumulative, and the incremental delta velocity is the value set on the DAP specialist display.

The attitude processor combines commanded attitude with the reference from the IMU to form attitude errors, which are sent to the ADI along with attitude. Body rates from flight control also are sent to the ADI.

STAR TRACKERS. The star trackers provide orbiter's orientation with respect to stars. The two star trackers are used in orbit to align the IMU's. The star trackers are located just forward of the crew compartment on the left side of the orbiter in a well on the extension of the navigation base on which the IMU's are mounted. They are slightly inclined off the vehicle negative Y and negative Z axes.

Each star tracker has a door which is closed during ascent and entry and opened in orbit. Each door rotation is driven by two electric motors. Limit switches stop the motors and drive a talkback when the door has reached limits of full travel, open or closed. Both doors have thermal protection to prevent heat leaks into the star tracker well during entry.

The star trackers are a strapped-down, wide field of veiw, image-dissector, electro-optical tracking device. Their major function is to search for, acquire, and track the 50 brightest navigation stars. By knowing the relationship of the star tracker to the orbiter and the location of the star in space, a line-of-sight vector from the orbiter to the star is defined. Two line-of-sight vectors define the orbiter inertial attitude. The star trackers align the IMU's and provide angular data from the orbiter to a target. Star tracker data is sent to the computers via the MDM's.

The difference between the inertial attitude defined by the star tracker and IMU is processed by software and results in IMU torquing angles. If the IMU gimbals are torqued or the matrix defining its orientation is recomputed, the effects of the IMU gyro drift are removed and the IMU is restored to its inertial attitude. If the IMU alignment is in error by more than 0.5 degree, the star tracker is unable to acquire and track stars correctly. This is because the angles the star tracker is given for searching are based on the current knowledge of the orbiter attitude, which is based on IMU gimbal angles. If that attitude is greatly in error, the star tracker may acquire and track the wrong star. In this case, the crew optical alignment sight must be used to re-align the IMU's to within 0.5 degree and then the star trackers used to re-align the IMU's more precisely.

The star tracker includes a light shade assembly and an electronics assembly mounted on top of the navigation base. The light shade assembly defines the field of veiw of the tracker (10 degrees square). It contains a shutter mechanism which can be opened manually by the crew using an entry on the CRT display. It may be opened and closed automatically by a bright object sensor. The bright object sensor reacts before a bright object such as the sun or moon could cause damage to the star tracker (the sensor has a larger field of view than the star tracker shutter).

The electronics assembly contains an image dissector tube mounted on the underside of the navigation base. The star tracker itself does not move. The field of view is scanned electronically. It may be commanded to scan the entire field of view or a smaller offset field of view (1 degree square) about a point defined by horizontal and vertical offsets. When an object of proper intensity and in the correct location is sensed, it is tracked. Star tracker outputs are the horizontal and vertical position within the field of view of the object being tracked and

 **Space Shuttle Spacecraft Systems**

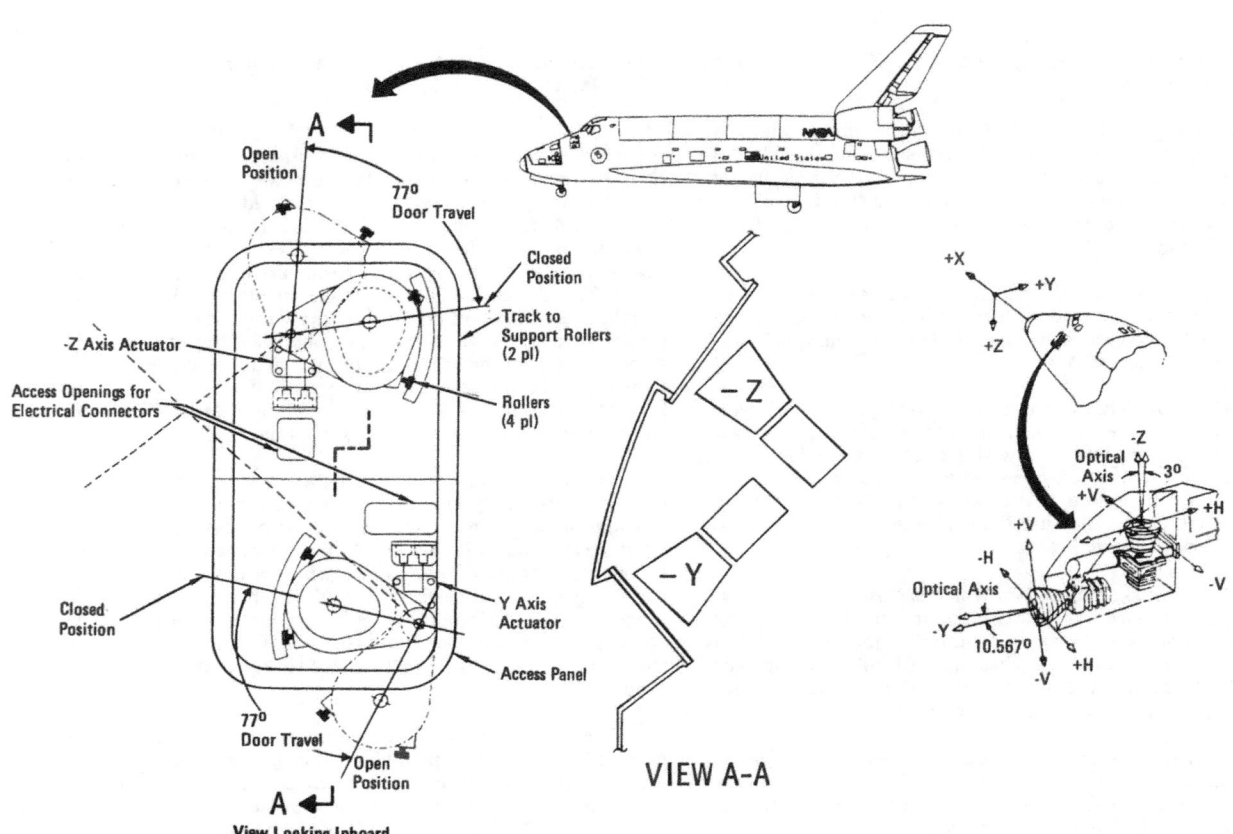

Star Trackers

Space Shuttle Spacecraft Systems

its intensity. The orbiter may have to be manually maneuvered to position the object in the star tracker field of view. If the flight crew cannot sight through the star tracker. They may use the COAS in its negative Z mount to verify the negative Z star tracker is tracking the correct star.

There is no redundancy management for the star tracker assemblies; they operate independently and either can do the whole task. They can be operated either singly or both at the same time.

The warmup time is 15 minutes. The door-open time with two electric motors is 60 seconds and with one motor is 120 seconds.

CREW OPTICAL ALIGNMENT SIGHT. This is used as a backup to the star trackers for aligning the IMU's. A secondary use of the COAS is as the primary optical docking device to measure range and rotational rates and allow crew members to align the vehicles and permit docking. The COAS also allows the flight crew to reassure themselves of proper attitude orientation during deorbit thrusting periods.

The COAS is an optical device with a reticle projected on a combining glass focused on infinity. A light bulb with variable brightness illuminates the reticle.

The COAS may be located in two locations. There are two mounts, one over the positive X commander's window, and one next to the aft flight deck right overhead negative Z window.

After mounting the COAS on desired location, the crew member must manually maneuver the orbiter until the selected star is in the field of view. The crew member maneuvers the orbiter so that the star will cross the center of the reticle. At the instant of the crossing, the crew member "makes a mark," which means he depresses the ATT REF (attitude reference) pushbutton. At the time of the mark, software stores the gimbal angles of the three IMU's. The mark can be taken again if it is felt the star was not centered as well as it could have been. When the crew member feels a good mark was taken, the software is notified to accept it. Good marks for two stars are required for an IMU alignment.

By knowing the star being sighted and the COAS location and mounting relationship in the orbiter, software can determine a line-of-sight vector from the COAS to the star in an inertial coordinate system. Line-of-sight vectors to two stars define the attitude of the orbiter in inertial space. This attitude can be compared to the attitude defined by the IMU's and, if the IMU's are in error, they can be realigned to the more correct orientation by the COAS sightings.

An additional COAS is used at the aft flight deck station to check the alignment of the payload bay doors.

The COAS requires 115-volt ac power for reticle illumination. The COAS is 241 x 152 x 109 millimeters (9.5 x 6.0 x 4.3 inches) and weighs 1.13 kilograms (2.5 pounds).

ORBITAL OPERATIONS. An orbital operations sequence/major mode is provided for orbital checkout of the orbiter systems used during entry. The activities are performed the day before deorbit and takes about 15 minutes. The system checkout is performed in two parts. The first part contains tests requiring one of the APU/hydraulic systems, the flight control system repositioning of the left and right main engines for entry, and cycling the aerosurfaces, hydraulic motors, and hydraulic switching valves. After the checkout, the APU is deactivated. The second part consists of a check of all the flight crew dedicated displays; self-test of the microwave scan beam landing system, TACAN, accelerometer assemblies, radar altimeter, rate gyro assemblies, and air data transducer assemblies; and

 **Space Shuttle Spacecraft Systems**

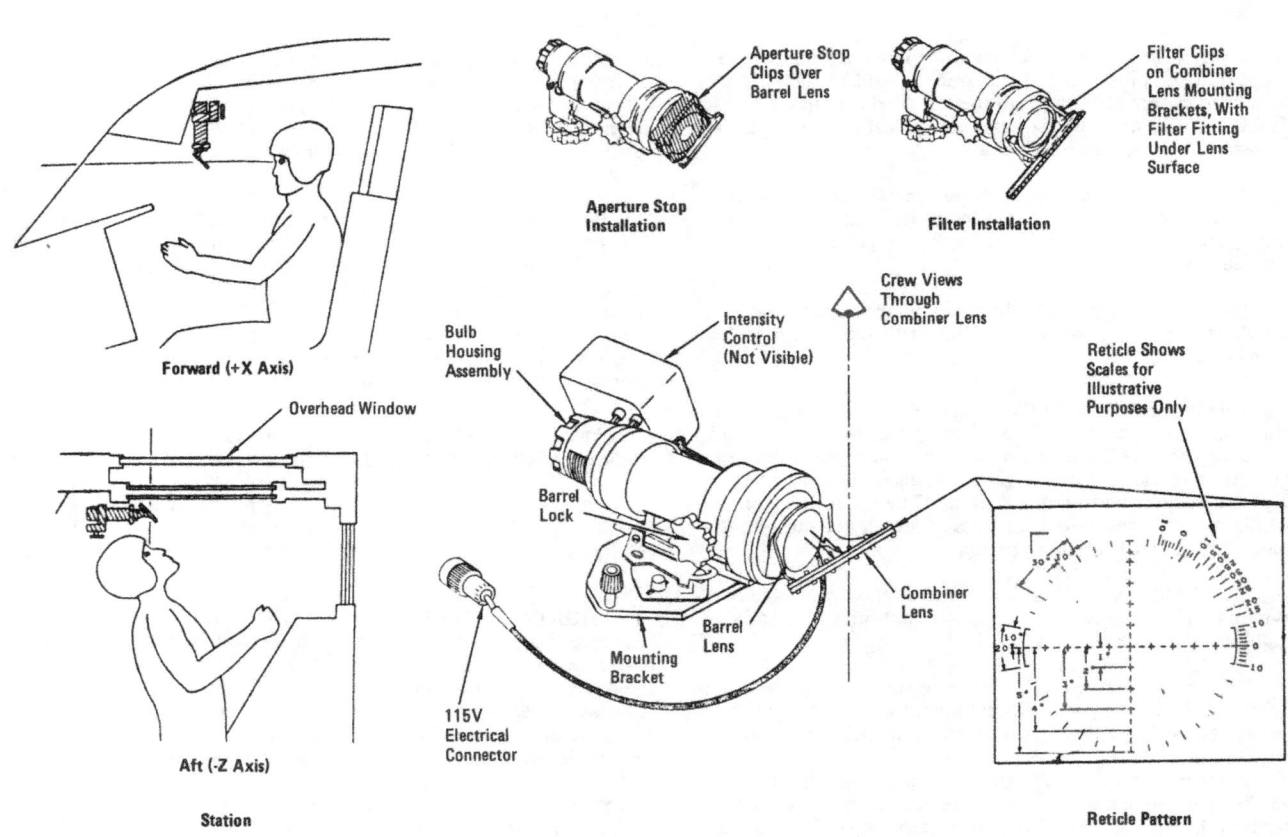

Crewman Optical Alignment Sight

Space Shuttle Spacecraft Systems

check of the flight crew controllers, rotation hand control, rudder pedal transducer assembly, speed brake, trim switch-panel, rotation hand control trim switches, speed brake takeover button, and mode/sequence pushbutton indicators.

DEORBIT. Deorbit guidance, navigation, and flight control software, through the transition DAP, provides maneuvering of the spacecraft to the OMS-1 and OMS-2 ignition attitude, OMS thrusting commands, OMS engine gimbaling for thrust vector control, and RCS thrusting commands in conjunction with the use of the orbital DAP control panels similar to orbit insertion.

The deorbit sequence reduces orbital velocity so that the orbiter enters the atmosphere for landing. The deorbit thrusting is nominally with two OMS engines and must establish the proper entry velocity and range conditions. It is possible to downmode to one OMS engine or plus X aft RCS jets for deorbit. The orbiter is positioned to a retrograde, tail-first thrusting attitude for deorbit.

Approximately four hours before deorbit, the ECLSS radiator bypass/flash evaporator system checkout is performed as the flash evaporator is used for cooling of the Freon-21 coolant loops when the ECLSS radiators are deactivated and the payload bay doors are closed. The high-load flash evaporator cools the ECLSS Freon-21 coolant loops until the ECLSS ammonia boilers are activated by the computers at approximately 42,672 meters (140,000 feet) altitude during entry. The orbiter IMU's are aligned, the star trackers deactivated,

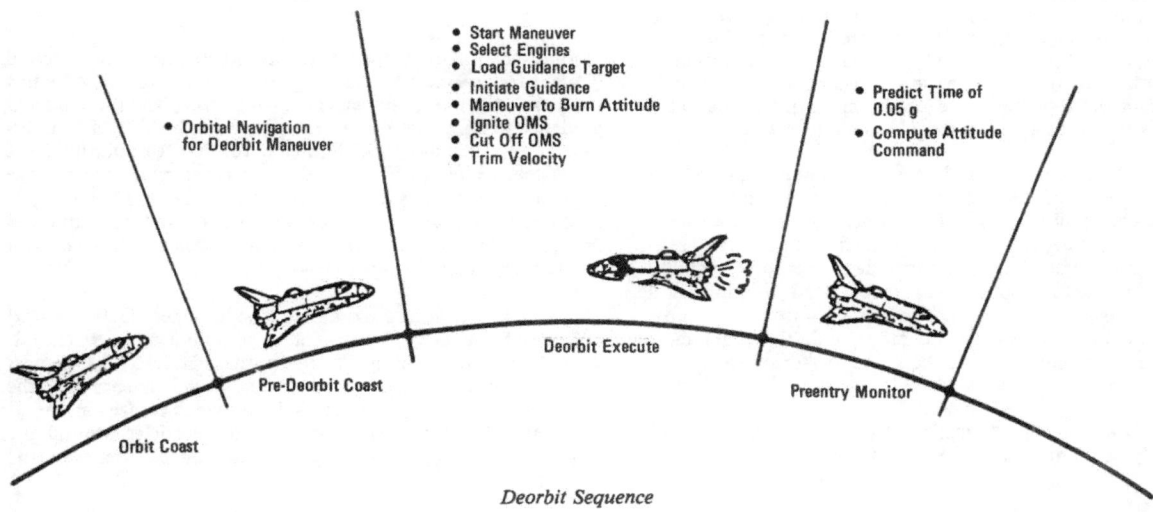

Deorbit Sequence

Space Shuttle Spacecraft Systems

and the star tracker doors closed. At about one hour before deorbit, the flight crew takes their seats and dons helmets. The spacecraft is then manually maneuvered utilizing the RCS jets to the deorbit attitude (retrograde). About 30 minutes before deorbit, the OMS preparations for the deorbit thrusting are accomplished. They consist of OMS thrust vector control gimbal checks, OMS system data checks, vent door closure and single APU start. The OMS deorbit thrusting consists of one OMS thrusting period. At the completion of the thrusting, the flight crew manually maneuvers the spacecraft to orient the nose first for the entry sequence utilizing the RCS jets. The forward RCS remaining propellants are dumped through the forward RCS engines, if required and the two remaining APU's are started and remain operating through entry and landing rollout along with the APU started before entry. The spacecraft hydraulic fluid system thermal conditioning also begins, if required.

ENTRY THROUGH LANDING. Guidance, navigation, and flight control software for the entry phase is initiated by the flight crew approximately five minutes before entry interface (EI), and the DAP processes the flight control laws in response to automatic or manual commands and feedback sensor signals that indicate the dynamic state of the spacecraft. The entry interface is the point at which aerodynamic forces are sensed.

RCS jets (pitch, roll, and yaw) are commanded and actuators for aerosurfaces (elevons, body flap, rudder/speed brake) are commanded once sufficient aerodynamic pressure is sensed on the surface to accomplish acceleration, rate, and attitude response. Commands would include turn control, stability augmentation, aerodynamic load relief, load limiting, and ground landing rollout control. The flight control system normally would be in an automatic mode with the flight crew primarily monitoring the operation and performance of the GN&C systems.

EI is at about 121,920 meters (400,000 feet) altitude, approximately 4,400 nautical miles (5,063 statute miles) from the

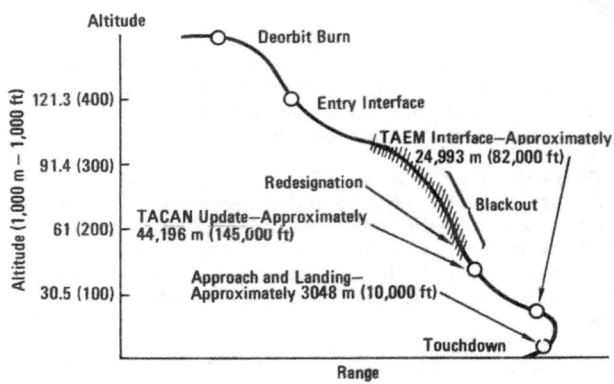

Entry Flight Profile

landing site, and at approximately 7,620 meters per second (25,000 feet per second) velocity. It continues to entry-TAEM (terminal area energy management) interface. The fundamental guidance requirement during entry is to reach the TAEM interface at approximately 25,298 meters (83,000 feet) altitude, 762 meters per second (2,500 feet per second) velocity, and on range 52 nautical miles (59 statute miles) from the landing runway and at a time interval between 34:45 minutes and 5:32 minutes before landing within a few degrees of tangency with the nearest heading alignment cylinder (HAC).

Before EI, at approximately 97,536 meters (320,000 feet) altitude, the spacecraft enters into a communications blackout. This lasts until an altitude of approximately 54,800 meters (180,000 feet). Between these altitudes, heat is generated as the spacecraft enters into the atmosphere, which ionizes atoms of air, which form a layer of ionized gas particles around the spacecraft. During this period, radio signals between the

Space Shuttle Spacecraft Systems

spacecraft and ground cannot penetrate this sheath of ionized particles, and radio communications is blocked for approximately 16 minutes.

A pre-entry phase begins at 121,920 meters (400,000 feet) altitude in which the spacecraft is maneuvered to zero degrees roll and yaw (wings level) and a predetermined angle of attack for entry. In the equatored landing mission the angle of attack is 40°, in polar orbit landing missions, it will be between 38° and 28°. The flight control system issues the commands to the roll, yaw, and pitch RCS jets for rate damping in attitude hold for entry into the earth's atmosphere until 0.176 g is sensed which corresponds to a dynamic pressure of 517 millimeters of mercury (mmHg) per meters squared (10 pounds per square foot), approximately the point where the aerosurfaces become active.

The forward RCS jets are inhibited at 121,920 meters (400,000 feet) altitude. The aft RCS jets maneuver the spacecraft until a dynamic pressure of 517 mmHg per meters squared (10 pounds per square foot) is sensed which is when the orbiter's ailerons become effective and the aft RCS roll jets are deactivated. At a dynamic pressure of 1,035 mmHg per meters squared (20 pounds per square foot), the orbiter's elevators become effective and the aft RCS pitch jets are deactivated. The orbiter's speed brake is used below Mach 10 to induce a more positive downward elevator trim deflection. At Mach 3.5, the rudder becomes activated and the aft RCS yaw yets are deactivated at 13,716 meters (45,000 feet).

In the entry phase, the RCS commands roll, pitch, and yaw. Lights on the commander's flight deck display and control Panel F6 are used to indicate the presence of an RCS command from the flight control system to the RCS jet selection logic; however, this does not indicate an actual RCS jet thrusting command. The minimum light-on duration is extended to allow the light to be seen even for minimum-impulse RCS jet thrusting commands. After the roll and pitch aft RCS jets are deactivated, the roll indicator lights are used to show that three or more yaw RCS jets have been requested. The pitch indicator lights are used to show elevon rate saturation.

Entry guidance must dissipate the tremendous amount of energy the orbiter possesses when it enters the earth's atmosphere ensuring the orbiter does not burn up (entry angle too steep) or skip out of the atmosphere (entry angle too shallow) and at TAEM interface the orbiter is capable of reaching the desired touchdown point.

During entry, energy is dissipated by the atmospheric drag on the orbiter's surface. Higher atmospheric drag levels enable faster energy dissipation with a steeper trajectory. Normally, the angle of attack and roll angle enable the atmospheric drag of any flight vehicle to be controlled. However, for the orbiter, angle of attack was rejected because it creates surface temperatures above the design specification. The angle of attack schedule used during entry is initialized-loaded as a function of

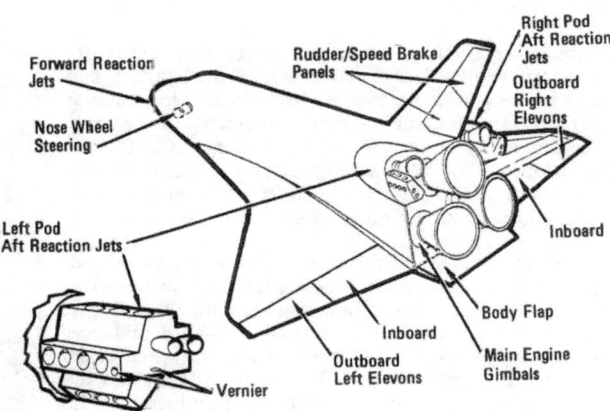

Flight Control Hardware Effector Elements

Space Shuttle Spacecraft Systems

relative velocity leaving roll angle for energy control. Increasing the roll angle decreases the vertical component of lift, causing a higher sink rate and energy dissipation rate. Increasing the roll rate does raise the surface temperature of the orbiter, but not nearly as drastically as an equal angle of attack command.

If the orbiter is low on energy (current range-to-go much greater than nominal at current velocity) entry guidance will command lower than nominal drag levels. If the orbiter has too much energy (current range-to-go much less than nominal at the current velocity) entry guidance will command higher than nominal drag levels to dissipate the extra energy.

Roll angle is used to control crossrange. Azimuth error is the angle between the plane containing the orbiter's position vector and the heading alignment cylinder tangency point and the plane containing the orbiter's position vector and velocity vector. When the azimuth error exceeds an initialized-loaded number, the orbiter's roll angle is reversed.

Thus descent, rate and down ranging are controlled by bank angles, the steeper the bank angle, the greater the descent rate and the greater the drag, conversely the minimum drag altitude is wings level. Cross-range is controlled by bank reversals.

The entry thermal control phase is designed to keep the backface temperatures within the design limits. A constant heating rate is established until below 5,791 meters per second (19,000 feet per second).

The equilibrium glide phase transitions the orbiter from the rapidly increasing drag levels of the temperature control phase to the constant drag level of the constant drag phase. The equilibrium glide flight is defined as flight in which the flight path angle, the angle between the local horizontal and the local velocity vector, remains constant. Equilibrium glide flight provides the maximum downrange capability. It lasts until the drag acceleration reaches 33 ft/second squared.

The constant drag phase begins at 33 ft/second squared. In the equatorial landing missions the angle of attack is initially 40 degrees but it begins to ramp down in this phase to approximately 36 degrees by the end of this phase, in polar orbit landing missions the angle of attack will be between 38° and 28°.

The transition phase is where the angle of attack continues to ramp down, reaching the approximate 14 degree angle of attack at the entry TAEM interface, approximately 25,298 meters (83,000 feet) altitude, 762 meters per second (2,500 feet per second), Mach 2.5 and 52 nautical miles (59 statute miles) from the landing runway. Control is then transferred to TAEM guidance.

It is noted, that during the aforementioned entry phases, the orbiter's roll commands keep the orbiter on the drag profile and control cross range.

Terminal area energy management guidance in entry begins at approximately 25,298 meters (83,000 feet) altitude and 762 meters per second (2,500 feet per second) and terminates at the A/L (approach and landing) guidance phase capture zone, which begins at approximately 3,048 meters (10,000 feet) altitude, at Mach 0.9. The spacecraft attains subsonic velocity at approximately 14,935 meters (49,000 feet) at a range of over 22 nautical miles (25 statute miles) from the landing site.

TAEM guidance steers the orbiter to the nearest of two heading alignment cylinders whose radius are 5,486 meters (18,000 feet), which are located tangent to and on either side of the runway centerline on the approach end. In TAEM guidance, excess energy is dissipated with an S turn and the speed brake can be utilized to modify drag, L/D (lift/drag) ratio, and the flight path angle in high energy conditions. This increases the

Space Shuttle Spacecraft Systems

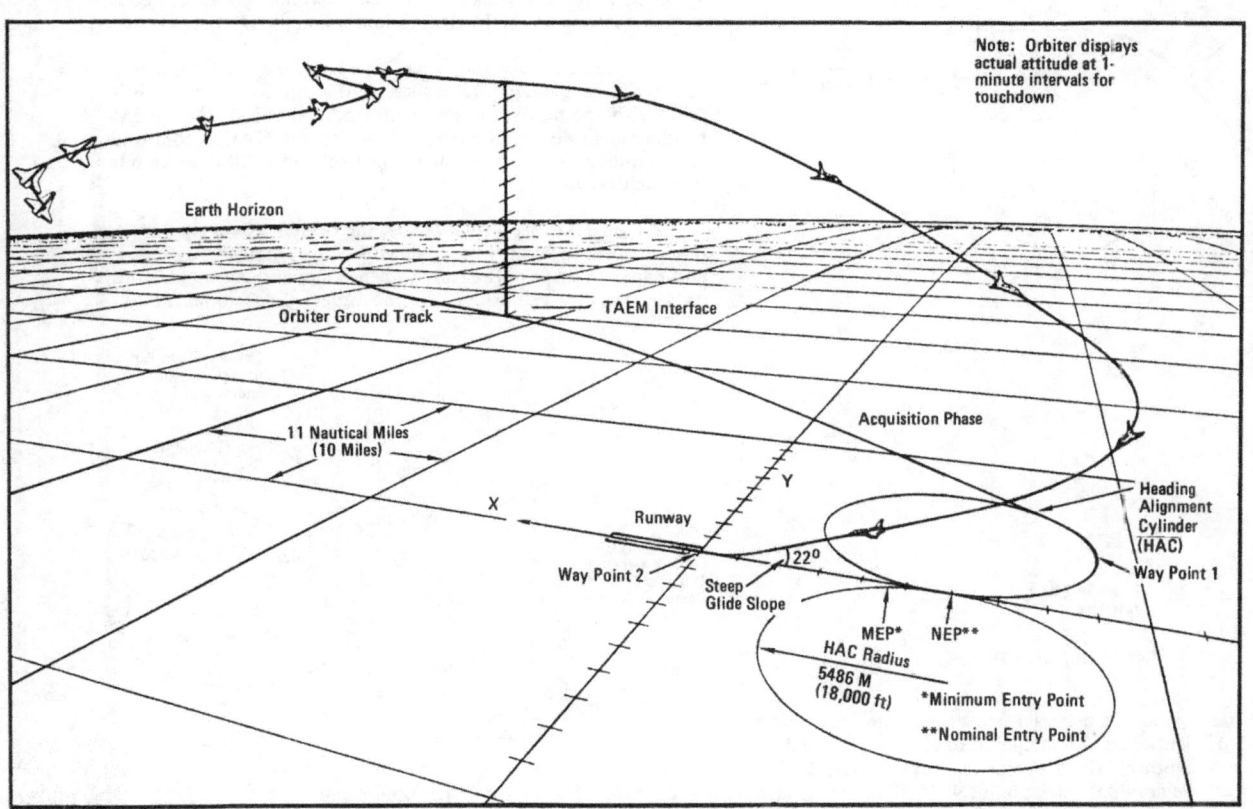

Entry Flight Profile

Space Shuttle Spacecraft Systems

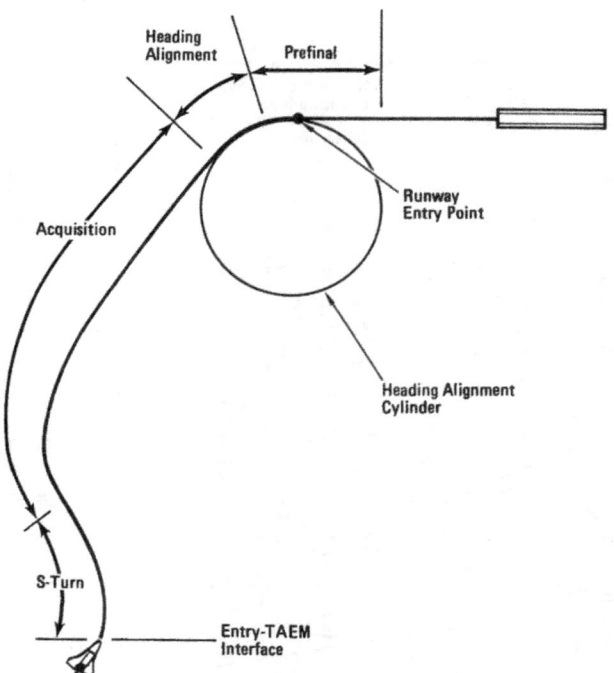

TAEM Guidance Phases

can be flown near the velocity for maximum lift over drag or wings level for the range stretch case, which moves the A/L guidance phase to the MEP (minimum entry point).

A TAEM acquisition, the orbiter is turned until it is aimed at a point tangent to the nearest HAC and continues until it reaches the point, WP-1 (waypoint one). At WP-1, the TAEM heading alignment phase begins in which the HAC is followed until landing runway alignment plus or minus 20 degrees has been achieved.

In the TAEM prefinal phase, the orbiter leaves the HAC, pitches down to acquire the steep glide slope, increases airspeed, and banks to acquire the runway centerline and continues until on the runway centerline, on the outer glide slope and on airspeed.

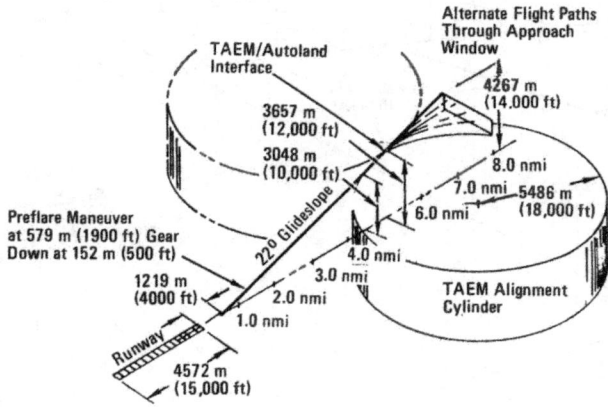

TAEM/Autoland

ground track range as the orbiter turns away from the nearest HAC until sufficient energy is dissipated to allow a normal A/L guidance phase capture, which begins at 3,048 meters (10,000 feet) altitude at the nominal entry point (NEP). The orbiter also

Space Shuttle Spacecraft Systems

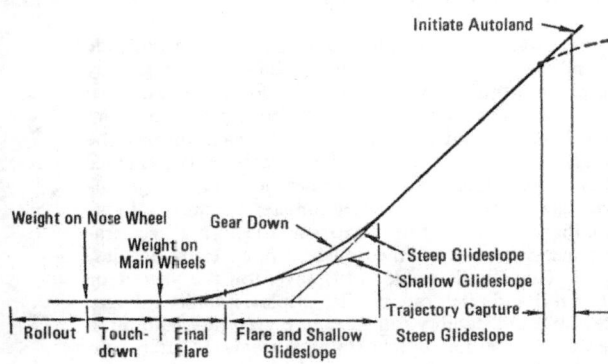

Autoland Profile

The A/L guidance phase begins with the completion of the TAEM prefinal phase and ends when the craft comes to a complete stop on the runway. The A/L interface airspeed requirement at 3,048 meters (10,000 feet) altitude is 290 plus or minus 12 knots, equivalent airspeed (EAS), 6.9 nautical miles (7.9 statute miles) from touchdown.

Autoland guidance is initiated at this point to guide the orbiter to the minus 20° glide slope (which is over seven times that of a commercial airliner's approach) aimed at a target approximately 0.86 nautical mile (one statute mile) in front of the runway. The descent rate in the later portion of TAEM and A/L is greater than 3,048 meters per minute (10,000 feet per minute — approximately 20 times higher rate of descent than a commercial airliner's standard three-degree instrument approach angle). The steep glide slope is tracked in azimuth and elevation and the speed brake is positioned as required.

At 533 meters (1,750 feet) above ground level a pre-flare maneuver is started to position the spacecraft for a 1.5° glide slope in preparation for landing with the speedbrake positioned as required. The flight crew deploys the landing gear at this point.

The final phase reduces the sink rate of the vehicle to less than 2.7 meters per second (9 feet per second). After the spacecraft crosses the runway threshold — WP-2 in the Autoland mode, navigation uses the radar altimeter vertical component of position in the state vector for guidance and navigation computations, this is from an altitude of 30 meters (100 feet) to touchdown. Touchdown occurs approximately 762 meters (2,500 feet) past the runway threshold at a speed of 184 to 196 knots (213 to 226 mph).

In the automatic mode, the orbiter is essentially like a missile; the flight crew monitors the instruments to verify that the vehicle is following the correct trajectory. The onboard computers execute the flight control laws (equations). If the vehicle diverges from the trajectory, the flight crew can take over at any time by switching to CSS. The orbiter can fly to a landing in the automatic mode (only landing gear extension and braking action on the runway are required by the flight crew).

The navigation system used from entry to landing consists of the IMU's and navigation aids (TACAN, air data system, MSBLS, and radar altimeter). The three IMU's maintain an inertial reference and provide delta velocities until MSBLS is acquired.

Navigation-derived air data — obtained after deployment of the two air data probes below Mach 3 — is needed from entry through landing as inputs to the guidance, flight control, and crew display. TACAN provides range and bearing measurements and is available at approximately 44,196 meters (145,000

Space Shuttle Spacecraft Systems

feet) altitude and will nominally accept the data into the state vector before 39,624 meters (130,000 feet) altitude. TACAN is used until MSBLS acquisition which provides range, azimuth, and elevation and is expected to occur at approximately 5,486 meters (18,000 feet) altitude. Radar altimeter data is available at approximately 2,743 meters (9,000 feet).

TACAN acquisition and operation is completely automatic. The crew is provided with the necessary controls and displays to evaluate the TACAN system performance and to take over if required. When the distance to the landing site is approximately 120 nautical miles (138 statute miles), the TACAN begins the navigation region of interrogating six navigation stations. As the spacecraft progresses, the distance to the remaining stations and the next nearest station is computed and the next nearest station will be selected automatically when the spacecraft is closer to it than it is to the previous locked-on station. Only one station is interrogated when the distance to the landing site is less than approximately 20 nautical miles (23 statute miles). Again, the TACAN's will automatically switch from the last locked-on navigation region station to begin searching for the landing site station. TACAN azimuth and range are provided on the CRT horizontal situation display. TACAN range and bearing cannot be used to produce a good estimate of the altitude position component, so navigation uses barometric altitude derived from the air data system probes.

MSBLS acquisition and operation is completely automatic; the flight crew provided with controls and displays to evaluate system performance and to take over if required. MSBLS acquisition is expected to occur at approximately 5,486 meters (18,000 feet) and approximately eight nautical miles (9.2 statute miles) from the runway. The range and azimuth measurements are provided by a ground antenna located at the end of the runway and to the left of the runway centerline. Elevation measurements are provided by a ground antenna to the left of the runway centerline approximately 800 meters (2,624 feet) from the runway threshold.

During entry, the commander and pilot ADI's (attitude direction indicators) becomes a two-axis ball displaying body roll and pitch attitudes with respect to local vertical/local horizontal. These are generated in the attitude processor from IMU data. The roll and pitch error needles each display the body roll and pitch attitude error with respect to entry guidance commands by using the bank guidance error and the angle-of-attack error generated from the accelerometer assemblies. In atmospheric flight, the roll attitude error and the normal acceleration error are displayed on the roll and pitch error needles, respectively. The sideslip angle is displayed on the yaw error needle. The roll and pitch rate needles display stability roll and body rates by using stability roll rate, rate gyro rate, and pitch rate. The yaw rate needle displays stability yaw rate. After main landing gear touchdown, the yaw error with respect to runway centerline and nose gear slapdown pitch rate error are displayed on the roll and pitch error needles. During rollout, the pitch error indicator indicates pitch error rate.

During entry, the commander and pilot HSI's (horizontal situation indicators) display a pictorial view of the spacecraft's location with respect to various navigation points. During entry, navigation attitude processor provides the inputs to the HSI until after leaving the communications blackout at approximately 44,196 meters (145,000 feet) altitude. Then TACAN is acquired and accepted for HSI inputs at approximately 39,624 meters (130,000 feet) altitude until MSBLS acquisition at approximately 5,486 meters (18,000 feet), and approximately 8 nautical miles (9.2 statute miles) from the runway.

When the approach mode and MLS source are selected for the commander and pilot HSI, data from the microwave scan beam landing system replaces TACAN data. MSBLS azimuth,

Space Shuttle Spacecraft Systems

elevation, and range are used from acquisition until runway threshold is reached and azimuth and range are used to control rollout.

At an altitude of 2,743 meters (9,000 feet), radar altitude one or two can be selected to measure the nearest terrain within the beamwidth of the altimeters. They provide this indication to the AVVI radar, altitude, meter display from 5,000 feet to 0.

Below Mach 3, the left and right air data system probes are deployed by the flight crew. This system senses air pressures related to the spacecraft movement through the atmosphere for updating the navigation state vector in altitude, guidance in steering and speed-brake command calculations, flight control for control law computations, and for display on the AMI's (alpha Mach indicators) and AVVI's (altitude, vertical velocity indicators).

The AMI's display essential flight parameters relative to the spacecraft's travel in the air mass such as angle of attack (alpha), acceleration, Mach/velocity, and knots equivalent airspeed. The source of data during the AMI's is determined by the position of the air data select switch. Prior to air data system probe deployment, the AMI's receive their inputs from the navigation altitude processor. When the air data probes are deployed, the left or right air data system provides the inputs to the AMI's except for the acceleration indicator, which remains on the navigation attitude processor.

The AVVI's display radar altitude, barometric altitude, altitude rate, and altitude acceleration. The data driving the AVVI's is determined by the air data select switch. When the air data probes are deployed, the right or left air data system provides the inputs to the AVVI's, except for the altitude acceleration indicator, which remains on the navigation attitude procsor, and the radar altitude, which are not operational until the orbiter descends to 1,524 meters (5,000 feet).

The three CRT's on the flight deck are used during entry to display entry trajectories, horizontal situation displays, guidance navigation, and control systems summaries.

When the orbiter is in atmospheric flight, it is flown by varying the forces it generates while moving through the atmosphere like any other aerodynamic vehicle. The forces are determined primarily by the speed and direction of the relative wind (the airstream is seen from the vehicle). The speed of the airstream is described as Mach number. The direction of the airspeed is described by the difference between the direction the vehicle is pointing (attitude) and the direction it is moving (velocity). It may be broken into two components — angle of attack (vertical component) and sideslip angle (horizontal component).

To rotate the orbiter in the atmosphere, aerodynamic control surfaces are deflected into the airstream. The orbiter has seven aerodynamic control surfaces. Four of these are on the trailing edge of the wing (two per wing) and are called elevons since they combine the effects of elevators and ailerons on ordinary airplanes. Deflecting the elevons up or down causes the vehicle to pitch up or down. If the right elevons are deflected up, while the left elevons are deflected down, the orbiter will roll to the right — that is, the right wing will fall while the left wing rises. The fifth control surface is the body flap, which is located on the rear lower portion of the aft fuselage and provides thermal protection for the three main engines during entry. During atmospheric flight, it provides pitch trim to reduce elevon deflections. The sixth and seventh control surfaces are the rudder/speed brake panels located on the aft portion of the vertical stabilizer. When both panels are deflected right or left, the spacecraft will yaw, moving the spacecraft nose right or left, thus acting as a rudder. If the panels are opened at the trailing edge, aerodynamic drag force increases and the spacecraft slows down. Thus, the opening of the panels is called a speed brake for speed control.

Space Shuttle Spacecraft Systems

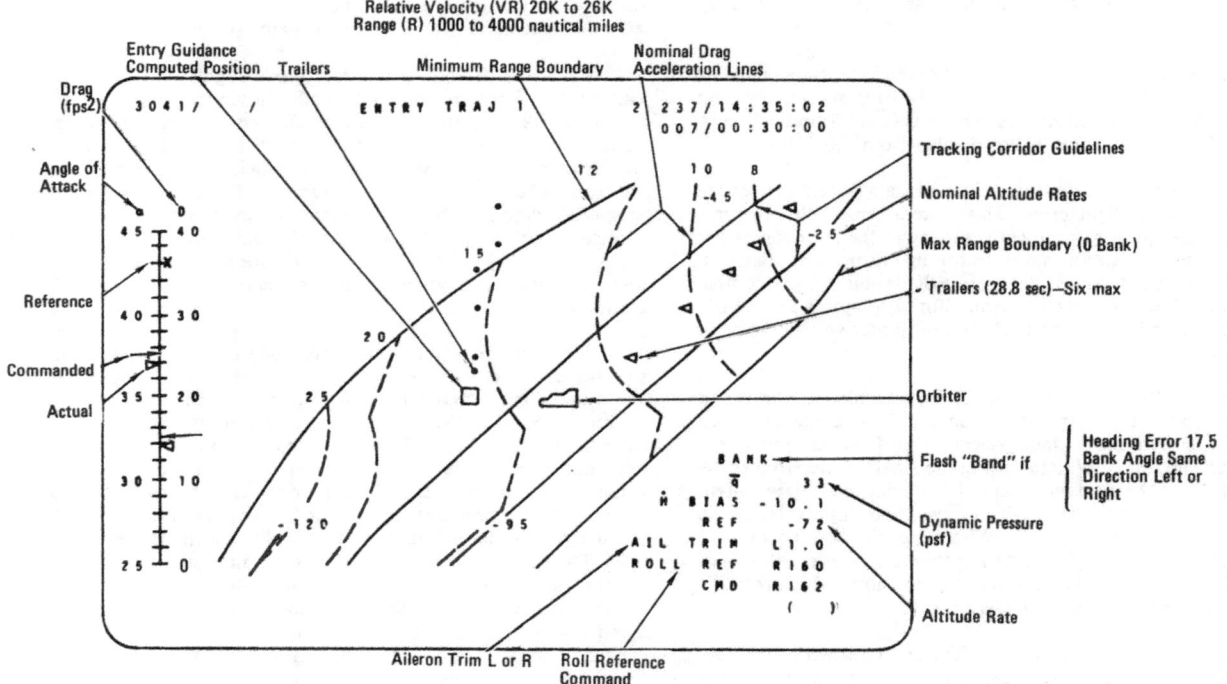

Entry Trajectory

Space Shuttle Spacecraft Systems

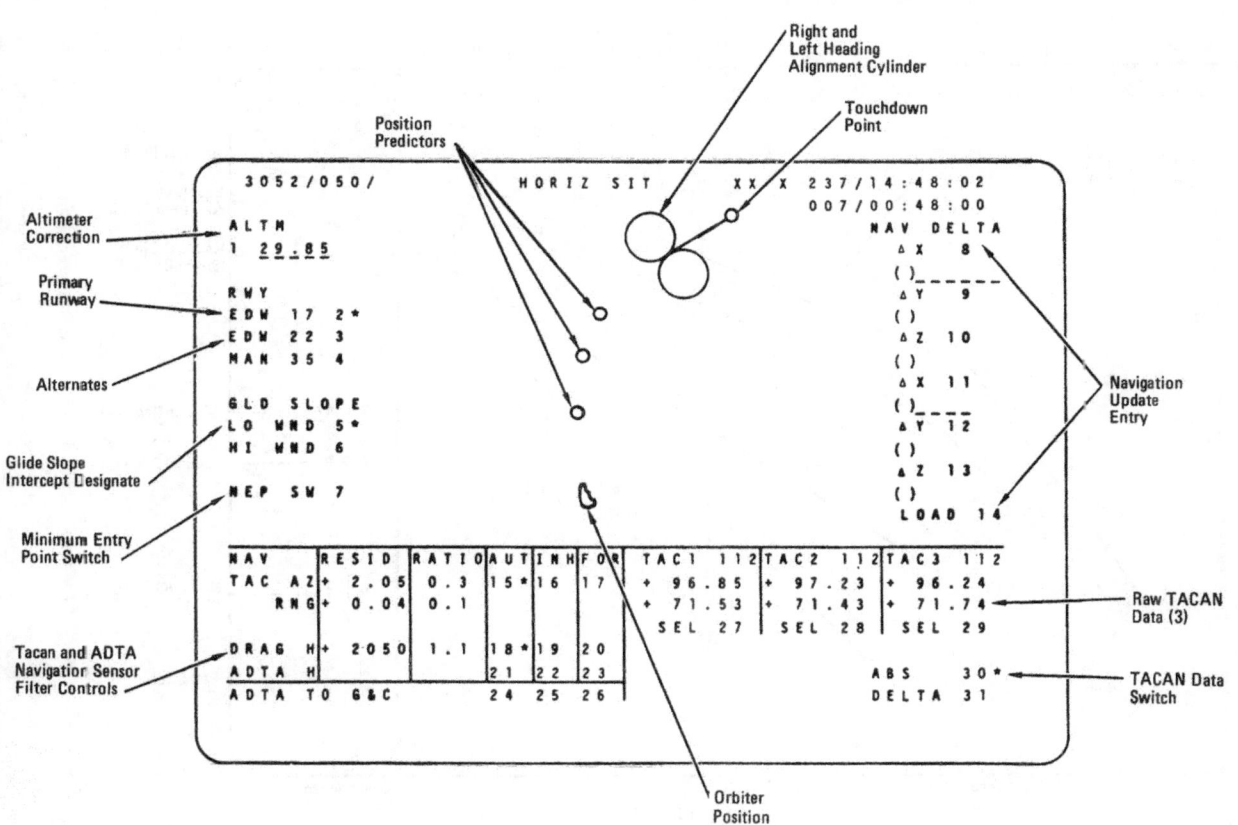

Horizontal Situation Display

Space Shuttle Spacecraft Systems

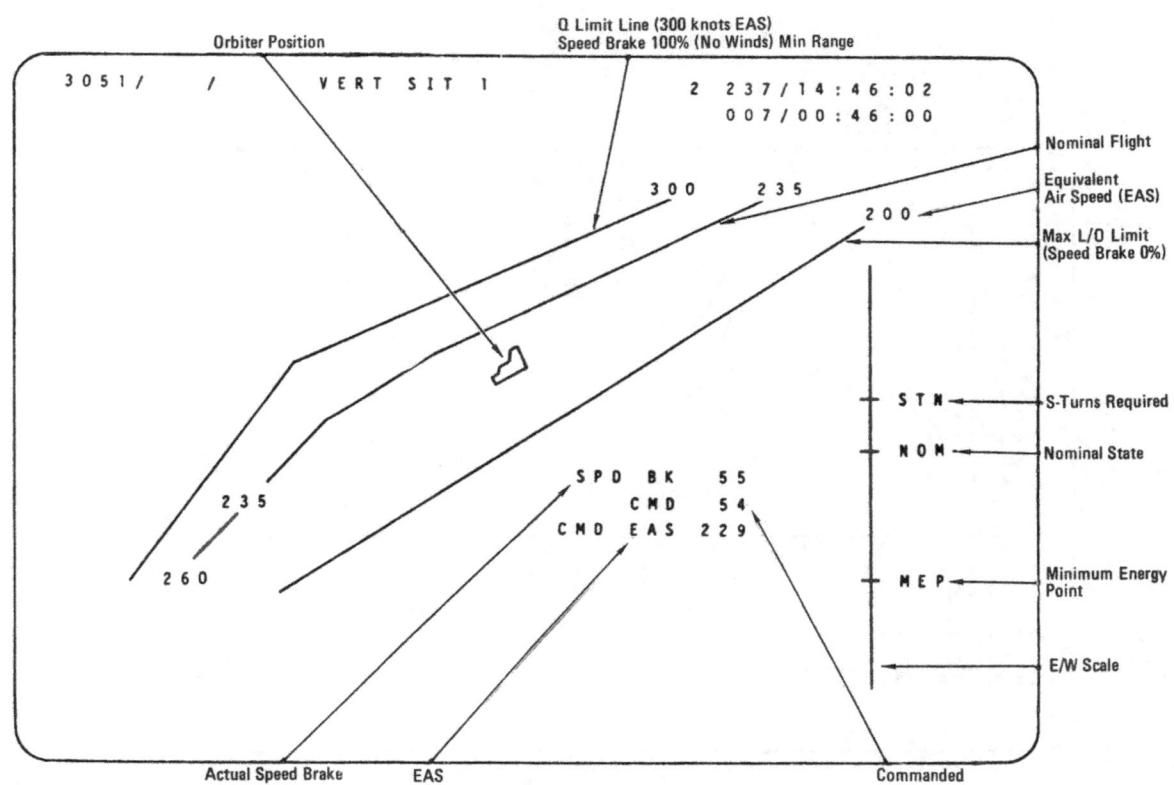

Vertical Situation Display

Space Shuttle Spacecraft Systems

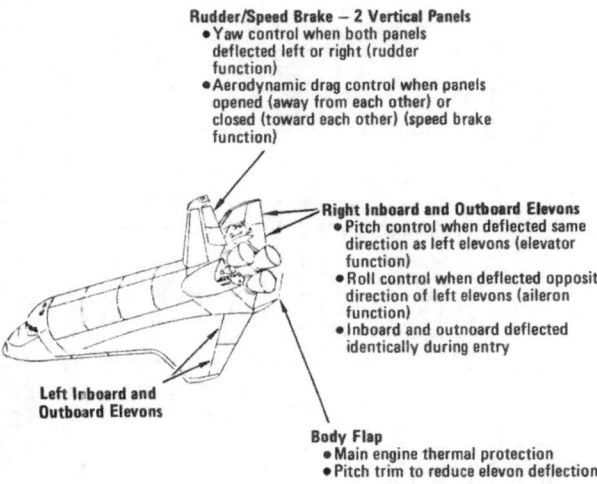

Aerodynamic Surfaces

On the flight deck display and control panel, between the commander and pilot stations on Panel F7, are the surface position indicators which display the position of each aerodynamic control surface. The positions of the aerodynamic surfaces are displayed on moving pointer displays. Elevon positions are shown in degrees from plus 20°, full down to minus 35°, full up. Body flap position is 0 to 100 percent. Rudder position is left 30°, right 30°. Aileron position is left 5°, right 5°. Speed brake position and command are 0 to 100 percent. An off flag is provided in each indicator to indicate power loss, erroneous input signals, or failure in any display channel.

The commander and pilot can select the flight control system modes of operation with pushbutton light indicators on the flight deck display and control glareshield/eyebrow panel. There are two modes — automatic and CSS (control stick steering). The flight crew can select separate modes for pitch and roll/yaw (roll and yaw must be in the same mode). The body flap and speed brake have an automatic and manual mode selection. Each pushbutton is a triple redundant, momentary contact, nonlatching switch.

Automatic pitch or automatic roll/yaw are selected by depressing the applicable pushbutton at either the commander or pilot station. When depressed, that pushbutton will illuminate at both stations. Automatic pitch provides automatic control in the pitch aixs and the automatic roll/yaw provides automatic control in the roll/yaw axes. During entry, the automatic mode uses the RCS jets while dynamic pressure permits the aerosurfaces to become effective; the aft RCS jets and spacecraft aerosurfaces are then used together until dynamic pressure becomes sufficient for aerosurface control only.

Control in the pitch axis is provided by the elevons, speed brake, and body flap. The elevons provide control to guidance normal acceleration commands, control of pitch rate during slapdown (landing) for nose wheel load protection, and static load relief after slapdown for main landing gear wheel and tire load protection. The speed brake provides control to guidance surface deflection (open/close, increase/decrease velocity) command. The body flap provides control to null elevon deflection. Pushbuttons at the commander and pilot flight deck display and control eyebrow panel illuminate to indicate automatic speed brake or body flap control.

Control in the roll and yaw axes is provided by the elevons and rudder. The elevons provide control to guidance bank angle command during TAEM and autoland and control to guidance wings-level command during flat turns 1.5 meters (5 feet) above touchdown. The rudder provides yaw stabilization during TAEM and autoland and control to guidance yaw rate command during flat turn and subsequent phases.

 **Space Shuttle Spacecraft Systems**

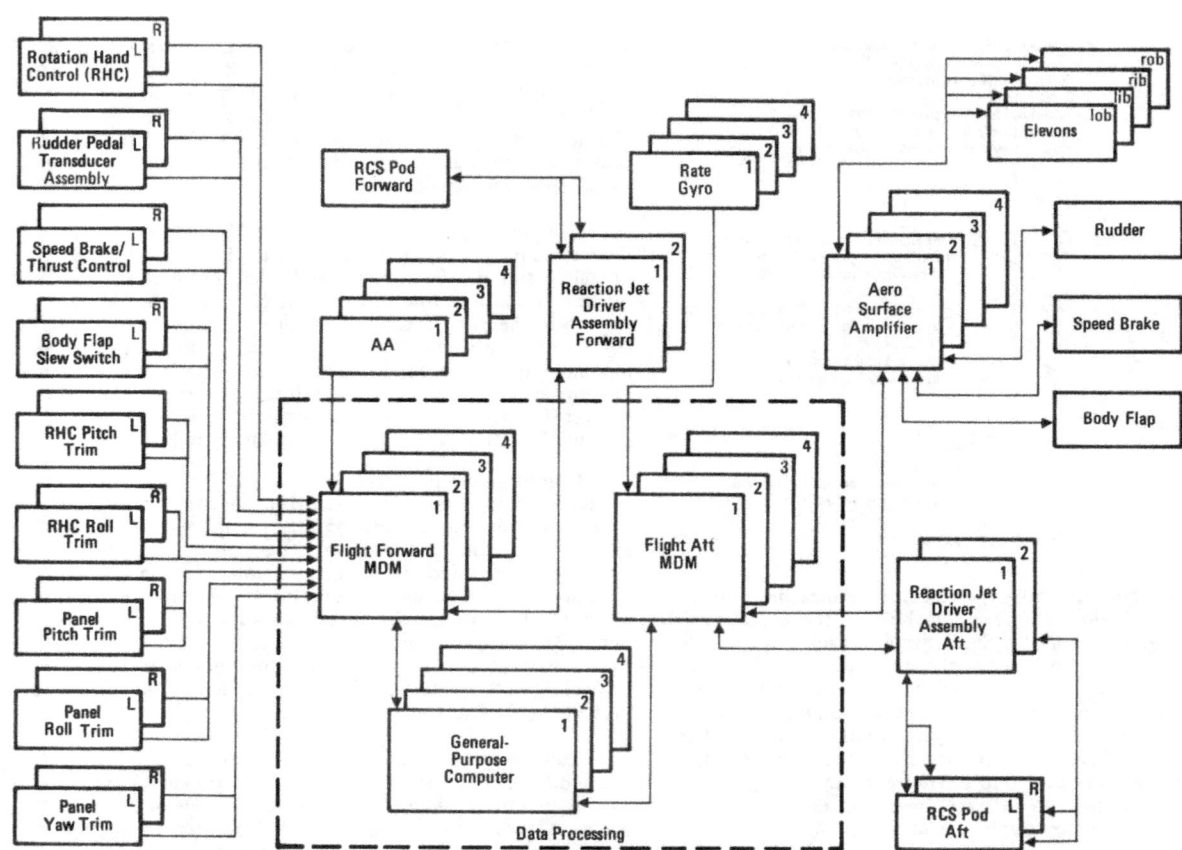

Flight Control System Hardware

Space Shuttle Spacecraft Systems

The three rate gyro assemblies of the flight control system measure and supply output data proportional to the orbiter's attitude rates about its three body axes. The three accelerometer assemblies of the flight control system measure and supply output data proportional to the orbiter's normal (vertical) and lateral (right/left) accelerations. The three rate gyro assemblies and three accelerometer assemblies are used in the flight control system for stability augmentation because of the orbiter's marginal stability in its pitch and yaw axes at subsonic speeds.

The three IMU's are an all-attitude stabilized platform that also measures and supplies output data proportional to the spacecraft's attitude (rotation) and acceleration (velocity). The three IMU's augment the flight control system's rate gyro assemblies and accelerometer assemblies.

The rate gyro assembly pitch rate (rotation) and the accelerometer assembly normal acceleration (velocity) are used to generate elevon (elevator) deflection commands. The rate gyro assembly yaw rate (rotation) and the accelerometer assembly lateral acceleration generate the rudder deflection required for directional stability. The rate gyro assembly roll rate (rotation) generates the elevon (aileron) deflection command required for lateral (roll) stability. The speed brake and body flap positions generate the elevon deflection required for trim near neutral to maximize roll effectiveness of the elevons.

When the orbiter is in the automatic pitch and roll/yaw mode, the flight crew manual control stick steering commands are inhibited. In the CSS mode, the crew flies the orbiter by deflecting the RHC and rudder pedals. The flight control system interprets the RHC motions as rate commands in pitch, roll, or yaw and controls the RCS jets and aerosurfaces. The larger the deflection, the larger the command. The flight control system compares these commands with inputs from rate gyros and accelerometers (what the vehicle is actually doing — motion sensors) and generates control signals to produce the desired rates.

If the flight crew releases the RHC, it will return to center, and the orbiter will maintain its present attitude (zero rates). The rudder pedals position the rudder during atmospheric flight; however, in actual use, because flight control software performs automatic turn coordination, the rudder pedals are not used until the wings are leveled prior to touchdown.

The CSS mode is similar to the automatic mode except the flight crew can issue three-axis commands, affecting spacecraft motion. They are augmented by the feedback from the same spacecraft motion sensors, except for the normal acceleration (velocity) accelerometer assemblies, to enhance control response and stability. CSS pitch or roll/yaw modes are selected by depressing a pushbutton at either the commander or pilots station. When depressed, that PBI will illuminate to show CSS control in that axis.

When the orbiter is in the CSS mode, flight crew inputs are provided by the commander or pilot RHC. The RHC commands are processed by the computers together with data from the motion sensors. The flight control module processes the flight control laws and provides commands to the flight control system, which positions the aerosurfaces in atmospheric flight.

Control in the roll/yaw axis is provided by the elevon (ailerons) and the rudder. The elevons provide augmented control to the RHC control. The rudder interface between the roll/yaw channel automatically positions the rudder for coordinated turns. A rudder pedal transducer assembly (RPTA) is provided at the commander and pilot's stations. The two rudder pedal assemblies are connected to their respective RPTA's. Because of the roll/yaw interface, rudder pedal use should not be required until just before touchdown. An artificial feel is provided in the rudder pedal assemblies. The RPTA commands are processed by the computers and the flight control module commands the flight control system to position the rudder.

# Space Shuttle Spacecraft Systems

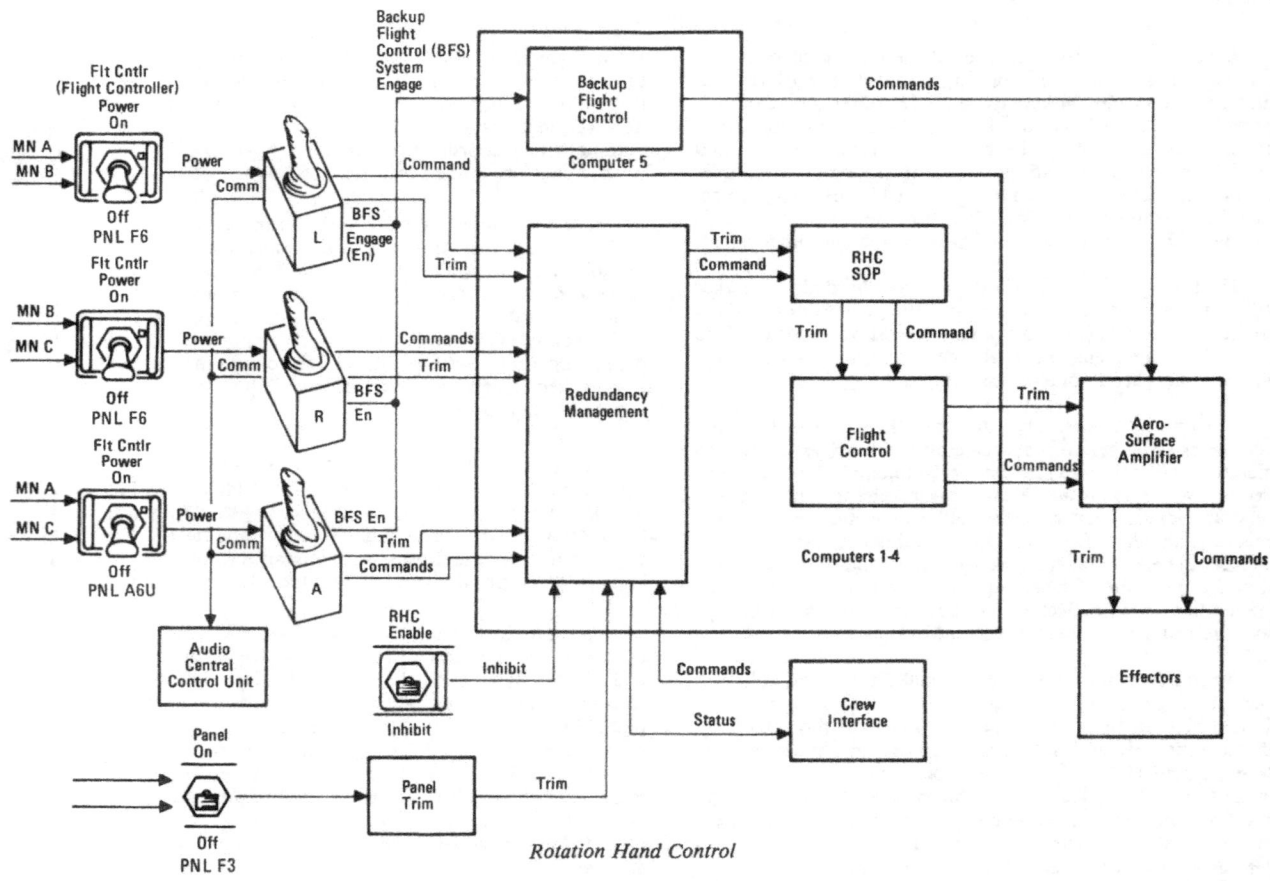

Rotation Hand Control

# Space Shuttle Spacecraft Systems

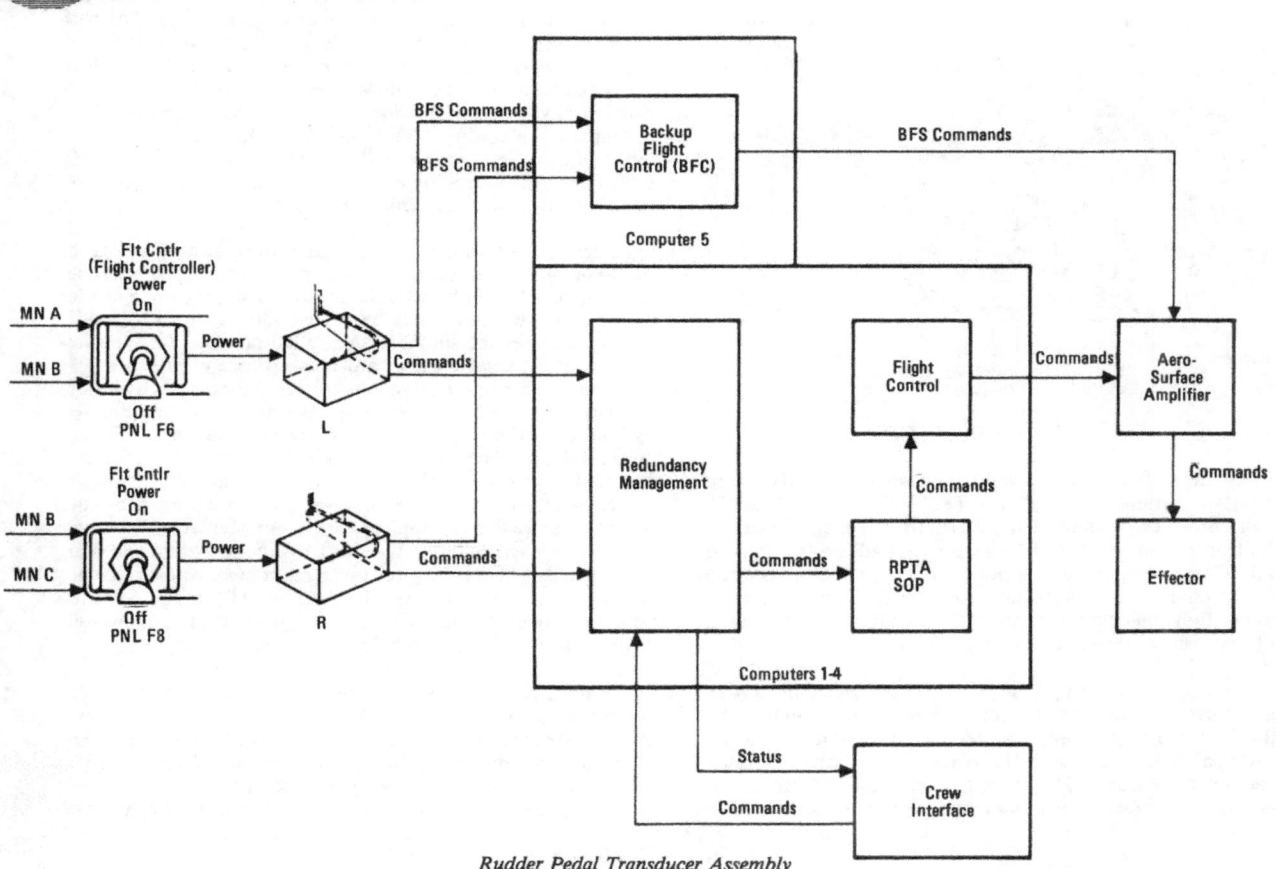

Rudder Pedal Transducer Assembly

Space Shuttle Spacecraft Systems

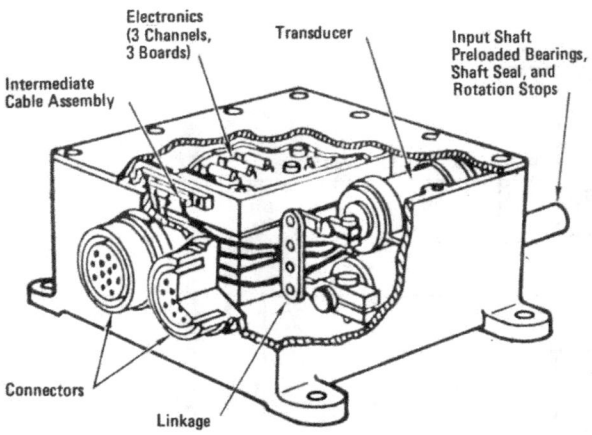

Rudder Pedal Transducer Assembly

In the CSS mode the command and pilot's RHC trim switch, in conjunction with the TRIM ENABLE/INHIBIT switch, activates or inhibits the RHC trim switch. When the RHC trim switch positioned forward or aft adds a trim rate to the RHC pitch command, positioning it left or right adds a roll trim. Redundancy management processes the two contacts in each position and enables or inhibits the use of the RHC trim in the flight control system software.

Aerosurfce servo amplifier's (ASA's) are electronic devices that receive the aerosurface commands during atmospheric flight from the flight control system software and electrically position hydraulic valves in the aerosurface actuators which cause the deflections. The ASA's receive position feedback from the aerosurface, which is summed with the position command to provide a servo loop closure for one of the four independent servo loops associated with the elevons, rudder, and speed brake. The body flap utilizes three servo loops. The path from an ASA to its servovalve in the actuators and the feedback sensor to an ASA is called a flight control channel; thus there are four flight control channels. The four ASA's are located in avionics bays 4, 5, and 6 in the orbiter aft fuselage. Each ASA is mounted on a coldplate and cooled by the Freon-21 coolant loops. Each ASA is 508 millimeters (20 inches) long, 162 millimeters (6.4 inches) high and 228 millimeters (9 inches) wide. Each weighs 13.69 kilograms (30.2 pounds).

Each of the four elevons, located on the trailing edge of the wings, has a servoactuator that positions it. Each servoactuator is supplied with hydraulic pressure from the three orbiter hydraulic systems. A switching valve controls which hydraulic system becomes the source of hydraulic pressure for that servoactuator. The valve allows a primary source of pressure (P1) to be supplied to that servoactuator. If the primary hydraulic pressure source fails, the switching valve allows the first standby hydraulic pressure (P2) to supply that servoactuator. Failure means a decrease of pressure between 77,625 mmHg (1,500 psi) and 62,100 mmHg (1,200 psi). If the first standby hydraulic source pressure fails, then the secondary standby hydraulic source pressure (P3) is supplied to that servoactuator. With two hydraulic system failures, the third hydraulic system will provide sufficient hydraulic pressure to operate all aerosurface actuators at a reduced rate. The yellow HYD PRESS caution/warning light would illuminate on the flight deck crew and display panel to indicate a hydraulic system failure.

Each elevon servoactuator receives four command signals, one from each of the four ASA's. Each servo channel consists of a two-stage servovalve that drives a modulating piston. Each of the four modulating pistons is summed on a common shaft, creating a force on the mechanical shaft to position a power spool that controls the flow of hydraulic fluid to the actuator

 # Space Shuttle Spacecraft Systems

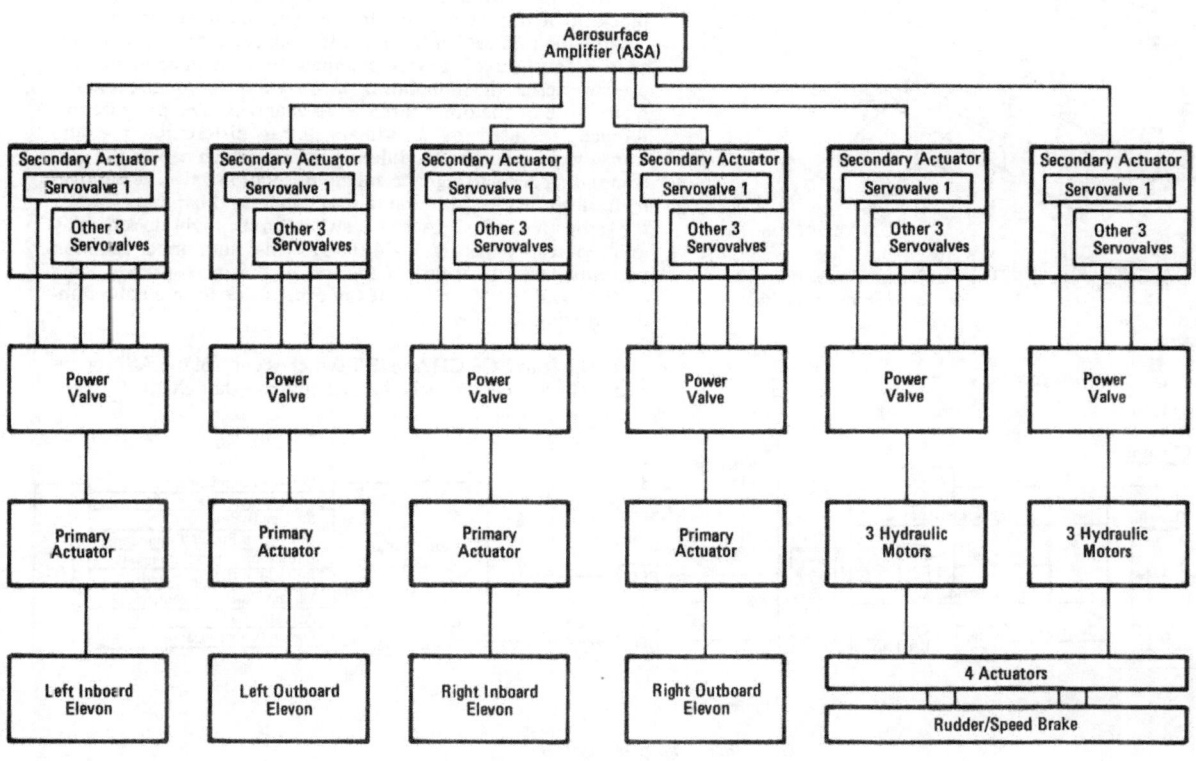

Aerosurface Amplifier

Space Shuttle Spacecraft Systems

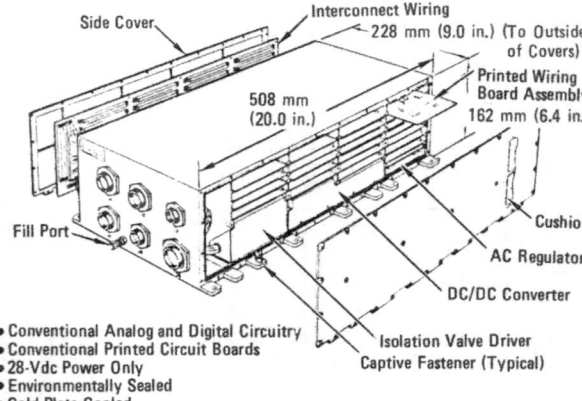

Aerosurface Servoamplifier

power ram, controlling the direction of ram movement and thus the elevon to the desired position. When the desired position is reached, the power spool positions the mechanical shaft to block the hydraulic pressure to the hydraulically operated ram, locking the ram at that position. If a malfunction occurs in one of the four servo control channels of a servoactuator, the pressure across the modulating piston will differ (delta pressure) from those of the other three servo channels. The pressure difference is sensed by a primary linear differential pressure transducer across the modulating piston, which causes the corresponding ASA to signal a solenoid isolation valve. It removes hydraulic pressure from the failed channel and bypasses it when the respective FCS CHANNEL switch on the flight deck display and control panel is in AUTO. This automatic function prevents excessive transient motion to that aerosurface. Such motion could result in loss of the orbiter due to manual redundancy being too slow.

The four FCS CHANNEL switches control the ASA channels. The switch for each channel controls that channel for the

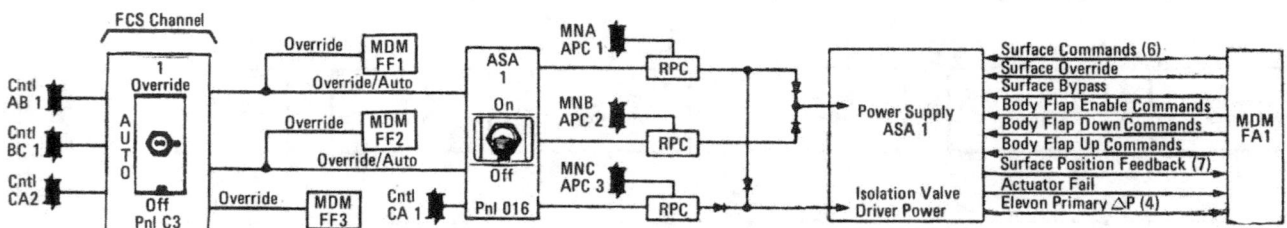

Aerosurface Amplifier Control

Space Shuttle Spacecraft Systems

elevons, rudder/speed brake, and body flap, except for channel 4, which has no body flap commands. When an RCS channel switch is positioned to OVERRIDE, it resets the channel that was bypassed in the automatic mode. The FCS channel switch OFF position bypasses that channel.

The FCS CHANNEL yellow caution/warning light will illuminate on the flight deck display and control panel to inform the flight crew of a failed channel. The red FCS SATURATION caution/warning light on the flight deck display and control panel will illuminate if one of the four elevons is greater than plus 15 degrees or less than minus 20 degrees.

In each elevon servoactuator ram, there are four linear ram position transducers and four linear ram secondary differential pressure transducers. The ram linear transducers provide position feedback to the corresponding servo loop in the ASA which is summed with the position command to close the servo loop. These are summed with the elevon ram linear secondary differential pressures which develop an electro-hydraulic valve drive current proportional to the error signal to position the ram. The maximum elevon deflection rate is 20 degrees per second.

The rudder/speed brake is located on the trailing edge of the orbiter's vertical stabilizer. The rudder/speed brake consists of upper and lower segments. One servoactuator positions both segments together as a rudder. Another servoactuator opens the segments at the flared end of the rudder to function as a speed brake.

The rudder and speed brake servoactuator receives four command signals from the four ASA's. Each of the four servo channels consists of a two-stage servovalve which functions similar to that of the elevons. The exception is that the power spool for the rudder controls the flow of hydraulic fluid to three rudder reversible hydraulic motors, and the power spool for the speed brake controls the flow of hydraulic fluid in the three speed brake hydraulic reversible motors. Each rudder hydraulic motor receives hydraulic pressure from only one of the orbiter's hydraulic systems and each speed brake hydraulic motor receives hydraulic pressure from only one of the orbiter's hydraulic systems. Each hydraulic motor has a hydraulic brake. When the motor is supplied with hydraulic pressure, the motor's brake is released. When the hydraulic pressure is blocked to that hydraulic motor, the hydraulic brake is applied holding that motor and the corresponding aerosurface at that position.

The three hydraulic motors provide the output to the rudder differential gearbox. The gearbox is connected to a mixer gearbox, which drives rotary shafts, which drive four rotary actuators, which position the rudder segments for rudder position.

The three speed brake hydraulic motors provide the power output to the speed brake differential gearbox. The differential gearbox is connected to the same mixer gearbox as that of the rudder, which drives rotary shafts, which drive the same four rotary actuators involved with the rudder. Within each of the four rotary actuators, are planetary gears which provide the blend of positioning the rudder and opening the rudder flared ends.

There are four rotary position transducers on the rudder differential gearbox output and one differential linear position transducer in each rudder servoactuator. The rotary position transducers provide position feedback to the corresponding servo loop in the ASA. The feedback is summed with the linear differential pressures that develop the electro-hydraulic valve drive current proportional to the error signal to position the rudder.

There are also four rotary position transducers on the speed brake differential gearbox output and one differential linear pressure transducer in each speed brake servoactuator.

Space Shuttle Spacecraft Systems

The rotary position transducers provide position feedback to the corresponding servo loop in the ASA. It is summed with the position command to close the servo loop. These are summed with the linear differential pressures that develop the electrohydraulic valve drive current proportional to the error signal to position the speed brake.

If a malfunction occurs in one of the four rudder servoactuator channels or one of the four speed brake servoactuator channels, the corresponding linear differential pressure transducer will cause the corresponding ASA to signal a solenoid isolation valve, which removes pressure from the failed channel and bypasses it, if that FCS channel switch is in AUTO. The FCS channel switch override and off position functions the same as for the elevons and the FCS channel caution/warning light. The HYD PRESS caution/warning light will indicate a hydraulic failure. The rudder deflection rate is a maximum of 14 degrees per second. The speed brake deflection rate is approximately 10 degrees per second. If two of the three hydraulic motors fail in the rudder or speed brake, the corresponding deflection rate is reduced approximately 50 percent.

Three servoactuators at the lower aft end of the fuselage are used to position the body flap. Each is supplied with hydraulic pressure from an orbiter hydraulic system. Each has a solenoid-operated enable valve controlled by one of the three ASA's (the fourth ASA is not used for the body flap commands). Each solenoid-operated enable valve supplies hydraulic pressure from one orbiter hydraulic system to a corresponding solenoid-operated pilot valve. Each pilot valve is controlled by one of the three ASA's. When the individual pilot valve receives a command signal from its corresponding ASA, it positions a common mechanical shaft in the control valve, which allows hydraulic pressure to be supplied to the hydraulic motors (normally one pilot valve is enabled and moves the other two). The hydraulic motors are reversible, allowing the body flap to be positioned up or down. The hydraulic brake associated with each hydraulic motor releases the hydraulic motor for rotation and, when the desired body flap position is reached, the control valves block the hydraulic pressure to the hydraulic motor and apply the hydraulic brake, holding that hydraulic motor at that position. Each hydraulic motor provides the power output to a differential gearbox, which drives a rotary shaft, and four rotary actuators, which position the body flap. The rotary position transducer associated with each rotary actuator provides position feedback to the ASA's; thus the fourth ASA is utilized for position feedback to the flight control system software.

The ASA's will isolate a body flap channel if FCS channel switches are in the automatic mode through the solenoid-operated enable valve if the corresponding solenoid-operated pilot valve malfunctions or the control valve associated with the pilot valve does not provide the proper response and allows the hydraulic pressure fluid to recirculate. The FCS channel switches and FCS channel caution/warning light would function the same as for the elevons. If the hydraulic system associated with the hydraulic motor fails, the remaining two hydraulic motors will position the body flap and the HYD PRESS caution/warning light will illuminate. The body flap deflection rate is approximately 4.5 degrees per second.

Manual control (CSS mode) in the pitch axis is provided by elevons, speed brake, and body flap. The elevons provide augmented control through the RHC pitch command. The speed brake can be switched to its manual mode at either commander or pilot stations by depressing a takeover switch on the SBTC handle. The takeover switch, when depressed, illuminates the SPD BK/THROT MAN pushbutton at the station depressed. Manual speed brake control can be transferred from one station to the other by activating the takeover switch. When the SBTC is at its forward setting, the speed brake is closed. Rotating the handle aft positions the speed brake at the desired position (open) and holds it. To regain automatic speed brake control, the pushbutton on the panel is depressed again and the

Space Shuttle Spacecraft Systems

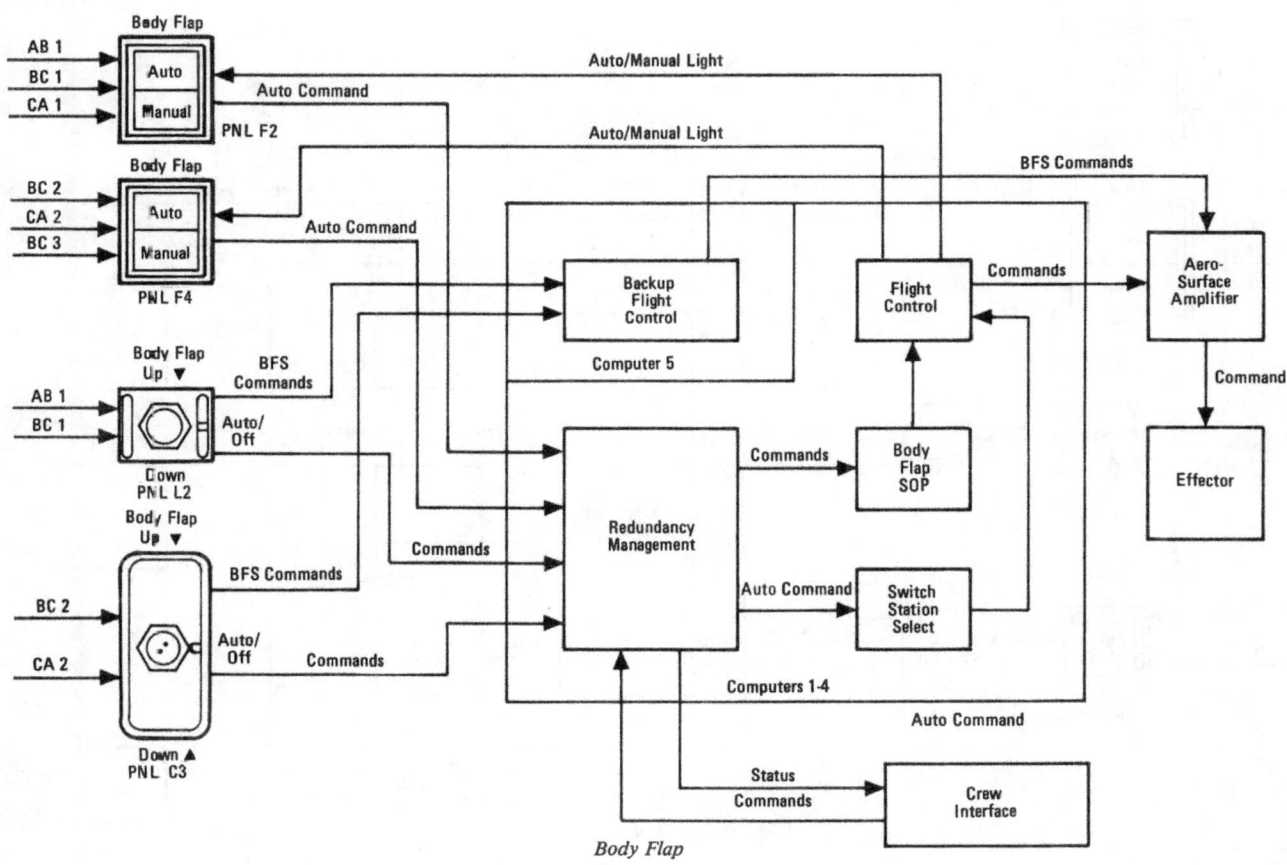

Body Flap

 **Space Shuttle Spacecraft Systems**

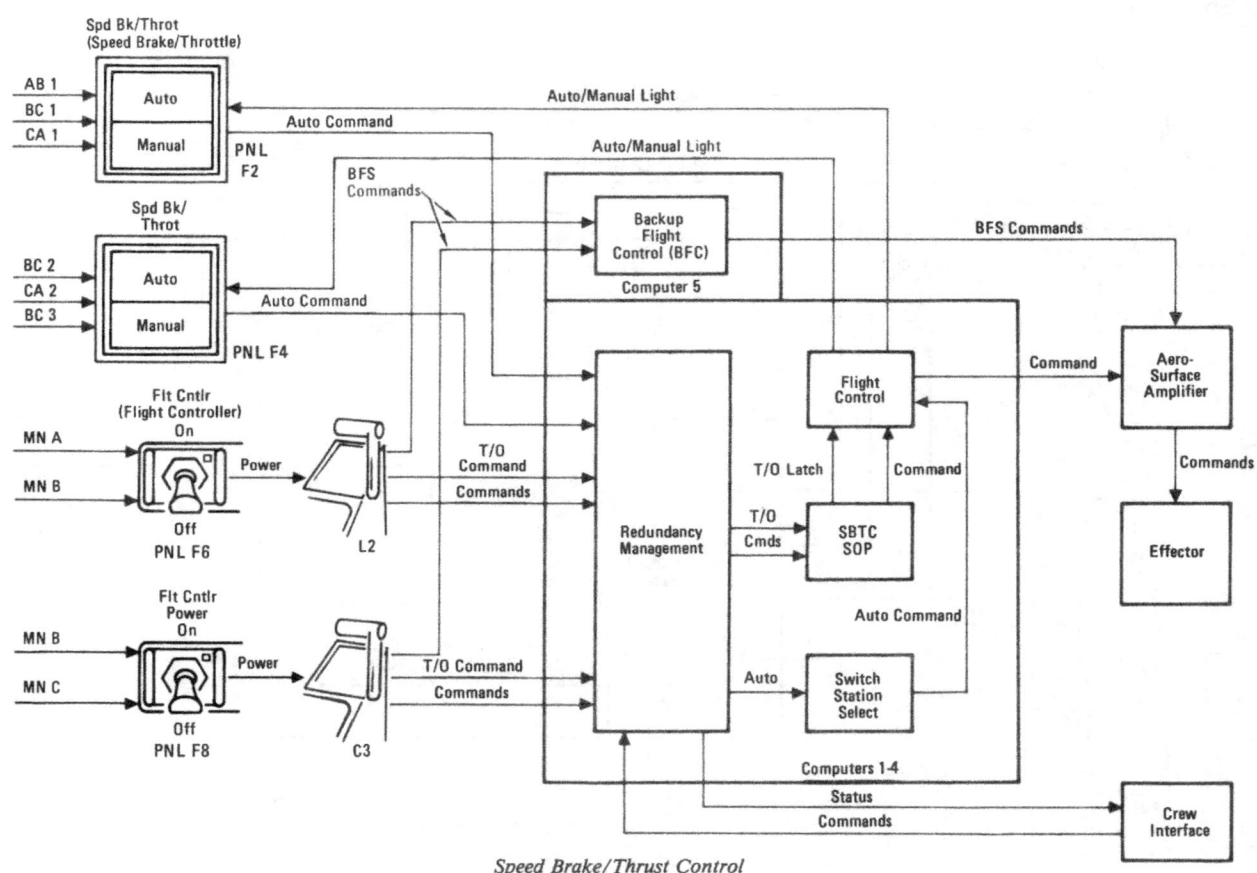

Speed Brake/Thrust Control

Space Shuttle Spacecraft Systems

AUTOMATIC (upper half) of the pushbutton will then be illuminated. In the manual mode, the speed brake commands are processed by the computers and the flight control module commands the flight control system to position the speed brake and hold it at the desired position. The body flap can be switched to its manual mode at PNL C3 by moving a toggle switch from AUTO/OFF to UP or DOWN for the desired body flap position. These are momentary switch positions: when released the switch returns to off. The BODY FLAP MAN (lower half) indicator will illuminate to indicate manual control of the body flap. To regain automatic body flap control, the pushbutton must be depressed again.

ORBITER VENT AND PURGE SYSTEM. The vent and purge system is controlled exclusively through software. There are no dedicated manual controls or displays on the crew compartment flight deck display and control panel.

The orbiter's vent and purge system comprises 18 active ports and is divided into six groups: left and right ports one and two, left and right port three, left and right port five, left and right ports four and seven, left and right port six, and left and right ports eight and nine.

All vent ports have a purge position with the exception of left and right vent ports three, four, five, and seven.

The vent and purge system is used for equalizing the pressure across the outer surface of the orbiter and to permit molecular venting of orbiter cavities and insulation blankets to achieve the required low internal blanket pressure. The purge position must maintain a positive pressure in the orbiter's payload bay area to prevent contamination and to vent any residue in the payload bay area while in the ground turnaround phase.

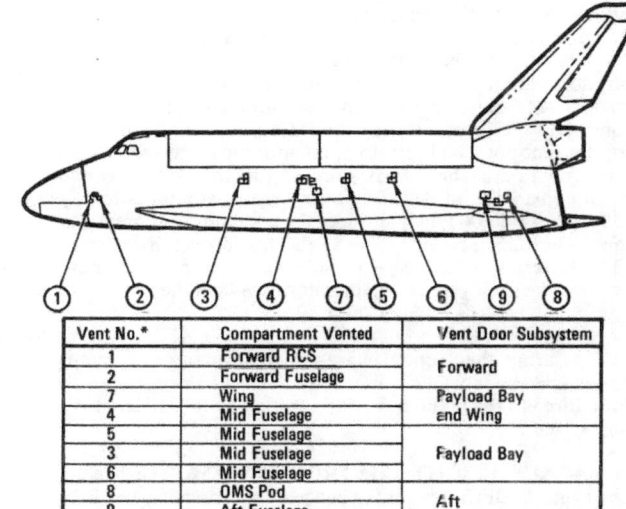

Orbiter Vent Doors

The sequencing of the active ports is by the software program in the onboard redundant set computers. The ports are cycled to open, close, or purge position as required in each mission phase. Positioning of the active ports is performed by the software based on mission time or mission events during ascent, entry, aborts, and by keyboard entry on the crew compartment flight deck display and control panel in orbit. Vent port status is displayed on the flight deck display and control panel CRT's in orbit.

Space Shuttle Spacecraft Systems

When a cue is received from the computer launch sequence to configure the vent ports to a launch configuration, vent ports one and two and eight and nine are commanded to the open position, and all other vent ports are closed. The status of the vent ports is transmitted to the computer launch sequence to determine that the vent ports have achieved the launch configuration within the specified time. The orbiter is launched with the vent ports in this configuration, and at T plus 10 seconds all vent doors are commanded open. In a nominal mission, the vent ports will remain open until the flight crew closes the ports with a keyboard entry on the flight deck display and control panel prior to deorbit. During entry, at approximately 24,384 meters (80,000 feet) altitude, the computers automatically open the fuselage vents. The vents should be fully open at 21,336 meters (70,000 feet) altitude and remain open until weight on the nose gear at landing, at which time they will go to purge.

If, during the launch phase—T minus ten to T minus zero—a launch abort has occurred, the vent door system is reconfigured to the prelaunch configuration by the launch processing system.

BACKUP FLIGHT CONTROL SYSTEM. The backup flight control (BFC) system is engaged by depressing the BFC switch located near the top of the commander or pilot's RHC or by depressing the BFC pushbutton on the commander or pilot's flight deck display and control eyebrow panels F2 and F4. The fifth computer is allocated as the backup flight control system. It has a separate independent software design and coding activity to protect against generic software failures in the primary flight control system computer set. It also provides unique functions such as non-critical monitoring and payload command and monitoring.

The BFC system is distinct from the primary flight system. The BFC system is used only in an extreme emergency. It utilizes control laws (equations) similar to the primary flight control system; however, it is a non-redundant mechanism. The BFC is constrained in a single-string flight control system to avoid generic problems that may exist with the primary computers or primary system software. It operates concurrently with the primary flight system, processing the same commands and sensor data; however, its output is inhibited. When engaged, the BFC system augments the CSS mode by using the commander's RHC. Thus there is no manual thrust vector control or manual throttling capability during first- and second-stage ascent and, during atmospheric flight, the speed brake would be positioned by using the SBTC and the body flap would be positioned manually.

The software of the BFC system is processed only for the commander's (left) side ADI, HSI, and RHC. The BFC system supplies attitude errors on the CRT trajectory display, whereas the primary flight system is supplying the attitude errors to the ADI's when in primary flight control, however, when the BFC system is engaged, the errors on the CRT are blanked.

The BFC can be disengaged by use of the BFC DISENGAGE switch at the commander's station PNL F6. It is used primarily for ground checkout.

GN&C contractors are: Rockwell International, Autonetics Group, Anaheim, CA (driver module controller, master event sequence controller, backup flight control system); Ball Brothers Research, Boulder, CO (star tracker); Rockwell International, Collins Radio Group, Cedar Rapids, IA (display driver unit, horizontal situation indicator); Honeywell Inc., St. Petersburg, FL (flight control system displays and controls, rotation hand control, translation hand control, accelerometer assembly, rudder pedal transducer assembly, speed brake/thrust control, forward and aft reaction jet and OMS drivers, aerosurface servo amplifier, ascent thrust vector control amplifier); IBM Corp., Federal Systems Division, Electronics Systems Center, Owego, NY (mass memory/multi-function cathode ray tube display unit, keyboard display electrical unit

Space Shuttle Spacecraft Systems

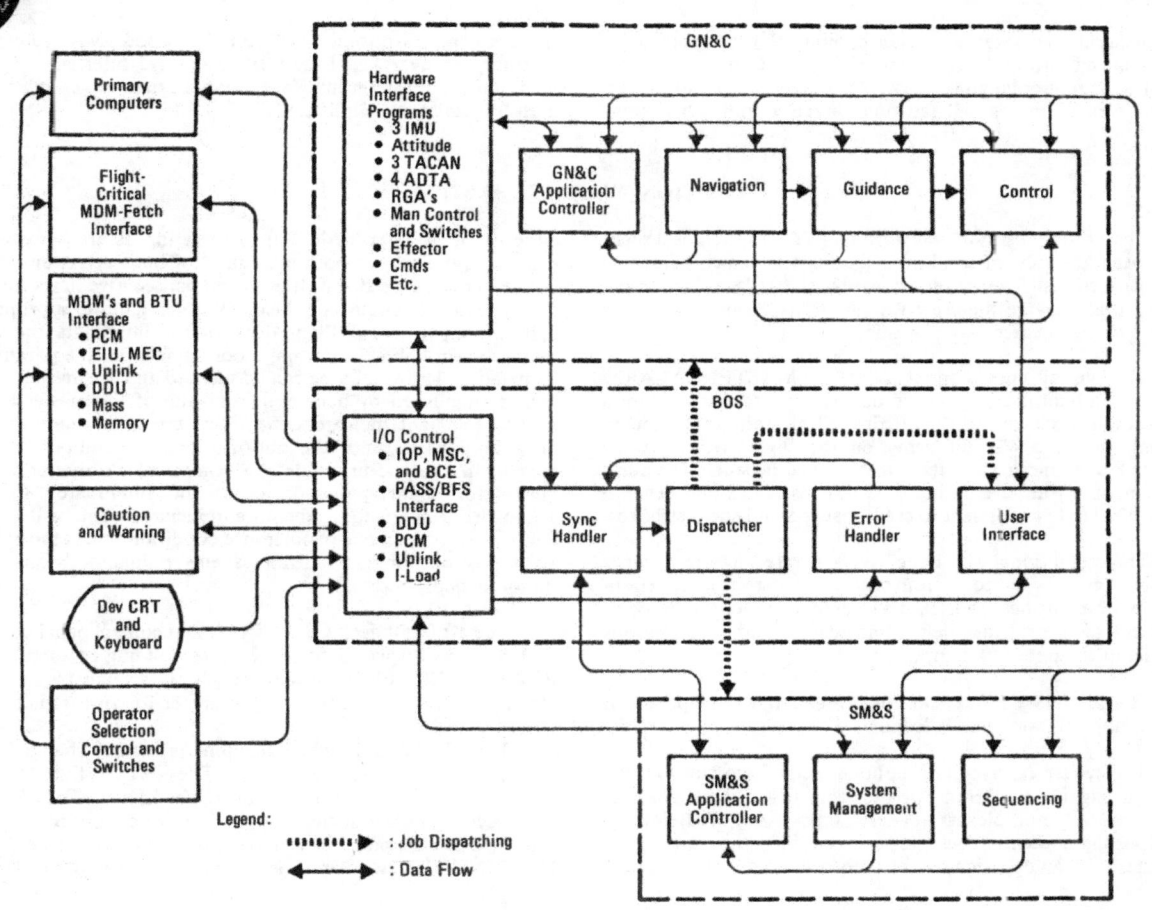

BFS Flow Diagram

Space Shuttle Spacecraft Systems

and general purpose computer-computer processor unit-input/output processor); Intermetrics Inc., Cambridge, MA (advance computer language, HAL/S, avionics software); Lear Siegler, Grand Rapids, MI (attitude direction indicator); Northrop Corp., Electronics Division, Norwood, MA (rate gyro assembly), Singer Kearfott, Little Falls, NJ (inertial measurement unit, multiplexer interface, adapter, data bus coupler, and data bus isolation amplifier).

CAUTION/WARNING SYSTEM

The primary caution and warning (C/W) system is designed to warn the crew of conditions that may adversely affect orbiter operations. The system consists of hardware and electronics that provide the crew with both visula and aural cues when a system exceeds set operating limits.

The visual cues consist of four MASTER ALARM pushbutton light indicators (two on the forward display panels of the flight deck, one at the aft flight deck station, and one in the mid-deck), a 40-light array on the flight deck forward display panel, and a 120-light array on the aft panel. The aural cue consists of an alternating tone that oscillates between 375 and 1,000 Hertz and is sent to crew headsets and speaker boxes.

Three additional aural cues are generated by the primary C/W system: a siren tone from the smoke detection system in the crew compartment, a Klaxon alarm tone from the crew compartment delta pressure/delta time sensor, and a continuous tone from the onboard computers.

The primary C/W system has three modes of operation: ascent, normal, and acknowledge.

The system receives 120 inputs directly from transducers, through signal conditioners, or from the flight forward multiplexer/demultiplexers. These inputs are fed into a multiplexing system. These inputs can be either analog or discretes; the analog signals are 0 to 5 volts dc; the discretes either 0, 5, or 28 volts dc. All of these inputs are designed to provide high, low, or both high and low limit detection. If the parameter has exceeded its limits eight consecutive times for 100 milliseconds, it will turn on the C/W tone, light the appropriate light on the forward flight deck panel, illuminate the four master alarm lights (in normal mode), and store the parameter in memory. The aural tone can be silenced and the master alarm light extinguished by depressing any one of the master alarm pushbutton light indicators; however, the C/W light will remain illuminated until the out-of-tolerance condition is corrected. In the ascent mode the commander's master alarm pushbutton light indicator does not illuminate. In the acknowledgement mode, the 40 annunciator lights will not illuminate during an out-of-tolerance condition unless the master alarm pushbutton light indicator at the commander's or pilot's station is depressed.

The aft flight deck C/W display and control panel consists of 120 status display lights, one corresponding to each input. PARAMETER SELECT thumbwheels on this panel are used to identify (by number) a specific parameter for further action.

The LIMIT SET switch grouping is used to change limits or to read out a parameter's limits. The VALUE thumbwheels are to select the lower or upper limits. The LIMIT UPPER position specifies the parameter's upper limit for change or read-out and the LOWER position specifies the parameter's lower limit. The FUNC SET switch position sets the limit (upper or lower)

Space Shuttle Spacecraft Systems

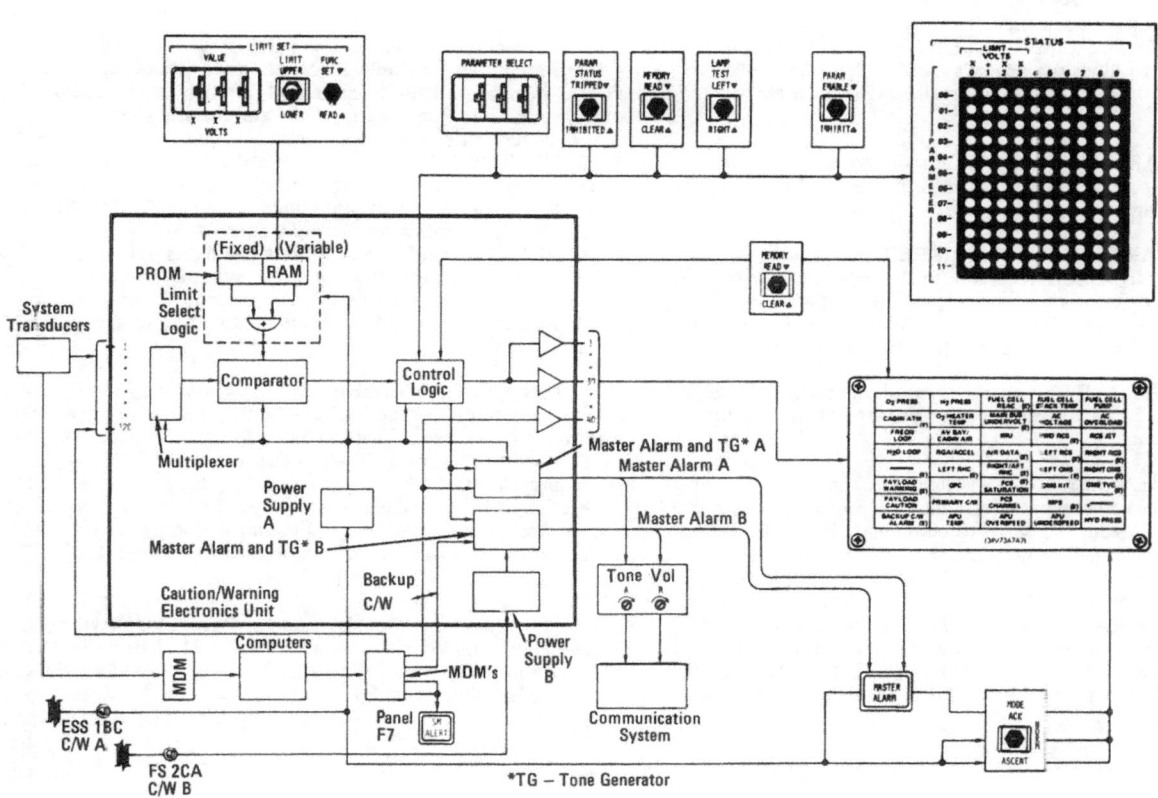

Caution and Warning System Hardware

Space Shuttle Spacecraft Systems

specified by the settings on the VALUE thumbwheels for the parameter on the PARAMETER SELECT thumbwheels. The READ position of the switch illuminates the C/W status lights under the STATUS LIMIT VALUE XXX VOLTS heading corresponding to that parameter's limit. The value read corresponds to the parameter's full-scale range on a scale of 0 to 5 volts dc.

The PARAMETER ENABLE switch position enables (activates) the parameter selected; if positioned to INHIBIT, the selected parameter is inactivated.

The PARAM STATUS TRIPPED switch illuminates all C/W status lights corresponding to the input parameters that are currently out of tolerance, regardless whether they are inhibited. The INHIBITED position illuminates C/W status lights corresponding to the parameters that have been inhibited.

The LAMP TEST switch tests the illumination in the C/W status lights; LEFT for the five left columns and RIGHT for the five right columns.

The MEMORY READ switch illuminates C/W status lights for all enabled parameters that have been out of tolerance since the last MEMORY CLEAR command. The CLEAR position clears the recall memory in the C/W electronics unit.

A CAUTION/WARNING READ switch on the forward C/W display and control panel illuminates lights on the forward C/W panel from enabled parameters that have been out of tolerance since the last MEMORY CLEAR command. The MEMORY CLEAR and LAMP TEST switches on the forward C/W display and control panel also have the same functions as those on the aft panel.

The backup caution and warning system is part of the systems management fault detection and annunciation, the GN&C system, and the backup flight system software (computer) programs. The backup C/W annunciation responds to the same conditions as the primary system. It illuminates BACKUP C/W light on the forward flight deck display and control panel, produces a continuous tone in the flight crew headsets and speaker boxes, illuminates the MASTER ALARM pushbutton light indicators, and produces a fault message on the CRT display. The MASTER ALARM lights operate in the same manner as in the primary C/W system. The computer keyboards can be used to change the lower and upper limit (not in all cases) or inhibit and re-enable parameters, similar to the primary C/W.

The systems management Alert program, which operates in a manner similar to the backup C/W system, is designed to inform the flight crew of a situation that may be leading up to a caution or warning, or a situation that may require additional procedures. When an Alert parameter exceeds its limits, the SM ALERT light is illuminated on the flight deck forward display and control console and a signal is sent to the primary C/W system to turn on the systems management tone and the CRT displays.

The contractors involved are: Aerospace Avionics, Bohemia, NY (annunciators), and Martin Marietta, Denver, CO (caution and warning electronics and caution and warning status display, limit module).

Space Shuttle Spacecraft Systems

SMOKE DETECTION AND FIRE SUPPRESSION

Smoke detection and fire suppression capabilities are provided in the crew cabin avionics bays, in the crew cabin, and the payload (Spacelab).

Ionization detection elements, which sense levels of smoke concentrations or rate of concentration change, trigger alarms and provide smoke-concentration-level intelligence to the performance-monitoring system on the crew cabin flight deck display and control panel. There are two ionization detection elements each in crew cabin avionics bays 1, 2, and 3A. There are three ionization detection elements in the crew cabin, one in the ECLSS cabin fan plenum outlet, which is located beneath the crew cabin mid deck floor, and one each in the crew cabin left and right flight deck return air ducts. There are also ionization detectors in Spacelab when it is in the payload bay.

Fire suppression in the crew cabin avionics bays 1, 2, and 3A is by remote-controlled fire extinguishing agents in each bay. For the remaining areas there are portable extinguishers.

The two smoke ionization detector elements (A and B) in crew cabin avionics bays 1, 2, or 3A will trip if the smoke concentration is greater than 2,000 plus or minus 200 micrograms per cubic meter (cubic foot) or if the rate of smoke increase is 22 micrograms per cubic meter (cubic foot) for eight consecutive counts in 20 seconds. The trip signal will illuminate the applicable AV BAY red SMOKE DETECTION A or B light on flight deck display and control Panel L1. A siren tone will sound in the crew members' headsets.

Avionics bays 1, 2, and 3A each have a Freon-1301 (bromotrifluoromethane) extinguishing agent bottle. To activate the bottle in the applicable bay, the applicable FIRE SUPPRESSION AV BAY switch on the flight deck display and control panel is positioned to ARM on Panel L1. The applicable FIRE SUPPRESSION AV BAY AGENT DISCH (discharge) guarded pushbutton while light indicator is depressed on the

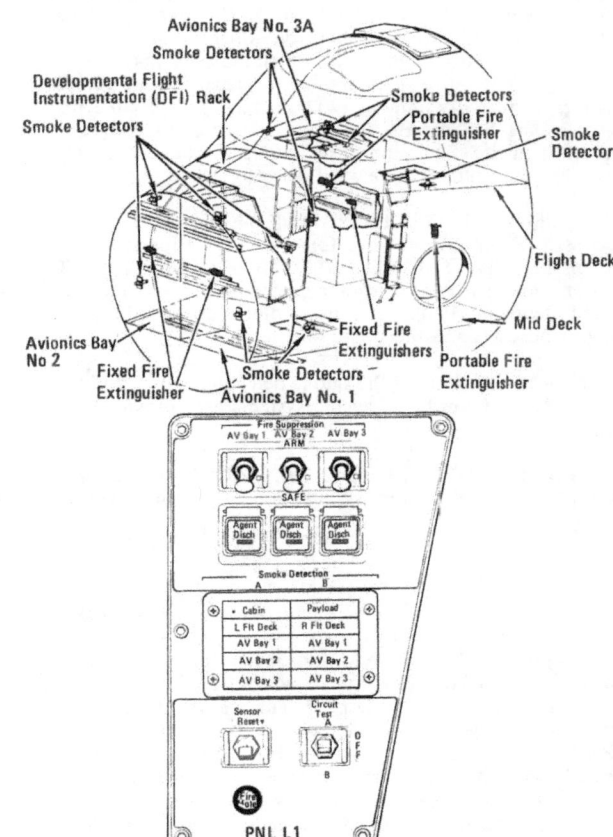

Smoke Detection and Fire Suppression System

Space Shuttle Spacecraft Systems

panel. The pushbutton white light indicator activates the corresponding pyro initiator controller (PIC), which drives a pyrotechnic valve to discharge the Freon-1301 fire extinguishing agent from the bottle. The pushbutton is held until the white light in the pushbutton illuminates, which indicates that fire extinguishing agent bottle pressure has decreased from 10,350 millimeters of mercury (mmHg) (200 psi at 21 °C (70 °F) to 3,105 plus or minus 517 mmHg (60 plus or minus 10 psi). The pushbutton white light indicator would illuminate if the pressure in that bottle had decayed prior to its use.

The SMOKE DETECTION switch on flight deck display and control Panel L1 provides for reset of a tripped smoke detection unit. The SMOKE DETECTION CIRCUIT TEST switch on Panel L1 tests the smoke detectors, the smoke detection red lights on Panel L1, the audible tone in the crew member's headset, and the siren.

Various parameters of the smoke detection system and the remote fire extinguishing agent system are provided to telemetry.

The SMOKE DETECTION CABIN red light on the flight deck display and control panel would illuminate from the smoke detection ionization element in the ECLSS cabin fan plenum; the L FLT DECK red FIRE DETECTION light would illuminate from the crew cabin left flight deck return air duct smoke ionization element, and the R FLT DECK red FIRE DETECTION light would illuminate from the right flight deck return air duct.

Portable hand-held fire extinguishers are available in the crew cabin. The extinguishing agent is Halon-1301 (monobromotrafluoromethane). Halon-1301 minimizes the major hazards of a conflagration: smoke, heat, oxygen depletion, and formation of pyrolysis products such as carbon monoxide.

Two of the fire extinguishers are located on the crew module cabin mid deck and two on the flight deck. The fire extinguisher nozzles can fire through fire hole ports in the display and control panels in the event of fire in back of the display and control panels.

The contractors involved with the smoke detection and fire suppression system are Brunswick Celesco, Costa Mesa, CA (smoke detectors and remote control fire extinguishing agent); J.L. Products, Gardena, CA (arming fire pushbutton); Metalcraft Inc., Baltimore, MD (portable fire extinguishers).

HEAD UP DISPLAY

In September 1979 the NASA directed Rockwell International's Shuttle Orbiter Division to add a Head UP Display (HUD) System to the Shuttle Orbiter. This was the result of approximately two years of study activity by the NASA and Rockwell with the support of several HUD subcontractors. Rockwell awarded a letter contract to Kaiser Electronics, San Jose, CA, in June, 1980 to provide a HUD system for the Orbiter. The program progressed on schedule and a breadboard hardware set was delivered to Rockwell in November, 1980 for use in the Avionics Development Laboratory for preliminary interface testing.

A head up display system allows an out of the window view while providing flight commands and information to the flight crew by superimposing this information on a transparent combiner in the out the window field of view. The baseline Orbiter,

Space Shuttle Spacecraft Systems

Head Up Display

like most commercial aircraft utilizes conventional electromechanical displays on a display panel beneath the glareshield which necessitates that the flight crew look down for information and then up for the out the window information. During critical flight phases, in particular the approach and landing case, this is not an easy task. In the Orbiter with its unique vehicle dynamics and approach trajectories this situation is even more critical.

Since the Orbiter is intended to be in service for several years, it was considered appropriate that the Orbiter be equipped with this system. In the study phase, it was determined that most recent military aircraft include HUD systems and that the airliners used by several European countries also contain HUD's. Additionally, it was apparent that the display portion of some existing HUD systems would lend themselves to installation in the Orbiter. So as to minimize development costs, the HUD system requirements for the Orbiter were patterned after existing hardware.

While the display portion of the Orbiter system could be similar to existing HUD systems, the drive electronics could not. The Orbiter Avionics is digital and since minimal impact to the Orbiter was paramount, the HUD drive electronics are designed to receive data from the Orbiter data buses. Most existing HUD drive electronics use analog data or a combination analog/digital interface. In the Orbiter system, the HUD drive electronics utilizes to the maximum extent possible the same data which drives the existing electromechanical display devices to minimize impact on the Orbiter software.

The Orbiter display device as designed by Kaiser Electronics uses a cathode ray tube (CRT) to create the image which is then projected through a series of lens on a combining glass which is very similar to a system they developed and produce for the Cobra Jet Aircraft. Certain Orbiter design requirements including vertical viewing angles, brightness and unique mounting requirements dictated some changes from the Cobra Jet configuration.

On the Orbiter, a HUD will be installed at each flight station (Commander and Pilot) with a HUD POWER ON/OFF switch at each flight station. Each HUD system is single string although connected to two data buses all redundancy is achieved, the fact that a system is installed at each station (station redundancy similar to that used for the present displays) and the fact that the existing displays can be used in the event of one HUD failure.

The HUD system was installed on the Orbiter in line in the production flow for OV-099, OV-103 and OV-104 and a kit supported installation of the HUD on OV-102. In addition, several other systems were built for test sites and simulators.

Space Shuttle Spacecraft Systems

A new improved display format is being developed for the head-up display which will further reduce the commander/pilot workload during approach and landing. This improved format will be implemented on future Space Shuttle missions.

The HUD is an electronic/optical device with two sets of combiner glasses located above the glareshield and in the direct line of sight of the Commander and the Pilot. Essential flight information for vehicle guidance and control during approach and landing is projected on the combiner glasses and is collimated at infinity.

In the example shown, the orbiter is in the final phase of the preflare maneuver with EAS (equivalent airspeed) = 280 kts (knots) (left scale), altitude = 500 feet (right scale), with orbiter heading (+) slightly to the left of runway centerline — which indicates a light crosswind from the left. The velocity vector ▷ -○- ◁ , is shown just crossing runway overrun. The guidance diamond is shown centered inside the velocity vector symbol. The flare triangles on the wing tips show that the pilot is precisely following the flare command. The lighted outline of the start of the runway zone can be seen at the top of the combiner. The HUD is capable of displaying speed brake command and position, discrete messages such as "Gear", and during rollout; deceleration and wing leveling parameters.

The images are generated by a small cathode ray tube (CRT) and passed through a series of lenses before being displayed to the flight crew on the combiners as lighted symbology. The transmissiveness of the combiner is such that the crew can look through them and see actual targets, i.e. runway, etc. Example: Assume the crew is conducting an instrument approach and is currently at 2,133 meters (7,000 feet) on the final approach course in a solid overcast, the base of which is at 1,524 meters (5,000 feet): The lighted outline of the runway would be displayed on the combiner. However when the orbiter exited the overcast at 1,524 meters (5,000 feet), the lighted outline of the runway would be superimposed on the real runway.

As the orbiter proceeds down the steep glideslope, the velocity vector is superimposed over the steep glideslope aim point. At preflare altitude, flare triangles move up to command the pullout. The pilot maintains the velocity vector symbol between the triangles. After a short period of stabilized flight on the shallow glideslope, the guidance diamond then commands a pitch-up until the nose is about 8 degrees above the horizon — which is essentially the touchdown attitude. After touchdown, during the rollout phase, the commander/pilot maintains the touchdown attitude, approximately plus 6 degrees theta (nose above the horizon) until 180 KEAS (knots equivalent airspeed), then a de-rotation maneuver is commanded.

The HUD is an excellent landing aid and is considered the primary pilot display during this phase. As the system matures, it is anticipated that the HUD will be used for star/land mark sightings as well as rendezvous with other orbiting vehicles, space platforms, etc.

Space Shuttle Spacecraft Systems

DISPLAYS AND CONTROLS

The displays and controls in the orbiter crew compartment enable the flight crew members to supervise, control, and monitor the Space Shuttle mission and vehicle. They include controllers, cathode ray tube (CRT) displays and keyboards, coding and conversion electronics for instruments and controllers, lighting, timing devices, and a caution and warning system.

The displays and controls are designed so that a crew of two can perform normal operations in all mission phases (except payload operations). They are designed to enable a safe return to earth from either commander's or pilot's seat; flight-critical displays and controls are accessible from the forward flight deck station from launch to orbital operations and from deorbit to landing rollout.

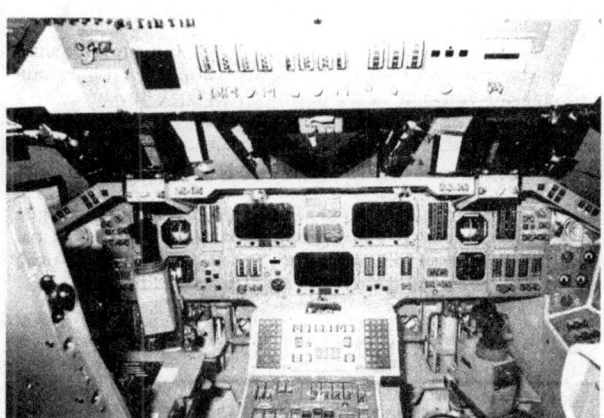

Forward Displays and Controls

Overhead Displays and Controls

 **Space Transportation System**

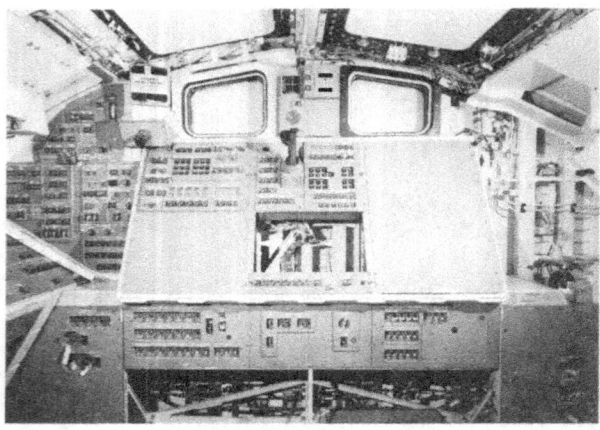

Aft Station Displays and Controls

All controls are protected against inadvertent activation. Toggle switches are protected by wicket guards, and lever lock switches are used wherever inadvertent action would be detrimental to flight operations or could damage equipment. Cover guards are used on switches where inadvertent actuation would be irreversible.

All displays and controls are provided with dimmable floodlighting in addition to integral meter lighting.

The contractors involved are: Abbott Transistor, Los Angeles, CA (transformers), Aerospace Avionics, Bohemia, NY (propellant quantity indicator and annunciators); Aiken Industries, Mechanical Product Division, Jackson, MI (thermal circuit breakers); Applied Resources, Fairfield, NJ (rotary switch), Bendix Corp., Teterboro, NJ (surface position, alpha-Mach, altitude, vertical velocity indicators), Bendix Corp., Davenport, IA (accelerometer indicator); Conrac Corp., West Caldwell, NJ (mission and event timer); Edison Electronics Division of McGraw Edison, Manchester, NH (digital select thumbwheels, toggle switches); Eldec Corp., Lynwood, WA (tape meter); Honeywell, Inc., St. Petersburg, FL (flight control system); IBM Corp., Federal Systems Division, Electronic Systems Center, Owego, NY (cathode ray tube display unit, computer keyboard), ILC Technology, Sunnyvale, CA (cabin interior and exterior lighting); J.L. Products, Gardena, CA (pushbutton switch), Lear Siegler, Grand Rapids, MI (attitude direction indicator); Martin Marietta, Denver CO (C/W status display, limit module), Weston Instruments, Newark, NY (event indicator, electrical indicator meter), Collins-Rockwell, Cedar Rapids, IA (display driver unit, horizontal situation indicator), U.S. Radium, Inc., Parisippany, NJ (integrally-illuminated panels), Betatronix, Hauppauge, NY (potentiometers).

 Space Shuttle Spacecraft Systems

OMITTED

Space Transportation System Payloads

PAYLOAD ASSIST MODULE (PAM)

The Payload Assist Module (formerly called the Spinning Solid Upper Stage — SSUS) is designed as a higher altitude booster of satellites deployed in near Earth orbit but operationally destined for higher altitudes.

The PAM-D is used to boost various satellites to geosynchronous orbit (35,887 kilometers — 22,300 miles) after deployment from the Space Shuttle spacecraft.

There are two versions of the PAM — the "D" which is utilized to launch lighter weight satellites and the "A" which is capable of launching satellites weighing up to 1,995 kilograms (4,400 pounds) into a 27-degree geosynchronous transfer orbit after being deployed from the Shuttle spacecraft's cargo bay.

The PAM-D is capable of launching satellite weights up to 1,247 kilograms (2,750 pounds) into a 27 degree geosynchronous orbit following deployment. A requirement for a 1,361 kilogram (3,000 pound) transfer orbit capability requires about a 10-percent increase in the PAM-D motor performance, which can be accomplished by adding more length to the motor case, but reducing the nozzle length the same amount to retain the overall stage length. The motor case extension is about 137 milimeters (5.4 inches). This uprating will require other changes, namely the strengthening and addition of cradle members so that the system structural dynamic frequency will avoid the Space Shuttle forcing frequencies.

The PAM-A and PAM-D have deployable (expendable) stage consisting of a spin stabilized solid rocket fueled motor (SRM), a payload attach fitting (PAF) to mate with the unmanned spacecraft, and the necessary timing, sequencing, power and control assemblies.

The reusable airborne support equipment (ASE) consists of the cradle structure for mounting the deployable system in the Space Shuttle orbiter payload bay, a spin system to provide the stabilizing rotation, a separation system to release and deploy the stage and unmanned spacecraft, and the necessary avionics to control, monitor, and power the system.

The PAM-A and PAM-D stages are supported through the spin table at the base of the motor and through restraints at the PAF. The forward restraints are retracted before deployment.

The PAM-D also provides a sunshield for thermal protection of the satellite when the Space Shuttle orbiter payload bay doors are open.

PAM-D AIRBORNE SUPPORT EQUIPMENT AND ORBITER INSTALLATION

The PAM-D Airborne Support Equipment (ASE) consists of all the reusable hardware elements that are required to mount, support, control, monitor, protect, and operate the PAM-D expendable hardware and unmanned spacecraft from liftoff to deployment from the Space Shuttle. It will also provide the same functions for the safing and return to the stage and spacecraft in case of an aborted mission. The ASE is designed to be as self-contained as possible, thereby minimizing dependence on orbiter or flight crew functions for its operation. The major ASE elements include the cradle for structural mounting and support, the spin table and drive system, the avionics system to control and monitor the ASE and the PAM-D vehicle and the thermal control system.

The cradle assembly provides a vertical structural mounting support for the PAM-D/unmanned spacecraft assembly in the orbiter payload bay. The nominal envelope for the PAM-D vertical installation provides a cylindrical volume 2,562 millimeters (100.88 inches) in height on the centerline and a

Space Transportation System Payloads

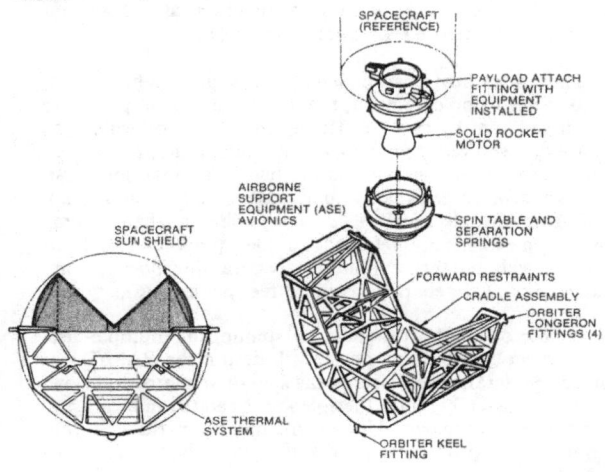

PAM-D System

PAM-D Sunshield Open

diameter of 2,184 millimeters (86 inches). The diameter limitation applies to all early unmanned spacecraft that require the capability to use the Delta launch vehicle as a backup to the Space Shuttle. After full transition to the Space Shuttle is complete, the unmanned spacecraft configuration may use the extra volume available within the Space Shuttle payload bay, a maximum diameter of 2,743 millimeters (108 inches) inside the cradle, 3,048 millimeters (120 inches) above the cradle. The cradle is 4.5 meters (15 feet) wide. The length of the cradle is 2,362 millimeters (93 inches) static and 2,438 millimeters (96 inches) dynamic. The open truss structure cradle is constructed of machined aluminum frame sections and chrome plated steel longeron and keel trunnions.

The spacecraft-to-cradle lateral loads are reacted by forward retractable retraction fittings between the payload attach fitting and cradle, which are driven by redundant dc electrical motors. After the reaction fittings are retracted, the spin table is free to spin the PAM unmanned spacecraft when commanded.

The spin table consists of three subsystems, spin, separation, and electrical interface. The spin subsystem consists of the spin table, the spin bearing, the rotating portion of the spin

Space Transportation System Payloads

PAM-D Sunshield Closed

table, a gear and gear support ring, two redundant drive motors, a despin braking device, and a rotational index and locking mechanism. The separation subsystem includes four compression springs mounted on the outside of the rotating spin table, each with an installed preload of 635 kilograms (1,400 pounds) and a Marman-type clamp band assembly.

The electrical interface subsystem is composed of a slip-ring assembly to carry electrical circuits for PAM-D and spacecraft across the rotating spin bearing. The electrical wiring from the slip ring terminates at electrical disconnects at the spin-table separation point. The slip-ring assembly is used to carry safety-critical command and monitor functions and those commands required before separation from the spin table.

The system provides a capability for spin rates between 45 and 100 rpm. Upon command, the spin table will be spun up to the nominal rpm by two electric motors, either of which can produce the required torque. When the spin table rpm has been verified and the proper point is reached in the parking orbit, redundant debris-free explosive bolt cutters are fired upon command from the electrical ASE to separate the band clamp (which is mechanically retained on the spin table) and the springs provide the thrust to attain a separation velocity of approximately 0.9 meters per second (3 feet per second).

In case of an abort mode after spinup, the multiple-disc-stack friction-type braking device will despin the PAM-D unmanned spacecraft assembly and the spin drive motor will slowly rotate the assembly until the solenoid-operated indexing and locking device is engaged. Upon confirmation by the ASE that the spin table is properly aligned and locked, the restraint pins will be re-engaged.

PAM-D MOUNTED THERMAL CONTROL SYSTEM

The PAM-D thermal control system is provided to alleviate severe thermal stresses on both the unmanned spacecraft and the PAM-D system.

The system consists of thermal blankets mounted on the cradle to provide thermal protection for the PAM-D system, and a passive sunshield mounted on the cradle to control the solar input to and heat loss from the payload when the orbiter payload bay doors are open.

Thermal blankets consisting of multilayered insulation mounted to the forward and aft sides of the cradle protect the

Space Transportation System Payloads

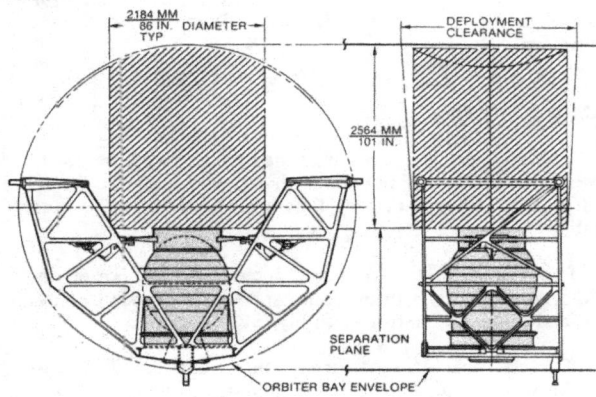

PAM-D Orbiter Vertical Installation

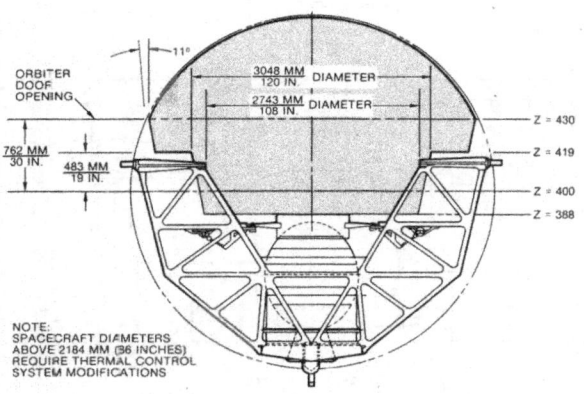

Maximum Spacecraft Envelope With STS PAM-D

PAM-D from thermal extremes. On the sides and the bottom, the orbiter payload bay liner protects the PAM-D from the environmental extremes.

A sunshield, consisting of multilayered, Mylar lightweight insulation supported on a tubular frame, mounts to the cradle and protects the unmanned spacecraft from environmental extremes. The sunshield panels on the sides are fixed and stationary. The portion of the shield covering the top of the unmanned spacecraft is a clamshell structure that remains closed to protect against thermal extremes when the orbiter payload

PAM-D/Telesat-F

Space Transportation System Payloads

bay doors are open. The sunshield resembles a two-piece baby buggy canopy. The clamshell is opened by redundant electric rotary actuators operating a control-cable system.

The sunshield required for the PAM-D growth will have a width adjustment capability to accommodate spacecraft up to 2,901 millimeters (115 inches) in diameter.

PAM-D VEHICLE CONFIGURATION

The PAM-D expendable vehicle hardware consists of a Thiokol Star-48 solid-fueled rocket motor, the payload attach fitting and its functional system. The Star-48 motor features a titanium case, an 89-percent solid propellant, a carbon-carbon throat insert, and a carbon-carbon exit cone. Maximum loading of propellant is 1,998 kilograms (4,405 pounds) with a nominal of 1,738 kilograms (3,833 pounds). The motor is 1,239 millimeters (48.8 inches) in diameter and is 1,828 millimeters (72 inches) long.

The payload attach fitting (PAF) structure is a machined forging and provides the subsystem mounting installations and mounts on the forward ring of the motor case. The two cradle reaction fittings provide structural support to the forward end of the PAM-D stage and unmanned spacecraft, and transmit loads to the ASE cradle structure. The forward interface of the PAF provides the spacecraft mounting and separation system. One steel band is preloaded to approximately 2,585 kilograms (5,700 pounds) and separation is achieved by redundant bolt cutters. Four separation springs, mounted inside the PAF provide the impetus for clear separation. The installed preload for each spring is approximately 90 kilograms (200 pounds) with a spring stroke of 133 millimeters (5.25 inches), providing a spacecraft separation velocity of about 0.9 meters per second (3 feet per second). The electrical interface connectors between the PAM-D and the spacecraft are mounted on brackets on opposite sides of the PAF. Other subsystems mounted on the PAF include the redundant safe-and-arm device for motor ignition, and telemetry components (if desired) and the S-band transmitter.

PAM-D AVIONICS

The electrical ASE minimizes the number of operations to be performed by the flight crew so that greater attention can be paid to monitoring functions that are critical to safety and reliability.

Flight crew control functions include system power on, SRM arming, deployment ordnance arming, emergency deployment and sequence control assembly (SCA) control.

The electrical ASE performs control and monitoring of restraint withdrawal, spin-table spin and deployment functions;

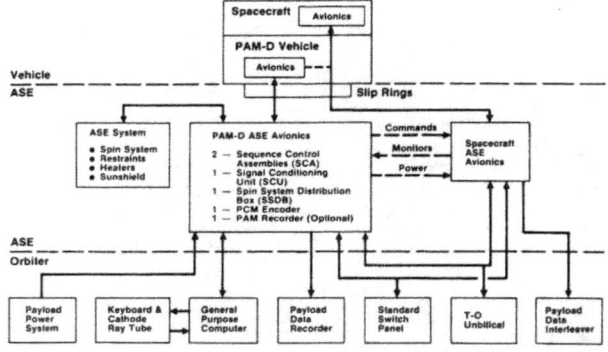

PAM-D Interfaces

Space Transportation System Payloads

arms (and disarms, if necessary) the SRM; controls and monitors the PAM-D vehicle electrical sequencing system (and telemetry system, when used); generates system status information for display to the flight crew (cathode ray tube) via the data lens and from the orbiter keyboard panel; and provides wiring to carry required spacecraft functions. And, as a mission option, it provides control and monitoring of spacecraft systems.

The Payload Assist Modules are designed and built by McDonnell Douglas Astronautics, Co., Huntington Beach, CA.

INERTIAL UPPER STAGE (IUS)

The Inertial Upper Stage (IUS) will be used with the Space Shuttle to transport NASA's Tracking and Data Relay Satellites (TDRS) and other NASA and Department of Defense satellites destined for much higher orbits than that of the Space Shuttle orbiter or for trajectories beyond earth orbit. For instance, many communications satellites are stationed at geosynchronous orbit, some 35,880 kilometers (22,300 statute miles) from earth.

The IUS was originally designed as a temporary stand-in for a reusable space tug and the vehicle was named the Interim Upper Stage. The word "Inertial" (signifying the satellites guidance system) later replaced "Interim" when it was seen that the IUS would be needed through the 1980's.

The IUS is being developed and built under contract to the Air Force Systems Command's Space Division. After two and a half years of competition, Boeing Aerospace Company, Seattle, WA, was selected in August 1976 to begin preliminary design of the IUS.

IUS is a two stage vehicle weighing approximately 14,742 kilograms (32,500 pounds). Each stage is a solid rocket motor and was selected over those of liquid fueled engines due to relative simplicity, high reliability, low cost and safety. Thus, the reference IUS-2 (two stage).

The IUS is 5.18 meters (17 feet) long and 2.8 meters (9.5 feet) in diameter. It consists of an aft skirt; an aft stage solid rocket motor containing 9,707 kilograms (21,400 pounds) of propellant and generating 95,187 Newtons (21,400 pounds) of thrust; an interstage; a forward stage solid rocket motor with 2,721 kilograms (6,000 pounds) of propellant generating 82,288 Newtons (18,500 pounds) of thrust, and an equipment section. The equipment support section contains the avionics which provide guidance, navigation, telemetry, command and data management, reaction control and electrical power. All mission-critical components of the avionics system are redundant, along with thrust vector actuators, reaction control thrusters, motor ignitor and pyrotechnic stage separation equipment to assure reliability of better than 98 percent.

AIRBORNE SUPPORT EQUIPMENT (ASE)

The IUS Airborne Support Equipment (ASE) is the mechanical, avionics, and structural equipment, located in the orbiter. The ASE supports and provides services to the IUS and the TDRS in the orbiter payload bay and provides positioning of the IUS/TDRS in an elevated position for final checkout prior to deployment from the orbiter.

The IUS ASE consists of the structure, batteries, electronics, and cabling to support the IUS vehicle/TDRS combination. These ASE subsystems enable the deployment of the combined vehicle, provide and/or distribute and control electrical power to the IUS and TDRS and communication paths between the IUS and/or TDRS and the orbiter.

Space Transportation System Payloads

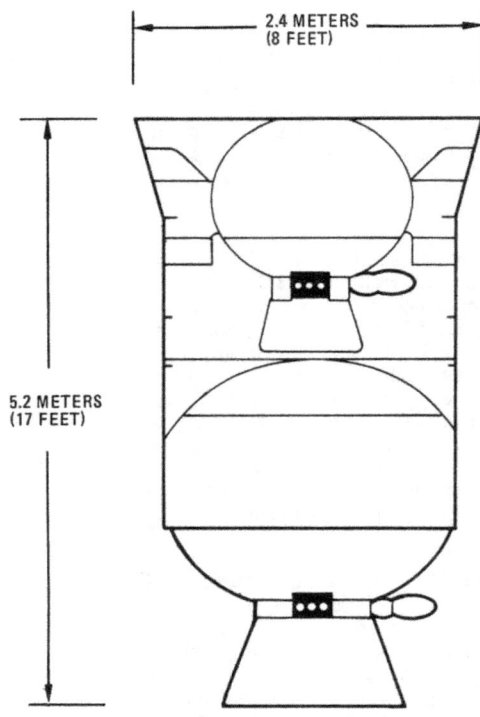

Inertial Upper Stage (IUS)-2

The ASE incorporates a low-response spreader beam and torsion-bar mechanism that reduces spacecraft dynamic loads to less than one-third of what would be experienced without this system. In addition, the forward ASE frame includes a hydraulic load leveler system to provide a balanced loading at the forward trunnion fittings.

The ASE data subsystem provides for the transfer of data and commands between the IUS/TDRS combination to the appropriate orbiter interface. Telemetry data includes TDRS data received over dedicated circuits via the IUS and TDRS telemetry streams. An interleaved stream is provided to the orbiter for transmission to the ground or transfer to ground support equipment.

The structural interfaces in the orbiter payload bay consist of six standard nondeployable attach fittings on each longeron which mate with the ASE aft and forward support frame trunnions, and two payload retention latch actuators at the forward ASE support frame. The IUS has a self-contained spring actuated deployment system which imparts a velocity to the IUS at release from the raised deployment attitude. Ducting from the orbiter purge system interfaces with the IUS at the forward ASE.

IUS STRUCTURE

The IUS structure is capable of transmitting all the loads generated internally and also by the cantilevered spacecraft during orbiter operations and the IUS free flight. In addition, the structure supports all the equipment and Solid Rocket Motor's (SRM's) within the IUS, and provides the mechanisms for IUS stage separation. The major structural assemblies of the two-stage IUS are the equipment support section, interstage, and aft skirt. The basic structure is made from aluminum skin-stringer construction with eight longerons and ring frames.

EQUIPMENT SUPPORT SECTION (ESS)

The Equipment Support Section (ESS) houses the majority of the avionics and control subsystems of the IUS. The top of

Space Transportation System Payloads

IUS/TDRS-A With Airborne Support Equipment in Payload Cannister Transporter

the ESS contains the 3 meter (10 feet) in diameter interface mounting ring and electrical interface connector segment for mating and integrating the TDRS with the IUS. Thermal isolation is provided by a multilayer insulation blanket across the interface between the IUS and spacecraft. All Line Replaceable Units (LRU's) mounted in the ESS can be removed and replaced via access doors even with the mated TDRS. This includes all equipment except the IUS Reaction Control System (RCS) tankage.

IUS AVIONICS SUBSYSTEM

The avionics subsystem consists of the telemetry, tracking, and command (TT&C) subsystem, guidance and navigation (G&N) subsystem, data management (DM) subsystem, thrust vector control (TVC) subsystem, and electrical power (EP) subsystem. This includes all the electronic and electrical hardware used to perform all computations, signal conditioning, data processing, and software formatting associated with navigation, guidance, control, data management and redundancy management. The IUS avionics subsystem also provides the communications between the orbiter and ground stations, in addition, it also provides for electrical power distribution.

Data management performs the computation, data processing and signal conditioning associated with guidance, navigation and control; safe-arm and ignition of the IUS two stage solid rocket motors (SRM's) and electro-explosive devices (EED's); command decoding and telemetry formatting; redundancy management; and issues spacecraft discretes. The data management subsystem consists of two computers, two signal conditioner units (SCU's) and a signal interface unit (SIU).

Modular, general-purpose computers use operational flight software to perform in-flight calculations and to intiate vehicle thrust and attitude control functions necessary to guide the IUS/satellite payload through a predetermined flight path to

Space Transportation System Payloads

a final orbit. A stored program including data known as the on-board digital data load, is loaded into the IUS flight computer memory from magnetic tape through the memory load unit (MLU) during pre-launch operations. Memory capacity is 65,536 (64K) 16-bit words.

The signal conditioner unit (SCU) provides the interface for commands and measurements between the IUS avionics computers and the IUS pyrotechnics, power, reaction control system (RCS), thrust vector control (TVC), telemetry, tracking, and command (TT&C), star scanner and the satellite payload. The SCU consists of two channels of signal conditioning and distribution for command and measurement functions. The two channels are designated channel A and B. Channel B is redundant to channel A for each measurement and command function.

The signal interface unit (SIU) performs buffering, switching, formatting and filtering of TT&C interface signals.

Communications and power control equipment are located and mounted at the orbiter aft flight deck payload station and

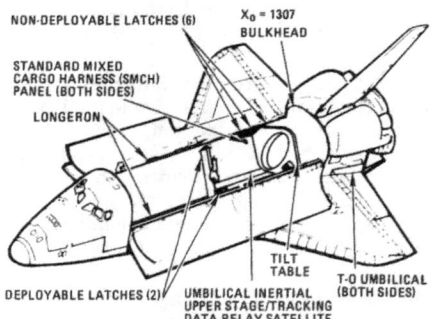

Inertial Upper Stage (IUS) Airborne Support Equipment (ASE)

operated in flight by the orbiter flight crew mission specialists. Electrical power and signal interfaces to the orbiter are located at the IUS equipment connectors. Cabling to the orbiter equipment is provided by the orbiter. In addition, the IUS provides dedicated hardwires from the TDRS through the IUS to an orbiter multiplexer/demultiplexer (MDM) for subsequent display on the orbiter cathode ray tube (CRT), parameters requiring observation and correction by the orbiter flight crew. This capability is provided until IUS ASE umbilical separation.

To support TDRS checkout or other IUS initiated functions, the IUS has the capability of issuing a maximum of eight discretes. These discretes may be initiated manually by the orbiter flight crew prior to deployment from the orbiter, or automatically by the IUS mission sequencing flight software after deployment. The discrete commands are generated in the IUS computer either as an event scheduling function (part of normal on-board automatic sequencing) or a command processing function initiated from an uplink command from the orbiter or Air Force Satellite Control Facilities (AFSCF) to alter the on-board event sequencing function and permit the discrete commands to be issued at any time in the mission.

During the ascent phase of the mision, the TDRS telemetry is interleaved with the IUS telemetry and is continually transmitted via the orbiter's S-Band FM downlink communications link. In addition the IUS/TDRS telemetry data is recorded on the orbiter's payload recorder. Telemetry transmission on the IUS RF link begins after the IUS/TDRS is tilted for deployment from the orbiter. TDRS telemetry data may be transmitted by the TDRS directly to the ground when in the orbiter payload bay with the orbiter payload bay doors open or during IUS/TDRS free flight.

IUS guidance and navigation consists of redundant strapped down IMU's and star tracker. The IMU consists of five rate integrating gyros, five accelerometers and associated

Space Transportation System Payloads

electronics. The IUS stellar-inertial guidance, navigation subsystem provides measurements of angular rates, linear accelerations and other sensor data to data management for appropriate processing by software resident in the computers. The electronics provides conditioned power, digital control, thermal control, synchronization, and the necessary computer interfaces for the inertial sensors. The electronics is configured to provide three fully independent channels of data to the computers. Two channels each, support two sets of sensors and the third channel supports one set. Data from all five gyro/accelerometer sets is sent simultaneously to both computers.

The guidance and navigation subsystem is calibrated and aligned on the launch pad. The navigation function is initialized at liftoff and data from the IMU is integrated in the navigation software to determine the current state vector. Before vehicle deployment, an attitude update maneuver is performed by the Orbiter, with the IUS star scanner scanning two stars separated by 60 to 120 degrees, to compensate for accumulated drift errors.

If, for any reason, the computer is powered-down prior to deployment, the navigation function is reinitialized by transferring Orbiter position, velocity, and attitude data to the IUS vehicle. Attitude updates are then performed as described above.

Star scanner attitude updates can also be performed after deployment. For example, an update can be used shortly before SRM-2 ignition to correct for attitude drift accumulated during the transfer orbit coast.

The IUS vehicle uses an explicit guidance algorithm (gamma guidance), to generate thrust steering commands, SRM ignition time and RCS vernier thrust cutoff time. Prior to each SRM ignition and each RCS vernier, the vehicle is oriented to a thrust attitude based on nominal performance of the remaining propulsion stages. During SRM burn, the current state vector determined from the navigation function is compared to the desired state vector, and commanded attitude is adjusted to compensate for the build up of position and velocity errors due to off-nominal SRM performance (thrust, I_{sp}). The primary purpose of the vernier thrust is to compensate for velocity errors resulting from SRM impulse and cutoff time dispersions. However, residual position errors remaining from the SRM thrusting and position errors introduced by impulse and cutoff time dispersions are also removed by the RCS.

Attitude control in response to guidance commands in provided by thrust vector control (TVC) during powered flight and by reaction control thrusters during coast. Measured attitude from the guidance and navigation subsystem is compared with guidance commands to generate error signals. During solid

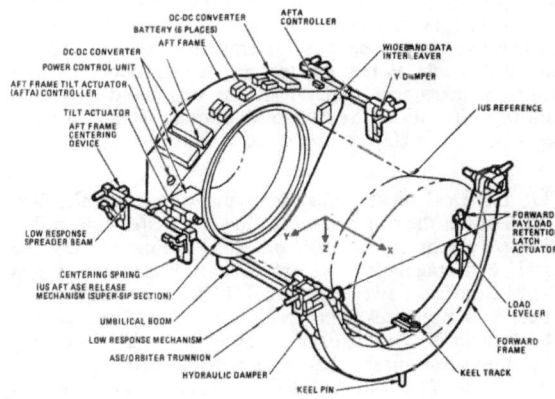

Inertial Upper Stage (IUS) Airborne Support Equipment (ASE)

Space Transportation System Payloads

motor thrusting, these error signals drive the motor nozzle actuator electronics in the TVC subsystem, the resulting nozzle deflections produce the desired attitude control torques in pitch and yaw. Roll control is maintained by the RCS roll-axis thrusters. During coast flight, the error signals are processed in the computer to generate RCS thruster commands to maintain vehicle attitude or to maneuver the vehicle. For attitude maneuvers, quarternion rotations are used.

Thrust vector control (TVC) provides the interface between the IUS guidance and navigation and the SRM gimbaled nozzle to accomplish powered-flight attitude control. Two complete electrically redundant channels are provided to minimize single point failure. The TVC subsystem consists of two controllers, four actuators and four potentiometers for each IUS SRM. Power is supplied through the signal conditioner unit (SCU) to the TVC controller which controls the actuators. The controller receives analog pitch and yaw commands, proportioned to desired nozzle angle, and converts them to pulse-width modulated voltages to power the actuator motors. The motor drives a ball screw which extends or retracts the actuator to position the SRM nozzle. Potentiometers provide servo loop closure and position instrumentation. A staging command from the SCU allows switching of the controller outputs from IUS stage one actuators to the IUS second stage actuators.

IUS electrical power subsystem consists of avionics batteries, IUS power distribution units, power transfer unit, utility batteries, pyrotechnic switching unit, IUS wiring harness and umbilical, and staging connectors. The IUS avionics system distributes electrical power to the IUS/TDRS interface connector for all mission phases from prelaunch to TDRS separation. The IUS system distributes orbiter power to the TDRS during ascent and on-orbit phases. The ASE provides batteries to supply power to the TDRS in the event of orbiter power interruption. A dedicated IUS/TDRS battery ensures uninterrupted

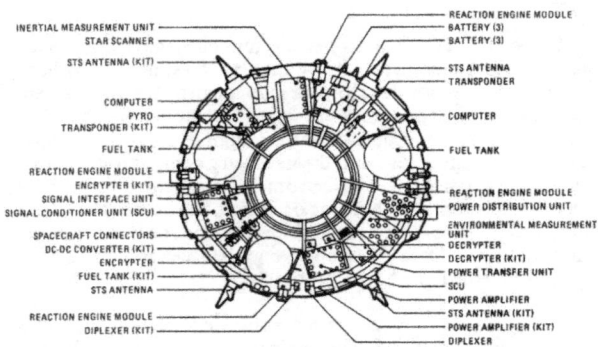

Inertial Upper Stage (IUS) Equipment Support Section

power to the TDRS from IUS deployment to TDRS separation. The IUS will also accomplish an IUS automatic power down if high temperature limits are experienced prior to opening of the orbiter payload bay doors. Dual buses ensure that no single power system failure can disable both A and B channels of avionics. For the IUS-two stage vehicle, four batteries (three avionics, one TDRS) are carried in the IUS first stage. Five batteries (two avionics, two utility, one TDRS) are provided to supply power for the IUS second stage, after staging. Redundant IUS switches transfer the power input between TDRS, GSE, ASE, and IUS battery sources.

It is noted that stage one to stage two IUS separation uses redundant low-shock ordnance devices that minimize the shock environment on the TDRS. The IUS provides and distributes ordnance power to the IUS/TDRS interface for firing TDRS ordnance devices in two groups of eight initiators, a prime group and a backup group. Four separation switches or breakwires provided by the TDRS are monitored by the IUS telemetry system to verify TDRS separation.

Space Transportation System Payloads

IUS SOLID ROCKET MOTORS (SRM's)

The IUS two stage vehicle incorporates a large solid rocket motor and a small solid rocket motor. These motors employ movable nozzles for thrust vector control (TVC). The nozzles are positioned by redundant electromechanical actuators permitting up to four degrees of steering on the large motor and seven degrees on the small motor. Kevlar filament wound cases with largely wound in aluminum end rings, provide high strength at minimum weight. The large motor is the longest thrusting duration solid rocket motor ever developed for space, having a thrusting duration of 145 seconds. Variations in user mission requirements are met by tailored propellant offloading. The small motor can be flown either with or without its extendible exit cone (EEC). The EEC provides an increase of 14.5 seconds in the delivered specific impulse (I_{sp}) of the small motor.

IUS REACTION CONTROL SYSTEM (RCS)

The IUS RCS is a hydrazine monopropellant positive expulsion system that controls the IUS/TDRS attitude during IUS coast periods, roll control during SRM thrustings, and the delta velocity impulses for accurate orbit injection. Valves and thrusters are redundant permitting continued operation with a minimum of one failure. The maximum thrusting time during an SRM thrusting period is 14 seconds.

The IUS baseline includes two RCS tanks with a capacity of 54 kilograms (120 pounds) of hydrazine each. A production option is available to add a third tank if required. To avoid TDRS contamination, the IUS has no forward facing thrusters. IUS users with spin stabilized spacecraft can use the RCS to provide spinup prior to spacecraft separation. The system is also used to provide the velocities for spacing between multiple spacecraft deployments and for a collision avoidance maneuver after TDRS separation.

Inertial Upper Stage Tracking Data Relay Satellite Deployment

The RCS is a sealed system which is serviced prior to TDRS mating. Propellant is isolated in the tanks with pyrotechnic squib operated valves which are not activated until IUS deployment from the orbiter. The tank and manifold safety factors are such that they are no safety constraints imposed on operations in the vicinity of the serviced tanks.

IUS TO TDRS INTERFACES

Physical attachment of the TDRS to IUS is provided by a maximum of eight attachment points. It provides substantial load carrying capability while minimizing thermal conditions across the interface. Separation between the TDRS and IUS is provided by the TDRS. The separation plane is forward of the IUS/TDRS attachment interface and the TDRS adapter will remain with the IUS after separation occurs.

 **Space Transportation System Payloads**

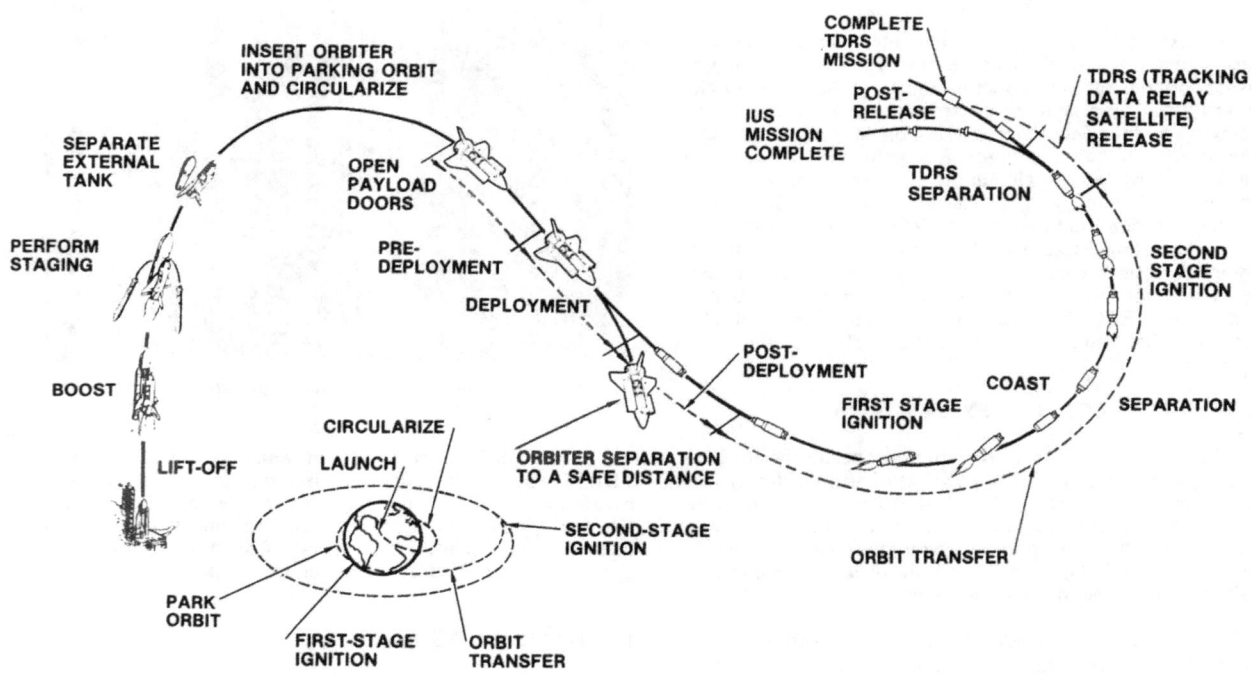

Sequence of Events for Typical Geosynchronous Mission

Space Transportation System Payloads

Power and data transmission to the spacecraft are provided by several IUS interface connectors. Access to these connectors can be provided on the spacecraft side of the interface plane, or through the access door on the IUS equipment bay.

The IUS provides a multilayer insulation blanket made up of aluminized Kapton with polyester net spacers with an aluminized Beta cloth outer layer across the 2,837 millimeter (111.7 inch) diameter IUS/TDRS interface. All IUS thermal blankets are vented toward and into the IUS cavity. All gases within the IUS cavity are vented to the orbiter payload bay. There is no gas flow between the TDRS and the IUS and the thermal blankets are grounded to the IUS structure.

FLIGHT SEQUENCE

After the orbiter payload bay doors are opened in earth orbit, the orbiter maintains a payload bay to earth attitude to fulfill payload thermal requirements and constraints except during those operations that require special attitudes (IUS star scan maneuvers, orbiter IMU alignments, RF communications and deployment operations.).

The IUS power is transferred to orbiter power and early predeployment checkout begins, followed by an IUS command link check and TDRS RF command check. The state vector is uplinked to the orbiter for orbiter trim maneuver(s) and the orbiter trim maneuver(s) are performed. The orbiter is oriented for IUS attitude initialization and downlink of the IUS attitude data is verified. The state vector is uplinked to the orbiter and transferred to the IUS and IUS predeployment checkout begins. The orbiter is then maneuvered to the IUS/TDRS deployment attitude.

The forward ASE payload retention latch actuator is released and the aft frame ASE electromechanical tilt actuator tilts the IUS/TDRS combination to 29 degrees. This extends the TDRS into space just outside the orbiter payload bay which allows direct checkout of the TDRS from earth. If a problem has developed within the TDRS, the IUS/TDRS can be restowed.

Prior to deployment, the TDRS electrical power source is switched from orbiter power to IUS internal power by the orbiter flight crew. Verification that the TDRS is on IUS internal power and that all IUS/TDRS predeployment operations have been successfully completed will be evaluated by data contained in the IUS and TDRS telemetry. IUS telemetry data is evaluated by the IUS Mission Control Center at Sunnyvale, CA, and the TDRS data by the TDRS Control Center Analysis of the telemetry will result in a GO/NO-GO decision for IUS/TDRS deployment from the orbiter.

When the orbiter flight crew is given a GO decision, the orbiter flight crew will activate the ordnance that separates the IUS/TDRS umbilical cables. The flight crew will then command the electromechanical tilt actuator to raise the tilt table to 58 degree deployment position. The Orbiter's RCS thrusters are inhibited and the Super* zip ordnance separation device is initiated which physically separates the IUS/TDRS combination from the tilt table and compressed springs provide the force to jettison the IUS/TDRS from the orbiter payload bay at approximately 0.12 meters per second (0.4 feet per second). The IUS/TDRS deployment is performed in the shadow of the orbiter or in earth eclipse. Approximately 14 minutes after IUS/TDRS deployment the orbiter Orbital Maneuvering System (OMS) engines are ignited to provide the orbiter with a separation maneuver from the IUS/TDRS.

The IUS/TDRS is now controlled by the IUS onboard computers. Approximately ten minutes after the IUS/TDRS are ejected from the orbiter, the IUS, RCS is enabled by the IUS onboard computers and all subsequent operations are similarly

Space Transportation System Payloads

sequenced by the IUS computers through TDRS separation and IUS deactivation.

Following RCS activation, the IUS will maneuver the TDRS to it's required thermal attitude and perform the required TDRS thermal control maneuver.

At approximately 46 minutes after IUS/TDRS ejection from the orbiter the SRM-1 ordnance inhibits are removed. It is noted that the belly of the orbiter is oriented towards the IUS/TDRS for window protection attitude from the IUS SRM-1 plume. The IUS will then recompute SRM-1 ignition time and maneuver to the proper attitude for the SRM-1 thrusting period. When the transfer orbit injection opportunity is reached, the IUS computer will enable and apply ordnance power, arm and safe arm devices and ignite the first stage SRM. The SRM-1 thrusting period is approximately 145 seconds to provide sufficient thrust for the orbit transfer phase of the geosynchronous mission. The IUS first stage and interstage are separated from the IUS second stage prior to reaching the apogee point of its trajectory.

During the coast phase the IUS is capable of performing the maneuvers required by the TDRS for thermal protection or communication reasons. If the TDRS requires improved accuracy, the IUS can perform a pre SRM-2 star scan attitude update.

At apogee, the second stage motor is ignited and its thrusting period is approximately 104 seconds providing the final injection to geosynchronous orbit. The IUS then supports TDRS separation and performs a final/collision/contamination avoidance maneuver before deactivating its subsystems.

Boeing's propulsion team member Chemical Systems Division of United Technologies, designed and tests the two solid rocket motors. Teamed with Boeing in the avionics area are TRW and Hamilton Standard Division of United Technologies. TRW provides software design and IUS telemetry, tracking and command system hardware. Hamilton Standard provides guidance system hardware support. Delco, under subcontract to Hamilton Standard, provides the avionics computer. Ball Aerospace Systems Division furnishes the star scanner.

In addition to the actual flight vehicles, Boeing is responsible for the development of ground support equipment for the checkout and handling of the IUS vehicles from factory to launch pad. Boeing also develops the airborne support equipment to support the IUS in the Space Shuttle and monitor it while it is in the Space Shuttle payload bay.

Under a separate contract, Boeing integrates the IUS with the various satellites and joins the satellite with the IUS, checks out the configuration and supports launch and mission control operations for both the Air Force and NASA.

The IUS is fabricated, assembled and tested at the Boeing Space Center in Kent, WA, south of Seattle. The first IUS to be used with the Space Shuttle and TDRS-A was shipped in June 1982 to Cape Canaveral, FL. Boeing is building eight IUS vehicles under its full-scale development contract with the Air Force which began in 1978. The Air Force expects to acquire six more IUS vehicles in 1983-1984.

TRACKING AND DATA RELAY SATELLITE SYSTEM (TDRSS)

A new era in space communications opened with the STS-6 mission with the deployment of the first Tracking and Data Relay Satellite (TDRS). This satellite, TDRS-A, is the first of three identical ones which are planned for the TDRS system. The TDRS system was developed following studies in the early 1970's which showed that a system of telecommunication

Space Transportation System Payloads

satellites operated from a single ground station could better support the projected scientific and application mission requirements and also halt the spiralling cost escalation of upgrading and operating a worldwide tracking and communications network of ground stations.

A second stage IUS booster failed to insert TDRS-A into its prescribed orbit. From April until mid-summer 1983 TRW and NASA engineers used small thrusters on the satellite, firing in short bursts to place TDRS-A into its circular orbit. Contact TRW Public Relations for details.

Six TDRS satellites are being built by TRW's Defense and Space Systems Group, Redondo Beach, CA, for Space Communictions Company (Spacecom) of Gathersburg, MD. Spacecom is jointly owned by Western Union and American Satellite Company, a partnership between Fairchild Industries and Continental Telephone Company. Spacecom owns and operates the TDRS system, which will consist of three multifunction communication satellites and the White Sands Ground Terminal (WSGT), New Mexico built jointly by the team of TRW, Harris Corporation and Spacecom.

One satellite will be stationed over the Pacific Ocean at the Equator southwest of Hawaii at 171 degrees West longitude and is referred to as TDRSS-West, one satellite will be stationed at the Equator over the northeast corner of Brazil at 41 degrees West longitude and is referred to as TDRSS-East and the remaining satellite is centrally located over the Equator at 79 degrees West which is referred to as in-orbit spare. These three satellites will comprise the space segment of the system. The in-orbit spare would be available for use in the event one of the operational satellite malfunctions, or to augment system capabilities during peak periods. The remaining three satellites will be available as flight ready spares.

NASA will lease TDRSS services from SPACECOM, for a 10 year period, under a contract awarded in December 1976.

When the TDRSS is fully operational (including the in-orbit spare), ground stations of the worldwide Spaceflight Tracking and Data Network (STDN) will be closed or consolidated resulting in savings in personnel, operating and maintenance costs with the exception of Merritt Island, FL, Ponce de Leon, FL, and Bermuda which remain open to support the launch of the Space Transportation System in addition to landing at the Kennedy Space Center, FL. Moreover, much of the equipment at the ground stations is almost 20 years old and inadequate to meet the demands of the Space Shuttle and today's advanced spacecraft.

Instead of the existing worldwide network of ground stations which can provide coverage up to only 20 percent of a satellite's or a spacecraft's orbit, limited to the brief periods when the satellite or spacecraft are within the sight of the tracking station. Each tracking station in the network can handle at most two satellites or spacecraft at one time and most stations can handle but one.

The TDRSS operational system can provide continuous global coverage of earth orbiting satellites above 1,200 kilometers (750 miles) up to an altitude of about 5,000 kilometers (3,100 miles). At lower altitudes there will be brief periods when satellites or spacecraft over the Indian Ocean near the Equator will be out of view. The TDRSS operational system will be able to provide almost full-time coverage not only for the Space Shuttle but up to 26 other near earth-orbiting satellites or spacecraft simultaneously.

Deep space probes and earth orbiting satellites above approximately 5,000 kilometers (3,100 miles) will use the three ground stations of the Deep Space Network (DSN) operated for NASA by the Jet Propulsion Laboratory, Pasadena, CA. The STDN stations that were co-located with the three DSN stations, Goldstone, CA, Madrid, Spain, and Canberra, Australia will be consolidated with the DSN.

 **Space Transportation System Payloads**

The TDRSS satellites will be deployed from the Space Shuttle spacecraft at an altitude of 283 kilometers (153 nautical miles) and the Inertial Upper Stage will provide the velocity to place the TDRS satellite at geosynchronous orbit above the Equator at an altitude of 35,880 kilometers (22,300 statute miles). At this altitude, because the speed of the satellite is the same as the rotational speed of earth, they remain "fixed" in orbit over one location.

The data acquired by the two TDRS satellites is relayed to a single centrally located ground terminal at NASA's White Sands Test Facility in New Mexico. From New Mexico, the raw data will be sent directly by domestic communications satellite (DOMSAT) to NASA control centers at Johnson Space Center, Houston, TX, for Space Shuttle operations and the Goddard Space Flight Center, Greenbelt, MD, which schedules TDRSS operations and controls a large number of unmanned satellites. To increase system reliability and availability, there will be no signal processing done onboard the TDRS satellites, they will act as repeaters, relaying signals to and from the ground stations or to and from user satellites or spacecraft. No user signal processing is done onboard the TDRS satellites.

The TDRSS communications capability extends across a wide spectrum that includes voice, television, analog and digital signals.

The highly automated ground station is located at NASA's White Sands Test Facility, New Mexico, and is owned and managed by Spacecom, which NASA also leases. The ground station provides a location at a longitude with a clear line-of-sight to the TDRS satellites and a location where rain conditions are very remote, as rain can interfere with the Ku-band uplink and downlink channels. It is one of the largest and most complex communication terminals ever built. All satellite or spacecraft transmissions are relayed by the TDRS satellites and funneled through the White Sands ground station. The most

Tracking Data Relay Satellite

Space Transportation System Payloads

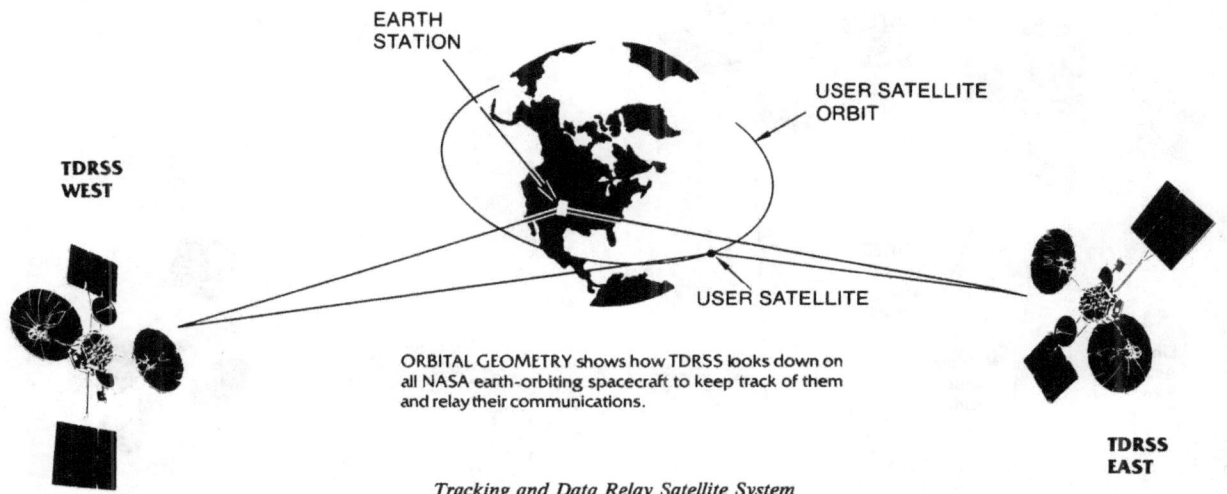

Tracking and Data Relay Satellite System

prominent features of the ground station are three 18 meter (59 feet) Ku-band antennas used to transmit and receive user traffic. Several other smaller antennas are used for S-band and Ku-band communications. NASA is developing a sophisticated operational control system to schedule the use of the system. These control facilities located at Goddard Space Flight Center and adjacent to the ground terminal at White Sands, will enable NASA to schedule the TDRSS support of each user and to distribute the user's data directly from White Sands to the user.

The two TDRS satellites are positioned 130 degrees apart at geosynchronous orbit — instead of the usual 180 degrees spacing. This 130 degree spacing reduces the ground station to one instead of two if the satellites were spaced at 180 degrees.

Initially the TDRSS will be used to support the Space Shuttle mission, Spacelab missions and the Landsat 4 earth resources Satellite program. The TDRSS operational system will provide data from Landsat 4 in near real time, thus eliminating the need to rely upon onboard tape recorders. DOMSAT satellites will be used to transmit Landsat 4 data from White Sands to the data processing facility at the Goddard Space Flight Center and subsequently to the Landsat data distribution center at the Earth Resources Observation System (EROS) Data Center at Sioux Falls, SD.

All TDRS satellites will be placed into geosynchronous orbit from the Space Shuttle spacecraft with an Inertial Upper Stage (IUS)-2 (two stage solid rocket motor) developed for the

 **Space Transportation System Payloads**

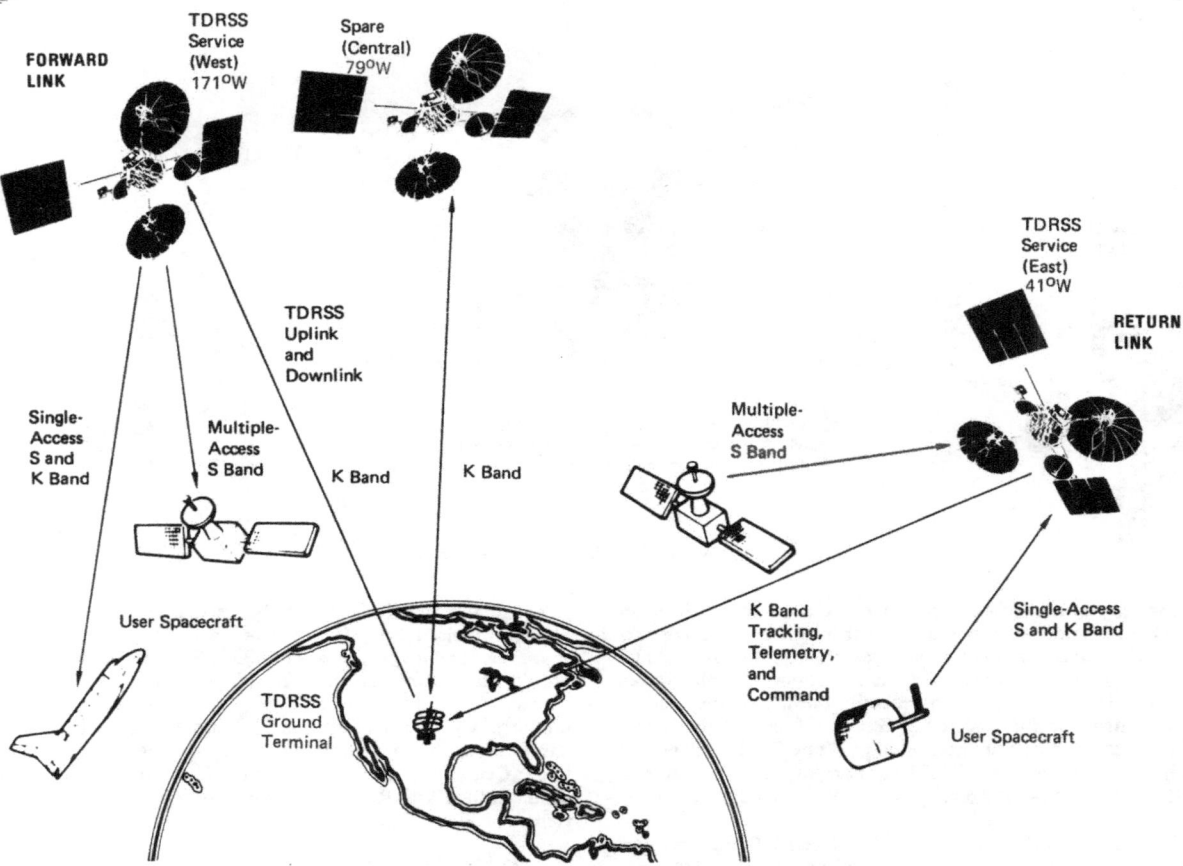

Linking Three Identical and Interchangeable Satellites With Earth Station

Space Transportation System Payloads

Air Force Space Division by Boeing Aerospace Company. The TDRS satellite attached to its IUS is jettisoned from the payload bay of the Space Shuttle Orbiter. The IUS is then under control of its own onboard computers. The Space Shuttle spacecraft is maneuvered to a safe distance from the IUS/TDRS. The IUS performs a series of preparatory maneuvers, then fires its first stage solid rocket motor which propels it towards its geosynchronous position. The Space Shuttle spacecraft flight crew will not see IUS ignition as the Space Shuttle spacecraft is positioned belly towards the IUS for Space Shuttle spacecraft window protection attitude from the IUS solid rocket motor plume. Between the IUS first and second stage firings, repeated earth pointing for TDRS commands and telemetry, plus thermal control maneuvers will take place. Injection into its final geosynchronous orbit is provided by the IUS second stage solid rocket motor.

Deployment of the TDRS satellite solar panels, C-band antenna and space ground link antenna occur prior to the TDRS satellite separation from the IUS. The IUS then separates from

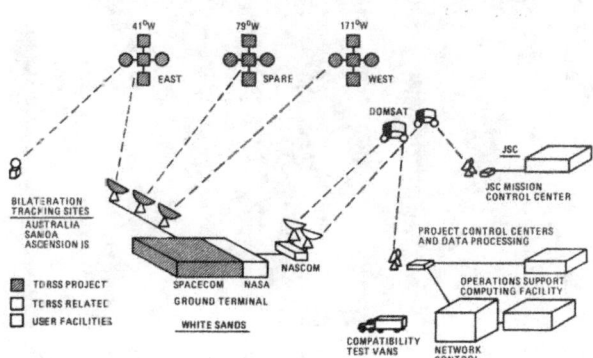

TDRSS System Elements

Tracking Data Relay Satellite Mating With Inertial Upper Stage

 **Space Transportation System Payloads**

the TDRS satellite when the TDRS satellite is in its final orbit and the IUS moves to a non-collision position.

The TDRS single access parabolic antennas deploy after separation from the IUS and subsequent to acquisition of the sun and earth by TDRS satellite sensors utilized for attitude control. Attitude and velocity adjustments place the TDRS satellite into its final geostationary position.

Three-axes stabilization onboard the TDRS satellite maintains attitude control. Body fixed momentum wheels in a vee configuration combine with body fixed antennas pointing constantly at earth while the TDRS satellite solar arrays track the sun. Monopropellant hydrazine thrusters are used for TDRS satellite positioning and north-south and east-west station keeping.

The TDRS satellite to date, are the largest privately owned telecommunications satellites ever built. Each satellite weighs nearly 2,268 kilograms (5,000 pounds) on orbit. The solar arrays on each satellite when deployed span more than 17 meters (57 feet) tip to tip. The two single-access high-gain parabolic antennas when deployed measure 4.9 meters (16 feet) each, in diameter and span 13 meters (42 feet) from tip to tip.

Each TDRS satellite is composed of three distinct modules; the equipment module, the communications payload module and the antenna module. The modular structure reduces the cost of individual design and construction efforts.

The equipment module housing the subsystems that operate the satellite and the communications service is located in the lower hexagon of the satellite. The attitude control subsystem stabilizes the satellite so that the antennas have the proper orientation toward the earth and the solar panels toward the sun. The electrical power subsystem consists of two solar panels that provide a 10 year life span of approximately 1,850 watts of power. Nickel cadmium batteries supply full power when the

Tracking Data Relay Satellite Solar Panels and Antenna Stowed

Space Transportation System Payloads

satellite is in the shadow of earth. The thermal control subsystem consists of surface coatings and controlled electric heaters.

The communications payload module on each satellite is composed of the electronic equipment and associated antennas required for linking the user spacecraft or satellite with the ground terminal. The receivers and transmitters are mounted in compartments on the back of the single-access antennas to reduce complexity and possible circuit losses.

The antenna module is composed of four antennas. For single-access services, each TDRS satellite has two dual feed S-band/Ku-band deployable parabolic antennas. These antennae are 4.9 meters (16 feet) in diameter, attached on two axes that can move horizontally or vertically to focus the beam on satellites or spacecraft below. These antennas are used primarily to relay communications to and from user satellites or spacecraft. The high-bit rate service made possible by these antennas is available to users on a time-shared basis. Each antenna simultaneously supports two user satellites or spacecraft (one at S-band and one at Ku-band), if both users are within the antenna bandwidth. The antennas primary reflector surface is a gold clad molybdenum wire mesh woven like cloth on the same type of machine used to make material for women's hosiery. When deployed 18.9 square meters (203 square feet) of mesh are stretched taughtly between 16 supporting tubular ribs by fine threadlike quartz cords like a glittering metallic spider web. The entire antenna structure, including the ribs, reflector surface, a duel frequency antenna feed and the deployment mechanisms needed to fold and unfold the structure like a parasol, weighs approximately 22 kilograms (50 pounds).

For multiple-access service, the multi-element S-band phased array of 30 helix antennas on each satellite is mounted on the satellite body. The multiple access (MA) forward link (between TDRS and the user satellite or spacecraft) transmits command data to the user satellite or spacecraft and in the return link the signal outputs from the array elements are sent separately to the White Sands Ground Terminal (WSGT) parallel processors. Signals from each helix antenna are received at the same frequency, frequency-division multiplexed into a single composite signal and transmitted to the ground. In the ground equipment, the signal is demultiplexed and distributed to 20 sets of beam forming equipment which allows discrimination of the 30 signals, to select the signals for individual users. The multiple access system uses 12 of the 30 helix antennas on each TDRS satellite to form three transmit beams (one from each TDRS satellite) in the direction of the users.

A fourth antenna, a 2 meter (6.5 feet) parabolic reflector, provides the prime link for relaying transmission to and from the ground terminal at Ku-band.

Each of the six Ku-band return service channels (two per TDRS satellite) have the capacity to receive up to 300 million bits-per-second of digital information. Receiving equipment at the White Sands Ground Terminal is provided to handle two channels simultaneously.

Thus, TDRSS will serve as a radio data relay, carrying voice, television, analog, and digital data signals. It will be the first telecommunications satellite to simultaneously offer three frequency band service: S-band, C-band, and high capacity Ku-band. The C-band transponders operate at 4-6 gigahertz and the Ku-band TDRS transponders operate at 12-14 gigahertz.

Automatic data processing equipment at the White Sands Ground Terminal aids in making user satellite tracking measurements, controls all communications equipment in the TDRS satellite and in the ground station, and collects system status data for transmission along with user satellite or spacecraft data to NASA.

Space Transportation System Payloads

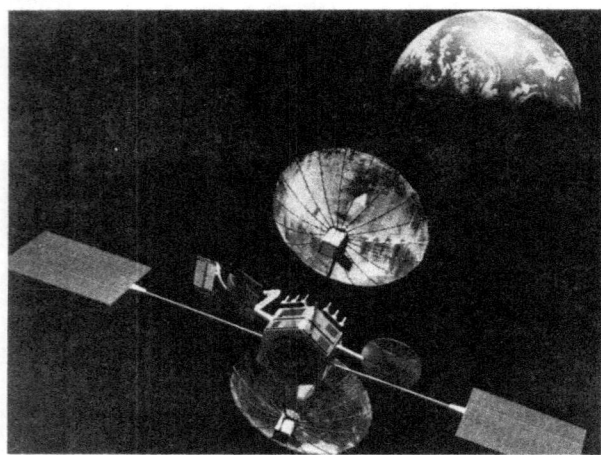

Tracking Data Relay Satellite on Station at Geosynchronous Orbit

Many command and control functions ordinarily found in the space segment of a system are performed by the ground station. The receive beam of the TDRS satellite multiple access phased-array antenna is formed and controlled by the ground station, as are the control and tracking functions of the TDRS satellite single access antennas.

The ground station software and computer component, with more than 900,000 machine language instructions controls the eventual three geosynchronous TDRS satellites and the 300 racks of ground station electronic equipment via a network of 10 computers.

The ground station is located on a nine-acre site at NASA's White Sands, New Mexico, Test Facility. The station includes the electronic equipment, three 18 meter (60 feet) dish antennas for Ku-band, a number of small antennas, and a multiprocessor computer network.

The ground station is owned and managerd by Space Communications Company and leased by NASA. Electronic hardware is jointly supplied by TRW and Harris Government Communications Division in Melbourne, FL. TRW performed integration and testing of the ground station and developed software for the TDRS system and integrates the software with the ground station and TDRS satellites, tying together the space and ground segment.

The TDRS-A satellite on STS-6 was positioned at East station, TDRS-B will be positioned at West station, and TDRS-C will be positioned at the Central station as backup. A three month checkout is scheduled for each satellite when it is on station.

The Space Shuttle Orbiter carries a 914 millimeter (36-inch) diameter orbiter Ku-band antenna stowed in the starboard forward portion of the orbiter payload bay and will be deployed after the orbiter is in orbit and the payload bay doors are open. The capability of installing a Ku-band antenna on the left-hand side is available. If the Ku-band antenna cannot be stowed, provisions are incorporated to jettison it so that the payload bay doors can be closed.

The orbiter Ku-band system operates in the Ku-band portion of the RF spectrum, which is 15,250 MHz to 17,250 MHz. The Ku-band provides a much higher gain signal with a smaller antenna than the S-band system. The S-band system can be used to communicate via the TDRS, but the low-data-rate mode must be used because of limited power since the S-band does not have a high enough signal gain to handle the high data rate. With Ku-band system, the higher data rates can be used.

One drawback of the Ku-band system is its narrow pencil beam, which makes it difficult for the antennas on the TDRS to

Space Transportation System Payloads

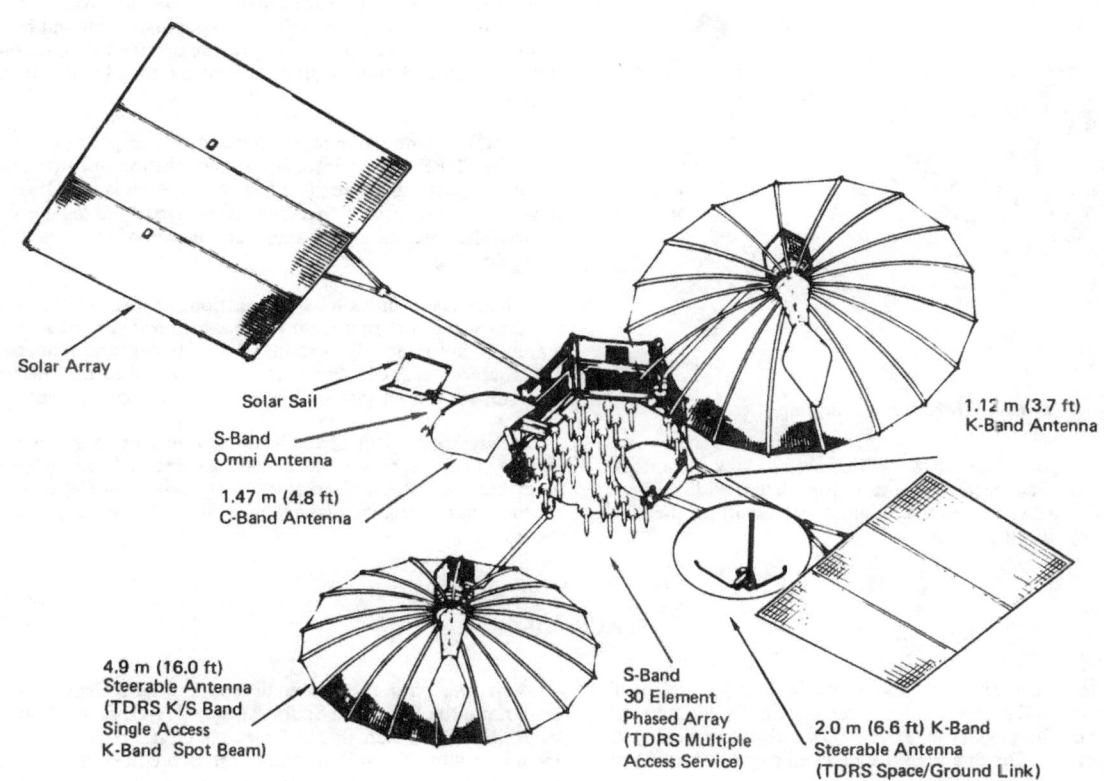

TDRS Satellite

Space Transportation System Payloads

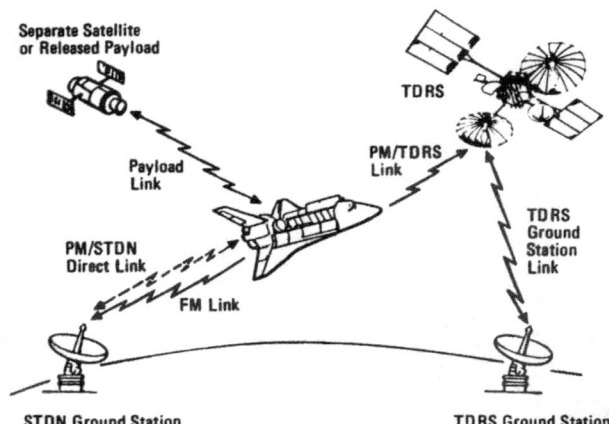

S-Band/TDRS Communications

lock on to the signal. The S-band will be used to lock the antenna into position first because it has a larger beam width. Once the S-band signal has locked the antenna into position, the Ku-band signal will be turned on.

The orbiter Ku-band system includes a rendezvous radar which will be used to skin-track satellites or payloads that are in orbit. This makes it easier for the orbiter to rendezvous with any satellite or payload in orbit. For large payloads that will be carried into orbit, one section at a time, the orbiter will rendezvous with the payload that is already in orbit to add on the next section.

The Ku-band antenna is gimbaled, which permits it to acquire the TDRS for communications acquisition or radar search for other space hardware. The Ku-band system is first given the general location of the space hardware from the orbiter computers. The antenna then makes a spiral scan of the area to pinpoint the target.

With communications acquisition, if the TDRS is not detected within the first eight degrees of spiral conical scan, the search is automatically expanded to 20 degrees. The entire TDRS search requires approximately three minutes. The scanning stops when an increase in the received signal is sensed.

Radar search for space hardware may use a wider spiral scan, up to 60 degrees. Objects may be detected by reflecting the radar beam off the surface of a target (passive mode) or by using the radar to trigger a transponder beam on the target (active mode).

SPACELAB

Spacelab is the manned laboratory built by a group of European nations. This laboratory, which normally will take up the majority of the payload bay. The Spacelab is not deployed free of the orbiter. The Spacelab is scheduled to make a number of flights aboard the Shuttle.

NASA, with the Marshall Space Flight Center as lead center and the European Space Agency (ESA), formerly known as ESRO (European Space Research Organization), signed a memorandum of understanding on September 24, 1973, in which ESA would design, develop, and test a space laboratory

Space Transportation System Payloads

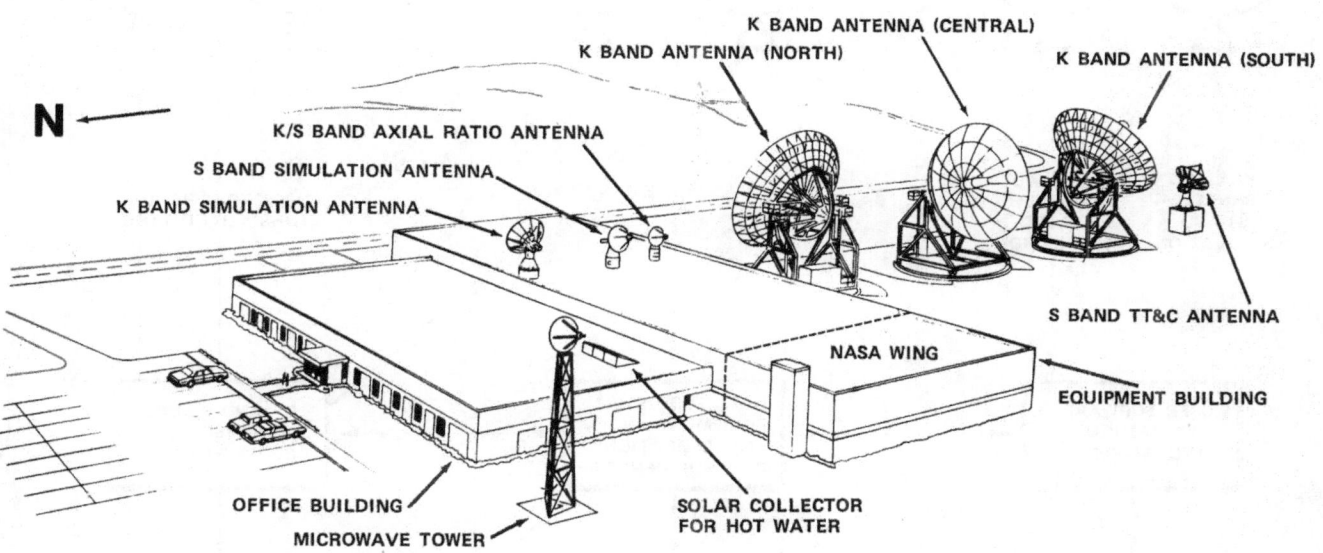

Tracking and Data Relay Satellite System Ground Station, White Sands, New Mexico

Space Transportation System Payloads

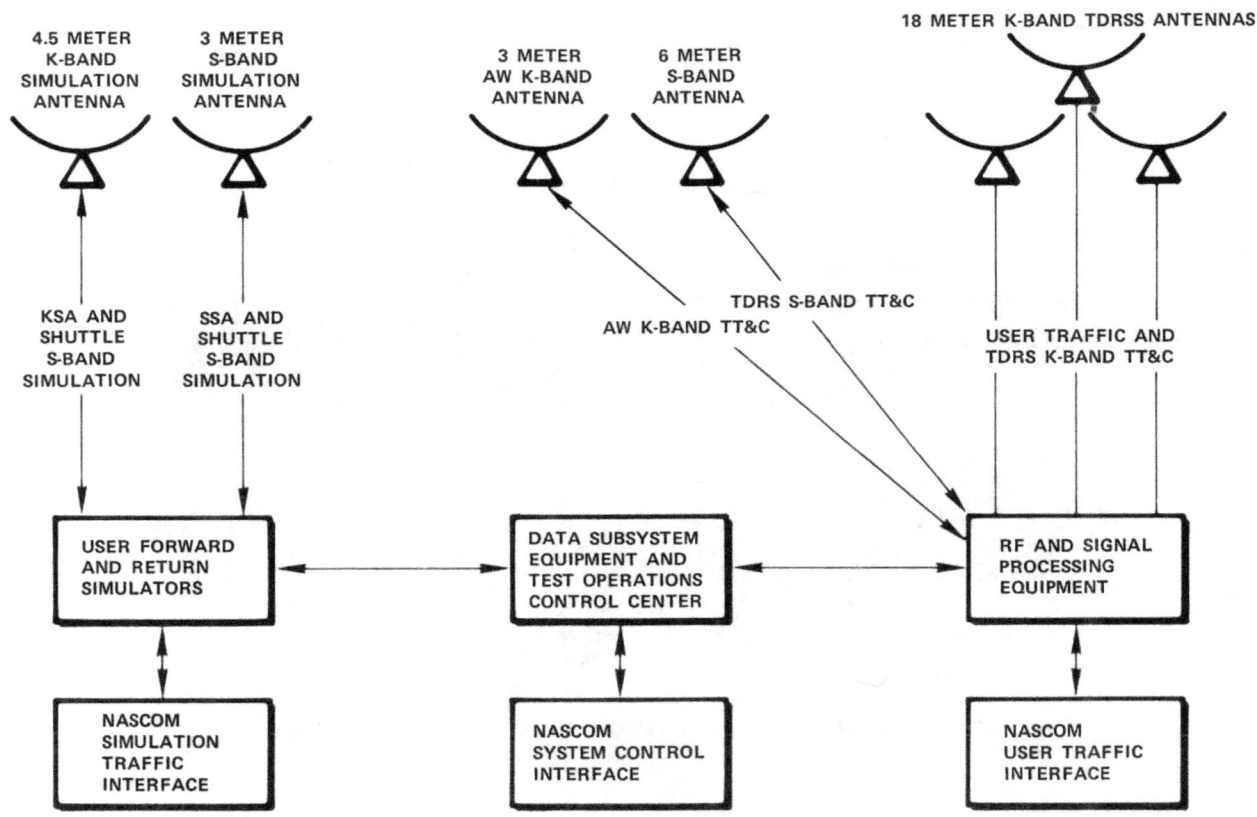

Tracking and Data Relay Satellite System Antenna

 Space Transportation System Payloads

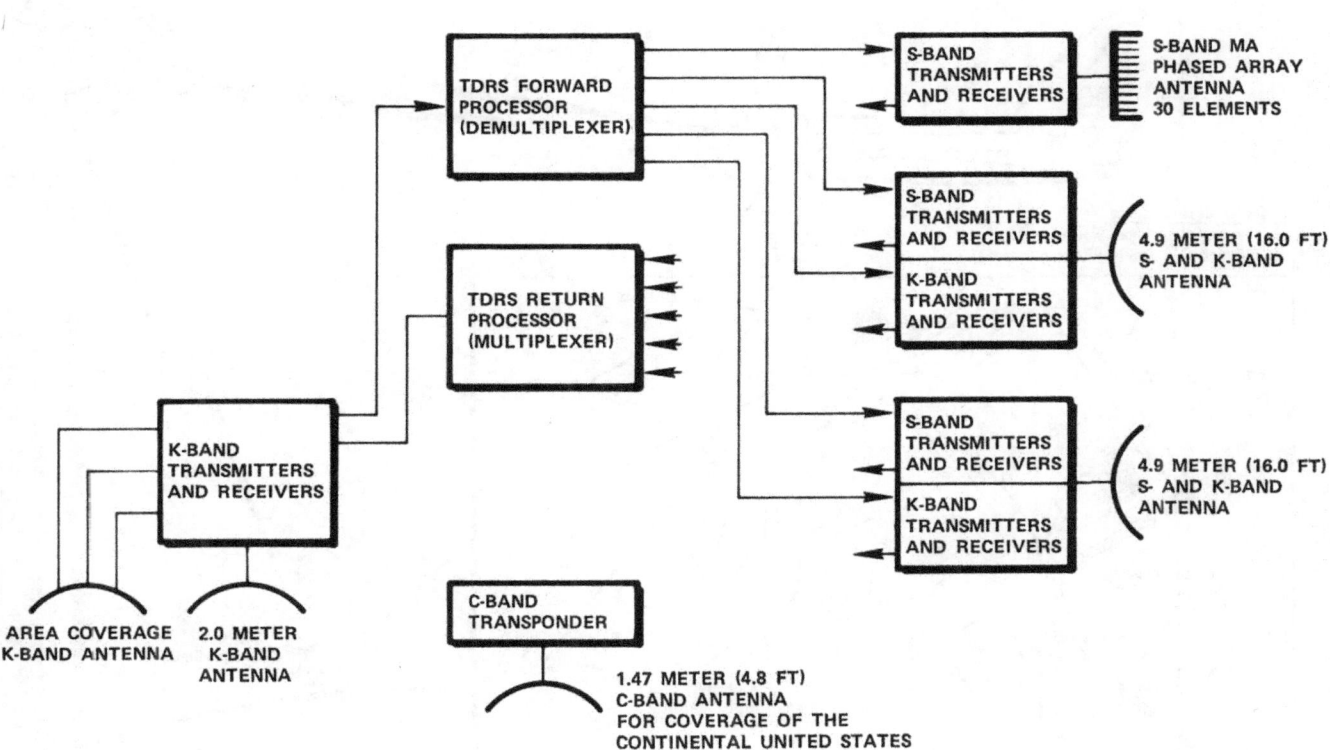

Tracking and Data Relay Satellite System Transmission and Receive System

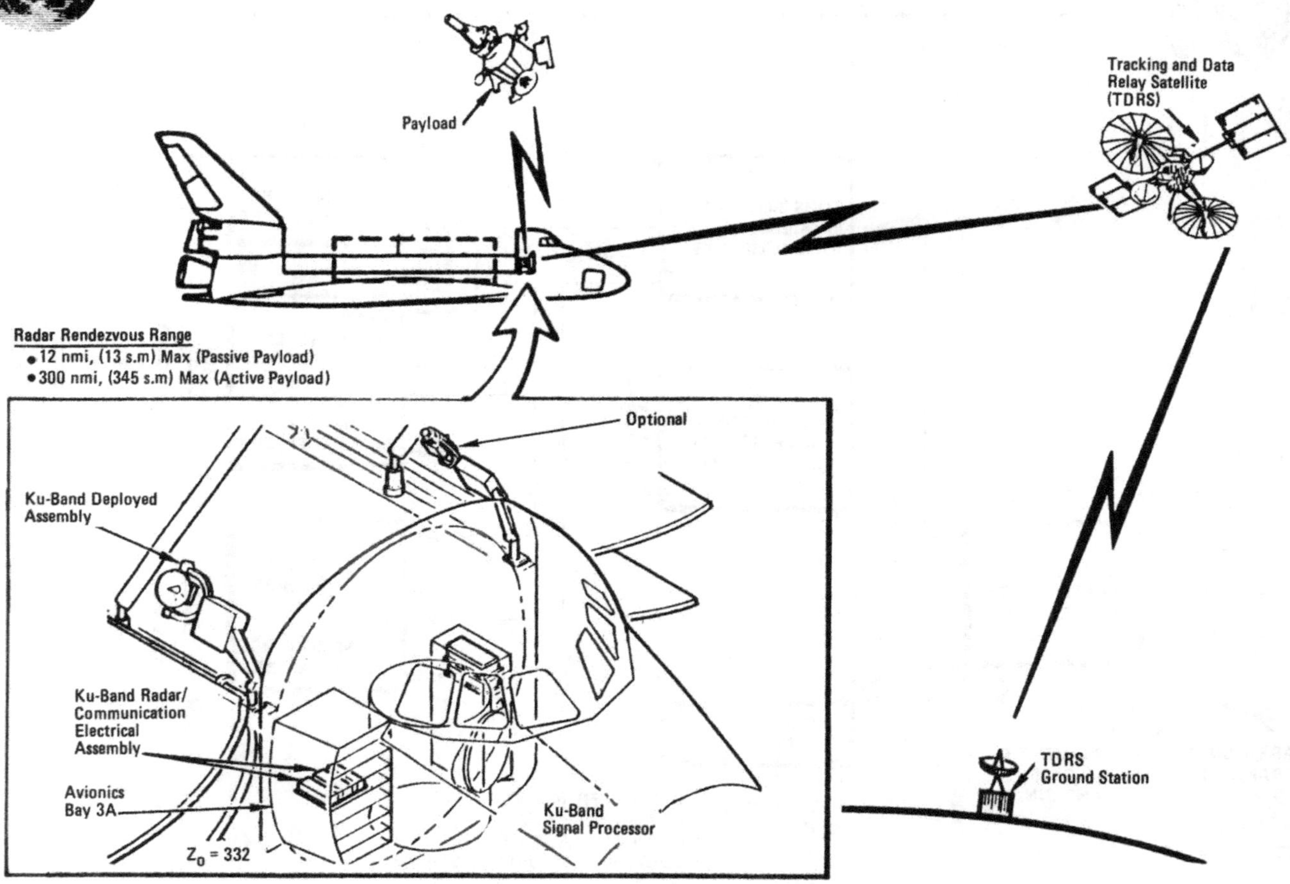

Space Transportation System Payloads

Radar Rendezvous Range
- 12 nmi, (13 s.m) Max (Passive Payload)
- 300 nmi, (345 s.m) Max (Active Payload)

Ku-Band Radar Communication System

Space Transportation System Payloads

to be flown in the cargo bay of the orbiter. ESA has 11 member nations: Belgium, Denmark, France, Ireland, Italy, the Netherlands, Spain, Sweden, Switzerland, United Kingdom, and West Germany. A twelfth, Austria, is an observer rather than a full member. All except Sweden are participating in the Spacelab program.

The development of Spacelab, construction, and delivery to NASA from ESA of one Spacelab flight unit, one engineering model and ground support equipment cost is about $800 million.

Spacelab is designed to have a lifetime of 50 missions or five years. Nominal mission duration of the Spacelab is seven days, but is designed so that missions up to 30 days can be completed by trading payload capability for consumables and a power extension package.

An industrial consortium headed by ERNO-VFW Fokker was named by ESA in June 1974 to build the Spacelab.

Spacelab is developed on a modular basis and can be varied to meet specific mission requirements. Its two principal components are the pressurized module, which provides a laboratory with a shirtsleeve working environment, and the open pallet that exposes materials and equipment to space. Each module is segmented, permitting additional flexibility.

The pressurized module or laboratory comes in two segments. One, called the core segment, contains supporting systems such as data processing equipment and utilities for both the pressurized modules and the pallets. It also has laboratory fixtures such as floor-mounted racks and work benches.

The second, called the experiment segment, is used to provide more working laboratory space. It contains only floor-mounted racks and benches. When only one segment is needed, the core segment is used.

Each pressurized segment is a cylinder 4.1 meters (13-1/2 feet) in diameter and 2.7 meters (9 feet) long. When both segments are assembled with end cones, their maximum outside length is 7 meters (23 feet). Both segments are covered with insulation. The segments are structurally attached to the orbiter by attach fittings.

Due to the orbiter center-of-gravity constraints, the Spacelab module can not lie at the very forward end of the orbiter payload bay. A tunnel is provided for crew equipment and transfer between the orbiter and the Spacelab module. The transfer tunnel is a flexible mounted cylindrical structure with an internal unobstructed dameter of 1,016 millimeters (40 inches) assembled in sections to allow length adjustment as required by different Spacelab configurations. In long the tunnel is 5.75 meters (18.88 feet) long and the short tunnel is 2.60 meters (8.72 feet) long. The McDonnell Douglas Huntington Beach, CA builds the tunnel.

The airlock, tunnel adapter, hatches, tunnel extension and tunnel permits the flight crew members to transfer from the spacecraft pressurized mid deck crew compartment into Spacelab in a pressurized shirt sleeve environment.

In addition, the airlock, tunnel adapter and hatches permit the EVA flight crew members to transfer from the airlock/tunnel adapter in the space suit assembly into the payload bay without depressurizing the spacecraft crew cabin and Spacelab.

It is noted, that if an EVA is required, no crew members will be in Spacelab.

Five unpressurized pallet segments are available. Each U-shaped pallet is 3 meters (10 feet) long, built by the British Aerospace Corp. under contract to ERNO (Zentralgesellschaft VFW-Fokker mbh) and the ESA (European Space Agency). Each pallet is not only a platform for mounting instrumentation but also can cool equipment, provide electrical power, and fur-

 **Space Transportation System Payloads**

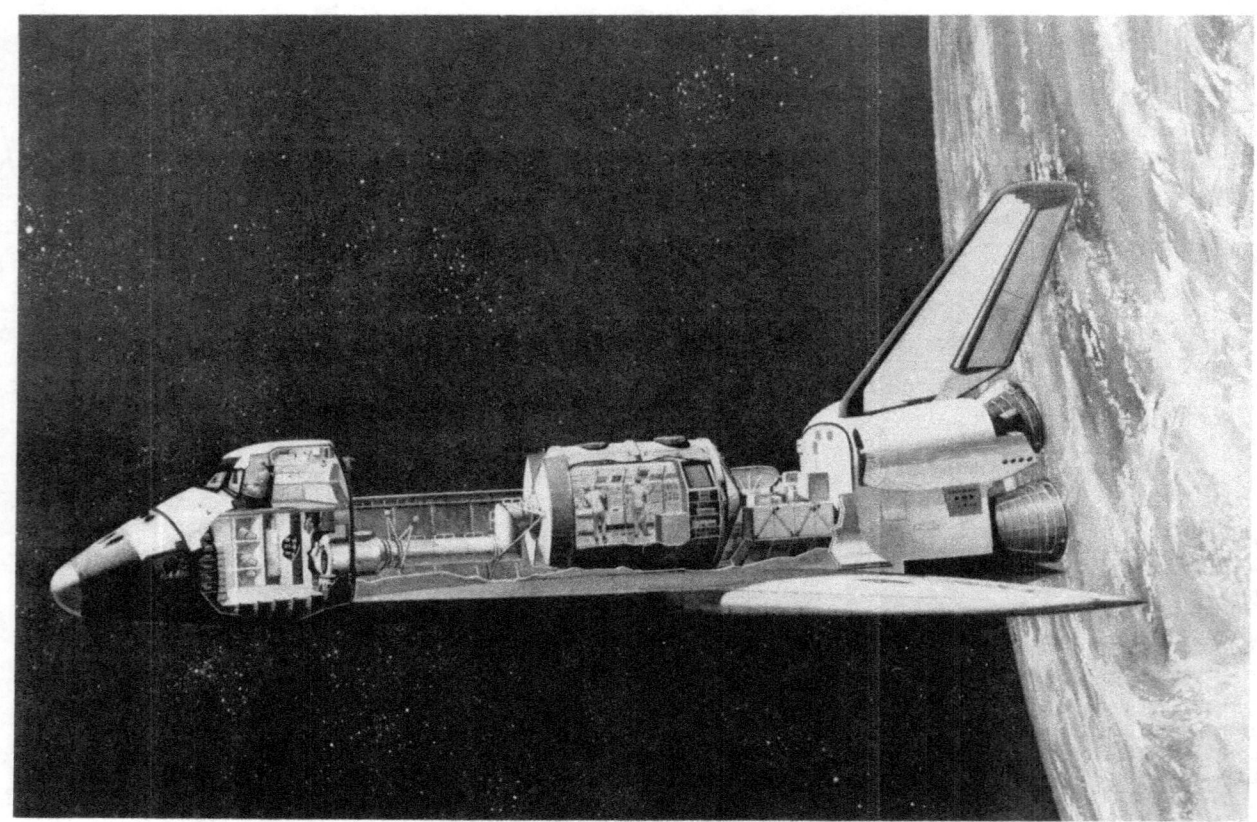

Spacelab

Space Transportation System Payloads

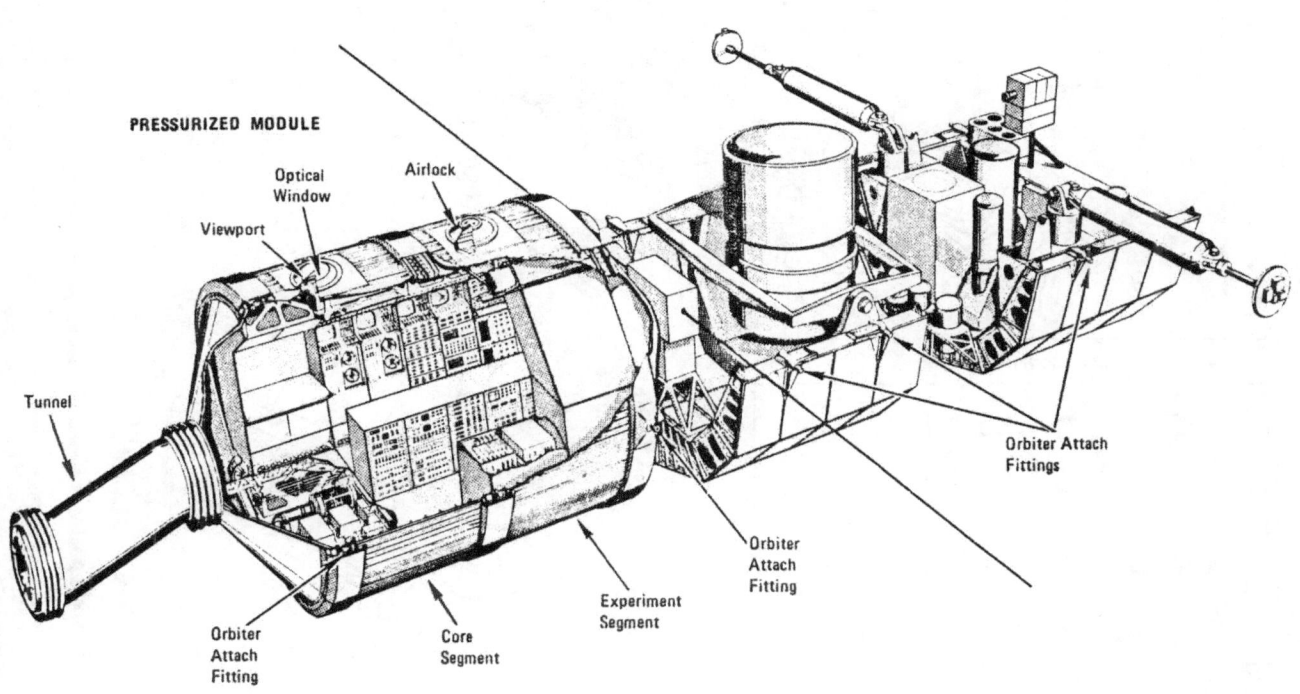

European Space Agency's Spacelab

 **Space Transportation System Payloads**

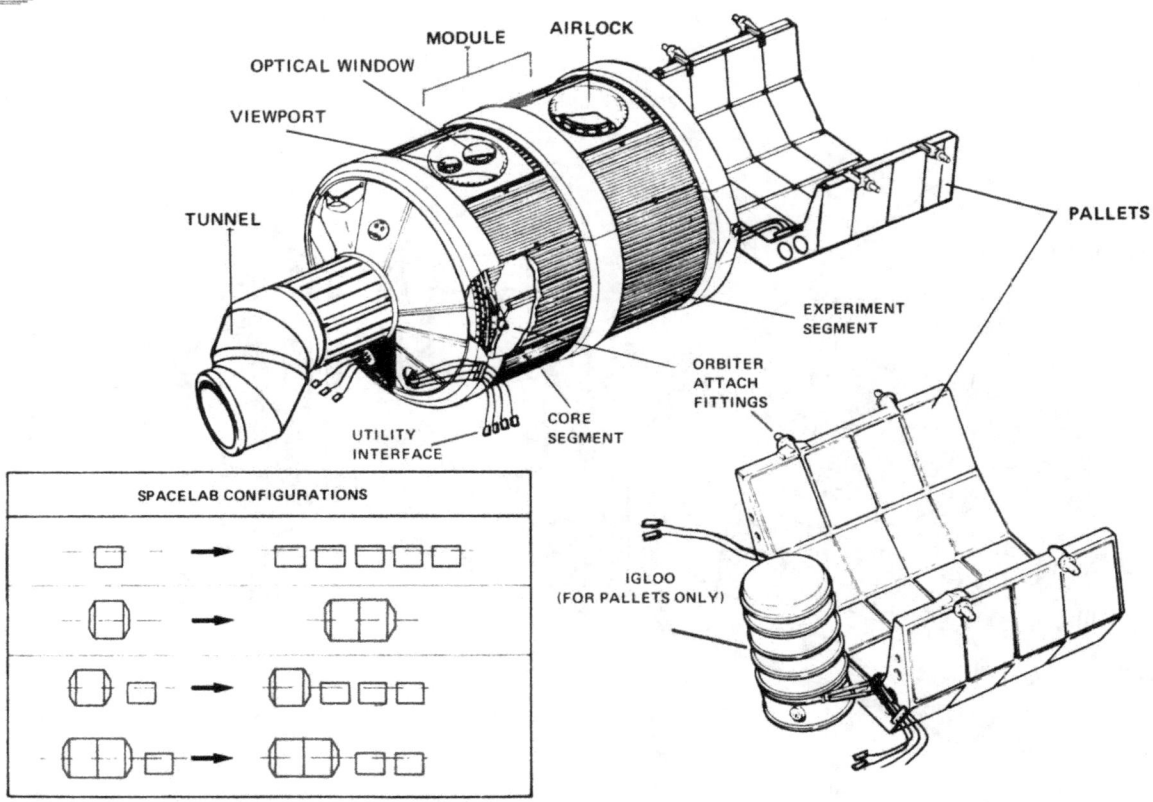

Spacelab External Design Features

Space Transportation System Payloads

nish connections for commanding and acquiring data from the experiments. When pallets only are used, Spacelab portions of the essential systems for supporting experiments (power, experiment control, data handling, and communication, etc.) are protected in a small pressurized and temperature-controlled housing called an igloo. Communications normally is an orbiter function on Spacelab pallet flights. The pallets are designed for large instruments, experiments requiring direct exposure to space, or those needing unobstructed or broad fields of view. Such equipment includes telescopes, antennas, and sensors such as radiometers and radars.

The U-shaped pallets are covered with aluminum honeycomb panels. A series of hard points attached to the main pallet structure are provided for mounting of heavy payload equipment. The pallet segments are mounted to the orbiter attach fittings.

Unlike the spacecraft, the activation of the Spacelab systems does not take place until on-orbit. This necessitates that the Spacelab system be powered up before ingress of the flight crew, which is accomplished via the spacecraft cathode ray tube (CRT) displays. The orbiter on-orbit general purpose computer (GPC) configuration will be, one GPC in GNC, one GPC in Systems Management (SM) and the other three off. Spacelab activation and deactivation will be accomplished under control of the orbiter SMGPC. Once the Spacelab systems are activated, the software functions of Spacelab are then handled by the Spacelab displays.

It is noted that the airlock, tunnel adapter, tunnel and Spacelab are pressurized prior to liftoff with an ambient atmosphere.

The experimenters aboard Spacelab are called payload specialists. They are nominated for flight by the organization sponsoring the payload. They are accepted, trained, and certified for flight by NASA. Their training includes zero-gravity exercises, simulation of operations and emergencies, briefings of flight plans, and space flight housekeeping.

From one to four payload specialists can be accommodated for each Spacelab flight. These specialists ride into space and and return to earth in the orbiter cabin, but they work in Spacelab from seven to 30 days and return to the orbiter's crew cabin when off duty.

More than 2,000 world scientists were represented in the responses to invitations to participate in the first Spacelab missions in the 1980's. Of these, NASA and ESA selected proposals representing 222 investigators from 15 countries. The countries are Austria, Belgium, Canada, Denmark, Federal Republic of Germany, France, Italy, India, Japan, the Netherlands, Norway, Spain, Switzerland, United Kingdom and United States. The investigations planned by the 222 scientists will be conducted in orbit by the Spacelab payload specialists who will be in contact with their colleagues on the ground.

Some of the experiments that may be flown about Spacelab include:

- Earth Surveys — These will gather a variety of information useful in transportation, urban planning, pollution control, farming, fishing, navigation, weather forecasting, and prospecting.

- Astronomy Observations — These are designed to add knowledge about our sun and its interactions with the earth's environment, to view transient events such as comets and novas, and to make observations of high-energy radiation. This radiation — gamma rays, X-rays, ultraviolet light — does not for the most part pass through the atmosphere and cannot be studies on

 **Space Transportation System Payloads**

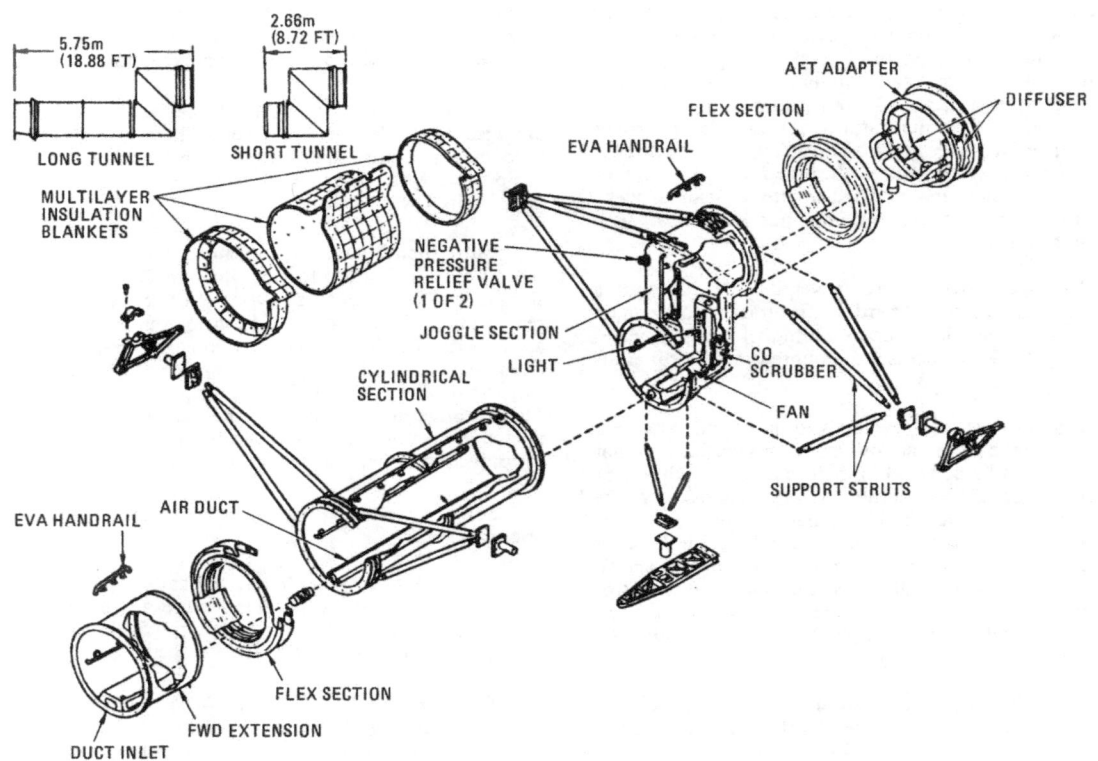

Spacelab transfer tunnel

Space Transportation System Payloads

earth. Locked in it are answers to many questions about the nature, origin, and evolution of celestial phenomena.

- Life Science — Studies of man and other living things in space have indicated significant metabolic changes resulting from the absence of gravity; continued studies are expected to increase the understanding of these changes and contribute to the advancement of medicine.

- Biomedicine — The gravity-free environment of space has demonstrated significant advantages in separating and purifying biological particles. Space processing provides increased opportunities for removal of impurities from vaccines that cause undesirable side effects and for isolating specific cells or antibodies for treatment of disease.

- Industrial Technology — Zero gravity lends itself to manufacturing of new alloys and other composite materials that are uniquely strong, lightweight, and temperature-resistant. Zero gravity has proved conducive to growing of very large crystals of high purity for use in electronics and producing pure glass-free of contamination for optical, electronic and laser uses.

SPACE TELESCOPE

Lockheed Missiles and Space Company, Incorporated, Sunnyvale, CA, designs and builds the Space Telescope support systems module which is 13 meters (43 feet) long, 4.2 meters (14 feet) in diameter — very similar in size to the 3,048 millimeter (120 inch) telescope at Lick Observatory — and houses optics, sensors, support system. They are also responsible for systems engineering, analysis, integration, verification of the complete assembly, in addition to providing support of ground and flight operations under contract from NASA. Perkin-Elmer Corporation's Optical Technology Division, Danbury, CO, is responsible for the Space Telescope optical telescope assembly, including the optics, focal plane structure, fine guidance sensor, and associated controls. NASA's Goddard Space Flight Center, Greenbelt, MD, is responsible for the scientific instruments, mission operations and data reduction. ESA provides one scientific instrument, solar arrays and participates in flight operations. NASA's Office of Space Science is responsible for overall direction of the Space Telescope System and NASA's Marshall Space Flight Center, Huntsville, AL, is responsible for overall project management.

The Space Telescope would be placed about 270 nautical miles (310 statute miles) above the Earth where it will be free of atmospheric interference, such as filtering, haze, twinkling and light pollution, that handicaps conventional telescopes such as the famous Hale telescope at Mount Palomar, CA, which is a 5,080 millimeter (200-inch) telescope. The Space Telescope 2,413 millimeter (95-inch) diameter optics will enable astronomers to detect objects 50 times fainter and view items with a clarity or resolution seven times better than ground-based observations for periods as long as 30 to 40 hours. A major technical challenge is to lock onto viewing targets and insure the telescope maintains a pointing stability of 0.007 arc seconds, which is roughly equivalent to a marksman in Boston zeroing in on a baseball in Los Angeles, CA.

In orbit the telesope consists of three parts: an Optical Telescope Assembly, Scientific Instruments, and a Support Systems Module.

Space Transportation System Payloads

Space Telescope Servicing

Space Telescope

Space Transportation System Payloads

The Optical Telescope Assembly comprises 2,413 millimeter (95-inch) glass reflecting telescope. A meteroid shield light baffles, and sunshade protects the optics.

The five Scientific Instruments (SI's) provide the means of converting the telescope images to useful scientific data. The SI's for the first flight include the wide field/planetary camera (Jet Propulsion Laboratory), faint object camera (ESA/Dormer), faint object spectrograph (Martin Marietta), high speed photometry (University of Wisconsin), and high resolution spectrograph (Ball Brothers). Astronomy (University of Texas) is performed using one of the five guidance sensors.

The Support Systems Module contains a very precise pointing and control system, communications system, thermal control system, and the power system. Electrical power is supplied by solar panels and batteries.

In operation, light enters the open end of the telescope, is projected by the primary mirror onto the smaller secondary mirror and from there is deflected to the scientific instruments for analysis. The Space Telescope will also make ultraviolet and infrared measurements not possible on Earth.

Celestial objects which exist under conditions of gravity, temperature, radiation and time that cannot be duplicated here on Earth will be studied, such as gaseous nebulae, dust clouds, variable stars, binary stars, novae, supernovae, pulsars, neutron stars, black holes, forming galaxies, exploding galaxies, and quasars. The nearby stars, solar system planets, comets, and the planetary systems of stars within 100 light years will also be subjected to Space Telescope scrutiny.

The detailed study of plasmas in space or of mysterious quasars for instance could unlock secrets of tremendous new energy sources. Closer observation of planets in the Earth's own solar system should also help us better understand the significance of changes now taking place on this planet. Such insights could be vital to preserving the world's environment or adapting to changes beyond our control.

The Space Telescope is planned to operate for 15 years on-orbit. It is capable of being serviced and repaired while it is in orbit and new instruments can be installed. It can also be returned to Earth for major repairs and improvements and then replaced into orbit.

GETAWAY SPECIAL

The getaway special (GAS), officially titled small self contained payloads (SSCPS) is offered by NASA to provide anyone who wishes the opportunity to fly a small experiment aboard the Space Shuttle.

Since the offer was first announced in the fall of 1976, more than 326 GAS reservations have been made by over 197 individuals and groups. Payload spares have been reserved by several foreign governments and individuals; United States industrialists, foundations, high schools, colleges and universities; professional societies, service clubs and many others. Although many reservations have been obtained by persons and groups having an obvious interest in space research, a large number of spaces have been reserved by persons and organizations entirely outside the space community.

There are no stringent requirements to qualify for space flight, but the payload must meet safety criteria and must have a scientific or technological objective. It is noted, that a person who wishes to fly items of a commemorative nature, such as medallions for later resale as "objects that have flown in space" would be refused.

 ## Space Transportation System Payloads

GAS requests must first be approved at NASA Headquarters, Washington, D.C. by the Director, Space Transportation Systems Utilization Office, Code OT6. It is at this point that requests for Space Shuttle space are screened for propriety, and scientific or technical aim. These requests must be accompanied or preceded by the payment of $500 earnest money.

Requests approved by the Space Transportation Systems Utilization Office are given a payload identification number and referred to the GAS Team at the Goddard Space Flight Center, Greenbelt, MD. The center has been designated the lead center or direct manager for the project.

The GAS Team screens the proposal for safety and provides advice and consultation for payload design. The GAS Team certifies that the proposed payload is safe, that it will not harm or interfere with the operations of the Space Shuttle, its crew, or other experiments on the flight. If any physical testing must be done on the payload to answer safety questions prior to the launch, the expense of these tests must be borne by the customer.

In flight, the crew will turn on and off up to three payload switches, but there will be no opportunity for flight crew monitoring of GAS experiments, or any form of in-flight servicing.

The cost of this unique service will depend on the size and weight of the experiment; getaway specials of 90 kilograms (200 pounds) and 0.14 cubic meter (5 cubic feet) may be flown at a cost of $10,000; 45 kilograms (100 pounds) and 0.07 cubic meter (2.5 cubic feet) for $3,000, and 27 kilograms (60 pounds) and 0.07 cubic meter (2.5 cubic feet) at $3,000. These prices remain fixed for the first three years of Space Shuttle operations.

The GAS container provides for internal pressure which can be varied from near vacuum to about one atmosphere. The bottom and sides of the container are always thermally insulated and the top may be insulated or not depending on the specific experiment, an opening lid or one with a window may be required. These may be offered as additional cost options.

The weight of the GAS container, experiment mounting plate and its attachment screws, and all hardware regularly supplied by NASA is not charged to the experimenter's weight allowance.

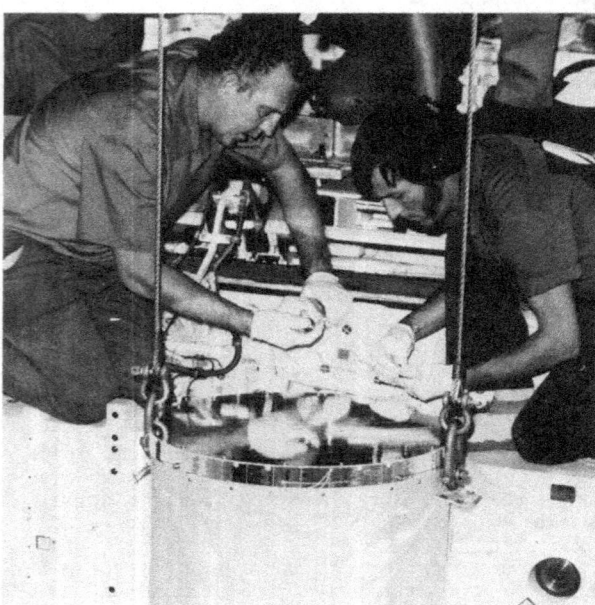
Getaway Special Container in Payload Bay

Space Transportation System Payloads

The GAS container is made of aluminum and the circular end plates are 15 millimeters (5/8 inch) thick aluminum. The bottom 76 millimeters (3 inches) of the container are reserved for NASA interface equipment such as command decoders and pressure regulating systems. The container is a pressure vessel capable of; evacuation prior to launch, or; evacuation during launch and repressurization during reentry, or; maintaining about one atmosphere pressure at all times; evacuation and repressurization during orbit as provided by the experimenter. The experimenters payload envelopes in the 0.14 cubic meter (5 cubic feet) container are 501 millimeters (19.95 inches) in diameter and 717 millimeters (28.25 inches) in length. The payload envelope in the 0.07 cubic meter (2.5 cubic feet) container are 501 millimeters (19.95 inches) in diameter and 358 millimeters (14.13 inches) in length.

The GAS program is managed by the Goddard Space Flight Center. Project manager is James S. Barrowman. Clarke Prouty, also of Goddard, is technical liaison officer, and queries can be addressed to him at Code 741, Goddard Space Flight Center, Greenbelt, MD, 20771. Program manager at NASA Headquarters, Washington, D.C. is Donna S. Miller.

Space Transportation System Background Information

ORBITER APPROACH AND LANDING TEST PROGRAM

The nine-month-long Approach and Landing Test (ALT) program in 1977 at NASA's Dryden Flight Research Center, Edwards Air Force Base, proved that America's newest spacecraft, the Enterprise (OV-101), is truly a "magnificent flying machine and its aerodynamic flight controls handled crisply and accurately."

Two NASA astronaut crews — Fred Haise and Gordon Fullerton and Joe Engle and Dick Truly — took turns flying the 68,040 kilograms (150,000-pound) spacecraft to free-flight landings and described it as "magnificient, precise, excellent, outstanding."

Enterprise Approaches Edwards AFB Runway in AZT Flight

The ALT program involved a total of 13 flights (five unmanned "captive," three manned captive, and five free flights), as well as various ground tests.

The ground tests included taxi tests of the mated 747-orbiter to determine structural loads and responses and assess the mated capability in ground handling and control characteristics up to flight takeoff speed. The taxi tests also validated 747 steering and braking with orbiter attached. A ground test of orbiter systems followed the unmanned captive tests. All orbiter systems were activated as they would be in atmospheric flight; this was the final preparation for the manned captive flight phase.

In the unmanned captive flights, the orbiter — its systems inert — was attached atop the 747 Shuttle carrier aircraft. These flights assessed the structural integrity and performance handling qualities of the mated craft.

The three manned captive flights which followed included an astronaut crew aboard the orbiter operating its flight control systems while the orbiter remained perched atop the 747. These flights were designed to exercise and evaluate all systems in the flight environment in preparation for the orbiter release (free) flights. They included flutter tests of the mated craft at low and high speed, a separation trajectory test, and a dress rehearsal for the first orbiter free flight.

In the five free flights the astronaut crew separated the spacecraft from the 747 and flew it back to a landing. These flights verified the orbiter pilot-guided approach and landing capability, demonstrated the orbiter subsonic terminal area energy management autoland approach capability, and verified the orbiter subsonic airworthiness, integrated system operation,

Space Transportation System Background Information

and selected subsystems in preparation for the first manned orbital flight. The flights demonstrated the orbiter's ability to approach and land safely with a minimum gross weight and using several center-of-gravity configurations.

For all of the captive flights and the first three free flights the Orbiter was outfitted with a tail cone covering its aft section to reduce aerodynamic drag and turbulence. The final two free flights were without the tail cone and the simulated engines — three main and two orbital maneuvering — were exposed aerodynamically.

The final phase of the ALT program prepared the spacecraft for four ferry flights. Fluid systems were drained and purged, the tail cone was re-installed, and elevon locks were installed. The forward attachment strut was replaced to lower the orbiter's cant from six to three degrees. This reduces drag to the mated vehicles during the ferry flights.

After the ferry flight tests, Orbiter 101 was returned to the NASA hangar at Dryden FRC and modified for vertical ground vibration tests at the Marshall Space Flight Center, Huntsville, AL, and then was ferried to Huntsville. At the Marshall Space Flight Center, the Enterprise was mated with the external tank and solid rocket boosters and subjected to vertical ground vibration tests. These tested the mated configuration's critical structural dynamic response modes, which were assessed against analytical math models used to design the various element interfaces.

The Enterprise then was ferried to the Kennedy Space Center, where it was mated with the external tank and solid rocket boosters and transported via the Mobile Launch Platform to Launch Complex 39A. At Launch Complex 39A, it served as a practice and launch complex fit check verification tool representing the flight vehicles.

Finally, the Enterprise was ferried back to Edwards AFB, CA, and returned via overland transport to Rockwell Palmdale final assembly facility. Certain components were refurbished for use on flight vehicles being assembled at Palmdale. On September 6, 1981, the *Enterprise* was returned overland to NASA;s Dryden Flight Research Center. In the May-June 1983 time period, *Enterprise* was ferried to the Paris, France air show as well as to Germany, Italy, England and Canada and returned to the Dryden Flight Research Center. It will eventually be used as a practice and fit check verification tool at Vandenberg AFB, CA.

SPACE SHUTTLE FLIGHT TEST PROGRAM

In preparation for the first flight into space of Orbiter 102 *(Columbia)*, tens of thousands of hours of tests and simulations were expended at government and contractor facilities throughout the nation to qualify all the structures, flight equipment, and computer programs software.

Major test programs included the 13 flights of the Approach and Landing Test (ALT) program at NASA's Dryden Flight Research Center, Edwards Air Force Base, CA. The main propulsion test article — consisting of the orbiter aft fuselage, three Space Shuttle main engines, external tank, and truss arrangement to simulate the mid fuselage — qualified the main propulsion system at the National Space Technology Laboratory in Mississippi. The solid rocket booster's solid-propellant rocket motor were qualified at Thiokol Chemical Corp. near Bringham City, UT. The orbiter reaction control and orbital maneuvering systems were qualified at NASA's White Sands Test Facility near Las Cruces, NM Flight simulation and avionics testing qualified the Space Shuttle avionics at JSC and at Rockwell's Space Systems Group in Downey, CA.

Space Transportation System Background Information

The orbiter full-scale structure was tested at Lockheed's facility in Palmdale, CA The orbiter vehicle (OV-101), external tank, and solid rocket boosters underwent a mated vertical ground vibration test program at MSFC. OV-101, external tank and solid rocket boosters. completed a fit check and checkout with the mobile launch platform and launch complex 39A at KSC.

The initial four flights of *Columbia* were launched from the Kennedy Space Center, FL, and verified the design and operational capability of the Space Shuttle and all of the ground-based monitoring, communications, and support systems. The first flight of *Columbia* was structured to minimize risks and complexity. The remaining test flights became progressively more complex and developed and demonstrated mission and payload capabilities. The spacecraft performance was above and beyond expectations in all aspects. The spacecraft has demonstrated that it also is a very stable platform for experiments in addition to providing a stable platform for deployment of satellites.

SHUTTLE CARRIER AIRCRAFT

The NASA Shuttle carrier aircraft (SCA) is a Boeing 747 (100 series) purchased from American Airlines on June 17, 1974. It was modified to ferry orbiters to and from various Shuttle facilities and to transport and release Orbiter 101 for the ALT program.

Modifications to the basic 747 aircraft included removal of interior equipment (passenger seats, galleys, etc.); changes to air conditioning ducts, electrical wiring, and plumbing; installation of higher-thrust engines (JT9D-7AHW) and the 747-200 series rudder ratio-charger; and alteration of the longitudinal trim system to permit two degrees more nose-down trim. Other changes included relocation and installation of antennas, addition of bulkheads and doublers in the fuselage main deck, addition of structural doublers and tip fins to the horizontal stabilizers to improve directional stability with the orbiter on top of the aircraft, and addition of one forward and two aft support assemblies for attachment of the orbiter. The modifications increased the basic weight of the aircraft by approximately 1,270 kilograms (2,800 pounds).

The Orbiter's mated location on the 747 was based on consideration of static stability and control, structural modifica-

Boeing 747 Shuttle Carrier

tion, weight, and performance. Center-of-gravity limits for the 747 with the orbiter mated are 15 percent of the 747's mean aerodynamic chord (MAC) for the forward limit and 33 percent MAC for the aft limit. Longitudinal stability is similar to that of

Space Transportation System Background Information

the basic 747; ballast must be added so that the center of gravity limits are not exceeded. The ballast is carried in standard 747 cargo containers in the forward cargo compartment. The mated configuration allows the 747 center of gravity to shift approximately 3 meters (10 feet) upward.

For the ferry flight configuration, the tail cone fairing is installed on the orbiter to decrease aerodynamic drag and buffet, and aerosurface control locks are added to the orbiter's elevons. The orbiter is unmanned and the orbiter systems inert. A bailout system also is installed in the 747.

Some modifications to the 747 SCA are removable. These include support struts for the orbiter, horizontal tip fans, and associated cabling and umbilicals.

SPACE SHUTTLE CHRONOLOGY

Date	Event
1972	
Aug. 9	Rockwell receives authority to proceed, Space Shuttle orbiter
1974	
June 4	Start structural assembly crew module (Orbiter 101)
July 17	Start long-lead fabrication (MPTA-098)
Aug. 26	Start structural assembly aft fuselage (Orbiter 101)
1975	
Jan. 6	Start long-lead fabrication aft fuselage (STA-099)
March 27	Mid fuselage on dock, Palmdale (Orbiter 101)
March 27	Start long-lead fabrication aft fuselage (Orbiter 102)
May 23	Wings on dock, Palmdale — less elevons, seals, and main gear doors (Orbiter 101)
May 27	Vertical stabilizer on dock, Palmdale (main fin box only) (Orbiter 101)
June 24	Start structural assembly (MPTA-098)
Aug. 25	Start final assembly and closeout system installation (Orbiter 101)
Sept. 5	Aft fuselage on dock, Palmdale (Orbiter 101)
Oct. 17	Space Shuttle main engine first mainstage test at NSTL
Oct. 31	Lower forward fuselage on dock, Palmdale (Orbiter 101)
Nov. 17	Start long-lead fabrication crew module (Orbiter 102)
Dec. 1	Upper forward fuselage on dock, Palmdale (Orbiter 101)
Dec. 20	First Space Shuttle main engine 60-second duration test, NSTL
Dec. 31	1/4-scale model ground vibration test facility construction complete, Downey, Building 288
1976	
Jan. 16	Crew module on dock, Palmdale (Orbiter 101)
Jan. 23	MPTA-098 truss on dock, Downey
Feb. 16	Start fabrication forward fuselage (STA-099)
March 3	Payload bay doors on dock, Palmdale (Orbiter 101)
March 12	Complete final assembly and closeout system installation, Palmdale (Orbiter 101)
March 15	Start functional checkout (Orbiter 101)

Space Transportation System Background Information

Date	Event
March 17	Complete pre-mate MPTA test structure, Downey, and deliver to Palmdale
April 2	Crew escape system test sled on dock, Downey
April 3	Complete assembly and deliver MPTA structure on dock, Lockheed Test Site, Palmdale
April 22	Body flap on dock, Palmdale (Orbiter 101)
May 3	Complete MPTA-098 proof load test set-up, Lockheed test site, Palmdale
June 14	Start aft fuselage assembly (STA-099)
June 24	Complete MPTA-098 proof load test, Lockheed test site and on dock, Palmdale
June 25	Complete functional checkout (Orbiter 101)
June 28	Start assembly crew module (Orbiter 102)
June 28	Start horizontal ground vibration tests and proof load tests (Orbiter 101)
June 29	MPTA-098 truss assembly, Palmdale, Bldg. 294 to Bldg. 295
June 30	Space Shuttle main engine dummy set on dock, Palmdale (Orbiter 101)
July 8	MPTA-098 on dock, Downey, without truss assembly
July 12	Start installation secondary structure (MPTA-098)
Aug. 2	Start final assembly forward fuselage (STA-099)
Aug. 2	Start carrier aircraft modification
Aug. 20	Complete horizontal ground vibration tests and proof load tests (Orbiter 101)
Aug. 23	Start Delta F modification (Orbiter 101)
Aug. 27	Reaction control system/orbital maneuvering system pods (simulated), approach and landing tests, on dock, Palmdale (Orbiter 101)
Aug.	Start overland roadway construction from Palmdale to DFRC
Sept. 10	Complete Delta F modifications (Orbiter 101)
Sept. 13	Start preparations for first rollout (Orbiter 101)
Sept. 13	Start assembly aft fuselage (Orbiter 102)
Sept. 17	Rollout (Orbiter 101)
Sept. 17	Complete on-stand construction, NSTL
Sept. 20	Start Delta F retest (Orbiter 101)
Oct. 1	Start final assembly, vertical stabilizer (STA-099)
Oct. 1	Start final assembly, wing (STA-099)
Oct. 15	Mid fuselage on dock, Palmdale (STA-099)
Oct. 26	Escape system test assembly sled ship from Downey to Holloman, NM
Oct. 29	Deliver ejection seats (Orbiter 101)
Oct. 29	Complete Delta F retest (Orbiter 101)
Oct. 31	Solid rocket booster 1/4-scale model (burnout configuration) on dock, Downey
Nov. 4	Complete 747 Shuttle carrier aircraft modification, rollout, Boeing
Nov. 18	Start escape system sled test, Holloman, NM
Nov. 26	Complete integrated checkout (Orbiter 101)
Nov. 28	Complete orbiter transporter strongback
Dec. 7	Tail cone fairing on dock, Palmdale
Dec. 10	Complete overland roadway construction, Palmdale to DFRC
Dec. 13	Start assembly upper forward fuselage (Orbiter 102)
Dec. 17	External tank 1/4-scale model on dock, Downey
1977	
Jan. 3	Start assembly vertical stabilizer (Orbiter 102)
Jan. 14	Complete post-checkout (Orbiter 101)
Jan. 14	Boeing 747 Shuttle carrier aircraft delivered to DFRC
Jan. 25	Complete aft fuselage assembly on dock, Palmdale (STA-099)
Jan. 28	Simulated crew module on dock, Palmdale (STA-099)
Jan. 31	Solid rocket booster 1/4-scale model (liftoff configuration) on dock, Downey

Space Transportation System Background Information

Date	Event
Jan. 31	Orbiter 101 transported overland from Palmdale to DFRC (36 miles)
Jan. 31	Mass simulated Space Shuttle Main Engines on dock, Palmdale (Orbiter 101)
Feb. 7	Orbiter 101 Shuttle carrier aircraft mate start
Feb. 10	Mid fuselage on dock, Palmdale (Orbiter 102)
Feb. 15	Complete Orbiter 101 Shuttle carrier aircraft mated ground vibration test and taxi tests
Feb. 18	Conduct first inert captive flight, DFRC (2 hr 5 min) (Orbiter 101)
Feb. 22	Conduct second inert captive flight, DFRC 3 hr 13 min) (Orbiter 101)
Feb. 25	Conduct third inert captive flight, DFRC (2 hr 28 min) (Orbiter 101)
Feb. 28	Conduct fourth inert captive flight, DFRC (2 hr 11 min) (Orbiter 101)
March 2	Conduct fifth inert captive flight, DFRC (1 hr 39 min) (Orbiter 101)
March 16	Wings on dock, Palmdale (STA-099)
March 21	Orbiter 1/4-scale model on dock, Downey
April 1	Lower forward fuselage on dock, Palmdale (STA-099)
April 6	Upper forward fuselage on dock, Palmdale (STA-099)
April 6	Vertical stabilizer on dock, Palmdale (STA-099)
May 20	Nose landing gear doors on dock, Palmdale (STA-099)
May 26	Aft payload bay doors on dock, Palmdale (STA-099)
May 27	Complete systems installation/final acceptance, MPTA-098, transport from Downey to Seal Beach
May 31	Solid rocket booster 1/4-scale model, burnout and maximum q configuration, on dock, Downey
May 31	Solid rocket booster 1/4-scale model, liftoff configuration, on dock, Downey
May 31	Body flap on dock, Palmdale (STA-099)
June 3	Ship MPTA-098 from Seal Beach to NSTL
June 7	Complete integrated checkout and hot fire ground test, DFRC (Orbiter 101)
June 18	Conduct first manned captive active flight, Orbiter 101 — 747, DFRC (55 min 46 sec)
June 23	Deliver first Space Shuttle main engine to NSTL (MPTA-098)
June 24	Deliver MPTA-098 to NSTL
June 28	Conduct second manned captive active flight, Orbiter 101 — Shuttle carrier aircraft, DFRC, 1 hr 2 min)
July 5	Start long-lead fabrication aft fuselage, Orbiter 103
July 8	Deliver second main engine to NSTL (MPTA-098)
July 14	Deliver third main engine to NSTL (MPTA-098)
July 18	Conduct two-minute firing of solid rocket booster at Brigham City, UT, Thiokol (2.4 million pounds of thrust)
July 22	Deliver forward payload bay doors, on dock, Palmdale (STA-099)
July 26	Conduct third manned captive active flight, Orbiter 101 — Shuttle carrier aircraft, DFRC (59 min, 50 sec)
Aug. 12	Conduct first free flight, ALT, tail cone on DFRC (5 min 21 sec) (Orbiter 101), Lakebed Runway 17
Aug. 26	Deliver wings on dock Palmdale (Orbiter 102)
Sept. 7	Lower forward fuselage on dock Palmdale (Orbiter 102)
Sept. 10	Deliver external tank MPTA-098 (Martin Marietta) to NSTL
Sept. 13	Conduct second free flight, ALT, tail cone on, DFRC (5 min 28 sec) (Orbiter 101), Lakebed Runway 17

Space Transportation System Background Information

Date	Event
Sept. 23	Conduct third free flight, ALT, tail cone on, DFRC (5 min 34 sec) (Orbiter 101), Lakebed Runway 15
Sept. 30	Complete mate vertical stabilizer, Palmdale (STA-099)
Oct. 12	Conduct fourth free flight, ALT, first tail cone off, DFRC (2 min 34 sec) (Orbiter 101), Lakebed Runway 17
Oct. 26	Conduct fifth free flight, ALT, final tail cone off, DFRC (2 min 1 sec) (Orbiter 101), Concrete Runway 04
Oct. 28	Lower forward fuselage on dock Palmdale (Orbiter 102)
Nov. 4	Start forward reaction control system development tests, White Sands Test Facility, NM
Nov. 4	Deliver aft fuselage on dock Palmdale (Orbiter 102)
Nov. 7	Forward reaction control system on dock Palmdale (STA-099)
Nov. 7	Start final assembly and closeout system installation Palmdale (Orbiter 102)
Nov. 15	First ferry flight test, DFRC (3 hr 21 min) (Orbiter 101)
Nov. 16	Second Ferry flight test, DFRC (4 hr 17 min) (Orbiter 101)
Nov. 17	Third ferry flight test, DFRC (4 hr 13 min) (Orbiter 101)
Nov. 18	Fourth ferry flight test, DFRC (3 hr 37 min) (Orbiter 101)
Dec. 9	Complete approach and landing flight tests including ferry flights (Orbiter 101)
Dec. 12	Start removal for mated vertical ground vibration test modification at DFRC (Orbiter 101)
Dec. 13	Complete propellant load testing NSTL (MPTA-098)

Date	Event
Dec. 31	Deliver Space Shuttle main engine envelope/electrical simulators, on dock Palmdale (Orbiter 102)
1978	
Jan. 10	Vertical stabilizer on dock Palmdale (Orbiter 102)
Jan. 18	Second solid rocket booster firing, Thiokol
Feb. 10	Complete final assembly STA-099, Palmdale
Feb. 14	STA-099 on dock, Lockheed facility, Palmdale
Feb. 17	Crew module on dock, Palmdale (Orbiter 102)
Feb. 24	Body flap on dock Palmdale (Orbiter 102)
March 3	Complete modification for mated vertical ground vibration test, DFRC (Orbiter 101)
March 6	Upper forward fuselage on dock, Palmdale (Orbiter 102)
March 10	Ferry Orbiter 101 atop Shuttle carrier aircraft from DFRC to Ellington Air Force Base, TX (approximately 3 hr 38 min)
March 13	Ferry Orbiter 101 atop Shuttle carrier aircraft from Ellington AFB to Marshall Space Flight Center, Huntsville, AL (approximately 2 hours)
March 19	Aft payload bay doors on dock, Palmdale (Orbiter 102)
March 31	External tank for mated vertical ground vibration tests arrives at MSFC from Martin Marietta, Michoud, LA
March 31	Operational readiness date, solid rocket booster refurbishment and subassembly, Kennedy Space Center, FL
April 14	Complete ground vibration test modification at MSFC; deliver Orbiter 101 for mated vertical ground vibration test

Space Transportation System Background Information

Date	Event
April 21	First static firing, MPTA-098, NSTL (2.5 seconds; stub nozzles)
April 23	Complete final assembly and closeout system installation, ready for power-on (Orbiter 102)
April 24	Start pre-combined systems test (Orbiter 102)
April 28	Forward payload bay doors on dock, Palmdale (Orbiter 102)
May 19	Second static firing MPTA-098, NSTL (15 seconds, 70% thrust; stub nozzles)
May 19	Start forward reaction control system thermal tests, WSTF
May 26	Start development test aft reaction control system, WSTF
May 26	Complete forward reaction control system structure (October 102)
May 26	Upper forward mate, Orbiter 102
May 30	Start test Orbiter 101/external tank mated vertical ground vibration test, MSFC
May 31	Loaded solid rocket boosters (2) arrive at MSFC for mated vertical ground vibration test
June 15	Third static firing, MPTA-098, NSTL (50 seconds, 90% thrust, last 5 seconds 70% thrust; stub nozzles)
July 3	Deliver left-hand orbital maneuvering system pod to WSTF
July 7	Fourth static firing, MPTA-098, NSTL (No. 2 engine 70% thrust for 90 seconds, simulated one engine out at 90 seconds, then No. 1 and No. 3 to 90% thrust — then to 70% thrust, then to 90% for 100 seconds; stub nozzles)
July 7	Complete mate forward and aft payload bay doors (Orbiter 102)
July 13	Complete forward reaction control system development tests, WSTF
July 13	Reconfigure from boost to launch, mated vertical ground vibration test, MSFC (Orbiter 101)
July 15	Deliver solid rocket boosters (2) empty to MSFC for mated vertical ground vibration test
July 21	First firing development test, orbital maneuvering system, WSTF
July 31	Operational readiness date, Orbiter Processing Facility Bay 1, Orbiter Landing Facility and Hypergolic Maintenance Facility, KSC
Aug. 11	Complete forward reaction control system (Orbiter 102)
Aug. 11	Complete test preparation STA-099, Lockheed facility, Palmdale
Aug. 11	Complete forward reaction control system thermal test, WSTF
Aug. 14	Start coefficient tests STA-099, Lockheed facility, Palmdale
Aug. 31	Operational readiness date, Vertical Assembly Building High Bay 3 and 4, KSC
Sept. 8	Start orbital maneuvering system left-hand development test, WSTF, NM
Sept. 20	Start launch configuration test liftoff configuration, Orbiter 101/external tank/solid rocket boosters mated vertical ground vibration test at MSFC
Sept. 20	Start acoustic test, forward reaction control system, WSTF
Sept. 25	Start pre-combined system test (Orbiter 102)
Sept. 29	Complete coefficient tests, STA-099, Lockheed facility, Palmdale
Sept. 30	Operational readiness date, mobile launch platform, KSC
Oct. 19	Third solid rocket booster firing, Thiokol
Nov. 11	Complete forward reaction control system acoustic test, WSTF

Space Transportation System Background Information

Date	Event
Nov. 15	Complete orbital maneuvering system development test, WSTF
Nov. 15	Complete aft reaction control system development test, WSTF
Nov. 30	Operational readiness date, Pad A, KSC
Dec. 9	Start orbital maneuvering system Phase I qualification tests, WSTF
Dec. 15	Start aft reaction control system Phase I qualification tests, WSTF
Dec. 15	Complete pre-combined system test, Palmdale (Orbiter 102)
1979	
Jan. 2	Start long-lead fabrication crew module (Orbiter 099)
Jan. 30	Start orbiter/external tank/solid rocket booster burnout mated vertical ground vibration test MSFC
Jan. 31	Deliver Shuttle carrier aircraft to DFRC for ferry operations
Jan. 31	Start left-hand orbital maneuvering system Phase I qualification test, WSTF
Jan. 31	Mission Control Center Houston/Goldstone Ready for operational flight test early operations
Feb. 3	Complete combined systems test, Palmdale (Orbiter 102)
Feb. 16	Airlock on dock, Palmdale (Orbiter 102)
Feb. 17	Fourth solid rocket booster firing, Thiokol
Feb. 26	Complete mated vertical ground vibration test program at MSFC (Orbiter 101)
Feb. 28	Operational readiness date, Shuttle landing site, DFRC (Edwards AFB) Runway 23, for first manned orbital flight
March 5	Complete post-checkout, Palmdale (Orbiter 102)
March 8	Complete closeout inspection, final acceptance, Palmdale (Orbiter 102)
March 8	Orbiter 102 transported overland from Palmdale to DFRC (38 miles)
March 9	Shuttle carrier aircraft/Orbiter 102 test flight at NASA DFRC
March 17	Space Shuttle main engine 2005, first flight engine delivered to NSTL for acceptance test firings
March 20	Ferry flight, Shuttle carrier aircraft/Orbiter 102 from DFRC to Biggs Army Air Base, El Paso, TX (3 hr 20 min)
March 22	Ferry flight, Shuttle carrier aircraft/Orbiter 102 from Biggs Army Air Base to Kelly AFB, San Antonio, TX (1 hr 39 min)
March 23	Ferry flight Shuttle carrier aircraft/Orbiter 102 from Kelly AFB to Eglin AFB, FL (2 hr 12 min)
March 24	Ferry flight Shuttle carrier aircraft/Orbiter 102 from Eglin AFB, to KSC (1 hr 33 min)
March 30	Space Shuttle main engine 2007, flight engine delivered to NSTL for acceptance test firing
April 6	Complete Phase I qualification tests, aft reaction control system, WSTF
April 10	Ferry flight, Shuttle carrier aircraft/Orbiter 101 from MSFC to KSC (1 hr 52 min)
April 16	Space Shuttle main engine 2006, flight engine delivered to NSTL, for acceptance test firing
April 18	Complete left-hand orbital maneuvering system Phase I qualification, WSTF
May 1	Orbiter 101/external tank/solid rocket boosters mated on mobile launch platform transported to Launch Complex 39A from VAB at KSC
May 4	Start forward reaction control system Phase I qualification tests, WSTF

 ## Space Transportation System Background Information

Date	Event
May 4	Fifth static firing, MPTA-098 NSTL, flight nozzles 1.5 seconds
May 10	Deliver right-hand orbital maneuvering system/reaction control system from McDonnell Douglas, St. Louis, to KSC (Orbiter 102)
May 15	Deliver left-hand orbital maneuvering system/reaction control system from McDonnell Douglas to KSC (Orbiter 102)
May 30	Deliver external tank used in mated vertical ground vibration test from MSFC to Martin Marietta for refurbishment
June 12	Fifth static firing, MPTA-098 NSTL, flight nozzles (54 seconds, early cutoff, accelerometer filters)
June 15	First solid rocket booster qualification firing, Thiokol, UT, 122 seconds; nozzle extension severed at end of run as in actual mission; full cycle gimbal
June 21	Start assembly crew module (Orbiter 099)
July 2	Six static firing, MPTA-098 NSTL, flight nozzles (19 seconds, early cutoff — main fuel valve rupture)
July 23	Orbiter 101, external tank, solid rocket boosters transported on mobile launch platform from Launch Complex 39A to VAB at KSC
Aug. 1	Start long-lead fabrication crew module (Orbiter 103)
Aug. 6	Complete limit test (STA 099), Lockheed facility, Palmdale
Aug. 10	Ferry flight, Shuttle carrier aircraft/Orbiter 101, KSC to Atlanta (1 hr 55 min)
Aug. 11	Ferry flight, Shuttle carrier aircraft/Orbiter 101, Atlanta to St. Louis (1 hr 50 min)
Aug. 12	Ferry flight, Shuttle carrier aircraft/Orbiter 101, St. Louis to Tulsa (1 hr 35 min)
Aug. 13	Ferry flight, Shuttle carrier aircraft/Orbiter 101, Tulsa airport to Denver (2 hr)
Aug. 14	Ferry flight, Shuttle carrier aircraft/Orbiter 101, Denver to Hill Air Force Base, Ogden, UT (1 hr 30 min)
Aug. 15	Ferry flight, Shuttle carrier aircraft/Orbiter 101, Ogden to Vandenberg Air Force Base (VAFB) (2 hr 20 min)
Aug. 16	Ferry flight, Shuttle carrier aircraft/Orbiter 101, VAFB to DFRC (1 hr 10 min)
Aug. 23	Orbiter 101/Shuttle carrier aircraft demate, DFRC
Aug. 27	Start long-lead fabrication, crew module (Orbiter 103)
Aug. 31	Complete OMS Phase II qualification tests, WSTF
Aug.	Second SRB qualification firing, Thiokol
Sept. 5	Complete forward RCS Phase I qualification tests, WSTF
Sept. 12	Start forward RCS Phase II qualification tests, WSTF
Sept. 21	Start aft RCS Phase II qualification tests, WSTF
Oct. 5	Complete setup and thermal tests (STA 099), Lockheed Facility, Palmdale
Oct. 24	Sixth static firing, MPTA-098, NSTL, flight nozzle (scrubbed, hydrogen detector oversensitive)
Oct. 30	Move Orbiter 101 from DFRC overland to Rockwell Palmdale facility (38 miles)
Nov. 2	Start OMS left-hand pod Phase II qualification, WSTF
Nov. 3	Complete Orbiter 102 APU hot fire tests, Orbiter Processing Facility, KSC
Nov. 4	MPTA-098 static firing, NSTL (10 seconds, flight nozzles, SSME LOX turbine, seal cavity pressure, high-cutoff steerhorn failure)

Space Transportation System Background Information

Date	Event
Nov. 7	Deliver STA-099 from Lockheed facility, Palmdale to Rockwell Palmdale for rework as second operational orbiter (redesignated OV-099)
Nov. 12	Complete qualification test OMS engine at WSTF
Dec. 7	Demate payload bay door, Palmdale, OV-099
Dec. 14	Complete demate body flap, Palmdale, OV-099
Dec. 16	Orbiter integrated test start Orbiter 102, KSC
Dec. 17	Sixth static firing, MPTA-098, NSTL, 554 seconds (340 seconds at 100% rated power level, then 90% at 385 seconds, 80% at 450 seconds, 70% at 505 seconds; 1 engine shutdown — other 2 continued at 70% until 554 seconds). POGO and gimbaling tests accomplished; stub nozzles
Dec. 21	Complete demate elevons, Palmdale, Orbiter 099
1980	
Jan. 14	Complete orbiter integrated test, Orbiter 102, KSC
Jan. 18	Vertical stabilizer on dock, Fairchild, NY, for rework, Orbiter 099
Jan. 25	Body flap on dock, Downey, Orbiter 099
Jan. 25	Payload bay doors on dock, Rockwell, Tulsa, OK, for rework, Orbiter 099
Jan. 28	Start instrumentation removal and prepare mid fuselage for modification, Palmdale, Orbiter 099
Feb. 1	Complete aft fuselage demate Palmdale, Orbiter 099 for rework on dock, Downey
Feb. 1	Elevons on dock, Grumman, NY, for rework, Orbiter 099
Feb. 1	Complete aft RCS, Phase II qualification tests, WSTF
Feb. 4	Start instrumentation removal and prepare wing for modification, Palmdale, Orbiter 099
Feb. 8	Demate forward RCS module, Palmdale, Orbiter 099
Feb. 14	Final qualification firing SRB, Thiokol, UT
Feb. 15	Complete demate upper forward fuselage, Palmdale, Orbiter 099
Feb. 18	Complete left-hand OMS Phase II qualification test, WSTF
Feb. 20	Complete forward RCS module qualification test, WSTF
Feb. 23	Upper forward fuselage on dock, Downey for rework, Orbiter 099
Feb. 28	Seventh static firing, MPTA-098, NSTL, 555 seconds (No. 2 engine planned shutdown at 520 seconds, throttle down to 70% from 100%), pogo and gimbaling tests; stub nozzles
March 3	Start detail fabrication, crew module, Orbiter 103
March 20	Eighth static firing, MPTA-098, NSTL, 539 seconds (started at 100%, two engines throttled to 70%, then up to 100%, two other engines throttled to 70%, then up to 100%; all three engines throttled to simulate 3-g mission profile, then to 70% and shutdown; simultaneous pogo and gimbal tests; stub nozzles
March 21	Forward RCS module on dock, Downey, for rework, Orbiter 099
March 31	Complete 1/4-scale model tests, Downey
April 16	Ninth static firing, MPTA-098, NSTL (No. 2 engine, 4.6 seconds, shutdown due to discharge. Overtemperature on high-pressure fuel turbo pump; No. 1 and No. 3 shutdown at 6 seconds; stub nozzles)
May 30	Complete preparation lower forward fuselage modification, Palmdale, Orbiter 099
May 30	Ninth static firing, MPTA-098, NSTL (46 seconds into run, throttled to 70% for 18

Space Transportation System Background Information

Date	Event
	seconds, then to 100%, to 95% at 375 seconds, to 85% at 400 seconds, 83% at 430 seconds, 75% at 460 seconds; No. 3 shutdown at 480 seconds, remaining two to 60% at 490 seconds; No. 2 shutdown at 565 seconds; No. 1 shutdown by engine cutoff sensors at 575 seconds). Pogo and gimbal tests; stub nozzles
June 1	Orbiter 102, Engine 2005 SSME fired for 520 seconds, NSTL
June 1	Start fabrication/assembly wings, Orbiter 103
June 5	Orbiter 102, Engine 2006 SSME fired for 520 seconds at NSTL
June 16	Orbiter 102, Engine 2007 SSME fired for 520 seconds at NSTL
June 20	Start fabrication lower forward fuselage, Orbiter 103
July 12	Tenth static firing, MPTA-098, NSTL, shutdown 105 seconds into firing due to burn through in Engine No. 3 fuel preburner, 102% thrust, flight nozzles
July 28	Start detail fabrication aft fuselage, Orbiter 104
Aug 3	Complete installation SSME's, Orbiter 102, KSC
Sept. 1	Start body flap modification Downey, Orbiter 099
Sept. 29	Start assembly crew module, Orbiter 103
Oct. 1	Start fabrication/assembly mid fuselage, Orbiter 103
Oct. 1	Start fabrication/assembly mid fuselage, Orbiter 104
Oct. 9 & 10	Removal of SSME's from Orbiter 102, KSC for modifications
Oct. 11 & 12	Installation of orbital maneuvering system/reaction control system pods, Orbiter 102, KSC
Nov. 3	Start initial systems installation crew module Downey, Orbiter 099
Nov. 3	External tank mated to solid rocket boosters in Vehicle Assembly Building, KSC, for STS-1
Nov. 3	Eleventh static firing MPTA-098, NSTL, shutdown 20 seconds into firing due to burn through in Engine No. 2 nozzle, 102% thrust, flight nozzles
Nov. 4	Structural integrity test aft fuselage, Orbiter 102, KSC
Nov. 5	External tank mated to solid rocket boosters at KSC
Nov. 8, 9 & 10	Reinstallation of SSME's, Orbiter 102 KSC
Nov. 10	Start assembly aft fuselage, Orbiter 103
Nov. 14	Complete modifications, Orbiter 102, KSC
Nov. 16	Complete thermal protection system installation, Orbiter 102, KSC
Nov. 21	Complete wing modification, Palmdale, Orbiter 099
Nov. 21	Complete modification lower forward fuselage, Palmdale, Orbiter 099
Nov. 24	Transfer Orbiter 102 from Orbiter Processing Facility to Vehicle Assembly Building, for STS-1, KSC
Nov. 26	Mating of Orbiter 102 to external tank and solid rocket boosters in Vehicle Assembly Building, for STS-1, KSC
Dec. 4	Space Shuttle vehicle power-up in Vehicle Assembly Building (Orbiter 102), KSC
Dec. 4	Eleventh static firing MPTA-098, NSTL stub nozzles, 591 seconds simulated flight profile, started at 100% then to 65% at 37 seconds, then to 102% at 65 seconds, then to 65% at 438

Space Transportation System Background Information

Date	Event
	seconds. Engine No. 2 shutdown at 442 seconds, remaining two to 65% at 508 seconds, then shutdown gimbal test
Dec. 8	Start initial system installation aft fuselage, Orbiter 103
Dec. 12	Complete rework aft fuselage, Downey, Orbiter 099
Dec.15-18	Space Shuttle interface test in Vehicle Assembly Building (Orbiter 102), KSC
Dec. 19	Start preparations for Space Shuttle rollout and ordnance installation (Orbiter 102), KSC
Dec. 29	Transfer Space Shuttle aboard mobile launch platform from Vehicle Assembly Building to Launch Complex 39A for STS-1 (Orbiter 102)
1981	
Jan. 5	Emergency egress test
Jan 17	Twelfth static firing MPTA-098, NSTL, flight nozzles, 625 seconds, 100% thrust, simulated abort mission profile, No. 1 engine shutdown at 239 seconds, remaining two shutdown at 625 seconds, POGO and gimbal tests. External tank test without anti-geyser line to verify feasibility of eventually removing it from later external tank versions.
Jan. 22	External tank LH_2 load, KSC, Orbiter 102
Jan. 23	Auxiliary power unit confidence run, Orbiter 102, KSC, each APU serially run for two minutes
Jan. 24	External tank LO_2 load, KSC, Orbiter 102
Jan. 29	Hypergolic load, reaction control system, orbital maneuvering system, KSC, Orbiter 102
Feb. 2	Start initial system installation forward reaction control system module, Downey, Orbiter 099
Feb. 2	Wet countdown demonstration test simulations, KSC, Orbiter 102
Feb. 4	Start series of countdown demonstration tests for flight readiness firing, Orbiter 102, KSC
Feb. 13	Left outboard elevon on dock, Palmdale, Orbiter 099
Feb. 20	Flight readiness firing (20 second firing of all three SSME's, Orbiter 102, KSC
Feb. 24, 25, & 26	Start mission verification tests, Oribter 102, KSC
March 1	Body flap from Downey to Palmdale, Orbiter 099
March 2	Start fabrication/assembly payload bay doors, Orbiter 103
March 14	Launch verification tests, Orbiter 102, KSC
March 27	Vertical stabilizer on dock, Palmdale, Orbiter 099
March 30	Elevon rework complete, on dock, Palmdale, Orbiter 099
April 12	Conduct STS-1, Orbiter 102, launch KSC, land Edwards AFB, CA, Runway 23 dry lakebed on April 14, 54 hour 21 minutes, 51 seconds mission, landed on Orbit 36
April 20	Ferry flight, Orbiter 102, from Edwards AFB, CA, to Tinker AFB, OK, 3-1/2 hour flight
April 20	Solid rocket booster stacking began on Mobile Launcher Platform for STS-2, KSC
April 21	Ferry flight, Orbiter 102, from Tinker AFB, OK, to KSC, 3 hours, 20 minute flight
April 22	External tank for STS-2 on dock, KSC
June 1	Start fabrication elevons, Orbiter 104
June 1	Start fabrication vertical stabilizer, Orbiter 103
June 1	Start fabrication/assembly wings, Orbiter 104
July 2	Upper forward fuselage on dock, Palmdale, Orbiter 099
July 10	Payload bay doors on dock, Palmdale, Orbiter 099

Space Transportation System Background Information

Date	Event
July 13	Start fabrication crew module, Orbiter 104
July 14	Crew module on dock, Palmdale, Orbiter 099
July 17	Complete body flap modification, Downey, Orbiter 099
July 21	Aft fuselage on dock, Palmdale, Orbiter 099
July 24	Body flap on dock, Palmdale, Orbiter 099
July 30	Began mating of External Tank to Solid Rocket Boosters on Mobile Launcher Platform for STS-2, KSC
Aug. 4	Transfer Orbiter 102 from Orbiter Processing Facility to Vehicle Assembly Building, KSC
Aug. 26	Transfer Space Shuttle aboard Mobile Launcher Platform from Vehicle Assembly Building to Launch Complex 39A for STS-2 KSC
Sept. 6	Transport OV-101 overland from Palmdale, CA, to NASA's Dryden Flight Research Center at Edwards AFB, CA
Oct. 3	External tanks for STS-3 on dock, KSC
Oct. 19	Start detailed fabrication/assembly, body flap, Orbiter 103
Oct. 19	Start detailed fabrication lower forward fuselage, Orbiter 104
Oct. 23	Complete airframe modifications, Palmdale, Orbiter 099
Oct. 26	Start initial systems installation, crew module, Downey, Orbiter 103
Oct. 26	Start initial subsystems test power-on, Orbiter 099
Nov. 2	Start initial subsystems test, Palmdale, Orbiter 099
Nov. 12	Conduct STS-2, Orbiter 102, launch KSC, land Edwards AFB, CA, Runway 23, lakebed, landed on Orbit 36, November 14, 54 hours, 24 minutes 4 seconds
Nov. 23	Start Solid Rocket Booster stacking on orbit Launcher Platform for STS-3, KSC
Nov. 23	Start assembly aft fuselage, Orbiter 104
Nov. 24	Ferry flight, Orbiter 102, from Edwards AFB, CA, to Bergstrom AFB, TX, 3-1/2 hour flight
Nov. 25	Ferry flight Orbiter 102, from Bergstrom AFB, TX to KSC, FL, 3 hour flight
Dec. 9	Start fabrication upper forward fuselage, Orbiter 104
Dec. 11	Spacelab-1, arrived KSC, FL
Dec. 19	Start mating of External Tank to Solid Rocket Boosters on Mobile Launcher Platform for STS-3 at KSC, FL
1982	
Jan. 4	Start initial system installation upper forward fuselage, Orbiter-103
Jan. 29	Complete initial subsystems test including Delta F. Palmdale, Orbiter 099
Jan. 31	Forward reaction control system module on dock Palmdale, Orbiter 099
Feb 3	Transfer of Orbiter 102 from Orbiter Processing Facility to Vehicle Assembly Building at KSC, FL for STS-3
Feb. 15	Start assembly crew module, Orbiter 104
Feb. 15	Right hand orbital maneuvering system/reaction control system pod on dock at Palmdale for thermal protection system installation, Orbiter 099
Feb. 16	Transfer of Space Shuttle (Orbiter 102) from Vehicle Assembly Building to Launch Complex 39-A for STS-3
Feb. 28	Complete main propulsion test program NSTL, MS

Space Transportation System Background Information

Date	Event
March 3	Left hand orbital maneuvering system/reaction control system pod on dock at Palmdale for thermal protection system installation, Orbiter-099.
March 16	Mid-fuselage on dock Palmdale, Orbiter 103
March 22	Conduct STS-3, Orbiter 102, launch KSC, FL, land White Sands Missile Range, NM, Lakebed Runway 17, landed on orbit 130, March 30, 192 hours (8 days) 6 minutes, 9 seconds duration
March 30	Elevons on dock Palmdale, Orbiter 103
April 6	Ferry flight, Orbiter 102 from White Sands Missile Range, NM, to Barksdale, AFB, LA to KSC, FL
April 7	Orbiter 102 at Orbiter Processing Facility KSC, FL
April 16	Complete mating of SRB's and ET for STS-4 in Vehicle Assembly Building, KSC, FL
April 16	Complete subsystems test, Palmdale, Orbiter 099
April 30	Wings on dock Palmdale, Orbiter 103
April 30	Lower forward fuselage on dock, Palmdale, Orbiter 103
April 30	Complete final acceptance test, Palmdale, Orbiter 099
May 16	Transfer of Orbiter 102 from Orbiter Processing Facility to Vehicle Assembly Building, KSC, FL for mating for STS-4
May 24	Start fabrication forward reaction control system module, Orbiter 104
May 25	Transfer of Space Shuttle (Orbiter 102) from Vehicle Assembly Building on Mobile Launcher Platform to Launch Complex 39-A for STS-4
June 4	Complete post checkout operations Palmdale, Orbiter 099
June 21	Complete configuration inspection Palmdale, Orbiter 099
June 27	Conduct STS-4, Orbiter 102, launch KSC, FL, land Edwards AFB, CA, concrete runway 22, land on orbit 113, July 4, 168 hours (7 days), 1 hour, 10 minutes 43 seconds duration
June 30	Orbiter 099 rollout Palmdale
July 1	Transport Orbiter 099 overland from Palmdale to Edwards AFB, CA, for ferry flight to KSC, FL
July 2	Deliver mission kits for STS-5 KSC, FL, Orbiter 102
July 4	Ferry flight Orbiter 099 from Edwards AFB, CA, to Ellington AFB, TX
July 5	Ferry flight Orbiter 099 from Ellington AFB, TX, to KSC, FL
July 7	Start assembly payload bay doors, Orbiter 104
July 14	Ferry flight Orbiter 102 from Edwards AFB, CA to Dyess AFB, TX
July 15	Ferry flight Orbiter 102 from Dyess AFB, TX, to KSC, FL
July 16	Upper forward fuselage on dock Palmdale, Orbiter 103
Aug. 5	Vertical stabilizer on dock Palmdale, Orbiter 103
Sept. 3	Orbital maneuvering system/reaction control system pods on dock KSC, FL, Orbiter 099
Sept. 3	Payload bay doors on dock Palmdale, Orbiter 103
Sept. 9	Transfer Orbiter 102 from Orbiter Processing Facility to Vehicle Assembly Building for mating for STS-5
Sept. 21	Transfer Space Shuttle (Orbiter 102) from Vehicle Assembly Building on Mobile Launcher Platform to Launch Complex 39A at KSC, FL, for STS-5

Space Transportation System Background Information

Date	Event
Oct. 4	Start fabrication/assembly vertical stabilizer, Orbiter 104
Oct. 15	Body flap on dock Palmdale, Orbiter 103
Nov. 11	Conduct STS-5, Orbiter 102, Launch KSC, FL, land Edwards AFB, CA, concrete runway 22 on orbit 82 Nov. 16, 120 hours (5 days) 2 hours, 15 minutes, 29 seconds duration
Nov. 12	Start fabrication orbital maneuvering system/reaction control system pods, Orbiter 104
Nov. 21	Ferry flight Orbiter 102 from Edwards AFB, CA, to Kelly AFB, TX
Nov. 22	Ferry flight Orbiter 102 from Kelly AFB, TX to KSC, FL
Nov. 23	Orbiter 102 at Orbiter Processing Facility at KSC, FL, for modifications to support Spacelab-1, STS-9
Nov. 23	Transfer Orbiter 099 from Orbiter Processing Facility to Vehicle Assembly Building at KSC, FL, for mating for STS-6
Nov. 30	Transfer Space Shuttle (Orbiter 099) from Vehicle Assembly Building on mobile launcher platform to Complex 39A at KSC, FL, for STS-6
Dec. 18	Flight readiness firing Orbiter 099, 20 seconds Launch Complex 39A, KSC, FL
Dec. 28	Crew module on dock Palmdale, Orbiter 103

1983

Date	Event
Jan. 11	Aft fuselage on dock Palmdale, Orbiter 103
Jan. 25	Second flight readiness firing Orbiter 099, 23 seconds Launch Complex 39A, KSC, FL
Feb. 25	Forward reaction control system on dock Palmdale, Orbiter 103
Feb. 25	Complete final assembly and closeout installation Palmdale, Orbiter 103

Date	Event
Feb. 28	Start initial subsystems test, power-on Palmdale, Oriber 103
Mar. 14	Start fabrication/assembly body flap, Orbiter 104
April 4	Conduct STS-6, first flight of Orbiter 099, launch KSC, FL Orbiter 099, land Edwards AFB, CA, concrete runway 22, April 9, land on orbit 81, 120 hours (5 days) 24 minutes, 31 seconds duration
April 14	Ferry flight Orbiter 099 Edwards AFB, CA, to Kelly AFB, TX
April 16	Ferry flight Orbiter 099 Kelly AFB, TX, to KSC, FL
April 17	Orbiter 099 at Orbiter Processing Facility
May 6	Mid-fuselage on dock Palmdale Orbiter 104
May 13	Complete initial subsystems testing including Delta-F, Orbiter 103
May 21	Orbiter 099 from Orbiter Processing Facility to Vehicle Assembly Building for STS-7
May 26	Orbiter 099 from Vehicle Assembly Building to Launch Complex 39A for STS-7
June 13	Wings on dock Palmdale, Orbiter 104
June 17	Elevons on dock Palmdale, Orbiter 104
June 18	Conduct STS-7 Orbiter 099, Launch KSC, FL, land Edwards AFB, CA, lakebed runway 15, June 24 on orbit 98, 144 hours (6 days) 25 minutes, 41 seconds duration
June 28	Ferry flight, Orbiter 099 Edwards AFB CA, to Kelly AFB, TX
June 28	Orbiter 102 at Orbiter Processing Facility for STS-9
June 29	Ferry flight, Orbiter 099 Kelly AFB, TX to KSC FL
June 30	Orbiter 099 at Orbiter Processing Facility for STS-8

Space Transportation System Background Information

Date	Event
July 26	Orbiter 099 from Orbiter Processing Facility to Vehicle Assembly Building for STS-8
July 29	Complete subsystems test, Palmdale, Orbiter 103
Aug. 1	OMS/RCS pods on dock, Palmdale, Orbiter 103
Aug. 2	Orbiter 099 from Vehicle Assembly Building to Launch Complex 39A for STS-8
Aug. 12	Complete final acceptance test, Palmdale, Orbiter 013
Aug. 19	Vertical stabilizer on dock Palmdale, Orbiter 104
Aug. 30	Conduct STS-8, Orbiter 099, launch KSC, FL, first night launch of Space Shuttle, land Edwards AFB, CA, concrete runway 22, Sept. 5 orbit 98 first night landing, 144 hours (6 days) 1 hour, 9 minutes, 33 seconds duration
Sept. 9	Ferry flight Orbiter 099 from Edwards AFB, CA, to Sheppard AFB, TX, to Kennedy Space Center, FL
Sept. 9	Complete post checkout, Orbiter 103, Palmdale
Sept. 11	Orbiter 099 at Orbiter Processing Facility, KSC, FL
Sept. 20	Orbiter 099 power down modification period begins at Orbiter Processing Facility, KSC, FL
Sept. 23	Orbiter 102 from Orbiter Processing Facility to Vehicle Assembly Building, KSC, FL for STS-9
Sept. 28	Orbiter 102 from Vehicle Assembly Building to Launch Complex 39 KSC, FL, for STS-9
Sept. 30	OMS/RCS pods on dock KSC, FL, for Orbiter 103
Oct. 16	Orbiter 103 rollout at Palmdale
Oct. 17	Orbiter 102 rollback from Launch Complex 39A to Vehicle Assembly Building, KSC, FL
Oct. 19	Orbiter 102 from Vehicle Assembly Building to Orbiter Processing Facility, KSC, FL
Oct. 28	SSME on dock KSC, FL, for Orbiter 103
Nov. 3	Orbiter 102 from Orbiter Processing Facility to Vehicle Assembly Building for remote for STS-9, KSC, FL
Nov. 5	Orbiter 103 overland transport from Palmdale to Edwards AFB, CA
Nov. 6	Ferry flight Orbiter 103 from Edwards AFB to Vandenberg AFB, CA
Nov. 7	Lower forward fuselage on deck, Palmdale, Orbiter 104
Nov. 8	Orbiter 102 mated from Vehicle Assembly Building to Launch Complex 39A for STS-9, KSC, FL
Nov. 8	Ferry flight Orbiter 103 from Vandenberg AFB, CA to Carswell AFB, TX
Nov. 9	Ferry flight Orbiter 103 from Carswell AFB, TX to KSC, FL
Nov. 15	Orbiter 103 modification period start, Orbiter Processing Facility, KSC, FL
Nov. 17	Orbiter 099 modification period retest begins at Orbiter Processing Facility, KSC, FL
Nov. 23	Orbiter 099 modification period retest ends at Orbiter Processing Facility, KSC, FL
Nov. 28	Conduct STS-9, Orbiter 102, launch KSC, FL, land Edwards AFB, CA, lakebed runway 17, Dec. 8, 1983, 240 hours (10 days) 7 hours, 48 minutes, 17 seconds duration
Dec. 2	Payload bay doors on dock, Palmdale Orbiter 104
Dec. 9	Orbiter 103 start Vehicle Assembly Building storage, KSC, FL
Dec. 11	Orbiter 099 Crew Equipment Interface Test (CEIT), Orbiter Processing Facility, KSC, FL
Dec. 14	Ferry flight Orbiter 102 from Edwards AFB to Biggs Army Air Base, El Paso, TX to Kelly Air Force Base, San Antonio, TX.

Space Transportation System Background Information

Date	Event
Dec. 15	Ferry flight Orbiter 102 from Kelly Air Force Base, San Antonio, TX to Eglin Air Force Base, FL, to KSC, FL.
Dec. 22	SSME on dock, KSC, FL, for Orbiter 103
1984	
Jan. 3	Orbiter 103 end Vehicle Assembly Building storage, transfer to Orbiter Processing Facility, KSC, FL
Jan. 5	SSME on dock KSC, FL, for Orbiter 103
Jan 6	Orbiter 099 transfer from Orbiter Processing Facility to Vehicle Assembly Building for mating for STS-11, KSC, FL
Jan. 6	Upper forward fuselage on dock, Palmdale, Orbiter 104
Jan. 11	Orbiter 099 mated transport from Vehicle Assembly Building to Launch Complex 39A, KSC, FL
Jan. 18	SSME on Dock KSC, FL, for Orbiter 103
Jan. 20	Body flap on dock Palmdale, Orbiter 104
Feb. 11	Orbiter 103 crew equipment interface test, Orbiter Processing Facility, KSC, FL
Feb. 11	Orbiter 104 aft fuselage on dock, Palmdale
March 2	Orbiter 104 crew compartment on dock, Palmdale
March 12	Orbiter 103 Orbiter Integrated Test (OIT), Orbiter Processing Facility, KSC, FL
April 3	Orbiter 103 transfer from Orbiter Processing Facility to Vehicle Assembly Building for mating, KSC, FL
April 4	Conduct STS-13, Orbiter 099 launch KSC, FL
April 6	Orbiter 104 Forward Reaction Control System on dock, Palmdale
May 1	Orbiter 103 flight readiness firing, LC-39A, KSC, FL
June 4	Conduct STS-14, Orbiter 103, launch KSC, FL

Rockwell Space Shuttle Management

DR. ROCCO PETRONE is president of the Space Transportation and Systems Group. Prior to this he was executive vice president of the Space Transportation and Systems Group. He provides Rockwell's technical expertise and leadership for engineering, design, fabrication and test requirements for the space vehicle systems. Prior to joining Rockwell in 1981, he was chairman and chief executive officer of the National Center for Resource Recovery since 1975. Petrone, a graduate of the United States Military Academy, was instrumental in development of rocket propulsion as the U.S. Army's Development officer at the Redstone Missile Department from 1952 through 1955. After four years as a member of the Army General staff, Dr. Petrone was assigned in 1960 to NASA's Kennedy Space Center as manager, Apollo program. In 1966 he was named director of Launch Operations for the Apollo flights through the first two lunar landings. He was then named director of the Apollo program. In 1973, Dr. Petrone was named director of the Marshall Space Flight Center, Huntsville, AL (which is on the site of Redstone Arsenal), and in 1974 he was named NASA's associate administrator until he resigned from NASA in 1975. Dr. Petrone whose achievements in meeting NASA's technological goals are recorded in "*Who's Who of Space.*"

WILLIAM C. STRATHERN is vice president and general manager of the Space Transportation and Systems Group, Shuttle Integration and Satellite Systems Division. He is responsible for the System and Cargo Integration and Space Operations for the transition of the Space Transportation System into an operational mode. This includes the responsibility of Rockwell Launch Operations at Kennedy Space Center and Vandenberg Air Force Base. Prior to this he was vice president of Business Development and Strategic Planning at Rockwell's Aerospace Operations. Before that, he was executive vice president of Rockwell's Space Systems Group. Strathern has held numerous executive positions with Rockwell in the more than 25 years he has been with the company. In 1973, he was President of Program Management and Marketing of the Government Telecommunications Division of Rockwell in Dallas, TX; and in 1977, he was Vice President and General Manager of the Collins Government Avionics Division in Cedar Rapids, IA. Strathern joined the Collins Division in 1955 following Air Force service during the Korean War, in which he was an electronics instructor at airborne and ground communications and navigation schools. First, he was an avionics field service engineer on Strategic Air Command programs and was later named Manager of Field Operations. In 1961, he was appointed Sales Manager for Government Marketing in Washington, D.C. Later, he became Director of Government Field Representation, responsible for all Rockwell Collins field representatives assigned to government projects. In 1968 Strathern was appointed Vice President of Government Sales and in 1972 was named Vice President of Government Operations in Washington, D.C. Strathern holds membership in the Institute of Electrical and Electronics Engineers, the National Contract Management Association, and the Radio Club of America. He belongs to the Air Force Association, the Association of the U.S. Army, the National Aviation Club, and the American Marketing Association.

SEYMOUR (SY) RUBENSTEIN is vice president and general manager of the Space Transportation and Systems Group, Shuttle Orbiter Division. He is responsible for the orbiter development and production. Prior to this, he was vice president and deputy program manager. He became vice president of engineering and chief engineer in 1977 after serving as associate chief program engineer for Space Shuttle avionics. Rubenstein transferred to the division from Rockwell's Autonetics Group in 1973 as director of Shuttle Avionics Systems. He received a BS in electrical engineering from the Massachusetts Institute of Technology and an MS in electrical engineering from New York University. He received an MBA from California State University at Fullerton. Rubenstein joined the Autonetics Division in 1961. In 1966 he joined the Strike Avionics Systems of Autonetics where he became chief of Digital Systems and later chief of Guidance Navigation Control. In 1970, he was named program manager for Information Systems of the company's NARISCO group.

EDWARD SMITH is vice president of the Space Transportation and Systems Group Engineering. He is responsible for the technical excellence of engineering throughout the Divisions. Prior to this, he was vice president of the orbiter development and production, responsible for the engineering of the orbiter vehicles, since June 1977. From 1976 to 1977, he was Space Shuttle orbiter vice president and program manager. From April 1974 to 1976, he was vice president of Engineering, serving as Space Shuttle chief engineer responsible for division wide engineering. He was Shuttle chief engineer from 1972 to 1974. With the company for more than 20 years, Smith came to the Space Division in 1966 as a senior project engineer on the Apollo program; he was named Apollo chief engineer in 1970. He received the NASA Public Service Award and the Certificate of Appreciation for his contributions to the Apollo program. Smith graduated from UCLA with a BS in mechanical engineering.

Management changes at Rockwell International's Space Shuttle Program Office will be updated as required.

Rockwell Space Shuttle Management

ROBERT G. MINOR is vice president and program manager Orbiter Operations Support, Space Transportation and Systems Group, Shuttle Orbiter Division. Prior to this he was chief engineer, Space Transportation and Systems Group, Shuttle Orbiter Division, engineering development. Previously, he was the assistant chief engineer, Space Transportation and Systems Group, engineering systems verification. He has held various positions during his 20 years in the company and developed an extensive technical and management background in avionics and project engineering. Minor has received the NASA Medal for Exceptional Engineering Achievement. He holds a Bachelor of Science degree in electrical engineering from Southern Methodist University and has done graduate work at UCLA. He is a member of the American Institute of Electrical Engineering.

ROLAND L. BENNER is vice president of the Space Shuttle Integration and Satellite Systems Division. He is responsible for Shuttle Integration and Payload/Cargo Integration and Space Operations, which includes Rockwell Launch Operations at the Kennedy Space Center and Vandenberg Air Force Base. Prior to this, he was director of Space Shuttle Systems Integration. He was assistant chief engineer for Space Shuttle Program Engineering from 1972 to 1973. Before that, Benner held the positions of assistant chief engineer, assistant program manager (spacecraft manager), and chief program engineer on Apollo. Benner also served as assistant project engineer and project engineer on the X-14 program. He received his BS in 1949 from Villanova University, Philadelphia, PA.

BENJAMIN M. BOYKIN is chief engineer, Space Transportation and Systems Group, Shuttle Orbiter Division. He was appointed to this position in January 1982 and is responsible for the Orbiter production program and operational flight support for Space Shuttle flights from Kennedy Space Center, FL and Vandenberg AFB, CA. He has held various positions in his 27 years in the company. He was director Space Shuttle engineering at Houston, Texas from 1974 to 1982; manager Space Shuttle project engineering at Downey, CA, from 1973 to 1974; manager Apollo project engineering at Downey, CA from 1968 to 1973; site project manager Apollo engineering from 1965 to 1968 at Houston; project engineer for the Apollo spacecraft-008 at Downey, CA, from 1963 to 1965 and other positions in Downey, CA from 1957 to 1963. He was a flight maintenance officer and pilot in the USAF from 1953 to 1957. He attended New Mexico State University, Las Cruces, NM, where he received a BSEE in 1953.

BRUCE A. GERSTNER is chief engineer of Space Transportation System Operations and Integration Engineering, Space Transportation and Systems Group, Shuttle Integration and Satellite Systems Division. He is responsible for the Shuttle Integration and Payload/Cargo Integration and Space Operations, including the responsibility for Rockwell Launch Operations at the Kennedy Space Center and Vandenberg Air Force Base. Prior to this, he was chief program engineer responsible for the administration of system integration contracts in engineering. Previously, he was chief project engineer of Integration and project manager in System Engineering. He attended the College of the City University of New York, where he received a BS degree in mechanical engineering in 1948.

Management changes at Rockwell International's Space Shuttle Program Office will be updated as required.

Rockwell Space Shuttle Management

DANIEL R. BROWN plant manager, Rockwell's Palmdale spacecraft assembly facility for the Space Transportation and Systems Group, Orbiter Division. He is responsible for assembly, test, and checkout of the spacecraft at Palmdale. Prior to this he was program director, Manufacturing and Test, for the Space Transportation and Systems Group, Shuttle Orbiter Division, responsible for production operations, test, and checkout of the spacecraft, and related facilities at Downey and Palmdale, as well as related traffic transportation activities. He has held key roles in building spacecraft, missiles, and aircraft. Brown joined Rockwell in 1969 as director of Manufacturing Operations and Test. Previously he was with Martin-Marietta Corporation, where he was a quality control engineer, system propulsion engineer, general supervisor of Manufacturing Installation and Checkout, and chief of Program and Production Control. During his service with Martin-Marietta, Brown held manufacturing assignments in Denver, CO, Orlando, FL, Vandenberg AFB, CA, and Little Rock, AR.

CORNELIUS MULLINS is plant manager, Manufacturing, for the Space Transportation and Systems Group, Shuttle Orbiter Division. He is responsible for the manufacturing of all parts and tools from the Space Shuttle Orbiter. Previously he was director of Manufacturing at Rockwell International's Rocketdyne Division from 1978 to 1982. He was director of Tests and Operations at Edwards Air Force Base during the Approach and Landing Test Program from 1976 to 1978 for the Space and Transportation Systems Group. Mullins was director of Launch Operations for the Space and Transportation Systems Group at the Kennedy Space Center from 1967 to 1976. Prior to joining Rockwell International in 1967, he was site manager for the Convair Atlas program at Florida. He served in the U.S. Army during the Korean conflict. He is a member of the National Management Association.

ROY H. BEAT is division director, Material for the Space Transportation and Systems Group, Shuttle Orbiter Division. He is responsible for all activities necessary to procure, store, and control all material required to produce an end product within cost and schedule limitations. Prior to this he was division director of Material and Subcontract Management in addition to director of Material, manager of Contracts/Pricing for Finance and Administration, director of Business Operations, manager of Contracts/Pricing on the Space Shuttle program, director of Business Management, director of Central Procurement, and director of Shuttle Major Subcontract Management. Beat attended both the University of Detroit and the University of Wichita. He is also a member of the National Management Association and the Purchasing Management Association.

Management changes at Rockwell International's Space Shuttle Program Office will be updated as required.

Rockwell Space Shuttle Management

RAY F. LARSON is project manager, Orbiter Production for the Space Transportation and Systems Group, Shuttle Orbiter Division. He is responsible for directing the design and supporting the fabrication of production orbiter vehicles, including the planning, organization, and integration of engineering activities. Prior to this, he was chief engineer Production Engineering for the Space Transportation Systems Group, Shuttle Orbiter Division. Larson served as vice president of Satellite Systems Advanced Programs from 1979 to 1980 and as vice president of Space Advanced Programs from 1976 to 1978. He was vice president and program engineer on the ASTP Command Service Module program from 1974 to 1976 after serving more than four years as the division's vice president for Assurance Management. Larson joined Space Division in 1954 and worked on missile programs and study projects until being assigned to the Apollo program in 1962. He later became an Apollo assistant program manager, responsible for management of division work on the spacecraft for the Apollo 9, 12, 13, 14, and 15 missions. He received the NASA Public Service Award in 1975 for his work on the ASTP program. He also serves as a consultant and chairman of the Subcommittee on Space Systems, NASA Space Systems Technology Advisory Committee, and as a consultant to the National Academy of Sciences on the Naval Studies Board Space Panel. Larson attended Wayne State College, NB, and earned a BS degree at the U.S. Naval Academy. He also took engineering courses at USC and did postgraduate work at the California Institute of Technology.

DONALD H. CARTER is program director, Material, for the Space Transportation and Systems Group, Shuttle Integration and Satellite Systems Division. He is responsible for managing all material activities for the Space Operations/Integration and Satellite Systems Division. Carter previously held key positions in material functions involving the Apollo and Space Shuttle programs. Carter attended UCLA, where he received a BA degree in accounting.

Management changes at Rockwell International's Space Shuttle Program Office will be updated as required.

Rockwell Space Shuttle Management

PAUL A. BARKER is Director Material for the Space Transportation and Systems Group, Shuttle Integration and Satellite Systems Division. He is responsible for the direction of Material Procurement. This includes all purchases required to be made except for major subcontracts. He has held this position since 1974. Prior to this position, he held the positions of Manager, General Supervisor, Supervisor, and Subcontract Administrator. Prior to joining Rockwell International he was assistant project manager at Pratt and Whitney Aircraft and subcontractor administrator at Raytheon.

Management changes at Rockwell International's Space Shuttle Program Office will be updated as required.

Space Transportation System Facilities

NATIONAL AERONAUTICS AND SPACE ADMINISTRATION (NASA)

Lyndon B. Johnson Space Center (JSC), Houston, TX
The Johnson Space Center is the lead center for overall Space Shuttle program development and is responsible for program control, overall systems engineering, and Space Shuttle systems integration and also is responsible for design and development of the orbiter vehicle and the Shuttle carrier aircraft.

The Johnson Space Center also provides the facilities for the Shuttle Avionics Integration Laboratory (SAIL), thermal-vacuum tests, thermal protection system (TPS) tests, simulators, trainers, and mockups.

Space Transportation System Facilities

NATIONAL AERONAUTICS AND SPACE ADMINISTRATION (NASA)

Marshall Space Flight Center (MSFC), Huntsville, AL
The Marshall Space Flight Center is responsible for managing the development of the solid rocket boosters, the Space Shuttle main engines, the external tank, and Spacelab.
　The mated (Orbiter 101, external tank and solid rocket booster) vertical ground vibration test was conducted at the Marshall Space Flight Center.

Space Transportation System Facilities

NATIONAL AERONAUTICS AND SPACE ADMINISTRATION (NASA)

Kennedy Space Center (KSC), FL

The Kennedy Space Center is responsible for the design and development of the Space Shuttle launch, landing, and refurbishment facilities and launch operations for launches in easterly azimuths (equatorial orbits).

The Vehicle Assembly Building (VAB), the Launch Control Center, the mobile launcher platforms, and Launch Pad 39A were modified to support the Space Shuttle vehicle and Launch Pad 39B will be modified to support it. The orbiter is ferried by the Shuttle carrier aircraft to KSC, the external tank is shipped from Michoud, LA, by barge to KSC and the solid rocket boosters are shipped via rail from Thiokol in Utah. The Space Shuttle is mated in the VAB and transported via the mobile launcher platform to the launch pad.

The Orbiter Processing Facility (OPF), located near the northwest corner of the VAB, will accommodate and process two orbiters simultaneously. A new landing facility for the orbiter is located approximately one-and-five-tenths mile north and west of the VAB. The landing runway is 4,572 meters (15,000 feet) long and 91 meters (300 feet) wide. The runway designation is 33 from southeast to northwest and 15 from northwest to southeast.

Space Transportation System Facilities

NATIONAL AERONAUTICS AND SPACE ADMINISTRATION (NASA)

National Space Technology Laboratory (NSTL), Bay St. Louis, MS .
 The National Space Technology Laboratory is utilized for the Space Shuttle main engine firings as well as the main propulsion test article (MPTA), external tank, and Space Shuttle main engine cluster test firings.

Space Transportation System Facilities

NATIONAL AERONAUTICS AND SPACE ADMINISTRATION (NASA)

Hugh L. Dryden Flight Research Center (DFRC), Edwards AFB, CA

The Dryden Flight Research Center was used to support the Orbiter 101 approach and landing test (ALT) program. The structure in the right foreground is used to mate and demate the orbiter from the Shuttle carrier aircraft. DFRC has been used as the landing site for the initial flights to date as well as an alternate landing site in future missions.

Space Transportation System Facilities

UNITED STATES AIR FORCE

Holloman Air Force Base (HAFB), NM

Holloman Air Force Base was used for the escape system sled tests. These tests involved the orbiter ejection panels and ejection seats that were provided for the commander and pilot in Orbiter 101 for the approach and landing test program and the initial four flight tests of Orbiter 102.

Space Transportation System Facilities

NATIONAL AERONAUTICS AND SPACE ADMINISTRATION (NASA)

White Sands Test Facility (WSTF), NM

The NASA White Sands Test Facility is used for the testing of the orbital maneuvering system/reaction control system engines and systems and the forward reaction control system module engines and systems.

The Northrup Strip at White Sands may be used as an alternate landing site.

Space Transportation System Facilities

UNITED STATES AIR FORCE

Vandenberg Air Force Base (VAFB), CA
Vandenberg Air Force Base will be used for Space Shuttle southern launches (polar and near-polar azimuths). The Space Shuttle facilities at Vandenberg AFB will be provided by the Department of Defense.

Space Transportation System Facilities

Rockwell International Corporation

The Space Transportation and Systems Group of Rockwell International's facilities at Downey, CA, was responsible to the Johnson Space Center for the design development, test and evaluation (DDT&E) of the Space Shuttle orbiter. The orbiter DDT&E called for the fabrication and testing of two orbiter spacecraft (Orbiter 101 and 102), a structural test article (STA) and a main propulsion test article (MPTA). In addition to the DDT&E contract, the group is responsible for the integration of the overall Space Transportation System. The group has converted the STA to an operational orbiter (Orbiter 099) in addition to building two additional orbiters (103 and 104). NASA has also awarded a structural spares contract to the group for the construction of an upper and lower forward fuselage, aft fuselage, a set of wings and a set of payload bay doors. The Downey facility provides engineering and manufacturing of the Space Shuttle orbiter crew module, forward fuselage and aft fuselage, and forward reaction control system module, as well as Shuttle integration.

The Downey facility also maintains a Flight Simulation Laboratory, an Avionics Development Laboratory (ADL), and Flight Control Hydraulics Laboratory (FCHL). These laboratories have supported the total orbiter system verification process by accomplishing early testing on breadboard/prototype hardware and preliminary releases of software provided by NASA. These tests provided the foundation for orbiter testing at the Palmdale facility and the final hardware/software verification testing at the Shuttle Avionics Integration Laboratory (SAIL) located at the Johnson Space Center (JSC) in Houston, TX. Both the Flight Simulation Laboratory and ADL start with system hardware design development, proceed into flight software evaluation, and end in a complete orbiter mission simulation using the "iron bird" of the FCHL.

Space Transportation System Facilities

The U.S. Air Force Plant No. 42 at Palmdale, CA (Site 1) provides Space Transportation and Systems Group with the facilities for the final assembly, test and checkout of the orbiters.

The Rocketdyne Division of Rockwell International, located in Canoga Park, CA, is responsible to the Marshall Space Flight Center for the design, development, manufacturing, testing, and assembly of the Space Shuttle main engines. The Rocketdyne facility at Santa Susana, CA, also is used for the Space Shuttle main engine test program.

Space Transportation System Facilities

Martin Marietta, Michoud, LA, Facility
Martin Marietta is responsible to the Marshall Space Flight Center and is using the Michoud facility in Louisiana for the design, development, manufacturing, testing, and assembly of the external tank.

 **Space Transportation System Facilities**

Thiokol Chemical Corporation, Brigham, City, UT
Thiokol is responsible to the Marshall Space Flight Center for the design, development, manufacturing, testing, and assembly of the solid rocket booster motors. The solid rocket boosters, which are separated during launch of the Space Transportation System, will descend to the ocean and be towed from the recovery area to a land return. The solid rocket booster motor casings will be returned to Thiokol for refurbishment.

Space Transportation System Facilities

McDonnell Douglas Astronautics, Huntington Beach, CA
McDonnell Douglas is responsible to the Marshall Space Flight Center for the design, development, manufacturing, testing, and assembly of the solid rocket booster structure.

 **Space Transportation System Facilities**

United Space Boosters, Incorporated, Sunnyvale, CA
United Space Boosters is responsible to the Marshall Space Flight Center for the solid rocket booster checkout, assembly, launch, and refurbishment except for the solid rocket booster motors.

Space Transportation System Glossary

AA	Accelerometer assembly
ABE	Arm-based electronics
ACCU	Audio control central unit
ACE	Automatic checkout equipment
ACK	Acknowledge
ADI	Attitude director indicator
ADS	Air data system
ADTA	Air data transducer assembly
AGL	Above ground level
AGS	Anti-gravity suit
A/L	Approach and landing
ALC	Aft load controller
ALT	Approach and landing test
AMC	Aft motor controller
AMI	Alpha Mach indicator
AOA	Abort once around
APC	Aft power controller
ARPCS	Atmospheric revitalization pressurization control system
ARS	Atmospheric revitalization system
ASA	Aerosurface servo amplifier
ASI	Augmented spark igniter
ASS	Airlock support system
ATO	Abort to orbit
AUX	Auxiliary
BFC(S)	Backup flight control (system)
CCA	Communications carrier assembly
CCTV	Closed circuit television
CCU	Crew communications unit
CCV	Chamber coolant valve
CDF	Confined detonating fuse
CDR	Commander
CI	Course (deviation display) invalid
COAS	Crew optical alignment sight
CPU	Control processor unit
CRT	Cathode ray tube
CSS	Control stick steering
C/W	Caution and warning
DA	Distribution assembly
DAP	Digital autopilot
DBC	Data bus coupler
DBIA	Data bus isolation amplifier
DBN	Data bus network
DDU	Display driver unit
D&C	Displays and Controls
DEU	Display electronics unit
DFI	Development flight instrumentation
DFRC	Dryden Flight Research Center
DK	Display keyboard
DME	Distance measuring equipment
DP/DT	Delta pressure/delta time
DPS	Data processing system
DSC	Dedicated signal conditioner
DU	Display unit
EAFB	Edwards Air Force Base
EAS	Equivalent airspeed
ECLSS	Environmental control and life support system
EES	Escape ejection suit
EI	Entry interface
EIU	Engine interface unit
EMU	Extravehicular mobility unit
EPDU	Electrical power distribution unit
EPS	Electrical power system
ESA	European Space Agency
ESRO	European Space Research Organization
ESS	Essential
ESVS	Escape suit ventilation system
ET	External tank or elapsed time
ETR	Eastern Test Range
EVA	Extravehicular activity

Space Transportation System Glossary

FC(S)	Flight control (system)
FCL	Freon coolant loop
FCOS	Flight computer operating system
FDM	Frequency division multiplexer
FES	Flash evaporator system
GH_2	Gaseous hydrogen
GN_2	Gaseous nitrogen
G&N	Guidance and navigation
GN&C	Guidance, navigation, and control
GMT	Greenwich mean time
GO_2	Gaseous oxygen
GPC	General purpose computer
GS	Glideslope
G/S	Guided steering
GSE	Ground support equipment
H_2	Hydrogen
HAC	Heading alignment cylinder
HAL/S	High order assembly language/Shuttle
HIU	Headset interface unit
H_2O	Water
HPFT	High-pressure fuel turbopump
HPOT	High-pressure oxygen turbopump
HRSI	High-temperature, reusable, surface insulation
HSI	Horizontal situation indicator
I	Initialized (computer program)
IEA	Integrated electronic assembly
IMU	Inertial measurement unit
IOP	Input-output processor
IUS	Inertial upper stage
IVA	Intra-vehicular activity
JSC	Johnson Space Center
KBU	Keyboard unit
LCA	Load control assembly
LDEF	Long-duration exposure facility
LH_2	Liquid hydrogen
LiOH	Lithium hydroxide
LO_2	Liquid oxygen
LPFT	Low-pressure fuel transducer
LPOT	Low-pressure oxygen transducer
LPS	Launch processing system
LRSI	Low-temperature, reusable, surface insulation
LVLH	Local vertical, local horizontal
LWHS	Lightweight headset
MAC	Mean aerodynamic chord
MCC	Main combustion chamber
MCC-H	Mission control center—Houston
MCDS	Multi-function CRT display system
MCIU	Manipulator controller interface unit
MDA	Motor drive amplifier
MDF	Mild detonating fuse
MDM	Multiplexer/demultiplexer
ME	Main engine
MECO	Main engine cutoff
MEP	Minimum entry point
MET	Mission elapsed time
MFV	Main fuel valve
MIA	Multiplex interface adapter
MLG	Main landing gear
MLP	Mobile launch platform
MLS	Microwave landing system
MM	Major mode or mass memory
MMC	Mid-motor controller
MMH	Monomethyl hydrazine
MMS	Multi-mission modular spacecraft
MMU	Master measurement unit
MOV	Main oxidizer valve
MPC	Mid-power control
MPL	Minimum power level

Space Transportation System Glossary

MPS	Main propulsion system
MPTA	Main propulsion test article
MSBLS	Microwave scan beam landing system
MSFC	Marshall Space Flight Center
MTU	Master timing unit
MTVC	Manual thrust vector control
N_2	Nitrogen
NASA	National Aeronautics and Space Administration
NEP	Nominal entry point
NH_3	Ammonia
NLG	Nose landing gear
N_2O_4	Nitrogen tetroxide
NSI	NASA standard initiator
NSP	Network signal processor
NSTL	National Space Technology Laboratories
O_2	Oxygen
OFI	Operational flight instrumentation
OI	Operational instrumentation
OMS	Orbital maneuvering system
OPF	Orbiter processing facility
OPOV	Oxidizer preburner oxidizer valve
OPS	Operational sequence
OV	Orbiter vehicle
PBI	Pushbutton indicator
PBS	Protective breathing system
PCA	Pneumatic control assembly
PCM	Pulse-code modulation
PCMMU	Pulse-code-modulation master unit
PDRS	Payload deployment and retrieval system
PDU	Power drive unit
PFCS	Primary flight control system
PRFLT	Preflight
PIC	Pyro initiator controller
PL	Payload
PLBD	Payload bay door
PLT	Pilot
PM	Phase modualtion
POS	Portable oxygen system
PRI	Primary
PROM	Programmable read-only memory
PRSD	Power reactant storage and distribution system
PSI	Pounds per square inch
PTC	Passive thermal control
RA	Radar altimeter
RAM	Random access memory
RCC	Reinforced carbon carbon
RCS	Reaction control system
RGA	Rate gyro assembly
RHC	Rotation hand controller
RJD	Reaction jet driver
RM	Redundancy management
RMS	Remote manipulation system
RPC	Remote power controller
RPL	Rated power level
RPTA	Rudder pedal transducer assembly
RS	Redundant set (computers)
RSS	Range safety system
RTLS	Return to launch site
SBTC	Speedbrake thrust control
SCA	Shuttle carrier aircraft
SCF	Satellite control facility
SEC	Secondary
SGLS	Space ground link system
SIP	Strain isolation pod
SM	System management
SMU	Speaker microphone unit
SOP	Software operating program
SPI	Surface position indicator
SRB	Solid rocket booster

Space Transportation System Glossary

SSME's	Space Shuttle main engines
SSUS	Solid spinning upper stage
ST	Star tracker
STA	Structural test article
STS	Space transportation system
TACAN	Tactical air navigation
TAEM	Terminal area energy management
TDRS	Tracking and Data Relay Satellite
THC	Translation hand control
TLM	Telemetry
TPS	Thermal protection system
TVC	Thrust vector control
T/W	Thrust to weight
UHF	Ultra-high frequency
VAB	Vehicle Assembly Building
Vdc	Volts direct current
VFM	View finder monitor
VIU	Video interface unit
W/B	Wide band
WBSC	Wide-band signal conditioner
WCLS	Water coolant loop subsystem
WONG	Weight on nose gear
WOW	Weight on wheels
WP	Way point
WSB	Water spray boiler
WSTF	White Sands Test Facility
WTR	Western Test Range

Space Transportation System Contractors

An experienced team of aerospace firms is working on the Space Shuttle System. Many of these firms, together with the system or component they produce, are listed in the following pages.

Contractor	Location	System/Component
Abbott Transistor	Los Angeles, CA	Transformer, displays and controls
Abex Corp., Aerospace Div.	Oxnard, CA	Pump, hydraulic, variable delivery
Aerocearic Castings	Bell Gardens, CA	Orbiter castings
Aerodyne Controls Corp.	Farmingdale, NY	Oxygen, hydrogen check valve (fuel cell and environmental control life support system) Valve, pressure relief (water) Orbiter castings
Aeroflex Laboratories	Plainview, NY	Line assembly vibration isolation mounts, feedline (main propulsion system)
Aerojet General, Aerojet Liquid Rocket Co.	Sacramento, CA	Orbital maneuvering system engines

Aeroquip Corp., Marman	Los Angeles, CA	For environmental control/life support system:
		  V-band coupling
		  Coupling No. 10T-bolt, 40 degree
		  Band clamp
		  Band clamp, light weight
		  Retaining strap
		  Forty degree flange
		  Flexible air duct

Aerospace Avionics	Bohemia, NY	Annunciator assembly (caution/warning system)
		General:
		  Annunciator display general requirements
		  Annunciator, performance monitoring
		  Annunciator, special
		  Annunciator, fire warning
		  Annunciator, assembly event sequence
		  Annunciator, assembly computer status
		  Annunciator, single event
		  Annunciator, control assembly
		  Indicator, quantity propellant
		  Floodlight, incandescent overhead
		  Light dimmer
		Battery charger and power supply (environmental control/life support system)

Space Transportation System Contractors

Contractor	Location	System/Component
Aerospace Research Associates	West Covina, CA	Cable attenuator assembly for crew escape system, Orbiters 101 and 102
Aiken Industries, Mechanical Products Div.	Jackson, MI	Circuit breakers, thermal
		Circuit breakers, three phase
Airco Inc.	Industry, CA	Cylinders, hydrogen and oxygen, Orbiter 101
Aircraft Engineering Corp.	Paramount, CA	Main propulsion test article structure and platform, orbiter overland transporter, strongback orbiter overland transporter, strongback for MSFC
Air Industries	Garden Grove, CA	Nuts and bolts
Aircraft Instruments Co.	Montgomery, PA	Indicator position (Orbiter 101)
AiResearch Manufacturing Co., Garrett Corp.	Torrance, CA	Air data transducer assembly and computer
		Safety valve, cabin air pressure
		Solenoid valve, shutoff, air
		Ground coolant unit (circulate Freon through orbiter heat exchanger during ground operations, checkout, preflight, postflight)
Airite Div., Sargent Industries	El Segundo, CA	Helium receiver (spherical), surge pressure relief of gaseous helium during actuation of valves and external tank disconnects (main propulsion system)
Allen Bradley	Milwaukee, WI	Potentiometer (controls and displays)
American Aerospace	Farmingdale, NY	Current, sensor ac, dc
		Current level detector
American Airlines	Tulsa, OK	Operations support and maintenance of Boeing 747 Shuttle carrier aircraft
Ametek Calmec	Pico Rivera, CA	2-inch shutoff valve, liquid hydrogen (main propulsion system)
		4-inch disconnect, liquid hydrogen, orbiter to tank recirculation and replenishment system (main propulsion system)
		2-inch disconnect, gaseous hydrogen/gaseous oxygen, orbiter to tank pressurization system (main propulsion system)
Ametek Straza	El Cajon, CA	Liquid hydrogen and liquid oxygen 8-inch fill and drain assembly (main propulsion system)
		Liquid hydrogen recirculation and replenishment 2-inch and 4-inch line assembly (main propulsion system)
		Gimbal joint pressure (main propulsion system)
Amex System	Lawndale, CA	Coax cable, high-temperature (flight instrumentation)

Space Transportation System Contractors

Contractor	Location	System/Component
AMI	Colorado Springs, CO	Operational flight crew and passenger seats
Anemostat Products	Scranton, PA	Cabin air diffuser
		Cabin air diffuser
Apex Mills	Los Angeles, CA	Separators - passive thermal control
Applied Resources	Fairfield, NJ	Switch, rotary
Arkwin Industries	Westbury, NY	Hydraulic reservoir, bootstrap
		Valve, 3-way, 2-position hydraulic control
		Oleophobic filter hydraulic canister
Arrowhead Products, Div. of Federal Mogul	Los Alamitos, CA	Space Shuttle main engine feedlines, 12- to 17-inch diameter liquid oxygen and liquid hydrogen (main propulsion system)
		Connector, flexible purge gas (main propulsion system)
		Environmental control/life support system:
		Coupling sleeve
		Duct flex air
		Coupling sleeve flex
		Clamp cushion flex sleeve
		Flexible connector
		Flexible connector
		Convoluted bellows long flexible connector drain system
Astech	Santa Ana, CA	Heat shield (main propulsion system)
Autonetics Group, Rockwell International	Anaheim, CA	Shuttle avionics test set—Avionics Development Laboratory, Downey
		Shuttle avionics test set—Johnson Space Center
		Driver module controller
		AC bus sensor
		Load control assemblies (forward No. 1-3, aft No. 1-3)
		Master event controller
		Backup flight control system, Orbiter 101, 102
Avco	Wilmington, MA	Ku-band antenna (microwave scan beam landing system)
		Ku waveguide assembly
	Nashville, TN	Manufacturing (crew module bulkheads)
	Lowell, MA	Elevon ablator
Aydin, Vector Div.	Newtown, PA	Wideband frequency division multiplexing unit

Space Transportation System Contractors

Contractor	Location	System/Component
Ball Brothers Research Corp.	Boulder, CO	Star tracker and light shade
		Active keel actuator
		Payload retention latch
Barry Controls	Burbank, CA	Vibration mount assembly kit
Beech Aircraft Corp., Boulder Division	Boulder, CO	Power reactant storage assembly
		Orbiter Freon coolant servicing unit
		Gaseous hydrogen and gaseous oxygen valve box unit (ground support equipment)
Bell and Howell	Pasadena, CA	Magnetic tape recorder
Bell Industries	Gardena, CA	Terminal boards, modular
Bendix Corp.	Sidney, NY	High density connector, data processing software
		Connector, electrical
	Teterboro, NJ	Indicator/surface position
		Indicator/alpha mach
		Indicator/altitude/vertical velocity
	Franklin, IN	Connector, triax, 93 ohm (electrical power distribution system)
	Davenport, IA	Accelerometer indicator (senses "g" forces in vehicles and displays on display & control panel) (backup)
Beldo Steel Corp.	Orlando, FL	Canister to carry payloads from checkout facilities to orbiter*
B.F. Goodrich Co.	Troy, OH	Main/nose landing gear wheel and main landing gear brake assembly
		Tires, main and nose gear
Bertea Corp.	Irvine, CA	Main landing gear uplock actuator, hydraulic
		Main landing gear strut actuator
		Nose landing gear uplock actuator
		External tank umbilical retractor actuator (retracts external tank feedline into orbiter at separation; composed of linear hydraulic actuators with integral control valves and locking devices)
Betatronix	Hauppauge, NY	Potentiometers
Boeing Aerospace Co.	Houston, TX	Sneak circuit analysis
	Seattle, WA	Carrier aircraft modification
		Tail cone fairing
		Load measurement system (3 load sensor cells and signal conditioner; used in ALT to measure and display to 747 and orbiter crews the flight loads during taxi, takeoff, cruise, separation, and landing)

*Separate contract from NASA

Space Transportation System Contractors

Contractor	Location	System/Component
Bomar/TIC	Newbury Park, CA	Variable transformer (displays and controls) (Orbiter 101)
Brunswick	Lincoln, NE	Filament wound tank, developmental program
		Pressure storage tanks for orbital maneuvering system, main propulsion system, reaction control system, atmospheric revitalization pressure control system
Brunswick Celesco	Costa Mesa, CA	Smoke detection
		Fire suppression system
Brunswick-Circle Seal	Anaheim, CA	Check valve (purge, vent, and drain)
		Check valve, water
		Check valve, 1-inch helium repressurization line (main propulsion system)
		Valve 3/8 inch relief liquid hydrogen, (main propulsion system)
		Engine isolation dual check valve (main propulsion system)
		Servicing check valve
		Water relief valve
Brunswick-Wintec	El Segundo, CA	Fuel line filter (auxiliary power unit)
		Filter, windshield (purge, vent, and drain)
		Filter, coolant return
		Filter assembly, cryo
		Filter, ammonia inline
		Filter, helium (main propulsion system)
		Filter, water
Bussman Div. of McGraw Edison	St. Louis, MO	Seal, bulkhead window conditioning system
		General-purpose fuse holder
		Fuse, general-purpose microminiature, axial load
		Fuse, dc limiter high current
		Fuse holder, dc, boltdown
Calspan	Buffalo, NY	Wind tunnel (hypersonic) tests, jet plume effects
		Ascent and entry rates
		Heat/pressure distribution
Carleton Controls	East Aurora, NY	Atmospheric pressure control system
		Valve, shutoff ram air inlet (Orbiter 101)
		Regulators, hydrogen and oxygen, cryo
		Airlock support component system

Space Transportation System Contractors

Contractor	Location	System/Component
Celesco Industries	Canoga Park, CA	Transducer pressure absolute, high level
		Discrete pressure transducer
Charles Stark Draper	Cambridge, MA	Software (guidance, navigation, and control)
Chem Tric	Rosemont, IL	Generator, silver/ion (environmental control/life support system—Orbiter 101)
Coast Metal Craft	Compton, CA	Metal flex hoses (main propulsion system)
		Bulkhead pen lines
		Flex line assembly propulsion
Collins Radio Group, Rockwell International	Cedar Rapids, IA	Display driver unit
		Horizontal situation indicator
Columbus Aircraft Div., Rockwell International	Columbus, OH	Body flap structure
		Manufacturing (detail parts and subassembly, forward and aft fuselage tooling, nose gear doors)
Communications Components	Costa Mesa, CA	Antenna, UHF (approach and landing test), Orbiter 101
Conrac Corp.	West Caldwell, NJ	Engine interface unit (main propulsion system)
		Mission timer
		Event timer
		Ground command interface logic box (enables crew to select either crew control or NASA ground control of 250 communications functions)
Consolidated Controls	El Segundo, CA	Fuel isolation valve (auxiliary power unit)
		Unidirectional/bidirectional shutoff valve (fuel cell, environmental control/life support system)
		Solenoid valve, hydrogen and oxygen, cryo
		High-pressure helium valves and low-pressure vernier engine manifold valves, dc (reaction control system)
		Valve, hydraulic and oxygen primary flow control pressure (main propulsion system)
		Pressurant flow control valve, hydrogen and oxygen flow from orbiter main engines for external tank main propulsion system
		Regulator, 750/20 psia, helium (main propulsion system)
		Regulator valve, 850 psig, helium (main propulsion system)
		Helium regulator pressure, 750 psia, (main propulsion system)
		Water solenoid latching valve

Space Transportation System Contractors

Contractor	Location	System/Component
Contractors Cargo	South Gate, CA	Tractor for overland transport, Palmdale to DFRC
Convair Aerospace Div., General Dynamics	San Diego, CA	Mid fuselage, including mid fuselage glove fairing
Corning Glass	Corning, NY	Windshield and windows and side hatch window
		Thermal protection system Macor machinable glass ceramic
Cox and Co.	New York, NY	Heater, water, relief valve, vent nozzle, and port
		Heater assembly tank, fuel and lube lines (auxiliary power unit)
		Heater, water boiler steam vent line
		Heater, oxygen purge lines
		Heater, strop, water relief valve
		Heater electrical set (reaction control system)
Crane Co., Hydro Aire	Burbank, CA	Main landing gear brake anti-skid hydraulic modules, transducers, control box, and wheel sensor
Crissair Inc.	El Segundo, CA	Check valve, hydraulic
		Flow restrictor, hydraulic
Curtiss Wright	Caldwell, NJ	Payload bay door actuation system (power drive units, rotary actuators, drive shaft, torque tubes, couplings)
		Radiator deploy actuator and latch mechanism
Datum Inc.	Anaheim, CA	Multi-channel, closed-loop structural test
Deutsch	Banning, CA	General-purpose connector, electrical
		Bulkhead feedthrough
Descent Controls Inc.	Costa Mesa, CA	Descent control and lines
Dynamics Corp.	Scranton, PA	Cabin diffuser
Eaton Corp. AIL Division	Farmingdale, NY	Microwave scan beam landing set navigation set
	Milwaukee, WI	Circuit breaker, remote control (Orbiter 101)
	Huntington, NY	Preamplifier assembly
Eckel Valve	San Fernando, CA	Check valves, purge (vertical stabilizer, forward reaction control system, and plenum)
Edcliff Instruments	Monrovia, CA	Position transducer, landing gear and rudder (Orbiter 101, ALT)
Edison Electronics Div., McGraw Edison	Manchester, NH	Digital select thumbwheel switch
		Toggle switches

Space Transportation System Contractors

Contractor	Location	System/Component
Eldec Corp.	Lynwood, WA	Dedicated signal conditioner (subsystem pressure, temperature, etc., to multiplexer/demultiplexer)
		Tape, meter
		Proximity switch (landing gear operation)
Electronics Association	West Long Branch, NJ	Analog computer system (Rockwell simulator)
Electronic Resources Inc., Tasker Industries	Los Angeles, CA	Coax cable, special, external temperature (communication tie, links)
Ellanef	Corona, NY	Hatch latch actuator
		Air data sensor probe actuator
		Star tracker door (consists of shutter-type door, electro-mechanical actuator, door supports)
		Manipulator retention latch actuator
Endevco	San Juan Capistrano, CA	Piezoelectric accelerometer (flight instrumentation, vibration-acoustic data)
		Acoustic pickup, piezoelectric (development flight instrumentation, acoustic data, Orbiter 101)
		Piezoelectric accelerometer (flight instrumentation, vibration-acoustic data)
ESA (European Space Agency)	Paris, France	Spacelab and U-shaped pallets*
Ex-Cell-O Corp., Div. of Cadillac Controls	Costa Mesa, CA	Hatch attenuator, main ingres/egress hatch
Explosive Technology	Fairfield, CA	Interseat energy transfer and sequencer, pyro (crew escape system, Orbiter 101s and 102)
		Severence system, pyro, ejection panel (Orbiters 101 and 102)
Fairchild Republic	Farmingdale, NY	Vertical tail
Fairchild Stratos	Manhattan Beach, CA	12-inch prevalves, shutoff propellant (main propulsion system)
		1-1/2 inch disconnect, liquid oxygen overboard bleed (main propulsion system)
		8-inch fill and drain valve, propellant (main propulsion system)
		Ammonia boiler subsystem (rejects heat during reentry)

*Memorandum of understanding signed by NASA and European Space Research Organization (ESRO), now the European Space Agency, effective 5/30/75

Space Transportation System Contractors

Contractor	Location	System/Component
		1-inch disconnect pneumatic, helium and gaseous nitrogen (main propulsion system)
		1-inch shutoff valve, liquid oxygen and liquid hydrogen relief (main propulsion system)
		Regulator, pressure, helium (series redundant), forward and aft (reaction control system)
		Cryogenic fluid and gas supply disconnects between orbiter power reactant storage and distribution system fill, drain, and vent lines and ground support equipment)
		Coupling, hypergolic servicing, nitrogen tetroxide/hydrazine (orbital maneuvering system/reaction control system)
		Cap, liquid hydrogen/liquid oxygen, 8-inch orbiter disconnect (main propulsion system)
		Coupling helium fill disconnect flight half
Ford Aerospace and Communications Corporation*	Palo Alto, CA	Ground communications system to support Space Shuttle at Vandenberg AFB
G&H Technology Co.	Santa Monica, CA	Connector, cryo
General Electric	Valley Forge, PA	Waste collector subsystem
	Waterford, NY	Thermal protection system room-temperature vulcanizing adhesive
George A. Fuller Co., Div. of Northrop Corp.	Chicago, IL	Fabrication and erection of mating-demating device for orbiter and 747 carrier aircraft at Dryden Flight Research Center**
Globe Albany	Auburn, ME	Nomex felt (thermal protection system)
Grimes Manufacturing	Urbana, OH	Floodlight, overhead, incandescent (Orbiter 101)
Grumman Corp.	Bethpage, NY	Wing (includes main landing gear doors, elevons, wing box glove)
Gulton Industries	Costa Mesa, CA	Accelerometer, linear low frequency (flight instrumentation, vibration-acoustic data)
		Differential pressure transducer, hydraulic actuators
		Transducer, pressure, pogo (main propulsion system)
		Transducer, sound pressure level
		Transducer, cabin acoustic
Hamilton Standard Div., United Technologies Corp.	Windsor Locks, CT	Freon coolant loop (includes fuel cell heat exchanger, Freon to water interchanger, Freon loop 1 and 2 inlet sensor, development instrumentation package, Freon pump package)

*Separate contract from USAF
**Separate contract from NASA

Space Transportation System Contractors

Contractor	Location	System/Component
		Atmospheric revitalization subsystem (includes cabin fan assembly and debris trap, CO_2 absorber and temperature control avionics cooling assembly, humidity control heat exchanger assembly, development instrumentation signal conditioner and avionics fan, secondary coolant pump and accumulator, primary coolant pump and accumulator)
		Water boiler, hydraulic thermal control unit
		Hydraulic cart (ground support equipment)
		Flash evaporator system (removes heat generated by orbiter systems from two coolant loops by turning water from fuel cells into steam and venting to space)
		Water management control panel
		Life support system for Space Shuttle extravehicular activity*
		Fabrication and field support of Space Shuttle EVA mobility unit suits*
Harris Corp., Electronics Systems Div.	Melbourne, FL Baltimore, MD	Pulse code modulation master unit Payload data interleaver
Hartman Electric Manufacturing Div. of A-T-O	Mansfield, OH	General-purpose power contactor
Haveg Industries, Inc.	Winooski, VT	Wire, general-purpose
Hayden Switch and Instrument	Waterbury, CT	Limit switches, hermetic seal (electrical power and distribution system)
Hexcell Aerospace	Dublin, CA	Attenuator pads (energy absorbers for inner and outer crew escape panels during deployment) (crew escape system, Orbiters 101 and 102)
Hi-Temp Insulation Inc.	Camarillo, CA	Hydraulic blanket insulation (hydraulic system) Fibrous bulk insulation-passive thermal control
Hoffman Electronics Corp. NavCom Systems Div.	El Monte, CA	TACAN (tactical air navigation)
Holloway Corp.	Titusville, FL	Construction of solid-rocket booster recovery and disassembly at Cape Canaveral Air Force Station*

Honeywell Inc.	St. Petersburg, FL	Flight control system (displays and controls):
		    Rotation hand control
		    Accelerometer assembly
		    Rudder pedal transducer assembly
		    Speed brake thrust control

*Separate contract from NASA

Space Transportation System Contractors

Contractor	Location	System/Component
		Translation hand control
		Forward and aft reaction jets and orbital maneuvering system drivers
		Aerosurface servo amplifier
		Ascent thrust vector control amplifier
	Minneapolis, MN	Radar altimeter
	McLean, VA	Central data processing for Space Shuttle launch processing system, Kennedy Space Center*
Hoover Electric	Los Angeles, CA	Umbilical door latch actuator
		Umbilical door actuator
		Umbilical door centerline latch actuator
		Orbiter payload door latch actuators (2 electromechanical rotary actuators to latch payload bay doors closed; one to extend and retract radiators and operate latches to hold radiators in retracted position)
		Actuator, payload bay door centerline latch
Hughes	El Segundo, CA	Ku-band radar/communication system, deployable antenna and electrical assembly
Hydraulic Research, Textron	Valencia, CA	Servo actuator, elevon-electro command, hydraulics, Orbiter 101
IBM Corp., Federal Systems Div., Electronics Systems Center	Owego, NY	Mass memory/multi-function cathode ray tube display unit, keyboard display electrical unit subsystem
		General-purpose computer processor unit and input-output processor
		Space Shuttle Mission Control Operations data processing complex at JSC*
		Orbiter computer programs*
		Shuttle data processing complex programming at JSC*
		Launch processing system at KSC*
	Gaithersburg, Md.	Space Shuttle orbiter data processing hardware maintenance*
ICI United States Inc.	Wilmington, DE	Thermal protection system alumina mat
ILC Technology	Sunnyvale, CA	Cabin interior lighting
		Orbiter floodlight system, payload bay floodlight, overhead docking floodlight, forward bulkhead floodlight, and floodlight electronic assemblies
Intermetrics Inc.	Cambridge, MA	Advanced computer programming language, HAL/S (high-order assembly language/Shuttle)
		Avionics software

*Separate contract from NASA

Space Transportation System Contractors

Contractor	Location	System/Component
ITT Cannon	Santa Ana, CA	Connector, power
		High-density connectors
		Rectangular connector
		Connector, coax TNC
		Connector, coax TNC bulkhead
		Connector, coax HN
		Connector, coax bulkhead
	Phoenix, AZ	Bulkhead feedthrough
J. C. Carter Co.	Costa Mesa, CA	Service coupling (auxiliary power unit)
Jet Electronics	Grand Rapids, MI	Indicator attitude, backup, Orbiter 101
J. L. Products	Gardena, CA	Crew compartment failure warning and corrective control:
		Arming fire switch, pushbutton
		Pushbutton switch
		Relay
Johns Manville	Manville, NJ	Advanced flexible reusable surface insulation quilted fabric blanket
J.P. Stevens Co.	Los Angeles, CA	Thermal protection system Quartz thread
Kaiser Electronics	San Jose, CA	Heads-up display Orbiter 102, STS-5, Orbiter 099 and subsequent
K-West	Westminster, CA	Wideband signal conditioner, accelerometer/acoustic
		Strain gage signal conditioner, stresses
		Ullage pressure signal conditioner for external tank (monitors and controls external tank ullage pressure, liquid oxygen and liquid hydrogen tanks)
		Differential pressure transducer and electronics, propellant head pressure in main feed and fill lines (main propulsion system)
Labarge	Santa Ana, CA	Wire, general-purpose
Leach Relay	Los Angeles, CA	General-purpose latching relay
		Relay, 10 amp
Lear Siegler	Grand Rapids, MI	Attitude direction indicator
	Elyria, OH	Hydraulic disconnect supply, 1/2-inch
		Coupling, heat exchanger test point (ground support equipment)
		Test point and nitrogen coupling
		Coupling, flight half
	Los Angeles, CA	Disconnect, flight half—air, gaseous nitrogen, purge

Space Transportation System Contractors

Contractor	Location	System/Component
Life Systems Inc.	Cleveland, OH	Separator, hydrogen/water unit
Lockheed-California Co.	Burbank, CA	Ejection seats, including drogue and personnel chute (crew escape system, Orbiters 101 and 102)
		Orbiter structural static and fatigue testing
Lockheed Electronics	White Sands, NM	Support services*
Lockheed Missiles and Space Co., Inc.	Sunnyvale, CA	Reusable surface insulation, high temperature and low temperature and high temperature reusable surface insulation
		Fibrous refractory composite insulation (FRCI) -8 and -12
		Space telescope support systems module* (14-ft. diameter, 43-ft. long cylindrical unit houses optics, sensors, and support systems; systems engineering, analysis, integration, verification of complete assembly and support of ground and flight operations
Los Angeles Aircraft Div., Rockwell International	Los Angeles, CA	Manufacturing of aft fuselage upper truss thrust structure
		Diffusion bonding
		Crew module panels
		1/4-scale ground vibration test model
		Tool fabrication
Magnavox	Ft. Wayne, IN	Receiver transmitter mount, UHF
		Receiver/transmitter, UHF
Malco Microdot Corp.	Pasadena, CA	Connector, coax bulkhead
Marquardt Co.	Van Nuys, CA	Reaction control system thrusters, 870-lb thrust
CCI Corp.		Reaction control system thrusters, 24-lb thrust
Marshall Space Flight Center (NASA)	Huntsville, AL	Solid rocket booster integration and final assembly
Martin Marietta	Denver, CO	Caution and warning electronics, status display, limit module
		Pyro initiator controller
		Reaction control system fluel and oxidizer tanks (forward and aft)
	New Orleans, LA	External Tank*
		Checkout, control, and monitor subsystem*

*Separate contract from NASA

Space Transportation System Contractors

Contractor	Location	System/Component
	Vandenberg Air Force Base, CA	Definition and planning of acquisition of ground systems to support Shuttle operations at VAFB**
Marvin Engineering	Inglewood, CA	Manufacturing (crew module skins and ejection panels)
McDonnell Douglas Astronautics Co.	St. Louis, MO	Orbital maneuvering system/reaction control system aft propulsion pod
	Huntington Beach, CA	Solid rocket booster structure*
		Assembly, checkout, launch operations, and refurbishment, solid rocket boosters*
		Spinning solid upper stage** (SSUS)
	Huntsville, AL	Spacelab integration
Megatek	Van Nuys, CA	Cryo seals, line flange (main propulsion system)
Menasco Manufacturing Co.	Burbank, CA	Main/nose landing gear shock struts and brace assembly
Merco Manufacturing Co.	Anaheim, CA	Manufacturing (crew module star tracker panels)
Metal Bellows Co.	Chatsworth, CA	Potable and waste water tanks
		Metal bellows assembly
		Flex metal tube, convoluted
		Flex metal tube assembly
		RCS flexible line assembly
		Propellant flex line assembly, OMS/RCS
Metalcraft Inc.	Baltimore, MD	Portable fire extinguishers
Micro Measurements	Romulus, MI	Strain gage
3M Co., Inc.	St. Paul, MN	Thermal protection system AB312 fibers
Modular Computer Systems	Ft. Lauderdale, FL	Data acquisition system (Rockwell laboratories)
		Central data subsystem of launch processing system (mini-computer)*
Moog, Inc.	East Aurora, NY	Main engine gimbal servo actuator
		Elevon servo actuators, Orbiter 102 and subsequent
Motorola	Scottsdale, AZ	Communications test set (ground support equipment)

*Separate contract from NASA
**Separate contract from Department of Defense

Space Transportation System Contractors

Contractor	Location	System/Component
Networks Electronics Corp., U.S. Bearings Div.	Chatsworth, CA	Hatch latch links, main hatch
Northrop Corp., Electronics Div.	Norwood, MA	Rate gyro assembly
OEA	Denver, CO	Thruster assembly, pyro, nose gear uplock release
		Guillotine assembly pyro crew escape system, Orbiters 101 and 102
		Thruster assembly ejection panel jettison, Orbiters 101 and 102
Parker Hannifin	Irvine, CA	17-inch disconnects, liquid hydrogen and liquid oxygen, orbiter to external tank feed system (main propulsion system)
		8-inch disconnect, liquid hydrogen and liquid oxygen, orbiter to ground fill and drain (main propulsion system)
		Accumulator, hydraulic
		Cryogenic pressure relief valve
		Orbital maneuvering system and reaction control system propellant isolation valve tank manifold interconnect lines, 64 valves on each orbiter, plus additional valves, payload delta V kit
		Valve, inline liquid hydrogen/liquid oxygen emergency relief (main propulsion system)
		AC motor valve, valve isolation, propellant (reaction control system)
		Hypergolic couplings
		Valve, pressure relief (reaction control system)
		Dual check valves, manually operated (main propulsion system)
		Valve, manually operated (orbital maneuvering system/reaction control system)
Perkin Elmer Corp., Optical Technology Div.	Danbury, CT	Space telescope assembly, fine pointing assembly and associated controls*
Pneu Devices	Goleta, CA	Shutoff valve, emergency thermal control, hydraulic
		Circulation pump, hydraulics, electric motor driven
Pneu Draulics	Montclair, CA	Priority valve reservoir primary, hydraulic
Porter Seal	Glendale, CA	Seal plate, air data sensor unit
Power Systems Division United Technologies Corp.	South Windsor, CT	Fuel cell power plant
Pressed Steel Tank Co.	Milwaukee, WI	Ammonia storage module tank, mid fuselage (Orbiter 101)

*Separate contract from NASA

Space Transportation System Contractors

Contractor	Location	System/Component
Pressure System Inc.	Los Angeles, CA	Hydrazine fuel tank, (auxiliary power unit)
Purolator Inc.	Newbury Park, CA	Hydraulic filter module assembly
		Helium fill connect, 5/8-inch
		Helium fill connect, 1/4-inch
		Hydrogen, oxygen fill and vent disconnects (EPS)
RDF Corp.	Hudson, NH	Temperature sensor/transducer, general
		Temperature resistance transducer (probe-type)
		Temperature resistance transducer (probe-type)
		Cryo temperature transducer
		Thermocouple reference junction (thermal protection system thermocouple signal conditioner)
		Transducer temperature tip
Regent Jack Manufacturing Co.	Downey, CA	Forward orbiter jacks
Remtech, Inc.	Huntsville, AL	Functional material
Resistoflex	Roseland, NJ	Hydraulic systems:
		Line connector, dynatube
		Line connector, 90° dynatube
		Line connector, tee dynatube
		Line connector, bulkhead dynatube
		Line connector, jam nut-type dynatube
		Line connector, female/male adapter dynatube
Rocketdyne Div., Rockwell International	Canoga Park, CA	Space Shuttle main engines*
		Valve quad check (orbital maneuvering system/reaction control system)
Rosemount Inc.	Eden Prairie, MN	Probe system, air data sensor
		Probe air data sensor, flight boom
		Sensor temperature probe
		Temperature transducer
		Sensor temperature surface, general
		Indicator, angle of attack (backup) (Orbiter 101)
		Indicator, angle of slideslip (backup) (Orbiter 101)

*Separate contract from NASA

Space Transportation System Contractors

Contractor	Location	System/Component
Radio Corp. of America, Astro-Electronics Div.	Princeton, NJ	Closed-circuit television, orbiter and remote manipulator systems*
R.V. Weatherford	Glendale, CA	Shunt
Santa Fe Textiles	Santa Ana, CA	Thermal protection system Inconel 750 wire spring and fabric sleeving
Scheldahl	Northfield, MN	Cover materials and inner layers - passive thermal control
Scott, Inc.	Downers Grove, IL	Main landing gear uplock release thruster actuator
Sealectro	Mamaroneck, NY	Connector, coaxial SMA series
Sentran Co.	Santa Barbara, CA	Vacuum sensors (located in upper and lower wing area, vertical stabilizer, and upper and lower sections of fuselage to measure aerodynamic pressure at reentry and compartment pressures)
Simmonds Precision Instruments	Vergennes, VT	Sensors and electronics, point level, liquid oxygen and liquid hydrogen (main propulsion system)
Simmonds Precision, Motion Controls Div.	Caldwell, NJ	Forward and aft vent doors (electrical, mechanical actuators, torque tubes, push/pull rods, bell cranks, and mechanism support for opeaning and closing vent/purge doors for all ground operations and flight operations)
Singer Kearfott	Little Falls, NJ	Inertial measurement unit
		Multiplexer interface adapter
		Data bus coupler
		Data bus isolation
		Maintenance, modification, and operations support, JSC simulation complex*
		Spacelab simulator*
Skipper and Co.	Cerritos, CA	Chemical processing (Rockwell facility)
Snap Tite	Northridge, CA	Coupling, quick-disconnect, high-pressure
Space Ordnance Systems Div., Trans Technology Corp.	Saugus, CA	Cartridge assembly detonator (frangible nut, tail cone separation, orbiter-carrier aircraft separation, Orbiter 101, and orbiter-external tank separation)
		Pressure cartridge
		Frangible nut, orbiter-external tank aft separation, 3/4-inch
		Sequencing assembly pyro, crew escape system, Orbiters 101 and 102
		Gas generator assembly, pyro crew escape system, Orbiters 101 and 102

*Separate contract from NASA

Space Transportation System Contractors

Contractor	Location	System/Component
		Initiator assembly, pyro panel jettison, Orbiters 101 and 102
		Booster cartridge, detonator assembly
		Cartridge, forward separation bolt, pyro, centering mechanism
Spar Aerospace	Toronto, Canada	Remote manipulator system for Space Shuttle Orbiter (arrangement between NASA and National Research Council of Canada, supported by RCA, Ltd., and CAE of Montreal, Canada)*
Spectran Instrument	La Habra, CA	Radiometer (thermal protection system)
		Calorimeter, standard
		Sensor, radiant heat flux
		Sensor, combination heat flux
		Sensor, temperature thermocouple (measures main engine plume temperature on thermal protection system)
Speedring Manufacturing Co.	Cullman, AL	Window retainers
Sperry Rand Corp., Flight Systems Div.	Phoenix, AZ	Multiplexer/demultiplexer
		Automatic landing
		Indicator, airspeed/mach, backup (4-inch round) (Orbiter 101)
		Indicator, altimeter, barometric (Orbiter 101)
SSP Products Inc.	Burbank, CA	Exhaust duct assembly (auxiliary power unit)
Statham Instruments	Oxnard, CA	Pressure transducers (low, medium, high), general systems
		Pressure transducer, cryo
Sterer Engineering & Manufacturing	Los Angeles, CA	Nose gear steering and damping system
		Valve, selector 3-way, solenoid-operated, landing gear uplock and control
		Shutoff valve, solenoid-operated, main engine hydraulic and hydraulic landing gear
Sterling Transformer Corp.	Brooklyn, NY	Transformer, power displays and controls, 115/26 volt (Orbiter 101)
Sundstrand Corp.	Rockford, IL	Auxiliary power unit
		Rudder-speed brake actuation unit and drive shaft
		Actuation unit, body flap
		Hydrogen recirculation pump assembly (main propulsion system)

*Separate contract from NASA

Space Transportation System Contractors

Contractor	Location	System/Component
Sundstrand Data Control	Redmond, WA	Heater thermostat (auxiliary power unit) Thermal switch
Symetrics	Canoga Park, CA	Hydraulic quick disconnects Fluid disconnect Fluid disconnects (environmental control life support system) Freon Quick disconnects, water boiler, water fill vent
Systron-Donner	Concord, CA	Accelerometer, angular, 3-axis (development flight instrumentation)

Subsidiary of Systron-Donner:
Seaton Wilson Inc.
	Burbank, CA	Water and coolant system:
		  Coupling, 2-inch, half, quick-disconnect female
		  Coupling, 2-inch, half, quick-disconnect male

Tavis Corp.	Mariposa, CA	Flow meter, Freon Transducer, pressure very low range
Tayco Engineering	Long Beach, CA	Fuel cell, excess water dump nozzle Urine, waste water, oxygen, and nitrogen waste dump
Teledynamics Div., Ambac Industries	Fort Washington, PA	S-band transmitter (development flight instrumentation) Frequency modulation S-band transceiver
Teledyne Kinetics	Solano Beach, CA	DC power contractor Limit switches, hermetic seal (electrical power and distribution)
Teledyne Thermatics	Elm City, NC	Cable coax prototype Wire, general-purpose Wire, TSP71 ohm Wire, T/C Wire, TFE insulated
Teledyne McCormack	Hollister, CA	Initiator assembly, pyro (crew escape system, Orbiters 101 and 102)
Telephonics Division, Instruments Systems Corp.	Huntington, NY	Orbiter audio distribution system (voice and tonal signals)
Thiokol Chemical Corp., Wastach Div.	Brigham City, UT	Solid rocket booster motors*

*Separate contract from NASA

Space Transportation System Contractors

Contractor	Location	System/Component
Times Wire and Cable	Wallingford, CT	Coax cable
		Coax cable
Titeflex Div.	Springfield, MA	Flex hose, low-pressure, windshield purge
		Flex line coolant loop (water coolant)
		Hose, high/low pressure (hydraulic system)
		Hose, swivel assembly (hydraulic system)
		Flexible stainless steel hose
Torrington Co.	Torrington, CT	Needle bearing (environmental control/life support system)
TRW Systems, Electronic Systems Div.	Redondo Beach, CA	S-band payload interrogator
		S-band network equipment
		Network signal processor
		Payload signal processor
		Materials processing in Spacelab program*
		Tracking data relay satellites (TDRSS)*
Transco Products	Venice, CA	S-band switch
Tulsa Div., Rockwell International	Tulsa, OK	Cargo bay doors
		Ground support equipment/parts off-load
United Space Boosters, Inc.	Sunnyvale, CA	Assembly, checkout, launch operations, and refurbishment of solid rocket boosters* including recovery ship operation*
U.S. Bearing	Chatsworth, CA	Spherical bearing, orbiter-external tank
U.S. Radium	Parsippany, NJ	Lighting panel overlay
Vacco Industries	El Monte, CA	Pressure relief valve, inline potable water
Vought Corp.	Dallas, TX	Leading edge structural subsystem and nose cap, reinforced carbon-carbon
		Radiator and flow control assembly system
		Manufacturing (crew module skins and bulkheads)
Waltham Precision Instruments	Waltham, MA	8-day clock, windup (Orbiter 101)
Watkins Johnson	Palo Alto, CA	Antennas:
		C-band radar altimeter
		UHF air traffic control voice

Space Transportation System Contractors

Contractor	Location	System/Component
		L-band TACAN (tactical air navigation)
		Antenna S-band (approach and landing test) Orbiter 101
		S-band, quad antenna
		S-band, hemi antenna
		S-band, payload antenna
Wavecom	Northridge, CA	S-band multiplexer (development flight instrumentation)
Westinghouse Electric Corp., Aerospace Electrical Div.	Lima, OH	Remote power controller
		Electrical system inverters, dc-ac
Westinghouse Electric Corp., Systems Development Div.	Baltimore, MD	Master timing unit
Weston Instruments	Newark, NJ	Event indicator
		Electrical indicator meter, round scale, vertical scale
Whittaker Corp.	North Hollywood, CA	Dump valve, manually operated, hydraulic accumulator (ground)
		Regulator, 750/20 psia, helium (main propulsion system)
Wright Components Inc.	Clifton Springs, NJ	2-way pneumatic solenoid valve (main propulsion system)
		3-way solenoid valve helium (main propulsion system)
		Valve, latching, solenoid-operated, hydraulic (hydraulic system and main propulsion system)
		Fuel pump seal cavity drain catch bottle relief (auxiliary power unit)
Xebec Corp.	Kansas City, MO	Automated circuit
Xerox Corp.	El Segundo, CA	Digital computer (Rockwell simulator)

*Separate contract from NASA

Shuttle Orbiter

Rockwell International's Space and Transportation Systems Group, Downey, CA, was responsible for design, development, test and evaluation (DDT&E) and production of the Shuttle spacecraft program under contract with the NASA's Johnson Space Center, TX. The basic contract (in three phases) called for completion of four spacecraft and associated equipment. In addition the NASA has ordered a structural spares contract for construction of the spacecraft's major structures.

Rockwell's Shuttle Integration and Satellite Systems Division has had payload and integration responsibilities on the Space Transportation Systems program. In addition the Division is under contract with the Department of Defense's Advanced Projects Agency (DRPA) and the U.S. Air Force for a number of programs including the Global Positioning System (GPS) as builder of the Navstar satellites and also the Teal Ruby sensor project and P80-1 spacecraft.

GPS Navstar

Apollo CSM

Saturn S-II

Skylab

ASTP

X-15

The Rockwell International Corporation, Downey, CA, facility was the prime contractor for design, development, fabrication, and test of the Apollo Command and Service Modules (spacecraft) used in the successful NASA Apollo lunar landing program, the three Skylab missions, the Apollo-Soyuz Test Project (ASTP), and also designed and fabricated the docking module and docking system used in ASTP.

Rockwell International's, Seal Beach, CA, facility designed and built the Saturn Second Stage (S-II) of the Saturn V launch vehicle used in the lunar landing program as well as the first flight of Skylab when it boosted the laboratory into earth orbit. In the mid-1950s, Rockwell International designed and built the three X-15 experimental aircraft used by NASA in the 1960s for space flight research.

Additional information may be obtained by contacting Rockwell International's Downey, CA, facility, Public Relations personnel:

R. E. Barton, Manager of Public Relations
Bus.: 213/922–1217
Home: 213/377–8962

W. F. Green, Technical Communications
Bus.: 213/922–2066
Home: 818/331–6773

R. V. Gordon, KSC Media Relations Manager
Bus.: 305/784–9825
Home: 305/784–3659

Or National Aeronautics and Space Administration (NASA) Public Information news centers:

Headquarters, Washington, D.C.
202/453–8400

Kennedy Space Center
305/867–2468

Johnson Space Center
713/483–5111

Marshall Space Flight Center
205/453–0034

Dryden Flight Research Facility
805/258–8381

Contacts for planning and information concerning payload use of the STS by potential customers should be made to Space Transportation Systems Operations Office, MO, NASA, Washington, D.C., 20546, telephone 202/755–2344 or Space Transportation Systems User Service Center, Space Shuttle Integration and Satellite Systems Division, Rockwell International, 12214 Lakewood Boulevard, Downey, CA, 90241, telephone 213/922–3344.

Metric Conversion Table

Multiply	By	To Obtain
Length:		
Inches	25.4	Millimeters (mm)
Feet	304.8	Millimeters (mm)
Yards	0.9144	Meters (m)
Miles	1.609	Kilometers (km)
Inches	2.54	Centimeters (cm)
Feet	0.3048	Meters (m)
Area:		
Square Inches	6.452	Centimeters Squared (cm^2)
Square Feet	0.0929	Meters Squared (m^2)
Square Yards	0.8361	Meters Squared (m^2)
1 Acre	0.4047	Hectares (ha)
Square Miles	2.59	Kilometers Squared (km^2)
Square Miles	259.0	Hectares (ha)
Volume:		
Cubic Inches	16.39	Cubic Centimeters (cm^3)
Cubic Feet	0.0283	Cubic Meters (m^3)
Cubic Yards	0.7646	Cubic Meters (m^3)
Fluid Ounces	29.57	Milliliters (ml)
Fluid Quarts	0.9464	Liters (L)
Gallons	3.785	Liters (L)
Fluid Ounces	0.0296	Liters (L)
Liters (L)	0.2642	Gallons
Mass:		
Ounces	28.35	Grams (g)
Pounds (lb)	0.4536	Kilograms (kg)
1 Ton (2000 pounds)	0.9072	Metric Tons (Tonnes - t)
Kilograms (kg)	2.205	Pounds

Multiply	By	To Obtain
Speed:		
Miles Per Hour (M/h)	0.447	Meters per Second (m/s)
	1.609	Kilometers per Hour (km/h)
Knots	1.852	Kilometers per Hour (km/h)
Distance:		
Nautical Miles (N.M.)	1.852	Kilometers (km)
Statute Miles (S.M.)	1.609	Kilometers (km)
Meters (m)	3.281	Feet
Kilometers (km)	3281	Feet
Kilometers (km)	0.6214	Statute Miles (S.M.)
Nautical Miles (N.M.)	1.1508	Statute Miles (S.M.)
Statute Miles (S.M.)	0.8689	Nautical Miles (N.M.)
Statute Miles (S.M.)	1760	Yards
Acceleration:		
Inches per Second Squared	2.54	Centimeters per Second Squared (cm^2/s)
Feet per Second	0.3048	Meters per Second (m/s)
Meters per Second (m/s)	2.237	Statute Miles per Hour
Feet per Second	0.6818	Statute Miles per Hour
Feet per Second	0.5925	Nautical Miles per Hour
Pressure:		
Pounds per Square Inch (PSI)	6.895	Kilonewtons per Square Meter (kilopascals - kPa)
Millimeters Mercury	133.32	Newtons per Square Meter (Pascals - Pa)
Pounds per Square Inch (PSI)	51.75	Millimeters of Mercury (mmHg)

Metric Conversion Table

Multiply	By	To Obtain
Force:		
Ounces Force	0.278	Newtons (N)
Pounds Force	4.448	Newtons (N)
Newtons	0.225	Pounds
Kilograms (kg)	9.807	Newtons (N)
Flow Rate:		
Cubic Feet per Minute	0.283	Cubic Meters per Second (m^3/s)
Gallons per Minute	3.7854	Liters per Minute (L/m)
Pounds Mass per Hour	0.4536	Kilograms per Hour (kg/hr)
Pounds Mass per Minute	0.4536	Kilograms per Minute (kg/m)
Pounds Mass per Second	0.4536	Kilograms per Second (kg/s)
Pounds Mass per Cubic Foot	16.02	Kilograms per Cubic Meter (kg/m^3)

Multiply	By	To Obtain
Power:		
British Thermal Units (Btu)	1.054	Kilojoules per Hour (kj/hr)
Brake Horsepower	0.7457	Kilowatts (kw)
Electric Horsepower	0.746	Kilowatts (kw)
Energy:		
Kilowatt Hours	3.60	Megajoules (mj)
Temperature:		
$\dfrac{°F - 32}{1.8}$		Degrees Celsius (°C)

lem
LUNAR EXCURSION MODULE

PROJECT APOLLO

LMA 790-1

NOW AVAILABLE!

FIRST MANNED LUNAR LANDING
FAMILIARIZATION MANUAL

GRUMMAN AIRCRAFT ENGINEERING CORPORATION • BETHPAGE, L. I., N. Y.

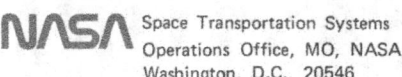 Space Transportation Systems Operations Office, MO, NASA, Washington, D.C., 20546

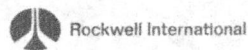 Rockwell International

Integration and Satellite Systems Division
12214 Lakewood Boulevard
Downey, CA, 90241

©2011 Periscope Film LLC All Rights Reserved
ISBN #978-1-935700-84-5
www.PeriscopeFilm.com